The Behavioral Biology and Conservation of Goitered Gazelles (*Gazella subgutturosa*)

David Blank

The Behavioral Biology and Conservation of Goitered Gazelles (*Gazella subgutturosa*)

Springer

David Blank
Research Center for Ecology and Environment of Central Asia
Bishkek, Kyrgyzstan

ISBN 978-3-031-83503-2 ISBN 978-3-031-83504-9 (eBook)
https://doi.org/10.1007/978-3-031-83504-9

This Springer imprint is published by the registered company Springer Nature Switzerland AG
The registered company address is: Gewerbestrasse 11, 6330 Cham, Switzerland

I dedicate this book to all my good friends who have helped me to create this monograph

Foreword

David Blank and I first met in January 1996. I had taken a year off from my job as a systems analyst in Seattle, Washington, to travel, and I signed up for several wildlife research projects through Earthwatch Institute expeditions. Dorcas gazelles in southern Israel were the reason I met David. He was the principal investigator on a project to gather data on the potential impact of the dorcas gazelle on the desert environment and its other species before the gazelles were to be released into the wild. It wasn't until the end of the project that I found out that David, a native of Kazakhstan, was learning both English and Hebrew at the same time. He had taught himself enough English to be able to read research papers, but he was working on speaking and writing two new languages simultaneously. With his native language of Russian, that meant he was not only using three languages, but he had to work in three different alphabets.

At the end of our two weeks in southern Israel, we volunteers moved on with our travels. David and I began corresponding via "snail mail" (Internet availability for both of us was still a few years away), and sometime later, I started providing him with research assistance, since I had a major university conveniently available and he, living and working in southern Israel, did not. At some point during the following years, I began reading through his research papers to assist him with his English. I have no specific training in writing or editing, other than my education (MBA) and job experience—and the fact that I read a lot, but with persistence, patience, and practice, we've developed a system that works. I've lost track of the number of papers he's written, but this is the second book we've worked on together.

As long as I've known David, he has been passionate about his work. At the same time as the project on dorcas gazelles, he was also gathering data on acacia gazelles, both of which once lived in southern Israel. When still in Kazakhstan and working on his Ph.D. in zoology, he made extensive, decade-long observations on goitered gazelles and has used his time since to thoroughly document and publish his findings. Goitered gazelles, too, were once abundant throughout the former Soviet Union, but unfortunately, that's no longer the case. Published data on previous goitered gazelle populations and similar species have ranged from non-existent to sporadic and very dated across the former Soviet republics. This book has given David the opportunity to pull together whatever research that has been published to date, including his own investigations, to provide as complete a source of documentation as possible on all aspects of what it means to be a goitered gazelle: history, evolution, distribution, habitat, ecology, morphology, taxonomy, behaviors, and everything else known about them.

It is the function of published research to provide a foundation on which to base further study of a particular subject. Having all existing data in one book is a good way to point out what's missing in the knowledge base, in this case, of a species of gazelle. Knowledge of a species is necessary to define and implement conservation efforts to protect that species. In order for goitered gazelles to survive in the wild, and hopefully thrive to repopulate their previous habitats, their needs require understanding and support. It is my hope, as someone who has worked on this book, that what David has created with this publication will be used as a basis for other researchers to further that understanding of goitered gazelles and work to develop support mechanisms for them to survive and expand as a species. I feel I've come to know these beautiful animals and would like to believe that they will be able to live a life of freedom in the wild.

I thank David for allowing me to participate in this effort. And I am honored to have been asked to write this Foreword for his book. He is my friend, and I am very proud of him and all of his lifelong accomplishments that have made this book a reality.

Tacoma, WA, USA
28.06.2023

Patricia A. Johnston

Preface

Dear reader!

This is a book that was conceived about 30 years ago when I was writing an article about goitered gazelles for "Mammalian Species" for the American Society of Mammologists. With a wonderful zoologist, friend, and fellow author named Dr. Steven C. Kingswood, we tried in that paper to touch on all aspects of goitered gazelle biology. But the format of that publication assumed a brief overview of the published literature with some of the data presented almost with tabular brevity. As a result, I decided to write an entire book, in which to collect everything published in both English and Russian about goitered gazelles. I, personally, have worked mainly in the field of goitered gazelle behavior and my study information has been published in a number of articles in various scientific journals. I wanted to collect and publish all information in one book. Unfortunately, almost everything that was published about the goitered gazelle was in Russian and inaccessible to an English reading audience. I translated all the main Russian publications into English in order to collect "under one roof" everything that is known about the goitered gazelle to create a kind of encyclopedia for this species.

At the beginning, I wanted to insert all data, even small observations, into the book so that the reader could evaluate and analyze this information directly. But then I realized that I was trying to translate books in their entirety, which would have created a multi-volume edition. Moreover, many data were repeated, or confirmed the same behaviors, many times. Therefore, I had to get rid of duplicate or irrelevant details. In addition, of all the publishers I contacted, only Springer, Elsevier, and Taylor and Francis provided me, as an author, with the freedom to use tables and graphs from my own articles, while other publishers would only sell me my own information for amounts that I could not pay out of pocket. Under those conditions, I could not include tables and graphs in the book from some of my own articles published by other publishers. Unfortunately, I did not know about this limitation when I sent my articles to publishers, or I would have been more selective and careful in choosing a publisher and a magazine. Consequently, any missing information is available for researching details in my published papers, and I have made links to these papers, where they can be found online in Google Scholar. Therefore, my original idea to collect everything in one book was not possible, and for this inconvenience, I apologize to my readers.

I have reviewed many articles by different authors and have gathered information from many books. I did not want to miss any interesting descriptions of goitered gazelle behavior, so I supplemented my chapters with any additional information I found. As a result, it turned out that some descriptions taken from different sources sometimes repeat each other, or in a few cases, contradict each other. After editing by my English editor and friend, Patricia Johnston, and after my own checking, many repetitions have been removed. It is quite possible that some of these repetitions still remain in the text, and for this, I also apologize.

In this book, I state my objections if I disagree with the statements of other authors, but I included those statements anyway, so the reader can see different perspectives. I was especially upset by the publications on the goitered gazelle taxonomy, where the authors presented the results of their genetic studies and offered dizzying conclusions about kinship with other gazelle species. At the same time, those authors were not at all confused by the contradictions of the results of their genetic studies with morphological data. They also did not try to explain the logic of their conclusions from a zoogeographic point of view. It is quite possible that I am wrong, since I do not work with genetic materials and analyze published materials. However, in my opinion, a critical approach is very important in order to check and recheck whether a hypothesis is right or wrong. Over time the truth will become clear, but a critical approach to one's own and others' conclusions should accompany a scientist throughout his or her professional path.

At the beginning, each chapter of this book was structured according to the format adopted for scientific articles, where there is an introduction, results, and discussion. However, later the text of most chapters was revised with the reduction of the paper format. But in any case, unlike published scientific articles, I do not provide research methodology or statistical analyses, because all this information is available in the published articles themselves, most of which are in English. As far as this approach worked or not is up to my dear readers to judge.

So, here is the result of my many years of work. I hope that my readers will find something of value in this book for further research and conservation. My opinion is not the ultimate truth, and the reader may agree or disagree with me, but the main thing I wanted to do with this publication is to give a new impetus to future research on the biology of the goitered gazelles and urge the scientific and non-scientific community to do everything possible to preserve this beautiful animal in the wild. If this is not done soon, then this once widespread gazelle will turn into an extinct species that can only be studied from fossils and books.

I wish my readers a pleasant and interesting reading experience.

With best wishes,

Bishkek, Kyrgyzstan David Blank

List of Animal Latin Names

Common names in alphabetical order (synonyms in parentheses) and Latin names of animal species involved in the discussion of the book.

Common name	Scientific Latin name	Common name	Scientific Latin name
Addax	Addax nasomaculatus	Impala	*Aepyceros melampus*
Alpine chamois	*Rupicapra rupicapra*	Imperial eagle	*Aquila heliaca*
Alpine ibex	*Capra ibex*	Indian hog deer	*Axis porcinus*
Alpine marmot	*Marmota marmota*	Jackal (Golden or common jackal)	*Canis aureus*
Anatolian leopard	*Panthera pardus tulliana*	Klipspringer	*Oreotragus oreotragus*
Aoudad	*Ammotragus lervia*	Kulan	*Equus hemionus*
Arabian gazelle	*Gazella arabica*	Leopard	*Panthera pardus*
Arabian sand gazelle (Reem)	*Gazella (subgutturosa) marica*	Lion	*Panthera leo*
Argali	*Ovis ammon*	Markhor	*Capra falconeri*
Asian mouflon	*Ovis vigney arcal*	Mongolian gazelle	*Procapra gutturosa*
Asiatic cheetah	*Acinonyx jubatus venaticus*	Moose	*Alces alces*
Asiatic ibex	*Capra sibirica*	Mongolian gazelle	*Procapra gutturosa*
Barasingha	*Cervus duvaucelii*	Mouflon sheep (European or Mediterranean mouflon)	*Ovis gmelini*
Barbary sheep	*Ammotragus lervia*	Kongoni	*Alcelaphus buselaphus*
Barren-ground caribou	*Rangifer tarandus arcticus*	Mountain gazelle	*Gazella gazella*
Beira	*Dorcatragus megalotis*	Mountain snow goats	*Oreamnos americanus*
Beisa (East African oryx)	*Oryx beisa*	Mule deer	*Odocoileus h. hemionus*
		Muntjac	*Muntiacus muntjak*
Bezoar wild goat	*Capra aegagrus*	Musk deer	*Moschus moschiferus*
Bighorn sheep	*Ovis canadensis*	Muskox	*Ovibos moschatus*

Common name	Scientific Latin name	Common name	Scientific Latin name
Bison	*Bison bison*	North American moose	*Alces alces andersoni*
Bharal	*Pseudois nayaur*	Nubian ibex	*Capra nubiana*
Black bear	*Ursus americanus*	Nyala (lowland nyala)	*Tragelaphus angasii*
Blackbuck (Indian blackbuck)	*Antilope cervicapra*	Oribi	*Ourebia ourebi*
Black-tailed deer (Columbian black-tailed deer)	*Odocoileus hemionus columbianus*	Oryx antelope	*Oryx gazella*
Black rhinos	*Diceros bicornis*	Oryx beisa	*Oryx beisa*
Black wildebeest	*Connochaetes gnou*	Pampas deer	*Ozotoceros bezoarticus*
Blesbok	*Damaliscus pygargus*	Peccary	*Tayassu pecari*
Blue duiker	*Cephalophus maxwelli*	Persian gazelle	*Gazella subgutturosa subgutturosa*
Blue wildebeest	*Connochaetes taurinus*	Plains zebra	*Equus quagga*
Bontebok	*Damaliscus pygargus*	Pronghorn	*Antilocapra americana*
Brown-necked raven	*Corvus ruficollis*	Przewalski's gazelles	*Procapra przewalskii*
Harnessed bushbuck (North bushbuck)	*Tragelaphus scriptus*	Pyrenean chamois (Apennine chamois)	*Rupicapra pyrenaica*
Caracal	*Felis caracal*	Raven (black-necked)	*Corvus ruficollis*
Caribou	*Rangifer tarandus*	Red deer	*Cervus elaphus*
Cattle	*Bos taurus*	Red-fronted gazelle	*Eudorcas rufifrons*
Chamois (Alpine chamois)	*Rupicapra rupicapra*	Red kangaroo	*Macropus rufus*
Cheetah	*Acinonyx jubatus*	Reedbuck	*Redunca arundinum, R. fulvorufula*
Chinese water deer	*Hydropotes inermis*	Reindeer	*Rangifer tarandus*
Chinkara (Indian gazelle)	*Gazella bennettii*		
Chiru (Tibetan antelope)	*Pantholops hodgsonii*	Roe deer (Common roe deer)	*Capreolus capreolus*
Chital	*Axes axes*	Ruppell's sand fox	*Vulpes rueppellii*
Collared peccary	*Tayassu tajacu*	Sable antelope	*Hippotragus niger*
Common fox (Red fox)	*Vulpes vulpes*	Saiga antelope	*Saiga tatarica*
Common duiker (bush duikers)	*Sylvicapra grimmia*	Sambar deer	*Cervus unicolor*
Corsac fox	*Vulpes corsac*	Saudi gazelle	*Gazella* (*dorcas*) *saudiya*
		Scimitar-horned oryx	*Oryx dammah*
Coyote	*Canis latrans*	Siberian musk deer	*Moschus moschiferus*
Cuvier's gazelle (Edmi)	*Gazella cuvieri*	Siberian roe deer	*Capreolus pygargus*
Dama gazelle	*Nanger dama*	Siberian ibex	*Capra sibirica*

Common name	Scientific Latin name	Common name	Scientific Latin name
Dibatag	*Ammodorcas clarkei*	Slender-horned gazelle (Rhim gazelle, African sand gazelle, Loder's gazelle)	*Gazella leptoceros*
		Snow leopard	*Uncia uncia*
Dik-dik	*Madoqua kirkii*	Soemmerring's gazelle	*Nanger soemmerringii*
Defassa waterbuck	*Kobus ellipsiprymnus defassa*	Southern reedbuck	Redunca arundinum
Domestic goat	*Capra hircus*	Spanish ibex (Iberian ibex)	*Capra pyrenaica*
Domestic sheep	*Ovis aries*	Spotted hyena	*Crocuta crocuta*
Dorcas gazelle	*Gazella dorcas*	Springbok	*Antidorcas marsupialis*
Duiker (forest duikers)	*Cephalophus*	Steenbok	*Raphicerus campestris*
Eland (Common eland)	*Taurotragus oryx*	Steppe eagle	*Aquila rapax*
Eland (Giant eland)	*Taurotragus derbianus*	Steppe ferret	*Mustela eversmanni*
Elk	*Cervus elaphus canadensis*	Striped hyena	*Hyena hyena*
Elephant seal	*Mirounga angustirostris*	Sundevall's jird	*Meriones crassus*
European bison	*Bison bonasus*	Suni	*Neotragus moschatus*
Fallow deer	*Dama dama*	Swamp deer	*Cervus duvaucelii*
Farasan gazelle	*Gazella gazella farasani*	Takin	*Budorcas taxicolor*
Feral dog	*Canis familiaris*	Thomson's gazelle	*Eudorcas thomsonii*
Feral goat	*Capra hircus*	Tibetan antelope	*Pantholops hodgsonii*
Four-horned antelope	*Tetracerus quadricornis*	Tibetan gazelle	*Gazella picticaudata*
Fox (red fox)	*Vulpes vulpes*	Tiger	*Panthera tigris*
Gemsbok (South African antelope)	*Oryx gazella*	Topi (sassaby, tiang, tsessebe)	*Damaliscus lunatus*
Gerenuk	*Litocranius walleri*	Korrigum (topi)	*Damaliscus lunatus korrigum*
Giraffe	*Giraffa camelopardalis*	Turanian tiger	*Panthera tigris virgata*
Goat	*Capra hircus*	Urial	*Ovis vignei*
Goitered gazelle	*Gazella subgutturosa*	Vole	*Microtus*
Golden eagle	*Aquila chrysaetos*	Wapiti	*Cervus elaphus canadensis*
Golden takin	*Budorcas taxicolor bedfordi*	Warthog	*Phacochoerus aethiopicus*
Goral	*Naemorhedus goral*	Waterbuck	*Kobus ellipsiprymnus*
Grant's gazelle	*Nanger granti*	White-tailed deer	*Odocoileus virginianus*

Common name	Scientific Latin name	Common name	Scientific Latin name
Greater kudu	*Tragelaphus strepsiceros*	Wild dog	*Lycaon pictus*
Grizzly bears	*Ursus arctos*	Wild goat	*Capra aegagrus*
Grysbok	*Raphicerus melanotis*	Wolf	*Canis lupus*
		Wood bison	*Bison bison*
Guanaco	*Lama guanicoe*	Woodland caribou	*Rangifer tarandus caribou*
Hamster	*Mesocricetus*	Yarkand gazelle	*Gazella subgutturosa yarkandensis*
Hartebeest	*Alcelaphus buselaphus*		
Hillier goitered gazelle	*Gazella subgutturosa hillieriana*		
Hyena	*Hyena hyena*		

Acknowledgments

I express my greatest gratitude to my friend and English editor Ms. Patricia Johnston, who has worked for many months to review the entire text of this book without any monetary reward. She has helped me with my English by editing many papers for me over many years. Also, thanks to Dr. Kathreen Ruckstuhl for her professional suggestion to the manuscript. Big thanks also go to Mark Pestov along with his entire team of NGO "Dront," as well as to director of the Ustyurt Nature Reserve Zhaskairat Nurmukhambetov and entire his team (especially Agyniaz Pulatov, Vladimir Terentiev) for the opportunity to use all their wonderful photos of goitered gazelles from the Southern Ustyurt Plateau (Kaplankyr) and Kyzylkum Desert (Kazakhstan). Because of these photos, my book is much more interesting and attractive for readers. Great thanks and appreciation to my German friend Ralf Pfeffer, who painted the picture of the male goitered gazelle especially for the cover page of my book; to my Austrian friend Petra Kaczensky for her excellent pictures of goitered gazelles and their habitats from Mongolia; and to my Chinese friend Xu Feng for his interesting pictures from Xinjiang; as well as to Askar Dautletbakov from Kyrgyz Academy of Sciences for his kind permission to give me opportunity to take pictures of goitered gazelle behaviors in the Issyk-Kul goitered gazelle nursery for acclimatization. Thank you very much to Azamat Osmonov for making goitered gazelle distribution maps for me and Ulanbek Kenzhetaev for his drawings of goitered gazelles and their behavioral postures made specifically for my book. I would also like to thank in particular my manager, Salamat Abdyzhapar uulu, and my entire team at the Research Center for Ecology and Environment of Central Asia (Bishkek) for their patience in granting me the time and opportunity to concentrate on writing this book for more than a year.

Many thanks to Lars Koemer and all undisclosed reviewers for believing in this project and supporting its publication. I am also very grateful to Springer Nature for publishing my book and to Springer, Elsevier, and Taylor and Francis for giving me free-use permissions to use my own data from previously published English papers.

About This Book

The book *Behavioral Biology of Goitered Gazelles* consists of 22 chapters. In Chap. 1, I collected all basic published information on the goitered gazelle biology. The following chapters are devoted mostly to behavioral ecology of goitered gazelles. And the final chapter describes the main unexplored phenomena. I included in the book a number of my published papers; though I did not keep a structure similar to articles and did not include the method section and a statistical analysis, because they are available in my published papers. I used in the book the ternary numeration of subchapters, and binary numeration of tables and figures, where the first number means belonging to a certain chapter, while second is sequence number of item.

Contents

About the Author

David Blank was born in Alma-Ata (Kazakhstan) in 1958. He studied in the Kazakh State University in Alma-Ata, Kazakhstan (1975–1980) and investigated *Capra sibirica* behavior in Tien-Shan Mountains. David got his Ph.D. degree in the Tel Aviv University for the theme "Social and mating behavior of Persian gazelle." He worked as a field zoologist in the Institute of Zoology, Kazakh Academy of Sciences (Alma-Ata), and observed goitered gazelle behavior in Kazakhstan (1981–1991). After that he worked as zoologist in the Israel Nature Reserve Authority (1993–2005). He participated in the UNDP project China Wetlands Biodiversity Conservation (2005–2007). He worked in the Hai Park Zoo in Israel. Later, David worked as an assistant visiting professor in the Xinjiang Institute of Ecology and Geography (Urumqi, Chinese Academy of Sciences—2010–2015). And now he works as ecologist in the Research Center of Ecology and Environment of Central Asia, Bishkek, Kyrgyzstan (2015–until now).

Basic Knowledge on Goitered Gazelle Biology

1

1.1 Description of the External Appearance

1.1.1 Signs in the Field

A relatively long black tail, vertically raised when alarmed and clearly visible against the background of a white rump patch (Fig. 1.1), always distinguishes a goitered gazelle from the similar-looking Mongolian gazelle (*Procapra gutturosa*), which sometimes shares habitat in places in Mongolia. In a calm state, the tail of the goitered gazelle is held down (Fig. 1.2). Goitered gazelles differ from the saiga by having a smaller, more elegant body. While galloping, the goitered gazelle runs with huge jumps, raising its head or holding it level with the body (Fig. 1.3), and not lowering it, as the saiga (*Saiga tatarica*) does when running (Heptner et al. 1961).

The black hooves are heart-shaped, narrow, and long, with a sharply pointed tip (Fig. 1.4; Sokolov et al. 1964). The hind hooves are slightly shorter than the front ones, as well as in most Bovidae. The peculiarity of the goitered gazelle's hooves is a relatively gentle steepness, both from the side and the front of the hoof; therefore, they look pointed (Sokolov et al. 1964). There is no longitudinal notch at the front end of the track, since the hooves almost do not move apart, even when resting on soft ground (Fig. 1.5). Gazelle droppings are small, dark pellets with sharp tips, about 10–15 mm long and 7–8 mm thick (Fig. 1.6). Goitered gazelle beds are usually protected by bushes (Fig. 1.7) or tall grasses and are round or oval holes, devoid of grass and carefully cleaned of rubble. There is always a pile of litter on the edge of the bed. In winter, goitered gazelles lie down on the bare ground, shoveling snow with their hooves. Trails to watering places are usually narrow and straight, with no bends. Trails are short in length and are usually clearly visible from no further than 100–200 m from the water. In herds where there are young, you can hear the quiet bleating of fawns and the voices of females resembling "moo" sounds. During the rut, males emit hoarse cries (Heptner et al. 1961).

D. Blank, *The Behavioral Biology and Conservation of Goitered Gazelles (Gazella subgutturosa)*, https://doi.org/10.1007/978-3-031-83504-9_1

Fig. 1.1 The long black tail is raised vertically in an alarmed gazelle, and it is visible at a distance against the background of a white rump-patch. Kalamali Ungulate Nature Reserve, Xinjiang, China, 15.12.2022. Photo by F. Xu with kind permission

Fig. 1.2 In the calm state, the tail of the goitered gazelle is lowered down. Issyk-Kul goitered gazelle nursery for re-acclimatization, 25.11.2022. Photo by D. Blank

Fig. 1.3 While running, the goitered gazelle jumps with strides, raising its head or keeping it at the same level as its body. Drawing by U. Kenzhetaev with kind permission

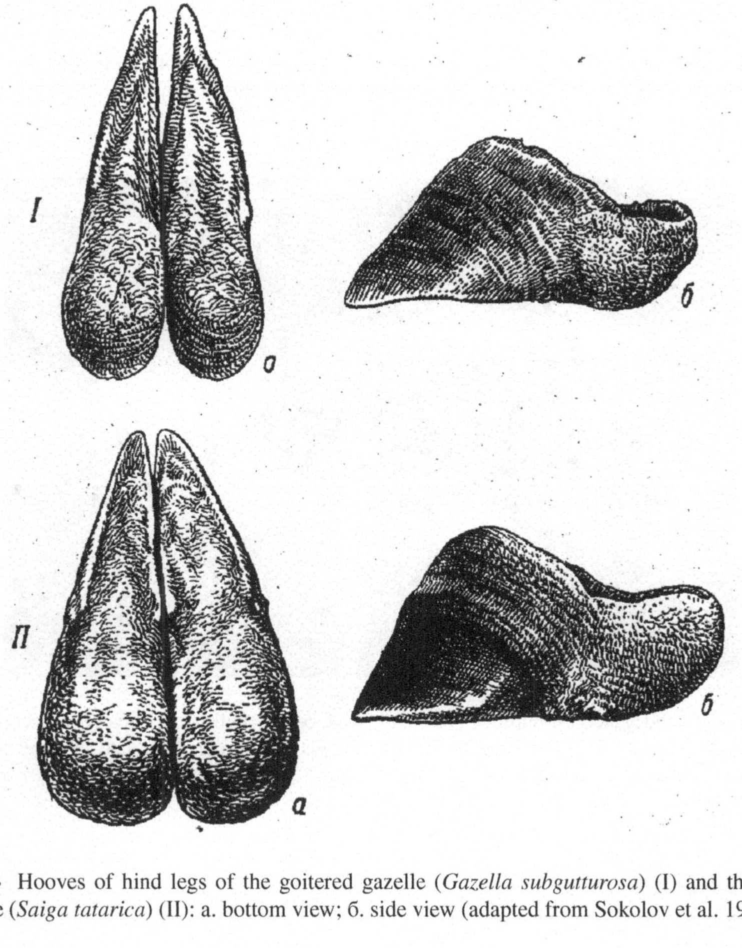

Fig. 1.4 Hooves of hind legs of the goitered gazelle (*Gazella subgutturosa*) (I) and the saiga antelope (*Saiga tatarica*) (II): a. bottom view; б. side view (adapted from Sokolov et al. 1964)

Fig. 1.5 Goitered gazelle hooves almost do not spread apart, even when resting on soft ground. Great Gobi B. 13.07.2008. Photo by P. Kaczensky with kind permission

Fig. 1.6 Goitered gazelle droppings are small, dark pellets with sharp tips. The gazelle droppings among kulan's (*Equus hemionus*) excrements. Kalamali Ungulate Nature Reserve. 23.07.2011. Photo by D. Blank

Fig. 1.7 Goitered gazelle resting sites are usually protected by shrubs. Issyk-Kul goitered gazelle nursery for re-acclimatization, 25.06. 2022. Photo by D. Blank

1.1.2 General Description

The goitered gazelle is a medium-sized species with a light, slim build and a short narrow body on tall slender legs. The dorsal pelage varies widely from light brown to various tints of gray, red, white, and yellow, while the ventral pelage is white (Fig. 1.8). Goitered gazelles are sexually dimorphic: males are larger and possess horns, while females are smaller and usually hornless. Only females of Arabian sand gazelles (*G. s. marica*) possess well-developed horns, albeit they are thinner than in males. The goitered gazelle male's total length is 94–126 cm, and the female's is 94–120 cm; the male's shoulder height is 58–79 cm and the female's is 56–76.5 cm; and the male's weight is 20–43 kg and the female's is 17.5–33.2 kg, respectively. The length of the body is equal to or slightly less than the height at the withers. The front legs are slightly shorter than the hind legs, which is why the height at the withers is usually slightly less than the height of the sacrum (Sokolov 1959). The Arabian sand gazelle has a significantly smaller body size than *G. s. subgutturosa* (Kingswood and Blank 1996).

The goitered gazelle's head is light and relatively small with a slightly build pointed muzzle. The upper lip is covered with hair and there is only a narrow strip of bare dark skin running from the nasal patch between the lower corners of the nostrils. The neck is relatively long and thin at 40% of the total body length. The larynx of both males and females is in the form of a small goiter that protrudes from

Fig. 1.8 Goitered gazelles are medium-sized ungulates, very slender and light built, with a straight back, thin and slender legs, a proportional thin, elongated neck and a high-set head. Adult female in the Kaplankyr, SW Kazakhstan. 03.06.2019. Photo of the camera-trap of the Ustyurt State Nature Reserve team with kind permission

the throats (Fig. 1.9); during the rut, the larynx of males increases significantly and becomes noticeable from a distance. The eyes are large, black, and convex and placed anterolaterally. The pupil of the eye is almost round, though with a slightly longitudinally oval; the iris is dark brown. The ears are quite long (11–16 cm), with pointed tips. When a goitered gazelle lifts its ears in response to danger, the ears do not stand parallel (like a horse), but at a significant angle, forming a triangle without a base (Fig. 1.10) (Sokolov 1959; Zhevnerov et al. 1983; Zhevnerov 1984). The tail is relatively long (longer than the ears) and tufted dorsodistally for 2/3 of its length with dark-brown or black hair. The preorbital, inguinal, carpal, and hoof glands are well-developed (Sokolov 1959).

In goitered gazelles, mostly males have long, black horns of various length, usually from 20 to 41 cm with 16–29 annular ridges. The lyrate horns can have different shapes, and some are widely spaced, while others have horns narrowly separated, but they are curved upward and backward with tips directed inward and slightly forward. The horns are always located on the front surface of the frontal bones, and their bases are brought close together (<15 mm apart). The transverse section of the horn is round, and occasionally there is a small anterior rib. The horns have transverse ring-shaped elevations or transverse bumps on the anterior side of the main half, more or less lyre-shaped, with the main part of each horn slightly curved forward (Fig. 1.11). The horn substance is black (Sokolov 1959; Zhevnerov et al. 1983). As an exception, there are horns in females, but they are rudimentary, irregularly shaped (crooked), and not more than 3–7 cm long (Fig. 1.12). Even if females

Fig. 1.9 Males (by sides) and a female (in middle) have a larynx in the form of a small goiter that protrudes from the throat. Issyk-Kul goitered gazelle nursery for re-acclimatization. 25.06.2022. Photo by Blank D.A

Fig. 1.10 In a dangerous situation, goitered gazelles raise their ears, which stand at a significant angle, forming a triangle with no base. Issyk-Kul goitered gazelle nursery for re-acclimatization. 25.06.2022. Photo by Blank D.A

Fig. 1.11 The male horns are quite thin and of medium length or rather long with transverse ring-shaped elevations. Issyk-Kul goitered gazelle nursery for re-acclimatization. 25.06.2022. Photo by Blank D.A

Fig. 1.12 In Central Asia, some goitered gazelle females have small rudimentary horns. A young female in the Issyk-Kul goitered gazelle nursery for re-acclimatization. 25.11.2022. Photo by Blank D.A

Fig. 1.13 Additional two small hooves in goitered gazelles are developed to a different degree and located higher than main hooves and back of every leg. Issyk-Kul goitered gazelle nursery for re-acclimatization. 25.06.2022. Photo by Blank D.A

do not have actual horns, their skulls always have well-defined bumps in the place of horns. In most goitered gazelle subspecies, females are hornless, with the exception of *G. s. marica*, where both sexes have horns (Kingswood and Blank 1996).

The limbs are slender. The length of the front hoof is 45–57 mm and 35–40 mm width, with a width two times less than their length; the rear hoof is 1–6 mm smaller. The height of the front hooves along the front edge is 28–35 mm, and the rear ones are usually 2–3 mm less or sometimes equal to the front ones. Small additional hooves, 9–12 cm long, are located high above the crumb of the hooves of the middle toes (Fig. 1.13) (Sokolov 1959). These hooves are poorly adapted to moving on snow, soft ground, or getting food from under the snow cover by raking away the snow. If ungulates (kulan, *Equus hemionus*; saiga, *Saiga tatarica*; argali, *Ovis ammon*) that can reach food from under the snow have an area of the front hooves almost two times larger than the rear ones, then in goitered gazelles they are almost equal. For example, in goitered gazelles of the Betpak-Dala Desert, the hoof area of the front leg of the male is 12.1 cm^2, while the rear hoof is 11.4 cm^2. It is very inconvenient and ineffective to rake up snow with a narrow hoof that has a sharp angle (Sludsky 1977), so the goitered gazelle is not adapted to living in areas with a deep snow cover. This fact is also supported by the large weight load of the body on the supporting surface of the hooves, equal to 491 g/cm^2 in males, 434 g/cm^2 in females, and 340 g/cm^2 in fawns. For comparison, a larger body-sized reindeer (*Rangifer tarandus*) has a weight load of only 140–180 g/cm^2 and they can survive in areas with deep winter snow (Sludskiy 1977).

1.1.3 Basic Vital Indices

Body mass: Male is 30.1 kg on average (range 24.2–36.0 kg); female is 26.7 kg on average (range 20.2–33.2 kg) (Maмbetzhumaev 1970; Prisiazhnuk and Soldatova 1985).

Reproductive activity duration: Male is 3.75 years (1.5–6 years); female is 5.75 years (0.5–11 years) (Zhevnerov et al. 1983).

Fecundity: On average 1.64 (range 0.88–1.8) (Sludskiy 1956a).

Jeiran is a Persian name for the goitered gazelle; Georgians call this animal a "Dzheirani"; Turkmens call it "zheren" or "duz keiigi"; Kyrgyz, Uzbek, and Kazakh people call it "karakuiruk" (black tail) or a common name used for a number of species "kiik"(wild goat) (Sludskiy 1956a); reem is an Arabian sand gazelle (Dunham et al. 2001); and in Farsi the gazelle is called ahu (Hemami and Groves 2001).

1.2 Taxonomy and Genetics

1.2.1 Problems with Goitered Gazelle Taxonomy

The goitered gazelle was first described by A.I. Güldenstaedt from a specimen extracted from northwestern Persia (Heptner et al. 1961). The goitered gazelle belongs to the order Artiodactyla (Owen 1848), suborder Ruminantia, infraorder Pecora, superfamily Bovoidea, family Bovidae (Gray 1821), subfamily Antilopinae, genus *Gazella* (De Blainville 1816), and subgenus *Trachelocele* (Ellerman and Morrison-Scott 1951). The goitered gazelle occupies a central position in the genus, although it has a number of ancestral features in its social structure and morphology. The absence of horns in females is a morphological primal characteristic. At the same time, the gazelle, as an inhabitant of deserts with a dense substrate, is a highly specialized form. This is expressed by carrying a very heavy load on the hooves, a small angle of the anterior wall of the hoof (Fig. 1.4), and an extreme elongation and thinning of the metatarsal and metacarpal bones (Vereshchagin 1939).

Until recently, four subspecies were recognized (Groves 1969, 1985; Corbet 1978; East 1992):

G. s. hillieriana (Heude 1894: 245) (synonyms: *mongolica* Heude, *reginae* Adlerberg; *sairensis* Lydekker): like the *G. s. yarkandensis*, even shorter-horned, and very pale sandy in coat color. Range: Mongolian Gobi, Dzungarian Xinjiang, and Caidam (Groves 1985).

G. s. subgutturosa (*Antilope dorcas* var. persica Gray, glacicornis Stroganov, seistanica Lydekker, and typical Lydekker are synonyms): fawns have a reddish hue coat; facial stripes are present, although inconspicuous; females usually hornless; head-to-tail length in males is 110–116 cm and in females 93–110 cm; height of males is 64–75 cm and in females 60–68 cm; range, from eastern Turkey, the Transcaucasian region, Iran, Pakistan, and former Soviet Central Asia; intermediate phenotypes between *G. s. marica* and *G. s. subgutturosa* occur in Iraq (Groves 1985).

G. s. yarkandensis Blanford, 1875: 112: sandy-yellow coat; facial stripes well-developed; male has rather short, stout horns; females always hornless; range, deserts of the Tarim Basin, Xinjiang (Groves 1985).

G. s. marica Thomas, 1897: 162: very light in color, nearly white; the face and body markings nearly invisible; females usually possess horns; head-to-tail length is 99 cm; weight of adult male is 17.5 kg; range, the Arabian Peninsula (Groves 1985).

G. s. subgutturosa: large subspecies; head-to-tail length up to 126 cm and weight up to 34 kg for males compared to significantly smaller *G. s. marica* (head-to-tail length up to 99 cm and 20 kg in males). These two subspecies intergrade in Kurdistan and the lower Tigris-Euphrates Valley of Iraq (Groves and Harrison 1967). Ellerman and Morrison-Scott (1951) defined *Gazella marica* as a subspecies of *G. leptoceros*, and Lange (1972), on the other hand, defined the North African *Gazella leptoceros* as a subspecies of *G. subgutturosa*. Range: Central Asia, Iran, Afghanistan, and Pakistan.

The detailed subspecies' taxonomy of goitered gazelles was expanded by Groves (1985) and is recognized by many researchers (Granjon et al. 1991; East 1992; Kingswood and Blank 1996; Abduriyim et al. 2018; Fadakar et al. 2020a). Until recently, the Arabian sand gazelle (*G. s. marica*) or "reem" was recognized as a subspecies of *G. subgutturosa*. This subspecies (*G. s. marica*) is found in open habitats of the Arabian Peninsula and ranges through Iraq, Jordan, and Syria into southern Turkey (Groves and Harrison 1967; Kingswood et al. 1996, 1997; Mallon and Kingswood 2001). Arabian sand gazelles have a pale body and white face. Females bear long slender horns, which can be too short or absent in some populations (Gürler et al. 2015). Adult *G. s. marica* males have a less obvious developed goiter than in *G. s. subgutturosa* (Groves and Harrison 1967). At the beginning, the Arabian sand gazelle was described as a full species (*G. marica* Thomas, 1897), and later it was combined together with the slender-horned gazelle (*Gazella leptoceros*) from northern Africa (Ellerman and Morrison-Scott 1951). More recently, *Gazella marica* has been considered a subspecies of *G. subgutturosa* based on morphological and karyological similarities (Groves and Harrison 1967; Kingswood et al. 1996, 1997).

Very striking differences have been found between the skull of *Gazella leptoceros*, from the Sahara, and skulls of *Gazella marica* (Groves 1969). The shape of the lambdoid suture is rectangular in *G. marica* and forms angles at either end of its upper surface. The lambdoid suture of *G. leptoceros*, in contrast, is much more rounded, without any angles. Also, the horns of *G. marica* rise close together, protruding from the skull at an angle, and following a lyrate form; in cross section they appear somewhat circular. In *G. leptoceros*, the horns are more laterally flattened, rise parallel to each other, and are almost straight throughout their entire length; the horn ridges are very weakly marked in contrast to *G. marica*. The breadth of the palate also distinguishes *G. leptoceros* from *G. marica* (Groves and Harrison 1967), while skulls of *G. subgutturosa* are very similar (Rostron 1972). In all those features wherein *G. marica* differs from *G. leptoceros*, *G. subgutturosa* does not. The skulls of *G. marica* are considerably smaller than *G. subgutturosa*, but their proportions are the same (Groves and Harrison 1967; Hayatgheib et al. 2011).

In addition, there is a distinctive difference in throat swelling which is seasonally very distinctive in *G. subgutturosa* and almost inconspicuous in *G. marica*. Interestingly, though, in captivity, *G. marica* individuals also exhibit throat swelling (Groves and Harrison 1967). Arabian sand gazelles possess a goiter-like swelling in the throat region of males, and similar to Persian goitered gazelles (*G. s. subgutturosa*), it becomes more prominent during the rutting season (Cunningham and Wronski 2011).

Regarding the presence of horns in females, *G. subgutturosa* females are hornless, while *G. marica* females have long horns. There is a geographic trend of an increasing occurrence of horns in females from east to west: goitered gazelle females in Mongolia are hornless, some females have short horns (50–70 mm length) in Central Asia, while *G. subgutturosa* females have much longer horns in Iran (209 mm long). Even within Iran, it appears that horn length is further increasing west of the Zagros, while hornless females also occur in the same area; and in Saudi Arabia, the horn length is typically 226 mm long in *G. s. marica*. Skulls from this area (west of Zagros) are also intermediate in size between *G. marica* and *G. subgutturosa* (Groves and Harrison 1967). The skulls from the lower Tigris-Euphrates Valley also have mixed features between *marica* and *subgutturosa*: the size is intermediate, and horned females occur very commonly, some of which have very long horns. Further north, in Kurdistan, there exists a phenotype the size of *marica*, but females are hornless. The coat patterns and color match those of *subgutturosa*. This area seems to be the zone of intergradation and great variability. Considering all arguments, Arabian sand gazelles should be called *Gazella subgutturosa marica* (Groves and Harrison 1967).

However, more recently, Wacher et al. (2011) ran genetic analyses on *G. s. marica* from the Arabian Peninsula and *G. s. subgutturosa* from east of the Euphrates-Tigris basin. His team analyzed the sequence variation of a mitochondrial marker (cytochrome b) of samples from wild and captive Arabian sand gazelles, Persian goitered gazelles, as well as from museum specimens. In light of their findings, Wacher et al. (2011) reconsidered the genetics of the sand gazelle and asserted that the former subspecies of goitered gazelle is worthy of species status. According to their results *G. s. marica* is much closer genetically to the North African *G. leptoceros* and *G. cuvieri* than to *G. subgutturosa*. However, these unexpected results are conflicting with comparisons of skull structure and horn formation of these species (Groves 1969). In addition, the *G. s. subgutturosa* samples used in the study were of unknown provenance and taken from captive specimens that might have inaccurately influenced Wacher et al.'s (2011) results. Specifically, different gazelle species, including *Gazella leptoceros*, were kept and interbred in Saudi Arabian enclosures, and were thus not purebred. A similar scenario happened with taxonomical investigations of *G. saudiya* and *G. bennetti*, where genetic materials from captive interbred individuals were used (Hammond et al. 2001). Further, Wacher et al.'s (2011) results are clearly conflicting with morphological comparisons of skulls and horns by Groves and Harrison (1967), and morphological cues are not less important than genetic material when distinguishing species or subspecies. Moreover, specimens from the region between the Euphrates-Tigris basin and the

Zagros Mountains of Iran had a mix of characteristics, and this zone may be considered a hybrid zone (Groves and Harrison 1967; Kingswood and Kumamoto 1988; Groves 1997; Al-Robaae and Kingswood 2001; Fadakar et al. 2020a). Finally, twinning is rarely found among gazelle species, except in *G. cuvieri* and *G. subgutturosa* (Zhevnerov et al. 1983; Furley 1986), while Arabian sand gazelles produce twins with approximately the same frequency as singletons, though Vesey-Fitzgerald (1952) described a higher proportion of twins among newborns. In contrast, twins for *G. leptoceros* have not been recorded at all (Furley 1986).

All this suggests that the results of the mtDNA study cannot be considered unambiguously, considering the morphological and physiological evidence. The stable zone of the hybrid population between *G. s. marica* and *G. s. subgutturosa* in the Tigris-Euphrates area and the Zagros Mountains (central Iran) (Groves 1996; Fadakar et al. 2020a) speaks to the biogeographical proximity of these gazelles and confirms the subspecies differences of these gazelles. In contrast, a biogeographical proximity of *G. s. marica* and African *Gazella leptoceros* and *G. cuvieri* does not exist. Moreover, *G. cuvieri* is very dark colored and lives in Atlas mountain habitats, while *G. s. marica* is quite light and lives in sandy deserts. In summary, I think that the results of the genetic study of *G. s. marica* and other subspecies of goitered gazelles demand further scrutiny and research to obtain representative specimens on which genotyping can be done. After that, genetic researchers need to integrate mtDNA typing and morphological and other criteria to definitively categorize the status of the gazelle species or subspecies and explain their locations from the biogeographical point of view.

A recent revision of the previously recognized subspecies by Groves and Grubb (2011) led to some subspecies being elevated to full species status based on predominantly morphological data (coloration, horn shape, and nasal bone shape), such as the Yarkand gazelle (*G. yarkandensis*) in China and Mongolia including *G. s. hillieriana*, *G. s. sairensis*, *G. s. reginae*, and *G. s. mongolica*; the Turkmen goitered gazelle (*G. gracilicornis*) in Turkmenistan, Kazakhstan, Tajikistan, and Uzbekistan; and the nominative subspecies, the Persian gazelle (*G. s. subgutturosa*) in Iran, Iraq, Azerbaijan, and Turkey (Fadakar et al. 2020a). But most of these species, however, were again based on a very small selection of individuals (Bärmann 2014). Furthermore, recent molecular studies show that differences between these new species are hardly detectable (Abduriyim et al. 2018), so a distinction on the species level might not be appropriate after all. Indeed, molecular data somewhat support the subspecies patterning within *G. subgutturosa* (Fadakar et al. 2020a). Abduriyim et al. (2018) analyzed the genetic diversity in mitochondrial DNA from 104 wild samples from the Xinjiang Uyghur Autonomous Region (XUAR) in northwestern China and reconstructed the phylogeny with additional sequences from across the species' range. He found two subspecies in Xinjiang, *G. s. hillieriana* in northern XUAR and *G. s. yarkadensis* in southern Xinjiang in China.

Most recently, Fadakar et al. (2020a) found three clades of goitered gazelles in Iran: the Middle Eastern, central Iranian, and Asiatic. However, geographic barriers and population connectivity strongly contradict the conclusions of these genetic investigations. For example, the Middle Eastern clade is found on both sides of the great Zagros Mountain range, despite it being impassable for goitered gazelles, and reaches the northern border with Turkmenistan, while the central Iranian clade mixes with the Middle Eastern clade in central Iran, and the Central Asian clade mixes with the Middle Eastern clade in northeastern Iran (Fadakar et al. 2020a). Such contradictions of genetic data and geographical distribution suggest that differences between these clades were overestimated, that genetic differences are insignificant between populations, and also that the same nominative subspecies (*Gazella subgutturosa subgutturosa*) lives both in Central Asia and Iran. Otherwise, it is difficult to explain the intermixture of different clades in different parts of Iran. In addition, quite recently Karami et al. (2002) stated that the presence of any other subspecies other than *G. s. subgutturosa* in Iran is doubtful, and craniometrical analysis of Iranian goitered gazelles by Hemami (1994) supports this conclusion, though, later, Hemami (2020) joined to the general tendency to create new species and subspecies (details below).

Based on the DNA analyses, Bärmann (2014) affirmed that new species proposed by Groves and Grubb (2011) belong to the same species *Gazella subgutturosa* and gave *G. s. marica* the status of a separate species, including *G. subgutturosa*, *G. bennetii*, and *G. marica*, for one clade of goitered gazelles. This is quite logical, but they also included two African gazelle species, *G. leptoceros* and *G. cuvieri*, in the goitered gazelle clade that completely contradicts zoogeographical data. Genetic materials for this species were taken from captive individuals from the San Diego Zoo and KKWRC (King Khalid Wildlife Research Center). In Saudi Arabia, many gazelle species from different localities of Africa and Asia interbreed without any records to document the results, which casts doubts on the purity of any genetic analyses. Thus, it is of paramount importance to obtain genetic materials from wild, undisturbed populations, if possible; otherwise DNA analyses and measurements of museum materials can be problematic.

New studies have created distribution maps based on the identification of habitat suitability and ecological niches of different subspecies of *G. subgutturosa* (*G. s. subgutturosa*, *G. s. yarkadensis*) and *G. s. marica* and their hybrid zones (Hemani et al. 2020). In my opinion, the ecological preferences of different subspecies of gazelle have not been studied well enough to distinguish the ecological niches of one subspecies from another. Secondly, following the erroneous claim of Groves and Grubb (2011), the names of subspecies do not correspond at all to their distribution. Third, determining the habitat suitability based on ten factors is a completely meaningless activity. It is much more productive to consider the habitats where one or another subspecies lives now or where it lived before. As a result of this mistaken approach, a map was presented of the distribution of the different gazelle subspecies (Hemami et al. 2020; Fig. 4), where *G. marica* appears in most of Iran, Afghanistan, Pakistan, and even in India and neither *G. marica* nor *G. subgutturosa* was found. It seems a more confusing representation of such a distribution is difficult to imagine and impossible to explain.

The purpose of this book is a detailed description of the former and modern study of this species. I cannot confirm the subspecific taxonomy of goitered gazelles but can only state that we currently have insufficient knowledge about this species' genetics, and urge scientists to continue these studies in order to bring full clarity to their subspecific taxonomy and distribution.

According to the zoogeographical point of view, I think that Arabian sand gazelles (*G. s. marica*) live in southern Iran, to the west of the Zagros Mountains, while *G. s. subgutturosa* inhabits the rest of the Iranian territory to the east of the Zagros Mountains; and morphological data support this assumption (Hayatgheib et al. 2011). The Zagros Mountains are an impassable barrier between two goitered gazelle subspecies. *G. s. subgutturosa* habitat also covers the entire area of Central Asia (Turkmenistan, Tajikistan, Uzbekistan, Kazakhstan). There are no geographical barriers within populations of Central Asia, as well as between the Central Asia and Iranian goitered gazelle populations that are east of the Zagros Mountains. As of quite recently, goitered gazelles freely migrate all over Central Asia and to Iran and back. *G. s. yarkadensis* lives only around the Takla Makan Desert, along the Tarim River, while *G. s. hillieriana* is in the Dzungarian Xinjiang and Mongolian Gobi (Fig. 1.14—Map). These subspecies are divided by the eastern Tien Shan Mountains. Apart from Saudi Arabia, *G. s. marica* inhabits Syria, Jordan, Iraq, and Turkey, forming hybrid populations in the eastern borders of its natural habitat. Genetic analysis of samples from the Turkey gazelles currently in the VPA (Valproic acid) carried out at Lisbon University has confirmed that they are genetically closer to *G. [s.] marica* (Carlos Fernandes in litt. 2012). Arabian sand gazelles also migrated within this area freely, because of a lack of geographical barriers between these countries. In Azerbaijan, there are two genetic clades, one from Turkey (apparently *G. s. marica*) and another from Iran (*G. s. subgutturosa*); found to the north from the Great Caucasus Mountains is *G. s. subgutturosa* (Fig. 1.14—Map) (Sorokin et al. 2011).

Arabian sand gazelles are a subspecies of goitered gazelles or already a separate species—it remains to be understood based on additional genetic investigations in the future. Meanwhile, in this book, I am sticking to older ideas about the taxonomy of this species and consider the Arabian sand gazelle (*G. s. marica*) as a subspecies of goitered gazelles (*G. subgutturosa*), at least temporarily, though I will try to separate information related to *G. s. marica* everywhere in the text.

1.2.2 Genetics

Goitered gazelles have 2n = 30–33 chromosomes, 26–28 metacentric and 0–5 acrocentric, with a submetacentric X and an acrocentric Y (Effron et al. 1976; Hsu and Benirschke 1977; Kingswood and Kumamoto 1988: Vassart et al. 1993). Variation in diploid number is the result of two chromosomal rearrangements, a putative centric fusion and an autosome-to-X translocation. The autosome-to-X translocation results in males having an unpaired autosome; the autosome involved in this rearrangement is the same one involved in the autosome-to-X translocations in *G. dama* and *G. soemmerringii* (Vassart et al., 1993). The centric fusion occurs in homozygous (females, 2n = 30; males, 2n = 31) and heterozygous (females, 2n = 31; males, 2n = 32) forms; individuals lacking the fusion are 2n = 32 (females) or 33 (males).

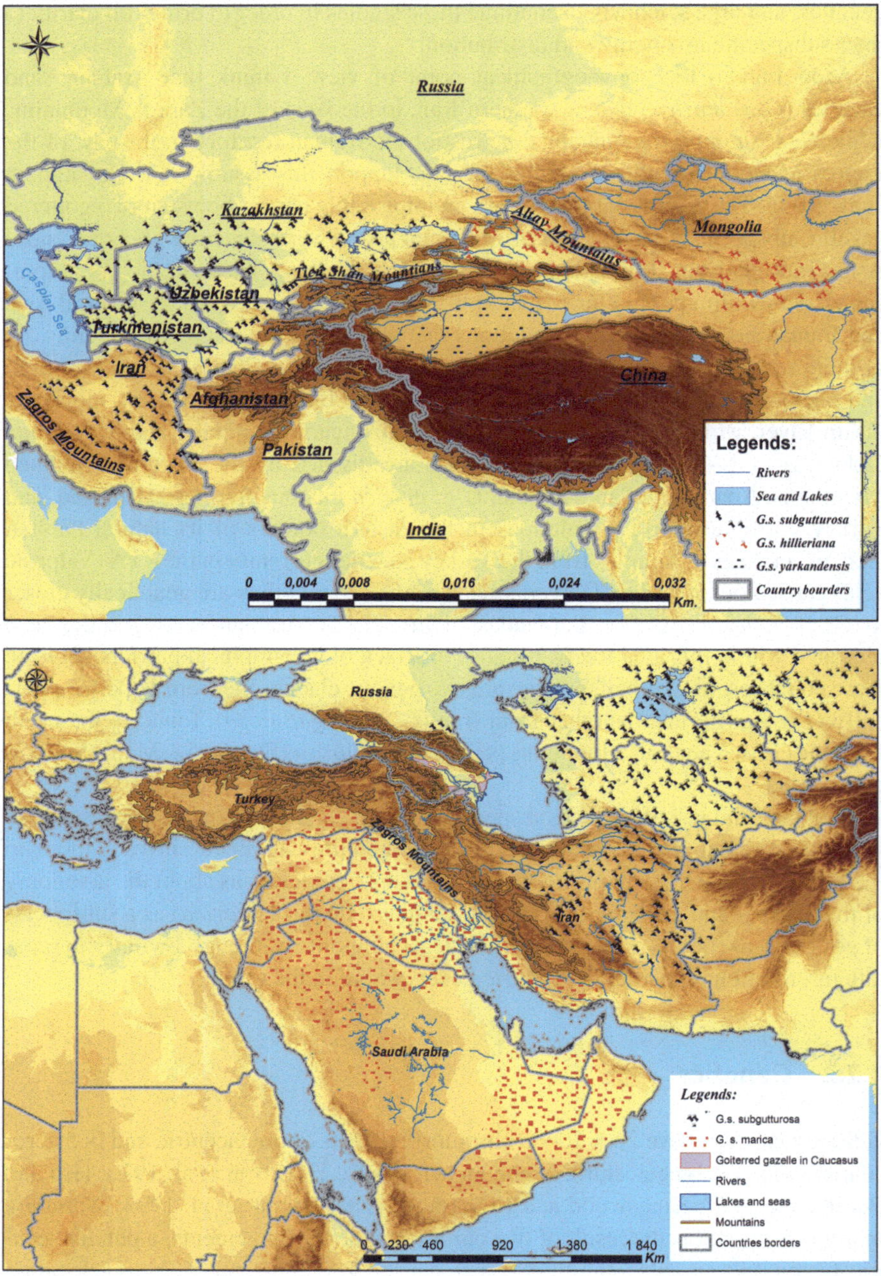

Fig. 1.14 Previous distribution of different subspecies of goitered gazelle according to zoogeographical points of view; taxonomical subspecies status of the goitered gazelle in the Caucasus is still undetermined (prepared by A. Osmonov with kind permission)

Specimens of putative *G. s. subgutturosa* are fusion homozygotes (2n = 30, 31); in *G. s. marica*, the fusion may be absent, or in homozygous forms (Kingswood and Kumamoto 1988; Vassart et al. 1993).

According to Granjon et al. (1991) and Vassart et al. (1993), the sand gazelles from Thumamah, Al Qassim, Jordan, and Qatar exhibit a variable number of chromosomes. The males have 31–33 chromosomes, while females have 30–32. A possible explanation for these differences includes (1) hybridization between *G. subgutturosa subgutturosa* and *G. s. marica*; (2) a geographic cline in chromosome numbers between northern and southern populations; or (3) natural chromosome polymorphism. Two wild males caught in northern Saudi Arabia exhibited 33 and 32 chromosomes (Hague and Smith 1996).

G-banding patterns of *G. s. subgutturosa* (2n = 30, 31) and *G. s. marica* (2n = 30, 31) are homologous (Kingswood and Kumamoto 1988). Most of the autosomes exhibit pericentromeric heterochromatin and a few have terminal G-bands; the Y and the short arm of the X are largely heterochromatic (Kingswood 1992). Neither centric fusion nor autosome-to-X translocation appears to reduce the meiotic fitness of male goitered gazelles (Kingswood et al. 1994). Individuals heterozygous for centric fusion are fertile and produce offspring when crossed with any of the cytotypes (Kingswood and Kumamoto 1988). The distribution of frequencies for different chromosome numbers conforms to the Hardy-Weinberg equilibrium (Vassart et al. 1993). *G. subgutturosa* has hybridized with *G. gazella* (*bennetti*), *G.g. gazella*, and *G. subgutturosa marica* (Gray 1972; Roberts 1977; Kingswood and Blank 1996). In a captive population of *G. s. marica*, the percentage of polymorphic loci (n = 20) and mean heterozygosity were 15% and 0.017, respectively (Granjon et al. 1991).

Analysis of the D-loop of mitochondrial DNA of goitered gazelle groups from Uzbekistan, Turkmenistan, and Azerbaijan provides evidence for the similarity of goitered gazelle populations from Turkmenistan and Uzbekistan. Haplotypes specific to animals from these territories are combined into one cluster. Populations from Ogurchinskii Island and Badkhyz Reserve in Turkmenistan are not genetically isolated. The population of the Shiranskaya steppe reserve from Azerbaijan is unique and strictly isolated from others. Analyses of the general sample including the data from NCBI on the goitered gazelles from Iran show that the sample of animals from Azerbaijan divides into two groups: one is cognate to a group from northwestern Iran, and the other is unique. Thus, the supposition on the isolation of the Transcaucasian population of gazelles from Central Asian populations based on analysis of morphological features (Heptner et al. 1961) agrees with the genetic data. The study data indicate that a part of the founding animals come from Iran. Nevertheless, at present, this isolated population is unique and requires additional protection (Sorokin et al. 2011).

G. subgutturosa can hybridize in captivity with *G. gazella gazella* and *G. leptoceros* (Gray 1972; Roberts 1977). Hybridization between *G. s. subgutturosa* females and *G.g. gazella* males resulted in sterile male offspring; the fertility of female offspring was not tested (Kingswood and Blank 1996). In the genus *Gazella*, the only record of hybridization in natural conditions is the introgression of mitochondrial

DNA between *G. marica* and *G. subgutturosa* in eastern Turkey (Murtskhvaladze et al. 2012). Morphologically, intermediate individuals, probably natural hybrids between the two nominal forms (Groves 1997; Groves and Harrison 1967; Mallon and Kingswood 2001), occur in a vast area between the Tigris-Euphrates Valley and the Zagros Mountains, but the only molecular study that included samples from this region (five captive individuals from the Rutba region in Iraq; Wacher et al. 2011) only detected mitochondrial sequences of *G. subgutturosa*. However recently, hybridization of two different species of gazelles, *G. bennettii* and *G. subgutturosa subgutturosa*, was found in natural conditions of central Iran. The hybrids have a nuclear genomic identity (based on two introns), habitat preference and phenotype of *G. bennettii*, but mitochondrial identity (based on cyt b) of *G. subgutturosa* (Fadakar et al. 2020b).

1.3 Morphology

1.3.1 Hair

Goitered gazelles have three types of hair that are differentiated in the coat: (1) protective, which are the coarsest and longest hairs located on the tail, in the area of the carpal and interdigital glands; (2) tactile or sensitive whiskers (vibrissae), which do not undergo change during molting; and (3) integumentary, or guard hairs, which cover the entire body of the animal and change during molting. Vibrissae are located on the upper and lower lip, lower eyelids, and eyebrows and have two types of coloring—white and black, with black ones usually being longer. The hair cover of the gazelle is quite thick (Table 1.1) and the hair covering the various parts of the body is of mixed lengths. A shorter and stiffer hair covers the muzzle, the legs below the wrist joint and hock joints, and the outer side of the auricles. On the inside of the auricles, the hair is arranged in separate longitudinal stripes that border four bare areas reaching half of the auricle (Fig. 1.15). The covered areas have a hair length of 15 to 20 mm and shorten at the top of the ear to 3–6 mm, becoming equal to the main hairline of the outer side of the ear (Zhevnerov 1984).

Table 1.1 The number of hairs per 1 cm^2 in adult male and female goitered gazelles (*G. s. subgutturosa*) in winter and summer on Barsakelmes Island, Aral Sea, Kazakhstan (adapted from Zhevnerov 1984)

	Winter				Summer
	Males ($n = 4$)		Females ($n = 2$)		
Body part	Range (min–max)	Mean	Range (min–max)	Mean	Male ($n = 1$)
Upper back (withers)	2340–3800	3070	2400–3500	2950	1200
Body flanks	2100–2300	2200	2250–2600	2420	1000
Belly	3270–4000	3730	3560–4050	3840	1500

Fig. 1.15 On the inside of the auricles, the hair is arranged in separate longitudinal stripes that border four bare areas reaching half of the auricle. Issyk-Kul goitered gazelle nursery for re-acclimatization, 25.06. 2022. Photo by D. Blank

The hair cover of the gazelle is relatively soft and elastic, but the ends of the hair are brittle. There is no clear division between the outer hair and undercoat. There are four orders of different lengths of the guard hairs on the back (Sokolov 1973). The hair does not form into groups and bundles but does grow in faintly discernible rows running across the body. These rows of hair can merge one into another or be broken by another row. The hair has a well-developed core layer (up to 88.7%), and the shape of the cuticle cells in the hair of all orders is similar. The first-order guard hairs generally have a slight expansion in the upper part of the rod. Across, the hair is bean-shaped—convex in front and slightly concave in back. The core runs all along the hair, except for the top and base. There are two to four cuticle cells across the hair and the hair rods are arched back and have no convolutions (Zhevnerov 1984).

In winter, the hair lengthens all over the body and the longest hair is on the back, sides, belly, groin area, and white rump patch (Fig. 1.16; Table 1.2); the udder, except for the nipples, is also covered with rather long soft hair (Sokolov 1959). In summer, the hair cover is less and shorter (Fig. 1.17), but the order of hair arrangement by density and length remains in the same sequence as in winter. Consequently, those parts of the gazelle's body that are more often exposed to wind, cold, and insolation are most well protected (Zhevnerov 1984). In this regard, the hair coat in adult gazelles is longer and denser on the back, in contrast in fawns' hair cover, which is longer and denser on the belly. This phenomenon is better for the thermal insulation of fawns' bellies because they lie down so much during their first days after birth and their bellies have longer and closer contact with the cold spring ground (Soldatova and Grazhdankin 1989). In addition, the hair cover of the gazelle

Fig. 1.16 In winter, the longest hair is on the back, sides, belly, in the groin area and a rump white patch. Issyk-Kul goitered gazelle nursery for re-acclimatization, 25.11.2022. Photo by D. Blank

has no undulation, so individual hairs do not cling to each other well and are unstable against blowing wind. The weak thermal insulation properties of their winter fur are also indicated by the fact that the snow melts on the ground under the gazelles as they lie to rest; therefore, goitered gazelles shovel snow before they bed down (Fig. 1.18) (Zhevnerov 1984).

1.3.2 Skin Glands

Goitered gazelles usually have two inguinal mammae, but occasionally four develop. When two pairs of mammae are present, the inguinal glands are connected with the posterior pair, indicating that the anterior pairs are supernumerary (Sokolov 1959). Goitered gazelles possess inguinal, carpal, interdigital (pedal), and preorbital glands (Pocock 1910; Sokolov 1959). Both males and females have one pair of inguinal glands in the form of narrow pockets about 3–5 cm deep, open laterally to the mammae. The two inguinal glands, with pouches extending 3–5 cm into vascular tissue above the mammae, open laterally to the mammae. Inguinal glands produce a yellow, waxy substance with a musky, aromatic odor. The carpals are composed of thickened skin and sebaceous components and are located beneath a knee-tuft of long, coarse, dark hair. These glands secrete yellow, greasy secretions to the touch. From the outside, these glands are distinguished by longer and coarser bundles of dark-colored hair compared to neighboring areas. The bases of the pedal gland hair

Table 1.2 The length of winter and summer hair coats (in mm) of adult male and female goitered gazelles (*G. s. subgutturosa*) from Barsakelmes Island, Aral Sea, Kazakhstan (adapted from Zhevnerov 1984)

Measurements (mm)	Winter Males (n = 5) Range (min–max)	Females (n = 4) Range (min–max)	Summer Males (n = 1)	Females (n = 1)
Muzzle top	4–5	4–5	3	3
Cheeks	25–30	20–30	8	8
Forehead	30–45	35–40	15	15
Ear outside Inside	3–6 15–20	4–5 15–20	4 17	4 16
Neck top Bottom	20–40 25–45	30–40 30–45	10 12	10 12
Body back	45–50	40–55	20	20
Body flanks	45–50	45–55	25	25
Belly middle Flank	40–45 65–70	50–65 65–70	15 25	15 25
Hip	35–45	40–45	10	10
Breast	50–60	55–60	20	20
Front legs: top Bottom Internal side of front legs: top Bottom	20–30 6–10 20–30 6–10	25–30 8–12 25–30 8–12	8 6 10 6	8 5 10 6
Hind legs: top Bottom Internal side of hind legs: top Bottom	40–50 15–20 40–50 15–20	35–50 15–20 35–50 15–20	15 7 15 10	15 8 15 10
Groin area	70–90	80–90	25	25
Rump patch: edge Center	65–70 30–40	65–75 30–40	10 5	10 5
Vibrissae on upper lip: black White On eyebrow Under eye	15–25 10–20 15–25 20–25	15–25 12–20 15–25 20–25	15 10 20 25	20 15 25 25
Eyelashes	10–15	10–15	10	10

are glued together by the secretions of the glands. All four legs have interdigital (carpal) glands that open on the front side of the headstock above the hooves with a longitudinal slit (Pocock 1910; Sokolov 1959). There are well-developed preorbital glands, consisting of a black integumental thickening formed around an invaginated pocket of skin; an elongated slit, located below and in front of each eye, excludes a black tarry substance (Pocock 1910). Females and males have preorbital glands, which are very large in males but are usually smaller in females (Walther 1984).

Fig. 1.17 In summer, the hair is relatively short all over the body, but the abdominal part is most well protected. Issyk-Kul goitered gazelle nursery for re-acclimatization, 25.06. 2022. Photo by D. Blank

Fig. 1.18 Male goitered gazelle tear the snow to the ground for a resting place. Kalamali Ungulate Nature Reserve, Xinjiang, China; 15.12. 2022. Photo by F. Xu with kind permission

Fig. 1.19 The body underside (belly and inner parts of legs) of goitered gazelles is purewhite. Young males (one-year old) in the feeding area. Issyk-Kul goitered gazelle nursery for reacclimatization, 25.06. 2022. Photo by D. Blank

Males occasionally open the lachrymal pits and display the black lining, which is very conspicuous on the light face (Kingswood and Blank 1996).

1.3.3 Color

Seasonal dimorphism in the color of the goitered gazelle is faintly expressed, with only some individual- and age-related color variability. In the winter, the color of the upper part of the body and neck is light brown or sandy gray. The general background of the coloring is presented by the pre-vertex yellow areas of hair, as well as individual white hair and the density of inclusions, which in different individuals has an individual character. The underside of the chest, neck, and back of the thighs on both sides of the tail (rump patch) are pure white (Fig. 1.19). Along the upper edge of the rump patch, there is a strip of dark-brown hair frames on both sides. The belly and inner parts of the legs are white. On the flanks of the body, along the border with the white color of the abdomen, almost at the level of the elbow and knee joints, a narrow strip of dark-brown hair stands out noticeably. A narrow strip of light-colored hair runs along the center of the lower side of the neck, which merges with white hair on the chest (Zhevnerov 1984).

The coloration of the outer hairs of the upper body is heterogeneous. On the lower part, the hair color at a length of 10 mm is almost white and, then, at a length of 15–20 mm, smoky. The apical part of the hair at a length of 2–5 mm is dark brown. In winter, the dark-colored tops of the hair break off, which leads to some lightening of the general background color of the upper body. On the front legs, a narrow strip of light sandy color runs down from the elbow joint, which expands

Fig. 1.20 The facial region of young goitered gazelles has many black spots and stripes that disappear with age. Female at the age of 20 months. Issyk-Kul goitered gazelle nursery for re-acclimatization, 24.11. 2022. Photo by D. Blank

lower to the wrist joints and acquires a light red color. In the area of the carpal glands, bundles of coarse tightly fitting hair with a length of 30–40 mm are noticeable, having two types of coloring—dark brown and gray. There are areas of dark brown hair on the interdigital glands and between hooves. The inner surface of the front legs from the elbow to the wrist joints has a white and smoky color, changing from the border of the wrist joints to gray with a brown tinge. The outside of the hind legs has a color from sandy to dark gray. In the thigh area, between the dark stripe that separates the white rump patch and the main color of the body, an illuminated triangular-shaped area stands out noticeably. The inner surface of the hind legs is white. The tail is a dark brown, caused by both black and dark brown hair (Zhevnerov 1984).

The gazelle has a fairly well-defined age-related coloration of the head and, above all, the facial region. In young individuals, the middle part of the forehead, neck, and tail from above are dark sandy and light brown. There are a lot of dark brown, almost black guard hairs on the upper part of the neck and all over the back, which makes the general background color of these parts of the body appear pockmarked. The most unusually colored part is the head, especially the front part of the muzzle, where black and white spots and lines create a complex pattern (Fig. 1.20). The lower lip is creamy white. A lighter stripe runs along the middle line of the nose, which ends with a black spot about 3–4 cm above the nostrils. Its size, color intensity, and the time of appearance and disappearance depend on the age of the gazelle. In newborns, the dark spot on the bridge of the nose stands out slightly and has blurred contours. The intensity of coloration increases rapidly during the first 2 months of life. By this time, the spot grows, becoming almost rectangular, up to 4 cm in size, and in this form, with minor variations, it persists until 8–9 months of

Fig. 1.21 Black spots and stripes on the facial region disappear with age. Female at the age of 2 years. Issyk-Kul goitered gazelle nursery for re-acclimatization, 25.06.2022. Photo by D. Blank

age. During the first spring molt, the spot decreases and brightens. By 18–19 months, it becomes a faintly noticeable dark ashy colored area of 1–2 cm in size. At 2.5–3 years of age, the spot disappears completely (Fig. 1.21). However, the disappearance of facial spots does not occur simultaneously in different individuals, and some gazelles gradually lose dark spots starting from the age of 1 year, while in other gazelles this process is delayed until the end of the second year of life. Therefore, Zhevnerov's statement (1984) that the presence of a large black spot on the muzzle above the nose is characteristic of all yearlings, and its absence suggests that this is an adult gazelle older than yearlings, is not entirely true, because the color of the muzzle also has individual variability and different individuals of the same age have different numbers and intensities of black stripes and spots (Fig. 1.22). Bands of dark-brown color stretch from the folds of the preorbital glands to the corners of the mouth, which also decrease and lighten with age. In older individuals, the color of the front part becomes white, except for darkened stripes in front of the eyes (Zhevnerov 1984). The older the individual, the paler the pattern on the muzzle: the head seems to turn gray and in older males, it is almost white (Ishunin 1961). The outer coloration of the ears is due to two types of differently colored hair—light sandy, which makes up the main hairline, and rarely occurring dark brown. The general background color of the outer side of the ears is sandy gray, and the inner side is white (Zhevnerov 1984). The tail is sharply bicolored: the main part of it is sandy brown on top, and the end is black or dark brown and sometimes carries a bunch of whitish hair. The hair covering the tail is stiff and longer than on the body (Heptner et al. 1961).

The summer coat color of the gazelles is somewhat brighter, and the border between dark- and light-colored areas stands out more clearly. The top of the head,

Fig. 1.22 Two females at the same age (19 months) have different intensities of black spots and stripes on the facial parts of their muzzles. Issyk-Kul goitered gazelle nursery for re-acclimatization, 24.11. 2022. Photo by D. Blank

neck, and body are bright sandy or light brown. On the sides of the body, above the dark-brown stripe passing at the level of the elbow and knee joints, a light ashy narrow stripe stands out. The outer side of the forelimbs is monochrome with the top of the body, while the inner side is light ashy and white. The hind limbs are light sandy on the outside and white on the inside. The light areas on the hips stand out more clearly in comparison with the winter colors (Zhevnerov 1984).

In newborn fawns, the upper part of the body is a continuous ashy or smoky color, and the belly and inner side of the limbs are white. The chest, in the area of the front legs, is pure white. On the sides of the body, a faintly pronounced red stripe is visible, which goes to the hips and borders the rump patch. Along the body ridge, especially in the area of the shoulders, a dark ashy strip 1–2 cm wide stands out noticeably, losing its color intensity at the level of the sacrum. The coloring of the head is a mosaic. The back of the head is dark ashy, interspersed with light sandy hair. The end of the muzzle is brown. There is a longitudinal smoky stripe 5–7 mm wide on the bridge of the nose. Dark blurry stripes stretch from the preorbital glands to the corners of the mouth. The cheeks have a light sandy color. The brush of the tail is black, and the rest of it is the same color as the top of the body (Zhevnerov 1984). On the back of the body, in the tail area, there is a small creamy white rump patch that does not enter the croup. The hooves are black, and the irises of the eyes are dark brown (Sludskiy 1956a).

Pelage coloration also varies geographically. *G. s. marica* varies from nearly white to light sandy yellow; *G. s. subgutturosa* is light brown dorsally with variable tones of gray, red, or yellow; and *G. s. hillieriana*, *G. s. reginae*, *G. s. seistanica*, and *G. s. yarkadensis* are light sandy yellow with less gray or red. Facial coloration and markings vary geographically and tend to whiten and fade with age (Groves 1985; Lydekke and Blaine 1914; Sokolov 1959; Zhevnerov et al. 1983). In *G. s. yarkadensis* and young *G. s. subgutturosa*, facial stripes are fairly distinct; *G. s. marica* has a white face that lacks markings (Kingswood and Blank 1996).

Roberts (1977) asserted that the light-colored pelage of gazelles is an important part of their overall adaptive strategy, decreasing heat absorption during hot summer days. A reduction of conductivity of two- to threefold for water vapor is observed in the pelage of goitered gazelles, as well, when a layer of still air appears between hairs restraining evaporation. Further, the more developed the hair layer, the greater the obstacle to evaporation and the lower the overall conductivity of the outer hair coating (Soldatova and Grazhdankin 1989).

1.3.4 Body Sizes

Gazella subgutturosa is generally larger and has a heavier build than other species of Asian gazelles. The size and weight of goitered gazelles from different regions of Kazakhstan are homogeneous (Tables 1.3, 1.4, and 1.5). There are similar sizes of goitered gazelles in Uzbekistan where their body length is 94–120 cm, body height is 67 to 80 cm, and the weight is 30.1 kg on average (range 24.2–36 kg) in males and 26.7 kg on average (range 20.2–32.2 kg) in females (Mambetzhumaev 1970; Prisiazhnuk and Soldatova 1985). Only the ears and tails are slightly longer in Uzbekistan than in Kazakhstan: ear length is between 12 and 16 cm and tail length

Table 1.3 Average (and range) of different body measurements (cm) and weights (kg) of adult male and female goitered gazelles (*G. s. subgutturosa*) from the Ustyurt Plateau, Western Kazakhstan (adapted from Mambetzhumaev 1970)

	Males ($n = 7$)		Females ($n = 9$)	
Indices	Range (min–max)	Mean	Range (min–max)	Mean
Body length	105–120	115	106–113	112
Head length	29–31	29.8	23–29	27.4
Tail length	12–20	16.6	10–20	16
Ear length	13–14	13.3	13–14	13.6
Slant body length	69–77	72.7	62–75	70
Shoulder height	64–79	71.2	59–73	68.8
Sacrum height	73–83.5	80	70–87.5	77.6
Chest circumference	72–79.5	76.5	71–78	75.2
Weight	24.2–29.6	27.2	20.2–33.2	24.2

Table 1.4 Average (and Range) of different body measurements (cm) and weights (kg) of adult male and female goitered gazelles (*G. s. subgutturosa*) from Barsakelmes Island, Aral Sea, Kazakhstan (adapted from Zhevnerov 1984)

	Males (n = 32)		Females (n = 10)	
Indices	Range (min–max)	Mean	Range (min–max)	Mean
Body length	107–126	118	101–119	108
Front leg length until elbow joint	46–52	47.0	44–51	46.0
Head length	27–31	29	25–29	27
Tail length	15–18	16	14–17	16
Ear length	13–15	14	15–16	14
Slant body length	62–76	70	57–74	69
Shoulder height	67–77	75	62–74	70
Sacrum height	71–84	82	68–81	75
Chest circumference	72–78	75	69–74	72
Weight	22–26	25	20–23	22
Hoof length front Hind	4.8–5.7 4.6–5.3	5.2 4.9	4.5–5.2 4.4–4.9	4.8 4.6
Hoof height front Hind	2.8–3.4 2.6–3.4	3.0 2.9	2.6–3.0 2.4–2.8	2.7 2.6
Hoof width front Hind	2.9–3.4 2.7–3.2	3.1 2.9	2.4–3.0 2.4–2.8	2.7 2.6

Table 1.5 Average (and range) of different body measurements (cm) and weights (kg) of adult male and female goitered gazelles (*G. s. subgutturosa*) from Betpak-Dala Desert, Central Kazakhstan (adapted from Sludskiy 1956a)

	Males (n = 18)		Females (n = 50)	
Indices	Range (min–max)	Mean	Range (min–max)	Mean
Body length	111–125	119	102–120	109
Head length	27–29	–	24–29	–
Tail length	16–21	–	10–23	–
Ear length	12–14	–	13–14	–
Slant body length	64–66	–	61–70	–
Shoulder height	69–79	77	68–76	71
Sacrum height	79–85	84	74–83	77
Chest circumference	71–81	–	67–75	–
Weight	27–33.8	30	21–33	24

is 17 to 20 cm. The sexual dimorphism in weight is also very small, and males are only slightly heavier than females in Uzbekistan (Mambetzhumaev 1970).

On the Barsakelmes Island, adult males weigh from 20 kg to 43 kg, on average, 36 kg (n = 15), and females from 18 to 28 kg, on average 24.2 kg (n = 14). Adult

Table 1.6 Monthly (May to October) weight changes in adult male and female goitered gazelles (*G. s. subgutturosa*) in the Betpak-Dala Desert, Central Kazakhstan (materials of 1949, 1955–1956—adapted from Sludskiy 1956a)

Months	Sex	Weights in kg		Number of individuals
		Min–max	Mean	
May	Males	29.9–33.1	31.2	5
	Females	21.0–33.0	27.0	58
June	Males	27.7–30.0	–	2
	Females	25.0	–	1
July	Males	–	–	–
	Females	20–25	23.3	6
August	Males	27.0–30.5	–	2
	Females	22.0–26.0	24.8	3
September	Males	–	–	–
	Females	24.0–31.0	26.5	4
October	Males	33.8	–	1
	Females	21.5–28.0	–	2
			Total	84

male gazelles from Betpak-Dala Desert have varying weights from 27 kg to 33.8 kg, and females vary from 21 to 33 kg, with females especially heavy in May when they are born. The weight of their embryos was often 5 to 6 kg. Seasonal weight changes are presented in Table 1.6. In the Ili Hollow (SE Kazakhstan), males weigh from 20 to 28 kg, for an average 25.3 kg ($n = 10$), and females from 20 to 26 kg, for an average 24.1 kg ($n = 11$), in January 21, 1949 (Sludskiy 1956a). In goitered gazelle, body weight loading on hooves is 491 g/cm^2 for males, 434 g/cm^2 for females, and 340 g/cm^2 for young, compared to 140–180 g/cm^2 for *Rangifer tarandus* male (Sokolov 1959; Sludskiy 1977; Zhevnerov 1984). This high load bearing on the hooves does not allow goitered gazelles to move over deep snow (Sludskiy 1963; Kingswood and Blank 1996).

As can be seen from Tables 1.3, 1.4, and 1.5, the individual variability of sizes in goitered gazelles can be quite large. The body length of males is often longer than that of females, but there are females with significantly longer body lengths than males. The difference between the main linear indicators of males and females is insignificant, which indicates a weakly expressed sexual dimorphism in body size among adults. Regarding weight indices, it should be noted that the live weight, especially of males, varies dramatically by season. The weight indicators are given for animals studied in December–March, so the absolute weight of both males and females turned out to be somewhat underestimated (Table 1.4). Normally, on the Barsakelmes Island, the average weight of males is 30 kg, and in females, it is slightly lower at 24 kg (four gazelles each) (Zhevnerov 1984). There are also no sharp differences in the proportions of the physique of males and females (Table 1.7). Comparing body measurements of gazelles from different mainland and island populations, only minor differences in linear sizes are observed, not exceeding the usual individual variability (Tables 1.3, 1.4, and 1.5).

Table 1.7 Indices of physique and ratio of measurements (%) of adult male and female goitered gazelle (*G. s. subgutturosa*) on Barsakelmes Island, Aral Sea, Kazakhstan (adapted from Zhevnerov 1984)

Indices	Ratio of measurements	Males	Females
Massiveness	Chest circumference × 100 Slant body length	107.0	103.8
Overgrowth	Sacrum height × 100 Shoulder height	104.5	105.0
Leggy	Leg height at elbow × 100 Slant body length	65.6	67.4
Relative head size	Head length × 100 Shoulder height	40.3	38.8
Format	Slant body length × 100 Shoulder height	97.0	98.5

The species *G. subgutturosa* is sexually dimorphic yet; males are larger in some extent and have longer horns than females. Means (and ranges) of external measurements (in mm) of adult males and females, respectively, are as follows: total body length, 1.158 (940–1260, *n* = 68) and 1097 (940–1200, *n* = 74); length of tail, 168.2 (120–210, *n* = 75) and 161.5 (100–230, *n* = 80); length of hind foot, 333.2 (232–350, *n* = 10) and 314.0 (270–335, *n* = 5); length of ear, 135.6 (116–150, *n* = 82) and 136.3 (120–160, *n* = 85); length of horn, 300.0 (203–400, *n* = 142) and 115.6 (33–226, *n* = 7); height at shoulder, 727.4 (580–790, *n* = 67) and 691.6 (560–765, *n* = 77); height at sacrum, 820 (710–840, *n* = 57) and 763 (680–875, *n* = 69); chest girth, 756.2 (700–810, *n* = 60) and 733.0 (670–780, *n* = 71); and body weight (kg), 27.4 (22.0–33.8, *n* = 60) and 23.2 (17.5–33.2, *n* = 72) (Groves and Harrison 1967; Harrison 1968; Harrison and Bates 1991; Klockenhoff 1969; Mowlavi 1978; Zhevnerov et al. 1983; Kingswood and Blank 1996).

Means (and ranges) of external measurements (in mm) of *G. s. marica* males indicate their smaller size (and greater horn development in females) and are as follows: total length, 974 (940–993, *n* = 3); length of tail, 148 (140–160, *n* = 3); length of hind foot, 286 (272–300, *n* = 2); length of ear, 125 (*n* = 1); length of horn (males), 286.8 (203–312, *n* = 29); length of horn (females), 148 (71–226, *n* = 5); and body weight, 20.2 kg (*n* = 6) (Groves and Harrison 1967; Harrison and Bates 1991; Rietkerk and Delima 1994). Slightly different results were found for *G. s. marica* for Turkey: horn length of adult males was 348.9 ± 0.92 mm (from 285 to 460 mm); body weight is 13.86 ± 0.76 kg (females, *n* = 54) and 19.39 ± 0.92 kg (males, *n* = 39) (Table 1.8; Gürler et al. 2015). In Saudi Arabia, *G. s. marica* has comparable weights: 17.4 kg (11–24 kg) for females and 19.6 kg (15–28 kg) for males (Wronski et al. 2010).

1.3.5 Skull

The skull of the goitered gazelle is relatively large, with widely spaced cheekbones. The skull has lachrymal vacuities, like those of antilocaprids (O'Gara 1978). The dental formula of goitered gazelles is I 0/3, C 0/1, P 3/3, M 3/3, total 32 (Kingswood

Table 1.8 Body sizes (cm) and weights (kg) from adult female and male goitered gazelles (*G. s. marica*) from Turkey, with associated statistical significance value for sex differences (P) (by Gürler et al. 2015—open access)

	Adult females ($n = 54$)		Adult males ($n = 39$)		
Measure	Range (min–max)	Mean ± SE	Range (min–max)	Mean ± SE	*P*
Body weight	11.1–15.3	13.86 ± 0.76	16.8–25.6	19.39 ± 0.92	<0.001
Chest circumference	55.0–64.0	59.50 ± 0.37	59.0–70.0	63.19 ± 0.36	<0.001
Shoulder height	52.0–64.0	58.39 ± 0.40	61.0–71.0	65.73 ± 0.56	<0.001
Ear length	13.0–15.0	13.86 ± 0.16	13.0–16.0	14.06 ± 0.20	>0.05
Foreleg length	35.0–50.0	43.31 ± 0.65	46.0–60.0	53.62 ± 0.77	<0.001
Hind leg length	49.0–65.0	55.86 ± 0.85	57.0–70.0	63.31 ± 0.62	<0.001
Neck circumference	21.0–25.0	23.42 ± 0.26	23.0–31.0	28.08 ± 0.53	<0.01
Head length	19.0–25.0	21.67 ± 0.24	20.0–26.0	22.69 ± 0.25	<0.05
Tail length	14.0–20.0	18.75 ± 0.18	15.0–22.0	19.54 ± 0.19	>0.05

and Blank 1996). Molariform teeth are hypsodont; each molar is bilobate and has a normal selenodont structure. The row of lower cheek teeth is nearly straight, the crowns narrow, and the medical cusps of molars prominent and trenchant (Harrison and Bates 1991). At birth, three incisors, one incisiform canine, and three deciduous cheek teeth are present on each side of the lower jaw; three premolars are present on each side of the upper jaw (Heptner et al. 1961; Legge and Rowley-Conwy 1987; Zhevnerov 1984). Two permanent teeth, m^1 and m^2, erupt during the first year; at about 14 months, m^3 erupts and p1, P^2, and P^3 have replaced the deciduous molars (Legge and Rowley-Conwy 1987). The age of goitered gazelles may be estimated between 1 month and 10 years according to tooth eruption and wear (details below; Binh 1963). Osteo-morphological characteristics of the skeleton of goitered gazelles are consistent with ten other species of gazelles (Peters 1989).

The nasal bones are wide and short; their front ends have a notch in the entire width of the bone; at the border of the nasal and frontal bones, the profile of the skull is straight or even concave; the preorbital pits are well-developed; the rods of the horns are massive; the auditory drums are very swollen. The skull has a relatively short preorbital part, the length of which from the anterior edge of the orbit to the end of the maxillary bones (prosthion) is less than 1.3-fold of the length of the posterior part of the skull from the posterior edge of the orbit to the extreme protrusion of the occipital condyles. The entrance to the nasal cavity is narrow; its width is not more than half the distance from prosthion to the anterior ends of the nasal bones. The nasal bones are relatively short and wide; their length is not more than 35% of the main length of the skull, and the greatest width is not less than 40% of its own length. The anterior ends of the nasal bones have cuttings for the entire width of the bone, and their lateral processes are only slightly inferior in length to the medial ones. At the border of the nasal and frontal bones, the profile of the skull is straight or even concave. The inner edges of the eye sockets have cut-outs above the lacrimal-frontal suture, and at their level, there is also the smallest width of the forehead between the eye sockets. The preocular pits on the skull are well-developed. The

Table 1.9 The skull sizes (in mm) of adult male and female goitered gazelles (*G. s. subgutturosa*) from different parts of their natural range in Kazakhstan (adapted from Zhevnerov et al. 1983)

	Ustyurt Plateau and Kyzylkum Desert				Barsakelmes Island			
	Males ($n = 37$)		Females ($n = 31$)		Males ($n = 23$)		Females ($n = 10$)	
Indices	Range (min–max)	Mean	Range (min–max)	Mean	Range (min–max)	Mean	Range (min–max)	Mean
Maximal length	200–225	208.1	183–208	199.1	196–222	210	180–217	199
Basic length	184–202	192.2	169–190	178.4	186–203	194	180–198	189
Maximal width	92–100	96.5	84–97	89.6	92–102	96	84–92	91
Interorbital width	51–60.5	56.2	46–55	51.4	56–65	60	52–59	56
Front length	141–156	147.9	136–150	143.4	124–157	150	134–145	139
Brain part length	84–97	90.3	76–86	81.5	87–98	92	77–81	79
Occipital width	–	–	–	–	70–78	74	68–73	71
Upper molars' length	–	–	–	–	63–70	67	63–66	65
Diastema length	–	–	–	–	27–37	35	30–38	34

horny processes are massive. The distance between the outer edges of the bases of the processes is greater than the interocular width of the forehead. The auditory capsules are very swollen. The distance from the lower edge of the bladder to the upper edge of the bony auditory canal is usually equal to or greater than the vertical diameter of the eye socket. The last premolar of the lower jaw (P_4) has a well-developed only antero-external tubercle of the tooth, so the anterior valley on its inner side remains open (Sokolov 1959).

The length of the skull of adult gazelles varies in males from 196 to 222 mm and in females from 180 to 217 mm. The greatest width of it in males is 92–102 and in females is 84–92 mm, which is slightly less than half of the total length of the skull. In different areas of the range, the geographical variability of the skull size of adults is quite significant (Table 1.9; Zhevnerov et al. 1983).

Individual fluctuations in the size of the skull of adult gazelles are quite significant (Table 1.10). To the greatest extent, the length of the skull and the facial part varies, while the individual variability of other signs of the skull is more moderate. A large range of variability in the length of the skull of the gazelle is also characteristic of mainland populations. The skull length of gazelles from the Betpak-Dala Desert in males is 172–224 mm and in females 182–208 mm (Sludskiy 1956a); in gazelles of the Ustyurt Plateau and the Kyzylkum Desert, these ranges are 200–225 mm and 183–208 mm, respectively (Table 1.9; Mambetzhumaev 1970).

Table 1.10 The range variability of the skull sizes of adult goitered gazelles (*G. s. subgutturosa*) in the Barsakelmes Island, Aral Sea, Kazakhstan (adapted from Zhevnerov 1984)

Indices, mm	Males	Difference	Females	Difference
Maximal length	196–222	26	180–217	37
Basic length	186–203	17	180–196	16
Maximal width	92–102	10	84–92	8
Interorbital width	56–65	9	52–59	7
Muzzle front part length	124–157	13	134–145	11
Brain part length	87–98	11	77–79	2
Length of upper teeth row	63–70	7	63–66	3

According to Mambetzhumaev (1970), anomalies are not uncommon in the skull of goitered gazelles (the appearance of dental rudiments on the surface of the diastema, the location of the interparietal bone between the parietal and frontal bones, the non-fusion of paired lobes of the parietal bone). Zhevnerov et al. (1983) found a dead adult male with a strong asymmetry of the skull, which consisted of a bend in the facial part to the right, which caused a shortening of the inter-jaw (*intermaxilla*) and upper maxillary bones on this side.

There are some differences in skull measurements in different subspecies. Means (and ranges) of greatest length of skull (in mm) of males and females, respectively, are the following: *G. s. hillieriana*, 203.1 (198–210, $n = 20$) and 201 (193–209, $n = 2$); *G. s. marica*, 181.2 (169–193, $n = 18$) and 169.3 (157.3–188, $n = 11$); *G. s. subgutturosa*, 208.5 (186.2–225, $n = 93$) and 196.3 (168–217, $n = 59$); and male *G. s. yarkadensis*, 214 (210–220, $n = 5$). Means (and ranges) of length of maxillary tooth row (in mm) of males and females, respectively, are as follows: *G. s. hillieriana*, 61 (55–65, $n = 28$) and 59.3 (56–64, $n = 3$); *G. s. marica*, 57.4 (52–61.2, $n = 34$) and 55.5 (52–59, $n = 10$); *G. s. subgutturosa*, 64.5 (54.5–71, $n = 73$) and 62.1 (54–66, $n = 31$); and male *G. s. yarkadensis*, 68.6 (64–73, $n = 7$) (Harrison 1968; Zhevnerov et al. 1983). Harrison and Bates (1991) reported means (and ranges) of skull measurements (in mm) of *G. s. marica* and *G. s. subgutturosa*, respectively, of unspecified sex as follows: greatest length of skull, 171.8 (151.6–191.8, $n = 16$) and 175.7 (160.2–186.2, n = 5), and length of maxillary tooth row, 57.1 (52.7–61.2, $n = 19$) and 56.1 (54.0–59.0, $n = 7$) (Kingswood and Blank 1996). Mean skull volumes in goitered gazelles from Turkey are 115.7 ± 2.4 cm^3 in males and 87.7 ± 0.9 cm^3 in females; the mean skull surface areas are 79.6 ± 8.5 cm^2 in males and 77.3 ± 1.18 cm^2 in females (Yılmaz et al. 2020).

1.3.6 Dental System

1.3.6.1 Description

Canines, as a rule, are absent in the goitered gazelles. One-month-old male and 2-month-old and 3-month-old females each were found to have a pair of rudimentary milk canines (Binh 1963). They were located on the sides of the anterior end of the upper jaw where the jawbone fuses with the mandibular bone. The second premolar (p^2) has one sharp outer vertex. The last two premolars are similar in appearance to molars and have two sharp outer vertices. The third premolar (p^3), unlike the fourth (p^4), has only one inner half-moon. Permanent premolars are more or less alike and single-lobed and have one sharp outer vertex. There are several depressions on the chewing surface of the newly erupted teeth—a trace of the roots of the milk premolars—the exception is the crown cup of P^4. The abrased chewing surfaces P^2 and P^3 take an irregular quadrangular shape and P^4 the shape of a Latin letter D with a convex side facing inward. Molars are hypsodont in shape and structure is generally alike. They have two blades in the form of the letter B, with the convex side facing inward. The widest molar in adults is M^2. The crown cups of the molars taper downward, forming a pattern that changes with age on the rubbing surface of the teeth. M^2 can be used to determine the age (Binh 1963).

The hooks of the milk teeth have the shape of symmetrical blades, slightly curved outward, and the remaining incisors and canines have the shape of irregular sickles with more curved inner edges and a ridge on the lingual side. In shape, permanent incisors and canines are like milk teeth, but they are larger and longer than the latter: the greatest length (with the root) of a milk tooth hook is 19 mm; the width and height of the crown are 9 mm ($n = 19$). For permanent hooks of incisors and canines, respectively, 24, 11, and 11.5 mm ($n = 38$). The neck of milk teeth hooks rises above the gum from the age of 9–10 months, while the neck of permanent hooks is only visible in old individuals. The dental canal entering the crown of the hook (milk and permanent teeth) expands, and the remaining incisors and canines remain narrow, as well as in the root of the tooth. The second and third milk premolars (p_2, p_3) are similar to permanent premolars. Their sharp tip is located in the middle part of the crown. The last milk premolar p_4 has three lobes, the convex side facing outward. Permanent premolars of the mandible are similar to each other, though they differ from each other in size: the rear ones are larger than the front ones. The second premolar P_2 has two noticeable grooves on the inner surface of the crown, P_3 and P_4; in addition, there is a depression in the posterior thickened part of the crown. In shape and structure, the lower molars are like an inverted image of the upper molars (except for the M_3, which has three blades). The posterior crown cups are deeper than the anterior ones; only the cup in the talonid of the third molar is smaller than the first two (Binh 1963).

1.3.7 Age Group Identification Using Teeth According to Binh (1963)

Up to 1 month of age, the teeth consist of milk incisors, canines, and premolars. In the depth of the alveoli of the upper and lower jaws, the tops of the forming crowns of the first molars ($M_1{}^1$) are visible. X-rays of the lower jaw would show that the crowns of the first molar M_1 and the tip of the anterior blade M_2 are visible, but there are no roots. The sutures between the four occipital bones are clearly visible (*basioccipitale*, *supraoccipitale*, *occipitalia lateralia*) and between *basioccipitale* and *basisphenoideum*. The intertribal and parietal bones are paired. The sagittal suture between the frontal bones is almost straight throughout, without teeth.

From 1 up to 2 months old, the milk premolars p^2 and p_4^4 begin to wear out. On the upper jaw, the tops of the anterior (in males) or the posterior (in females) half-moons M^1 cut through the bone. The M^2 crown begins to be laid. On the lower jaw, the tops of the anterior half-moons of the first molar M_1 cut through the gum. The X-rays would show that the tops of the anterior and posterior blades M_2 have formed. Molars are without roots. The parietal bones begin to fuse along the midline. Intertribal bones started to merge (in females) and have merged (in males). The *basioccipitale* and *occipitalia lateralia* begin to merge, so only traces of their seams remain. Teeth begin to form in the middle part of the frontal suture.

From 2 up to 3 months old, the tops of the anterior half-moons of the first upper molar M^1 have cut through, and the anterior and posterior half-moons M_1 and M^1 roots appear in the lower jaw. The crowns of the M_2^2 that have not yet erupted are growing, and the *basioccipitale* and *occipitalia lateralia*, as well as the parietal and interparietal bones, have merged (though sometimes the trace of the seam between the parietal bones in females is still unclear).

From 3 up to 4 months, all milk premolars are abrased. The posterior half-moons of the first upper molar M^1, whose roots are forming, have cut through the gum. The tip of the second upper molar M^2 reaches the upper edge of the bony alveoli. The tip of the anterior blade M_2 is cut through the bone. The X-ray image would show that M_2 does not yet have roots and the alveoli of the third lower layer of M_3 are beginning to form. In the male, the frontal suture has teeth in both the middle and posterior parts.

From 4 up to 5 months of age, the alveoli of permanent hooks begin to form I_1. The posterior lobes of the first upper molars M^1 are abrased. The rear blades of M_2 were cut through the bone. The tops of the anterior and posterior lobes of the third lower molar M_3 are laid down. In males, the dentation of the false suture is getting more complicated in the middle part.

From 5 up to 6 months, the top of the crown of the permanent hook of I_1 is visible on the X-ray. The tops are cut through the bone only in front of their anterior (in females) and/or posterior (in males) half-moons M_2. Crowns of the third molar M^3 begin to form. The front half-moons of M_2 have cut through the gum. The roots of M_2 and the tops of the talonid of the third lower molar M_3 are formed in the alveoli. The sutures of the females are the same as in the previous age group. Males begin to fuse *occipitalia lateralia*, as well as *supraoccipitale* and *occipitalia lateralia* (only traces of seams between them are visible). In females, the false suture has teeth in the middle and rear parts.

From 6 up to 7 months of age, the crowns of the hooks (I_1) that have not yet erupted grow upward; crowns of the second and third upper molars (M^2, M^3) grow upward, as well. The anterior lobe of the second upper molar M^2 and the posterior lobe of the second lower molar M_2 have cut through the gum.

From 7 up to 8 months of age, the chewing surface of the front blade p_4 is smoothed. Alveoli of permanent premolars appear between the roots of milk premolars. The tops of the posterior lobe of the second upper molar M^2 have cut through the gum, while the posterior lobe of the second lower molar M_2 is already beginning to wear off. The tip of the formed crown of the third lower molar M_3 is located at the level of the upper edge of the bone alveoli.

From 9 up to 10 months of age, the crown of the hooks has finally formed I_1; the alveoli of the second incisors begin to form I_2. The milk premolars p_2, p_3, and p_4 are strongly erased. The anterior lobe of the third lower molar M_3 has cut through the bone. In females, the *occipitalis lateralia* and also *supraoccipitale* and *occipitalis lateralia* begin to fuse (only traces of seams between them are visible). The serration of the frontal suture in the middle part is getting more complex.

From 12 up to 13 months of age, the root of the hook (I_1) that has not yet erupted is formed, and the crown and neck of the second hook I_2, alveoli of the third incisor I_3, and permanent canine C start to form. The crowns of the permanent premolars of both jaws are laid down. The anterior lobes of the third molars M_3^3 in the upper jaw cut only through the bone and in the lower one through the gum. The root of the front blade M_3 is formed. In males, the dentation of the frontal suture is complicated in the middle and posterior parts of the suture.

From 13 up to 14 months of age, the crown of the incised hook I_1 reaches 2/3 of what will be its full height. The root of the second incisor is formed I_2; the crown of the third incisor I_3 and canines C and the roots of P_4^4 are formed. The anterior lobe of the third upper molar M^2 and the posterior lobe of the lower molar M_3 were cut through the gum. The seam between the *basisphenoideum* and the *basisoccipitale* are almost overgrown on female skulls.

From 14 up to 15 months of age, the hooks I_1 are fully developed. The root of the third incisor is formed I_3, and the crown and neck of canine C are formed. The chewing surface of milk premolars (p^2, p^3, and p^4) is smoothed (i.e., traces of crown cups disappear). Between the roots of the milk premolars, the tops of the permanent premolars (P^2, P^3, P^4) cut through the bone are now visible. The roots of all permanent premolars are laid down. The posterior M^3 blades cut through the gum, and the M_3 talonid cuts through the bone. The roots of the third upper molar M^3 appear. There remains only a trace of a seam between the *basioccipitale* and *basisphenoideum*.

From 15 up to 16 months of age, permanent incisors and premolars reach 2/3 of the total height, and permanent canines cut through the bone in females. In males, permanent incisors I_2 and I_3 and permanent canines will have erupted only in one out of three cases, and permanent premolars cut through the bone. Talonid M_3 cuts through the gum but did not reach the level of the front blades. The seam between the *basisoccipitale* and *basisphenoideum* is completely overgrown on the skulls of females, while males still retain its trace.

From 16 up to 18 months of age, the crowns of the permanent incisors reach a normal height; the permanent canines have cut through the bone. Permanent premolars reach the level of molars or even begin to wear off (in females). The rear blades of the M_2^3 begin to wear out. The root of the second blade M_3 is laid down.

From 18 up to 19 months of age, the hooks begin to wear off. The canines reach the height of the incisors. Permanent premolars begin to wear out (except for P_2 and P_3).

From 2 up to 2.5 years of age, the second incisors (I_2) begin to wear off, as well as the permanent premolars (P_2 and P_3) and talonid M_3.

From 2.5 up to 3 years, I_1 and I_2 start to erase and protrude from the alveoli of the neck P^2 and P^3. The back wall of the M^3 crown begins to wear out. The trace of the crown cup on the rubbing surface of M_1 is shorter than in the previous age group.

From 3 up to 3.5 years old, the dentin shines through under the thinned layer of enamel on the hooks I_1. The second incisors, I_2, and canines, C, begin to wear out not only the tops but the ridges along the lingual surface of the teeth. P^2 is strongly abrased. Necks of P_2^2, P_3^3, and P^4 protrude from the alveoli. The crown cups of the posterior blade M^3 and the talonid M_3 are narrowed.

From 4.5 up to 5.5 years old, dentin shines through under the enamel of the third incisors I_3. Ridges on the lingual surface I_2 and I_3 and the canines are erased in both the upper and lower parts of the crowns. The rubbing surface of P^2 is almost smoothed. The trace of the crown cup of the front blade M_1 is in the form of an irregular speck.

From 5.5 up to 6.5 years, dental asterisks appear (a trace of the dental canal on the rubbing surface)—in the form of yellowish dashes on the rubbing surfaces of the hooks (I_1). The chewing surface of P^3 is almost smoothed. Trace of the crown cup M_1 appears in the form of a dash. The rubbing surface of the front blade M_1 is almost smoothed. The trace of the crown cup of talonid M_3 is in the form of a short dash.

From 6.5 up to 7.5 years, the chewing surface of P^3 is smoothed. Traces of crown cups M^1 are in the form of irregular spots. The rubbing surface of M_1 and the talonid M_3 is smoothed.

From 7.5 up to 8.5 years, the trace of the crown cup on P^4 is in the form of an irregular speck. The chewing surface M^1 and the front blade M_2 are almost or completely smoothed.

From 8.5 up to 9.5 years of age, the chewing surface of I_1 and the dental asterisk on it take an irregular triangular shape. The chewing surface M_1 and the front blade M_2 are smoothed. The trace of the crown cup of the front blade M_3 is in the form of a speck.

In more than 10 years old, the hooks are worn down to the gums. Dental asterisks are in the form of an irregular rounded spot. I_2 and I_3 and canine are worn out. Chewing surface of P_4, M_2^2, and M_3 is smoothed out and traces of crown cups disappear.

Wearing Out of Permanent Teeth

Upper Jaw. The second premolar P^2 is smoothed out first (depressions on the rubbing surface disappear) 2.5–3 years after its cutting (at the age of 4.5–5 years), then P^3 at the age of 6.5–7.5 years, and the last P^4—at the age of more than 10 years. Molars that have not yet begun to wear out have four vertices each. The inner vertices are erased faster than the outer ones so that the strongly abrased ones have only two outer vertices. The shape of the crown calyx of molars changes with age, as the calyx becomes narrower and shorter, and it eventually disappears. This happens on M^1 at the age of 8–9 years and on M^2—at the age of more than 10 years (Binh 1963).

Lower Jaw. The abrasion is noticeable first on the hooks I_1, then on internal incisors I_2, on fangs C, and, finally, on the outskirts I_2. In this case, the labial side is abrasing faster than the lingual side, and as a result, the initial slope of the lingual side of the incisor crown decreases, and their tops become flattened. When the enamel layer is worn away, dentin is shown on the rubbing surface of the hooks. The rubbing surfaces of weakly erased hooks are narrow and elongated; as the incisor is erased, they become wider and shorter to an almost rounded shape. A change in the shape of the dental asterisk is characteristic. The asterisk on the hooks initially looks like a transverse dark yellowish stripe and then rounded. The wear of the remaining incisors and canines develops in approximately the same way; only the shape of the dental asterisk is rounded and then almost does not change (Binh 1963).

Unlike the upper premolars, the fourth premolar P_4 is first erased on the lower jaw, then the third P_3, and finally the second P_2. As the premolars wear out, the shape of the rubbing surface becomes relatively wider and less rough, as the grooves on the inside of the crown are crushed. In basic terms, the lower molars are worn out in the same way as the upper ones; the trace of the crown cup on the rubbing surface of the anterior blade always disappears earlier than on the posterior blade, with the exception of the third lower molar M_3, where first of all, the trace of the crown cup of the talonid disappears. The rubbing surface of the first lower molar M_1 begins to smooth out in individuals aged 6.5–7.5 years (Binh 1963).

Below are some conclusions according to Binh (1963):

1. Crowns of permanent incisors and canines appear at the age of 5–13 months; roots are formed at 11–15 months. The change of milk incisors and canines to permanent teeth occurs at the age of 13–18 months. The first are replaced hooks in I_1 and then simultaneously in the middle and outer incisors I_2 and I_3. This happens 1–2 weeks later in males than in females.
2. Permanent premolar crowns are laid down at the age of 11–12 months; roots are formed at 13–16 months. The change of milk premolars to permanent ones begins from the 14th month and ends by the 18th month (males are 3–4 weeks later than in females). In the upper jaw, the teeth are replaced almost simultaneously and, in the lower jaw, first the posterior p_4 and then the anterior p_3 and p_2.
3. Crowns M_1 are laid down 1–2 months before birth, crown M_2 at birth, and crown M_3 at the age of 3–4 months. The crowns of the corresponding upper molars are laid down 1 month later than the lower ones. Molar roots form 3–4 weeks after the eruption of these teeth through the gum. In the lower jaw, the first molar cuts through the gum at the age of 1 month, the second from the age of 5 months, and the third from the age of 11 months. The talonid erupts by the age of 16 months but reaches the level of a number of molars only by the age of 2 years. The upper molars usually appear 3–4 weeks later than the lower ones. With age, the dental index, especially the third molars M_3^3, decreases markedly.
4. When erasing permanent incisors and canines, the enamel wears out and dentin appears; then the dental plaque becomes noticeable, which changes its shape on the front incisors as it wears out. Among the permanent premolars, P^2 is the first to smooth out at the age of 4.5–5.5 years. The shape of the crown calyx of molars changes with age, becoming narrower and shorter and eventually disappearing (the tooth is smoothed). Among molars, M_1 is the first to be smoothed at the age of 6.5–7.5 years.

Fig. 1.23 Adult males of goitered gazelles have lyre-shaped horns with ends facing forward-inside A male in the enclosure of the Issyk-Kul goitered gazelle nursery for re-acclimatization, 25.11. 2022. Photo by D. Blank

5. The timing of the eruption of molars, the change of incisors, canines, and premolars are most suitable for determining the age of a goitered gazelle skull up to 18 months; then age estimates can be guided by the degree of wear of the teeth, mainly the molars.

1.3.8 Horns

The horns, which are usually present in males, are blackish with lyre-shaped curves, converging at the base and diverging backward and to the side, with the ends facing each other and forward (Fig. 1.23). The surface of the horn covers consists of a larger number of transverse undulating rings, more strongly developed on the anterior side; at the base of the horns, the rings are located close to each other and in the middle part to the tops more rarely. Adult gazelles have up to 26–28 such rings (more often 16–20). The curved ends of the horns are smooth for 7–9 cm from the tips. The horn length of an adult male is typically between 27 and 41 cm in Uzbekistan (Ishunin 1961) and 39–42 cm in Tajikistan; the diameter at the base is 38 to 40 × 28 to 31 mm ($n = 76$; Stroganov 1956). Individual variability in the shape and size of the horns is stronger than body and skull parameters (Table 1.11). Individuals with the typical lyre-shaped horns are observed rarely on Barsakelmes Island. Of 70 skulls studied, 52 (74.3%) were asymmetrical. This was most often

Table 1.11 Horn sizes (cm) of adult males of goitered gazelles (n = 70) in the Barsakelmes Island, Aral Sea, Kazakhstan (adapted from Zhevnerov 1984)

Indices	Horn length along curve	Circumstance in horn base	Distance between horns		Ring number on horns
			At base	At the end	
Minimal	26	11	0.7	2.5	11
Maximal	40	13	1.5	22.4	20
Average	30.6	11.7	0.9	7.1	15

Fig. 1.24 Some goitered gazelle males have different curvature of their horns as a result of unfavorable heredity, and they do not have a regular lyre shape. A young male (1 year) in the Issyk-Kul goitered gazelle nursery for re-acclimatization, 05.05.2024. Photo by D. Blank

expressed by the crossing of the horns at the ends and in the difference in their length. The prevalence of this asymmetry is obviously a consequence of the small number of the first breeding males brought to the Barsakelmes Island resulting in a small gene pool (Zhevnerov 1984). The similar phenomenon is observed in goitered gazelle males of the Issyk-Kul Breeding Center (Fig. 1.24), which got the gazelles from Bukhara nursery, the goitered gazelle population of which also had a small gene pool.

Horn development in females apparently increases from Mongolia and China to the Arabian Peninsula. Females of *G. s. hillieriana* and *G. s. yarkandensis* are hornless (Mongolia, China); females of *G. s. subgutturosa* (Central Asia) are hornless or have short horns (usually <70 mm) (Fig. 1.25); and females of *G. s. marica* (Saudi Arabia) have well-developed horns (Kingswood and Blank 1996; Zhevnerov 1984).

Fig. 1.25 Goitered gazelle females are usually hornless, though sometimes they have one or two short horns (less than 70 mm). A female with her fawn in the Kyzylkum Desert. 18.07.2017. Photo by Zh. Nurmukhambetov with kind permission

Sludskiy (1956a) claims that in Kazakhstan horned females make up 5.5% of all females. Three such females were found in the northern Kyzylkum Desert during 1961–1962 with horns of 1.5, 10, and 3 cm; in the latter case, one horn was curved, like in a chamois where it is bent forward. The horns of other females also differed having irregular shapes. Two of the females had fawns. Two other females with horns of 3 and 5 cm were seen in Aryskum Sands (northeastern Aral Sea region) in 1956 and 1958. Horned females are not a rarity in Turkmenistan, and horns up to 10 cm were sometimes noted (thin and irregular in shape). In Badkhyz (southern Turkmenistan), the percentage of horned females corresponds to that in Kazakhstan. For the region of the lower Murgab River, 20% to 25% of females had horns according to information from the 1930s (Lobachev and Smirin 1970).

The mean curved horn lengths for the collected (in Mahazat Al-Sayd Protected Area, Saudi Arabia) adult skulls (males older than 18 months, females older than 12 months—*G. s. marica*) were 29.05 ± 3.53 cm (n = 168) and 20.84 ± 3.21 cm (n = 121), and the maximum horn lengths were 36.5 cm and 30.5 cm, for males and females, respectively. Horn length obtained from skulls of adult males (*G. s. marica*) differed significantly and was greatly shorter compared to males of the same age group held at the King Khalid Wildlife Research Center in Saudi Arabia. The same is true for adult females with horns that are noticeably longer in the captive breeding population (Cunningham and Wronski 2011).

1.3.9 Skeleton

The gazelle has a rather massive skeleton, more massive than, for example, the saiga. Males, in contrast to females, have a heavier axial skeleton, possibly due to the presence of horns. The skeleton of the limbs with their belts in the goitered gazelles is generally quite well-developed and quite massive, which is to be expected of a high-legged and short-bodied ungulate (Sokolov et al. 1964).

In the axial skeleton, it is necessary to first note not only the absolutely but also relatively short skull size of the goitered gazelle. This observation is essential in determining the physique of the animal and in the position of its center of gravity. The correlative development of the spine is undoubtedly connected with the type of movement of the animal. The number of vertebrae in each of the main divisions of the goitered gazelle is similar to other ungulates: cervical, 7; thoracic, 23; lumbar, 6; and sacral, 4. However, the relative length of the same sections in different species may differ significantly. The goitered gazelle has a relatively long neck and short thoracic and lumbar parts (Table 1.12; Sokolov et al. 1964). In addition, the pelvic cavity is significantly different in males and females (sexual dimorphism) in terms of the conjugate diameter, vertical diameter, and the intermediate transversal diameter that is related to the birthing function of females (Demircioglu et al. 2021).

Regarding limbs and their belts, it should be noted first that goitered gazelles are high-legged animals. The hind limbs are better developed in both weight and structure compared to the front as in most other terrestrial mammalian species. However, the goitered gazelle has a rather pronounced disproportion between the front and hind legs, especially in terms of weight. Comparison of the weight of the forelimbs and hind limbs with the weight of the entire skeleton also confirms that the weight of the skeleton of the hind limb in the goitered gazelle is noticeably greater. Correspondingly, the muscle weight of the hind legs is 1.5 times heavier than the muscles of the front legs. The development of the large muscles of the posterior limb of the goitered gazelle corresponds to the large longitudinal and transverse dimensions of its bones in that limb. The goitered gazelle has distinctive features in the ratio of the lengths of individual limb sections. The metacarpal bone is the longest in the forelimb, and the metatarsal bone is longer than the femur, which of all the long tubular bones of the hind limb turns out to be the shortest. The posterior toe is shorter than the anterior, while the disproportion in the goitered gazelle is only slightly expressed, which is due to the general elongation of the hind limbs, especially the distal parts (Sokolov et al. 1964).

In the postembryonic period, the development of the skeleton in goitered gazelles occurs quite quickly. The final ossification of the skeleton, in particular, the fusion of the epiphyses with the diaphyses and the disappearance of the epiphyseal sutures, occurs in a 2-year-old gazelle, and the corresponding sutures on the ulna, femoral sciatic, and tibia bones disappear, whereas in saiga, for example, this happens at the age of 4–5 years. The goitered gazelle has a fairly rapid ossification of the vertebrae: by the age of 2 years, the vertebrae of the sacrum and the sutures on the fifth and sixth lumbar vertebrae are almost completely merged; in saiga, this happens at the age of 5–7 years (Sokolov et al. 1964). The more detailed description of limb bones

Table 1.12 The mean and range of body proportions and length in centimeters of adult male and female goitered gazelle (*G. s. subgutturosa*) skeleton (adapted from Sokolov et al. 1964)

Indices	Males (n = 5)		Females (n = 8)	
	Mean	Range (min–max)	Mean	Range (min–max)
Main length of the skull/vertebral length	22.7	22.4–23.1	22.2	21.6–22.7
Spinal column parts/vertebral column length:				
Neck	32.8	32.0–33.6	33.4	32.7–33.9
Chest	36.2	35.9–37.0	36.3	35.9–37.1
Loin	21.7	21.0–22.3	21.7	21.3–22.4
Sacrum	9.3	8.8–10.0	8.5	8.0–8.8
Front leg length/vertebral column length	58.8	58.1–59.5	58.7	57.0–60.8
Hind leg length/vertebral column length	72.6	71.5–74.2	71.8	70.0–74.3
Front leg length/hind leg length	80.8	80.0–82.1	81.6	81.1–83.1
Length of front leg parts/front leg length:				
Shoulder	28.8	27.4–29.5	29.0	28.5–29.8
Arm bone	27.2	26.7–27.6	27.2	26.8–27.7
Radial bone	34.6	34.1–35.0	34.0	33.5–34.9
Metacarpus	38.1	37.4–38.7	38.8	38.3–39.9
Reward finger length/front finger length	92.6	92.4–92.8	91.1	90.2–92.4
Length of hind leg parts/hind leg length:				
Hip	30.0	29.7–30.4	29.9	29.7–30.2
Tibia	38.0	37.5–38.4	37.9	37.6–38.5
Metatarsal	31.9	31.5–32.8	32.2	31.8–32.5
Finger	13.1	12.9–13.4	12.9	12.6–13.3
Metacarpus diaphysis width/metacarpus length	6.6	6.2–7.1	6.2	5.9–6.4
Metatarsal diaphysis width/metatarsal length	5.8	5.5–6.2	5.7	5.4–6.0
Metatarsal diaphysis width/metacarpus diaphysis width	92.0	88.3–94.9	94.4	90.9–96.5
Arm bone diameter/hip bone diameter	83.3	80.5–86.9	82.4	78.2–90.2
Skeleton weight/body weight	7.47	–	5.7	4.8–6.2
Vertebrate column weight/entire skeleton	56.1	–	43.6	40.7–46.4
Leg bones' weight/entire skeleton	43.9	–	56.3	53.3–59.3
Front leg bone weight/hind leg weight	72.5	65.0–77.1	68.4	65.5–71.7
Front leg bone/entire skeleton	6.3	–	7.8	7.5–8.1
Hind leg bone/entire skeleton	8.6	–	11.4	10.7–12.2

of *G. s. marica* from Turkey was done with the help of computer tomography (Demircioglu and Gezer Ince 2020).

1.3.10 Muscles

The general topography of the muscles of the goitered gazelle does not differ from that of other antelope species. Differences in musculature are revealed only when determining the weight and size and making a more detailed comparison of the structure of individual muscles. One conditional indicator of the overall development of the musculature can be its relative weight as a percentage of the total body weight. This index reaches 62.8% in a male and 60.7% (range is 59.3–62.0%) in a female. Despite the high degree of variability of this indicator, the goitered gazelle's musculature on average makes up a larger percentage of the whole-body weight compared to other antelopes of the same size (e.g., saiga). The relative weight of the muscles of the limbs of the goitered gazelle is also quite high and amounts to 13.1% of the body weight. The extensor muscles of the back are also better developed in the gazelle (2.5% of body weight) and weight twice as much as, for example, with saiga. The weight of the muscles of the hind limbs of the goitered gazelle is almost one and a half times greater than the weight of the muscles of the forelimbs (Sokolov et al. 1964).

The goitered gazelle has especially strongly developed muscles of the anterior girdle (52.1% of the total muscle weight). Especially powerful in this group of muscles are *m. brachiocephalicus*, *m. trapezius*, *m. serratus*, and *m. pectoralis*, which are involved in holding a certain position and moving the head, neck, and free forelimb (Sokolov et al. 1964).

1.3.11 Internal Organs

The absolute weight of the internal organs of males is greater than that of females, and the relative weight is somewhat less. Extreme indicators of the size of the intestinal sections of males and females overlap each other (Table 1.13) (Zhevnerov 1984).

According to Mambetzhumaev (1970), weights and sizes of adult males' internal organs from southern Ustyurt Plateau are as follows: lung weight, 700 g; heart weight, 223 g; liver, 400 g; kidneys, 142 g; spleen, 59 g; length of small intestine, 1240 cm; large intestine, 650 cm; and cecum, 111 cm.

An examined liver was dark brown in color, rectangular shaped, and located on the right side of the thoracic part of the abdomen. It measured 19 cm, 11 cm, and 2.5 cm in length, width, and depth, respectively, with a weight of 275.40 ± 0.21 g. It is divided into left, right, caudate, and quadrate lobes in addition to caudate and papillary processes. It is supplied by the hepatic artery, which is divided into the right, left, and middle branches. The gallbladder was a narrow, elongated pear-shaped organ attached to the cystic fossa of the liver. The biliary system consisted of the right and left hepatic ducts, the common hepatic duct, the cystic duct, and

Table 1.13 Sizes (in millimeters) and weights (in grams) of internal organs in adult male and female goitered gazelles (*G. s. subgutturosa*) of the Barsakelmes Island, Aral Sea, Kazakhstan (adapted from Zhevnerov 1984)

	Males ($n = 15$)		Females ($n = 7$)	
Indices	Range (min–max)	Mean	Range (min–max)	Mean
Weight, g				
Heart	246–278	261	220–250	223
Lungs	460–612	529	425–515	473
Liver	692–809	715	485–800	672
Kidneys	180–250	205	172–250	197
Spleen	42–68	56	38–63	49
Weight of organs relative to body weight, %				
Heart	0.6–0.7	0.7	0.8–0.9	0.8
Lungs	1.0–1.6	1.3	1.7–1.9	1.8
Liver	1.7–2.1	1.9	1.9–2.7	2.2
Kidneys	0.4–0.6	0.5	0.7–0.9	0.7
Spleen	0.1–0.2	0.1	0.1–0.2	0.1
Length, mm				
Small intestine	1250–1475	1395	1252–1430	1354
Large intestine	600–690	650	602–681	625
Cecum	25–35	30	27–32	29

common bile duct. The goitered gazella liver is similar to that of small ruminants but different in its blood supply (Dawood and Khamas 2017).

The small intestine is composed of three segments: the duodenum, jejunum, and ileum, but with no anatomical demarcation line isolated between them. It is situated in the most right caudal area of the abdominal cavity which starts in the stomach and ends at the cecum. The mesentery allows for an extraordinary level of mobility to the small intestine and maintains the fixation of the small intestine within the midline of the body. The ratio of small intestine to the total body length is about nine times larger (Hamza and Siwan 2017).

1.3.12 Vomeronasal Organ

The vomeronasal organ is symmetrical tubular organ that includes two parts: conducting and olfactory parts. Olfactory parts involve the vomeronasal duct. In general, the histological features of the vomeronasal organ of indigenous gazelles are similar to those of both small and large ruminants except that the incisive duct is completely lined with respiratory epithelium. The significant diameter of the duct of the vomeronasal organ was recorded at the incisive duct and at the rostral portion of the vomeronasal duct, compared to other (middle and caudal) portions of the duct. The significant epithelial height is evident at the rostral and middle portions of the vomeronasal duct in comparison with caudal portions. The vomeronasal duct shows an undulating respiratory epithelium, which indicates that the distension of the vomeronasal duct depends on the distension of the epithelium of the lateral wall. Both neuroepithelium and respiratory epithelium extend into the caudal portion of the vomeronasal organ, which is an important feature. Finally, the Jacobson glands in indigenous gazelle are a mucous-tubuloalveolar type (Abood and Hussein 2017, 2018).

1.3.13 Thyroid Gland

The thyroid gland is one of the most important endocrine glands which secretes thyroglobulin hormones, thyroxin, triiodothyronine, and calcitonin hormones (Khaleel and Salih 2017). These hormones play a central role in regulating metabolic activities of the body: fetus development in mammals, regulation of nutrients, calcium absorption and calorigenesis, normal reproductive functions, and milk and fiber production in domestic animals. The thyroid gland and thyroid hormones also play a crucial role in the nervous, immune, reproductive, and other systems, as well as development and functions related to mammalian adaptations to changes in the environmental conditions (Khaleel and Salih 2017).

The histological structure of the thyroid gland in the goitered gazelle is covered by a connective tissue capsule which sends out trabeculae that divide the glandular parenchyma into irregular lobules which make up an aggregation of various shapes and sizes of follicles that line the predominantly simple cuboidal and simple columnar epithelium. Parafollicular cells are present in epithelium and in follicular tissue, and myoepithelial cells are also present in the gazelle thyroid. The colloid material found in the follicular lumen is PAS (periodic acid-Schiff) positive. The isthmus is glandular in this species. The histomorphometrical analysis revealed there is a significant importance at $p < 0.05$ between left and right lobes with the isthmus in goitered gazelles (Khaleel and Salih, 2017).

1.3.14 Blood

The blood count results of goitered gazelles ($n = 11–62$) as reviewed by Kingswood and Blank (1996) are as follows:

- 5600 white blood cells/mm3 (range, 4300–6900)
- 9.8 million erythrocytes/mm3 (range, 4300–6900)
- 9.8 million erythrocytes/mm3 (range, 7.7–11.2 million)
- Hemoglobin of 16.2 g/100 cm3 (range, 44.5–17.8)
- Hematocrit of 47.7% packed-cell volume (range, 44.5–51.5)
- Mean corpuscular volume of 50.5 μm3 (range, 46.5–66.5)
- Mean corpuscular hemoglobin content of 17.1 pg (range, 16.7–21.9)
- Mean corpuscular hemoglobin concentration of 32.2% (range, 33.0–34.6)
- 4600 segmented neutrophils/mm3 (range, 100–5900)
- 2300 lymphocytes/mm3 (range, 2000–2700)

- 227 monocytes/mm3 (range, 40–280)
- 143 eosinophils/mm3 (range, 142–143)

Chemical immobilization of captive goitered gazelles apparently caused a decrease of hemoglobin from ca. 15.8 to 13.1 g/100 cm^3 and hematocrit from 47.8 to 40.0% packed-cell volume (Furley 1986). Decreases of hemoglobin and hematocrit were explained by an increase in heart rate caused by handling stress and the drug etorphine. The heart rate of a sub-adult female fluctuated between ca. 106 and 131 beats/min during a 22 min period (Furley 1986). Heart rate, respiratory rate, and certain hematological values decreased during anesthesia with xylazine/ketamine (Rietkerk and Delima 1994). The cardiac index of adult males is 6.56–8.37, of adult females 9.31–10.0, and of juveniles 9.04–11.59 (Heptner et al. 1988).

A study of the distribution of acyl carnitine (AC) and free carnitine (FC) in the erythrocytes of Arabian sand gazelle (*G. s. marica*) blood found more acyl carnitine than in whole blood and plasma and the ratio of AC/FC was significantly higher than in cats, dogs, and humans, but it was comparable to the ratio in camels, which have also significantly higher ratio to other species. The high level of AC found in the Arabian sand gazelle erythrocyte, whole blood, and plasma and also the higher AC/FC ratio in this species could be a unique adaptive mechanism which may be common for many desert animals and which may provide a necessary source of energy to various tissues during scarcity of food and water for long periods (Al-Eissa and Alhomida 1997).

1.3.15 Young Growth and Development

1.3.15.1 Newborn Sizes and Weights and Their Development with Age

A month after the end of the rut (January), embryos can already be seen with the naked eye in the uteruses of pregnant females. In December 1945, the uterus was swollen in ten adult females on the island of Barsakelmes, but the embryos were not yet visible (Ismagilov 1954). A female, investigated on January 8, 1946, had two embryos that were 2.2 cm long and weighed 0.7 g each. Two embryos found in a female hunted on February 13, 1946, weighed 40 g and 45 g; the body length of one of them was 13.5 cm. In the female, hunted on March 12, 1946, the embryos already weighed 400 g. Their body length was 30 cm, and the height at the withers was 18.5 cm. A female, who had one embryo weighing 665 g, was killed on April 15, 1946 (Ismagilov 1954). In the females studied by Sludskiy (1956a) from May 6 to May 15, 1955, in the central part of Betpak-Dala, the weight of the embryos ranged from 800 to 2900 g. In young females who had only one embryo, their weight varied from 800 g up to 1900 g, and body length varied from 36 cm to 42 cm (three individuals each), which seems typical. They were still hairless, but there were already well-developed brushes of long hair on the wrist joints (calluses). The hooves only began to harden and darken at the very top. In adult females with 2 embryos, their weight ranged from 2200 g to 2900 g (22 individuals each); and the fetuses' body length was between 48 cm and 60 cm (Sludskiy 1956a).

Table 1.14 Sizes (cm) and weights (kg) of goitered gazelle newborns (*G. s. subgutturosa*) in the Central Betpak-Dala Desert, Central Kazakhstan, in May (adapted from Sludskiy 1956a)

Sizes and weight	May 10, female	May 13, male	May 13, female	May 21, male	May 21, female
Body weight	3.1	2.7	2.8	2.9	2.6
Length of body	55	57	55	58	–
Head	16	15	15	16	–
Corpus	25	28	27	–	–
Tail	10	10	9.5	11	–
Ear	8	9	8.5	9.0	–
Oblique length of the body	30	30	30	33	–
Height at:					
Withers	40	39	40	41	–
Sacrum	42	42	43	43	–
Front leg	29	30	30	30	–
Hock joint	22	22	22	23	–
Metacarpus	20.5	19	20	–	–
Chest girth	31	31	30	–	–
Belly girth	–	24	26	–	–

The goitered gazelle newborns from Kyzylkum Desert have body length of 45–47 cm (Kim 1953) or 53–59 cm; the sacrum height is around 43 cm and their weight is 2.6 kg (Mambetzhumaev 1970).

It is interesting to note that in large fetuses of gazelles extracted in May in Betpak-Dala Desert, the weight of males was from 2.3 to 3.3 kg ($n = 12$) and females from 2.2 to 3.0 kg ($n = 17$). In females who had two embryos, the male weighed more than the female by 100–200 g. Based on this data, it can be assumed that in Betpak-Dala, goitered gazelle newborns are somewhat stronger than in other habitats (Sludskiy 1956a). For example, a newborn was found in the Ili Valley, which weighed only 1.9 kg and had a body length of 53 cm (Antipin 1941).

Fawns of goitered gazelles are born well-developed. They are covered with a thick coat of wool. In the first hours of life, they are already able to stand on their feet. Sexual differences in size and body weight in newborns are not clearly visible and the extreme values of the corresponding measurements overlap each other (Tables 1.14 and 1.15). In captivity, the weights of newborns are not different compared to wild fawns: males from twins weigh between 1900–2400 g and 2400–2600 g; single male weighs 3100 g; females from twins weigh between 2350–3420 g and 2100–2085 g. Thus, the weight of newborns fluctuates between 1900 g and 2600 g in twins to more than 3000 g in singletons. Twins weigh less in primiparous females, but in subsequent years the weight of newborns increases in the same females (Pereladova and Pereladov 1986). The weight of newborn males is on average 7.6% of that of adult animals and the females' ~ 10%; body length compared to adults are 47% and 49%, respectively. Newborns differ sharply from adults with their long legs and small absolute size of the body (Table 1.16). Due to

Table 1.15 Means and ranges of sizes (cm) and weights (kg) of goitered gazelle newborns (*G. s. subgutturosa*) on Barsakelmes Island, Aral Sea, Kazakhstan, for males and female fawns, respectively (adapted from Zhevnerov 1984)

Body size and weight	Males ($n = 38$)		Females ($n = 43$)	
	Range (min–max)	Mean	Range (min–max)	Mean
Weight	2.1–3.0	2.4	2.0–2.6	2.3
Body length	49.0–57.5	53.5	48.0–58.0	53.0
Corpus length	24.7–30.5	27.7	25.0–30.0	27.8
Head length	15.0–16.9	15.4	14.2–17.0	15.5
Ear length	7.5–8.4	8.1	8.0–8.5	8.3
Tail length	8.2–10.9	9.6	8.5–11.0	9.5
Oblique length of the body	28.3–33.0	30.2	28.2–32.7	30.0
Height at the withers	38.0–43.0	40.0	38.0–43.8	40.2
Height in the sacrum	41.0–46.0	43.8	42.0–46.0	44.5
Chest circumference	28.5–34.0	32.0	28.0–34.8	31.1

Table 1.16 Changes in body proportions of goitered gazelle fawns from birth to adulthood

	Indices			
Age	Leggy	Format	Massiveness	Overgrowth
Newborn	103	69	80	107
4 days	103	69	71	106
10 days	103	69	80	104
1 month	95	89	89	106
3 months	90	91	96	103
6 months	73	95	104	106
10 months	68	97	109	104
1 year	59	96	106	103
1.5 years	62	97	105	105
Adults	65	97	107	105

Leggy = height of the leg at the elbow × 100/oblique length of the trunk; format = oblique length of the trunk × 100/shoulder height; massiveness = chest girth × 100/oblique length of the trunk; overgrowth = sacrum height × 100/shoulder height) (adapted from Zhevnerov 1984)

this ratio of body parts, fawns are forced to spread their front legs wide at their first grazing attempts. This disparity is smoothed out by the age of 1 month, when forage contributes significantly to their diet (Zhevnerov 1984).

At birth, fawns of *G. s. marica* have an average body weight of 1.86 kg ($n = 27$) (Frazier and Hunt 1994; Lindsay and Wood 1992) or more exactly 1.84 ± 0.02 kg for female fawns ($n = 22$) and 1.94 ± 0.04 kg for male fawns ($n = 25$), though statistical differences between weights of sexes at birth are absent (Gürler et al. 2015); female fawns of *G. s. subgutturosa* have a weight on average of 2.2 kg ($n = 51$) and

Table 1.17 Birth weights (kg) of newborn (1–3 days old) *G. s. marica* (?) fawns from Turkey by Gürler et al. (2015)—open access

	Females			Males			All	
Year	*N*	Range (min–max)	Mean ± SE	*N*	Range (min–max)	Mean ± SE	*N*	Mean ± SE
2008	5	1.56–2.04	1.85 ± 0.04	6	1.58–2.52	2.03 ± 0.09	11	1.946 ± 0.06
2009	8	1.62–2.06	1.84 ± 0.04	7	1.60–2.21	1.91 ± 0.06	15	1.875 ± 0.04
2010	9	1.69–2.01	1.83 ± 0.03	12	1.70–2.12	1.92 ± 0.04	21	1.877 ± 0.03
All	22	1.56–2.06	1.84 ± 0.02	25	1.58–2.52	1.95 ± 0.04	47	1.899 ± 0.02

male fawns on average of 2.4 kg (n = 38) (Frazier and Hunt 1994). Birth weights ranged between 1.71 and 1.95 kg in twins (n = 10) and 1.58 and 2.52 kg in singletons (n = 37). The difference between the mean birth weights of twins (1.824 ± 0.75 kg) and singletons (1.919 ± 0.024 kg) was statistically significant (Gürler et al. 2015).

Although the weight differences are not great between the sexes in *G. s. marica*, they are still significantly different indicating sexual dimorphism together with males having longer horns and thicker necks and standing taller at the shoulder than females (Table 1.17; Gürler et al. 2015). Growth within the first 6 months is very similar between the sexes with an overall rapid growth observed within the first 18 months for males and females. Body weight levels off at around 4 years for females and 6 years for males under captive conditions which is similar to figures presented for *G. subgutturosa* (Jamsheed 1976; Zhevnerov et al. 1983). Mean weight for adult *G. subgutturosa marica* is similar to that published by Kingswood and Blank (1996; 20.2 kg) and Habibi et al. (1997) (17.7–23.8 kg for males and 15.3–18.4 kg for females). Maximum body weight established for both sexes is slightly heavier than reported by Kingswood and Blank (1996) and Habibi et al. (1997) but expected from animals in captivity (Cunningham et al. 2011).

Significant differences are observed between horn length—straight and curved—for males and females with juvenile males showing a rapid increase in horn length within the first 6 months of growth compared with females. Although horn development is continuous throughout life, the rapid growth phase levels off between 3 and 5 years for males and females which is similar for *G. subgutturosa* in Iran (Jamsheed 1976). The 18% of females (n = 41) with horn abnormalities (deformities and breakages) are disconcerting which either result from inbreeding—which requires immediate further investigation—or handling-related damages, especially from holding crates, during capture operations when they were young, i.e., when horns are still flexible, with female horns being more susceptible to damage. This is in contrast to 8% (n = 289) of abnormalities observed (from skulls collected) for female sand gazelles in a free-ranging population—Mahazat as-Sayd Protected Area—in central western Saudi Arabia (Cunningham 2009). However, according to Habibi (1991), horn damage in females is common as a result of the length and slender nature of the horns.

G. s. marica fawns born late in the year (September or October) are heavier than those born early in the year (March, April, or May), and younger mothers generally give birth to heavier offspring. Moreover, males are heavier than females, and singletons are substantially heavier than twins, whereas the difference in weight at birth between sexes is smaller in twins than singletons (Riesch et al. 2013). A mother's birth type (i.e., whether she was born as a singleton or twin) or her weight at birth again has no impact on her offspring's weight at birth. When investigating sex-specific twin-offspring weight at birth, no statistically significant effects of a mother's birth type on male or female offspring were found (Riesch et al. 2013).

The fawns are born sighted and covered with thick hair, and after an hour the strongest of them can already stand on their feet and run over short distances. Incisors of newborns are fully erupted, white and sharp, eight pieces; two medium ones are especially large and are separated to the sides. The "triangular" space between them is probably an adaptation for suckling (Zhevnerov et al. 1983). The weight of newborn gazelles in the Betpak-Dala Desert ranges between 2.7 and 3.1 kg (Sokolov 1959). In the Betpak-Dala Desert, the weight of five newborn gazelles found by Sludskiy (1956a, b) ranged from 2.6 to 3.1 kg, and the body length was from 55 to 58 cm. On the island of Barsakelmes, the weight of newborns ranges from 2.0 to 3.0 kg; and the body length is from 48 to 58 cm. Sexual differences in size and body weight in newborns are not pronounced (Table 1.15; Zhevnerov et al. 1983).

The color of the hair of a newborn fawn on the head, neck, back, and top of the tail is light brown. According to this basic background, there are a lot of dark brown, almost black guard hairs on the upper part of the neck and back, and the overall color tone is pockmarked. There is no dark pattern characteristic of semi-adult animals on the muzzle. The lower lip is almost white. The chest at the front legs is pure white. The upper sides are light brown, gradually lightening to light toward the chest and belly. The groin area is pure white, and the rump patches are white. The terminal half of the tail is dark brown above, almost black. The hooves are black. The irises of the eyes are dark brown. This color is very in harmony with the background of the clay-gravelly desert, and it is difficult to notice a hidden fawn as it lies motionless. Sometimes he can be found betrayed only by its eyes, which are usually not closed (Zhevnerov et al. 1983).

The intensity of growth of newborns during the first month of life is characterized by significant unevenness. The greatest leap in the development of all parts of the body occurs in the first 10 days and is about 50% of the monthly increase (Tables 1.18, 1.19, 1.20, and 1.21). Biologically, this is beneficial, since the period of passive (hiding) response to predators of fawns is reduced (Sokolov 1959). At the beginning, daily weight gain ranges from 30–80 g to 300 g per day; the average daily weight gain during the first 7 to 9 days after birth is 18 to 250 g. After the ninth day, the daily weight gain stabilizes, and the range of fluctuations decreases from 180 to 250 g, on average, to 190 to 200 g per day and is even in the twins. In a single fawn, the gain is stable almost from the very beginning and amounts to 180–240 g (an average of 192 g) per day. By the age of 1 month, the fawns have reached a

Table 1.18 Mean and standard deviation (SD) of body measurements for different male and female age classes of *G. subgutturosa marica* as determined from the KKWRC living collection in Saudi Arabia (by Cunningham et al. 2011—License Number 5475801447513)

		Age (days)	Weight (kg)	Horn, straight (cm)	Horn, cured (cm)	Tip to tip dist. (cm)	Neck (cm)	Shoulder height (cm)	Forehead/ horn ratio	Ear (cm)
Juvenile male <6 months	Mean	73.6	8.9	3.9	4.0	3.9	19.6	53.3	3.4	11.8
	SD	17.5	1.8	1.4	1.5	0.7	1.5	3.2	3.2	0.6
Juvenile female <6 months	Mean	86.6	9.2	3.4	3.5	4.1	17.9	53.0	4.3	11.8
	SD	28.8	2.4	1.7	1.8	0.7	1.1	2.2	3.6	0.7
Sub-adult male (6–18 months)	Mean	369.0	17.4	19.9	21.8	4.3	26.2	65.1	0.6	–
	SD	35.4	2.3	2.5	2.5	2.4	1.6	1.9	0.1	13.5 0.4
Sub-adult female (6–18 months)	Mean	257.3	15.8	20.2	21.5	9.4	21.7	60.5	0.6	12.7
	SD	40.8	4.3	8.3	8.6	5.7	1.8	2.4	0.3	0.8
Adult male (>18 months)	Mean	905.6	19.5	28.8	33.2	11.7	29.0	64.9	0.4	13.3
	SD	433.2	3.0	2.3	3.2	3.9	3.3	2.5	0.1	0.5
Adult female (>18 months)	Mean	1428.6	16.6	20.4	21.4	7.9	22.1	61.8	0.6	12.9
	SD	1155.0	2.7	2.9	3.1	4.7	1.4	2.2	0.1	0.6

weight of 6.5 to 7 kg and by the age of 6 months 16–18 kg (Pereladova and Pereladov 1986).

The average increase in all body parts in the first month of life is significantly higher than in the months that follow. The increase in body length during the first month reaches 38%, height at the withers 26%, in the sacrum 20%, and chest circumference 39%. In the following months, these growth rates decrease, with the exception of body length, which continues to increase rapidly until the age of 6 months (Table 1.16). Young gazelles gain weight quite fast. If a 1-day-old fawn weighs between 1.9 and 3.3 kg (average of 2.4 kg, $n = 16$), then after a month (4–5 weeks), its weight reaches 5 to 9 kg, or about 1/4 of the weight of an adult animal. For example, a young female examined in Betpak-Dala on June 25, 1954, at approximately 1 month of age weighed 7.1 kg and had a body length of 70 cm and a height at the withers of 49 cm, and a male measured in June 26, also about 1 month of age, weighed 9.5 kg and had a body length of 80 cm and a height at the withers of 55 cm (Zhevnerov 1984).

A fawn reaches 7–13.5 kg at 10–13 weeks (Heptner et al. 1961), and they can reach half the height of adults at this age. The horns of males begin to grow at the age of about a month (Sokolov 1959). At the age of 3 months, the average weight of fawns is 13.5 kg. Four-month-old fawns already weigh from 14 kg to 18 kg. Thus, a young female, measured on September 18, 1949, weighed 14 kg, and a male from September 19, 1954 weighed 17.5 kg. A female examined in September 28 weighed the same amount, but as early as August 7, 1949, a young male weighed 18 kg. At the age of 6 months, the males weighed between 15 and 20 kg. And at 7 to 8 months, the body length of the arrivals approaches the smallest limits of adult females. For

Table 1.19 Development of goitered gazelle fawns in the Barsakelmes Island (mean indices), compared to that of adult males and females. Measurements are given in kilograms and centimeters, respectively (adapted from Zhevnerov et al. 1983)

Sizes (cm), weight (kg)	Goitered gazelles' age									Adults	
	Newborns ($n = 38$)	4 days ($n = 2$)	10 days ($n = 1$)	1 month ($n = 1$)	3 months ($n = 2$)	6 months ($n = 2$)	10 months ($n = 7$)	12 months ($n = 1$)	18 months ($n = 1$)	Males ($n = 22$)	Females ($n = 10$)
Weight	2.4	3.0	4.5	8.8	11.8	14.7	13.0	21.0	23.5	30.0	24.0
Length of											
body	53.5	57.0	61.5	69.0	76.0	96.0	102	103	108	118	108.0
Head	15.4	17.0	–	20.0	21.0	25.0	25.0	25.5	26.5	29.0	27.0
Ear	8.1	9.5	10.5	12.0	12.0	13.8	15.0	14.0	14.5	14.0	14.0
Tail	9.6	11.0	11.0	14.0	15.0	16.0	18.0	17.0	16.0	16.0	16.0
Wither height	40.0	46.0	48.0	52.0	55.0	60.0	62.5	64.0	68.0	75.0	70.0
Sacrum height	43.8	49.0	50.0	54.0	57.0	63.0	64.0	67.0	72.0	82.0	75.0
Chest girth	32.0	34.0	39.0	44.0	53.0	65.0	69.0	70.0	72.0	75.0	72.0
Horn length	–	–	–	1.0	2.5	6.5	12.0	17.0	24.0	30.6	–

Table 1.20 Development of goitered gazelle fawns in the Betpak-Dala Desert (weights in kg and measurements in cm), compared to adult size measurements (adapted from Sludskiy 1956a)

	Goitered gazelles' age									
	Newborn	1 Month	2.5 months	3 months	4 months	6 months	1 year	1.5 year	Adults	
Sizes (cm), weight (kg)	($n = 16$)	($n = 3$)	(n–1)	($n = 5$)	($n = 4$)	($n = 1$)	($n = 1$)	($n = 15$)	Males ($n = 18$)	Females ($n = 50$)
Weight	1.9–3.3 (2.4)	7.1–9.5	12.0	(13.5)	14–18	17.5–19.5	24	(20.4)	20–43 (30)	18–33 (24)
Length of										
Body	53–58 (54.3)	68–70	72	(74)	(92)	–	102	(102)	111–125 (119)	102–120 (109)
Head	15–16	20	–	–	23	–	25	–	27–29	24–29
Corpus	25–28	40	–	–	47	–	51	–	57–58	(26)
Tail	8–11	12–15	15	(16.5)	18.5	–	21	(18)	16–21 (18)	55–60 (56)
Ear	8–9	11	11.5	(12.5)	15	–	15	(13)	12–14 (13)	10–23 (17) 13–14 (13)
Oblique length of body	30–33	44	–	–	55	–	74	–	64–66	61–70 (67)
Wither height	36–41	49–55	53	(55)	61	–	–	(69)	69–79 (77)	68–76 (71.4)
Sacrum height	42–45	52–61	–	(58)	69	–	67	(74)	79–85 (81)	74–83 (77.4)
Front leg height	29–30	36–88	–	–	42	–	46	–	44–46	44–47 (45)
Hock joint height	22–23	24.5	–	–	30	–	33	–	30–31	30–31 (30.5)
Metacarpus height	19–20	22	–	–	–	–	28	–	27–28	26–27
Chest girth	30–31	44	–	–	65	–	70	–	72–81	67–75
Belly girth	24–26	36	–	–	–	–	69	–	67–68	74–76
Horn length on curve	–	1–1.5	–	–	6	–	17	8.5–30 (17.3)	33–34 (33.5)	2.5–5

Table 1.21 Growth dynamics of goitered gazelles (monthly measurements in cm and kg, respectively) (adapted from Zhevnerov 1984)

	Age in months												
Measurement	Newborn	1	2	3	4	5	6	7	8	9	10	11	12
Length:													
Body	50	69	75	79.5	93.7	97	–	101	102	–	103	104.5	105
Head	15	18	21	22.5	23	24.5	25	25.5	26	102.5	6.5	27	27
Ear	8	11	12	12.5	13	13.7	14	14	14.5	–	15	15	15
Tail	8.5	11	–	14	16	17.5	18	19	19.5	19.5	20	20	20
Height at:													
Shoulder	38	48	53	56	57	57	58	60	61	–	63	64	64.5
Sacrum	41	49.5	54.5	58	–	61	–	63.5	64	65	–	66	67.5
Chest girth	31.5	43	48.5	54	60	–	65	68	68.5	69	69	70	70
Slanting body length	29	39.5	45	50	–	55	57.5	59	60	62.5	63	63.5	63.5
Weight, kg	2.1	6.8	9.6	12.5	14.9	–	18.4	19.6	20.7	–	21.6	22.3	23

example, on October 9, 1949, two young gazelles weighed 17.2 and 19.5 kg. The fawns weighed 16.5–20 kg at the age of 6–8 months. A 1-year-old gazelle weighed 20.9 kg (Antipin 1941), and 1.5-year-olds averaged 20.4 kg (n = 15; Ismagilov 1954). A 1-year-old male, studied by Sludskiy (1956a) on May 13, 1956, in the central part of Betpak-Dala Desert, weighed 22.4 kg, and two 1-year-old females in May 24 in the same place weighed 22.4 kg and 16.2 kg. In addition, the ratio of different body parts in growing gazelles also changes (Table 1.16). Young reach approximate adult weight at 18 to 19 months, while weight changes already have seasonal characteristics (Zhevnerov et al. 1983).

In Iran, external measurements (in mm) of females (n = 3) at 3, 6, and 12 months of age, respectively, are as follows: total length, 1000, 1010, and 1040; length of tail, 170, 170, and 190; length of hind foot, 270, 320, and 320; length of ear, 120, 130, and 130; height at shoulder, 520, 650, and 650; and chest girth, 470, 600, and 610 (Mowlavi 1978).

In the third or fourth month, there is a slight decrease in growth rates. An increase in height at the withers and sacrum, for example, occurs by only 5–7% and chest circumference by 12%. However, the increase in body length is still faster than the rest of the body. At 7–8 months of age, the body length index of the adolescents, as a rule, approaches the extreme indicators of adult females (Table 1.19; Zhevnerov 1984).

In the next 3 months of life, falling in the winter period, with a lack of a full availability of feed and the intensive consumption of fat deposits, the growth of profits slows down significantly compared to the summer–autumn period. During the winter period, the body length increases by an average of only 4.5 cm. The height at the withers and sacrum increases even more slowly. Subsequently, in the second summer period of life, growth proceeds more slowly than in the first. The size of adult animals is reached at the age of 19–20 months (Zhevnerov 1984), though gazelles continue to grow later, until about the age of 2, and possibly longer, but this growth is much slower than in the first half of the year. Young gazelles at the age of 1–2 years are often difficult to distinguish in size from adults. Only on the muzzle of the young is a dark-brown lateral stripe that passes through the eye, which is almost invisible in adult gazelles. In spring, in May, young gazelles differ from adults by a brownish coat color, whereas in old gazelles it is grayer (Sludskiy 1956a). Males reach their near maximum weight (ca. 30 kg) by 3–4 years of age and continue to grow until their fifth year; females reach full growth (ca. 25 kg) by 2–3 years (Jamsheed 1976; Zhevnerov et al. 1983). Horns have a true lyrate shape by 2–2.5 years (Zhevnerov 1984). Horn growth after 5 years of age is negligible, although slight thickening of horns continues (Jamsheed 1976). Thus, three periods are markedly different in the postembryonic development of the goitered gazelle: the period of intensive growth, which lasts from birth to 6–7 months of age; the period of slowed growth during the winter period; and the period of completion of growth that begins in spring and lasts about 8 months (Zhevnerov 1984).

The growth and development of young animals is directly dependent on the feeding conditions in a year. In years of severe drought, the growth lag in young gazelles begins to manifest itself at 3–4 months, with the end of the lactation period. At 7 to

8 months old, the young's linear body length indicators, for example, are 3–5 cm less than normal and the weight lag is even more significant. In the dry winter of 1970, the live weight of fawns in December averaged 11 kg ($n = 4$), which is actually typical for individuals of 3 months of age developing in favorable conditions. Generally, fawns artificially fed were similar in growth rates to individuals developing in the wild and even surpassed them in some indices (Tables 1.18 and 1.19). These differences were especially evident during the winter period. While in wild individuals, growth slowed down quite sharply in the winter months, in captive individuals, thanks to regular feeding, growth occurred more smoothly. In the process of growth and development of young gazelles, the proportions of their physique also change. In newborns, the index of leggedness significantly prevails over the indices of format and massiveness (Table 1.15). With age, due to the uneven growth of body parts, the ratio of body shape indices becomes reversed and the indices of the format and massiveness increase, and leggedness decreases. In 6- to 7-month-old individuals, the body proportions are close to those of adult gazelles (Tables 1.16 and 1.19; Zhevnerov 1984).

Table 1.22 shows the change in weight and linear parameters of some internal organs of the goitered gazelle during growth. The greatest increase in the absolute weight of internal organs occurs in the first month of life of newborns: the weights of the heart increase, for example, by more than three times, and lungs and liver by 1.2 and 2.4 times, respectively. The size of the intestine increases even more significantly. In fawns at the age of 1 month, the length of the small and large intestines is 80% and 50% of that in adult animals, which is associated with the consumption of a significant amount of plant food. The relative weight of internal organs changes slightly with age. At the age of 1.5 years, the weight and size of the internal organs of the gazelle are close to that of adult individuals (Zhevnerov 1984).

Longevity averages 6 years in the wild; most males live less than average but females can live 8–12 years (Zhevnerov et al. 1983). Longevity in captivity is at

Table 1.22 The changes of some internal organs during young development (adapted from Zhevnerov 1984)

	Young age					
Indices	Newborn	4 days	1 month	6 months	12 months	1.5 years
Weight, g						
Heart	18	31	71.5	90	190	210
Lungs	52	61	102	127	319	387
Liver	49	79	120	165	592	710
Spleen	4	8	–	19	33	57
Length, cm						
Small intestine	680	700	1012	1110	1280	1250
Colon	130	164	200	500	560	600
Cecum	9	17	18	22	25	27
Number of specimens	5	2	1	2	1	1

least 15 years and 8 months (Jones 1993). There is likely mistaken report of a male that lived to the age of 20 years (Kingswood and Blank 1996).

1.3.15.2 Age-Related Variability of Limbs and Skull

The growth and fusion of limb bones, especially calcaneum or heel bones, are quite significant, showing clear signs of growth. One sign is that the body of the bone lengthens and thickens considerably during the first year of life. Another sign is that at about 14 months, the process (a small extension to which a tendon is attached), which begins as a separate piece of bone, then fuses to the main structure. In newborn gazelles, one group of heel bones is in the process of being unfused. In yearlings, a second cluster has the process unfused, but they are almost as large as the adult bones. Adult gazelles have large heel bones with the process fully fused (Legge and Rowley-Conwy 1987).

Age-related changes in the proportions of the goitered gazelle skull are quite significant. The skull of a newborn differs from the skull of an adult with a relatively short facial region and a large brain box. With age, the muzzle lengthens, and the brain part shortens (Table 1.23; Zhevnerov 1984).

Over the period of postembryonic development, the absolute length of the skull and facial and cerebral parts increases by 1.6–1.8 times. The proportions of the facial region change little with age. For example, the relative width of the nasal opening, the interorbital, and the width along the back walls of the eye sockets almost do not change. The proportions of the brain region change significantly with age: for example, the relative width of the brain box decreases from 42.6% to 30.5% (Table 1.24). The most intensive growth of all parts of the skull proper is observed during the first 3 months after birth. The total length of the skull during this period increases by 33%, whereas in the future, up to 1 year of age, it only increases by 19.3%. After 3 months of age, the intensity of skull growth decreases markedly (Table 1.24). So, over the next 2 months, the increase in most indicators was 4 to 15 times less. A relatively smooth increase in all parts of the skull is observed from the 5th to the 10th month and from the 10th to the 15th month. At the age of 1.5 years, the size of the skull in the main indicators is close to that of adults (Tables 1.23 and 1.24; Zhevnerov 1984).

1.3.15.3 Horn Development

Newborn males have only funnel-shaped curls of wool in place of future horns. The horns become noticeable at the age of 1 month. Intensive growth of bone processes begins at 3 months (Table 1.25). Horns in males begin to break through as early as the 12th day after birth and become noticeable at the age of 1 month, when the horn covers reach a length of about 1 cm and 2–3 weeks later reach 2 cm in length and after 2 months up to 3 cm. In young males, the horns are first devoid of transverse rollers and the ends are bent inward (Fig. 1.26). Only at the age of 5 months (in September) are the first rings seen, with the next ones appearing after 2 months. Rings are formed at the base of the horns, pushing the previous ring up. The rings are transverse folds of the epidermis, which has not yet had time to harden, but soon they become keratinized. The rings formed in autumn and winter lie close to each

Table 1.23 Age-related changes in goitered gazelle fawn skull measurements

Fawn age	Total length	Front part length	Brain part length	Maximal width	Interorbital width	Brain chamber width	Nose hole width
Newborn							
1	117	81	36	55	35	51	15
2	100	69.2	30.7	47.0	29.9	42.7	12.7
4 days							
1	121	84	37	59	37	51	15
2	100	69.4	30.5	48.7	30.5	42.1	12.4
3 months							
1	161	116	45	74	43	54	18
2	100	72.0	28.0	43.4	26.7	33.5	11.2
5 months							
1	164	120	47	77	43	55	20
2	100	78.9	26.2	47.0	26.2	33.5	12.2
10 months							
1	182	132	50	85	48	58	24
2	100	70.8	27.3	46.6	26.3	31.8	13.2
15 months							
1	194	101	53	90	50	60	25
2	100	72.6	27.3	46.4	25.7	30.8	12.6
1.5 years							
1	208	146	60	95	57	62	25
2	100	71.1	28.8	45	27.4	29.8	12.0
Adult male							
1	220.1	150	92	96.9	60.6	64	25
2	100	71.4	42.8	46.1	28.4	30.8	11.9
Adult female							
1	199.3	149.3	79	91	56	62	25
2	100	69.8	32.1	45.6	28.1	31.1	12.6

Legend: 1, absolute number; 2, proportion (%) of skull length (adapted from Zhevnerov 1984)

other; the rings are significantly narrower in spring and summer and are formed three times faster than in the autumn–winter (Ishunin 1961).

The age of an individual male can be judged by the horns' length. In fawns at the age of 1 month, the horns are 1 to 1.5 cm long, in a 4-month-old 6 cm (on the curve), in a 1.5 -year-old from 8.5 to 30 (17.3) cm, and in a fully grown adult from 33 to 43 (38.5) cm. In male yearlings, the number of rings is between 6 and 8 (Fig. 1.27), and in males at the age of 1.5 years, the number of rings on the horns does not exceed 11 (Fig. 1.28) (and sometimes they are smooth); in adults, there are 12 to 25 rings or 18 rings on average. The number of rings on the horns of gazelles of the same age can be different, so it is impossible to judge the age of a male by ring number. In fawns, the horns grow by about 2 cm without rings. At 5 months, in September (often earlier), the first ring appears and in November the second ring. During this

Table 1.24 Age-related changes of skull measurements in goitered gazelle fawns during ontogeny in % (adapted from Zhevnerov 1984)

	Fawn age					
Indices	1–4 days	4 days–3 months	3–5 months	5–10 months	10–15 months	15–18 months
Total length	3.4	33.0	1.8	11	6.5	7.2
Front part length	3.7	37.3	3.4	10	6.8	4.9
Brain part length	2.7	21.6	4.4	6.3	6.0	13.2
Maximal width	7.2	25.4	4.0	10.3	5.8	5.5
Interorbital width	5.7	16.2	–	11.6	4.1	14.0
Brain chamber width	–	5.8	1.8	5.4	3.4	3.3
Nose hole width	–	20.0	11.1	20.0	4.1	–

Table 1.25 Development of horns in goitered gazelles (adapted from Zhevnerov 1984)

	Age of goitered gazelle in months					
Indices	3	6	10	12	18	36
Horn length (in straight line)	1.0	3.9	6.5	7.9	9.0	14.0
Distance between horns:						
At the base	2.5	2.0	1.8	1.5	1.5	1.0
At the ends	4.0	5.5	6.0	7.5	8.5	12.5
Girth of the horns at the base	3.5	5.5	7.3	8.5	10.0	11.5

Fig. 1.26 In young males, the horns are first devoid of transverse rollers and their horn ends are bent inward. Goitered gazelle male of 3-month-old in the Southern Ustyurt, SW Kazakhstan. 22.08.2022. Photo of the camera-trap of the Ustyurt State Nature Reserve team with kind permission

Fig. 1.27 At the age of one year, males have 6-8 rings on their horns. Male-yearling in the Ustyurt Plateau area, SW Kazakhstan. 05.06.2019. Photo of the camera-trap of the Ustyurt State Nature Reserve team with kind permission

Fig. 1.28 In adult males, the number of horn rings does not exceed 11. A goitered gazelle male at the age of 1.5 years. Issyk-Kul goitered gazelle nursery for re-acclimatization, 25.11.2022. Photo by D. Blank

Fig. 1.29 At the age of 5-6 months, male horns of goitered gazelles have a hooked shape, the ends of which are directed towards each other and they are strongly curved outward in the middle parts. Drawing of U. Kenzhetaev with kind permission

period, the horns gradually acquire the concavity of the middle part outward, and their ends bend inward, toward each other (Fig. 1.29). The distance between the base of the horns in young individuals is not only relatively but also absolutely greater than in adults, which is due to the thickening of the horn itself during growth and a change in the shape of the bone stump (Zhevnerov et al. 1983).

At 6–7 months of age, the ends of the horns facing inward go behind each other; one to two rings begin to vaguely show. Relatively rapid growth of horns occurs from birth to 6 months of age and from 1 year to 1.5 years. Apparently, there is a difference in the rate of growth and development of horns, depending on external conditions and on individual characteristics—in different parts of the goitered gazelle habitats, the horns of young have different sizes (Figs. 1.30 and 1.31). By the end of the spring period, the horns become equal to or slightly more than half the length of the horns of an adult gazelle. They diverge wider apart, but their ends are still facing inward, toward each other (Fig. 1.32). In gazelles aged 15–18 months, the upper layers of the horny substance between the rings constantly exfoliate and peel off. In individuals older than 1.5 years, this process ends. This makes it possible to distinguish the horns of young from the horns of fully adult individuals since they are already almost equal in size (up to 26 cm). By the age of 2–2.5 years, the horns take the correct lyre-like shape, and their growth, as a rule, ends (Fig. 1.33) (Zhevnerov et al. 1983).

Fig. 1.30 There is a difference in the rates of growth and development of horns of goitered gazelle young under different environmental conditions. A male young at the age of seven months has relatively longer horns. Kalamali Ungulate Nature Reserve, Xinjiang, China, 15.12.2022. Photo by F. Xu with kind permission

Fig. 1.31 A male fawn at the age of ten months (stands behind the female) has relatively shorter horns. South Gobi Region., Mongolia. 23.03.2014. Photo by P. Kaczensky with kind permission

Fig. 1.32 A male-yearling at the age of 12-13 months. Its horns become a slightly longer than half of an adult male, but their ends are still facing towards each other. Great Gobi B, Mongolia. 01.07.2012. Photo by P. Kaczensky with kind permission

Fig. 1.33 IBy the age of 2 years, the male's horns take the correct lyre-shape. Goitered gazelle male older than 2.5 years. Ustyurt Plateau, Southern Chink (Kaplankyr). 18.10.2019. Photo of the camera-trap of the Ustyurt State Nature Reserve team with kind permission

1.4 Habitat

1.4.1 Central Asia

The goitered gazelle is a typical representative of the fauna found in arid landscapes. Its range consists of almost all types of deserts and semi-deserts of Asia (Heptner et al. 1961). Gazelles occur in flat or rolling areas, along foothills with broken terrain, and mountain valleys and plateaus. They avoid rocky cliffs, terrain devoid of gullies and ravines, thick, woody vegetation, and land used for cultivation or intensive livestock grazing (Roberts 1977; Neronov and Bobrov 1990). Goitered gazelles, however, will graze at the edge of cultivated land throughout much of their range (Kingswood and Blank 1996). The northern distribution of goitered gazelles is limited by the difficulty of foraging where the snow depth exceeds 10–15 cm and the inadequacy of pelage that does not provide effective insulation from wind and bitter cold (Sludskiy 1963; Zhevnerov 1984). In winter, they inhabit windy, snow-free areas, finding shelter from wind in deep wadis, gorges of low mountains, or thickets (Blank 1990). The main distribution areas of the goitered gazelle are open plains, but unlike the other two Asian antelopes (the Mongolian gazelle *Procapra gutturosa* and the saiga antelope *Saiga tatarica*), this species not only tolerates the rugged terrain, but even prefers semi-closed areas of dry riverbeds and foothills (Sokolov 1959; Zhevnerov 1975).

In western and southern Central Asian regions, the favorite habitats of goitered gazelles are fixed hummocky sands with sparse vegetation and sparse saxaul (*Haloxylon* spp.) forests (Kostin 1955; Zhevnerov et al. 1983). Gazelles live only in sandy deserts fixed by vegetation cover. But their highest preference is areas with dense soil of clay and gravel deserts (Zhevnerov et al. 1983). They clearly avoid poorly anchored sand dunes, as well as open areas with heavy stony soil (Kostin 1955). Favorite locations are fixed bumpy sands with sparse shrubby vegetation, as well as rare saxaul growth, which provides protection from both heat in the summer and bad weather in the winter. However, goitered gazelles avoid both dense shrubby thickets of shrubs and saxaul. In summer, goitered gazelles visit salt marshes (Kostin 1955). In eastern Azerbaijan, in summer, the goitered gazelles stay in stony places of calcareous rocks, where, due to the presence of more diverse vegetation, they are better provided with fresh green food. Areas of sandy deserts often serve as habitats in winter, when the soil becomes more compacted (Vereshchagin 1939). Mainly during deep snows of winter, gazelles gather in significant numbers in the sands and saxaul thickets of the Kyzylkums. In particularly unfavorable years due to snow, they migrate to this desert far from the north (Ishunin 1961). The goitered gazelle is not adapted to moving through deep snow cover and getting to the food under it, so in winter they stay only in areas where the highest average snow cover is 10 cm or less; therefore, gazelles appear only in summer in semi-deserts that accumulate more snow in winter and avoid the steppe zone during winters with its deep snow cover (Zhevnerov et al. 1983).

In Transcaucasia and Central Asia, gazelles inhabit not only open plains but also gentle foothills, as well as mountain valleys and plateaus with elevations of 2500 to

3000 m ASL. In the Ketmen Range of northern Tien Shan, several groups of goitered gazelles were observed higher than the line of mountainous coniferous forests, at an elevation of around 3000 m ASL (Zhevnerov et al. 1983). In winter they also live in the Boguty Mountains (the eastern spurs of the Zaiiliskiy Alatau, Tien Shan) on sun-exposed slopes without snow (Sokolov 1939). In the spurs of the Dzungarian Alatau, goitered gazelles live as high as 1500 m ASL (Antipin 1941), and in Mongolia and the Iranian Highlands, they are found even at altitudes of about 2000 m ASL (Dementyev 1935; Bannikov 1954). The limit of vertical distribution is not set by the height of the terrain itself, but by the topography and the nature of the vegetation associated with it. Goitered gazelles carefully avoid high mountains and rocky areas, even low mountains, because they are not adapted to moving through steep and highly rugged terrain (Sokolov 1959).

In the northern areas of their natural habitat, where the snow cover is higher than 10 cm, goitered gazelles migrate south for the winter or descend from the foothills to the lowlands. In the spring–summer period, the gazelles go far north into the semi-deserts and climb quite high into the mountains (Zhevnerov 1984). Due to this ecological feature of this species, the northern part of its range is divided into areas of summer and winter quarters. The gazelles appear in their summer habitats in May and stay there until September–October. In Kazakhstan, the areas of the goitered gazelle's summer habitats are located between the Emba River and southern Ustyurt, in the Daryalyk Desert (north of the Syr Darya River), in the interfluve of the Sarysu and Chu rivers (Sludskiy 1956a, b; Sludskiy 1963).

Dominant vegetation of goitered gazelle habitats includes *Anabasis*, *Artemisia*, *Arthrophyton*, *Astragalus*, *Calamagrostis*, *Haloxylon*, *Eurotia*, *Nanophyton*, *Peganum*, *Phragmites*, *Poaceae*, *Salsola*, *Stipa*, *Stipagrostis*, *Suaeda*, and *Ziziphus* (Kingswood and Blank 1996). Biomass of forage species in the plant community is 340 kg/ha (Mowlavi 1978).

1.4.2 Kazakhstan

1.4.2.1 Ustyurt Plateau

In the Ustyurt Plateau (Fig. 1.34), the main locations for goitered gazelles, at least in the summer, are extensive depressions with takyr-like soils (takyr is a flat clay depression in the arid zone of Central Asia, periodically flooded with water after spring snowmelt and heavy rains; Fig. 1.35). The gazelles are attracted not by the features of the relief and soils in these areas, but rather by the presence of juicy green vegetation, which is more abundant and persists for a longer time in depressions that allow the goitered gazelles to exist without watering longer than in other sites (Fig. 1.36). In the second half of summer, temporal watering holes (puddles) and most of the vegetation dry up, and then the goitered gazelles need to find operating watering holes, which are very rarely found along the Ustyurt Chink (plateau cliff). However, goitered gazelles cannot survive in this region without water and willingly visit available water sources (Fig. 1.37), although often the water is quite salty, and in some places, it is still bitter from an excess of salts. In contrast, on the

Fig. 1.34 In the Ustyurt Plateau, goitered gazelles are found mainly in the extensive depressions along the chink (plateau cliff) in SW Kazakhstan. 26.04.2011. Photo by M. Pestov with kind permission

Fig. 1.35 Takyrs are flat clay depressions typical in the arid zone of Central Asia; they are periodically flooded with water after spring snowmelt and heavy rains. A takyr with the horns of a goitered gazelle male near the Ustyurt Plateau, SW Kazakhstan. 22.05.2017. Photo by M. Pestov with kind permission

Fig. 1.36 Even in a dry autumn, vegetation is more varied in depressions, and *Salsola* species and other plants from the Family Chenopodioideae play an important role in the autumn diet of goitered gazelles. A five-month-old male with his twin-female in a valley along the Ustyurt Plateau (far in the background), W Kazakhstan. 11.09.2016. Photo by M. Pestov with kind permission

Fig. 1.37 Goitered gazelles often visit rarely found water sources during dry summer and autumn. Four goitered gazelle males seen during drinking water in the southern area of the Ustyurt Chink (Kaplankyr), in SW Kazakhstan. 30.08.2019. Photo taken by the camera-trap of the Ustyurt State Nature Reserve team with the kind permission

elevated parts of the Ustyurt Plateau, goitered gazelles are absent or appear very rarely, because of the absence of juicy vegetation and watering holes (Minervin 1944).

1.4.2.2 Buzachi Peninsula

Here, the goitered gazelles prefer small-sized solonetz soils; solonetz is a type of soil characterized by a large amount of sodium in the soil absorbing a complex of illuvial deposits; unlike salt marshes, solonetz soil contain water-soluble salts at some depth and not in the uppermost horizon, and about half of all the individual gazelles encountered were observed among the leveled sands, closer to the northern peninsula. The other half of gazelles were found in the central parts of the peninsula. Here gazelles preferred bumpy sands to flat sands. The number of gazelles decreased from west to east, from the Caspian Sea into the peninsula. There are other watering holes on the peninsula besides the sea and these watering holes are located on salt flats, which attract gazelles. In spring, the entire Ustyurt Plateau and Mangyshlak Peninsula are flooded with water. By the end of spring and the beginning of summer, most of these reservoirs, which are used by gazelles and other wild ungulates as watering holes, dry up. The water lasts the longest on soras (shallow, drainless salt marshes in closed depressions) and salt marshes, where it apparently evaporates last. This gives a certain advantage to the gazelles living on the Buzachi Peninsula, which is surrounded on all sides by soras and kultuks (solonchak that is created by a drying sea), in terms of water availability. This is especially important for females after giving birth in May–June. Obviously, such an abundance of water, although to varying degrees salty, lengthens the life of plants (Blank 1991).

1.4.2.3 Barsakelmes Island

General Characteristics of Habitat The Barsakelmes Island is located in the northern part of the Aral Sea (45 N and 59 E) and in its climate, vegetation, relief, and soil represent a characteristic zonal section of the northern moderate deserts of Eurasia. The length of the island is 28 km; its greatest width in the western part is about 11.5 km, with an average width of about 7.2 km and an area of 184 km^2 (Zhevnerov 1984).

A clay-salt desert occupies the main massif of the Barsakelmes Island (Fig. 1.38); on three sides it is bounded by bumpy sands (Fig. 1.39) and in the southwest by cliffs. According to the relief, it is divided into two parts: the southern elevated plateau and the northern lower part. Two main types of goitered gazelle habitats can be observed: bumpy fixed sands with thickets of shrubs and a clay-salt desert. Goitered gazelles inhabit both in the flat parts of the Barsakelmes Island and into the barkhan area, which is covered by shrubs, preferring the latter to a greater extent. Goitered gazelles concentrate (80%) in barkhan sands (7% of the entire island) in most of seasons, except during the rut when only 12.5% inhabit the barkhan sands and most rutting males with females stay in the open plains (Zhevnerov 1975). Bumpy fixed sands have a well-defined ridge character. The sandy ridges stretch parallel to the seashore and sometimes form separate sandy mounds. The height of ridges and mounds varies, the highest reach 5–7 m. They are separated from the sea by a coastal

Fig. 1.38 The clay-salt desert occupies the main massif of the Barsakelmes Island. Barsakelmes Island. 04.05.2007. Photo by A. Kovshar with kind permission

Fig. 1.39 The bumpy sand dunes are located on three sides of the Barsakelmes Island. 06.05.2007. Photo by A. Kovshar with kind permission

flat strip with a width of 200–500 m. In the western part of the island, the width of the bumpy sands reaches 2 km, gradually narrows to the east, and in some places is only about 200 m. The lengths between the sandy ridges have different widths and are usually very long. In some places they are interspersed with salt marshes. The largest of them, formed in the 1970s, is a salt marsh on the site of a salt lake, 5 km long and 500 m wide. Temporary freshwater reservoirs are located in two places: on the border of the sand zone and the reduced plain (Zhevnerov 1984).

The southern half of the island is a slightly hilly, gradually lowering plateau. Its height above the level of the Aral Sea is 40–70 m. Along the southern border of the plateau and the very edge of the cliff to the sea stretches a sandy ridge overgrown with saxaul. In the southeastern part, there is a whole system of ravines that merge into each other and descend to the sea. The ravines reach 3–4 m deep and have rather steep overgrown slopes. In the western and eastern parts of the plateau, there are two more systems of ravines. There is no fertile soil layer here, and instead saline tertiary clays rich in crystalline gypsum lie on the surface. Ravines and spurs of various sizes diverge here in different directions. Their slopes are devoid of vegetation and quite steep. The northern lowered half of the clay-salt marsh desert is an undulating plain, where there are a lot of valleys interspersed with gentle hills. Some of them do not reach the sandy zone of the northern coast and are closed by the rises of the clay plain with takyrs in the depressions. At the end of other valleys reaching the sand zone, plump salt marshes develop.

The vegetation of the sandy strip is similar to the sandy deserts surrounding the Aral Sea (Demchenko 1950). Shrubs are represented by saxaul (*Haloxylon* sp.), juzgun (*Calligonum* sp.), and tamarix (*Tamarix* sp.). The lower tier is dominated by *Ephedra* sp., *Atraphaxis* sp., and *Nitraria schoberi*. Of the herbs, the most typical are wheatgrass (*Elytrigia repens*), mortuk (*Eremopyrum* sp.), wild rye (*Secale sylvestre*), bonfire (*Bromus* sp.), and *Aeluropus littoralis*. In general, the vegetation of ridge sands is quite diverse and varies in different areas. This is due to the degree of salinity of the soil, its moisture content, and terrain features. The most diverse and lush vegetation is in the northwest and west of the sandy strip. The most extensive areas overgrown with saxaul are also located here. Despite the insignificance of the occupied area (9%), the most diverse vegetation of the island is developed in the bumpy sands. Of the 24 plant species most eaten by the goitered gazelle, 18 (75%) are found here.

The vegetation of the clay desert is mainly represented by wormwood-biyurgun (*Artemisia*-*Anabasis salsa*), biyurgun, and wheatgrass (*Elytrigia* sp.) associations. The wormwood-biyurgun association is the predominant grouping. It occupies elevated parts of the relief, spreading throughout the plateau and plain parts. Saxaul is found in individual bushes, forming only insignificant sparse groves along the slopes of the gullies, as well as along the southern edge of the elevated plateau. In the eastern part of the plain and in the center of the plateau, there are areas of *Caragana* thickets. Along the steep southern edge stretches a strip of bushes of boyalych (*Salsola arbusculaeformis*), *Atraphaxis* sp., and saxaul. On flat areas and along the bottom of the ravines, there are numerous salt marshes. In the spring,

Table 1.26 The relative frequency of encountering of goitered gazelle in main types of desert habitats in the Barsakelmes Island (1970–1975), % (adapted from Zhevnerov 1984)

Months	Fixed sands ($n = 2550$)	Clay-solonchak ($n = 2445$)	Months	Fixed sands ($n = 2550$)	Clay-solonchak ($n = 2445$)
January	13.1	6.8	August	8.1	5.6
February	12.5	3.9	September	11.6	3.8
March	9.2	4.5	October	10.6	5.2
April	4.7	8.8	November	6.0	18.8
May	4.5	12.2	December	6.9	16.6
June	5.4	8.3			
July	7.4	5.5	Total	100.0	100.0

ephemeras appear, from which mortuk (*Eremopyrum* sp.), bluegrass (*Poa* sp.), and *Bromus* sp. form distinctive curtains in some places.

The nature of the relief and the distribution of vegetation allow the gazelles to use the entire territory of the island all year round. However, the intensity of use of some habitats varies significantly throughout the year (Table 1.26). The greatest number of goitered gazelles in ridge sands is observed in the autumn period and from January to March, but the number of gazelles is noticeably reduced here during the rutting season (November–December). Spring and the beginning of summer are characterized by the average occupancy of the ridge sands by goitered gazelles and their more uniform distribution throughout the island. This period coincides with the mass appearance of green vegetation and birthing. The distribution of gazelles on some sites of these complexes is also different. In the fixed sands, extended areas with sparse shrubby thickets and large open spaces between ridges of dunes are more populated, as well as areas with plump salt marshes that are used as recreation areas in the hot months of the year. In the clay-saline desert, gazelles prefer areas with rugged terrain. There is not a single biotope on the island that would be used equally intensively throughout the year. This is mainly due to the seasonal change of forage and its availability, as well as the biological cycles of the gazelles. In addition, the biotopic placement of gazelles has a well-defined daily cycle (Zhevnerov 1984).

Seasonal Changes of Habitat When considering the main features related to the placement of the gazelles in the territory, three periods can be identified: winter, spring, and summer–autumn. This division does not completely coincide with phenological phenomena in nature, but is interconnected primarily with the specifics of nutrition and morphological features of the gazelles, which determine their territorial behavior. In winter-early spring (January–March), their main habitat is bumpy sands with shrubby thickets. They account for 65.9% of the total number of gazelles found during this period. At the same time, the determining factor of placement is the state of the snow cover (Zhevnerov 1984).

Under the influence of frequent strong winds, snow condenses quickly and usually in January on a snowy plain reaches a density of 0.20–0.27 g/cm^3 and on an elevated plateau reaches a density of 0.30–0.33 g/cm^3. Therefore, foraging on the flat areas of the island becomes difficult to access by the end of January, and gazelles begin to concentrate in the fixed sands, where the snow is looser and shrubby vegetation prevails. In the clay-salt marsh desert, gazelles are found mainly in the near-barkhan zone and in low-snow areas of the reduced plain. Goitered gazelles appear extremely rarely on the elevated plateau, and in some winters (1971/72), they are not observed here until the snow cover has melted (Zhevnerov 1984).

Goitered gazelles are very sensitive to low air temperatures and wind, which significantly limits their distribution in winter. Gazelles appear mainly on windless days on low-snow flat areas. Even a short-term lull is often used for grazing on pastures. During such periods, their appearance in open areas increases markedly, but open areas are not a permanent habitat and are used only for grazing. For example, on January 26, 1972, with a wind force of 10 m/s and an air temperature of −13°C, all 23 gazelles encountered along a 5-kilometer observation route were in the zone of bumpy sands, and the gazelles grazed on the leeward side of the dunes. On another January day, in sunny weather with an air temperature of −20 °C, out of 55 gazelles seen along a 10-kilometer route, 40 individuals fed on the near-barkhan areas of *Artemisia-Anabasis* plains, while the other 15 animals were located in the fixed sands. In warm winters with little snow, the gazelles use the island's territory more evenly, with the bulk of them concentrated in the fixed sands only during periods of adverse weather (Zhevnerov 1984).

In the spring–early-summer period (April–June), after the snow cover melts and before the beginning of the mass vegetation of ephemera, gazelles mostly still adhere to winter habitats. Most often they are found in fixed sands and shrubby thickets. However, during this period, the activity of the gazelles increases in search of the first green vegetation, with their locations noticeably shifting as gazelles concentrate on the southern slopes of hills, depressions, and inter-row spaces of settled sands. In early spring, they are common on the slopes of the plateau, freed from snow, on which green shoots begin to appear; in the flat part, they are much less common. In years with a good autumn vegetation of plants immediately after the snow melts, gazelles have the opportunity to eat green shoots of wormwood (*Artemisia terrae-albae*) and cereals; therefore, they are distributed dispersed throughout the island (Zhevnerov 1984).

In April, with the beginning of the mass vegetation of ephemera and ephemeroids, the gazelles have a rather sharp change of habitat. They shift to the near-barkhan zone and to the flat pastures of the clay-salt desert. The number of encounters in the ridge sands decreases two to three times compared to the winter period. During the birthing period (May), the population density of gazelles here is further reduced due to the dispersal of pregnant females on the plateau and the lowered plain. In the clay-saline desert, the occurrence of gazelles at this time increases to 62.7%. However, such a significant increase in the number of gazelle sightings in this biotope during this period is characteristic only in years with good spring vegetation of plants. In dry years, gazelles are more attracted to the dune and dune zones. The birthing sites are also shifting. Most females move for childbirth to areas directly adjacent to shrubby thickets and fixed sands (Zhevnerov 1984).

With the drying of ephemera and ephemeroids and the beginning of flowering of shrubs and semi-shrubs, which usually occurs at the end of May to the first half of June, the gazelles have a characteristic alternation of grazing places. The need for juicy forages is partially compensated for due to the use of the green mass of shrubs, so gazelles are regularly found in the fixed sands mainly in the early morning and evening hours. As the air temperature increases, gazelles move to the borders of the sands, to places blown by the wind, and go out to flat areas. The occurrence of gazelles in certain biotopes is determined by the cyclicity of their daily activity. There is no constant concentration in one habitat (Zhevnerov 1984).

The summer–autumn period (July–October) does not significantly redistribute gazelles by biotope, but their occurrence in places of spring concentration is noticeably reduced. Usually, from the end of June, their distribution over the territory becomes characteristic of the summer–autumn period. Starting from July, the hottest and driest time of the year, gazelles most fully use the green mass and fruits of shrubs and dwarf shrubs, which are associated with more frequent visits to the fixed sands and adjacent areas. Another determining factor in the concentration of gazelles here is the proximity of the sea. When the water dries up in fresh temporary reservoirs, which usually happens in July, the sea remains the only source of drinking water for gazelles. In addition, the coastal part is a convenient place for the rest of the gazelles in the hot time of the day. The clay-salt marsh desert in this respect becomes a less attractive habitat. With the beginning of budding and flowering of wormwood (*Artemisia terrae-albae*) and biyurgun (*Anabasis salsa*), the amount of consumption by gazelles drops sharply. On these pastures, gazelles use mainly dry remnants of cereals and ephemera, as well as partially perennial parts of wormwood. Here the gazelles feed mainly in the evening hours and stay overnight, and in the morning, they move to the areas of ridge sands, vividly showing the daily cyclicity in the change of habitats (Zhevnerov 1984).

In September, with a decrease in air temperature and an increase in precipitation, secondary vegetation begins in cereals and sagebrush. Gazelles are more common in the hollows and depressions of the lowered plain, as well as along the southern and southwestern steep edges of the elevated plateau on the wheatgrass (*Elytrigia* sp.) site. However, the autumn vegetation of plants does not occur annually, and in some years, it is poorly expressed or completely absent. Significant redistribution of gazelles across the territory occurs in November–December, which is associated with the rut. Usually during this period, the main population of mature animals goes to the flat areas. Young gazelles and adult animals that do not participate in reproduction are mainly kept in the ridge sands. Gradually, during the rut, weakened males join them.

1.4.2.4 Betpak-Dala Desert

In the Betpak-Dala Desert in May, it was more often possible to meet goitered gazelles on clay-rocky areas with rare trees of saxaul, *Spirianthus*, and other shrubs and woody plants. In contrast to the saiga that lives in the same areas, the presence

Fig. 1.40 During the spring rains and the abundance of meltwater residues in the soil after the melting of the snow cover, deserts and semi-deserts turn into flowering meadows with rich and diverse vegetation. The Betpak-Dala Desert (Central Kazakhstan). 02.05. 2009. Photo by O. Belyalov with kind permission

of rare bushes is an indispensable condition for attractive places for goitered gazelles. In the bushes, the gazelles not only escape from the heat, rest during the day, and hide the fawns, but in case of danger, they also run away there, hiding from predators (Sokolov 1959).

The habitat of gazelles in the western half of the Betpak-Dala Desert is a clay plain overgrown with sea wormwoods and white-earth wormwoods (*Artemisia maritima* and *A. terrae-albae*) and *Salsola arbuscula*. In the south of this desert, sparse thickets of black saxaul (*Haloxylon aphyllum*) are quite extensive, and to the north, in the lowlands, there are caragana, meadowsweet, and an endemic tall shrub *Spiraeanthus schrenckianus*. Ferules of two species with bright green foliage grow luxuriantly on the outskirts of the thickets of shrubs in spring (Fig. 1.40). The largest part of the Betpak-Dala has a steep relief; the smaller one is a shallow-hilly ground. In Betpak-Dala, gazelles migrate in April from wintering places from Muyunkum Desert and the foothills of the Karatau Mountains. In the southern and central parts of Betpak-Dala, they give births; some of the animals remain for the summer in the area of the few springs and water pits here, and the others go north to the semi-desert with its more diverse vegetation of sagebrush and cereals. By the winter, almost all gazelles go to the south and southwest for wintering for around 500 kilometers. Goitered gazelles have the similar habitats in the deserts to south of Balkhash Lake (Taukum and Saryishikotrau) and Alakol Basin (Zhevnerov et al. 1983).

Fig. 1.41 Many sandy areas of the Kazakhstan are covered with white saxaul trees (*Haloxylon persicum*) in Kazakhstan's Kyzylkum Desert. 14.02.2023. Photo by M. Pestov with kind permission

Fig. 1.42 Kyzylkym is the largest sand desert in Central Asia. Low sandy mounds in the northern part of the desert. Kyzylkum Desert, Kazakhstan. 14.02.2023. Photo by M. Pestov with kind permission

1.4.2.5 Sand Deserts

There are many salt marshes in the vast fixed bumpy or ridge sands where the goitered gazelle lives (Aral Karakum, Aryskum, Kyzylkum, Sarysu and Muyunkums, Taukum, Saryishikotrau): from vegetation-free takyrs to plump soil, covered with salt patches. Low sandy mounds (up to 10 m) are overgrown with *Calligonum*, *Astragalus*, *Tamarix*, white saxaul (*Haloxylon persicum*; Fig. 1.41), sandy acacia (*Ammodendron* spp.), *Atraphaxis frutescens* and grasses desert sedge (*Carex pachystylis*), sagebrush, and various grasses (Fig. 1.42 —Kyzylkum). In the wide ancient riverbeds, extensive "forests" of black saxaul (*Haloxylon aphyllum*) are common with a height of up to 4 m and a trunk thickness of up to 30 cm, as well as turanga trees (*Populus diversifolia*) (Zhevnerov et al. 1983). In the northern parts of the sandy Muyunkum Desert, the goitered gazelle inhabits a sandy loam hilly plain covered with black saxaul and bumpy sands, where these trees find favorable conditions for itself during all seasons of the year (Pole and Karpov 1986).

In more southern sands (Kyzylkums, Muyunkums, Taukum, and Saryishikotrau), snow cover, as a rule, is low and often it melts or evaporates on sunny slopes. Among the hills and near the shrubs, antelopes escape from the winds in winter, as well as in the saxaul forests and turanga groves. Near rivers (Zhanadarya, Syrdarya, Sarysu, Chu, Ili, Karatal, Aksu), some antelopes live all year round. In summer, the animals go to the rivers' channels and floodplain lakes to drink, crossing tugai (riparian forest) from jigda (*Elaeagnus angustifolia*), turanga, and grebenshchik (*Tamarix* spp.), where in winter they feed and shelter from the winds. Throughout the year, these gazelles were found in wide low-snow valleys, such as Ili, Syugatinskaya, and Jalanashskaya at an altitude of 600–800 m above sea level, with their gravelly or clay soils and a low sparse cover of sagebrush and tipchak (*Festuca valesiaca*) (Zhevnerov et al. 1983).

1.4.2.6 Ili Hollow

Ili Hollow is a vast tectonic depression located in the SE Kazakhstan. Mountain hills (up to 1817 m ASL) are situated within the hollow, with slightly inclined planes of 3–7° slope (Fig. 1.43). The hollow is represented by a gravel desert, which is scarred with a thick net of dry riverbeds, karst craters, and depressions, and intermingled with clusters of small hills and plateaus (Fig. 1.44). The major zone in the central part of the hollow is occupied by the Ili River, with a few sandy dunes along banks (Fig. 1.45). Three types of vegetation are found here: (1) brush deserts

Fig. 1.43 Ili Hollow is slightly inclined plane with a 3-7 degree slope toward Tian Shan mountain spurs (SE Kazakhstan). 30.03.2018. Photo by D. Blank

Fig. 1.44 The Ili Hollow is represented by a gravel desert, which is scarred with a thick net of dry river beds, intermingled with clusters of small hills and plateaus. Katu-Tau Mountains. 31.03.2018. Photo by D. Blank

Fig. 1.45 A sandy mountain that sing during windy weather, is the only large sandy massif located along the Ili River. Ili Hollow. 13.10.2005. Photo by A. Kovshar. with kind permission

Fig. 1.46 An abundance of shrubs is typical for the desert of the northern type. Gravel desert in the Ili Hollow. 30.03.2018. Photo by D. Blank

(Fig. 1.46), the common components of which are *Nanophyton erinaceum*, *Arthrophytum iliense*, *Salsola arbusculaeformis*, *Atraphaxis frutescens*, *Eurotia ceratoides*, *Caragana balchaschensis*, *Salsola orientalis*, and *Haloxylon aphyllum*; (2) extremely arid areas, where only *Nanophyton erinaceum* is noted (and there are many plots without any vegetation) (Fig. 1.47); and (3) waterside vegetation along the floodplains of the Ili River, as well as of the oases in the central region of the hollow. There, low riparian forests are common, which consist of *Populus diversifolia*, *Berberis iliensis*, *Lonicera iliensis*, *Halimodendron halodendron*, *Nitraria sibirica*, and *Lasiagrostis splendens*.

Thus, in Kazakhstan, the goitered gazelle lives mainly in deserts with hard soils, especially during warm seasons. It is readily found in places with rough terrain along the plumes of mountains and in the melkosopochnik (hilly area), as well as in semi-closed biotopes of saxaul forests, thickets of *Tamarix*, and sparse riparian forests (tugai). This to a certain extent distinguishes the goitered gazelle from saiga antelopes, Mongolian gazelles, and kulans that are more inclined to open spaces with a flattened relief (Zhevnerov et al. 1983).

1.4.3 Uzbekistan

In Uzbekistan, goitered gazelles prefer flat and hilly deserts, shrub and dwarf-shrub deserts, semi-deserts and steppe areas, as well as grass-*Salsola* deserts. In spring, they enter the desolate foothills and climb the mountains up to 2000 to 3000 meters

Fig. 1.47 Extremely arid areas of gravel desert in the Ili Hollow (SE Kazakhstan). 13.08.2008. Photo by O. Belyalov with kind permission

ASL along the river valleys, preferring terrain with hard soils (gravelly or clay). The biotopes used are shrubby, semishrubby, grass-*Salsola* deserts and semi-deserts, fixed sands, and sparse *Haloxylon* thickets, willingly staying in places with rough terrain. To escape from cold winds and snow in winter, the gazelles stay in steep chinks of the plateau and hide in gorges and ravines; they also enter the outskirts of the tugai forests, *Haloxylon* thickets, *Tamarix* thickets, hollows, and lowland depressions. Females choose semi-closed biotopes for birthing, such as hollows, cliffs, slopes of beams, depressions covered with bushes, and sparse *Haloxylon* thickets.

In the Kyzylkum Desert in the spring, goitered gazelles stay in the sandy shrub-grassy desert, where juicy ephemera and ephemeroids form the basis of the vegetation cover and serve as the main food source. The fact that the goitered gazelles are confined to the sandy desert is connected not only with the development of ephemera but also with the presence of secluded places for birthing in the overgrown bumpy sands with thickets of shrubs. Mounds with a height of 3–8 m stretch in different directions and, connecting in places, form hollows with a diameter of 100–150 m. *Haloxylon*, *Ammodendon*, and *Calligonum* bushes grow on the slopes of the hills. The herbaceous cover forms are viviparous *Poa*, *Carex deserti*, *Aristida pennata*, and other grasses, while *Salsola* and *Artemisia* are found along the bottom of the basins (Mambetzhumaev 1954).

From the end of May to the beginning of June, due to the burning of succulent grasses and the drying of rain puddles in the takyr depressions, the bulk of the gazelles move to desert areas with permanent water sources. In the central Kyzylkum Desert, ungulates gather in the vicinity of salty and fresh springs, where vegetation on sandy-clay plains is mainly represented by wormwood, shrubby *Salsola* (*Salsola rigida*, *S. arbusculaeformis*), harmala (*Peganum harmala*), and herbaceous *Salsola* (*Halimocnemis*) (Mambetzhumaev 1954). On purely sandy islands, unburned *Poa*, *Aristida pennata* hummocks, *Ammodendron*, and a large umbrella of *Ferula* stand out. Gazelles also gather in the foothill plains covered with steppe vegetation, where the landscape is created by thickets of *Artemisia* and *Stipa* sp. In places where salty springs protrude, *Salsola* dominates. In November–December, after the rains, the goitered gazelles migrate to the sandy plains with adjacent water paddles. At this time, the *Poa bulbosa* and *Carex deserti*, growing on the sands, which the gazelles feed on, begin to turn green. In the salt marsh basin of Sorkor, goitered gazelles, who had stayed in tamarisk thickets near a salty stream, moved the next day after a heavy rain (November 21, 1953) to a sandy bumpy plain with puddles of water. Here they stay until the puddles of water are drying up. The same phenomenon is observed in other places. Thus, in the autumn–winter and early spring periods, when there is a lot of precipitation and puddles form in the desert, and the grass becomes moist and soft, gazelles wander through the Kyzylkum Desert in search of food, willing to move quite far from water sources (Ishunin 1961).

1.4.4 Turkmenistan

In Turkmenistan, the conditionally indigenous pastures are weakly modified with the prevalence of *Carex*-ephemeral associations and *Artemisia* and various herbs in some places. The projective vegetative cover in spring is equal to 70% and productivity can reach 5 center/ha. In these habitats, goitered gazelle females give birth. The preferable landscape variability is quite high: elevated plateau (700–800 m ASL), gentle undulating ridges (up to 900 m ASL), pistachio-sparse woodlands with a heavily divided relief, a line of deep drainless depressions with vast solonchaks, and dunes. Badkhyz Upland pastures are best in Turkmenistan, where most of the ground is covered by vegetation with a predominance of the cereal Poaceae, ephemeras, and large grasses, where the pasture productivity can reach 6 center/ha (Kuznetzov 1986).

1.4.5 Tajikistan

In Tajikistan, goitered gazelles stay on the gentle slopes of Karatau, Sarsariak, and other mountains along the lower reaches of the Vakhsh and Panj rivers. In the Dangara Depression, gazelles are more often found on the semi-desert slopes of the neighboring low ridges rather than on the plains. In some areas, they even migrate for the summer from open deserts to the foothills, attracted by the presence of green xerophytic vegetation and the lower number of bloodsucking insects.

1.4.6 Iran

Iran forms a large part of the Iranian plateau. Seven desert plains and depressions give the landscape a completely different appearance: Dasht-e Kavir in central Iran, Dasht-e Lut Desert, Sistan and Jazmurian depressions in the southeast, Khuzestan plain in the southwest, Moghan steppe in the northwest, and the Turkman-Sahra steppe in the northeast (Zehzad et al. 2002).

Iran's climate is mainly arid or semiarid, except the northern coastal areas and parts of western Iran. The climate is extremely continental with hot and dry summer and very cold winter particularly in inland areas. Apart from the coastal areas, the temperature in Iran is characterized by relatively large annual range from about −22 °C to +26 °C. The rainy period in most of the country is from November to May followed by a dry period between May and October with rare precipitation. The average annual rainfall of the country is about 240 mm with maximum amounts in the Caspian Sea plains and Alborz and Zagros slopes with more than 1800 and 480 mm, respectively. The Dasht-e Kavir and Dasht-e Lut deserts are the driest parts with less than 150 mm annual precipitation (Heshmati 2007). The Dasht-e Kavir is located in the north of Iran, and it is presented by vast saline desert, and Dasht-e Lut is in the south, which is presented largely by a sand and gravel desert, which is one of the hottest deserts in the world (Heshmati 2007).

According to plant geography, Iran is divided into three distinct phytogeographic regions: the Irano-Turanian region, the Nubo-Sindian region in the south, and the Hyrcanian region in the north. The largest of them is the Irano-Turanian region, which occupies more than two-thirds of the land surface (Zehzad et al. 2002). The central Iranian sector hosts the most typical vegetation of Iran's steppe and desert regions. Dwarf scrub vegetation is common in large areas of the interior of Iran and is very diverse and rich in species; in nonsaline areas, a variant with many thorn-cushions (Zohary 1973) is formed. Under extremely arid conditions, a very open variant of the dwarf shrublands appears, also characteristic of large areas of the Iranian interior; the dominant species are sagebrush (*Artemisia siberi*, *Astragalus gossypinus*, and others) (Frey and Probst 1986). In areas receiving over 100 mm of rain, other genera such as *Pteropyrum*, *Zygophyllum*, and *Amygdalus* can also be found. With regard to vegetation of the sand deserts in the interior regions of Iran, among the more characteristic genera are *Ephedra*, *Calligonum*, *Heliotropium*, and others (Frey and Probst 1986). Many species here are highly specialized as psammophytes; these sand-adapted species are estimated to make up one third to one half of the total number of species in the sand deserts of Iran and Afghanistan (Freitag 1986). Iranian deserts also have a striking number of *Tamarix* species; they have been reported to occur on the margins of the more sandy and gravelly parts of the Dasht-e Lut (Breckle 1983). Halophytic communities of varying composition are found on the margins of the undrained salt pans of the central Iranian region, such as the Dasht-e Kavir. In such areas, clays and sand soils have a high surface salt content due to insufficient water and high summer evaporation. Characteristic genera and species include *Aellenia* spp., *Halocnemum strobilaceum*, *Haloxylon* spp.,

Salsola spp., and others (Frey and Probst 1986). The inner parts of the salt pans have almost no vegetation.

For centuries pasturelands in Afghanistan and Iran have been overburdened with domestic grazing which has led to the complete destruction of their vegetative cover all over vast areas. The entire territory of these countries presents the modified pasture of second and third orders with small projective cover (10–15%) and sparse annual plants on pastures of low productivity (less than 0.4–0.8 center/ha) (Kuznetzov 1986).

In northeastern Iran, goitered gazelles prefer a combination of hilly plains with flat plains. In hilly plains, including scattered hills and mountains, dry riverbeds, and depressions, they stay mainly during nights, while flat plains are used mostly for pasturing during daytime. Dry riverbeds are used as a pasture less often, and plains covered by *Tamarix* thickets are not used for grazing due to food scarcity; the gazelles pass through these areas traveling to agricultural fields, where they graze through summer and autumn nights (Farhadinia et al. 2009).

In the Golestan National Park, goitered gazelles live at an elevation of 1100 to 1300 m ASL, where the average annual precipitation is 400 mm. The average annual temperature varies between 11.5 °C and 17.5 °C, with an absolute minimum of −25 °C and a maximum of +45 °C. The main plant coverage includes thick shrubs, evergreen, and annual vegetation, belonging to Iran-Turanian elements. The most often found among these are several *Artemisia* species. The canopy ground cover consists of 58% shrub species, 25% grasses, and 17% forbs. The main shrubs include *Artemisia herba-alba*, *Salsola rigida*, *Eurotia ceratoides*, *Salsola vermiculata*, and *Anabasis aphylla*. Among grasses *Stipa barbata*, *Poa bulbosa*, and *Eremopyrum bonaepartis* are dominated. Over ten different forbs are present in the park, including annual forbs, such as *Alyssum strigosum* and *Astragalus* spp., which are dominant and commonly used by goitered gazelles (Rad et al. 2015).

1.4.7 Mongolia

The Mongolian Gobi Desert is an important refuge for goitered gazelles and is dominated by plains in the east and rolling hills in the west, with an elevation range of 1000 to 2840 m ASL. Open water is distributed unevenly, with dry western and southwestern parts (Fig. 1. 48). The Gobi climate is quite harsh and continental, with annual temperatures around 1.0 °C and temperature extremes varying from −43 °C in winter to +35 °C in summer. The average annual rainfall is 96 mm with a peak during summer. The average snow cover lasts 97 days. Rain and snowfalls are highly variable from year to year, and season to season, and in areas with non-equilibrium dynamics (Fernandez-Gimenez and Allen-Diaz 1999).

The species are found in the deserts and desert steppes of the Central Asian dry and arid regions in Mongolia and sometimes in mountain steppes. Suitable habitats include the Gobi and low desert mountain foothills, rolling hills, mountainous valleys, around marshy depressions and ponds, and flat plains that are distributed by *Anabasis brevifolia*, *Reaumuria songarica*, *Salsola passerina*, *Sympegma Regelii*,

Fig. 1.48 Open water is limited and distributed unevenly in arid regions, so these areas are very important for the goitered gazelles' watering during the hottest periods of the year. Great Gobi B. 18.06.2012. Photo by P. Kaczensky with kind permission

Stipa spp., *Caragana* spp., *Allium* spp., *Artemisia* spp., and *Nitraria* spp. and shrubbery rocky desert and desert steppe habitats.

In Mongolia, gazelles are more often found in the feather-grass-onion-grass semi-desert, as well as in *Salsola* deserts (Bannikov 1954). The results of the surveys of gazelles in Mongolia showed that the optimal conditions for gazelles are located within real deserts (Fig. 1.49), as well as the desolate steppes on the plains or low intermountain valleys (Bugaev 1989; Fig. 1.50). However, throughout the entire range, the habitats of the goitered gazelles are quite diverse and are determined primarily by the seasonal changes in forage, its availability, as well as protective landscape features (Zhevnerov 1984). Desert areas are widely dominated by Chenopodiaceae (Fig. 1.51), such as *Haloxylon ammodendron* and *Anabasis brevifolia*. The steppe areas are dominated by Asteraceae, such as *Artemisia* and *Ajania*, and Poaceae like *Stipa* and *Ptilagrostis* (von Wehrden et al. 2006). Trees are rare and only found in places where underground water comes closer to the surface (Fig. 1.52). *Ulmus pumila* dominant wide shingle, natural springs, oases, salty lakes, and ponds play an important role in existence within the habitats providing important water access. Additionally, *Haloxylon ammodendron* is found in depression areas (Lkhagvasure et al. 2011).

Fig. 1.49 The optimal environmental conditions for goitered gazelles in Mongolia are located within the real deserts. South Gobi region. Summer, 2003. Photo by P. Kaczensky

Fig. 1.50 In the desolate steppes of the plains or low intermountain valleys, the forage is more abundant than in the extremely arid areas. Great Gobi B, Mongolia. Photo by P. Kaczensky with kind permission

Fig. 1.51 Desert areas are widely dominated by plants from Fem. Chenopodiaceae which is a part of Fem. Amaranthaceae accroding to new taxonomy. Great Gobi B, Mongolia. 02.10.2015. Photo by P. Kaczensky with kind permission

Fig. 1.52 Trees are rarely found in goitered gazelle habitats and they are located in places where the underground water is closer to the surface. South Gobi region. 18.10.2015. Photo by P. Kaczensky with kind permission

1.4.8 China (Dzungarian Basin, Xinjiang)

The climate is continental with long cold winters (October to early April) and short hot summers (mid-May to September). Most precipitation is noted during winter, spring, and early summer. The yearly precipitation is around 190 mm. The average temperature of January is −24.3 °C, with an absolute minimum of −45 °C and maximum of +1.4 °C; the average temperature in July is +20.5 °C with an absolute maximum of +38.4 °C and minimum of +6.0 °C. The annual average temperature is +1.99 °C, which is lower compared to deserts of Central Asia but higher than in northern deserts of Mongolia. The most significant aspect of this region is the exceptional changes in temperatures over the period of a single day, when it is possible to experience frost at dawn and very hot temperatures by midday (up to +30 °C). There is no permanent surface water, and only a dozen permanent water springs and some artificial watering holes that can be used by wildlife in the region (Xu et al. 2012).

The landscape of Dzungarian Basin in Xinjiang is dominated by plains in the north and south (Fig. 1.53), sand dunes in the west, and the rolling hills of Mt. Kalamaili in the center (Fig. 1.54). Altitudes range from 600 m to 1464 m ASL. Vegetation is sparse in the region with dominant plants including *Allium polyrhizum*, *Anabasis salsa*, *Artemisia desertorum*, *Atraphaxis frutescens*, *Calligonum mongolicum*, *Ceratocarpus arenarius*, *Ceratoides latens*, *Haloxylon ammodendron*, *Reaumuria songarica*, and *Stipa glareosa*. Ephemeral plants, such

Fig. 1.53 The landscape of Dzungarian Xinjiang is dominated by plains. Kalamali Ungulate Nature Reserve, Xinjiang, China. 21.07.2010. Photo by D. Blank

Fig. 1.54 Rolling hills are located only in the center of the Kalamali Ungulate Nature Reserve, Xinjiang, China. 06.09.2006. Photo by D. Blank

as *Alyssum linifolium*, *Astragalus*, *Cancrinia discoidea*, *Chorispora tenella*, *Eremurus inderiensis*, *Erysimum flavum*, *Lappula*, *Scorzonera*, *Sonchus*, and *Sterigmostemum matthioloides*, are abundant and common in spring and early summer under patches of shrubs (Xu et al. 2012).

1.5 Number and Distribution

In the past, goitered gazelles were widely distributed in Near, Middle, and Central Asia: Iran, Afghanistan, part of Pakistan, Dzungaria and the Gobi Desert in Mongolia, the northern part of Tibet, Alashan, and Ordos in China. Within the countries of the former USSR, the gazelles occupied an almost continuous distribution area covering deserts and semi-deserts, as well as the foothills of Central Asia and Eastern Transcaucasia (Afanasiev et al. 1953; Sokolov 1959). Goitered gazelles have lost as much as 80% of their former range globally, but this species is still categorized as Vulnerable (Mallon 2008), with a decreasing population trend attributed primarily to exploitation, habitat degradation, and human disturbance (Clark et al. 2006). In most areas of their former habitat, the goitered gazelles are now very sporadically distributed, with only separate areas of this species preserved in Kazakhstan: Mangyshlak Peninsula, Ustyurt Plateau, and Karakum, Kyzylkum, Betpak-Dala, and southern Balkhash deserts. In Turkmenistan, Uzbekistan, and Tajikistan, the gazelle has become rare everywhere. Density population changes to

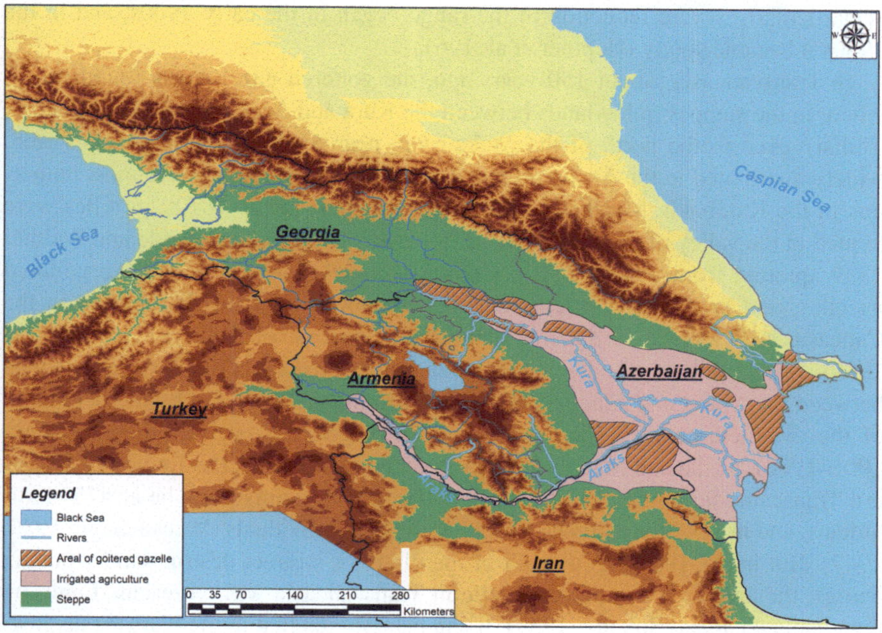

Fig. 1.55 The distribution of goitered gazelles in the Caucasus in the early 20th century (prepared by A. Osmonov with kind permission)

a wide extent are from 0.02 to 4.21 gazelles/km^2 (Blank 1990; Livov 1979; Mowlavi 1978; Schaller 1998; Thouless et al. 1991). Only in the semi-captive conditions of the Bukhara nursery does the population density reach a higher density of 16–24 individuals/km^2 (Marmazinskaya 1996).

1.5.1 Transcaucasia

Previously goitered gazelles were distributed over all the steppe and semi-desert zones in Transcaucasia (Fig. 1.55). The area encompassing the range of the goitered gazelles in the Kura-Araks lowland of Eastern Transcaucasia is completely isolated from the rest of the distribution area of this species, and this isolation probably occurred in prehistoric times. The existence of very ancient irrigated agriculture on the plains of Eastern Transcaucasia (from the fourth to eighth century BC), in the absence of early sources of information about the spread of the goitered gazelles, does not make it possible to restore the outlines of the area here before human influence altered it. However, the areas occupied by the goitered gazelles at the end of the nineteenth century were obviously close to those in the early epochs, and the habitat began to decline sharply only in the twentieth century (Vereshchagin 1939).

A particularly severe reduction in the range began in the early 1900s, and in the 1930s it became spotty (Heptner et al. 1961).

In Transcaucasia, about 150 years ago, the goitered gazelle was found everywhere in the steppes in lowlands between the Kura and Araks rivers (from Baku to Tbilisi), even to the north of the Araks river, near Avchala. These gazelles were widely distributed in the Apsheron Peninsula (Caspian Sea) and were most numerous in the Karabakh, Mulskaya, and Mugan steppes. In the 1870s, gazelles were hunted in the valley of the Kura River, 30–40 km below Tbilisi (Radde and Walther 1889, quoting from Dinnik 1910). Currently, none of these places have goitered gazelles, and in the early 1920s, the western border of their distribution in the Caucasus was located in the Iori Plateau (60 km east of Tbilisi) (Vereshchagin 1939). In the 1830s, gazelles were found in the territory of the Erivan province between the Ararat and Alagez mountains (Nordmann 1840), but by the beginning of the twentieth century, there was no trace of goitered gazelles there (Satunin 1905–1906). At the end of the 1930s, the total number of gazelles living in the eastern Transcaucasia in an area of about 8000 km^2 was estimated to be at 5750 individuals, and in 1940 the estimate was 5500 to 6000 individuals (Vereshchagin 1939, 1947). But in the 1930s, as a result of the direct or indirect destruction by human beings, the area of the goitered gazelle in Transcaucasia was represented only by eight isolated areas (Sokolov 1959). It can be assumed that in the Kura-Araks lowlands, there was a single area isolated from the Iranian plateau closest to it. Only later, by the 1920s, did it split into a number of small parts, the number of which reached ten by the mid-1930s (Vereshchagin 1939, 1947).

In the following years, there was a further progressive reduction of the range and a decrease in the number of gazelles as a result of not only direct extermination by poachers but also the development of steppes and semi-deserts for agriculture and grazing livestock (Sokolov 1959). Until the 1940s, goitered gazelles disappeared completely in Georgia and Armenia, and goitered gazelles could be found only in small, isolated areas of Azerbaijan (Arabuli 1970; Zhevnerov et al. 1983). In June of 1960, an aerial survey was made all over the area between the Kura and Araks rivers, and a total of 131 goitered gazelles were found only in the Milkaya, Karabakh, and Shirvan steppes, an area of 14.3 thousand km^2 (Safarov 1961). Soon, the number decreased up to 77 individuals during the 1960s. After that survey, measures were taken for the protection of goitered gazelles against poachers. As a result, there were 450 individuals in 1969 and 3600 gazelles in 1980; 4000 gazelles were found in 1982 and around 5000 gazelles in 1983 (Shchadilov 1986).

At present, the goitered gazelles' natural habitat is reduced to five separate populations. The main population is located in the southeast Shirvan steppes in the Shirvan State Reserve/Bandovan Sanctuary. Latest estimates put the current population at 3000 to 4500 individuals. The other populations are in the Bozdag Range and adjacent plains to the south of the Mingechevir Reservoir (100–150 gazelles), the eastern end of the Apsheron Peninsula (30–40 individuals), and a part of the Gobustan Plateau (an unknown number but certainly very few) and the descendants of a small, introduced population on Bulla Island in the Caspian Sea (15 gazelles). Gazelles also occur sporadically in the Murgab steppe, along the left bank of the

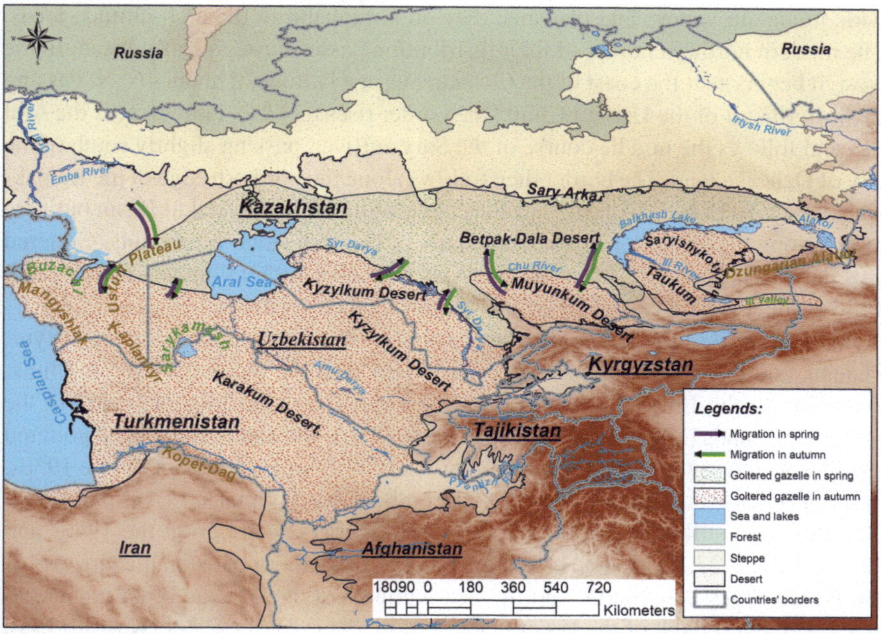

Fig. 1.56 The former (1940s) distribution and seasonal migrations of goitered gazelles in Central Asia (prepared by A. Osmonov with kind permission)

lower Kura and possibly in the Ajinour steppe, northeast of the Mingechevir Reservoir. The largest and most stable population is on the Shirvan steppe, and this numbered 11,400 in 1989 and 8900 in 1990. The latest data suggests that the total number dropped to around 4500 in 1994, 90% of which occurred in the Shirvan Nature Reserve. The second largest population occurs in the region of the Mingechevir Reservoir. The remaining small populations are doomed to gradual loss unless special measures are taken, such as ensuring effective protection from hunting and reinforcing their populations by translocation of gazelles from Shirvan (Shchadilov and Hadjiev 2001). Now, the total number of goitered gazelles in Azerbaijan is estimated at more than 8000 in 2019 (Sarukhanova 2021).

1.5.2 Kazakhstan

Goitered gazelles once occupied an unbroken range throughout the semi-desert and desert zones of Kazakhstan, as well as the foothills of Tien Shan Mountains (Fig. 1.56). The goitered gazelle population was estimated to be more than 200,000 in the 1930s but declined steadily until only about 10,000 remained by the end of the 1970s (Afanasiev et al. 1953; Sludskiy 1977; Zhevnerov et al. 1983). Goitered gazelles can live at heights of up to 1500 to 3000 ASL and are found in wide mountain valleys and plateaus; they do not inhabit in the high-altitude areas. The modern northern border of their distribution passes through Kazakhstan. In the west, it begins near the coast of the Caspian Sea at a latitude of about 46° N. Passing to the northeast of the Ustyurt Chink, the border rises to 47° N latitude near the Aral Sea and follows the middle course of the Sarysu River, passing slightly south of the city of Dzhezkazgan. The border then follows along the northern edge of the Betpak-Dala Desert to Lake Balkhash. Gazelles are also found in the Alakul Basin but very rarely along the western and southern spurs of the Tarbagatai Mountains. Goitered gazelles lived also in the Zaisan Basin (Afanasiev et al. 1953; Flint and Prisyazhnuk 1986), where they disappeared by the 1890s (Khakhlov 1928).

In the middle of the last century, this species was quite numerous on the Ustyurt Plateau, and their distribution went even further than the Northern Ustyurt Chink, extending to the Emba River and even the Ural River, and in some years to the Volga-Ural sands (Eversman 1850; Karelin 1883). At the end of the 1920s, goitered gazelles were very common in the southwestern Ustyurt Plateau, but by the 1930s, these gazelles were already "in a small number" on this plateau (Antipin 1941). Gazelles were still common throughout the Ustyurt Plateau during the 1940s and 1950s (Table 1.27; Bazhanov 1951), especially the southern parts on the territory of the Barsakelmes depression, and the near-chink (Ustyurt Plateau cliff) tugai wetlands on the left bank of the lower reaches of the Amu-Darya River (Kostin 1955), and on the Mangyshlak Peninsula. During a survey between 1955 and 1957 in Mangyshlak and the western Ustyurt Plateau, 11,301 gazelles were noted during a trip of 12,524 km (Table 1.27; Molodoskiy 1975). Until the 1950s, a total of about 100,000 gazelles lived on the Mangyshlak Peninsula and the Ustyurt Plateau (Sludskiy 1977). And during the 1960 spring migration of these animals, herds of up to 2.5 thousand gazelles were observed in the Mangyshlak region (Solomatin 1966).

The goitered gazelle is a common species in the Aral Karakum Sands (128 gazelles per 5 km in 1948; Bazhanov 1951), in the Syr Darya River valley, and in the entire Kyzylkum area between the Syr Darya and Amu-Darya rivers (Bazhanov 1945). Gazelles are still common in the Betpak-Dala Desert, in the Muyunkum Sands, and in the vast territory of the Sary-Arka steppe to the north of the Betpak-Dala Desert (Kulpin 1954). They are found in the northern Balkhash region but are very rare (Antipin 1941). In Semirechye, these gazelles were most often found in the Ili River basin (left bank—Sugaty-Boguty mountains) and the foothills of the Dzungarian Alatau (right bank—Altyn-Emel). Sporadically, gazelles were found in the Alakol Basin (Shnitnikov 1936).

In 1929, nine goitered gazelles were taken to the Barsakelmes Island in the Aral Sea, where they were not found previously. The gazelles were very well acclimated to this island and began to rapidly increase their numbers up to 4650 individuals in 1948. However, during the severe winter of 1949, the goitered gazelle population was dramatically reduced up to 64 individuals. Since then, the number of gazelles has not grown, and in 1957, their number did not exceed 20–30 gazelles (Sokolov 1959). In the autumn of 1960, an additional group of six gazelles was taken to Barsakelmes Island. They adapted well to the new conditions, and the population was around 200 gazelles during the 1970s (Zhevnerov 1984).

Table 1.27 The number and density of goitered gazelles in various regions of western Kazakhstan

Regions	Date	Number of individuals	Distance, km	Density of gazelles per 1 km	Author
Buzachi Peninsula, Caspian Sea	Spring–summer, 1941	10–15 gazelles per group	–	Common	Bazhanov (1951)
	Autumn–winter, 1941	20–30 gazelles per group	–	Common	Bazhanov (1951)
	June, 1964	5	20	0.25	Bekenov and Savinov (1975)
	June 12, 1975	1	54	0.02	Bekenov and Savinov (1975)
Ustyurt Plateau, W Kazakhstan	1940–1941	–	–	Common	Bazhanov (1951)
	June 26, 1941, car trip	500	150	3.33	Bazhanov (1951)
Ustyurt Plateau, W chink	December 25, 1964, car trip	100	25	4.00	Bekenov and Savinov (1975)
	November 10–21, 1966	7	50	0.14	Bekenov and Savinov (1975)
	April 1974; car trip	1	400	0.0025	Bekenov and Savinov (1975)
	May 8–9, 1975, car trip	3	145	0.02	Bekenov and Savinov (1975)
	May–June, 1980, car trip	50	700	0.07	Savinov
W chink	Nov. 1964 Apr. 1965	43 8	48 20	1.12 0.40	Mambetzhumaev (1970)
Ustyurt	1964	500	200	0.40	Solomatin (1966)
Ustyurt Plateau, N chink	September 1–October 14, 1950–1952	3797	2600	1.46	Sludskiy (1977)
Ustyurt Plateau, Actubinsk oblast	September 19–21, 1950	1109	40	27.73	Sludskiy (1977)
Mangyshlak, Ustyurt Plateau	October 1941	100,000	100	Estimation	Sludskiy (1977)
	1955–1957, car trip	11,301	12,524	0.90	Molodoskiy (1975)
	1960, spring migration	2500 gazelles per group	–	–	Solomatin (1966)
	1964	200	500	0.04	Solomatin (1966)
	Nov.–Dec. 1971	0	500	0	Bekenov and Savinov (1975)
	May–June 1980	50	400 km^2	0.12	Savinov

(continued)

Table 1.27 (continued)

Regions	Date	Number of individuals	Distance, km	Density of gazelles per 1 km	Author
N Aral Sea Sam Sands	July 12, 1948, by foot	128	5	25.6	Kostin (1955)
N Aral Sea Barsuki Sands	1950s	30	100	0.3	Zhevnerov et al. (1983)
	1971–1972	39	40,000	0.001	Zhevnerov et al. (1983)
Barsakelmes hollow, around	1944–1953	–	–	Common	Kostin (1955)

In January 1960, an aerial survey of 9500 km was made of goitered gazelles all over Kazakhstan. The survey defined the total wintering grounds of the goitered gazelles at 20620 km^2; the average population density fluctuated from 0.01 up to 0.82 gazelles per 1 km^2 in different deserts, and the total number of gazelles was estimated in 9300. The highest density of the gazelle population was in the Kyzylkum Desert, with an estimated number of 5700, and in the Muyunkum Desert, with estimated numbers of 2900 gazelles (Table 1.28; Sludskiy and Shubin 1963). The next aerial survey of goitered gazelles was made on February 27 through March 30, 1971, and again all over Kazakhstan; the total survey length was 17,000 km with only 24 gazelles in total found across all deserts. In the spring of 1974, during an aerial survey of 2450 km covering the Kyzylkum and Betpak-Dala deserts, only 11 goitered gazelles were found. These surveys demonstrated that the number of goitered gazelles has continuously decreased all over Kazakhstan starting from the 1940s and 1950s; during the 1970s, the total number of goitered gazelles in Kazakhstan was estimated at a few thousand. The main cause of this phenomenon was heavy poaching and periodical severe winters (Zhevnerov et al. 1983).

Due to the ecological needs of this gazelle, the northern part of its range is divided into areas for summer and winter. The gazelle appears in the summer habitats in May and stays there until September–October. In Kazakhstan, the areas of goitered gazelle summer habitats are located between the Emba River and the southern Ustyurt Plateau, in the Daryalyk Desert (north of the Syr Darya River), and in the interfluve of the Sarysu and Chu rivers (Sludskiy 1956b, 1963).

During 1982 through 1985 in Kazakhstan, an aerial survey of the number of goitered gazelle populations was made in the vast desert areas of the Ili Hollow, around Balkhash Lake, and the Muyunkum and Kyzylkum deserts. The observations were made from an AN-2 aircraft (Muyunkum and Kyzylkum deserts) and a Mi-8 helicopter (the rest of the areas) by five observers: two on each side of the aircraft and one ahead (plotting the course). Flights were conducted on parallel routes every 5 km in the Ili Valley and every 20 km in the Saryishikotrau Desert and every 10 km in the other areas. The flight altitude was 50–70 m, the speed was

Table 1.28 The number and density of goitered gazelles in various regions of central and eastern Kazakhstan

Regions	Date	Number of individuals	Distance, km	Density of gazelles per 1 km	Author
Kyzylkum Desert	1952	20,000–25,000	–	0.5 per 1 km^2	Krivosheev et al. (1960)
	May 1953	18–20	Per 5 km	3.6–4	Mambetzhumaev (1970)
	June 1953	10	Per 20 km	0.50	
	January 1960; aerial trip	71	100	0.71	Sludskiy and Shubin (1963)
	January 1960; aerial trip	5700		Entire area estimation	Sludskiy and Shubin (1963)
	Autumn, 1967, car trip	5	1000	0.0005	Rapoport et al. (1975)
	1970–1972, car trip	2	1000	0.0002	Rapoport et al. (1975)
	September 21–28, 1977, car trip	33	314	0.11	Zhevnerov et al. (1983)
	April 7, 1978, car trip	4	87	0.05	Zhevnerov et al. (1983)
Betpak-Dala Desert	October 6–13, 1949, hiking	500	10	50	Zhevnerov et al. (1983)
	July 18–August 23, 1951; car trip	121	781	0.15	Sludskiy (1962)
	May, 1955, car trip	447	1620	0.28	Sludskiy (1962)
	May 6–28, 1955, car trip	413	2285	0.18	Fadeev (1973)
	January 1960; aerial survey	187	710	0.26	Sludskiy and Shubin (1963)
	April 22, 1965	21	70	0.3	Fadeev (1973)
	April 22, 1965	12	16	0.75	Fadeev (1973)
	April 23, 1965	29	73	0.4	Fadeev 1973
	December 1972	26	500	0.05	Fadeev (1973)

(continued)

Table 1.28 (continued)

Regions	Date	Number of individuals	Distance, km	Density of gazelles per 1 km	Author
Chuluck-espe, 2000 km^2	March 1976	5	260	0.02, estimated number—40	Fadeev (1986)
Dzhambul Mountain, 1200 km^2	March 1976	10	170	0.06, estimated number—70	
Taukum Desert, 11,200 km^2	March 1976	29	650	0.04, estimated number—450	Fadeev (1986)
	March 1978	14	400	0.03, estimated number—310	
	March 1982	16	400	0.04, estimated number - 450	
Muyunkum Desert, 32,000 km^2	January 1960; aerial survey	74	90	0.82	Sludskiy and Shubin (1963)
	January 1960; aerial survey	2900	37,500 km^2	Entire area estimation	Sludskiy and Shubin (1963)
	November 1–5, 1980	14	106	0.13	Zhevnerov et al. (1983)
	March 1972	28	610	0.04, estimated number—1280	Fadeev (1986)
	March 1974	27	850	0.03, estimated number—1000	
	April 1977	21	640	0.03–950	
	March 1979	26	450	0.05, estimated number—1600	
Southern near-Balkhash	January 1960	27	50	0.54	Sludskiy and Shubin (1963)
Ili Hollow, right bank	October 1943	400		Survey	Sludskiy (1977)
	October 1944	4120		Survey	Sludskiy (1977)
	October 1945	5200		Survey	Sludskiy (1977)
	October 1946	6200		Survey	Sludskiy (1977)
	1970s	Not more than 30 individuals			Zhiriakov (1976)
Ili Hollow, Charyn-Chilik interfluve	1942	14,000		Survey	Sludskiy (1977)
	November 20–23, 1955	8	360	0.02	Sludskiy (1977)
	1970s			Rarely	Sludskiy (1977)

Table 1.29 The number and population density of goitered gazelles in aerial survey areas in Kazakhstan (adapted from Kovshar and Blank 1986)

Area	Survey time	Encountered gazelles	Route length, km	Density per 1 km^2	Area size, km^2	Estimated number
Kapchagai Preserve	25.11.1983	709	356	2.67	950	3695
Aktau, Katutau	25.11.1983	39	208	0.23	600	120
Ili left bank	09.01. 1985	94	333	0.30	1300	360
Taukum	26.11–01.12.1983	220	890	0.22	7500	1650
Saryishikotrau	26.11–01.12.1983	100	1386	0.07	21,130	1400
Muyunkum	15.12.1983–29.01.1984	201	3655	0.10	24,800	2480
Kyzylkum	02–09.10.1984	273	12,000	0.04	102,300	4100
Total		1636	18,828	–	158,580	13,805

160 km/h, and the accounting lane was 500 m from each side. This was the maximum distance and a special check showed that almost all gazelles were counted in the lane no further than 300 m from the route line. The windows in the helicopter were open during the survey. All gazelle encounters were recorded in registration protocol indicating the exact time of observation. Aerial surveys from the Mi-8 helicopter covered a total area of 15,360 km^2, within which 2123 gazelles were counted on routes with a length of 4911 km. From the AN-2 aircraft, 17,475 km of routes was laid out in an area of 127,100 km^2 and 475 gazelles were registered (Table 1.29; Blank and Kovshar 1986).

Kapchagai Hunting Preserve (Altyn-Emel National Park at Present) The highest population density of goitered gazelles in Kazakhstan during 1982–1983 was observed on the right bank of the Ili River in the territory of the Kapchagai Reserve. Here, surveys were carried out three times: on December 9, 1982; November 25, 1983; and December 7, 1983. The maximum density in some parts of the preserve was 9.6 individuals per 1 km^2, and the minimum was 0.3 individuals per 1 km^2, with an average of 2.67 gazelles (November 25, 1983). The number of goitered gazelles later fluctuated widely in the Kapchagai Hunting Preserve (Table 1.30; Kovshar and Blank 1986).

The distribution of gazelles in the territory of the Kapchagai Hunting Preserve shows that intensive economic development (plowing of land, grazing of livestock) in its western part led to the displacement of goitered gazelles, with not a single individual found there during the survey period. Automobile routes from 1981 through 1987 and stationary observations showed that in the central part of the hunting preserve near the Chulak-Matai mountains, the population density of goitered gazelles remained the highest throughout the year. It was in this territory that five flocks of sheep were placed for the winter, which of course affected the feeding base of the gazelles and, accordingly, their distribution and number. Cattle grazing was

Table 1.30 The number of goitered gazelles in the Kapchagai Hunting Preserve (the average density is 3.2–3.9 individuals per 1 km^2) (adapted from Bekenov et al. 1999)

Survey time	Survey kind	Estimated number	Source
1988	Auto-survey	2500	Bekenov et al. (1999)
1989		3000	
1990		3000	
August 1991	Aerosurvey	5625	
1992	Auto-survey	6200	Park's information
1993		5100	Park's information
August 1994		5423	
February 1995		6097	
March 1996		7280	
April 1997		4800	
March 1998		4083	

not allowed in this area because most of the gazelles were birthing their young here. It should be emphasized that the main clusters of gazelles on the preserve were located outside the protected part of it (Kovshar and Blank 1986).

During a car survey to the east of the Kapchagai Preserve, four main but small groupings of goitered gazelles were found (Boguty Mountains, Charyn River, Bakhtykurai, Panfilov Karakum) with 30–40 goitered gazelles living in each location. Sugaty Valley has no goitered gazelles at all. From the foothills of the Big Boguty, gazelles were displaced every winter to the interior of these mountains, since all watering holes were occupied by domestic flocks of sheep. The Panfilov Karakum sands were surrounded on all sides by agricultural lands, which has minimized gazelle migrations. Frequent visits to these places by hydrogeologists, border guards, and local peoples for the purpose of poaching, as well as a well-developed network of roads through the sands, make the existence of the goitered gazelles in this area problematic in the near future (Kovshar and Blank 1986).

Taukum Sands Gazelles were rarely found in the extreme southeastern and northwestern parts of this desert, where the population density did not exceed 0.01 individuals/km^2. These areas were excluded when extrapolating data to cover the entire accounting area. The highest density of gazelles was found in the more central desert areas, where the density reached 0.71 individuals/km^2, with an average of 0.22 individual/km^2 (Table 1.29). Almost all goitered gazelles were concentrated in the internal parts of the desert, in areas with large-bumpy sands (Kovshar and Blank 1986). Other authors (Pole and Karpov 1986) also found that the gazelles' maximal population density was in the central parts of the Taukum Desert, to the east from the Kolshengel-Topar highway.

The Desert of Saryishikotrau The dry channels of the Bakanas and the southern and central sections of the Ili-Karatal interfluve were surveyed. The average population density of the goitered gazelles in this desert was half as many as in the Taukum

Desert (Table 1.29). In Saryishikotrau, gazelles were concentrated in the saxaul forests or along the dry beds of the Bakanas (former Ili River) (Kovshar and Blank 1986). In the internal areas of the desert, the goitered gazelles were rarely found (Pole and Karpov 1986). In the Saryishikotrau Desert (right bank of the Ili River), the population density was significantly lower than in Taukum (left bank of the Ili River; Table 1.29). Such a significant difference in the population density of goitered gazelles is explained by less favorable habitat conditions in Saryishikotrau (snow on a significantly sized area in winter) and more favorable conditions for poaching, since Taukum has bumpy sands and Saryishikotrau has ridge sands with numerous takyr areas. Here, the distribution of gazelles varies significantly by season: in winter and spring, they are found in the interior of the desert, and then in summer and autumn, they migrate toward Lake Balkhash (Kovshar and Blank 1986).

Muyunkum Desert During the 1950s, there was a fairly high population in the desert, as in the whole of Kazakhstan. A mass death of gazelles occurred in this area after an unusually snowy winter in 1968–1969. In the spring of 1969, there were one to two dead gazelles for every 25 km of the survey route, with the remains of old males predominating. The reason for this mass death was the inaccessibility of basic winter food. With a deep snow cover, the lack of forage was not compensated for by the presence of shrubs and saxauls. Over the following 2 years, their numbers dropped to one individual per 100 km along the survey route. Since 1970, the number of goitered gazelles in the Muyunkum Desert has fluctuated quite significantly over the past 12 years from 1 to 24 individuals per 100 km of the route. By 1974, the density of gazelles reached its original number, and in 1978, the density significantly exceeded the long-term indicators (24 individuals per 100 km). The maximum number of gazelles in this area was in 1978 and the minimum number was in the winter of 1970/1971 and 1982; the average was about four gazelles per 100 km of the route. In the Muyunkum Desert, the total number of gazelles was estimated at about 350 to 400 individuals in 1980. In the next 3 years, there was a decline, and in the autumn of 1982, only traces of single individuals and small groups were found in Muyunkums (Pole and Karpov 1986).

During the aerial survey of 1983–1984 (Table 1.29), the highest population density of goitered gazelles was found in dense saxaul forests south of the village of Furmanovka (0.25 individuals/km^2). It was here that more than 100 gazelles were seen gathering in December 1983. To the south of this saxaul forest, where there were thickets with *Calligonum* or places devoid of bushes, gazelles practically did not occur. In the south of the Muyunkum sands, goitered gazelles began to meet only at the longitude of the city of Dzhambul, where a group of 14 individuals was noted. Another site with a relatively high density of gazelles (0.12 individuals/km^2) was found in the central Muyunkum. Here, gazelles were concentrated in the coarse-grained sands, and gatherings were rare to the north and south of the center (Kovshar and Blank 1986).

In all parts of the Muyunkum Desert, gazelles were found in relatively dense saxaul forests or heavily rugged terrain, where they could be protected from cold winter winds and hide from poachers. In the summer, the goitered gazelle was often found in Muyunkum, as indicated by ground surveys and interview data. So, at the southern edge of these sands on the border with the valley of the Talas River, an automobile survey with a length of 100 km from May 20 to 27, 1983, found gazelles three times with a total number of seven gazelles: two single females, a group of two adults and two young, and one more single female, leading away from a fawn. A local inspector observed 17 goitered gazelles in 1 night and detained poachers with 5 dead gazelles (Kovshar and Blank 1986).

Goitered gazelles are much more numerous at the northern edge of the Muyunkum Desert, bordering the valley of the Chu River. In summer, the gazelles are somewhat dispersed in the central areas of the desert, but periodically return to their northern edge, where the salt sites are located. Some shepherds, going to the north for the summer to Sary-Arka, left pieces of salt for goitered gazelles at the koshar (premises for the night keeping of sheep) and wintering grounds, where sometimes up to ten gazelles gathered. After a sudden heavy rain in July 1983, goitered gazelle groups of one to three individuals each were observed on the highway, where they drank water from puddles on the asphalt 50–100 m from moving cars (Kovshar and Blank 1986).

Goitered gazelles probably can roam between desert habitats of Muyunkum, Betpak-Dala, and Taukum during their spring and autumn migrations, the directions and timing of which coincide with the migrations of saiga, though the distance of travel is significantly less than saiga (Pole and Karpov 1986). The gazelle population was positively influenced by an increase in the number of artesian, self-emptying wells, which led to an increase in the number of permanent watering places in the west of Muyunkum Desert and in the Eastern Kyzylkum Desert. Unfortunately, these changes also increased the pressure from poaching in areas of ungulate concentration. Thus, in 1980–1981, the state hunting inspectorate bodies registered 14 cases of illegal poaching of these gazelles in the Dzhambul oblast and 58 cases in the Shymkent regions of Kazakhstan (Kovshar and Blank 1986).

Betpak-Dala Desert In the north of the Chu River, in the Betpak-Dala Desert, the goitered gazelles were so rare during the mid-1980s that it made no sense to conduct aerial surveys in this desert. In the Betpak-Dala, goitered gazelles have been preserved only in its southeastern and the southwestern parts of the desert (Table 1.31). Moreover, gazelles live here only in summer and in winter migrate to the Chu River and cross it to join the Muyunkum population (Kovshar and Blank 1986).

Kyzylkum Desert The minimal population density of goitered gazelles in Kazakhstan was in the Kyzylkum Desert. Since this desert was located very far from administrative centers, poaching was more prevalent here than any other regions (Table 1.29). A car survey of goitered gazelles in the Kyzylkum Desert (April 19–May 7, 1984) also demonstrated the low population density (5 individuals per 806 km of the survey route, 6.2 gazelles per 1000 km) compared to other populations, except the Betpak-Dala Desert (Tables 1.29 and 1.30). All gazelles encountered

Table 1.31 The number and population density of goitered gazelles in the Betpak-Dala Desert during the 1980s (adapted from Kovshar and Blank 1986)

Region	Dates	Number of individuals	Distance, km	Density of gazelles per 1000 km	Author
Western Betpak-Dala Desert					
NW from Juvantube Village	May 4–10, 1981	0	200	0	Kovshar and Blank (1986)
Sarysu River, left bank	June 20–30, 1983	1 female+juv	1035	1 individual per 1000 km	Kovshar and Blank (1986)
Chulak-Espe borehole	May–June 1983	2 individuals	–	–	Local people
Katynkum, Sasykchenel	May 1984	0	475	0	Kovshar and Blank (1986)
Central Betpak-Dala Desert					
Kogashi-Koktal well	June 9–25, 1984	0	1123	0	Kovshar and Blank (1986)
Betpak-Dala Station	Before 1980	7	–	–	Kovshar and Blank (1986)
Eastern Betpak-Dala Desert					
Furmanovka Village	1963	Groups, 300 individuals each			Kovshar and Blank (1986)
Furmanovka Village	Winter 1981	20–30	–	–	
Andasai Preserve	July 25–31, 1982	4	1840	2.72 gazelles per 1000 km	
	April 27–May 17, 1983	1 male	2000	0.5 gazelles per 1000 km	–
40 km to north of Furmanovka Village	May 29–June 9, 1984	1 female	900	1.11 gazelles per 1000 km	–
Entire Betpak-Dala Desert					
Betpak-Dala Desert	1980s	9	4600 km	1.96 gazelles per 1000 km	Kovshar and Blank (1986)

in Kyzylkums stayed near the edge of the sands. Goitered gazelles still inhabit the Kyzylkum, but in summer they are scattered throughout the sands and only in September come out on the clay plains. Poaching of gazelles is highly developed here, since there are no saigas in this area. A particular danger, in addition to hunting

Table 1.32 Results of the aerosurvey (aircraft AN-2) and number of goitered gazelles in various areas of the Ustyurt Plateau and Mangyshlak Peninsula

Region	Number of encountered gazelles	Route length, km	Density of gazelles within route, ind/km^2	Plot area, km^2	Estimated number, individuals
Buzachi Peninsula	764	1200	1.27	13,500	17,145
Karatau and Aktau mountains	49	280	0.35	3600	1260
Eastern Karatau	17	76	0.45	320	144
Sam sands	18	325	0.11	3000	330
Bostankum sands	16	160	0.20	1300	260
Karynjaryk sands	10	160	0.13	1800	234
Kaskyrjol	6	66	0.18	360	65
Western and southern chink zone	68	970	0.15	2000	300
Total					19,108

Five counters, 2 on each side of plane and one in the pilot cabin, survey routes per 15 km; flight height of 30–50 m; survey area of 500 m; flight speed of 170–180 km/h (adapted from Blank 1991)

by car headlights, was the poaching of gazelles by shepherds at watering holes with the help of traps (Kovshar and Blank 1986).

The total number of gazelles in the surveyed areas of Kazakhstan was estimated at 12,000 individuals. Despite the fact that the main area of the goitered gazelles in Central Asia is located in the territory of Kazakhstan, there is not a single reserve in the republic in which a significant population of gazelles is protected (Kovshar and Blank 1986).

In August 10–28, 1989, an aerial survey of goitered gazelles was made in Western Kazakhstan (Ustyurt Plateau and Buzachi and Mangyshlak peninsulas) (Blank 1991). The results of this survey are presented in Table 1.32.

Buzachi Peninsula Goitered gazelles preferred the northern and central parts of the Buzachi Peninsula and were much less common in the southern part. In addition, the number of gazelles decreased from west to east, from the Caspian Sea into the peninsula. In spring, the Ustyurt Plateau and Mangyshlak Peninsula are flooded with water, which gradually dries out, and by June only individual puddles remain. The water lasts the longest on a large area of salt marshes, where the water evaporates last. This gives a certain advantage to the goitered gazelles on the Buzachi Peninsula, which is surrounded on all sides by sors (shallow drainless salt marsh with salt crust) and salt marshes. Obviously, such an abundance of water, although salty, lengthens the vegetation of plants. Another reason for the high density of goitered gazelles is the small number of cattle, for which the abundance of sors creates unfavorable conditions for grazing. Cattle do not graze at all on the salt flats of the northern part of the Buzachi Peninsula. Wintering of livestock is located mainly in the southern area of the peninsula, less often in the central sandy massifs. But agricultural work, the so-called haymaking, is carried out in the southern and eastern parts, which are detrimental to vegetation, animals, and soil cover. In fact, this is the uprooting of salt pans, which not only completely deprives the land of vegetation cover but also loosens the soil. Gypsum soil cover is mainly widespread here, the mechanical damage of which, during such haymaking, leads to its complete and irreversible destruction causing surface erosion. After such events, according to local residents, a significant number of goitered gazelles die. Therefore, the goitered gazelle number is lower in these parts (Blank 1991).

In addition, the habitat of gazelles such as salt pans, sores, and kultuks (different kinds of salt marshes), which are completely inaccessible or accessible only for a short time for transports, limits poaching here to certain accessible areas and times. And another point, in my opinion the most important reason for the growth of the gazelle population, is the insignificant industrial development in the exploration and development of minerals on the Buzachi Peninsula. In the 1960s and 1970s, there were many geologists on the peninsula and the number of gazelles was then kept at the level of several hundred. Now there are no geologists there and the number of gazelles has grown to more than 17,000 (Table 1.32; Blank 1991).

The Eastern Part of Karatau In this part, cut off from the rest of the mountains by railways and highways, the population density of goitered gazelles is noticeably higher than in other parts of western Kazakhstan (Table 1.32). Gazelles are kept here both in the foothill and the chink areas, and in the mountains themselves, although there were noticeably fewer in the inner parts of the mountains. In all cases, gazelles were observed near watering places, while the height and degree of roughness of the terrain had little effect on their distribution. Often gazelles met in the same places as mouflons (*Ovis vignei arkal*). The economic development of these mountains is greater than the territory of the Buzachi Peninsula. There are several large settlements located here, which means there are more people and transport and a number of gas and oil pipelines, with railways and highways, are passing through. The largest city of western Kazakhstan, Aktau (formerly Shevchenko), is located nearby (within 90–100 km). The preservation of a fairly large number of goitered gazelles here is explained by the extreme roughness of relief and comparative inaccessibility for poaching and transportation. In the west of the ridge, there is a nature reserve where rangers are constantly on duty, keeping almost the entire ridge under protection. In addition, this goitered gazelle population is associated with the Buzachi Peninsula population, and the gazelles have the possibility for migrations between these areas (Blank 1991).

Sands of Sam This is one of the largest sandy massifs of the Ustyurt Plateau with an area of 3000 km^2, located south of the Northern Chink. There are flat areas here, overgrown with rare *Salsola*, and small-bumped sands with dense saxaul thickets. All the goitered gazelles were found in the bumpy sands among the saxaul thickets. To the north and south of the sandy massif, there are salt marshes and sores, where, unlike the Buzachi Peninsula, not a single goitered gazelle was encountered. Apparently, there was no water for drinking here at that time of the year. According to the previous survey data, previously, there were significant numbers of goitered gazelles living here, but subsequently, the sands became winter pastures for livestock. Shepherd houses were quite often located here every 10–15 km, which indicated a large number of cattle in these sands, and hence people with dogs who poached were taking advantage of the remoteness of these areas. In the warm season, geological expeditions often came to this area, and anti-plague detachments were stationed twice a year in spring and autumn. All this together with the accessibility of the area for vehicles made these sands attractive to all poachers. The hunting inspectorate visited the sands only occasionally because of the remoteness from the nearest cities, so poaching in these places had no real restrictions. Therefore, the lowest population density of goitered gazelles was observed here (Table 1.32; Blank 1991).

The Sands of Bostankum These sands are located 15–20 km northeast of the town of Novy Uzen. In Bostamkum, goitered gazelles were found in the southeastern parts of the massif, which is covered with saxaul thickets and various *Salsola* species. In the southern areas, the sands become flat. The sands are overgrown with various types of cereals, and almost no shrubs are observed. All the gazelles were observed here even though there was a fairly well-developed network of wintering pastures for domestic cattle. From the south, the highway from Novy Uzen approaches the Bostankum sands, with two settlements on either side of these sands. In addition, an oil pipeline approaches the sands from the south. Despite all this, the population density of the gazelles was noticeably higher than in the sands of Sam and Karynzharyk, which are significantly more remote from the territories developed by humans but probably better protected by hunting inspectors (Blank 1991).

Karynzharyk Sands The massif is located to the southeast of the city of Novy Uzen and is a sandy shallow-hummocky massif with an area of more than 1800 km^2, squeezed on one side by a chink and on the other delimited by salt marshes of 12 km wide; these sands stretch from north to south for almost 100 km. The eastern edge of these sands and almost all of the salt marshes are part of the Ustyurt Nature Reserve. The lowest population density of the goitered gazelles was here after the sands of Sam (Table 1.32). At the same time, the southern, narrower part of the sands is not likely inhabited by gazelles, since not a single gazelle was encountered there. The low population density of gazelles here is explained, in addition to the reasons mentioned above for other sands, by migrations toward the Western Chink and Mount Karamaya, where there are many watering places (Blank 1991).

Kaskyrzhol and Western Chink Mountains Suitable habitat for the goitered gazelles is not the chink itself or the mountains, but the nearby plains. Usually these are flat sandy massifs in combination with sores and salt marshes and less often flattened smooth mounds of sedimentary rocks and nodules of bedrock. The Karamaya Mountain located here has a layered structure and is almost completely devoid of vegetation. Rare shrubs and herbaceous species appear only around the water sites at the foot of the mountain (Blank 1991).

The population density of goitered gazelles in the near-chink zone is slightly higher compared to the sandy massifs of Sam and Karynzharyk, which in my opinion is not due to the preferences of the gazelles for this area, but rather due to lower accessibility to transportation of this zone, and therefore to poachers. Chinks attract gazelles because of watering holes and shade provided by the many niches and caves. In terms of fodder, chinks are often poorer locations than the sands. During the survey, it was not a single gazelle had been seen on the Ustyurt Plateau itself for the entire 8000-km-long route, although occasionally, gazelles were observed near the chink cliff of the plateau. About 20, 000 gazelles live in the area of the Ustyurt Plateau and the Buzachi Peninsula. The largest number is located on the Buzachi Peninsula (more than 17, 000 individuals) and in the Aktau and Karatau mountains and Mangyshlak Peninsula (about 1400 individuals). The remaining 1200 goitered gazelles are distributed over large sandy massifs, Sam, Bostankum, and Kyrynzharyk sands, as well as a narrow near-chink zone. In the rest of the territory of western Kazakhstan, goitered gazelles are not found (Blank 1991).

The slow increase of gazelles and other wild ungulates in the region was due to the industrial development from oil and gas fields, along with the associated pollution, physical disturbances, and the complete degradation of habitats. Also, geological and industrial work attracts a large number of people to previously deserted areas, which in turn leads to an intense increase in poaching. Obviously, for this reason, goitered gazelles disappeared from the sandy massifs to the north of the Aral Sea. Back in the middle of the twentieth century, gazelles were found here in considerable numbers, reaching as far north as the Emba River and even the Ural River (Blank 1991).

In general, the primary reasons for the decline in the number of goitered gazelles in western Kazakhstan were poorly controlled poaching, due to the remoteness of these areas from administrative centers, and intensive economic activity, which was accompanied by the development of previously arid and deserted areas. Special harm to wild ungulates was caused by the so-called haymaking, which is in fact the complete uprooting of natural vegetation. The lack of native forage then leads to the disappearance of goitered gazelles. Moreover, the vegetation in these places can only be restored with difficulty, and traces of haymaking are clearly visible up to 5–7 years later (Blank 1991).

The government harvested gazelle meat intensely during the period from 1942 through 1950; for example, in 1944, soldiers shot 14,000 gazelles in the Ili Depression alone (Sludskiy 1963; Zhevnerov et al. 1983). In addition to the legal harvest, poaching was completely uncontrolled. As a result, gazelle numbers decreased rapidly, and they disappeared from most areas of their habitat. During the 1940s and 1950s, there was a rapid opening up of new lands to industrial development. Following constant expansion of the area under cultivation, the available rangeland decreased, and the number of domestic livestock increased. As a result of overgrazing, as much as 63–64% of the pastures became degraded (Kurochkina et al. 1986). Goitered gazelles are also susceptible to the effects of winter dzhuts (mass ungulate death due to deep snow cover) that occur every 7–8 years on average and cause periodic mass mortality usually due to starvation (Bekenov et al. 2001).

Since the 1930s, the goitered gazelle range has decreased considerably and has become fragmented into several large populations and many small ones. Populations in the Ili Basin; Taukum, Muyunkum, and Kyzylkum deserts; and Ustyurt-Mangyshlak have all become isolated (Blank 1991). Currently, the most important areas are the Buzachi and Mangyshlak peninsulas on the eastern shore of the Caspian Sea and the Ili Valley in eastern Kazakhstan. In the mid-1930s, estimated numbers were 200,000, of which as many as 100,000 inhabited in the Ustyurt Plateau and Mangyshlak Peninsula areas. A sharp decline followed and, by the end of the 1970s, fewer than 10,000 remained (Zhevnerov et al. 1983). During the end of the 1980s, the goitered gazelle numbers were estimated at 30,000 to 50,000, of which almost half lived on the Buzachi Peninsula and in other parts of the Mangyshlak (Mangistau) District, an area encompassing only 1.5% of their original natural habitat (Blank 1991). During the 1980s and 1990s, the second largest population was found in the Altyn-Emel National Park in the Ili Valley. During the summer, 5000 to 6000 gazelles can be found here (compared to 2000 to 3000 in the mid-1980s) and in winter their numbers can increase to 10,000. Two main populations in Mangistau and Altyn-Emel National Park are currently increasing slowly, but the smaller fragmented populations have decreasing trends (Bekenov et al. 2001).

1.5.3 Uzbekistan

In the plains and foothill regions of Turkmenistan and Uzbekistan, the goitered gazelle is distributed almost everywhere, with the exception of densely populated areas and areas with intensive agriculture development (Fig. 1.35) (Sokolov 1959). The first information about goitered gazelles in Uzbekistan belongs to Rychkov (1762). Eversman (1850) reported that the Persian antelope (goitered gazelle) was found in abundance in the steppes near the cities of Khiva and Bukhara, where gazelles were often tamed and kept in the courtyards of rich people for decoration and fun. Information about the comparative numbers there, however, is quite limited. Goitered gazelles are found in the Fergana Valley and in a number of regions and districts of Uzbekistan. This gazelle has inhabited the Fergana Valley since ancient times. Back in the 1940s, the goitered gazelle was found throughout the flat part of the republic and was quite numerous in some places. Over the last 40–60 years, the number and range of these gazelles have decreased sharply, especially during the 1970s; in some areas, they have disappeared completely, such as in the Khorezm region (Ishunin 1961; Flint and Prisyazhnuk 1986).

Ustyurt Plateau (Uzbekistan Part) Until the middle of the nineteenth century, Ustyurt remained a region of relatively untouched land and undisturbed animals due to its remoteness, desolation, and the very small number of vehicles in the first post-war years. In 1950 through 1955, only a few expeditions and survey parties visited this region, and shepherds and hunters rarely encroached further than 40–50 km from the developed zone of the Amu-Darya oasis. In 1956–1960, the number of people in the Ustyurt Plateau increased significantly, as survey work increased and drilling began in a number of places. After the 1960s, there was a period of intensive development of the plateau. In connection with the construction of gas pipelines and the laying of the railway, a large number of people and a lot of construction equipment and vehicles were concentrated here. In a short time, large settlements were erected, a water supply was laid, and many gas compressor stations and radio relay points were built. During these years, state farms and collective farms of the republic were also established, creating pastures for sheep and camels (Sabilaev and Koptsev 1986).

Naturally, the movement of vehicles has sharply increased over time. The Ustyurt Plateau was cut up into a dense network of roads, along which construction materials, fuel, food, and other goods were transported. From about 1971–1973, regular traffic between the republics of Central Asia and Russia began through the center of the plateau along the railway line. The economic development has had a negative impact on the flora and fauna of the desert, and the lack of proper controls and impunity from the laws that existed contributed to mass poaching, which led to a sharp decline of a number of rare ungulates, with gazelles suffering the most (Sabilaev and Koptsev 1986).

In the pre-war years, significant herds of gazelles were common everywhere on the Ustyurt Plateau. The total gazelle population in 1950 in the Karakalpak part of Ustyurt (part of Uzbekistan) exceeded 6000 individuals. From survey route records in 1956–1959, the population density of gazelles was no more than 60–70 individuals per 1000 km^2, and the total population did not exceed 4500 individuals. In 1963–1965, there were only 35 to 40 individuals per 1000 km^2, and the total number decreased to 2500 gazelles. By 1970, gazelles on Ustyurt had become a rarity. Most had been destroyed by poachers, and the survivors turned out to be frightened and displaced from open spaces to marginal areas and depressions covered with poor vegetation. The total number of gazelles by this time in Ustyurt hardly exceeded 1500 individuals with an average density of 20 to 25 individuals per 1000 km^2 (Sabilaev and Koptsev 1986).

The largest remaining populations of goitered gazelles are found in the north of the country, one in the southern Ustyurt Plateau and another in the Kyzylkum Desert to the southeast of the Aral Sea. A survey of Ustyurt in 1981–1984 showed that almost a third of it (2.4 million hectares) was missing any sign of goitered gazelles. In the southern and northeastern parts of Ustyurt (3.7 million hectares), this species is extremely rare, with 3.5 individuals per 1000 km^2. Only on a territory of 600,000 hectares in the zone of the Sam Sands and Mataikum sands and in the extreme northeast part and near the Sarykamysh depression are goitered gazelles still more or less common (45–50 individuals per 1000 km^2). In general, the Karakalpak population of gazelles does not exceed 400–500 individuals. Thus, in the Uzbekistan part of Ustyurt Plateau, the number of gazelles has decreased by more than 10–12 times just during the last 30–35 years (Sabilaev and Koptsev 1986).

Kyzylkum Desert (Within Uzbekistan) The overall picture of the decline in the population as a whole was similar to the situation described for the Ustyurt Plateau. Due to the proximity to oases, the marginal areas of these sands have been used for grazing, for harvesting firewood, and for constant hunting since ancient times. Until the 1950s, the penetration of people into the desert was restrained by the same reasons as in Ustyurt. However, due to the significant number of wells, this part of the territory was used by people much more intensively. The active development of the Kyzylkums occurred in 1956–1960s and continued later when numerous motor transports began to arrive in the republic, and numerous artesian wells were drilled in the sands and numerous livestock settlements were built in remote areas of pastures (Sabilaev and Koptsev 1986).

Until 1960, gazelles in the desert were common everywhere (Table 1.33); they constantly stayed on the edges of the sands and often went deep into the oases. In 1954–1959, the average population density of gazelles in northern Karakalpakstan (autonomous republic of Uzbekistan), with border areas of Kazakhstan in Kyzylkums, was 300 to 400 individuals per 1000 km^2 (Sabilaev and Koptsev 1986). A similar density was observed in the gazelle population in the Kazakh part of Kyzylkum (Krivosheev et al. 1952). In the southern part of Karakalpakstan, there were significantly fewer gazelles during this period (80 to 100 individuals per 1000 km^2). The total number of gazelles in Kyzylkums within Uzbekistan was about 12,000–14,000 individuals (Sabilaev and Koptsev 1986). The appearance of automobiles and poaching led to goitered gazelle numbers decreasing by more than three times during the second half of the 1960s. In the period 1983–1984, goitered gazelles were forced out of areas where they had been very common and have been rarely observed in the 3 million hectare area ever since (Sabilaev and Koptsev 1986).

The main number of goitered gazelles has been preserved only in very limited areas near the Aral Sea and the Kyzylkum Desert. The last documented assessment defined the number of gazelles at not exceeding 40 to 50 individuals per 1000 km^2 and with a total population estimated at ca. 1600 to 2000 individuals (Sabilaev and Koptsev 1986).

In general, the state of the number of goitered gazelles in Karakalpakstan should be recognized as catastrophic. To preserve the remaining number of goitered gazelles, it is urgently necessary to take strict conservation measures and to restore the numbers by organizing two reserves with a fairly extensive territory in Kyzylkum

Table 1.33 Results of surveys of goitered gazelle numbers during the last 50 years in Uzbekistan (individuals per 1000 km^2) (adapted from Marmazinskaya and Mardonov 2001, 2004)

Survey location	Population number	Year	Observed number	Authors
Ustyurt	Common	1950s	128 per 4 km	Kostin (1955)
Ustyurt, Karakalpak part	Common	June 1957	Common	Mambetzhumaev (1970)
	Common	1961	60 per 2 h walking	
Eastern Ustyurt	Rarely noted	1965	1 per 50 km^2	
Eastern and southern Ustyurt		1967–1968	2–3 per hundred km or absent	
Karakalpak part	6000	1950		Sabilaev and Koptsev (1986)
Other Ustyurt parts	4500	1956–1959	65–70	
	2500	1963–1965	35–40	
	1500	1970	20–25	
	400–500	1982–1984	35–50	
Northern Kyzylkum	12,000–14,000	1954–1959	80–400	Sabilaev and Koptsev (1986)
	1600–2000	1983–1984	40–50	
Central Kyzylkum Kukayaz Hollow		1979	300–400	Tikhomolov personal inf.
Bukhantau		1979	12–14 individuals	Filatov per. inf.
Kuldzhuktau		1997	150–180 individuals	Soldatova and Salimov (1997)
Amu-Darya region		1997	100–120 individuals	
Southwestern Kyzylkum Kimerikum		1997	50–80 individuals	
Sundukli		1997	80–100 individuals	
Karnabchulskiy Bukhara nursery		1997	500 and 712 individuals	
Karnabchulskiy preserve			Unknown	
Kashkadarya		1983	70 individuals	Red Data Book of Uzbekistan, 1983
Sukhadarya			Unknown	

(southeastern Aral Sea region) and Ustyurt Plateau (in the Sam and Mataikum sand zone) (Sabilaev and Koptsev 1986).

The fragmentation of the area into separate groupings is observed throughout Uzbekistan, and there is the steady tendency to further reduce these areas, continuing to decrease the number of isolated groups and populations (Flint and Prisyazhnuk 1986). Over the past 30–40 years, the number of gazelles on the Ustyurt Plateau has decreased by more than 10–12 times and in Kyzylkum by 12–13 times (Sabilaev and Koptsev 1986). Currently, seven subpopulations have been identified: Ustyurt,

Table 1.34 The estimated carrying capacity of different subpopulations in Uzbekistan based on 5.7–7.0 ha per individual (adapted from Mardonov 1993; Marmazinskaya and Mardonov 2004)

Population	Estimated area (km²)	Range (km²)
Ustyurt	65,000	60,000–70,000
Northern Kyzylkum	27,000	20,000–30,000
Central Kyzylkum	40,000	35,000–45,000
Southwestern Kyzylkum	6000	6000–10,000
Kharnabchulckiy	2000	1800–2300
Kashkadarinskaya	3900	3500–4000
Sukhadarinskaya	1600	1400–1800
Total	145,500 km²	127,700–163,100

north Kyzylkum, central Kyzylkum and southwestern Kyzylkum, Karnabchul, Kashkadarya, and Surkhandarya. Natural resettlement is difficult due to the strong isolation of populations and strong anthropogenic pressures (Tables 1.33 and 1.34).

Over the past 15–20 years, no specific studies have been conducted on the spread of the goitered gazelle in Uzbekistan, and I do not have any data on the current status of the subspecies, though the total estimated number is 3000 individuals in Uzbekistan at present. In order to develop effective measures for its preservation, it is necessary to determine the current location of the existing subpopulations and the presence of connections between them, if there are any, and to identify the causes of isolation, the ways of seasonal movements, and their migrations. Special attention should be paid to the definition of seasonal and year-round habitats, as well as habitats where gazelles occur sporadically. The following studies should be carried out for each subpopulation: the study of the number and density, sex and age composition, spatial and social structure, and habitat analysis (forage, water resources, the degree of anthropogenic pressure, the number of livestock and interactions). Full details of former numbers are unavailable, but the 1983 estimate is believed to represent a decline of 10–12 times from the levels in the 1950s. Many examples of local declines and extinctions have been reported (Marmazinskaya and Mardonov 2001, 2004).

In the Bukhara nursery, the number of the semi-captive population reached its maximum maintainable value (1224 individuals in 1989) and then began to decrease (to 628 individuals in 1993). The result of the population reaching its maximum allowable density was a high percentage of males in the population compared to females (39.9% to 56.6%), a decrease in the number of births (54.7% to 72.5% of females had just one fawn), as well as a high percentage of barren females (24.8% to 43.4%) (Marmazinskaya and Mardonov 2001, 2004).

1.5.4 Turkmenistan

Back in the 1880s, goitered gazelles were very numerous in southern Turkmenistan, in particular between the Murghab and Tejen rivers (Radde and Walther 1889). At the beginning of the twentieth century, goitered gazelles inhabited the entire flat part and foothills with wide valleys in central Turkmenistan. The range of the species at that time occupied an area of about 370,000 km^2 or about 76.4% of the current territory of the republic, and the total population was estimated at 250,000–300,000 individuals (Rustamov et al. 1986; Zhevnerov 1984). At the beginning of the 1930s, a survey in the central areas of Turkmenistan put the average population density at 2–4.3 gazelles per 1 km^2.

In the Murghab Valley, these gazelles are still common, but their former numbers are now unthinkable (Flerov 1935). They were common also in southwestern Turkmenistan but there are only a few animals left there. The number was decreasing every year, which was associated with increased hunting from vehicles (Dementyev 1945, 1951). Twenty years ago, and perhaps even now, the goitered gazelle was an ordinary animal seen in the strip of mountain slopes of the southwestern Kopetdag Range (Flerov and Gromov 1935). It was found everywhere, although in small numbers, around the Uzboy Bay and in central Karakum Desert (Dementyev 1951).

In the 1930s, the goitered gazelle was common in the strip of mountain slopes of the southwestern Kopetdag, in the basin of the Sumbara and Chandyr rivers. There were especially many goitered gazelles in the very south of the republic on the Badkhyz and Karabil hills. In June 1942, there were from 1.3 to 6 gazelles per 1 km along the routes of various valleys of Badkhyz. In November 1944, 20 herds of gazelles of 100 to 150 individuals each were seen in the Badkhyz Reserve. Systematic poaching in all areas of Badkhyz has catastrophically reduced the number of these gazelles. If in 1956–1957 it was possible to find groups of 100 to 200 individuals in Badkhyz in winter, then in 1964–1965 herds of even 20 gazelles became rare. At the end of the 1960s, only 300 to 400 gazelles survived on an area of 8000 km^2; then, thanks to increased protections, their number increased to about 1000 individuals by the summer of 1972 (Gorelov 1972).

During the 1960s and 1970s, the number of gazelles in Turkmenistan decreased dramatically. In the early 1940s, the total number of gazelles in the republic was estimated at more than 100,000 individuals, but by the 1960s, there were found only about 15,000 individuals. In 1959–1961, gazelles were still found on an area of 283,000 km^2, while in the rest of the area (87,000 km^2 or 23.5%), they had been exterminated. By 1974, a small number of gazelles were still preserved in the north of the republic (along the Unguz) and in the northwest on the rough terrain near the chink of the Kaplankyr, in the central parts of the Karakum Desert, as well as in the uplands of Badkhyz and Karabil. According to the most cautious estimates, the total number of gazelles during the 1970s in Turkmenistan was no more than 4000 to 5000 individuals, and their range had decreased by 70% to 80% (Zhevnerov 1984). Now the area of the goitered gazelles occupies only 3% of the territory of the republic, and full-fledged populations have been preserved only in two reserves. In the rest of the territory, gazelles are found in rare, scattered groups (Rustamov et al. 1986).

Due to the improvement of environmental measures by the early 1980s, a slight increase in the number of goitered gazelles in Turkmenistan was noted, when the

Table 1.35 The number and density of goitered gazelle in Western Turkmenistan in 1983–1984 (hiking and bicycle routes) (adapted from Gorbunov and Zarkhidze 1986)

	Territory of permanent residence			Territory of temporal staying			
Region	Area, km^2	Individuals per 100 km route	Total number	Area, km^2	Individuals per 100 km route	Total number	Population density, per 10 km^2
Krasnovodsk Peninsula	750	5.8	60	1000	0.3	10	0.4
Zauzboiskiy district	1100	8.0	100	2900	1.0	10	0.3
S Ustyurt	6000	6.0	200	12,000	0.5	50	0.14
Kora-Bogaz-Gol	100	12.0	12	100	3.0	3	0.75
Kurendag foothills	200	8.0	20	100	2.5	5	0.8

population reached up to 10,000 individuals. Extremely high air temperatures and drought in the summer of 1983 caused a mass death of goitered gazelles in various regions of the republic, which led to a new decrease in their numbers to 6,000 individuals. Cases of deaths of gazelles in the summer of 1983 are mainly associated with insufficient drinking water. Previously functioning shallow reservoirs dried up, and the remaining watering holes sharply decreased their water amounts, and the consumption of such water by gazelles led to poisoning. For goitered gazelle survival, it is necessary, first of all, to ensure the availability of guaranteed year-round watering holes in areas inhabited by large populations by using all available opportunities and means, up to the delivery of water to permanent watering holes (Rustamov et al. 1986).

Over the past 10 years, fairly complete records have been kept of gazelles in western Turkmenistan, where up to 80% to 90% of surviving gazelles are concentrated. A survey accounted for gazelles found along hiking (20 km) and bicycling (50 km) routes. During the counting of the gazelles, 2 to 3 days were spent on each allocated site, with four to six hiking and bicycling routes explored. As can be seen from the results (Table 1.35), the number of gazelles in six locations was about 390 individuals, and the total area of their habitats reached 8000 km^2. The average population density of gazelles was 0.5 individuals per 10 km^2. The sporadic distribution of the species covers an area of 16,000 km^2. According to approximate estimates, 80 gazelles lived within this territory. Thus, in the mid-1980s, the number of gazelles in the surveyed areas of western Turkmenistan and southern Ustyurt Plateau was about 470 individuals on an area of 24,000 km^2. Considering migrating groups, about 500 to 600 gazelles probably lived in this area. In the Sarykamysh Lake area, the population density was quite high compared to other parts of Turkmenistan (Table 1.36; Chernov 1986).

In the southern regions of Turkmenistan, a similar decrease in distribution and number of goitered gazelles was observed because of human economic activities. As a result, the constant population split into several small populations with

Table 1.36 Number of goitered gazelles in the area of the Sarykamysh Lake (vehicle and aerial surveys) (adapted from Chernov 1986)

Region	Survey method	Survey time	Gazelle number per 100 km of route	Population density per 1 km^2
Southern shore	Air route	September 1983	15	0.7
Southern shore	Vehicle route	October 1983	15	0.3
Eastern shore	Vehicle route	June 1985	50	–
Northern, western, and southern shore	Air route	August 1985	12	0.6
Southern and western shore	Air route	October	27	0.7

numbers of not more than 200 to 300 individuals each, and in early 1985 their total number was not more than 2000 individuals. The most distinctive population of goitered gazelles is located in the Badkhyz Nature Reserve which covers 800 km^2, which has variable environmental conditions that help the goitered gazelles stay here all year round and support a relatively high population density: 6000 gazelles at the end of the 1970s. In 1983, because of intensive poaching and poisoning of spring water, the concentration of salts due to severe drought increased, and the number of gazelles fell to 2.1 to 2.3 thousand individuals by the spring of 1985. To promote the survival of these gazelles, it will be necessary to improve protections from poaching, enlarge the area of the nature reserve, and prepare artificial watering places for wildlife (Kuznetzov 1986).

Goitered gazelle distribution once covered 76% of Turkmenistan, but their distribution is now down to only 3%. The population declined from an estimated 100,000 at the beginning of the 1940s to 15,000 in the 1960s and 4000 to 5600 in 1993. The main area of current distribution lies in the south: in Badkhyz (between the Tejen and Murgab rivers, *ca.* 2000 gazelles) and Karabil (between the Murgab River and the Karakum Canal, 1000–2000 individuals). Gazelles also occur in Kaplankyr in the north (probably 300–500); northwest Turkmenistan (<100); eastern Karakum Desert (probably 100–300); the Kopetdag foothills (<100); and the right bank of the Amu-Darya River (a few animals). Gazelles are extremely rare in the central Karakum Desert (no data, but very few), and they are completely absent from southwest Turkmenistan. There is an introduced population on Ogurchinsk Island in the Aral Sea (400 gazelles in 1994), and most of the remaining animals are found in protected areas (Gorelov 2001).

1.5.5 Tajikistan

The spread of goitered gazelles in Tajikistan has greatly decreased, especially over the 50 years from 1935 to 1985. Before this, goitered gazelles inhabited all semi-desert and desert areas along the upper reaches of the Amu Darya River and the lower reaches of the Panj, Vakhsh, and Kafirnigan rivers. However, they were most common on the Dangara Plateau. They were still preserved in the lower reaches of the Panj River in the Tigrovaya Balka Nature Reserve, where about 50 gazelles were kept on an area of 6000 hectares in September 1970. The gazelle population was estimated at about 200 in 1986 (including 10–20 in the north, 90 to 120 in the Tigerbulk Reserve, and 50 to 80 in the foothills of the southwest) and 100 in 1989. Possibly, the goitered gazelles can still be found, albeit rarely, along the low mountain ranges of Panjisky Karatau, Tergalitau, and Arktau and in the mountains surrounding the Beshkent Valley (Sokov 1986, 1989).

Currently, the goitered gazelles in Tajikistan are on the verge of complete extinction. By now, the distribution of all populations has acquired an isolated character. The maximum area of goitered gazelle distribution is only 300–350 km^2 at present and within 150–200 km^2 of this, occurrence is only sporadic (Sokov 1986, 1989). Currently only 70–80 gazelles survive in Tajikistan (Abdusalyamov 2001).

1.5.6 Kyrgyzstan

During the 1930s, goitered gazelles lived all over the Issyk-Kul Basin, and here their range merged with the range in southeastern Kazakhstan. During the 1950s, this species was observed only along the northern coast of Issyk-Kul, and it has been preserved only on the lake's western tip, between the Chu River and Lake Issyk-Kul, and did not go further than the village of Rybachy (Zhevnerov 1984). In this area, 20 goitered gazelles were still observed in 1959. During a survey in 1966, only a few singles were found in the area of the Issyk-Kul Basin. In 1967, ten gazelles were found in the southwestern Kyrgyzstan (Kanigui Mountains; Yanushevich et al. 1972). In the collections of the Zoological Institute of the Academy of Sciences of the former USSR, there is a specimen of a goitered gazelle hunted in 1913 on the southwestern shore of the Issyk-Kul Lake (Agalen tract).

During the 1960s–1980s in Kyrgyzstan, there were only two places left where goitered gazelles were distributed: the Issyk-Kul Basin and the western slopes of the Turkestan Ridge (Aizin 1986). In Issyk-Kul, the goitered gazelles inhabit a very limited area of about 20,000 hectares with typically desert vegetation. The number of gazelles in this area was determined on the basis of survey records obtained from the air, on horseback, and on foot: 29 gazelles in 1962, 7 in 1963, 17 in 1979, 15 in 1981, and 10 in 1985. In southern Kyrgyzstan, in the foothills of the Turkestan Ridge, the total number was estimated at 50 individuals in mid-1985. The main cause of the goitered gazelle disappearance from Kyrgyzstan is poaching and economic human activities. Recommendations for survival include a preserve in the Issyk-Kul Basin on an area of 12,000 ha; biotechnical processes, especially establishing watering places for wildlife; improvements of protections from poachers and dogs; yearly surveys of goitered gazelle populations; and encouragement for support of wildlife among local societies (Aizin 1986).

Table 1.37 Distribution of goitered gazelle in the Trans-Altai Gobi, Mongolia (1980–1982) (adapted from Bugaev 1989)

Region name	Average population density per 1000 ha	
	Summer–autumn	Winter–spring
Shivetulanskiy	0.23	–
Otgonusskiy	0.21	0.57
Endriginskiy	0.33	0.15
Burynkhyiarskiy	0.10	0.10
Nomigobiyskiy	–	0.10
Shargyngobiyskiy	0.10	0.05
Atas-Chingizskiy	0.01	–
Central Gobi-Tianshanskiy	–	–
Tsaganbogdinskiy	–	0.14
Priakhingoliskiy	0.25	–
Total number	For the area of four million ha	800–1000 individuals

1.5.7 Turkey

According to various travelers, goitered gazelles were once plentiful all over Anatolia, while their numbers decreased quickly, and by the 1980s, these gazelles were found only in limited areas. Kumerloeve (1969) estimated the goitered gazelle population in the Ceylanpinar area between 1500 and 2000 individuals. At the end of the 1970s, the number had decreased to 200 gazelles (Kasparek 1986). Erkan (1978) believed that the main cause for the decrease in gazelle numbers was agricultural chemicals. In Ceylanpinar in 1977, a breeding farm was founded for reintroduction of goitered gazelles to the wild. During the 1980s, around 80 goitered gazelles were kept in enclosures in Ceylanpinar (Kasparek 1986). Currently, the number of *G. s. marica* is 317 individuals in the wild and 1530 gazelles captive in the enclosure (Durmuş 2010).

1.5.8 Iran

Goitered gazelles occur from the Mond Preserved Area in Bushehr Province of southwest Iran, north to Sehrin in Zanjan Province, east to Khorasan, and south to the northern Seistan Basin. The species also occurs on four islands in the Persian Gulf. It has disappeared from the far northwest of Iran, the Caspian Sea coast, and the southeast. A national survey of its distribution in 1994 found approximately 50 populations of goitered gazelle, numbering from 5 to 1000 to 1500 individuals. There are currently estimated to be more than 9045 goitered gazelles within protected areas. At least 1622 are estimated to occur outside protected areas, of which *ca.* 1100 are found on Kharg Island (Hemami and Groves 2001). During the last 3 years, numbers of gazelles in some protected areas have shown a significantly increasing trend. In central Iran (Ghameshlou National Park), for example, 5170 goitered gazelles were noted during a survey in 2007–2009 (Hosseini-Zavarei et al. 2010).

1.5.9 Afghanistan

Most of Afghanistan lies between 600 m and 3050 m ASL and can be divided into three regions: the northern plains (30% of the entire territory), the southwestern plateau, and the central highlands, reaching 50% of the entire area of the country (Habibi 2001). Due to the civil war, information on the current status of antelopes is very limited and difficult to obtain. They were once widespread in the steppes and semi-deserts of the western and southwestern regions of the country, and Aitchison (1889) describes a distribution extending from Quetta to Kandahar and Herat. Its numbers were drastically reduced with the construction of irrigation projects in its range in the Helmand and Kandahar Provinces and the building of the Kandahar-Herat Highway. The Street Expedition in 1965 collected and observed the species in Grishk and Kalat (Hassinger 1968). Naumann and Nogge (1973) observed goitered gazelles in the Regestan and Margo Deserts. In northern Afghanistan, the goitered gazelle was sparsely distributed in the Khulm Plain and parts of Hairatan Desert, and young gazelles collected from these areas were seen for sale in Kabul in 1973 (Habibi 1977). In the mid-1970s, a small group of gazelles was seen in the Hamadan Desert by officials of the Herat Livestock Company. By the late 1970s, the species was reduced to a small portion of its former range, and the continuation of war may have further threatened its survival in the deserts of Afghanistan. A goitered gazelle estimate of numbers is not available (Habibi 2001).

1.5.10 Mongolia

In Mongolia, the spread of goitered gazelles covers the southern part of the country, the Gobi Desert, the valley between the Khangai Ridge and the Mongolian Altai, and the basin of Large Lakes up to Lake Ubsu-Nur, penetrating north to 46° N (Fig. 1.57—Map) (Sokolov 1959). Goitered gazelles have a wide distribution across the semi-desert and desert zones of western, central, and southern Mongolia. Although their range and numbers decreased by 30% during the 1940s–1960s, they remain widespread. In recent years, a small increase in numbers has been noted and they have reappeared in some parts of their former range. Within the Great Gobi Reserve, with an area of 5.3 million hectares, their range occupies a third of the entire Mongolian area and is located in the southwest of Mongolia. In the entire Dzungarian Gobi, the total number of goitered gazelles is estimated at 13,000 individuals (Bugaev 1989). A high density of gazelles is known in the desolate steppes and deserts in the Trans-Altai Gobi (south of the Altai Gobi - Table 1.35), where the number is estimated at 5.5 thousand (Sokolov et al. 1978). There were a total of 60,000 goitered gazelles in Mongolia in 1990 (Amgalan 2000). In the Great Gobi B,

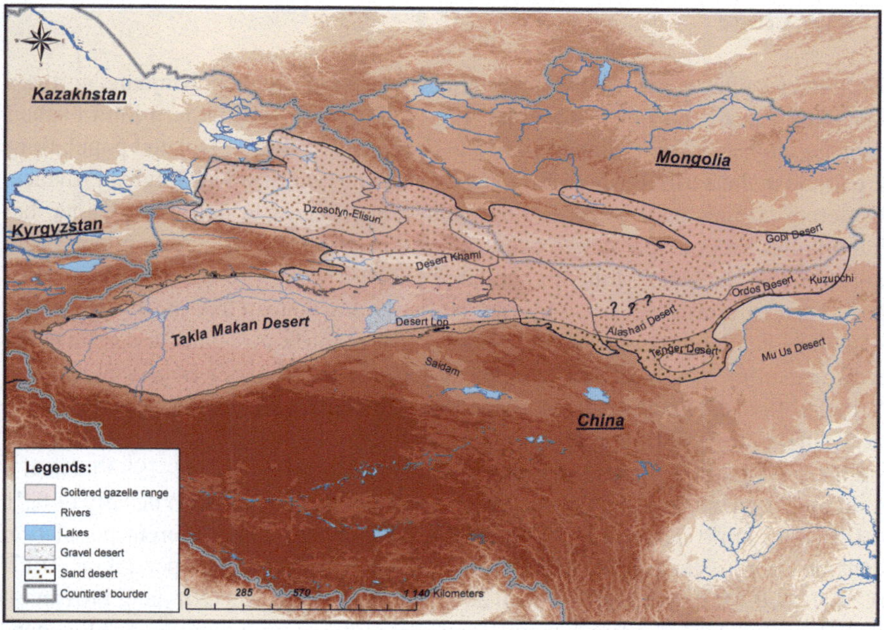

Fig. 1.57 The distribution of goitered gazelles (bounded by a solid line) in China and Mongolia (prepared by A. Osmonov with kind permission)

goitered gazelles were counted through 33 surveys, covering the area of 10,383 km^2, for a total number ranging from 2564 up to 10,766 gazelles (Kaczensky et al. 2015). In another survey in 2010, 5909 goitered gazelles were in Great Gobi B in the study area of 11,000 km^2 (Ransom et al. 2012). Aerial and auto-surveys of the goitered gazelle population conducted between 2012 and 2015 in the Mongolian Southern Gobi found an estimated total of 28,462 individuals (range 21,326 to 37,987 gazelles) (Buuveibaatar et al. 2017).

1.5.11 Russia

In December 2019–January 2020, two animals were observed, independently from each other up north of the Russian-Mongolian border near the Solovievsk railway station, Zabaikalsky Krai. One of them, an adult male, was photographed on January 11, 2020 (ca. 49.93°N 112.867°E), on a sod field with well-preserved dead grass in a forb-feather steppe. In the following days, the gazelle was observed several times on the Mongolian territory, down south of the Russian-Mongolian border (Kirilyuk 2021).

1.5.12 China

Goitered gazelles occur widely in western and northern China, in Xinjiang, northern Qinghai, Gansu, and Inner Mongolia (Fig. 1.57—Map). The eastern limit of their distribution lies at approximately 115°N (Heptner et al. 1961). Isolated populations are found in the Tarim Basin in Xinjiang and the Qaidam Basin in Qinghai Province. Goitered gazelles used to be an important game species in northwest China, but hunting has become rare due to overexploitation. Estimates of current numbers are not available. In Inner Mongolia, goitered gazelles were formerly abundant. They have been exterminated, or almost so, on the Ordos Plateau and most other areas, with only small numbers surviving, mainly close to the border with Mongolia in the west and in the southern and western margins of the Tengger Desert (Wang and Schaller 1996). The isolated population in the Qaidam Basin was reported to be declining and possibly in danger of extinction (Cai et al. 1990). Surveys near the southern edge of the Zuangker Basin in northern Xinjiang in 1992–1993 located a total of 221 gazelles along 19 transect routes (818 km), with a mean density of 0.71 (±0.17) per km^2 (Gao and Ji 1996). In the 2000s, there were different population densities in the Xinjiang Province from 0.17 up to 0.83 gazelles per 1 km^2 (0.17, 0.24, 0.36, 0.71, 0.8, 0.83) (Sun et al. 2002). In the Hami Basin (Xinjiang Province), in November of 2004, various populations were fond of high, middle, and low densities. The high-density population was 3.33 ± 0.24 ind/km^2 with an estimated number of 12,431 ± 896 individuals; the middle density was 0.95 ± 0.055 ind/km^2 with an estimated number of 21,651 ± 1254 gazelles; and the low density was 0.056 ± 0.042 ind/km^2 with an estimated number of 3026 ± 2270 individuals. The total population size in the Hami Basin was estimated at 37108 ± 4420 individuals (Gao and Yao 2006).

1.5.13 Saudi Arabia (Arabian Sand Gazelles)

According to Philby (1933) and Thesiger (1994), sand gazelles were once found throughout most of the Arabian Peninsula, including the central Rub al Khali (Empty Quarter), and were even observed as abundant in northern Saudi Arabia (Harrison and Bates 1991) (Fig. 1.58—Map). Vesey-Fitzgerald (1952) reports the Arabian sand gazelle presence from northern to central Saudi Arabia, being especially abundant on the Bisaita plain and Taisiya Plateau. Sand gazelles extend southward into the tropic, where they have been observed with other gazelle species together. Grimwood (1962) states that these gazelles stay in the sands of the southern Rub al Khali, and do not move far from these sands. Philby (1933) and Vesey-Fitzgerald (1952) state their estimates being in the thousands and in groups of 50–100 individuals.

The number of free-ranging Arabian sand gazelles (reem) has declined dramatically in the northern protected areas of their habitat (Harrat al-Harrah and Ak-Khunfah) and they are now possibly extinct. There are still some gazelles in the

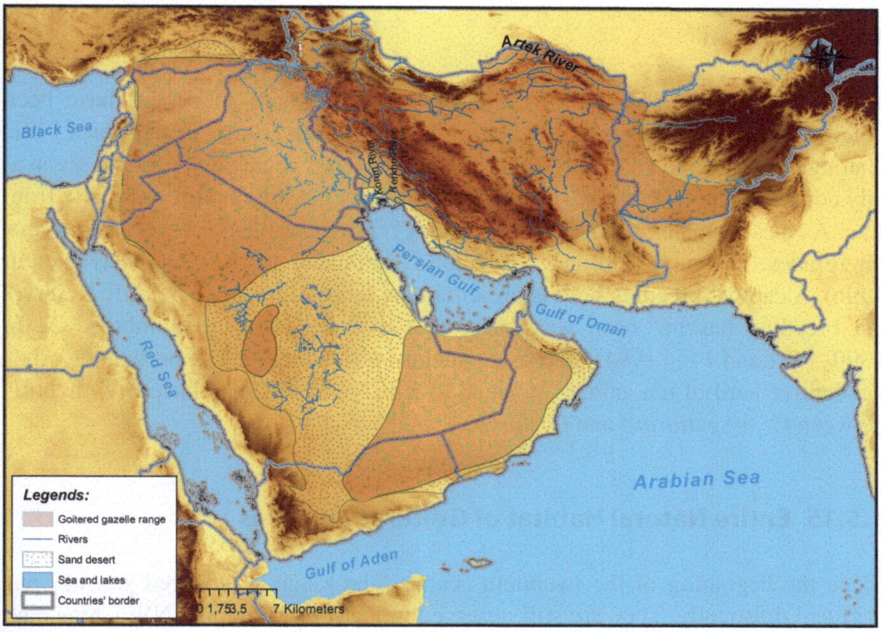

Fig. 1.58 The former distribution of goitered gazelle in the Arabian Peninsula, Iraq and Iran (prepared by A. Osmonov with kind permission)

southeastern Rub al Khali along the Saudi-Omani border, representing mainly a resident population in Oman. In contrast, gazelle numbers in the only fenced protected areas in central western Saudi Arabia (Mahazat as-Sayd) fluctuate widely according to food availability. Sand gazelle numbers in the southern protected area (Uruq Bani Ma'arid) remain changeable but relatively stable. Large numbers of sand gazelles are thought to be kept in private collections throughout Saudi Arabia (in the vicinity of 2000 to 5000 individuals), although with inconsistent husbandry methods and the potential for inbreeding and possibilities for crossbreeding, compromising the conservation value of these stocks (Cunningham and Wacher 2009).

Most recently, the total population number was estimated to be fewer than 10,000 mature individuals. Among them are four populations living in protected areas in Saudi Arabia with total population estimates ranging between 2650 and 3050 animals: Al-Khunfah, Harrat al-Harrah, Mahazat as-Sayd, and Uruq Bani Ma'arid (Cunningham and Wacher 2009).

In addition to Saudi Arabia, Arabian sand gazelles are found in small numbers in Oman and the coastal and offshore islands along the Arabian Gulf (e.g., United Arab Emirates (UAE), Qatar, and Bahrain), although the purity of the Gulf populations is questionable making the Saudi Arabian population even more internationally important (Thouless et al. 1991).

1.5.14 Jordan

Sand gazelles are known from the eastern desert, where specimens have been obtained from El Katrane and 160 km southeast of Hibar (Harrison and Bates 1991; Hatough-Bouran and Disi 1991). The goitered gazelles occurring in Jordan are usually considered to be *G. s. marica*, but goitered gazelles are imported illegally from Iraq. As these females are hornless, they may represent either *G. s. subgutturosa* or intergrades. The current Jordan population is estimated at <100 (Quemsiyeh et al. 1996). A captive-breeding herd is present in the fenced Shaumari Wildlife Reserve, but breeding has not been very successful to date. The herd numbered 22 in 1990–1991 and 14 in 1994. Moreover, Quemsiyeh et al. (1996) report that the original captive herd of ten animals donated to RSCN in 1978 may have been hybrids between *G. subgutturosa* and *G. dorcas*.

1.5.15 Entire Natural Habitat of Goitered Gazelles

Since the beginning of the twentieth century, the range of goitered gazelles has declined drastically. They are still observed in areas of Mongolia, NW China, and the former Soviet Central Asia, but elsewhere they have declined and occur primarily as isolated remnant populations in remote areas or protected reserves (Fig. 1.59,

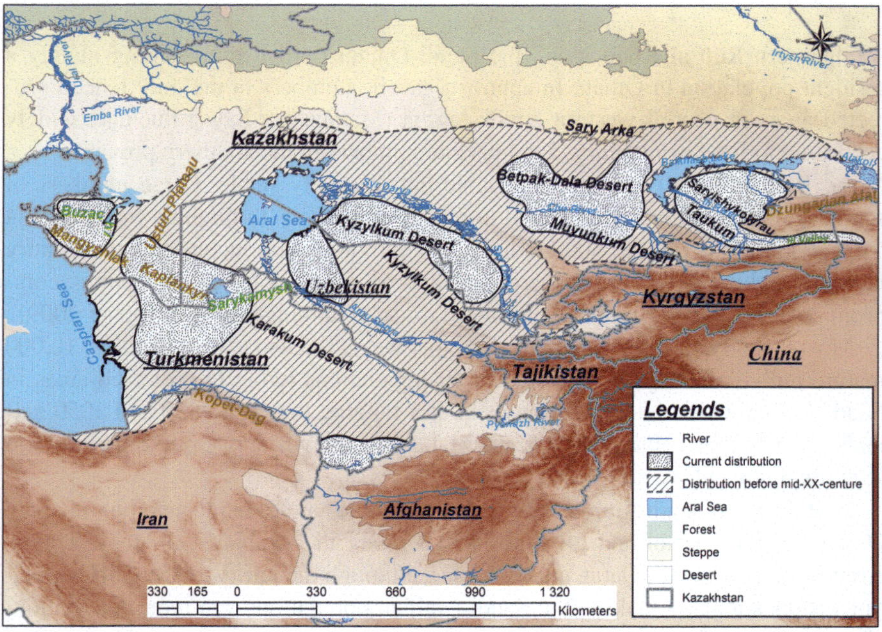

Fig. 1.59 Natural habitat of goitered gazelles decreased dramatically in Central Asia during twentieth century (prepared by A. Osmonov with kind permission)

Zhevnerov et al. 1983; East 1992). In Iran, goitered gazelles are widespread and increasing in protected areas outside the central deserts (East 1992). Goitered gazelles occur from Mond Preserved Area in Bushehr Province of southwest Iran, north to Sehrin in Zanjan Province, east to Khorasan, and south to the northern Seistan Basin. The species also occurs on four islands in the Persian Gulf. It has disappeared from the far northwest of Iran, the Caspian Sea coast, and the southeast. A national survey of its distribution in 1994 found approximately 50 populations of goitered gazelles, numbering from 5 to 1000–1500 (Hemami and Groves 2001). In the Sorkh-e-Hesar National Park, for example, in 91.68 km^2, an estimated population of 56–107 gazelles was found with a population density of 0.62–1.17 ind/km^2 (Jahandizi et al. 2015). There is currently an estimated >9045 goitered gazelles within protected areas. At least 1622 are estimated to occur outside protected areas, of which *ca.* 1100 are found on Kharg Island in the Persian Gulf (Iran) (Hemami and Groves 2001). In Iraq the population of *G. s. marica* is estimated at 1000 individuals and *G. s. subgutturosa* is estimated to be 500 individuals (Al-Robaae and Kingswood 2001).

Goitered gazelles are probably extinct in Georgia (Flint and Prisyazhnuk 1986), Kuwait, Syria, UAE, Qatar, and Yemen and nearly extinct in Jordan (<100 individuals of *G. s. marica*) (East 1992). A few small populations are said to survive in western and southern Afghanistan (Hamadan, Margo, and Registan) and Pakistan (Maslakh Game Reserve and northern Baluchistan) (East 1992). Larger populations occur in southeastern Turkey (300 individuals), northern Saudi Arabia (from 50–100 up to several thousand individuals; exact numbers are absent), and the Rub al Khali and Wahiba Sands of Oman and Yemen (no data if they still survive) (East 1992; Thouless et al. 1991). Gazelles on Kharg Island in the Persian Gulf are believed to be *G. subgutturosa* (Karami and Groves 1992). Populations on Barsakelmes in the Aral Sea and Ogurchinskiy in the Caspian Sea were introduced (Heptner et al. 1961; Kuznetsov 1986), and populations on the islands of Bahrain (250 individuals) and other areas in the United Arab Emirates also may have been introduced (East 1992; Gallagher and Harrison 1975).

In elevation, *G. subgutturosa* is found below 1000 m in Afghanistan (Habibi 1977), from sea level to about 2100 m in Iran (Lay 1967), and from 1050 up to 2100 m in Pakistan (Roberts 1977). They occur at altitudes of 2700 to 3000 m in the Qaidam Basin in China (Cai et al. 1990), but it can live in mountains from 3000 m to 3500 m during warm months in Kazakhstan, Mongolia, and China (Heptner et al. 1988; Kingswood and Blank 1996).

In total there were several hundred thousand goitered gazelles in Kazakhstan and Middle Asia during the 1930s and 1940s (Heptner et al. 1961). During these years, goitered gazelle hunting supplied a significant amount of meat and served as a serious boost to the economy of the people of desert areas. For example, in Transcaucasia, up to 2,000 gazelles were hunted per year; and in Central Asia, especially in Turkmenistan, several tens of thousands were taken per year (Heptner et al. 1961). And during the World War II, the Alma-Ata Nature Reserve supplied the army with several thousand goitered gazelles in 1942–1943 (Zverev 1948). For thousands of years, the goitered gazelle has served as the most important target of hunting in the deserts. But during the 1940s and 1950s, advancements in technology have allowed human beings to make significant changes in the desert zones, which have led to a substantial reduction in the natural range of the gazelle. In addition, unusually harsh winters and malicious poaching and extermination of gazelles from hunters in cars quickly led to the cataclysmic decline in gazelle numbers, and they have disappeared in many areas where until not that long ago they were numerous (Zhevnerov et al. 1983).

In the south of its range, goitered gazelles were widely distributed across deserts from the Arabian Peninsula in the west to Balochistan (Pakistan) in the east. In Afghanistan, the goitered gazelle is found everywhere in suitable habitats, reaching the border with Pakistan in the south. In Pakistan, it is found on the Quetta-Pishin Highlands, penetrating east to the Suleiman Mountains or almost to the Indus River (Zhevnerov 1984).

From areas where the snow cover is higher than 10 cm, this species migrates south for the winter or descends from the foothills to the lowlands. In the spring–summer period, the gazelles go far north into the semi-deserts and climb quite high into the mountains (Zhevnerov et al. 1983).

1.5.16 Goitered Gazelle Estimated Numbers

In about 1900, an estimated one million goitered gazelles inhabited Soviet Central Asia (Sludskiy 1977). Subsequently, they have declined or have been extirpated in many parts of their range as a result of overhunting, overgrazing by domestic livestock, and development of agriculture. Currently, populations are estimated to number about 100,000 in Mongolia (Zhirnov and Lushchekina 1992); 35,000 in Kazakhstan (Blank 1990); 8000 in Turkmenistan; 5000 in Azerbaijan; 5000 in Uzbekistan; maybe a few individuals in Kyrgyzstan; 1200 in Saudi Arabia; several hundred in Tajikistan; and 300 in Turkey (Flint and Prisyazhnuk 1986; Ogurlu 1992; Kingswood and Blank 1996). Thus, the total number of gazelles in Central Asia and Transcaucasia was estimated in the mid-1980s to be about from 30,000 to 55,000 individuals (Flint and Prisyazhnuk 1986). In Iran the current number of goitered gazelles is estimated at around 10,000 individuals; Afghanistan and Pakistan may have a few hundred each. The number of *G.s marica* may be estimated at a few thousand individuals in the Arabian Peninsula and Iraq in total (Mallon and Kingswood 2001). Thus, the total number can be estimated at 65,000–70,000 for *G.g. subgutturosa* and <2000–3000 for *G.g. marica.*

Elsewhere, goitered gazelles occur primarily in smaller populations, which are vulnerable to further reductions because of their isolation (Blank 1991; East 1992). There are more than 1000 on the various islands in Bahrain and other areas in the United Arab Emirates, but some of these are introduced populations probably originating from mixed releases of *G. s. marica* and *G. s. subgutturosa* (East 1992). The captive population numbers are estimated to be at least 529. In Afghanistan, China, the former Soviet Central Asia, and Turkey, *G. subgutturosa* is rated from vulnerable to extinct (Cai et al., 1990; Habibi 1977; Neronov and Bobrov 1990; Ogurlu

1992), and *G. s. marica* is considered endangered (Groombridge 1993). Protected areas with populations of goitered gazelles exist in Mongolia, China, Kazakhstan, Tajikistan, Uzbekistan, Turkmenistan, Iran, Turkey, Jordan, Saudi Arabia, United Arab Emirates, and Oman (Neronov and Bobro 1990; Ogurlu 1992). Reintroductions are being conducted in Saudi Arabia to develop additional reserve populations (Launay and Launay 1992).

By the mid-1980s, the northern distribution border of the goitered gazelle in Kazakhstan had shifted strongly to the south. Goitered gazelles have disappeared in most of the sands around the Aral Sea and almost the entire territory of the Betpak-Dala Desert. In general, within Central Asia, the gazelle still lives on an area of 300,000 km^2. At the same time, more than half of this area falls on the deserts of Kazakhstan (Blank and Kovshar 1986; Blank 1991) and about a quarter in Karakalpakstan. The distribution area of goitered gazelles in Transcaucasia is equal to 0.3% of their former species' range and in Tajikistan about 0.1%. The range of this species is steadily shrinking and breaking up into large and small populations. In most of the areas, there is still an exchange of individuals between populations, but many populations are already completely isolated from each other. Such isolated populations exist in Uzbekistan, Turkmenistan, Tajikistan, and Kazakhstan.

1.5.17 An Example of an Aerial Survey of Goitered Gazelles in Azerbaijan in 1982–1983

This survey was done with an MI-2 helicopter. The flights were carried out at an altitude of 50 m at a speed of about 60 km/h. These parameters for the flights were determined to be the most favorable, considering environmental and behavioral conditions for goitered gazelles. Despite the relatively low altitude, the helicopter did not bother the gazelles very much: some gazelles did not even get up from their beds or run away (and usually just a short distance and then stopped). At the same time, even resting individuals were well identified from this height. With a low flight speed, it was possible to accurately estimate the number of gazelles in herds of 15 to 20 individuals; in larger groups the estimate error was about 10%. Throughout the surveyed territory, 25 routes (tracks) with a total length of about 350 km were laid evenly on the map every 1 km. The main landmarks were the coast of the Caspian Sea and the highway, between which flights were conducted in a west to east direction. The parallel routes and the exact path of the helicopter to the desired points were provided by three cars, two of which were moving along the highway and one along the seashore. The location for the helicopter's approach to the next line route was determined by a standing car. Counting occurred over 2 days (about 6 flight hours). On the first day, the count ended on a route laid along the channel, which was a natural boundary that the gazelles could not overcome at night. Thus, the entire territory of the nature reserve and the preserve was covered by air routes (Khakhin et al. 1986).

The counting was conducted by a group of counters consisting of five people: a navigator who controlled the altitude and speed and helped the pilot to maintain a given direction and two counters on each side, one of which counted the gazelles forward and the other backward. During the counting, the counters, independently of each other, recorded the time of the beginning and end of the route, the time of sighting gazelles and their number, and the distance of the gazelles from the helicopter route. The width of the search stripe was 1000 m (500 m from each side). An important point was the identification of the remoteness of the sighting of gazelles from the helicopter course and the exclusion of gazelles encountered beyond the search path. For this purpose, the portholes in the helicopter cabin were marked with horizontal marks using a special template identifying 0–50 m, 50–100 m, 100–200 m, 200–300 m, 300–400 m, and 400–500 m distances from the flight path, in which the gazelles were encountered. The histogram of the frequency distribution showed a relatively uniform distribution within the search bands. A smaller number of gazelles were recorded in the 400–500 m band, which is due to the relative difficulty of detecting them, especially animals lying down (Khakhin et al. 1986).

It is well known that it is easier to count animals when they are more or less evenly distributed across the territory in small groups or one at a time. This was the case for a survey conducted in October 1982, when about 90% of the gazelle population was in groups convenient for counting. Another survey was in mid-November 1983, when the animals began to concentrate and about 35% of all gazelles were kept in a relatively small area in large groups (more than 50 individuals), forming significant clusters. The calculation of numbers in such groupings cannot be visually accurate. It follows from this that it is better to conduct air counting in October or earlier, before large groups of goitered gazelles begin to form (Khakhin et al. 1986).

According to the result of this aerial survey, the number of gazelles in the Shirvan Reserve and the Byandovan Preserve (Azerbaijan) in 1982–1983 can be estimated at not less than 4 to 4.5 individuals in this area (Khakhin et al. 1986).

1.6 Migration and Movements

1.6.1 Migrations in Central Asia

In most of their environment, goitered gazelles make spring and autumn migrations. As a result of spring migrations, goitered gazelles leave wintering grounds with relatively warmer weather and appear in areas with a wetter and cooler climate in spring and summer, where there are the most favorable feeding conditions. In their summer grounds, the gazelles stay through the birthing and rearing of their young. Goitered gazelles can stay in these areas until late autumn, but with the frosts and snowfalls that result in the loss of suitable forage availability, they gather into large groups and migrate to the southern deserts of their rutting grounds with a warmer and snowless environment (Zhevnerov et al. 1983). Autumn and spring migrations are well expressed, mainly in the northern parts of the gazelle natural habitats, and are associated with the presence of deep snow cover. In Central Asia, goitered gazelles historically migrated in autumn and spring between the southern edges of the steppes having a deep snow cover and the deserts in the south (Fig. 1.56); during summer dry periods, they may also be nomadic in search of water (Antipin 1941; Heptner et al. 1961).

In the flat parts of Central Asia, with snowfall—the Mangyshlak Peninsula, the deserts of the Northern Aral Sea, and Betpak-Dala Desert—most of the goitered gazelles migrate to the south and end up in the sands south of the Chu and Syr Darya rivers for the winter (Fig. 1.56). However, winter movements on the Ustyurt and especially on the Sarysu River are usually poorly expressed, and in mild winters, the goitered gazelles remain for the winter in areas along the river (Sludskiy 1953; Kostin 1955; Mambetzhumaev 1954). In the autumn–winter period in Mongolia, the increase in herding is associated with seasonal migrations, which are most pronounced in the northern parts of the area. Here, there is a shift of goitered gazelles from the zone of deep snow cover to more southern snowless areas of the deserts, and gazelle migrations to foothill areas, where there are better pastures, characterized by little snow (Bugaev 1989). In the southern parts of the goitered gazelle habitats—in Turkmenistan, Uzbekistan, Tajikistan, and southern Xinjiang in China—there are limited migrations of gazelles, due to the absence of snow or an insignificant snow cover (Heptner et al. 1961). However in the recent past, and perhaps at present, there were mass migrations of the gazelles from Turkmenistan to Iran and Afghanistan through the Parapamiz pass between the Amu Darya and Tejen rivers (Dementyev 1935). In addition, in the spring of 1919, a herd of gazelles of 10,000 individuals was observed in Iran, escaping in a continuous mass. It was the migration of gazelles from their wintering areas in the sandy deserts, which are waterless during the hot season, to their summer areas with juicy forage and watering holes in the foothills of the Kopet-Dag (Dementyev 1935).

Having emerged as a desert species, the gazelle, having small narrow hooves and a large weight load on the supporting surface of the limbs, is not adapted to moving on a deep, loose snow cover and getting food from under it. Therefore, its distribution in winter does not go anywhere beyond the boundaries of areas with an average decadal snow cover height above 10 cm. From areas where the snow cover is higher, the gazelles migrate south for the winter or descend from the foothills to the lowlands. In the spring–summer period, the gazelles go far north into the semi-deserts and climb quite high into the mountains (Zhevnerov 1984). Due to this ecological feature, the northern part of the gazelle's range is divided into areas of summer and winter zones (Fig. 1.56). The gazelle appears in the summer habitats in May and stays there until September–October (Sludskiy 1956b, 1963).

During dry periods in Kazakhstan, goitered gazelles stay in interior desert areas while they have temporary water sources and juicy vegetation, but when everything dries up, they become nomadic and start to search for water, migrating to large rivers and lakes (Zhevnerov et al. 1983). In NW Kazakhstan, summer areas for goitered gazelles are located between the Emba River and north Ustyurt Plateau. In late autumn or early winter, from the Buzachi and Mangyshlak peninsulas (Caspian Sea), as well as from the northwestern part of Ustyurt, most of the gazelles migrate previously to the south to the Kara-Bogaz-Gol Bay and the Krasnovodsk region, where annually in December–January, there are significantly more gazelles there than in summer. These migrations have now stopped because of fences constructed along the entire Kazakh-Turkmen borders (Pestov et al. 2018). However, inside Turkmenistan, on the Krasnovodsk Peninsula, gazelles have retained their local scale migrations and from June to September move from drier areas to wetter ones, where there are watering holes and succulent vegetation grows. In dry years, these movements begin earlier and their scale is more significant (Gorbunov 1983).

In June–July, goitered gazelles stay in the NE Ustyurt (Sam Sands) covered by shrubs of sand acacia (*Ammodendron* ssp.), *Calligonum* spp., and *Astragalus* spp., as well as a saxaul forest (*Haloxylon persicum*), that stretch to the Barsakelmes Hollow, where it is possible to find drinking water in takyrs. In October–November, goitered gazelles from the Sam Sands migrate far to the south by two migrating routes, to the Sarykamysh Hollow and the Kaplankyr Chink (Turkmenistan), covering 450 to 700 km. In winter, escaping from cold winds and snow, many gazelles stay in the chink zone in southern Ustyurt, hiding in gorges and ravines, as well as near the Amu Darya on the outskirts of tugai forests. They especially gathered near the Amu Darya in severe and snowy winters (Zhevnerov et al. 1983).

Other summer areas are located in the Daryalyk Desert (lying to the north of the Syr Darya River—Fig. 1.56), in the interfluve of Sarysu-Chu rivers. For wintering, goitered gazelles migrate for the winter from beyond the Syr Darya River to the Kyzylkum Desert, and before the ice breaks in early spring, they cross the river on their opposite migration (Heptner et al. 1961). By winter, all the gazelles that during the summer stay in the semi-desert of the Sary-Arka Valley and almost all the animals from the Betpak-Dala Desert migrate beyond the Chu River to the Chui Muyunkum Desert. Goitered gazelles avoid loose sands, so they are absent or rare in the depths of the vast deserts such as the Karakum (Zhevnerov et al. 1983).

From the Karsakpay region, the gazelles move in autumn to the lower reaches of the Irgiz and Turgay rivers, where the winters are not as snowy. From the flat part of the Ili-Karatal interfluve in the southern Balkhash region, the gazelles annually migrate in mass to the Kalkan-Matai Valley in Mynbulak (the southwestern spurs of the Dzungarian Alatau) for the winter (Dereviagin 1947). A mass annual migration of gazelles is also known in the autumn from the Betpak-Dala Desert to the south, to the Muyunkum sands, and in the spring, they migrate back in the opposite direction (Fig. 1.56; Kulpin 1954). The length of these migrations reaches hundreds of kilometers and is associated with the better protection in winter in the Muyunkum Desert (bumpy terrain, saxaul thickets). Migrations to the south in Betpak-Dala begins with the fall of rain and snow (late November–December), with the Chu River crossed only on ice (Sokolov 1959).

Inside of the Chui Muyunkum Desert, the goitered gazelles move in the spring from the saxaul forests in the north of the desert to its southern areas, devoid of trees. Here in the area of lakes, they spend the whole summer. In autumn, with the arrival of sheep flocks to the area, they return to the saxaul thickets (*Haloxylon aphylla*) where they stay all winter. During aerial surveys conducted in the winter of 1983–1984, all 200 gazelles noted here were seen in the saxaul thickets of the northern part of the Muyunkum Desert, while there was not a single individual in the southern part. In an automobile survey in August 1986 (about 250 km), no gazelles were found in the saxaul thickets; they appeared only to the south of the saxaul thickets. A much smaller group of animals migrates in the spring from the Chui Muyunkum Desert to the north, to the Betpak-Dala Desert, in the southeastern (Saksauldala district) and southwestern parts (Chulak- Espe district). There are no gazelles in other parts of the Betpak-Dala (Fadeev 1986). In general, the times of the autumn and spring migrations are quite variable (from September to November in autumn and February to March in spring) and depend on the location and specific environmental conditions at that particular time of year (Heptner et al. 1961).

In winter in the Taukum Desert (to the south of the Ili River; Fig. 1.56), gazelles are concentrated in the central and southeastern parts of the desert, to the east of the village of Kolshengel (Kovshar and Blank 1986). There are coarse-grained sands here and relatively fewer shepherd wintering grounds. In spring, with the departure of the winter livestock, the gazelles disperse throughout the desert and are more often found along its edges than in the center. As green vegetation disappears, the gazelles shift more prominently toward watering holes located along the edge of the desert and toward the Ili River delta; they are also noted on islands in the delta itself. In September 1986, I examined three areas in the Taukum Desert. I discovered that the population density of the goitered gazelles in the Taukum decreased as it moved from west to east. At night, the gazelles moved from the sands to the plains adjacent to the desert for grazing and watering, and they returned back to the sands in the morning. This is especially typical for the dry period of the year, when almost all vegetation disappears (July–September). Another group of gazelles went further to the south, to the Chui-Ili Mountains, where they remained. Now such migrations are impossible due to the network of fenced pastures located throughout the plains along the sands. In October–November, with the arrival of livestock in the Taukum Desert, and after a snowfall, these gazelles returned to the central areas of the desert (Blank 1990).

In the spring in the Saryishikotrau Desert (to the south of Balkhash Lake; Fig. 1.56), gazelles are also dispersed quite widely. With the drying of temporary water sources (puddles on takyr-like grounds), they shift to the Ili River delta. At this time, they are found in the southern and northern areas of the desert and very rarely in the central and eastern parts (Pole and Karpov 1986). In winter, the animals are concentrated in the well-protected areas of the dry beds and saxaul thickets in the area of the Bakanas village (Kovshar and Blank 1986).

In Mongolia, population movement and migrations are dependent on seasonal and pastureland conditions. In autumn and winter time, herds of Mongolian populations join each other constituting groups of several dozens up to a hundred (Lkhagvasure et al. 2011). The gazelles regularly migrate from Mongolia into China from October onward and return to Mongolia in late February (Wang et al. 1997). This migration may have led to a larger population in the Xinjiang area (near Mongolia), with a winter population of 19,677 individuals, while in other seasons it runs between 6000 and 14,000 individuals.

In a number of places, migrations running in the opposite directions are known. In the southeast of Kazakhstan, a significant number of pregnant females in the spring climb into the mountains of the southwestern spurs of the Dzungarian Alatau, where they birth and rear their young; for feeding, they descend to the adjacent valleys. Local, and sometimes distant, irregular migrations can occur due to the vegetation loss, the drying up of water sources, the onset of ice, the fall of deep snow, or the displacement of the gazelles by flocks of domestic sheep (Sokolov 1959).

Apart from a deep snow cover, the main reason for migration is the unequal supply of forage and surface water in different seasons. Summer migrations are associated with water and pasture availability (Heptner et al. 1961). These migrations represent the movement of the bulk of the gazelles from desert vegetation that has disappeared to more juicy pastures or closer to water sources. In summer, gazelles visit pastures that are located within 10–15 km from watering places. In the southeastern Karakum, gazelles gather for the summer on the banks of the Murgab, Kushka, and Tejen rivers in late June and July. In southern Mangyshlak and Ustyurt, goitered gazelles are concentrated near the coast of the Caspian Sea (Heptner et al. 1961). In other regions of southern Turkmenistan and Kazakhstan, most of the gazelles go to the foothills in summer, where pastures are juicier and there are springs (Heptner et al. 1961). In Kyzylkums and Ustyurt, migrations of gazelles are also associated with the disappearance of temporary watering holes and nomadic livestock (Zhevnerov et al. 1983). Only during the years of summer droughts are migrations in Mongolia and China similar to those that are common in Central Asia. At the same time, in Mongolia gazelles are more often concentrated not near rivers and lakes, but in the foothills (Bannikov 1954).

Nonperiodic migrations of goitered gazelles occur during natural disasters, most often when heavy snow falls or ice covers the snow on the ground (Sludskiy 1953; Mambetzhumaev 1970). Gazelles cannot get food from under deep snow, so they will leave an area that receives a deep snow cover. This species is also not capable of long-term migrations over long distances, like saigas, and they die in very large numbers. Besides ice and snow, nonperiodic migrations can be caused by droughts, steppe fires, and other natural disasters. In southern Turkmenistan, where for years the grasses were very thick and tall (mostly *Poa* spp.), there were large fires covering many thousands of hectares, forcing the goitered gazelles to vacate huge areas (Heptner et al. 1961).

Arabian sand gazelles (*Gazella subgutturosa marica*) are viewed as highly nomadic ungulates that travel vast distances for suitable forage following winter rains (Harrison and Bates 1991; Thouless et al. 1991) or display lengthy seasonal movements (Kingdon 1990). Sand gazelles migrated north out of the desert during the spring. On reaching the moister environment near the Euphrates Valley, the pregnant females give birth. The herds spend a few months in their northern territory before heading south again in July (Legge and Rowley-Conwy 1987). Now these migrations have stopped, as the number of these gazelles has decreased so much that their presence in the wild in the north of Saudi Arabia is very questionable (Nader 1989; Cunningham and Wacher 2009).

In the past, with the fall of snow, goitered gazelles migrated in large groups from the northern borders of their natural habitat (steppes and semi-deserts) far into the southern deserts (Fig. 1.56). There were also distinctive summer migrations associated with the drying up of green vegetation and temporary watering holes (Sludskiy

1953; Heptner et al. 1961; Zhevnerov et al. 1983). During the 1960s, as a result of the extermination of the gazelle population in the north, the northern border of their range has shifted to the south, and autumn migrations have been minimized, becoming local and limited (to no more than 50–60 km). However, in areas of their current range that were once the gazelles' wintering grounds, due to the drying up of temporary spring reservoirs together with livestock migrations, the gazelles are forced to undertake autumn–winter and summer migrations. However, long-distance migrations are made only occasionally, and only in unfavorable years (Sludskiy 1963).

1.6.2 Movements and Distribution in the Ili Hollow

In the territory of the Ili Hollow, the goitered gazelles in March–April move from the eastern regions of the hollow to the western areas, where they stay until the fall. Here, with ephemeral-wormwood associations, plants begin to grow much earlier in spring. Sometimes gazelles seek out crops and occasionally graze on harvested fields in the western part of the hollow (September–October). They also cross the mountains to the fields located on the slopes with a northern exposure. Undoubtedly, some of the ungulates also go farther into the Saryishikotrau Desert. With the onset of cold weather (October–November) and the arrival of shepherds to the western part of the hollow, goitered gazelles go back to the east. In the western part of the hollow, I never observed more than 10–15 individuals during autumn–winter over a 5-year period (1982–1986) (Blank 1990).

The distribution and movements of gazelles in summer are mainly associated with watering holes. In the dry hot months (July–August), they generally gather near springs and along the Kapchagai Reservoir. The spring and summer of 1987, however, had unusually rainy months and puddles formed everywhere, so there was no need to go to watering holes; therefore, the gazelles stayed away from the reservoirs near the mountains in the western and northern parts of the territory (Blank 1990). Goitered gazelles show similar behavior after releasing in the dry habitats of Azerbaijan: when water sources are limited in summer, they move closer to watering places (lakes and rivers) and usually cover 2–5 km per day and as much as 10–15 km in some days (Askerov et al. 2021).

There is a certain difference in the distribution of males and females, although at any time of the year, they can be found everywhere. In spring, during and after birthing of the females, their young mostly stay relatively close (within 5–7 km) to watering holes. At first mothers leave their fawns in specific hiding places, but do not go far away. So, the proximity of a watering hole allows her to leave her young for just a short time, which is why females are concentrated in the southern and eastern foothills of Chulak Mountains, where the preferred rugged landscape is combined with nearby watering holes. On the strip of plains along mountains (30 km long and 2 km wide), 180 gazelles and 220 fawns were observed, which accounted for 35% of all females and 50% of gazelles in the reserve. Later, at the age of 1.5 months, the females led their fawns to watering holes and left them

Table 1.38 Frequency of encounters in the southern foothills of the Katutau and Aktau mountains (adapted from Blank 1990)

Area	Month	Route length, km	Gazelle number	Encounter frequency ind/km^2
Katutau	November	28	17	0.61
	December	25	102	4.08
	May	46	76	1.87
	June	81	128	1.58
Katutau-Aktau valley	November	6	27	4.50
	December	6	34	5.67
	May	6	5	0.83
	June	10	21	2.10
Aktau	November	2	9	4.50
	December	3	9	3.00
	May	43	10	0.23
	June	28	12	0.43

already in the immediate vicinity of the water source. From then on, the fawns would drink without being escorted by their mothers (Blank 1990).

In the eastern part of the Ili Hollow, goitered gazelles migrate farther to the east toward the Katutau-Aktau mountains (right bank of the Ili Hollow). This happens after a snowfall in late autumn, as water sources during the summer–autumn period are limited here. At the beginning of November, when there was no snow yet, the population density of gazelles in the gravelly southern foothills of the Katutau Mountains was low. Only in a small area south of the mountains, around Aktau, a lot of animals gather in an area that has extensive takyrs that preserve water for a long time. In December, with the fall of snow, the migration starts from the western areas of the Ili Hollow, causing their numbers in the foothills of Karatau to increase dramatically (Table 1.38). Gazelles spend the whole winter, spring, and early summer here, but as the temporary watering holes dry up, they move back to the west. In rainy months, and June of 1987 was a good example, the population density of gazelles at the retaining rainwater takyrs in the Katutau-Aktau valley temporarily increased due to their arrival from the adjacent gravelly foothills of Katutau. But with the drying of the last puddles, the gazelles go to the west. Isolated individuals stay in the Katutau-Aktau valley, where they drink from the Ili River to eliminate traversing a distance of 20–25 km to reach other water sources (Blank 1990).

In the Boguty Mountains (left bank of the Ili Hollow), goitered gazelles move to the interior of the mountains in summer and in winter are pushed back by shepherds and flocks to the foothills and peripheral slopes. At the beginning of May 1987, on an automobile route 148 km along the Boguty foothills, 324 gazelles were counted (0.43 individual/km^2, with an accounting band of 4 km), but in July along a 114 km route in the same area, 112 gazelles were counted (0.20 individual/km^2), indicating that the number of animals decreased by more than half due to migration. In addition, some of the gazelles from these mountains go to the plains to the west of the

Boguty Mountains for the winter and return to the mountains in spring. On November 15, 1986, on a walking route of 6 km, 15 gazelles were seen, while on May of 1987, along a 23 km route, only 2 individuals were found (Blank 1990).

1.7 General Activity

1.7.1 General Statements

The daily cycle of goitered gazelle behavior depends on specific conditions: the time of year, weather patterns, and the state of the food supply (Heptner et al. 1961; Sludskiy 1977; Zhevnerov 1984). The gazelle, like most ungulates, demonstrates polyphasic activity. In summer, gazelle activity has two well-defined peaks, in the mornings and evenings, with a distinct break during the hottest hours of the day. This midday break, when the gazelles are resting, lasts from 9:00 to 10:00 until 16:00 to 17:00 (Fig. 1.60) and helps the animals to retain bodily moisture. In some areas, such as the Great Gobi B in Mongolia, goitered gazelles demonstrate one long morning peak of activity and a short evening peak (Ransom et al. 2012) while apparently spending most of the night on their beds (Zhevnerov 1984). In winter, goitered gazelles graze throughout the day with only a short break around noon (Fig. 1.60). Goitered gazelles are mainly diurnal, but on bright, moonlit nights, especially if they were not able to graze quietly during daytime hours, they go out to feed, even though it is their usual time to rest (Sokolov 1959).

Over a year, the times and durations between grazing and resting in the daily activity cycle change considerably over months. In March, feeding activity starts relatively late at 08:00 (sunrise) then varies insignificantly at a low level to around 13:00, when many goitered gazelles are seen grazing everywhere until 19:00. Thereafter, most gazelles disappear into the broken terrain. Sunrise is earlier in April and gazelle activity starts 2 h earlier than in March, with gazelles demonstrating almost the same active feeding periods as during morning and evening, with some decline in their foraging level midday (Blank 2022a).

In the first half of May, gazelles graze from sunrise at 7:00 to 12:00 and from 16:00, when the middle day heat lessens, to sunset at 20:00 with three peaks: 8:00 to 9:00, 16:00 to 17:00, and 19:00 to 20:00. In the hottest part of the day, they lie quietly and masticate their cuds. In the second half of May, the early evening period of 16:00 to 17:00 disappears, the morning period shifts 2 h earlier, and the daytime rest period becomes 3 h longer. In the middle of the day, there is no complete cessation of movement or grazing (Fig. 1.60), which is most likely due to the needs of mothers feeding fawns. The total duration of daily activity decreases by 1 h at the expense of the early-evening hours when the heat has not yet decreased. All these changes occur as daylight and daytime temperatures are increasing (Blank 1990).

The maximal duration of goitered gazelle activity is observed at the end of June, when their activity lasts 11 h compared to the 7 h at the beginning of June (Blank 1990). Zhevnerov et al. (1983) found that goitered gazelles graze from 11 to 13 h and rest 6–9 h, while movements take another 1.5–2 h of a 24 h day. Goitered

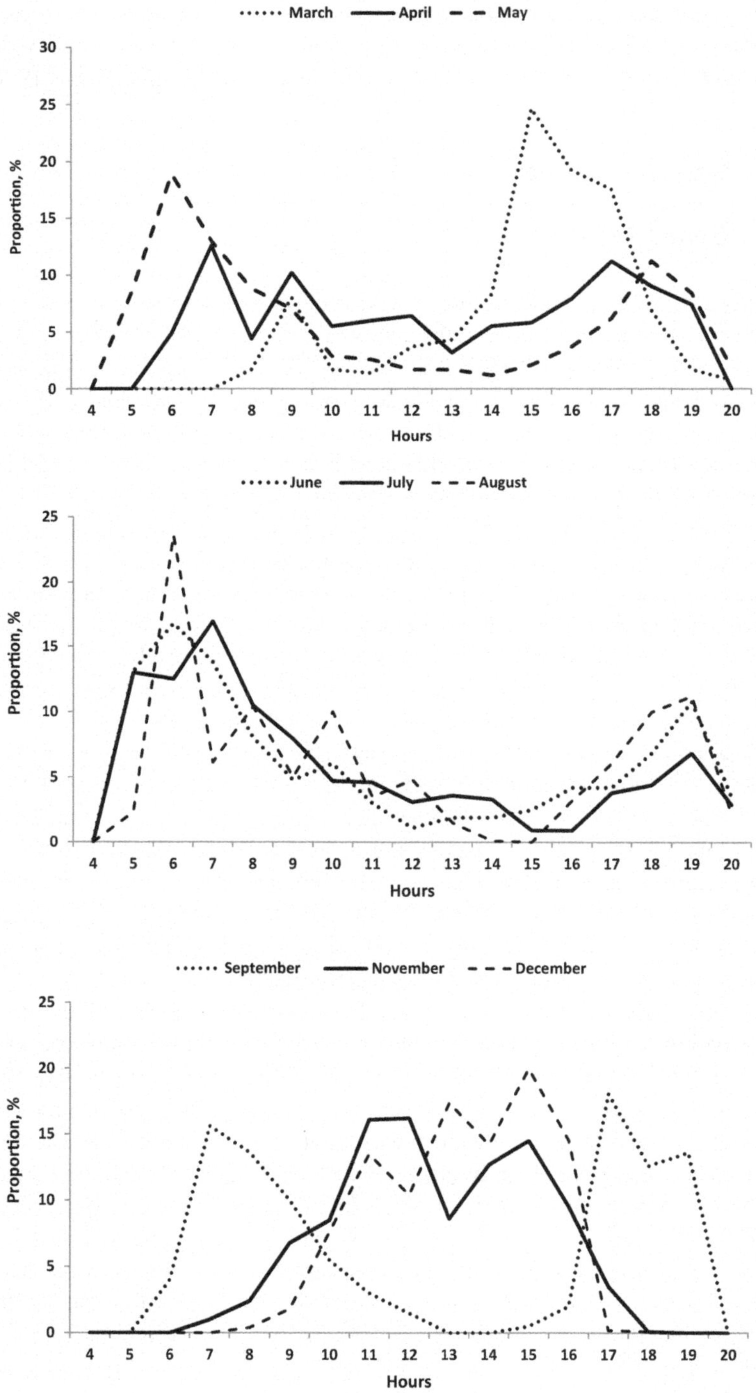

Fig. 1.60 Proportions of active goitered gazelles during daytime (per hour) in different months (by Blank 2022b)

gazelles graze from sunrise until 7:00 to 7:30 and then take a short rest for 30–40 min, before grazing again until 10:00. After that, the gazelles rest under the shade of trees and shrubs, in gullies or under precipices, or sometimes in elevated windy areas (Zhevnerov et al. 1983). The duration of their midday activity break during this time is reduced to 5 h (Fig. 1.60), though a reversal in this shift in activity-rest processes has already begun by the end of June. In my opinion, there are two reasons for this: first, at this time of year, the vegetation dries up, precipitation decreases, and day-time temperatures rise, causing a significant increase in the number of gazelles going to drink water during the day (Blank 1990); and second, overall activity is strongly influenced by the diurnal rhythms of females with fawns, whose activity shifts toward noon. Specifically, during this period of rearing young, mothers spend noticeably more time with their fawns than in May (Blank 1985).

In early July, the morning peak shifts by another 2 h toward midday. Gazelles stay longer in their night pastures at watering holes and graze less on the way to their daytime resting places. In the evening, the duration of movements is also reduced. At this time, the diurnal rhythm of females with fawns does not significantly affect the general activity budget, since almost all females lead their fawns from the foothills to watering holes. In the second half of November, the goitered gazelles demonstrate a winter diurnal lifestyle, when daytime rest decreases and feeding time increases due to the reduction of daylight hours and the deterioration of pastures (Zhevnerov et al. 1983). Although at this time gazelles are more active in the morning and evening than in the afternoon, this is not as pronounced as in summer. The total duration of grazing increases to 9 h. In December, this process continues, resulting in gazelles becoming most active in the middle of the day when it is warmer (Fig. 1.60) and resting during the cold morning and evening hours. In March, the daily rhythm is still on a winter schedule with gazelles most active in the middle of a day and afternoon (Blank 1990).

The activity of females with fawns differs markedly from the regular routine, especially immediately after giving birth (Tables 1.39 and 1.40). They do not have a noticeable increase in the duration of grazing, as noted by Zhevnerov (1984). However, mothers are more often active during the day, feeding their fawns two–four times and spending from 0.5 to 1.5 h with them (Blank and Yang 2015). When a female has two fawns, which she feeds separately for the first 3–4 days (sometimes up to 7 days), she has to interrupt grazing and resting for feedings up to eight times a day (Blank and Yang 2015). In the daytime activity budget of a mother, the suckling of the fawns takes a significant amount of time (10.9% to 23.9%, average 14.2%, $n = 669$) (Table 1.40) as well as reduces the duration of the mother's daily

Table 1.39 Daily budget of females without fawns (numerator is case number and denominator is proportion, %) (adapted from Blank 1990)

Date	Feeding	Rest	Movements	Vigilance	Scratching	Total
April 4	126/27.7	291/64.0	38/8.8	–	–	455/100
April 10	120/92.3	–	4	5	2	131/100
May 6	161/53.3	87/28.8	15/5.0	31/10.2	8/2.6	302/100
Average	136/57.8	126/30.9	19/5.5	12/4.4	3/1.4	293/100

Table 1.40 The daily budget of females with fawns (in May) (numerator, observed cases; denominator, proportion) (adapted from Blank 1990)

Date	Feeding	Rest	Suckling	Movement	Vigilance	Scratch	Ur./def.	Sexual pattern	Total
24	322	268	199	24	7	1	1	–	32
	38.7	32.2	23.9	4.1	0.9	0.1	0.1		100
26	207	204	86	155	24	1	1	13	691
	30.0	29.4	12.4	22.5	3.5	0.1	0.1	1.9	100
27	399	142	88	32	6	2	1	5	675
	59.2	21.1	13.0	4.7	0.9	0.3	0.1	0.7	100
28	295	95	103	137	1	–	1	–	635
	46.4	15.0	16.7	21.7	0.1		0.1		100
29	213	257	75	137	6	–	–	–	688
	31.0	37.4	10.9	19.9	0.8				100
30	318	99	118	107	4	–	1	–	647
	49.2	15.3	18.2	16.5	0.6		0.2		100
Average	286	159	95	114	8	1	1	4	458
	43.3	23.7	14.2	17.0	1.1	0.1	0.1	0.5	100

rest. In April, rest periods for pregnant females occupy 30.9% of their daily time budgets, while after giving birth, the average is 23.7% ($n = 1065$). Transitions from nursing one fawn to another without grazing also requires more time (17.0%) compared to movements in April without grazing that use only 5.5% ($n = 57$). In addition to female activity in April and early May, a fairly large part of all gazelle activity, such as scratching, is in response to discomfort associated with intense molting (1.5% to 2.6%), though this is later minimized (0.1% to 0.3%) (Table 1.39; Blank 2022b).

During the first days after giving birth, rainy weather does not affect so much the activities of females compared to a clear not rainy day, when grazing duration lasts 41.7% of the daily budget. As an example, on a rainy day on May 24, 1985, when it rained all day with only small interruptions, goitered gazelle mothers were active all day and grazed 38.7% of the 14 h of observations (Table 1.40; Blank 1990). Furthermore, Zhevnerov (1984) confirmed that females devote the maximum amount of time to feeding fawns in bad weather for all the days of observations.

There are many differences in the daily activity of fawns from that of adults. In their first 3–4 days, fawns remain lying down and hidden, rising only to suckle. On their fifth to sixth day, they begin to get up on their own, moving from one hiding site to another. At the age of 1 month, they are briefly grazing, both with and without their mothers, though still in the hiding phase of their development. It is not until their third month at the end of July that fawns begin to accompany their mothers constantly and start demonstrating activities similar to adults (Blank 2022a).

The activity of males in spring differs from that in autumn and winter. In winter, outside the rut, gazelle males graze all day, sometimes with a short break at midday. Their nights are spent on their beds, which are most often located in land depressions (Heptner et al. 1961). During the rut (November and December), however, the

Table 1.41 The daily budget of males (in April) (numerator, observed cases; denominator, proportion) (adapted from Blank 1990)

Date	Feeding	Rest	Scratching	Movement	Sexual patterns	Vigilance	Marking	Aggression	Total
5	198 71.7	77 28.3	–	–	–	–	–	–	275 100
10	120 92.3	–	2 1.5	4 3.1	–	4 3.1	–	–	130 100
	251 66.9	106 28.3	7 1.9	1 0.3	–	5 1.3	1 0.3	4 1.0	375 100
11	56 33.3	69 41.1	5 3.0	35 20.8	1 0.6	1 0.6	1 0.6	–	168 100
	108 31.6	185 54.2	5 1.5	28 8.2	2 0.6	4 1.2	3 0.9	6 1.8	341 100
12	110 97.3	–	1 0.9	–	–	1 0.9	1 0.9	–	113 100
21	107 59.4	62 34.5	6 3.3	–	–	5 2.8	–	–	180 100
	34 18.3	134 72.1	1 0.5	8 4.3	–	8 4.3	1 0.5	–	186 100
22	182 80.9	–	4 1.8	3 1.3	27 12.0	3 1.3	1 0.4	5 2.3	225 100
	44 48.9	38 42.2	1 1.1	6 6.7	–	1 1.1	–	–	93 100
28	27 16.9	90 55.3	1 0.6	41 25.6	–	–	1 0.6	–	160 100
29	133 88.7	1 0.7	3 2.0	7 4.6	–	5 3.3	1 0.7	–	150 100
Average	114 57.0	64 32.0	3 1.5	11 5.5	3 1.5	2 1.0	2 1.0	1 0.5	200 100

portion of their grazing time is reduced by more than two times, when it ranges from 11% to 38% (average 21.2%) of the daily time budget, while in April the range is 16.9% to 93% (average 57.0%), depending on the time of the day (Tables 1.41 and 1.42). Rest time in both cases is approximately the same: 27.8% during the autumn rut and 32.0% in April. It takes a considerable amount of time for a male to move from one end of his rutting territory to the other when courting females or chasing other males due to territorial aggression. Courtship itself takes relatively little time, 7.8% to 13.6% (average 11.1%). Thus, grazing time for a male during the rutting season is reduced not because of courtship, but because of the need to protect his territory from the invasion of neighbors or male bachelors. For this reason, the proportion of time for territorial boundary inspections is noticeable (Table 1.42). But all this concerns territorial males. The daily cycle of bachelor males is characterized by general activity during this period (Table 1.41) and consists almost exclusively of grazing and resting. Courting females takes them a little time for non-territorial males, since they are not specifically looking for females; instead, non-rutting males

Table 1.42 The daily budget of males during the rut (in November and December) (numerator, observed cases; denominator, proportion) (adapted from Blank 1990)

Date	Feeding	Snow eating	Rest	Scratching	Movement	Sexual pattern	Vigilance	Marking	Aggression	Total
Nov. 20	42 11.7	4 1.1	117 32.5	–	121 33.6	28 7.8	5 1.4	5 1.4	38 10.5	360 100
Nov. 25	128 37.7	19 5.6	67 19.8	–	55 16.2	29 8.6	13 3.8	6 1.8	22 6.5	339 100
Nov. 27	105 23.7	28 6.3	83 18.7	–	103 23.8	46 10.3	10 2.2	13 2.9	54 12.1	445 100
Dec. 2	84 15.0	–	139 24.8	–	159 28.4	76 13.6	1 0.2	8 1.4	93 16.6	560 100
Dec. 12	71 21.9	–	158 48.6	–	46 14.2	44 13.5	2 0.6	1 0.3	1 0.3	325 100
Average	86 21.2	10 2.5	113 27.8	–	97 23.9	45 11.1	6 1.5	7 1.7	42 10.3	403 100

just follow females during shared movements from watering holes to their daytime resting sites and back (Blank 2022a).

1.7.2 Activity Behavior in an Arid Environment

Blank (2022a)—License Number: 5478100079243

Foraging behavior is regulated by both intrinsic and extrinsic factors promoting ungulates to satisfy their daily energetic needs while also making time for other activities (Hamel and Côté 2008). The daily activity of ungulates consists of mostly of periods of foraging and resting (Belovsky and Slade 1986). The ratio of grazing and resting changes under the impact of various extrinsic factors, such as temperature. Goitered gazelles live in various kinds of arid conditions, including the harshest summers with very high temperatures combined with minimal levels of rainfall and high insolation. In such hard environments, ungulates need to adapt in order to maintain an almost constant body temperature with only small fluctuations and minimal water loss (Williamson and Delima 1991). Ungulates have different mechanisms to acclimate to the desert environment, but behavioral mechanisms are the most common because of their minimal energetic cost (Cain et al. 2006).

Similar to other ungulate species, one of the most common behavioral adaptations in summer is to decrease foraging and other activities during the hottest midday hours while being more active in the mornings and evenings (Belovsky and Slade 1986; du Toit and Yetman 2005). The reverse activity shift is seen in winter, when ungulates show minimal activity in the cold mornings and evenings and instead forage in midday when it is significantly warmer (Arnold 1981; Arnold et al. 2004). I examined the activity fluctuations in goitered gazelles across seasons and found that temperature is the strongest factor impacting the activity allocation between hours and seasons, changing time proportions between activity and rest over daytime hours and seasons (Blank 2022a; Peng et al. 2024). During the hot summer months, the goitered gazelles' highest foraging activities are mostly in the cooler mornings and evenings, while minimal movements and grazing occur during the hot midday. If the days are exceptionally hot, then goitered gazelles often graze much longer at nights, with the frequency of nighttime grazing fluctuating with the daytime temperatures (Abaturov et al. 1998). Saigas (*Saiga tatarica*) also demonstrate this behavior: when day temperatures reach upward of 42 °C in July and August, saiga antelopes decrease their daytime grazing to a minimum with active grazing continuing only when the temperature drops to 30 °C, often only in the evening, where they extend their grazing time to a total of 20% of a 24-hour day (Larionov et al. 2010). The black wildebeest (*Connochaetes gnou*; Maloney et al. 2005), reedbucks (*Redunca fulvorufula*; Irby 1981), steenbok (*Raphicerus campestris*), and impala (*Aepyceros melampus*) (du Toit and Yetman 2005) also decrease their grazing during hot days and increase feeding during nights. Moreover, large-bodied ungulates are much less affected by ambient temperature and can demonstrate more activities during hot middays (eland *Taurotragus oryx*, oryx *Oryx beisa*, buffalo *Syncerus caffer*) (Lewis 1977) compared to small- and medium-sized ungulates (du Toit and Yetman 2005). In temperate mountain environments, with relatively cool temperatures in the middle of the day, the Pyrenean chamois (*Rupicapra pyrenaica*), for example, can demonstrate a polymodal rhythm of activity and graze almost the entire day (almost 70% of the daytime), changing from grazing to resting many times in a row without dropping activities in the midday (Aldezabal et al. 1999). The same polycyclic daily rhythm during summers is typical for Arctic species, such as the wild reindeer (*Rangifer tarandus*), for which high midday temperatures are usually absent. Because the Arctic also has sunlight 24 h a day during the summer months, Arctic species can shift between grazing and resting during the entire 24 h period (Colman et al. 2001).

The reverser activity is demonstrated during cold months, when goitered gazelles are most active at midday with the more comfortable windless and warmer conditions (Blank 2022a). It follows, then, that, similar to other ungulate species, goitered gazelles optimize their foraging and resting periods at times that minimize the energy costs associated with the maintenance of their bodies' thermal balance and water turnover (Belovsky 1981).

Additionally, goitered gazelles demonstrate longer activity in the mornings during summer than in the evenings, since mornings tend to be the coolest time of day (Blank 2020). Thomson's gazelles (*Eudorcas thomsonii*) also have this same preference of morning hours for grazing over evenings (Rautiainen 2015). During seasonal weather changes between cold and hot (April into summer and September into winter), the difference between the duration of morning and evening activities is almost absent, while only in March are afternoon and evening grazing preferable for goitered gazelles compared to morning activities (Fig. 1.60). This unusual distribution of activities during March can be explained by strong winds that are common on spring mornings. Due to the poor adaptation of the gazelles' coats to protect them from strong winds and low temperatures, they have imperfect thermoregulation (Soldatova and Grazhdankin 1989) and prefer to rest inside sheltered areas, such as deep gullies and dry riverbeds that protect them from strong winds (Blank 2020).

Apart from temperature, the availability and quality of forage also have an impact on foraging activity. In general, ungulates need to graze longer time during autumn and winter, when forage amounts and quality are low, and graze for a shorter time in spring and early summer when forage is abundant and its quality is high (Pérez-Barbería and Nores 1996). Mountain ungulate species, being generally intermediate feeders, have an abundance of high-quality forage in mid-summer that allows them to decrease their forage duration. In autumn and winter, though, they have only low-quality food and become more bulk grazers, increasing their feeding duration (Pérez-Barbería and Nores 1996; Mysterud 1998; Hamel and Côté 2008). In addition, semi-desert ungulates, such as saiga antelopes and domestic sheep, also increase their grazing time with limited availability of forage plants on pastures (Larionov et al. 2010). I tested these changes in grazing behavior for goitered gazelles in desert habitat conditions and found that grazing duration does not depend on forage availability or quality. This variation in goitered gazelles is because this species is generally an intermediate feeder but can switch to being a highly selective feeder with the ability to find enough high-quality food in any season (Mysterud 1998; du Toit and Owen-Smith 1989; du Toit and Yetman 2005). Grass and herb cover are not significant in goitered gazelle habitats and only available for a limited time period in spring (2–3 weeks per year). Therefore, the main content of the gazelle diet consists of leaves, flowers, and shoots from shrubs (Zhevnerov et al. 1983; Blank 1990).

In addition, goitered gazelles move constantly while feeding and do not remain in one place for long. This behavior, in turn, allows goitered gazelles to be highly selective feeders (concentrated selectors) and collect enough high-quality food most months of the year (Blank 1990). The saiga antelopes also move continuously during grazing and only bite off the upper parts of plants, likely making them concentrated selectors, as well (Larionov et al. 2010). By contrast, bulk grazers, such as black wildebeests (*Connochaetes gnou*) and African buffalo (*Syncerus caffer*), living in conditions of constant abundance of water and forage, demonstrate a mostly sedentary lifestyle and stay in one place for long periods of time (Jarman 1974; Hofmann 1989). In mountain conditions of forage and water abundance, mountain species also exhibit sedentary behavior (Aldezabal et al. 1999).

Allometry of internal organs, body sizes, and stomach specialization also have an impact on the characteristics of feeding behavior (Underwood 1982; Van Soest 1996). Small ungulates were found to have a higher metabolic rate and therefore a need for higher-quality food than larger-sized animals (du Toit and Yetman 2005); larger animals have the ability to forage in the heat of midday due to thermal inertia which also gives them more foraging periods a day (Caboń-Raczyńska et al. 1987; Bruno and Lovari 1989; Zeng and Song 2001). Of course, large ungulates also need greater amounts of food per day compared to small ones. Some caprids and cervids have a big size difference between sexes, where males are larger than females, sometimes by several times (Pelletier and Festa-Bianchet 2004; Neuhaus and Ruckstuhl 2002). Goitered gazelles are also sexually dimorphic, with males having larger bodies (around 30%) compared to females. In my research, I initially assumed that this difference between males and females in goitered gazelles would mean that the larger males would demonstrate longer feeding times and shorter resting periods than the smaller females. However, my data does not support this hypothesis (Blank 2022a).

These results for goitered gazelles can be explained by an insignificant difference in body size between sexes, which relates to any insignificant difference in activity (Blank 2020). Another reason also may be due to a difference in forage behavior, when sexes may have the same activity duration but different bite rate, bite size, standing duration, or stepping rate (Abaturov et al. 1996; Ginnett and Demment 1997; Neuhaus and Ruckstuhl 2002). Saiga antelopes, for example, will change their stepping rate or decrease movement during grazing to enlarge their forage intake (Larionov and Nikonova 2006). Some ungulate species also have shown that sexual dimorphism in body mass has not impacted their activity budget (Ginnett and Demment 1997; Hamel and Côté 2008).

Pregnant and lactating females of many ungulate species (*Oreotragus oreotragus* (Norton 1981); *Oreamnos americanus* (Hamel and Côté 2008); *Cervus elaphus*

(Landete-Castillejos et al. 2000); *Ovis canadensis* (Ruckstuhl and Festa-Bianchet 1998); *Capra ibex* (Neuhaus and Ruckstuhl 2002); *Bison bison* (Komers et al. 1993)) demonstrate a meaningful extension in foraging duration compared to the period before birthing and rearing young. This difference can be related to the fact that pregnancy and lactation demand greater energy; therefore mothers are forced to stay longer on pastures to increase their food intake (Ruckstuhl and Festa-Bianchet 1998; Hamel and Côté 2008), especially during the first weeks after giving birth, when energy demands are especially high (Landete-Castillejos et al. 2000). My research data, however, does not show this phenomenon to be true for goitered gazelle mothers, with pregnant females and mothers with young not grazing any longer than in other seasons (Blank 2022a). This is because goitered gazelles are a hiding species, and immediately after giving birth, mothers are very busy with suckling fawns, behaviors related to approaching and constantly monitoring fawn's hiding places, and training their young (Blank 2017, 2022a; Blank et al. 2015). All of these activities demand a great deal of effort, making increased grazing time difficult. Goitered gazelle mothers, however, can increase their food intake by taking more bites of food compared to nonlactating females, which is similar to *Oryx gazella* females (Ruckstuhl and Neuhaus 2009), or they can increase the duration of night grazing, when they leave their fawns in hiding for the entire night (Blank 1985). Saiga antelope females exhibit similar behavior, grazing more at night, especially during the hottest weather (Larionov et al. 2010).

Finally, growing young need additional energy for development and growth until they reach adulthood, so they must forage longer compared to mature individuals with relatively lower energy demands (Owen-Smith 1998; Neuhaus and Ruckstuhl 2002). This statement, however, is only suitable for goitered gazelle fawns starting from August through November, not for June and July (Blank 2020). This, perhaps, is logical for goitered gazelles, since in June the young feed mainly on milk and plant-based forage is noticeably less important in their diet. Their feeding duration increases continuously during both June and July, but they still graze less time than adults. By August, when their mother's milk is mostly unavailable, they graze as much as their mothers, or maybe even a little longer than some adults (Blank 2022a); at this point in their development, they join the herd and follow their mothers constantly (Blank 1985). Staying in the group, fawns coordinate their activities with the adults, starting and stopping grazing together (Ruckstuhl 1999), though they may have different bite rates, steps, and vigilance frequencies that can affect their forage intake (Ruckstuhl and Neuhaus 2002). In addition, fawns, being smaller in size, may be more selective and choose mostly higher-quality food when grazing (du Toit and Yetman 2005). Similar to goitered gazelle fawns, young of some other ungulate species do not increase their feeding duration compared to adults (Hamel and Côté 2008).

So, I can conclude that external factors, such as temperature, have the strongest influence on daily feeding activity over seasons, while body size and the birthing and development of young have a more limited impact on feeding duration and the daily schedule of goitered gazelles, though some changes can be seen in the feeding behavior of females due to the hiding behavior of their fawns.

1.8 Predators

Wolves (*Canis lupus*) are the main predators of goitered gazelles (Sludskiy 1977; Zhevnerov 1984; Heptner et al. 1988), especially when a deep snow cover increases their vulnerability to pursuit (Jamsheed 1976). Historically, tigers (*Panthera tigris*) were known to hunt gazelles at watering holes (Heptner et al. 1988). In Turkmenistan, they were likely preyed upon by cheetahs (*Acinonyx jubatus*; Heptner et al. 1988), although Mowlavi (1978) did not find evidence that cheetahs affected goitered gazelle in the Khosh Yeilagh Wildlife Refuge in Iran. Young gazelles, particularly neonates, fall prey to a variety of predators, such as foxes (*Vulpes vulpes*, *V. rueppellii*), feral dogs (*Canis familiaris*), caracals (*Felis caracal*), imperial eagles (*Aquila heliaca*), and ravens (*Corvus ruficollis*; Heptner et al. 1988; Rietkerk et al. 1992; Sludskiy 1977; Zhevnerov 1984; Hague and Smith 1996).

The influence of the predator is reflected in the gender and age structure of the goitered gazelle. Fawns and older individuals are the most likely victims of predators, and among adult individuals, almost exclusively males became victims of the wolf (Hosseini-Zavarei et al. 2010). One of the reasons for such selectivity of predators may be due to the different levels of protective behavior displayed by adult females, fawns, and adult males. Adult females show a protective reaction to any potential danger more often than adult males, such that adult female gazelles are more vigilant than males. And females observe the terrain much longer and with greater frequency, while the gazelle adult males show mainly social vigilance, watching their relatives, especially during rutting season. Adolescents and yearlings show the lowest level of risk assessments, most likely because information about the potential risk of a threat is not as useful for them, due to their lack of sufficient experience to recognize predators and avoid them. Instead of relying on their own experience, young animals follow and repeat the behavior of adults. As a result, young animals of many ungulate species, including goitered gazelles, usually bear the highest losses from predators, slightly fewer losses in adult males, with adult females showing the lowest loss rate (Blank 2019).

Predators, along with other environmental factors, are also an important factor influencing the social structure of grazing ungulates (Roberts 1996), including the size of groups of chinkara gazelles (*Gazella bennettii*) in Iran (Akbari et al. 2015). Goitered gazelles, however, prefer to change the vigilance rate in response to predator pressure than increase their group size (Blank 2018).

1.8.1 Wolves

The wolf is one of the most dangerous predators of the goitered gazelles over almost their entire range, from the Caucasus in the west to Mongolia and China in the east. On the Ustyurt Plateau, wolves constantly roam behind herds of goitered gazelles, moving with them in autumn to the Mangyshlak Peninsula to the south (Kostin 1955). In snowy harsh winters near the Syr Darya River in the area of the city of Kyzyl-Orda, the arrival of goitered gazelles to the river valley from the north is usually observed in December to January and is always accompanied by the appearance of a large number of wolves following them. From the end of December to the beginning of April, wolves are usually absent in the northern half of the Betpak-Dala Desert, since there are no wild ungulates there then and rodents, another food source, hibernate. Only in low-snow winters, when some goitered gazelles are wintering in the north part of this desert, do wolves also stay there (winters 1952/1953 and 1954/1955). In late spring and summer, wolves will stay in the north of the Betpak-Dala Desert as long as gazelles are still there. In May 1955, an automobile survey in this desert area showed that along a route of 1620 km with a viewing width of 2 km in each direction, a total of 21 wolves were seen or 3 predators per 100 km^2. On the same survey, 1165 saigas and 447 gazelles were also encountered. Thus, there was an average of 55 saigas and 21 goitered gazelles also present, for a total 76 adult antelopes per wolf (Sludskiy 1962).

In the spring diet of the wolf, saiga and goitered gazelle, taken together, were found in 44.8% of all the studied data gathered over a 4-year period. In those areas where the saiga is absent, but the goitered gazelle is abundant, the gazelles become the main target of wolf hunting. For example, on the southern chink of the Ustyurt Plateau, in the excrements from 123 wolves, the gazelle remains were found in 51.2% of cases, while Asian mouflon (*Ovis vigney arkal*) was only found in 14.6% (Sludskiy 1962). In the Betpak-Dala Desert, the remains of goitered gazelles found in excrements were 22.4%, while in areas without saiga antelopes, the goitered gazelle remains in excrements increased up to 57.2% (Sludskiy 1977). In some nature reserves of Iran, remains of goitered gazelles in the wolf excrements fluctuated between 22% and 31% (Farhadinia et al. 2009; Hosseini-Zavarei et al. 2010).

The initial running speed of a goitered gazelle living in the Betpak-Dala Desert is about 60 km/h. In exceptional cases, females can run for a short distance at a speed of up to 65 km/h. In the Syugatinsky Valley, the goitered gazelle runs at a speed of 56 to 60 km/h, and only once on a short stretch, its running speed reached 64.5 km/h (Sludskiy 1962). According to Kostin (1955), gazelles can run at a speed of 60 km/h on the Ustyurt Plateau. Obviously, all plain antelopes, including goitered gazelles, can run faster than wolves. Only in exceptional cases (e.g., pregnancy) can some wolves catch up with a gazelle. Since wolves are not usually able to hunt gazelles by chasing, they have adapted their attacks using other techniques. In summer to autumn, wolves are especially successful in hunting goitered gazelles by waiting for them at watering places. Wolves hunt ungulates on pastures, hiding in bushes, ravines, and river valleys, and also accompany them during migrations, ambushing the gazelles on trails and roads. With the formation of packs, wolves can catch gazelles through forming roundups and pens. This method of hunting is especially widespread in late autumn and winter. During roundups, wolves in some cases can drive an antelope into water, on ice, into deep snow, or on thin salt or thin ice. During and after the rut, weakened males are caught in their individual territories, where the tired males often fall asleep. At the same time, they will actively catch single males. Finally, in May, during the period of mass birthing, wolves are actively looking for hidden fawns (Sludskiy 1962).

The drier the summer, the more gazelles die at watering holes from wolf predation, since there are fewer reservoirs and both gazelles and wolves concentrate near remaining water. Frequent attacks by wolves from thickets bordering water sources have led gazelles to avoid drinking in places where the water is surrounded with bushes in which wolves can hide. All known gazelle watering holes on lakes lying in the valley of the Chu, Ili, and Sarysu rivers, as well as Balkhash Lake, have always been located on completely open, low-lying banks, devoid of thickets, and hundreds of meters away from reeds (*Phragmites communis*) and *Tamarix* thickets (Sludskiy 1962).

In Turkmenistan, where there are few watering holes, goitered gazelles seeking water sometimes will approach somewhere along a river, even from a high bank, having to go down steep slopes to the water; gazelles from deserts try to walk along logs and gullies facing the river to reach the water. Trails appear along such gullies at some distance from the rivers, made by both goitered gazelles and kulans. Gazelles, preferring open spaces, are sometimes forced to pass through thickets of bushes when going for water in the floodplains of some desert rivers, such as the Amu-Darya, Syr Darya, Ili, and many others. Gazelles in Turkmenistan also prefer open areas for watering with shrubs several tens of meters away; but in some places, these gazelles, too, are forced to drink where thickets of shrubs come close to the shore (Sludskiy 1962).

If the gazelles are not very thirsty, they will carefully approach a watering hole, even if it is located in a completely open place, studying the surroundings for a long time, apparently pretending that they are not interested in water; some will run away, then return, and so on several times until they are convinced that there is no danger. In Betpak-Dala Desert, they will graze near the watering hole for as much as a few hours until they go to drink. Young antelopes are the first to go to water, and adult males are the last. A female gazelle with young walks very quickly to a watering hole, then slows until she finally stops, freezes, and listens while scanning the area. Sometimes the female pretends to graze, tearing off stems to eat. Gazelles will spend almost an hour, or possibly more, walking the last distance (300 m in one case) to reach the water. The same cautiousness is seen in the goitered gazelles in Turkmenistan when approaching water. They carefully look and listen from afar, sometimes standing immobile for half an hour. If they notice something, they usually become alert again and stop their approach. Sometimes quite a long time will pass before they calm down; and quite often, gazelles leave the water immediately and do not return (Sludskiy 1962).

Formation of large herds when watering in the summer–autumn period is not known in the goitered gazelle. In Kazakhstan, these gazelles usually come to drink alone, in families, or in small groups of up to ten individuals. In July in Turkmenistan, it is noted that gazelles most often go to water alone or in groups of two to four individuals. Apparently, deserts of the southern type with poor herbaceous vegetation and watering places do not create conditions for the formation of large herds of goitered gazelles (Sludskiy 1962).

Unlike saigas, goitered gazelles do not avoid thickets all year round. For example, in spring and summer, they are constantly found in sparse saxaul or thickets of shrubs, where they often hide in case of danger. The gazelles living in the thickets are much more cautious than saigas and have a number of morphological and behavioral features, such as when not grazing gazelles keep their heads high, while saigas leave their heads down and cannot lift them as high as gazelles (Sludskiy 1962).

Wolves cannot successfully chase goitered gazelles on a deep but loose snow cover, since the length of the front legs to the chest of wolves is about the same as the length of gazelles. A completely different picture is observed in winter, when there is a lot of snow, but it is covered with a fragile ice cover. Comparing the data on the weight load of goitered gazelles and wolves, it can be seen that, unlike gazelles, a wolf does not fall through the ice crust, while gazelles do fall through, making it very difficult or even impossible for them to run quickly to escape chasing predators. In addition, fleeing pursuit and falling through, ungulates can cut the skin on the metatarsals of the legs, resulting in their running speed falling and their gradual inability to run (Sludskiy 1962).

Wolves, having a weight load on the trail much less than the goitered gazelles, chase them not only in winter through deep snow covered by an ice crust but also in spring through muddy marshes and takyrs. The gazelles, exhausted by the lack of food in winter, and females pregnant in the spring, cannot quickly run through softened snow or soil, occasionally getting badly stuck, so wolves hunting them try to drive them into salt marshes to trap them. In early spring, some gazelles are wet up to half their bodies and covered with mud. Gazelles are instinctively afraid of marshy places, and if driven to a viscous salt marsh, they will run along the shore or try to break back, but not run into the swamp. Gazelles are also afraid of loose sands and always try to run around them. Wolves are also known to chase antelopes into a roundup and then drive them into the water or onto the ice (Sludskiy 1962).

Wolves, attacking goitered gazelles and other antelopes, always kill them with the same technique. They grab the victim with their teeth by the side in front of the groin and, with the first hit, usually inflict a large wound with a diameter of 10–15 cm, through which the intestines fall out. Sometimes there are several such wounds. If the predator cannot immediately grab the antelope by the side, he bites it up to several times on the rear end and thighs and, when the victim's running speed decreases, inflicts a decisive bite in the side. The wolf's grip is so strong that it easily pulls out large pieces of meat. An antelope with a wound on its side often runs for several tens of meters, dragging its intestines and bleeding until it falls. All the victims noted that had been killed by wolves had one or more wounds on the side, while the throat and top of the neck, as well as the head, were generally without damage. There are cases when wolves, attacking an antelope, do not have time to inflict a large wound on its side, and the victim escapes. Only in rare cases will a wolf be able to knock down an antelope by biting the rear or thighs and then kill it by grabbing it by the throat (Sludskiy 1962).

Wolves, having killed a goitered gazelle during a period when predators lack food, quickly devour it, eating and pulling away the meat. The prey is eaten almost entirely, along with bones, skin, and hooves. Usually, all that remains is a little wool, the contents of the stomach, horns with part of the cranial crest, and occasionally one to two pieces of a jaw with teeth or a tubular bone. Sometimes there are no such remnants, and the fact that wolves kill and eat antelopes can be judged only by traces of blood and a small amount of wool. Even after eating the body, the horns are left, which are usually quickly found by other wolves or foxes that gnaw on the horn covers. Predators like to gnaw on the soft horny cover so often that there are many horns lying around with chewed ends (Sludskiy 1962).

A wolf can eat no more than 3 kg of meat at a time, and most often noticeably smaller amounts (1–1.5 kg). Not being particularly voracious, a predator often keeps food reserves, hiding excess prey in secluded places or burying it. Sometimes the wolf, after overeating, will leave the prey, regurgitate the meat, and, after burying it, return to the carcass to finish the remains. In some areas, from August to March inclusive, the wolf mainly hunts and kills antelopes every 2 days or so; therefore for this specified time period (approximately 240 days), a wolf will kill about 80 antelopes. The wolf also hunts these animals in other months; in the spring and summer, ungulate remains are found in about 40% of all cases of the wolf's diet. Thus, he eats antelope meat for 48 days out of 120 days (4 months) and will have to kill 16 more antelopes, for a total of about 96 antelopes of different ages. Theoretical calculations about the number of antelopes killed by one wolf are very approximate, but it can be assumed that each predator kills several dozen of these animals during a year. For example, 1 wolf using the frozen surface of Aral Sea, arrived to Barsakelmes Island where the predator killed 34 gazelles and 1 saiga during the period from March to June. A wolf that ran across the ice to the island in 1945 killed or wounded eight gazelles in one night. In the gazelle population, the mortality by the end of September is 38.9% of the number of fawns born. So, it is possible to say with confidence that the bulk of the gazelle fawns are killed by wolves (Sludskiy 1962).

1.8.2 Common Fox

Adult goitered gazelles are not afraid of foxes at all, and a fox will sometimes pass them from a distance of 2–5 m without disturbing the gazelles. Remains of goitered gazelles were found in 3.5% ($n = 61$) of fox excrement collected in the Betpak-Dala Desert in spring and summer of 1950–1952; in 1956, the amount found was 3.5% ($n = 114$), and in 1957, 7.8% ($n = 64$), with most of the remains belonging to fawns. In the spring and summer, the fox feeds mainly on rodents in Betpak-Dala Desert and less often birds, reptiles, and insects. In autumn and winter, foxes eagerly eat the corpses of antelopes killed by wolves. Thus, in the Betpak-Dala Desert (October 19 to December 25, 1957), the remains of antelopes, sometimes almost only fat, weighing up to 200 g of the stomach contents, were found in 12 stomachs of 25 foxes examined (Sludskiy 1962).

The fox attacks the goitered gazelle fawns as well. There are reports that foxes cause significant damage to the population of gazelle in Azerbaijan and Turkmenistan. A case from the Syugatinsky Valley (Ili Basin) described finding 32 legs of goitered gazelle fawns near 1 fox burrow, which could belong to as many as 8 individuals, but it remains unknown whether this was the result of hunting for live newborns or the fox had scavenged the remains of dead fawns. In Uzbekistan and Tajikistan, the remains of young goitered gazelles have also been found repeatedly near fox dens. Even if it will be assumed that a fox rarely attacks antelope fawns, given the minimum number of foxes (several tens of thousands) and if each kills one fawn, then newborn antelopes would die from foxes in large numbers. And this causes serious harm to the goitered gazelle population, as well as other antelope species (Sludskiy 1962).

1.8.3 Corsac Fox

A small weak corsac fox is unlikely to attack antelope fawns. Nevertheless, the goitered gazelle occupies a prominent place in the diet of this predator. For example, in spring–summer experiments with corsacs, collected in the central part of the Betpak-Dala Desert, antelope remains were encountered regularly. It seems reasonable to assume that the presence of antelope remains in corsac excrement should be attributed to carrion, willingly eaten by this fox. With the beginning of winter and the establishment of a snow cover in some areas, the corsac almost completely switches to feeding on victims that have fallen and killed by wolves, as its summer foods become unavailable (through hibernation, migration, and becoming sedentary in burrows). With the establishment of a snow cover, corsac foxes experience a serious lack of food, since gerbils are the only rodents that remain active, though, as sedentary species, almost never come out of their burrows. Therefore, corsac foxes migrate in winter following the herds of ungulates, feeding on the remains of antelope carcasses that die from wolves or cold or exhaustion. Thus, wolves follow migrating ungulates, and corsac foxes follow the wolves, as both survive from the migrating ungulate herds (Sludskiy 1962).

1.8.4 Ruppell's Sand Fox

These foxes were observed climbing the fences of the quarantine pens holding young gazelles on several occasions and scavenging dead fawns. There is no direct evidence that fawns have been killed by these foxes; and in the wild, there are no recordings of predation of Arabian sand gazelles by them either (Hague and Smith 1996).

1.8.5 Wild Cats

Of the large wild cats, predation of gazelles can be attributed primarily to the Turanian tiger (*Panthera tigris virgata*), which previously was widely distributed in the Transcaucasia, in the south of Kazakhstan, and in all the republics of Central Asia. Turanian tigers often hunted goitered gazelles in the lower reaches of the Syr Darya River, and various authors reported about the tigers attacking gazelles while drinking water by the Ili River. The remains of gazelles torn apart by tigers were once found quite often in the tugai thickets where the tigers waited for gazelles to drink. Often the tiger hunts in winter, when the gazelles entered the floodplains. Judging by the remains of the gazelles, the predator ate its prey whole, with skin and bones, and remnants of gazelles were often found in a tiger den. Later during the early twentieth century, Turanian tigers were completely exterminated by people completely everywhere in Central Asia and Caucasus. In Turkmenistan (Kopet-Dag Mountains and foothills), an Anatolian leopard (*Panthera pardus tulliana*) occasionally attacks goitered gazelles. Even a snow leopard (*Uncia uncia*) hunts goitered gazelles that sometimes go high into the mountains. For example, the snow leopard is a predator that lives in the southwestern spurs of the Dzungarian Alatau, where the elevation barely reaches 600 m ASL; they often attack gazelles that enter this area in significant numbers. In the Altai Gobi in Mongolia, snow leopards descend into the valleys and catch goitered gazelles going to drink there (Sludskiy 1962).

In Ustyurt Plateau, one of the main predators of the goitered gazelle was the Asiatic cheetah (*Acinonyx jubatus*). The goitered gazelle was considered the most frequent prey of the cheetah during all seasons of the year. Out of several cases of encounters with this predator in Turkmenistan, two gazelles were observed to have just been killed (winter 1940–1941 and May 20, 1948). Cheetahs were often observed during their hunting for gazelles in the Badkhyz Upland (August 10, 1950, and June 2, 1953). Outside of the feeding period of fawns, the cheetah constantly followed the migrating herds of goitered gazelles and stayed in places where they had mass concentrations. Within the former Soviet Union, the distribution of the cheetah depended on the distribution of gazelles, since this predator was primarily associated with this species. In other parts of the cheetah's range, where the goitered gazelle was absent, it follows other antelope species (Sludskiy 1962).

The cheetah hunts goitered gazelles in the following way: noticing prey, it crouches to the ground and begins to crawl toward it, using the slightest irregularity of the soil, bushes, and stones for shelter. The cheetah hides from the wind that can carry its scent to its prey. Within 10–150 m from its intended target, the cheetah jumps up and rushes to its prey with lightning speed. Catching up to the gazelle, the cheetah knocks it down with a paw swipe, then grabs it by the throat with his teeth, and strangles it until the gazelle stops breathing. The cheetah is considered the fastest of mammals and for a short distance (up to 500 m), it is able to run at a speed of 84 km/h, running 91 m in 4.5 s (Pocock 1939). Other authors write that this predator has run even faster at up to 104–112 km/h (Bourliére 1951). The initial speed of the goitered gazelle rarely reaches 65 km/h. Therefore, few gazelles manage to escape from a cheetah chasing them, but if the predator does not catch its prey quickly, it, after running 400 to 500 m, abandons the pursuit, since the cheetah is not adapted to long runs. Sometimes a cheetah waits for its prey at a watering hole and, when a gazelle approaches, it jumps up and rushes at it from ambush. Often this predator hunts in pairs or families consisting of three to five individuals. So, in the Badkhyz Uplands in the autumn of 1949, a hunting group of cheetahs looking for gazelles was observed. Two of the group lay down in hiding, and a third approached the gazelles that were grazing 200 to 300 m away from the hidden animals (Sludskiy 1962).

After killing a goitered gazelle, the cheetah eats part of its internal organs and licks the blood but eats little meat. In zoos, a cheetah eats 2.77 kg of meat with bones per day (in spring and summer) and 3.33 kg (in winter). Unlike other wild cat species, it usually does not return to the remains of its prey and does not eat carrion. Given that the cheetah mainly feeds only on fresh meat, it attacks its victims more often than other predators. Constantly roaming after migrating goitered gazelles, each predator causes significant damage to the gazelle numbers. However, almost everywhere in Central Asia, this species disappeared after 1953 (Sludskiy 1962), though a few Asiatic cheetahs are still observed in Iran (Farhadinia, 2004).

There are descriptions in the literature that young gazelles in Turkmenistan are attacked by a rather large cat—the caracal (*Felis caracal*; Ognev 1935). This cat's hunting of antelopes of small species is known also in Africa. Within the former Soviet Union, the caracal is distributed mainly in Turkmenistan, from where individuals penetrate into western Kazakhstan to the Ustyurt and the southern part of Karakalpakstan. The caracal is very rarely observed and kills an insignificant number of goitered gazelles (Sludskiy 1962).

1.8.6 Various Carnivore Species

Goitered gazelle fawns occasionally fall prey to striped hyena (*Hyaena hyaena*) and even small predators such as the steppe ferret (*Mustela eversmanni*) (Sludskiy 1962).

1.8.7 Shepherd Dogs

Shepherd dogs and stray dogs are among the most dangerous enemies of goitered gazelles and other antelopes, especially fawns in the first days of life. Since the end of the 1940s, in connection with the intense development of animal husbandry in the deserts of Central Asia, since the end of the 1940s for animal husbandry, a huge number of cattle, numbering in the millions of head, have been driven through the deserts. These cattle drives usually occur during the months of April and May, which coincide with the birthing period in all antelope species; the routes of these domestic animals often pass through the places for birthing of gazelles and saigas. With each flock of sheep numbering between 400 and 800 heads, there are commonly 1–4 dogs. Very often, shepherds do not feed their dogs, and instead leave them to feed themselves. They look around for flocks, catching and destroying various animals, including birds, egg clutches, and antelope fawns until the fawns get stronger and start to run away from predators instead of hide. Often shepherds, for their entertainment and the training of their dogs, leave their herds and direct two or three dogs to chase goitered gazelle and saiga fawns. At first, while the dogs are hungry, they will eat any fawns they catch, but when full, the killing becomes sport, and the meat is left untouched. On May 20, 1955, in the central part of the

Betpak-Dala Desert, where a mass birthing of saigas was taking place, shepherd dogs crushed all the fawns that they came across, and a shepherd on a camel had eight dead fawns killed by dogs. Sludskiy (1962) found a well where there were 14 dead fawns that had been left by shepherds of 2 passing flocks. Apparently, only in the Betpak-Dala Desert, during the cattle drives through the antelope birthing sites, shepherd dogs can kill thousands of young antelopes. In the spring of 1955, about 550,000 heads, or approximately 1100 sheep flocks, were driven through this desert. With an average of two dogs for each flock, about 2200 dogs crossed the desert and birthing grounds. Then, if each dog killed several fawns a day, the total number of pointlessly killed fawns is a seriously high number after each cattle run. Many fawns are killed by herding dogs in other parts of Kazakhstan, in Kalmykia, as well as in Mongolia (Sludskiy 1962).

Due to the mass death of antelope fawns from dogs, female antelopes are very afraid of shepherds riding camels or horses, as they are usually accompanied by dogs. As soon as shepherds on horseback appear on the horizon, the females begin to raise their young, and together, gathering in large groups with other mothers and young, they run into the desert for a distance of 3–4 km. Even antelopes grazing 2–2.5 km from the road run away. As a rule, dogs cannot catch up with an adult antelope, although the Tazy dog (breed of dog) sometimes will manage to catch an isolated gazelle that strayed from the herd during the pursuit. The pursuit of the ungulate herd does not bring any results. Antelopes of all kinds, when danger appears, will first of all run into the herd, and then the herd runs away (Sludskiy 1962).

During dzhuts (mass death from hunger), when antelopes weaken and there is a fragile ice cover that cannot withstand their weight, deaths from dogs increase. During these periods, dogs can easily catch up with antelopes and kill them in large numbers. During the 1945–1946 dzhut in the lower reaches of the Ili River, individual villagers, using their dogs, exterminated more than a hundred weakened goitered gazelles each, and 1 dog killed as many as 300 fawns. During the 1949–1950 dzhut in the Syr Darya River valley and in 1950–1951 in the lower reaches of the Chu River, the dogs went hunting without people and killed 11 males at one time and continued to chase weakened antelopes to satisfy their hunting instinct (Sludskiy 1962).

Stray dogs directly hunt gazelles, as a result of which they can leave their habitats, as shown in Iran (Nayeri et al. 2022).

1.8.8 Golden Eagle (*Aquila chrysaetos*)

The predators of antelopes include eagles of large species. This feathered predator is found at nesting sites or wintering grounds throughout the goitered gazelles' range, although its numbers are low everywhere. According to Zarudny in Eastern Iran, the golden eagle of the southern subspecies often catches young goitered gazelles and once even caught a kulan colt. Bones of young gazelles were also found at a golden eagle's nest site in the low mountain rocks in Kyzylkum Desert. The golden eagle also hunts young gazelles in Turkmenistan. In the central part of the Ustyurt Plateau, a torn corpse of an adult gazelle was found near a golden eagle's nest with fledgling and chicks (Kostin 1955). Apparently, the bird brought some of the remains of an antelope to the nest, since the golden eagle cannot lift an adult gazelle into the air. Golden eagles, after special training by humans, will catch adult gazelles, and large golden eagle females also catch wolves. Looking for prey, the golden eagle hovers high in the air, slowly circling its hunting area. Noticing a victim, the bird folds its wings and begins to dive down, or alternatively, it first descends lower and then rushes at the victim. Having flown up to the antelope, the golden eagle stretches out its powerful legs and grabs it behind its back with one claw and its head with the other; then having knocked the antelope to the ground, it begins to tear its neck with its beak or pecks at its eyes. Sitting on its prey, the eagle does not fold its wings, but tries to maintain balance with its wings. Judging by the diet of hunting golden eagles, this raptor eats up to 1 kg of meat per day, while at the same time, it can do without food for 4–5 weeks in a row and still be active (Sludskiy 1962).

Steppe eagle (*Aquila rapax*). The steppe eagle also eats young antelopes, but this relatively small eagle feeds mainly on the carrion of fallen fawns and rarely attacks live ones. Female gazelles, if they are close to their hidden fawns, try to protect them by jumping up and trying to butt the feathered predator with their heads, or rise up on their hind legs and try to hit the flying bird with their front hooves (Sludskiy 1962).

The reflex of protection from the attack of birds of prey is present in adult gazelles. When tracking antelopes from an airplane, the gazelles do not run away from it, but direct their forehead up and forward, and wait for the approach of the plane. When it flies up to the gazelles, they jump up to hit it with their heads and horns. Such a reaction to the plane has been observed repeatedly in different parts of Kazakhstan. Obviously, the gazelles take the plane as a bird of prey and behave the same way as when large eagles attack them (Sludskiy 1962).

1.8.9 Raven

Brown-necked ravens (*Corvus ruficollis*) were observed harassing goitered gazelle neonates on four occasions, and it is possible that ravens contributed to three of eight deaths (Hague and Smith 1996).

1.8.10 Interrelationship Between Predators and Antelopes

From a detailed acquaintance with the enemies of the goitered gazelles, it is known that everywhere and for most of the year they are especially often attacked by wolves. One wolf destroys several dozen antelopes over a year, while all wolves kill hundreds of thousands. Other large predators have been exterminated over time and no longer hunt antelopes. Small predators also can cause harm to the antelope population but only during the first days of a fawn's life. It is also important to note that in deserts, semi-deserts, and steppes, the successful survival of wolf litters usually depends not on the abundance of ungulates, but on the number of rodents, which are the main food for pups. It is the abundance of rodents, not antelopes, that limits the increase in the number of wolves based on their surviving young. In addition, the survival of wolf pups in the deserts also depends to a greater extent on the preservation of watering holes available to them. In dry years in the deserts, some of the pups die, due to a lack of water, resulting in a decrease in the number of these predators. Following the goitered gazelle rut, mainly adult males, weakened by rutting activity and the lack of winter food, die from wolf attacks (Sludskiy 1962).

Goitered gazelles have excellent vision and hearing and can easily spot danger at a distance of up to 2 km and have been noted to react on sounds from a camera shutter at a distance of 50–60 m. According to their reaction to potential threats, goitered gazelles are more cautious than fearful. Keeping an absolutely motionless posture, a gazelle may stand for 5–8 min, observing the danger from behind a shrub. Going to the Aral Sea for water, goitered gazelles rarely return to the trail and, instead, after walking some distance along the seashore, turn into the sands. From open areas (seashore, solonchak), goitered gazelles enter the sands or bushy sites and fixed sands very carefully. The slightest suspicion of movement makes the gazelles to turn back (Zhevnerov 1984).

1.8.11 Human

When a car approaches gazelles, they flee by running in a chain formation that scatters to the sides 100–150 m in the front of the car. Others gazelles often join the fleeing herd. It is characteristic of gazelles to run across the path directly in front of the car, but in places of intense pursuit, the gazelles developed a different behavior and stopped running across the road. Wounded and surrounded by people, the gazelles often lie down and try to hide, tightly fitting their bodies to the ground (Sokolov 1959).

In Caucasus, when goitered gazelles are in a prolonged pursuit by hunters on horseback, they will change their open habitat for a closed, which is not typical for them. They then stay in tugai shrubs and reed thickets for as long as 2–3 days. When pursued by a car, goitered gazelles stay in solonchaks and areas with rough terrain (ravines, hummocks) that are impassable for a car (Vereshchagin 1959).

How invisible the gazelle is lying on the ground depends on the terrain where they are hiding. Due to their sandy-red color, they can almost merge with the surrounding background in most environments. When fleeing, the raised tail makes the bright white patch near the tail even more noticeable, which probably serves as a good reference point for animals when running in groups at night (Sokolov 1959).

In 1959, Sokolov (1959) reported that he pursued a young female and a pregnant adult female by car in the Betpak-Dala Desert. The young female was able to easily cross in front of the car and escape, while the pregnant female paralleled the car for 5 min at a speed of 60 km per hour, before also crossing the car's path and entering the bushes. Another time on a takyr, a female with a broken front leg successfully competed with a car traveling at a speed of 60 km per hour, and she escaped in the thickets of shrubs, where the car could not follow at high speed (Sokolov 1959).

Gazelles can run when necessary, at a speed of 60 km per hour for 20–90 min; however, they get tired very quickly (Kostin 1955).

The gazelle is a very cautious and sensitive animal, whose eyesight, sense of smell, and hearing are perfectly developed for their survival. In places in the open areas, where they are often chased, gazelles rarely let a person get closer than 1.5–2 km. When they are not frightened, they are still distrustful but can show signs of curiosity. There have been cases when it was possible for a human to approach them by 50 to 60 steps in plain sight (Shnitnikov 1936).

In the Betpak-Dala Desert, the goitered gazelles rarely let a truck with people approach closer than 300–400 m, but they also do not show panicky fear. Once a female came up to within 500 m of our camp and stared intensely at us for a long time, until a shot drove her away. Moving away from the car, gazelles usually run at a slow, calm trot. Even after an unsuccessful shot of a hunter, after making several jumps, they very often stop (giving the hunter the opportunity to shoot again) before running away; from time to time, they will stop and turn their heads in the direction from which the danger appeared (Sokolov 1959).

In reacting to danger in the fixed sands, gazelles notice a suspicious object, for instance; then they run for some distance and stop behind a bush or an overgrown hillock to consider the threat. They can stay in a completely motionless position for quite a long time (10–15 min). If they cannot determine the degree of danger from their shelter, they return from the side to improve the view of the terrain. And only after determining if there is a real threat, they will either run away quickly, weaving between the ridges, or stay in place where there is no danger (Zhevnerov 1984).

In the Miandasht Wildlife Refuge (Iran), it is shown that the predator living here, the cheetah, forces the gazelle to adhere to the safest open areas, where they can detect the predator at long distances. At the same time, these open areas are located next to hilly areas, which contribute to reducing the likelihood of becoming a victim of poachers since they can hide from their persecution (Farhadinia et al. 2009). That is, the distribution over the territory can also be directly affected by human persecution in the form of poaching. For example, an increase in the risk of poaching in open habitats is shown for the Arabian gazelle in Saudi Arabia (Attum 2007) and for the gazelles in the Southern Gobi (Buuveibaatar et al. 2016).

In Central Asia and Caucasus, livestock return from their mountain summer pastures to the foothills and plains. Filling the pastures on the plains by herds of domestic sheep, cattle, and camels starts gradually from October, and at the beginning, the appearance of the livestock leads to the redistribution of the goitered gazelles, which seek pastures free from livestock, though these pastures are worse than others. But with an increase in the number of pastoral camps and livestock herds, goitered gazelles are sometimes forced to graze near sheep herds and even interspersed with domestic ungulates. This closeness is always temporary and unstable (Vereshchagin 1959). In addition to being exterminated by poachers from cars and shepherds, gazelles are being displaced due to habitat destruction and the appearance of viscous soil due to the plowing of the steppes and development of irrigation. However, gazelles still exist safely among the rainfed grain crops on the Ajinour plateau. In May–June, they are kept there even in the high barley and wheat, since the bogara soil remains dense enough for gazelles to escape from predators (Vereshchagin 1959).

1.9 Parasites

1.9.1 General Remarks

In Kazakhstan, 23 species of helminths have been found in goitered gazelles: 1 trematode, 5 cestodes, and 17 nematodes (Boev et al. 1962; Baitursinov 2008b). All the helminths noted in gazelles can also be found in farm animals. There are seven species of ticks parasitizing on gazelles. Of 34 species of gastrointestinal nematodes found in goitered gazelles all over the natural habitat, 29 have also been found in domestic animals, indicating some association between goitered gazelles and livestock (Boev et al. 1962; Eslami et al. 1981). Eslami et al. (1981) proposed that wild canids were reservoir hosts for larval cestodes, while the absence of trematode parasites may be due to the absence of their intermediate hosts in desert areas inhabited by gazelles (Kingswood and Blank 1996). Parasitic infections probably do not cause significant mortality in goitered gazelles (Zhevnerov et al. 1983); of all gazelles examined by Eslami et al. (1981), 65 individuals were infested with at least 1 helminth species, but only 1 of the hosts was in poor physical condition. *Eimeria rheemi*, a coccidian protozoan believed endemic to *G. subgutturosa*, was found to be pathogenic to captive young (2–4 months of age), causing mild to severe mucoid diarrhea but treatable with sulfonamide (Hussein and Mohammed 1992).

1.9.2 Ectoparasites

The goitered gazelle is the host of a number of endo- and ectoparasites. Of the ectoparasites, two species of subcutaneous gadflies are most common: *Pavlovskiata subgutturosae* and *Crivellia corinnae* (Grunin 1949, 1956). Ticks have been found parasitizing goitered gazelles. Among other ectoparasites, lice *Linognathus tibialis* and fluff-eaters *Damalinia appendiculata* have been found on goitered gazelles, as well as other ungulates, in particular African antelopes (Galuzo 1953; Sokolov 1959).

Seven species of ticks have been found on goitered gazelles, including *Rhipicephalus turanicus*, *Rh. rossicus*, *Rh. pumilio*, *Rh. schulzei*, *Hyalomma asiaticum*, *H. plumbeum*, and *H. scupense*. According to Zhevnerov et al. (1983) data, however, ticks are rarely observed: in a thorough inspection of 65 gazelles hunted in May through June in the Betpak-Dala Desert, only one to six specimens of ticks (five belonged to *H. asiaticum* and one belonged to *Rh. schulzei*) were found on six animals (8.1%) and more often just one or two specimens. Kostin (1955) also speaks about a minor tick infection of goitered gazelles. So, apparently, the gazelles in Kazakhstan play an insignificant role as tick carriers (Zhevnerov et al. 1983).

Goitered gazelles can be heavily infected with the two previously mentioned types of skin gadflies, the subcutaneous larvae of which are especially often found on the white rump patch (Grunin 1949; Kostin 1955). Larvae of *Pavlovskiata subgutturosae* have been found on goitered gazelles in most regions of Kazakhstan (except for the Betpak-Dala Desert), as well as in Uzbekistan, Tajikistan, and Turkmenistan. In Kazakhstan, the gazelle gadflies are marked in winter, spring, and summer. Pupation of them, seemingly, occurs in end of summer and autumn. Sometimes these larvae are numerous—from 15 to 180 specimens—and they greatly exhaust animals (Kostin 1955). In the Uzbekistan's Nuratinskiy Nature Reserve, 2 goitered gazelles were examined and found to be heavily infected by cutaneous gadflies (*Pavlovskiata subguttutosae*): 101 larvae on 1 adult male (August 20, 1947) and a 2-year-old male with 129 larvae (August 23). On the level of hip and upper body parts of the left hind leg were 53 larvae, with 35 on the right side and others spread all over the body back from the rump to the shoulders (Ishunin 1961).

In summer and early autumn, gazelles are haunted by horseflies, mosquitoes, midges, wood lice, and flies. Gazelles go to drink in the evening, and at night or dawn, when there are no horseflies, and spend their days in the desert. In the Betpak-Dala Desert and in the lower reaches of the Chu River, bloodsucking flies (*Melophagus antilopae*) often parasitize gazelles. In May and June of various years, Zhevnerov et al. (1983) found 10–70 specimens per each gazelle or 32.4% to 41.2% of the animals examined. Flies concentrate on the croup, especially on the white rump patch, and less on the chest (Zhevnerov et al. 1983).

The scabies mite (*Sarcoptes scabiei*) was found on free-ranging goitered gazelles in Iran (Bazargani et al. 2007). The captured animals presented with an alopecic pruritic skin disease with signs of crusted skin lesions and numerous small nodules which first appeared on the lips or nostrils and then extended toward the eyelids, around the ears, and, in some cases, over the entire face, neck, trunk, and legs. Skin over the affected area gradually became bald, thick, and hard, being dry and doughy to the touch, with serous fluid or sometimes blood oozing from the lesions which had a severe malodor. Skin scrapings confirmed the presence of the mite *S. scabiei*. Histopathology of lesions demonstrated marked acanthosis, hyperkeratosis, and parakeratosis. Microscopical examination also revealed all stages of *S. scabiei*, which were located mainly in the stratum corneum and also in the stratum granulosum. During the capture and sampling of the gazelles, four people ranging in age from 25 to 62 years were exposed to scabies. Two relatives of one of these people were also affected by familial contact. Clinical signs appeared within 9 days of exposure, and they developed several pruritic erythematous papules with intense itching (Bazargani et al. 2007).

1.9.3 Endoparasites

Among the endoparasites, about two dozen helminths have been found in goitered gazelles (Sokolova and Bondarev 1948; Sokolov 1959). The largest number of these parasites was found in the stomach and intestines, less in the liver (e.g., *Fasciola hepatica*), in the abdominal cavity, and in the subcutaneous tissue. The vast majority of helminth species found in the gazelle were common with those found in domestic ruminants; however, there are four types that are specific only for goitered gazelles. The frequency of the occurrence of endoparasites can be judged by the fact that after complete helminthological autopsies, not one of the ten dissected individuals was free of parasites (Sokolova and Bondarev 1948).

Regarding endoparasites, 23 species of helminths have been found in the gazelles in Kazakhstan (Boev et al., 1962), including *Fasciola hepatica*, *Coenurus cerebralis*, *Cysticercus tenuicollis*, *Ostertagiella circumcincta*, and *Trichostrongylus* sp. Sometimes they parasitize people, but only two are dangerous for people—*Multiceps multiceps* and *Taenia* sp. (= *Cysticercus* sp.). In October and November 1947, eight species of helminths were found in nine gazelles on Barsakelmes Island (Sokolova and Boev 1950). A comparison of studies of mainland and island goitered gazelles showed that significant changes have occurred in the helminth fauna of the gazelles during the gazelles' life on the Barsakelmes Island. Over 25 years, the island gazelles have lost four helminth species, *Taenia hydatigena*, *Nematodirus oiratianus*, *N. mauritanicus*, and *Skrjabinodera saiga*, and gained two: *Ostertagiella occidentalis* and *Marshallagia mongolica*. Island gazelles have lost four species of helminths, *Taenia hydatigena* (larvae), *Nematodirus oiratianus*, *N. mauritanicus*, and *Skrjabinodera saiga*, and acquired two—*Ostertagiella occidentalis* and *Marshallagia mongolica*. These changes in the fauna of parasitic worms are caused by changes in the natural and climatic factors of the island. In the 1960s, due to the fall in the level of the Aral Sea, salty reservoirs dried up on the island and reed beds disappeared, and the area of fresh temporary reservoirs significantly decreased, as well. Favorable conditions for the life of many species of bloodsucking insects, and possibly their intermediate hosts, have disappeared (Zhevnerov 1984). In addition, island gazelles no longer had contacts with domestic livestock, many of which were likely important for supporting nematode numbers within the goitered gazelle population.

These changes may be related to the drying of the Aral Sea as well, affecting the biotopes of goitered gazelles. A mass death of gazelles from helminths is not known, which can partly be explained by their weakly expressed herding behavior, when goitered gazelles form mainly small-sized herds widely distributed within the area. In addition, goitered gazelles live mostly in deserts, in dry environments within which not many helminth species can survive successfully (Zhevnerov et al. 1983).

In 1974 and 1975, four gazelles were examined through a complete helminthological autopsy, and six helminths were found, including *Skrjabinema ovis*, *Oslertagiella occidentalis*, *Marshallagia marshalli*, *M. mongolica*, *Nematodircus gazellae*, and *Nematodirella gazelle* (Table 1.43). Nematodes parasitized the gastrointestinal tract where rennet is produced, the cecum, and small and large intestines (Zhevnerov 1984).

Skrjabinema ovis was found in all four gazelles and only in the large intestine. In addition to goitered gazelles, this nematode species is widespread among sheep, goats, argali, and saiga in all natural zones of Kazakhstan (Boev et al. 1962).

Marshallagia marshalli are rennet nematodes, occasionally penetrating into the small intestine. Seven hundred and seventy two specimens were found in the cecum of all four gazelles. Individuals of this species occupy one of the leading places in the helminthiasis of the gastrointestinal tract. The maximum number found in a single goitered gazelle is 476 specimens. It should be noted that there were many immature individuals with a high invasion intensity. This type of nematodes can

Table 1.43 Helminth infestation of goitered gazelles in the Barsakelmes Island (numerator, invasion intensity (range); denominator, average) (adapted from Zhevnerov 1984)

	1949–1950		1973–1974	
Helminth species	Spreading, %	Intensity, number	Spreading, %	Intensity, number
Taenia hydatigena (larvae)	16.6	1	–	–
Skrjabinodera saiga	33.3	*1–8* 4	–	–
Skrjabinema ovis	100	*4–220* 52	100	*10–35* 22
Ostertagiella occidentalis	–	–	25	20
Marshallagia marshalli	66.6	*16–37* 25	100	*40–476* 193
M. mongolica	–	–	75	*6–105* 61
Nematodirus gazellae	16.6	102	100	*35–818* 375
N. mauritanicus	66.6	*1–560* 147	–	–
N. oiratianus	16.6	11	–	–
N. gazelli	16.6	11	50	*40–200* 120

cause serious harm to gazelles and is almost always a component of helminthiasis of small domestic cattle (Sokolova and Lavrov 1956).

Marshallagia mongolica was identified in all examined gazelles. Localization most often was in the rennet, less often in the small intestines.

Nematodirus gazellae was found in all studied gazelles. There were both mature males and females, as well as larval forms. This species is particular to gazelles but is also found in saigas and sheep.

Nematodirella gazelli was found in the small intestine in two gazelles from the Barsakelmes Island. The nematodes were almost all sexually mature. Only a few were in the larval state (Boev et al. 1962).

In Kazakhstan, helminths of the family Taeniidae and especially larvae of *Taenia hydatigena* (1–12 specimens per individual) were found frequently in domestic and wild ungulates, including *Gazella subgutturosa*. Nematodes of the family Trichostrongylidae, especially the genera *Trichostrongylus* (four species), *Ostertagiella* (six species), *Nematodirus* (nine species), *Marshallagia* (two species), and *Nematodirella* (two species), were also found. No specific helminths were found in goitered gazelles, and all four helminths noted in goitered gazelles were also found in moose and other ungulate species. A nematode *Camelostrongylus mentulatus* that is a specific parasite to camels was unexpectedly found also in the goitered gazelle (Baitursinov 2008a).

Compared with wild ruminants, the infection of domestic sheep with 31 species of helminths was relatively high both in terms of the extensiveness and intensity of invasion. The research results suggest that wild and domestic ungulates are subject to helminthic diseases and are constantly exchanging helminths with each other. Diseases such as echinococcosis, cenurosis, and tenuicolic (or small interstinal) cysticercosis are widespread in sheep, and there is no doubt that these domestic animals are the source of these diseases in wild ungulates. On the other hand, the role of most wild ungulates, including gazelles, in the spread of helminthic diseases to livestock is insignificant. Firstly, the number of wild ungulates is very limited and they live in hard-to-reach limited areas of pastures and do not make long-term migrations. And, secondly, these wild ruminants have fewer helminth fauna compared to domestic animals (Baitursinov 2008a).

Many types of helminths with more or less wide coverage of wild animals are recognized as particularly pathogenic species. In the conditions of Kazakhstan, first of all, these are pathogens of larval tenidoses, the infection of which in some years takes on a significant character. Localized in parenchymal organs, parasite cysts have a negative effect on the general well-being and endurance of wild ruminant hooves. The family Anoplocephalidae is also very pathogenic for wild animals, the prevalence of which has high indicators. It should be assumed that for dicrocelia (*Dicrocoelium lanceatum*), *Parabronema scrjabini*, *Setaria*, *Scriabinema*, *Habertia*, gastrointestinal strongyles (especially *Marshallagia*, *Haemonchus*, and nematodes), trichophiles, and many other helminths are also dangerous to the health of ungulates within the conditions of their natural habitats. Helminths become particularly dangerous in late winter to early spring, when the body's resistance is reduced, and can play a significant role during dzhutes (mass death because of hunger) in reducing the number of wild animals (Baitursinov 2008a).

To combat helminthiasis in wild animals, it is necessary to reduce their contact with livestock. One of these methods is fenced pastures for domestic ungulates. However, a number of wild ungulates make seasonal migrations in Kazakhstan, and encountering barbed wire fences on their way can be detrimental to survival as some immediately die entangled in the barbed wire fences, and some die later, not finding suitable seasonal pastures and watering holes. Therefore, fenced pastures cause significant harm to wild ungulate populations, fragmenting and reducing their numbers. The existence of fenced pastures for domestic ungulates and the inhabitation of wild ungulates are incompatible in the same territory (Baitursinov 2008a).

It is necessary to organize watering holes exclusively for wild ungulates. Damp biotopes near water sources are places for the maturation of eggs and larvae of a number of helminths in the summer. Wild ungulates do not drink water alongside livestock but visit watering holes after the domestic ungulates go away. Nevertheless, the watering holes become a place of contact and exchange of infestations dispersed by drinking ungulates. The higher the load on the watering hole, the more intense the infection rate of wild ungulates and the infection with many helminths. This point is an especially important factor for goitered gazelles, which lack watering holes in most parts of their habitats during hot summers, and all water sources are used by domestic livestock. Therefore, the creation of reservoirs suitable for watering animals is an important problem, not only for domestic ungulates but also for wild ungulates, particularly for goitered gazelles. An increase in the number of organized watering holes (e.g., artesian wells and other water sources) contributes to the growth of disunity of herds, which leads to an overly dispersed population of invasive elements and becomes one of the important factors in the recovery of wild and domestic ungulates from parasitic infection (Baitursinov 2008a).

In the summer and autumn periods of the year, vegetation with anthelmintic properties makes up a significant share of the diet of wild ungulates. For example, wormwoods (*Artemisia*) and ferules (*Ferula*) are found in the food masses of saiga antelope, and an ecological feature of the boar is eating *Peganum harmala*, *Aconitum*, and *Anemone*, which are active agents against parasitic helminths. Previously, local people drove their sheep to the *Artemisia* and *Ferula* pastures before wintering to help rid them of helminths (Baitursinov 2008a).

Predatory mammals also play an important role in the circulation of parasitic helminths to wild ungulates. For example, wolves, foxes, and domestic dogs are the final hosts to these parasites through their hunting and help to disperse the infestation to both wild and domestic ungulates, thus making it necessary to control the number of wolves, as well as domestic dogs. It is also necessary to establish a strict legal restriction on the number of dogs in private farming and animal husbandry and other forms of economic activity of shepherds, and all livestock owners should be obliged to carry out mandatory deworming of their dogs (Baitursinov 2008a).

The cohabitation of the goitered gazelle with livestock largely determines its helminth burden. For example, in the territory of Karakalpakstan (Uzbekistan), where all sectors of animal husbandry are developed, from 32 to 37 species of helminths from the classes of trematodes, nematodes, and cestodes were counted in goitered gazelles, and a significant number of these parasitic species are also common in livestock. The great species similarity of the helminthofauna is associated with the common habitat of the gazelle and livestock (Akramova et al. 2020). Helminthological investigations were made for young and adult goitered gazelles in southern Turkmenistan and Uzbekistan (Table 1.44). They found six species of nematodes and one species of cestodes in the goitered gazelles examined and each individual gazelle was infested with two to four types of helminths. The average intensity of the total infestation was 755 parasites per individual with a variation from 316 to 1035. Some parasites were found in isolated instances, while the greatest invasion intensity was by nematodes. In Uzbekistan, schistosomiasis (parasitic flatworms called schistosomes) was found in goitered gazelles, with the urinary tract or the intestines infected. The disease is spread by contact with fresh water contaminated with these parasites (Azimov et al. 2020).

Until 1983, livestock grazed alongside with goitered gazelles in the Bukhara nursery on shared territory. The helminthological situation in the nursery was complicated by the maintenance of other wild ungulates (kulans, saigas, and Bukhara deer (*Cervus elaphus bactrianus*)) and shared grazing with sheep, goats, donkeys, cows, and a horse. The number of grazed cattle steadily increased and by the end of 1981 exceeded 100 heads. Moreover, the cattle roamed over almost all the territory of the nursery. In May 1981, coprological material was collected from the captive gazelles that roamed freely through the park and from goats and sheep of the personal households of the service personnel. Study samples were taken from 46 animals for analysis (Table 1.45; Nazarova and Prisyazhnuk 1986).

Table 1.44 Helminths of goitered gazelles from northern Kopetdag (Turkmenistan) and Bukhara nursery (Uzbekistan) (adapted from Nazarova and Prisyazhnuk 1986)

Helminth species	Location in body	Number of parasites and host's sex and age				
		Northern Kopetdag				Bukhara
		Male, 6 months	Female, 6 months	Female, 6 months	Male, 6 months	Adult male
Moniezia expansa	Small intestine	–	–	–	1	–
Skrjabinema ovis	Large intestine	27	219	–	186	–
Setaria transcaucasica	Venous sinus	1	–	–	–	–
Setaria sp.	Body cavity	–	–	–	–	1
Marshallagia mongolica	Rennet	–	–	126	–	–
Marshallagia sp.	Large intestine		–	–	–	1
Nematodirus oiratianus	Rennet, small intestine	700	474	977	129	–
Nematodirus mauritanicus	Small intestine	–	–	–	–	1032
Trichocephalus sp.						
Total		728	693	1003	316	1035

Table 1.45 Results of the coprological analyses of goitered gazelles, goats, and sheep from the area of the Bukhara nursery (Uzbekistan) (adapted from Nazarova and Prisyazhnuk 1986)

Investigated gazelles	It was found per 1 g excrements				
	Eggs			Larvae	
	Strongyloidiasis	Nematoda	*Chabertia ovina*	*Dictyocaulus filaria*	*Protostrongylus*
Average of 13 samples—gazelles of open area	3.2	2.8	2.0	–	0.4
Goitered gazelles from enclosure					
Female no. 1	4.0	2.0	–	–	–
Female no. 2	3.0	4.0	1.0	–	–
Male	5.0	1.0	–	–	3.0
Domestic ungulates, 15 samples each					
Goats	3.0	2.5	2.0	1.8	4.0
Sheep	4.5	4.0	1.5	2.0	2.5

The relatively high incidence of some larval cestodes in wild sheep and gazelle indicates the presence of canine reservoir hosts, which could be from either the sheep dogs or wild predator canines. The prevalence of *Cysticercus ovis* in gazelles (6.25%), in wild sheep (0.4%), and in domestic sheep (1.4%) (Aminoseheh 1961) is of particular interest, though, since *Taenia ovis* has not been recorded in dogs in Iran. It can be concluded, therefore, that the definitive hosts of this parasite are probably wild carnivores whose geographic distribution is similar to the habitat of the gazelle.

Although all animals examined were found to be infested with one or more helminths, including nematodes, their weight and general appearance showed that they were in good physical condition. Only one gazelle, which was heavily infested with *Avitellina centripunctata* (300 grams), was in poor condition and emaciated (Eslami et al. 1981).

The following parasitic species were found in Iran: *Marshallagia marshalli*, *Camelostrongylus mentulatus*, *Ostertagia occidentalis*, *O. circumcincta*, *Trichostrongylus colubriformis*, *T. vitrinus*, *Nematodirus gazellae*, *N. oiratianus*, *N. spathiger*, *N. archari*, *N. abnormalis*, *N. filicollis*, *N. sugatini*, *Nematodirella cameli*, *N. longissimespiculata*, *N. antilocaprae*, *N. gazelli*, *Skrjabinema ovis*, *Trichuris ovis*, *T. gazellae*, *T. skrjabini*, and *T. discolor*. Among the species found, *Marshallagia marshalli*, *Nematodirus* spp., and *Nematodirella longissimespiculata* were the most prevalent helminths (Eslami et al. 1980).

In comparison with Uzbekistan, the helminth fauna of Mongolian goitered gazelles is relatively poor. One of the most extensive studies of the helminth fauna of wild and domestic ruminants of Mongolia has shown that 12 species of helminths belonging to only 1 class of nematodes are found in goitered gazelles. The following helminth faunas were found: *Trichuris ovis*, *Skrjabinema ovis*, *Scrajbinema dserani*, *Parabronema skrjabini*, *Setaria* sp. (found in the chest and abdominal cavities), *Marshallagia mongolica*, *M. dentispicularis*, *M. shumakovichi*, *Ostertagia dahurica*, *Trichostrongylus probolurus*, *Nematodirus oiratianus*, and *Nematodirella longissimespiculata* (Sharhuu and Sharkhuu 2004).

1.9.4 Helminthofauna Exchange Between Wild and Domestic Ungulates

Depending on the intensity of land use, the biocenotic relationship between wild and domestic ungulates varies greatly in different regions. Wild ungulates often use the same pastures as domestic animals, especially in the areas where livestock are moved from pasture to pasture. These movements and shared pastures open the possibility of animals to exchange parasites, including helminths, between wild and domestic animals. As a result of various studies, 73 species of helminths were found in wild ungulates in Kazakhstan, including 4 species of trematodes, 8 of cestodes, 60 of nematodes, and 1 of *Acanthocephalus*, many of which are also characteristically found in domestic ruminants (Baitursinov 2008b). Naturally, with historically long-term cohabitation on common pastures, there is a flow and exchange of parasitic invasions from domestic animals to wild ungulates and vice versa. The coefficient of similarity of the composition of parasitic helminths of wild ungulates and sheep according to research data is given for a number of ungulate species. According to these results, goitered gazelles have a quite low coefficient of similarity with domestic sheep (only 6%), compared to saiga antelope (up to 51%) and caprids (*Ovis ammon* (43%) or *Capra sibirica* (21%)). The research results confirm the previously expressed opinion that in the community of helminthofauna between wild ungulates, systematic kinship is of paramount importance, followed by ecological ties (Baitursinov 2008b).

Helminths of 23 species have been identified in goitered gazelles in Kazakhstan. In general, there is a tendency where the pressure of parasites from domestic ungulates is more prevalent. And there is no doubt that livestock husbandry is the leading factor determining the formation of helminths in wild ungulates. The species composition of goitered gazelle helminths is closer to ruminant domestic ungulates, although the index similarity is much lower in the gazelle than in the caprids. Here, again, the systematic remoteness of gazelles from domestic ungulates affects their parasitic load in comparison with wild argali, ibexes, and urials (*Ovis vignei cycloceros*) (Baitursinov 2008b).

1.10 Paleontology

Fossils from the tribe Antilopini appeared in Eurasia ca. 17–16 million years ago (Miocene) (Vrba 1985). Later, throughout the Pliocene species from the subfamily Antilopinae were known only in Europe and Asia, where they were well-developed quite well, occupying a prominent place among other Bovidae. In Africa, in the location of its current greatest diversity and abundance, Antilopinae have been known only since the Pleistocene and did not appear here until the end of the Pliocene. And there have been none in Europe since the Pleistocene. Thus, Antilopinae should be considered as a group of Eurasian origin that has moved to Africa relatively recently. Apparently, the formation of the Antilopinae should be attributed to the end of the Miocene, since in the lower Pliocene, it was already clearly marked while the original group Antilopinae was not yet found. The subfamily includes a number of fossil genera; however, they all belong to modern groups. More richly represented in the fossil state is Antilopini (eight genera from Europe, Asia, Africa); from Neotragini only one Pleistocene African genus is known, and the forms of modern genera are known from the Pleistocene of Africa. Thus, everything said above regarding the history of the group concerns mainly Antilopini, and Neotragini should be considered as a relatively new, purely African branch of the subfamily (Heptner et al. 1961). Quaternary finds of the goitered gazelles lie within the modern range. In the fauna of the Paleolithic site in western Transbaikalia, probably, not one gazelle was found, but another species of the same genus was (Gromov 1948).

The paleontological history of the Gazellinae can be traced relatively deeply, but, as with most of the groups of Bovidae, the form that could be considered ancestral to all subsequent members of the group is unknown. The genus *Gazella* is the most ancient of all modern Bovidae. Already in the Miocene-Pliocene, forms appeared that showed all the essential features of modern gazelles (Sokolov 1959). There is no exact information about the center of origin and the ways of settlement, although Sokolov (1959) considers the area of the Asian continent lying to the west of the Himalayas to be the center of the origin of gazelles. *Gazella* is one the oldest of bovid genera (15–10 million years ago). First fossils were recorded from the early Miocene in Libya (Savage and Russel 1983) and they probably developed in the Near East. During the middle Miocene, *Gazella* appears firstly in Europe and Asia (Savage and Russel 1983). Genus *Gazella* belongs to the Miocene period: Fragments of horn rods and teeth of gazelles, partially similar to the Pliocene *Gazella deperdita* Gerv., were found in the Caucasus in the Sarmat deposits and in the middle Pliocene sandstones of southern Baku, as well as in the lower Quaternary thickness of the Taman semi-peninsula. Radiation levels found in *Gazella* apparently coincided with the expansion into more open habitats in the late Miocene and early Pliocene in Europe and Asia and later in Africa (Vereshchagin 1959; Savage and Russel 1983). The widest distribution of *Gazella* probably occurred in the Pliocene when it reached England (Heptner et al. 1988).

The divergence times between *Gazella* and *Gazella subgutturosa* were estimated from 1.08 to approximately 2.5 million years ago according to the cyt b molecular clock (Tancheng et al. 2016). Due to the unexplored osteological differences between the species of the subgenus *Gazella* s. str., the exact time of the appearance on the geological timeline of *G. subgutturosa* is difficult to determine. But, judging by a number of finds in Europe and China (Boule and Teilhard 1928; Bogachev 1938), close to modern forms lived as early as the second half of the Pliocene. The remains of this species are known from sediments from the Quaternary period, except for the areas of modern habitat, and then only in the fauna of the paleontological sites of Trans-Baikal area (Gromov 1948; Sokolov 1959). The remains of real modern goitered gazelles are known from central Iran. The remains of fossil gazelles in the former USSR are known from the Mesolithic layers of cave canopies near the cities of Dzhebel and Krasnovodsk and from the deposits of the Holocene (Vereshchagin 1959); the oldest remains from the Lower Anthropogene were found in the Pavlodar, Kostanay, and Alma-Ata regions (Kozhamkulova 1969). Goitered gazelles are known to have occurred in Georgia since the early Holocene when the climate became more arid (Vekua and Lordkipanidze 2008).

Gazella subgutturosa has existed in Central Asia since the late Pliocene and early Pleistocene (Sokolov 1959; Gromova and Baranova 1981) and was found in the Pleistocene fauna of Kazakhstan, Uzbekistan, and Tajikistan (Kozhamkulova 1981). In the territory of Uzbekistan, fossil remains of goitered gazelles were found in Kaunchi during excavations of a cave city, in the upper layers with remnants from the iron culture dating from 900–800 to 500–300 years BC. Remains were also found in the settlement of Shuralisai, 6 km from Kaunchi, where in the fourth to sixth centuries AD there were ordinary settled human settlements (Ishunin 1961). In the Fergana Valley, during the excavations of a Chust site from the Bronze Age, horns and bones of gazelles were also discovered (Malenkov 1958).

Remains of gazelles which were closely related to goitered gazelles were noted in Lebanon, Israel, and Syria within Paleolithic layers from Saint-Acheul until the late Neolithic. There were no goitered gazelles in the Transcaucasia either in the Middle or in the Upper Pleistocene. Reliable remains of goitered gazelles in the Caucasus are known only from the Holocene layers of the historical epoch of the Pre-Caucasus and Transcaucasia. Especially numerous are the remains of gazelles in the layers of ancient Baku, where they make up 14.5% of the total number of bones and 39% of the bones of sheep and goats (Vereshchagin 1959). Quaternary fossils of *G. subgutturosa* have been found within the limits of its present distribution (Heptner et al. 1988).

The absence of goitered gazelles in the Middle Pleistocene fauna of the Caucasus suggests that this species appeared here relatively recently—in the postglacial time—replacing the saiga in the current territory of Azerbaijan. Most likely, the goitered gazelle penetrated directly from the south through the ridges of northwestern Iran. Such an expansion of the area became possible with the onset of the dry and warm phase of the Holocene. It is also possible that gazelles may be imported to the Kura lowlands by ancient nomadic tribes who spent the summer in the highlands of northern Iran, as well as the ones similar to nomadic tribes during recent historic periods (Shahsevans—Turkic Oghuz group of Azerbaijan people) in the winter in the Ugan steppe on the plain. The persecution of goitered gazelles by the horsemen of the Medes, Albanians, Persians, Arabs, and Mongols was also probably a reason for the settlement of these gazelles on the plateaus and their penetration into isolated valleys (Vereshchagin 1959).

1.11 Hunting History

Archaeological findings of bone fragments indicate that people have hunted goitered gazelles since ancient times, and the oldest remains of these hunts have been found in Eastern Asia (4000 BP) and the Middle East (9000–11,000 BP), where the gazelle meat was the most crucial source of the animal protein food for the Mesolithic and Neolithic people. In some ancient human sites, up to 80% of gazelle bones were found, which indicated that gazelles were the most frequent object of hunting by ancient people in open arid areas. Humans hunted goitered gazelles mainly just after the birthing season (late April into early May), because remains of offspring bones were common in the sites of ancient humans (Legge and Rowley-Conwy 1987). In western Turkmenistan, paleontological and archaeological studies have established that hunting for ungulates was the most critical way of existence for the people of ancient settlements. In the Neolithic settlement of Jeytun, bezoar goats (*Capra aegagrus*) and urials (*Ovis vignei*) provided 70% of meat products, and goitered gazelles cover 11.3% (Tsalkin 1956; Sapozhenkov 1969).

Later on, ancient people tamed several species of wild ungulates and began to breed them as domestic animals for getting milk and meat from them and became much less dependent on the results of hunting. Agriculture and animal breeding of domestic ungulates started to develop very quickly in the Neolithic period, and archaeological data confirmed that from that time (3000 BC), almost 80% of bone fragments in the ancient human settlements belonged to domestic animals (Grebenukov 2008). As a consequence, they used less meat of wild ungulates, including goitered gazelles (Dinesman et al. 1986; Legge and Rowley-Conwy 1987). Despite this, people continued hunting, gradually improving the methods and tools of hunting and achieving better and better results of their hunts. Finally, people learned to build huge pens ("desert kites") with high stone walls, which were arranged so that people could drive a large number of wild ungulates inside of enclosed parts of these pens and thereby provide themselves with meat for a long time. These hunting kites have been known since the Neolithic, and hunters actively used them up to the early 1900s (Mendelssohn 1974; Legge and Rowley-Conwy 1987; Heptner et al. 1988). These kite enclosures are known in Syria, Jordan, Saudi Arabia, and the Sinai Peninsula. Nowadays, similar methods of mass corralling and killing wild ungulates from the prehistory of American Indians living on the vast plains of the North American continent have become known. Using different methods, they directed herds of reindeer (*Rangifer tarandus*), bison (*Bison bison*), and pronghorns (*Antilocapra americana*) inside of large traps (Legge and Rowley-Conwy 1987).

Modern satellite imaging methods have made it possible to find hunting kites in many other places, and there were many times more of them than previously thought (Betts and Burke 2015). More than 5200 hunting kites have been found in the vast plains of Central and Southwestern Asia, the Caucasus, and many countries of the Mediterranean and the Middle East (Svizzero and Tisdell 2018). Some authors believe that all these large kites could be used not only for hunting but also for keeping large herds of semidomesticated ungulates, as the first stage of complete domestication of them. People could keep these ungulates, herding them inside of kites during part of the time, and periodically release them to the outdoor pastures as a free-ranging livestock (Echallier and Braemer 1995). Such "free-range" ungulate pasturing was very common in ancient communities and is possible to see in Central Asia even now (Svizzero and Tisdell 2018). However, though hunting of wild ungulates or using desert kites as game traps (mostly gazelles) was continued, people could use the kites for herding of wild ungulates, as well. Among the main prey species hunted using enclosure traps were defined wild oxen (*Bos* sp.) and bison, wild goats (*Capra* sp.) and sheep (*Ovis* sp.), ostriches (*Struthio camelus*) and wild camels (*Camelus dromedarius*), oryxes (*Oryx* sp.), onagers (*Equus hemionus*), and several *Gazella* species. These animals are suitable for hunting with kites, because they behave in a similar way: they are all predominantly gregarious, and when threatened, they escape, gathering in larger herds and running in the same direction along the familiar pathways. These preys have valuable meat or other raw materials of economic benefit to people (Svizzero and Tisdell 2018).

In all cases, the catching of ungulates in large enclosures occurred in the same way: the ungulate herds are being led by drovers toward kite entrance or a camouflaged trap, where hunters catch them easily. This method depends on three factors: open country, large herds of prey animals, and enough hunters. Structures such as kites were seen by European explorer in actual exploitation during the twentieth century. The hunting kites can have a variety of designs and sizes, but all of them have some general principles. First of all, in the general design, there is a central enclosure, which can be of different sizes, with a diameter from several meters to one and a half hundred meters. A tight admission with diverging sides of the stony fence is typical for such enclosure; the fence can stretch for several kilometers across the plain. The walls of the kites can be built from big flat stony slabs, while others represent piles of boulders (Legge and Rowley-Conwy 1987).

According to professional archaeologists and amateur collectors of historical artifacts, the kites were used in different ways depending on the building materials and location. The mostly basalt stones as a result of cooled and petrified lava, located in the eastern and northern areas of Jordan, were preferable building blocks for kite walls. The round niches with small windows were arranged in the stone walls, in which hunters hid to be invisible to ungulates, and they trap and shoot with bows and arrows every individual that happens to be close to the hunters. In contrast, in other areas (e.g., in Syria), where the ancient lava is absent, people used stones from other materials. In addition, outside of the main enclosure were often pitfalls; people scare gazelles that are in the enclosure in such a way that the animals run toward the pits and fall into them, breaking their legs. Injured gazelles were easy for killing and meat was processed for long-term storage, for which it was dried and salted (Legge and Rowley-Conwy 1987).

The invention of bows and arrows during the Neolithic period made hunting more successful for even such fast and caution animals as goitered gazelles (Semenov and Korobkova 1983). Archaeological evidence suggests that gazelles were one of the most frequent targets of ancient hunters in vast arid areas of Asia. In the territory of Central Asia, including Kyrgyzstan and Uzbekistan, there are many prehistoric finds of the goitered gazelle remains in ancient human settlements (Iron Age and Early Bronze Age) (Rodrigues Soares et al. 2022). Later on, humans continued hunting goitered gazelles at all times. During the period of Golden Horde, gazelles were driven into the massive corral leading them toward hidden hunters, who struck all the animals that came to them, shooting by bows and arrows. This method of hunting was the most popular and widely used for supplying Mongol army with meat and allowed warriors to train (Smirnov 1965).

During current times, hunting goitered gazelles was very popular until recently among the peoples of Central Asia; they hunted with the help of dogs and falcons and did it on horseback and camel (Heptner et al. 1961; Roberts 1977). The introduction of current firearms and automobiles during World War I significantly raised the susceptibility of these gazelles to hunting, leading to a substantial reduction in their populations in the Middle East and the former Soviet Union (Heptner et al. 1961; Thouless et al. 1991). From the 1880s to the 1940s, industrial-scale hunting of goitered gazelles for their meat and hides was carried out in Central Asia. Approximately 4000 to 10,000 gazelles were harvested annually. And such kind of hunting still occurs in Mongolia (Sludskiy 1963; Smirnov 1965; Sokolov 1990). The skin of the gazelle is used to make bags for storing water and milk. It is also occasionally used for making shoes (Heptner et al. 1961). In the Caucasus, Central Asia, and the Middle East, local people traditionally keep captured fawns at home as pets; they are easily tamed in captivity (Mendelssohn 1974; Vereshchagin 1967; Heptner et al. 1961). In captivity, goiter gazelles can be bred manually to reduce neonatal mortality. Interestingly, tame fawns seem to be as suitable for reintroduction projects as mother-reared gazelles (Lindsay and Wood 1992).

Goitered gazelles are valuable primarily for meat. Gazelle meat is considered delicious, with its fat, and sometimes considered equal to or better than saiga antelope meat. Morrison-Scott (1939) remarked that Bedouins prefer the meat of goitered gazelles over that of other gazelles. With the effective organization of hunting and protection from extermination, the gazelle can become a useful wildlife resource of the desert and semi-desert zone and an important additional source of meat production (Sokolov 1959). Even in the pre-war years and the years of the World War II, attempts were made to plan for harvesting of gazelle meat (Zverev 1948; Shnitnikov 1936).

Suede can be made from the skin of the goitered gazelle, as well as clothes and summer shoes. But the main use of the skin is the production of containers for keeping and transporting water, ayran, and kumis. Containers made from gazelle skins are light and elastic and do not give the liquids an unpleasant smell. Skin buckets are made from the skins of young gazelles, which are convenient when traveling. The horns of males can serve as ornaments (Sokolov 1959).

When gazelles were numerous, the ways of hunting them were diverse. Hiding while hunting was almost impossible due to the open habitats and the high sensitivity of the animals. In some areas, hunters approached the gazelle by hiding behind a camel, which did not inspire much suspicion to the gazelle. One of the most common and effective ways to hunt these animals is to wait for them and shoot them at their watering holes and trails (Sokolov 1959). The habit of gazelles to go to their watering places in certain ways provide an opportunity to their capture using traps placed on the trails (Kolenikov 1952).

Local people also hunt gazelles with the help of Central Asian greyhounds (Tazy). In the southeastern Karakum Desert, a Tazy caught 15 gazelles in 1 year. Since even the best greyhound can rarely catch up with a gazelle, the hunter usually used replacement dogs. For this purpose, the hunt was organized in such a way that the animal, tired of being chased, was directed to a reserve group of dogs, which easily caught up with it (Dementyev 1935).

In some places in Central Asia, and especially in India and Afghanistan, there is still a method of hunting with the help of birds of prey (golden eagles, large hawks, and falcons), specially kept and trained for this purpose. Having gotten as closer as possible on horseback to the gazelle, the hunter releases his bird to catch up to and grab and hold the animal until the hunter arrives. The training of large saker falcons and hawks includes teaching the raptors to beat the prey animal in the head and try to peck its eyes, since even a large saker falcon is not able to keep ahold of a goitered gazelle that was grabbed by the back (Sokolov 1959).

Of group methods for hunting gazelles, the most common is hunting with a "catching up." Having found a group of animals, the hunters are divided into two parties. One hides with their guns behind some kind of cover, and the other very carefully and slowly tries to get around the gazelles from the opposite side and then direct them toward the hidden shooters. Another, rather barbaric method of hunting is also described, where a funnel-shaped corral for gazelles is arranged with a wide entrance and a narrowed exit, the sides of which are made up of piles of earth that frighten the animals. A group of riders with shouts drives the animals to the wide side of the corral. Once in the funnel, the gazelles rush to the narrow exit at the opposite end and run into sharp dry reeds placed behind it in a transverse row with their ends toward the running animals. Undoubtedly, many animals died uselessly with this method (Sokolov 1959).

No less barbaric, a hunting method developed in Central Asia and Transcaucasia over the past decades (1950) has hunters pursuing and shooting from a car. Animals chased to the point of death are shot almost at point-blank range or often killed with knives or crushed by the wheels of the car. A large number of surviving, but exhausted, animals die later as a result of overworking their heart muscles. Hunting from cars is prohibited everywhere, but, unfortunately, in some places it is still practiced almost openly (Sokolov 1959).

Another way to hunt from a car is a night hunt with a headlight. Its application is based on the premise that animals blinded by a stream of bright light directed at them (a car or motorcycle headlight is usually adapted for this purpose) become paralyzed by it and lets a shooter come very close to them out of the darkness. In complete darkness (moonlight does not support successful hunting), a driver, a farrier, and a shooter drive to a location known to attract gazelles. The headlights are turned off and a headlight reflector is held by the shooter's partner. The car moves at a low speed, and the farrier smoothly moves the headlight reflector around, trying to catch the glow of a gazelles' eyes in its field of view (Sokolov 1959).

It's a very spectacular sight to see a group of gazelles caught in a beam of light. At first, you notice several pairs of eyes burning with a phosphorescent shine. Then you identify the silhouettes of the animals. As if fascinated, they look at the light source approaching them, trying, but not often being able to tear themselves away from it. In extreme cases, the gazelles remain motionless until a shot rings out, but usually they try to escape from the light and sometimes succeed. The farrier, noticing the animals, begins to slightly shake the light to dazzle the animals even more. With a conditional light signal, he indicates to the driver the direction in which the car should move, and when it should stop, while trying not to lose sight of the gazelles. When the car stops, the driver continues to play with the light, while the shooter, who is in the dark, quietly jumps out of the car with a gun and runs sideways toward the animals. The shooter is able to get very close, for an accurate shot. In addition to the abundance of gazelles, the success of hunting with a headlight depends primarily on the skills of the farrier to detect the animals and keep them in the light until the car and the shooter approach. A car is also of great importance. A silent car or a light truck is best suited for this hunt. A heavy, rattling car scares away most of the gazelles long before they are affected by the paralyzing effects of the light (Sokolov 1959).

Goitered gazelles are very resistant to succumbing to injury. Even when shot from a three-line rifle with a telescopic sight, a lot of injured gazelles unfortunately get away. After being shot in the leg, one female ran on three legs at a speed of

60 km per hour. With a peritoneal aorta wound, another female ran about 100 m. Another female that was pierced through by a bullet from the crotch to the base of the neck ran about 10 m before falling down. An even more striking case occurred with a male: after several shots, he continued to move with his entrails hanging from the rupture in his abdominal wall and escaped, since it was impossible to catch up with him either on foot or at the slow speed of the car due to the terrain conditions (thickets of bushes, soil irregularities), not allowing the rapid movement of the car (Sokolov 1959).

A gazelle caught when it is young easily adapts to people, has a gentle disposition, and becomes quite tame. There are numerous cases of people keeping tamed gazelles in their homes. However, gazelle reproduction did not occur when they were kept in a human home setting and therefore they had no practical significance for domestication (Sokolov 1959). Nevertheless, the pastoralists of Transcaucasia even now are constantly catching and keeping young gazelles near their caravans. In Syria, Palestine, Lebanon, and generally the countries of the Near and Middle East, newborn gazelles were caught by the thousands annually for domestication and cultivation for meat. Osama ibn Munkiz described such a hunt (cited by Vereshchagin 1959): "When gazelles bring newborns, hunters go out on foot and take those fawns that were born on the night, the previous night or 2-3 nights earlier. Three thousand newborns a day were gathered at the Jabar fortress." Adult gazelles were also once sent to leaders and kings as gifts, often over very long distances (Vereshchagin 1959).

References

Abaturov BD, Kuznetsov GI, Magomedov MRD, Petelin DA, Kassaie F (1996) Assessment of the forage supply of wild ungulates in the dry season in the meadow pastures of Ethiopia. Zool Zhurnal 75(3):439–450. (in Russian, English abstract)

Abaturov BD, Petritshev BI, Kolesnikov MP (1998) The seasonal dynamics of forage resources and feeding of the saiga antelope on the natural pasture in the semi-desert. Adv Modern Biol 118(5):564–584. (in Russian)

Abduriyim S, Zibibulla G, Eli S, Ismayil Z, Halik M (2018) Phylogeny and genetic structure of the goitered gazelle (Artiodactyla, Bovidae) in North-Western China indicated by the hypervariable mitochondrial control region. Syst Biodivers 16:527–537

Abdusalyamov IA (2001) Chapter 29. Tajikistan. In: Mallon DP, Kingswood SC (eds) Antelopes. Part 4. North Africa, the Middle East, and Asia. Global survey and regional plans. SSC Antelope Specialist Group/IUCN, Gland/Cambridge, pp 148–149

Abood DA, Hussein ZM (2017) Histological features of vomer nasal organ in indigenous gazelle (*Gazella subgutturosa*). J Entomol Zool Stud 5(4):598–604

Abood DA, Hussein ZM (2018) Histological and histochemical features for Jacobson's glands in male *Gazella subgutturosa*. Indian J Nat Sci 8:13107–13114

Afanasiev AV, Bazhanov VS, Karelov MN, Sludskiy AA, Strautman EN (1953) Goitered gazelle. In: Afanasiev AV (ed) Animal of Kazakhstan. Publication of Kazakh Academy of Sciences, Alma-Ata, pp 484–488. (in Russian)

Aitchison JET (1889) The zoology of the afghan delimitation commission. Trans Linnaean Soc London Serial II 5:53–144

Aizin BM (1986) Distribution, number and perspectives of goitered gazelle conservation in Kyrgyzstan. In: Flint VE, Prisyazhnuk VE (eds) Conservation and perspectives of the number restoration of goitered gazelle in USSR. All-Union Scientific Institute of Nature Conservation and Nature Reserves, Moscow, pp 39–41. (in Russian)

Akbari H, Moradi HV, Rezaie HR, Baghestani N (2015) Seasonal changes in group size and composition of Chinkara (*Gazella bennettii shikarii*) (Mammalia: Bovidae) in Central Iran. Italian J Zool 82(4):609–615

Akramova FD, Sh TM, Shakarbaev UA, Rakhmonova LA, Azimov DA, Erkinova LU (2020) Ecological analysis of the helminth Fauna of wild artiodactyls (Mammalia: Artiodactyla) in Karakalpakstan. Rus J Paras 14(4):11–23. (in Russian)

Aldezabal A, Garin I, García-González R (1999) Activity rhythms and the influence of some environmental variables on summer ungulate behaviour in Ordesa-Monte Perdido National Park. Pirineos 153:145–157

Al-Eissa MS, Alhomida AS (1997) A study of the distribution of total, free, short-chain acyl and long- chain acyl carnitine in whole blood and plasma of Arabian sand gazelles (*Gazella subgutturosa marica*). Comp Haematol Int 1:65–69

Al-Robaae K, Kingswood SC (2001) Iraq. In: Mallon DP, Kingswood SC (eds) Antelopes. Part 4: North Africa the Middle East and Asia. Global survey and regional action plans. SSC Antelope Specialist Group and IUCN, Gland and Cambridge, pp 88–92

Amgalan L (2000) Conservation and ecology of the goitered gazelle (*Gazella subgutturosa*) of Mongolia". PhD dissertation (in Mongolian)

Aminoseheh J (1961) The presence of *Cysticercus tenuicollis* and C. *ovis* in sheep and goats in Tehran Abattoir. Vet Med Diss, Tehran University (in Persian)

Antipin VM (1941) Mammals of Kazakhstan. In: Ungulates, vol 3. Publication OGIZ, Alma-Ata, pp 47–54. (in Russian)

Arabuli AB (1970) Distribution and number of some hunting mammals in the eastern Georgia. Zool Zhurnal Moscow 49(3):418–421. (in Russian)

Arnold GW (1981) Grazing behaviour. In: Morley FHW (ed) Grazing Animals. Elsevier, Amsterdam, pp 79–102

Arnold W, Ruf T, Reimoser S, Tataruch F, Onderscheka K (2004) Nocturnal hypometabolism as an overwintering strategy of red deer (Cervus elaphus). Am J Physiol Regul Integr Comp Physiol 286:R174–R181

Askerov E, Mallon DP, Zazanashvili N, Kochiashvili V, Sarukhanova S, Muradov A (2021) An attempt to restore the population of the goitered gazelle (*Gazella subgutturosa*) in its indigenous range in the Caucasus (Mammalia: Bovidae). Zool Middle East 67(3):189–197

Attum O (2007) Can landscape use be among the factors that potentially make some ungulates species more difficult to conserve? J Arid Environ 69(3):410–419

Azimov DA, Norkobilov BT, Akramova FD, Shakarbaev UA, Mirzayeva AU, Rakhmanova LA (2020) Schistosomiasis of animals in Uzbekistan. Eur J Molecular Clin Med 7(1):3401–3408

Baitursinov KK (2008a) Ecological bases of helminthiasis prevention of wild ungulates in Kazakhstan. Russian Parasitological J 1:1–7. (in Russian)

Baitursinov KK (2008b) Factors of formation of the helmintofauna community in wild and domestic ungulates of Kazakhstan. Russian Parasitological J 4:5–12. (in Russian)

Bannikov AG (1954) Persian gazelle. In: Heptner VG (ed) Mammals of Mongolian People Republic. Publication of USSR Academy of Sciences, Moscow-Leningrad, pp 250–258. (in Russian)

Bärmann EV (2014) The evolution of body size, horn shape and social behaviour state in crowns Antilopini—an ancestral character state analysis. Zitteliana B 32:185–186

Bazargani T, Hallan JA, Nabian S, Rahbari S (2007) Sarcoptic mange of gazelle (*Gazella subgutturosa*) and its medical importance in Iran. Parasitol Res 101:1517–1520

Bazhanov VS (1945) Materials on ecology and zoogeography of ungulates. Proc Kazakh Acad Sci Series Zool 5:75–76. (in Russian)

Bazhanov VS (1951) New data for the theriological fauna of eastern Ustyurt. Proc Kazakh Acad Sci Series zoological 10:43–44. (in Russian)

Bekenov A, Savinov EF (1975) The number of goitered gazelle in the Usturt and Mangyshlak. In: Ungulates of fauna USSR, Nauka Publication, Moscow, pp. 261–263 (in Russian)

Bekenov AB, Blank DA, Grachev YA, Plakhov KN (2001) Chapter 26. Kazakhstan. In: Mallon DP, Kingswood SC (eds) Antelopes. Global survey and regional plans. Part 4. North Africa, the Middle East, and Asia. SSC Antelope Specialist Group, IUCN, Gland, Cambridge, pp 134–140
Belovsky GE (1981) Optimal activity times and habitat choice of moose. Oecologia 48:22–30
Belovsky GE, Slade JB (1986) Time budget of grassland herbivores: body size similarities. Oecologia 70:53–62
Betts A, Burke D (2015) Desert kites in Jordan—a new appraisal. Arab Archael Apigraphy 26:74–94
Binh DTS (1963) Age variability of the dental system in *Gazella subgutturosa* Guldenstaedt, 1780. Zool Zhurnal 42(6):932–943. (in Russian, English abstract)
Blank DA (1985) Peculiarities of social and reproductive behaviour of *Gazella subgutturosa* in the Ili Valley. Zool Zhurnal 64:1059–1070. (in Russian, English abstract)
Blank DA (1990) Persian gazelle. In: Kovshar AF (ed) Rare animals of desert regions. Nauka Press of Kazakh SSR, USSR, Alma-Ata, pp 56–80. (in Russian)
Blank DA (1991) The number and distribution of goitered gazelle in the Ustyurt plateau and Mangyshlak peninsula. In: Kovshar AF (ed) Rare species of birds and animals of Kazakhstan. Gylym Press, Alma-Ata, pp 268–273. (in Russian)
Blank DA (2017) Antipredator tactics are largely maternally controlled in goitered gazelle, a hider ungulate. Behav Proc 136:28–35
Blank DA (2018) Vigilance, staring and escape running in antipredator behavior of goitered gazelle. Behav Proc 157:408–416
Blank DA (2019) Diversity of risk assessment pattern in antipredator behavior of goitered gazelles. Behav Proc 158:211–218
Blank DA (2020) Using microclimate of arid landscape as a resource in goitered gazelle comfort behavior. J Arid Environ 180:104201
Blank DA (2022a) Activity behavior of goitered gazelles in an arid environment. Behav Proc 204:104804
Blank DA (2022b) Grooming behavior in goitered gazelles: the programmed versus stimulus-driven hypothesis. Ethol Ecol Evol 35:1–21
Blank DA, Kovshar AF (1986) The number of goitered gazelle populations in deserts of southern and South-Eastern Kazakhstan. In: Flint VE, Prisyazhnuk VE (eds) Conservation and perspectives of the number restoration of goitered gazelle in USSR. All-Union Scientific Institute of Nature Conservation and Nature Reserves, Moscow, pp 8–12. (in Russian)
Blank D, Yang W (2015) Strategy of goitered gazelle suckling behavior and its adaptation to the environment. Mamm Res 60(4):393–401
Blank DA, Ruckstuhl K, Yang W (2015) Antipredator strategy of female goitered gazelles (*Gazella subgutturosa* Guld., 1780) with hiding fawn. Behav Proc 119:44–49
Boev SN, Sokolova IB, Panin VYA (1962) Helminths of ungulate animal of Kazakhstan, vol 1,2. Academy Sciences of KazSSR, Alma-Ata. (in Russian)
Bogachev VV (1938) Zoogeography issues of Caucasus and its geological history. Proc Azerbaijan Depart Acad Sci USSR 10:89–90
Boule M, Teilhard P (1928) Arch. De l'inst. Paleontol. Hurn., 4: 66. (cited by Sokolov 1959)
Bourliére F (1951) Vie et Moeuers d. Mammiferes, 12:112
Breckle SW (1983) Temperate deserts and semi-deserts of Afghanistan and Iran. Ecosystems World 5:271–319
Bruno E, Lovari S (1989) Foraging behaviour of adult female Apennine chamois in relation to seasonal variation in food supply. Acta Theriol 34(37):513–523
Bugaev KE (1989) Large mammals of the trans-Altai Gobi (ecology and conservation). Ph.D. thesis. Moscow State University, Moscow
Buuveibaatar B, Mueller T, Strindberg S, Leimgruber P, Kaczensky P, Fuller TK (2016) Human activities negatively impact distribution of ungulates in the Mongolian Gobi. Biol Conserv 203:168–175

Buuveibaatar B, Strindberg S, Kaczensky P, Payne J, Chimeddorj B, Naranbaatar G, Amarsaikhan S et al (2017) Mongolian Gobi supports the world's largest populations of khulan *Equus hemionus* and goitered gazelles *Gazella subgutturosa*. Oryx 51(4):639–647

Caboń-Raczyńska K, Krasinska M, Krasinski ZA, Wojcik JM (1987) Rhythm of daily activity and behavior of European bison in the Bialowieza Forest in the period without snow cover. Acta Theriol 32: 21–31.

Cai G, Liu Y, O'Gara BW (1990) Observations of large mammals in the Qaidam Basin and its peripheral mountainous areas in the People's Republic of China. Can J Zool 68:2021–2024

Cain JW, Krausman PR, Rosenstock SS, Turner JC (2006) Mechanisms of thermoregulation and water balance in desert ungulates. Wildl Soc Bull 34(3):570–581

Carlos Fernandes in litt. (2012) – cited by Mallon D, Woodfine T, Gilbert T, Restoration of Goitered Gazelles to the Iori Plateau–Vashlovani Landscape, Georgia

Chernov VU (1986) The number and distribution of goitered gazelles in the Sarykamysh Hollow. In: Flint VE, Prisyazhnuk VE (eds) Conservation and perspectives of the number restoration of goitered gazelle in USSR. All-Union Scientific Institute of Nature Conservation and Nature Reserves, Moscow, p 28–31 (in Russian)

Clark EL, Mukhbat J, Dulamtseren S, Baillie JEM, Batsaikhan N, Samiya R, Stubbe M (2006) Mongolian red list of mammals. Zoological Society of London, London

Colman JE, Pedersen C, Hjermann DØ, Holand Ø, Moe SR, Reimers E (2001) Twenty-four-hour feeding and lying patterns of wild reindeer *Rangifer tarandus tarandus* in summer. Can J Zool 79(12):2168–2175

Corbet GB (1978) The mammals of the Palaearctic region: a taxonomic review. In: British museum (natural history), London

Cunningham PL (2009) Some horn measurement comparisons for reem between Mahazat as-Sayd and Thumamah. Unpublished quarterly report—January to March 2009. King Khalid wildlife research Centre, Thumamah

Cunningham PL, Wacher T (2009) Changes in the distribution, abundance and status of Arabian sand gazelle (*Gazella subgutturosa marica*) in Saudi Arabia: a review. Mammalia 73:203–210

Cunningham PL, Wronski T (2011) Sex ratios of Arabian sand gazelle *Gazella subgutturosa marica* Thomas, 1897 in the Mahazat as Sayd protected area, Saudi Arabia. Mammalia 75:243–248

Cunningham PL, Sandoka M, Wronski T (2011) Some morphological characteristics of Arabian sand gazelle *Gazella subgutturosa marica* and its implications for management. Eur J Wildl Res 57(4):865–872

Dawood MS, Khamas MJ (2017) Anatomical features of the liver, gallbladder and biliary duct system of indigenous gazelle (*Gazella subgutturosa*). J Entomol Zool Stud 5(6):2200–2205

De Blainville HMD (1816) Sur plusiers especes d'animaux mammiferes. Bulletin des Sciences Philomathique, Paris, 3:73–82

Demchenko LA (1950) Vegetation of the Barsakelmes Island as a food base for ungulates. Proceedings of theBarsakelmes Nature Reserve, issue 1. Publication of Academy of Sciences of KazSSR: Alma-Ata, 104 pp. (in Russian)

Dementyev GP (1935) Mongolian gazelle, goitered gazelle, saiga antelope. KOIZ Publication, Moscow-Leningrad, pp 15–30. (in Russian)

Dementyev GP (1945) Scientific notes of the Moscow University 83: 76–78 (in Russian)

Dementyev GP (1951) Proceedings of the Turkmenistan branch of the USSR Academy of Sciences 1: 58–59 (in Russian)

Demircioglu I, Gezer Ince N (2020) Three-dimensional modelling of computed tomography images of limb bones in gazelles (*Gazella subgutturosa*). Anat Histol Embryol 49(6):695–707

Demircioglu I, Yilmaz B, Gündemir O, Dayan MO (2021) A three-dimensional pelvimetric assessment on pelvic cavity of gazelle (*Gazella subgutturosa*) by computed tomography. Anat Histol Embriol 50(1):43–49

Dereviagin PYA (1947) On seasonal migrations of goitered gazelle from southern pre-Balkhash deserts into inter-mountain valley Kalkan-Matay-Mynbulack. Bull Acad Sci of Kazakh SSR 4(25):33–38. (in Russian)

Dinesman LG, Knyazev AV, Bold G (1986) On the history of hunting and cattle-breading in Mongolia. Bull Moscow Naturalists Depart Biol 91:31–37. (in Russian, English summary)
Dinnik NYA (1910) Animal of Caucasus, part 1, cetaceans and ungulates. Western Caucasus Depart Imperial Russian Geographical Society (Tiflis) 27(1):215–220. (in Russian)
du Toit JT, Owen-Smith N (1989) Body size, population metabolism, and habitat specialization among large African herbivores. Am Nat 133:736–740
du Toit JT, Yetman CA (2005) Effects of body size on the diurnal activity budgets of African browsing ruminants. Oecologia 143:317–325
Dunham KM, Williamson DT, Joubert E (2001) Saudi Arabia. In: Mallon DP, Kingswood SC (eds) Antelopes, part 4: North Africa, the Middle East, and Asia. Global survey and action plans. SSC Antelope Specialist Group. IUCN, Gland and Cambridge, pp 55–62
Durmuş M (2010) Determination of home range and habitat selection of gazelles (*Gazella subgutturosa*) by GPS telemetry in Şanliurfa. Master degree thesis. Middle East Technical University, Ankara
East R (1992) Conservation status of antelopes in Asia and the Middle East, part 1. Species 19:23–25
Echallier J-C, Braemer F (1995) Nature et fonctions des "desert kites", données et hypothèses nouvelles. Paléorient 21(1):35–63
Effron M, Bogart MH, Kumamoto AT, Benirschke K (1976) Chromosome studies in the mammalian subfamily Antilopinae. Genetica 46:419–444
Ellerman JR, Morrison-Scott TCS (1951) Checklist of palearctic and Indian mammals, 1758 to 1946. Trustees of the British Museum, London
Erkan F (1978) Ceylanpinar Seyahati. – Av Dergisi, 11(5):14–15, Ankara
Eslami A, Rahbari S, Nikbin S (1980) Gastro-intestinal nematodes of gazelle, *Gazella subgutturosa*, in Iran. Vet Parasitol 7(1):75–78
Eslami A, Rahbari S, Meydani M (1981) Cestodes and trematodes of wild sheep, *Ovis ammon orientalis*, and goitered gazelle, *Gazella subgutturosa* in Iran. Vet Parasitol 8(1):99–101
Eversman TA (1850) Natural history of the Orenburg region. Part 2. Kazan (in Russian)
Fadakar D, Barmann EV, Lerp H, Mirzakhah M, Nasari MN, Rezaei HR (2020a) Diversification and subspecies patterning of the goitered gazelle (*Gazella subgutturosa*) in Iran. Ecol Evol 10:5877–5991
Fadakar D, Malekian M, Hemami MR, Lerp H, Rezaei HR, Bärmann EV (2020b) Repeated hybridization of two closely related gazelle species (*Gazella bennettii* and *Gazella subgutturosa*) in Central Iran. Ecol Evol 10(20):11372–11386
Fadeev VA (1973) Distribution and number of goitered gazelles in the Betpak-Dala Desert. In: Rare species of mammals of the USSR fauna. Vol. 2. Nauka Publication: Moscow, pp. 56–58 (in Russian)
Fadeev VA (1986) The number of goitered gazelle in the Taukum and Muyunkum deserts (according results of aero-survey). In: Kovshar, A.F. (ed.) Rare animals of Kazakhstan. Nauka Press of Kazakh SSR: Alma-Ata, 21–23 pp (in Russian)
Farhadinia M (2004) The last stronghold: cheetah in Iran. Cat News 40:11–14
Farhadinia MS, Esfandabad BS, Karami M, Hosseini-Zavarei F, Absalan H, Nezami B (2009) Goitered gazelle, *Gazella subgutturosa*: its habitat preference and conservation needs in Miandasht wildlife refuge, North-Eastern Iran (Mammalia: Artiodactyla). Zool Middle East 46(1):9–18
Fernandez-Gimenez ME, Allen-Diaz B (1999) Testing a non-equilibrium model of rangeland vegetation dynamics in Mongolia. J Appl Ecol 36:871–885
Flerov KK (1935) Persian gazelle. In: Pavlovskiy EN (ed) Wild animals of Tajikistan their life and significance for man. Academy of Sciences of the USSR, Moscow, pp 109–112. (in Russian)
Flerov KK, Gromov IM (1935) Proceedings of the Tajik Comprehensive Expedition of 1932 10:281–284. (in Russian)
Flint VE, Prisyazhnuk VE (1986) Population status and perspectives of goitered gazelle restoration in the USSR. In: Flint VE, Prisyazhnuk VE (eds) Conservation and perspectives of the number

restoration of goitered gazelle in USSR. All-Union Scientific Institute of Nature Conservation and Nature Reserves, Moscow, pp 3–8. (in Russian)

Frazier H, Hunt K (eds) (1994) Zoo infant development notebook. Am Assoc Zoo Keepers, Topeka, Kansas 2:921–1755

Freitag H (1986) Notes on the distribution, climate and flora of the sand deserts of Iran and Afghanistan. Proc Royal Soc Edinburgh Sect B Biol Sci 89:135–146

Frey W, Probst W (1986) A synopsis of the vegetation in Iran. In: Kurschner H (ed) Contribution to the vegetation of Southwest Asia, pp 9–43

Furley CW (1986) Reproductive parameters of African gazelles: gestation, first fertile mating, first parturition and twinning. Afr J Ecol 24:121–128

Gallagher M, Harrison DL (1975) The terrestrial mammals of Bahrain. J Bombay Natur Hist Soc 72:407–421

Galuzo IG (1953) Blood-sucking ticks of Kazakhstan. Acad Sci, KazSSR Alma-Ata, Kazakhstan 5. (in Russian)

Gao W, Ji L (1996) An analysis of the nutrient composition of five species of sandy plants growing in west inner Mongonia. J Inner Mongolia Inst Argric Anim Husband 17:23–28

Gao X, Yao J (2006) Study on the geography distribution and population of Gazella subgutturosa in the Hami Basin, Xinjiang in early winter. Arid Land Geograp 29(2):213–218. (in Chinese, English abstract)

Ginnett TF, Demment MW (1997) Sex differences in giraffe foraging behavior at two spatial scales. Oecologia 110(2):291–300

Gorbuniv AB, Zarkhidze (1986) Multi-year changes of number, current distribution and perspectives of goitered gazelle population restoring I western Turkmenistan. In: Flint, V.E. and Prisyazhnuk, V.E., (eds.), Conservation and perspectives of the number restoration of goitered gazelle in USSR. All-Union Scientific Institute of Nature Conservation and Nature Reserves, Moscow, pp. 22–28 (in Russian)

Gorbunov AV (1983) About ecology and behavior of goitered gazelles in the Krasnovodsk peninsula. Proc Turkmen Acad Sci Series Biol Sci 3:66–68. (in Russian)

Gorelov YK (1972) Breeding of goitered gazelle (*Gazella subgutturosa* Güldenstaedt) and problem of restoring of its number in the Badkhyz (SE Turkmenistan). In: Theriology, vol 1. Nauka of Siberian Department of Academy of Sciences of USSR, Novosibirsk, pp 420–424. (in Russian)

Gorelov YK (2001) Chapter 30. Turkmenistan. In: Mallon DP, Kingswood SC (eds) Antelopes. Part 4. North Africa, the Middle East, and Asia. Global survey and regional plans. SSC Antelope Specialist Group, IUCN, Gland and Cambridge, pp 150–154

Granjon M, Vassart M, Greth A, Cribibiu EP (1991) Genetic study of sand gazelles (*Gazella subgutturosa marica*) from Saudi Arabia. Zeitschrift für Saügetierkunde 56:169–176

Gray AP (1972) Mammalian hybrids. A check-list with bibliography, 2nd edn. Commonwealth Agricultural Bureaux, Farnham Royal

Gray JE (1821) On the natural arrangement of vertebrose animals. London Medical Repository, 15(1):296–310

Grebenukov VI (2008) The problem of the Western-Kazakhstan variant Keltimirov. Culture Bull Nizhne-Vartovskogo State Univ 1:1–6. (in Russian)

Grimwood I (1962) Operation oryx. Oryx 6:308–334

Gromov VI (1948) Paleontological and archeological justification of the continental sediment stratigraphy of the quaternary period on the territory of the USSR. Proc Institute Geol Sci 64:1–521. (in Russian)

Gromova IM, Baranova GI (1981) The mammals' catalogue of USSR. Pliocene-present time. Academy of Sciences of the USSR, Leningrad. (in Russian)

Groves CP (1969) On the smaller gazelles of the genus Gazella de Blainville, 1816. Zeitschrift für Säugetierkunde 34:38–60

Groves CP (1985) An introduction to the gazelles. Chinkara 1(1):4–16

Groves CP (1996) Taxonomic diversity in Arabian gazelles: the state of the art. In: Greth A, Magin C, Ancrenaz M (eds) Conservation of Arabian gazelles. NCWCD, Riyadh, pp 8–39

Groves CP (1997) Taxonomy of Arabian gazelles. In: Habibi K, Abu-Zinada A, Nader I (eds) The gazelles of Arabia. National Commission for Wildlife Conservation and Development, Riyadh, pp 24–51
Groves CP, Grubb P (2011) Ungulate taxonomy. JHU Press, Baltimore
Groves CP, Harrison DL (1967) The taxonomy of the gazelles (genus *Gazella*) of Arabia. J Zool 152:381–387
Grunin KY (1949) New genera of subcutaneous gadflies of Asiatic antelopes. Rep USSR Acad Sci 64(4):603–606. (in Russian)
Grunin KY (1956) Subcutaneous gadflies (Diptera, Hyperdermatidae) of goitered gazelles (*Gazella subgutturosa* Gmel.). Entomol Rev 35(3):716–722. (in Russian)
Gürler Ş, Bozkaya F, Özüt D, Durmuş M (2015) Some morphological characteristics and neonatal weights of reintroduced gazelle (*Gazella subgutturosa*) in Turkey. Turkish J Zool 39(3):458–466
Habibi K (1977) The mammals of Afghanistan: their distribution and status.FO:DP/AFG/74/016. Field document no.1. FAO, Rome
Habibi K (1991) Arabian gazelles. National Commission for Wildlife Conservation and Development, Riyadh
Habibi K (2001) Chapter 22. Afghanistan. In: Mallon DP, Kingswood SC (eds) Antelopes. Part 4. North Africa, the Middle East, and Asia. Global survey and regional plans. SSC Antelope Specialist Group, IUCN, Gland and Cambridge, pp 119–121
Habibi K, Abuzinada AH, Nader IA (1997) The gazelles of Arabia. National Commission for Wildlife Conservation and Development, Riyadh
Hague MN, Smith TR (1996) Reintroduction of Arabian sand gazelle *Gazella subgutturosa marica* in Saudi Arabia. Biol Conserv 76:203–207
Hamel S, Côté SD (2008) Trade-off in activity budget in an alpine ungulate: contrasting lactating and nonlactating females. Anim Behav 75:217–227
Hamza LO, Siwan NA (2017) Morphological features of the small intestine in the adult indigenous gazelle (*Gazella subgutturosa*). Int J Sci Nat 8(2):223–229
Harrison DL (1968) The mammals of Arabia, vol 2. Ernest Benn, London, pp 193–381
Harrison GL, Bates PJJ (1991) The mammals of Arabia. Harrison Zoological Museum, Sevenoaks
Hassinger JD (1968) A survey of mammals of Afghanistan resulting from the 1965 street expedition (excluding bats). Fieldiana (Zoology) 60:1–195
Hatough-Bouran A, Disi AM (1991) History, distribution and conservation of large mammals and their habitats in Jordan. Environ Conserv 18(19–32):44
Hayatgheib D, Karami M, Farahmand H, Mehrabani YH, Farhadinia MS (2011) Morphological study and comparison of western and eastern goitered gazelle (*Gazella subgutturosa*: Gueldenstaedt, 1780) populations in Iran. Int J Environ Res 5(1):225–232
Hemami MR (1994) Taxonomic status and distribution of Iranian gazelles. P.D. Dissertation. University of Tehran
Hemami MR, Groves CP (2001) Chapter 21. Iran. In: Mallons DP, Kingswood SC (eds) Antelopes. Part 4. North Africa, the Middle East, and Asia. Global survey and regional plans. SSC Antelope Specialist Group, IUCN, Gland and Cambridge, pp 114–118
Hemami MR, Khosravi R, Groves C, Ahmadi M (2020) Morphological diversity and ecological niche divergence in goitered and sand gazelles. Ecol Evol 10(20):11535–11548
Heptner G, Nasimovich AA, Bannikov AG (1961) Mammals of Soviet Union, Vol.1. Artiodactyla and Perissodactyla. Vysshaya Shkola, Moscow
Heptner VG, Nasimovich AA, Bannikov AG (1988) Mammals of the Soviet Union. Artiodactyla and Perissodactyla, vol 1. Smithsonian Institution Libraries and The National Science Foundation, pp 1–1147. (translation from Russian)
Heshmati GA (2007) Vegetation characteristics of four ecological zones of Iran. Int J Plant Production 1(2):25–224
Heude PM (1894) Etudes odontologiques, etc. Memoires concernant l'histoire naturelle de l'Empire chinois par des peres de la Compagnie de Jesus. Impr. de la Mission Catholique, Chang-Hai, 2(3-4):117–247

Hofmann RR (1989) Evolutionary steps of ecophysiological adaptation and diversification of ruminants: a comparative view of their digestive system. Oecologia 78:443–457

Hosseini-Zavarei F, Farhadinia MS, Hemami MR, Karami M, Daniali R, Omidi M (2010) Sex-age structure of bovids in Ghameshlou, Central Iran: (Mammalia: Bovidae). Zool Middle East 51(1):3–8

Hsu TC, Benirschke K (1977) *Gazella subgutturosa* (Persian gazelle) 2n = 31 male, 30 female. An atlas of mammalian chrosomosomes. Vol. 10, Folio 504. Springer-Verlag, New York: 213-216.

Hussein HS, Mohammed OB (1992) Eimeria rheemi sp. n. (Apicomplexa, Eimeriidae) from the Arabian sand gazelle, *Gazella subgutturosa marica* (Artiodactyla, Bovidae) in Saudi Arabia. J Helminthol Soc Wash 59(2):190–194

Irby LR (1981) Mountain reedbuck activity patterns in the Loskop dam nature reserve. South Afr J Wildl Res 11(4):115–120

Ismagilov MI (1954) The goitered gazelle ecology on the Barsakelmes Island. The Bulletin of the Moscow Society of Naturalists, Biol Department 60(6):3–9

Ishunin GI (1961) Goitered gazelle. In: Zakhidov TZ (ed) Fauna of Uzbek SSR, mammals (carnivores and ungulates), vol 3. Publication of the Academy of Sciences of Uzbek SSR, Tashkent, pp 165–174. (in Russian)

Jahandizi EK, Kaboli M, Karami M, Soufi M (2015) The assessment of carrying-capacity of Persian gazelle in Sorkh-E-Hesar National Park, Iran. J Earth Environ Health Sci 1(2):66

Jamsheed R (1976) Big game animals of Iran (Persia). Rashid Jamsheed:1–122

Jarman RJ (1974) The social organization of antelope in relation to their ecology. Behaviour 48:215–267

Jones ML (1993) Longevity of ungulates in captivity. Int Zoo Yearb 32:159–169

Kaczensky P, Ganbaatar O, Altansukh N, Enksaikhan N, Kramer-Schadt S (2015) Monitoring of khulans and goitered gazelles in the Mongolian Gobi – potential and limitations of ground-based line transects. Open Ecol J 8:92–110

Karami M, Groves CP (1992) A mammal species new for Iran: Gazella gazella (Artiodactyla, Bovidae). In: Proceedings of the International Symposium "Ongulés/Ungulates 91", Toulouse, September 2-6, 1991. F. Spitz, G. Janeau, G. Gonzalez and S. Aulagnier, eds]. Société Française pour l'Étude et la Protection des Mammifères, Paris, Institut de Recherche sur les Grands Mammifères. Paris, Toulouse: 339–342

Karami M, Hemami MR, Groves CP (2002) Taxonomic, distributional and ecological data on gazelles in Iran. Zoology in the Middle East. 26(1):29–36

Karelin GS (1883) The trip along the Caspian Sea. Notes of the Russian geographical society for general geography, S-Peterburg, p 10

Kasparek M (1986) On the historical distribution and present situation of gazelles, *Gazella* spp., in Turkey. Zool Middle East 1:11–15

Khakhin GV, Tschadilov UM, Kuzmin IF (1986) The survey experience of goitered gazelles in the protected areas of Azerbaijan. In: Flint VE, Prisyazhnuk VE (eds) Conservation and perspectives of the number restoration of goitered gazelle in USSR. All-Union Scientific Institute of Nature Conservation and Nature Reserves, Moscow, p 49–59 (in Russian)

Khakhlov VA (1928) Zaisan Basin and Tarbagatai, Tomsk, p 1–9 (in Russian)

Khaleel IM, Salih AAM (2017) Comparative histomorphological and histochemical study of thyroid gland in adult indigenous gazelle (*Gazella subgutturosa*) and sheep (*Ovis aries*). J Entomol Zoology Stud 5(6):1236–1241

Kim TA (1953) Mammals of northern and western Kyzylkum Desert (mainly agricultural areas). Dissertation for the degree of candidate of biological sciences. Moscow University, Moscow

Kingdon J (1990) Arabian mammals: a natural history. Government Press, Ministry of Information, Isa Town

Kingswood SC (1992) Meiosis in chromosomally heteromorphic goitered gazelle. *Gazella subgutturosa*. M.S. thesis. Southwest Texas State University, San Marcos. 84 pp

Kingswood SC, Blank DA (1996) Gazella subgutturosa. Mamm Species 518:1–10

Kingswood SC, Kumamoto AT (1988) Research and management of Arabian sand gazelle in the USA. In: Dixon A, Jones D (eds) Conservation and biology of desert antelopes. Christopher Helm, London, pp 212–226

Kingswood SC, Kumamoto AT, Sudman PD, Fletcher KC, Greenbaum IF (1994) Meiosis in chromosomally heteromorphic goitered gazelle, *Gazella subgutturosa* (Artiodactyla, Bovidae). Chromosom Res 2:37–46

Kingswood SC, Vassart M, Williamson D (1996) Systematics of *Gazella subgutturosa*: implications for conservation in the Arabian peninsula. In: Greth A, Magin C, Ancrenaz M (eds) Conservation of Arabian gazelles. National Commission for Wildlife Conservation and Development, Riyadh, pp 104–111

Kingswood SC, Rebholz WER, Vassart M, Kumamoto AT (1997) Cytogenetics and molecular genetics of Arabian gazelles: chromosomes, proteins and nucleic acids. In: Habibi K, Abu-Zinada A, Nader I (eds) The gazelles of Arabia. National Commission for Wildlife Conservation and Development, Riyadh, pp 159–191

Kirilyuk VE (2021) Rapidly increasing migratory activity of Mongolian gazelle (*Procapra gutturosa*) and the sightings of Goitered gazelle (*Gazella subgutturosa*) in Transbaikalia as an alarm. Russian J Theriol 20(1):25–30

Klockenhoff H (1969) Über die Kropfgazellen, Gazella subgutturosa (Güuldenstaedt), 1780, Afghanistan und ihre Haltung im Zoologischen Garten Kabul. Freunde des Kolner Zoo 3:91–96

Kolenikov II (1952) About vertebrates of the southern Ustyurt plateau and their economic significance. Proceedings of the central Asian University 32. Biol Sci 2:13–26. (in Russian)

Komers PE, Messier F, Gates CC (1993) Group structure in wood bison: nutritional and reproductive determinants. Can J Zool 71:1367–1371

Kostin VP (1955) Materials on ecology and commercial value of goitered gazelle. Proc Institute Zool Acad Sci Kazakh SSR 4:67–82. (in Russian)

Kovshar AF, Blank DA (1986) On the number of goitered gazelle in the southern regions of Kazakhstan. In: Kovshar AF (ed) Rare animals of Kazakhstan. Nauka Press of Kazakh SSR, Alma-Ata, pp 23–30. (in Russian)

Kozhamkulova BS (1969) Anthropogenic fossil teriofauna of Kazakhstan. Nauka Press of Kazakh SSR, Alma-Ata. 149 pp. (in Russian)

Kozhamkulova BS (1981) The late-Cainozoic [sic] ungulates of Kazakhstan. Nauka Press of Kazakh SSR, Alma-Ata. 145 pp. (in Russian)

Kulpin PI (1954) Goitered gazelles and saiga antelopes in the Betpak-Dala Desert. Priroda (Nature), 4:102–104

Kumerloeve H (1969) Bemerkungen zum Gazellen Vorkommen im südöstlichen Kleinasien. Z. Saugetierkd. 34, 113–20

Kurochkina L Ya, Osmanova LT, Karibaeva KN (1986) Forage plants of Kazakhstan deserts. Nauka of KazSSR: Alma-Ata. 208 pp. (in Russian)

Kuznetzov VI (1986) Goitered gazelles in Badkhyz. In: Flint VE, Prisyazhnuk VE (eds) Conservation and perspectives of the number restoration of goitered gazelle in USSR. All-Union Scientific Institute of Nature Conservation and Nature Reserves, Moscow, pp 31–37. (in Russian)

Landete-Castillejos T, Garcia A, Garde J, Gallego L (2000) Milk intake and production curves and allosuckling in captive Iberian red deer, *Cervus elaphus hispanicus*. Anim Behav 60:679–687

Lange J (1972) Studien an Gazellenschadeln. Ein Beitrag zur Systematik der klineren Gazellen, Gazella (De Blainville, 1816). Säugetierkundliche Mitteilungen 20:193–249

Larionov KO, Nikonova OA (2006) The daily activity of tame saiga antelopes. In: Materials of international conference "problems of population ecology of animals", dedicated to the memory of academician Shilov IA. Tomsk State University, Tomsk, pp 495–496

Larionov KO, Nikonova OA, Abaturov BD (2010) The daily activity of the saiga antelope and domestic sheep on the semi-desert pasture in the northern Caspian Sea region. Arid Ecosyst 16(5):111–120. (in Russian)

Launay F, Launay C (1992) Daily activity and social organization of the goitered gazelle (gazella subgutturosa marica) Ongules/ Ungulates 91:373–377

Lay DM (1967) A study of the mammals of Iran resulting from the street expeditions of 1962- 63. Fieldiana: Zool 54:1–282

Legge AJ, Rowley-Conwy PA (1987) Gazelle killing in Stone Age Syria. Sci Am 257(2):76–83

Lewis JG (1977) Game domestication for animal production in Kenya: activity patterns of eland, oryx, buffalo and zebu cattle. J Agricul Sci 89(3):551–563

Lindsay N, Wood J (1992) Hand-rearing three species of gazelle, *Gazella* spp., in the Kingdom of Saudi Arabia. Int Zoo Yearb 31:250–255

Lkhagvasure B, Chimeddorj B., Sanjmyatav D (2011) Barriers to migrations: case study in Mongolia. Report: Ulaanbaatar, pp 1–33

Lobachev VS, Smirin VM (1970) Horns in females of the saiga and Persian gazelle. Zoological Zhurnal 49:1421–1423. (in Russian, English summary)

Livov IA (1979) Ecology of the dzheiran antelope (Gazella subgutturosa) of eastern Mongolia. Soviet Journal of Ecology, 10:154–155 (in Russian)

Lydekke R, Blaine G (1914) Catalogue of the ungulate mammals in the British museum (natural history), vol 3. Trustees of the British Museum, London, pp 1–283

Malenkov II (1958) On the goitered gazelle distribution and jackal in the Fergana Valley. Zool Zhurnal, Moscow 37(4):629. (in Russian)

Mallon DP (2008) *Gazella subgutturosa*. The IUCN Red List of Threatened Species 2008: e.T8976A12945246. 10.2305/IUCN.UK.2008.RLTS.T876A12945246.en. Accessed 3 Dec 2015

Mallon DP, Kingswood SC (2001) Antelopes. Part 4. North Africa, the Middle East, and Asia. Global survey and regional plans. The IUCN/SSC Antelope Specialist Group, IUCN, Gland and Cambridge

Maloney SK, Moss G, Cartmell T, Mitchell D (2005) Alteration in diel activity patterns as a thermoregulatory strategy in black wildebeest (*Connochaetes gnou*). J Comp Physiol A 191:1055–1064

Mambetzhumaev AM (1954) The goitered gazelle biology (*Gazella subgutturosa* Güld.). Dissertation for the degree of candidate of biological sciences. Institute of Zoology of Uzbek Academy of Sciences, Tashkent

Mambetzhumaev AM (1970) Goitered gazelle. FAN publication, Tashkent, pp 1–196. (in Russian)

Mardonov BK (1993) Impact of goitered gazelles on the vegetation cover of desert. Dissertation for the degree of candidate of biological sciences. Institute of Geography of the Russian Academy of Sciences, Moscow, pp 1–26. (in Russian)

Marmazinskaya NV (1996) Territorial and marking behavior of goitered gazelle *Gazella subgutturosa* (Artiodactyla, Bovidae) in the Bukhara ecological center. Zoological Zhurnal, Moscow 75(11):1737–1751. (in Russian, English summary)

Marmazinskaya NV, Mardonov BK (2001) Chapter 27. Uzbekistan. In: Mallon DP, Kingswood SC (eds) Antelopes. Part 4. North Africa, the Middle East, and Asia. Global survey and regional plans. The IUCN/SSC Antelope Specialist Group, IUCN, Glands and Cambridge, pp 141–145

Marmazinskaya NV, Mardonov BK (2004) The current status of populations and conservation goitered gazelle (*Gazella subgutturosa subgutturosa* Güldenstaedt, 1780) in Uzbekistan. Proc Nat Reserves Uzbekistan (4–5) Tashkent:124–133

Mendelssohn H (1974) The development of the populations of gazelles in Israel and their behavioral adaptations. Int Union Conservation Nat Nat Resour Publications New Series 24:722–743

Minervin VN (1944) On the watering-feeding regime of the goitered gazelle and domestic sheep in desert. Proc Turkmenistan Filial USSR Acad Sci 5:227–241. (in Russian)

Molodoskiy AV (1975) Distribution of goitered gazelle in the Mangyshlak and Usturt in 1955-1957. In: Sokolov VE (ed) Ungulates of USSR Fauna. Nauka of Acad Nauk USSR, Moscow, pp 1–352

Morrison-Scott TCS (1939) Some Arabian mammals collected by Mr. H. J.B., Philby, C.I.E. Novitates Zoologicae 41:181–211

Mowlavi M (1978) Ecological studies of the goitered gazelle (*Gazella subgutturosa*) in Khosh Yeilagh Wildlife Refuge, Iran. M.S. thesis. Michigan State University, East Lansing, p 40

Murtskhvaladze M, Gurielidze Z, Kopaliani N, Tarkhnishvili D (2012) Gene introgression between *Gazella subgutturosa* and *G. Marica*: limitations of maternal inheritance analysis for species identification with conservation purposes. Acta Theriol 57(4):383–386

Mysterud A (1998) The relative roles of body size and feeding type on activity time of temperate ruminants. Oecologia 113:442–446

Nader IA (1989) Rare and endangered mammals of Saudi Arabia. In: Abuzinada AH, Goriup PD, Nade IA (eds) Wildlife conservation and development in Saudi Arabia. NCWCD, Riyadh, pp 220–233

Naumann C, Nogge G (1973) Die Grossäuger Afghanistans. Zeitschrift des Kölner Zoo 16:79–93

Nayeri D, Mohammadi A, Qashqaei AT, Vanak AT, Gompper ME (2022) Free-ranging dogs as a potential threat to Iranian mammals. Oryx 56(3):383–389

Nazarova NS, Prisyazhnuk VE (1986) Prevention of helminthiasis during breeding and introduction of goitered gazelles. In: Flint VE, Prisyazhnuk VE (eds) Conservation and perspectives of the number restoration of goitered gazelle in USSR. All-Union Scientific Institute of Nature Conservation and Nature Reserves, Moscow, pp 84–89. (in Russian)

Neronov VM, Borbrov VV (1990) Conservation of rare mammals in deserts of USSR. Arid Land Newsletter 30:15–20

Neuhaus P, Ruckstuhl KE (2002) Foraging behaviour in Alpine ibex (*Capra ibex*): consequences of reproductive status, body size, age and sex. Ethol Ecol Evol 14:373–381

Nordmann (1840) Fauna pontique, Paris, pp. 1–61

Norton PM (1981) Activity patterns of klipspringers in two areas of the Cape Province. S Afr J Wildl Res 11(4):126–134

O'Gara BW (1978) Antilocapra americana. Mamm Spec 90:1–7

Ognev SI (1935) Animals of USSR and adjacent countries, vol 3. USSR Academy of Sciences, Moscow-Leningrad. (in Russian)

Ogurlu I (1992) Wild ungulates in Turkey. Ongulés/Ungulates 91:575–577

Owen R (1848) Description of teeth and portions of jaws of two extinct Anthracotherioid quadrupeds (Hyopotamus vectianus and Hyop. bovinus) discovered by the Marchioness of Hastings in the Eocene deposits on the NW coast of the Isle of Wight: with an attempt to develop Cuvier's idea of the Classification of Pachyderms by the number of their toes: Quarterly Journal of the Geological Society of London, 4: 103–141

Owen-Smith N (1998) How high ambient temperature affects the daily activity and foraging time of a subtropical ungulate, the greater kudu (*Tragelaphus strepsiceros*). J Zool 246:183–192

Pelletier F, Festa-Bianchet M (2004) Effects of body mass, age, dominance and parasite load on foraging time of bighorn rams, *Ovis canadensis*. Behav Ecol Sociobiol 56:546–551

Peng D, Zhang Z, Chen J, Meng D, Liang Y, Hu T, Teng L, Liu Z (2024) The activity patterns and grouping characteristics of the remaining goitered gazelle (*Gazella subgutturosa*) in an isolated habitat of Western China. Animals 14(16):2338

Pereladova OV, Pereladov SV (1986) The goitered gazelle breeding in nursery enclosures of the Syunt-Khasardak nature reserve. In: Flint VE, Prisyazhnuk VE (eds) Conservation and perspectives of the number restoration of goitered gazelle in USSR. All-Union Scientific Institute of Nature Conservation and Nature Reserves, Moscow, pp 49–59. (in Russian)

Pérez-Barbería FJ, Nores C (1996) Grazing activity of breeding and non-breeding female Cantabrian chamois (*Rupicapra pyrenaica parva*). Ethol Ecol Evol 8:353–363

Pestov MV, Dieterich T, Teretyev VA, Nurmuhambetov ZE, Muhashov AT, Denisov DA (2018) Problems and solutions regarding the border fence that mitigates the migration of ungulates on the territory of the Mangystau oblast of the Republic of Kazakhstan. Selevinia 26:92–98. (on Russian, English abstract)

Peters J (1989) Osteomorphological features of the appendicular skeleton of gazelles, genus Gazella Blainville 1816, bohor reedbuck, redunca redunca (Pallas, 1767) and bushbuck, Tragelaphus scriptus (Pallas, 1766). Anatomia, Histologia, Embryologia. 1989 Jun;18(2):97–113

Philby HSJB (1933) The empty quarter. Constable and Co. Publisher, London, p 576

Pocock RI (1910) On the specialized cutaneous glands of ruminants. Proc Zool Soc London 80(4):840–845. Oxford, UK: Blackwell Publishing Ltd

Pocock RJ (1939) The fauna of British India including Ceylon and Burma. Vol. 1. Mammalia. London

Pole SB, Karpov AA (1986) On the goitered gazelle number in the Muyunkum, Taukum, and adjacent areas. In: Kovshar AF (ed) Rare animals of Kazakhstan. Nauka Press of Kazakh SSR, Alma-Ata, pp 19–21. (in Russian)

Prisiazhnuk VE, Soldatova NV (1985) Forming of exterior indices of goitered gazelle in the ontogenesis (Bukhara nursery). In: Flint EV (ed) Ecological features of animal life conservation, Moscow, pp 84–89. (in Russian)

Quemsiyeh MB, Amr ZS, Budari AM (1996) Status and conservation of Artiodactyla (Mammalia) in Jordan. Mammalia 60:417–430

Rad EB, Mesdaghi M, Ahmad N, Abdullah M (2015) Nutritional quality and quantity of available forages relative to demand: a case study of the goitered gazelles of the Golestan National Park, Iran. Rangelands 37(2):68–80

Radde G, Walther A (1889) Zool. Jahrb. Abt Sys 4:1060–1064

Ransom JI, Kaczensky P, Lubow BC, Ganbaatar O, Altansukh N (2012) A collaborative approach for estimating terrestrial wildlife abundance. Biol Conserv 153:219–226

Rapoport LP, Subbotin UI, Karpov AA (1975) The goitered gazelle number in the deserts of Southern Kazakhstan. In: Ungulates of the USSR fauna. Nauka Publication, Moscow, pp. 275–276 (in Russian)

Rautiainen H (2015) Foraging and anti-predation behavior of Thomson's gazelles (*Gazella thomsoni*) and Grant's gazelles (*Gazella granti*) at a waterhole. Unpublished report

Riesch R, Martin RA, Lerp H, Plath M, Wronski T (2013) Size and sex matter: reproductive biology and determinants of offspring survival in *Gazella marica*. Biol J Linn Soc 110(1):116–127

Rietkerk FE, Delima EC (1994) Clinical and haematological changes in gazelles during xylazine/ketamine anaesthesia and following reversal with RX-821002A

Rietkerk FE, Lindsay N, Tatwany H, Mubarak S, Badri O, Williamson DT (1992) Population dynamics of a captive herd of Arabian sand gazelles. Ongulés/Ungulates 91:379–382

Roberts TJ (1977) The mammals of Pakistan. Ernest Benn, London

Roberts G (1996) Why individual vigilance declines as group size increases. Anim Behav 51(5):1077–1086

Rodrigues Soares AE, Boroffka N, Schröder O, Sverchkov L, Benecke N, Günther T (2022) Ancient DNA from a 2700-year-old goitered gazelle (*Gazella subgutturosa*) supports gazelle hunting in iron age Central Asia. R Soc Open Sci 9(6). https://doi.org/10.1098/rsos.220104

Rostron J (1972) A multivariate statistical study of skull measurements of fie taxa of gazelles. Biol J Linn Sci 4:1–14

Ruckstuhl KE (1999) To synchronize or not to synchronize: a dilemma for young bighorn males? Behaviour 136:805–818

Ruckstuhl KE, Festa-Bianchet M (1998) Do reproductive status and lamb gender affect the foraging behavior of bighorn ewes? Ethology 104(11):941–954

Ruckstuhl KE, Neuhaus P (2002) Sexual segregation in ungulates: a comparative test of three hypotheses. Biol Rev Camb Philos Soc 77(1):77–96

Ruckstuhl KE, Neuhaus P (2009) Activity budgets and sociality in a monomorphic ungulate: the African oryx (Oryx gazella). Can J Zool 87:165–174

Rustamov AK, Atamuradov KI, Sopyev OS, Kurbanov OP, Koshunov MA (1986) Goitered gazelles in Turkmenistan: population status, resettlement and breeding? Conservation measures. In: Flint VE, Prisyazhnuk VE (eds) Conservation and perspectives of the number restoration of goitered gazelle in USSR. All-Union Scientific Institute of Nature Conservation and Nature Reserves, Moscow, pp 17–21. (in Russian)

Rychkov P (1762) Topography Orenburg or detail description of the Orenburg Province. Saint-Petersburg

Sabilaev AC, Koptsev LA (1986) Past and present number of goitered gazelle in the Karakalpak part of the Ustyurt plateau and Kyzylkum Desert. In: Flint VE, Prisyazhnuk VE (eds) Conservation and perspectives of the number restoration of goitered gazelle in USSR. All-Union Scientific Institute of Nature Conservation and Nature Reserves, Moscow, pp 12–16. (in Russian)

Safarov MA (1961) Results of airplane survey of goitered gazelle number in Azerbaijan. Proceedings of Academy of Sciences of Azerb SSR, Biological series 1:56–63 (in Russian)
Sapozhenkov YF (1969) Essays on the history of hunting and commercial animals of Turkmenistan. Bull Moscow Soc Naturalists Biol Depart 54(2):33–41. (in Russian)
Sarukhanova S (2021) The study level of the goitered gazelle (Artiodactyla, Bovidae) in Azerbaijan. Bull Sci Practice 7(9):42–58. (in Russian)
Satunin RA (1905–1906) Mammals of the Talysh and Mugani. Proc Caucasus Museum 2(3–4):212–213 (in Russian)
Savage DE, Russel DE (1983) Mammalian paleo-faunas of the world. Addison-Wesley Publishing Company, London
Schaller GB (1998) Wildlife of the Tibetan steppe. University of Chicago Press
Semenov SA, Korobkova GF (1983) The Technology of the Ancient Production Facilities: Mesolithic-Eniolithic. Nauka, Leningrad, p 256. (in Russian)
Sharhuu G, Sharkhuu T (2004) The helminth fauna of wild and domestic ruminants in Mongolia—a review. Eur J Wildl Res 50:150–156
Shchadilov YM (1986) Goitered gazelle in Azerbaijan. In: Flint VE, Prisyazhnuk VE (eds) Conservation and perspectives of the number restoration of goitered gazelle in USSR. All-Union Scientific Institute of Nature Conservation and Nature Reserves, Moscow, pp 41–43. (in Russian)
Shchadilov YM, Hadjiev EM (2001) Chapter 24. Azerbaijan. In: Mallon DP, Kingswood SC (eds) Antelopes. Part 4. North Africa, the Middle East, and Asia. Global survey and regional plans. The IUCN/SSC Antelope Specialist Group, IUCN, Glands and Cambridge, pp 129–131
Shnitnikov VN (1936) Mammals of Semerechie. Publication of USSR Academy of Sciences, Moscow-Leningrad, pp 127–128. (in Russian)
Sludskiy AA (1953) The mass death of wild ungulates in the Kazakhstan deserts and its impact on the animal numbers. Proceedings of the institute of zoology, vol 2. Publication of Academy of Sciences KazSSR, Alma-Ata, pp 3–30. (in Russian)
Sludskiy AA (1956a) Breeding of goitered gazelle. Proceedings of institute of zoology, Kazakh Academy of Sciences, vol 6. Nauka Press of Kazakh SSR, Alma-Ata, pp 78–108. (in Russian)
Sludskiy AA (1956b) Fauna of Kazakhstan and its conservation. Proceedings of the institute of zoology Academy of Sciences of Kazakh SSR, vol 6. Nauka Press of Kazakh SSR, Alma-Ata, pp 3–27. (in Russian)
Sludskiy AA (1962) Interrelation between predators and prey. Proceedings of the institute of zoology. Publication of Academy of Sciences KazSSR 17:24–143. (in Russian)
Sludskiy AA (1963) The mass death of wild ungulates in steppes and deserts of Europe and Asia. Proceedings of the Institute of Zoology, Academy of Sciences of the Kazakh SSR 20:5–88. (in Russian)
Sludskiy AA (1977) Goitered gazelle. In: Rare animals of USSR. Publication of Lesnaya Promyshlennost, Moscow, pp 28–61. (in Russian)
Sludskiy AA, Shubin IG (1963) Airplane survey of saiga, goitered gazelle and these ungulate resources in the Kazakhstan deserts. In: The fauna resources of game animals in USSR and their accounting. Publication of the Academy of Sciences, USSR, Moscow: 84-91 (in Russian)
Smirnov YA (1965) The influence of hunt on trapper's fauna of Kazakhstan. Kainar, Alma-Ata. (in Russian)
Sokolov II (1959) *Gazella subgutturosa* Guldenstaedt—Goitered gazelle. In: Pavlovskiy EN (ed) Fauna USSR, mammals: ungulates (orders Perissodactyla and Artiodactyla). Publication of the Academy of Sciences USSR, Moscow, pp 308–331. (in Russian)
Sokolov VE (1973) The skin cover of mammals. Nauka Press: Moscow, 487 pp. (in Russian)
Sokolov V (1990) Management of gazelles in Mongolia. Gnusletter 9(2):12
Sokolov II, Klebanova EA, Sokolov AS (1964) Morphology-functional features of locomotion organs in saiga and goitered gazelle. Proceedings of the zoological institute, vol 33, morphology of vertebrate animals. Nauka Press, Moscow-Leningrad, pp 319–348. (in Russian)

Sokolov VE, Dulamtseren S, Kholtolgu N, Orlov VN (1978) Ungulate rare species in great Gobian reserve (Mongolia), current status and perspectives. In: Geography and dynamics of flora and fauna in Mongolian People Republic, vol 10. Nauka Press, Moscow, pp 112–124. (in Russian)

Sokolova IB, Bondarev VI (1948) Proceedings of the Kazakh Academy of Sciences. Parasitology series 6:110–112. (in Russian)

Sokolova IB, Lavrov LI (1956) Fauna of helminths in domestic and wild ungulates in the Caspian Sea region. Proc Institute Zool Acad Sci KazSSR 5:105–111. (in Russian)

Sokov AI (1986) Distribution, number and population status in goitered gazelle in Tajikistan. In: Flint VE, Prisyazhnuk VE (eds) Conservation and perspectives of the number restoration of goitered gazelle in USSR. All-Union Scientific Institute of Nature Conservation and Nature Reserves, Moscow, pp 37–39. (in Russian)

Sokov AI (1989) Condition of rare ungulates and methods for protecting them in Tajikistan. In: Sokolov VE (ed) Ecology, morphology, exploitation and protection of wild ungulates. Academy of Sciences, Moscow, pp 79–81. (in Russian)

Sokolova IB, Boev SN (1950) Materials on helminth fauna of wild ungulates of Kazakhstan. Proceedings of Academy of Sciences of KazSSR, Parasitology series, 8.:226–233 (in Russian)

Soldatova NV, Grazhdankin AV (1989) Evaporative water losses in goitered gazelle (*Gazella subgutturosa*). Zool Zhurnal (Moscow) 68(4):102–109. (In Russian)

Soldatova NV, Salimov KhV (1997) The current status, number, distribution and conservation strategy of goitered gazelle population in the Bukhara area. In: Proceedings of nature reserves of Uzbekistan. Issue 2, Tashkent: 84–92 (in Russian)

Solomatin AO (1966) Trip in the Usturt and Magyshlak. Hunt Hunting Bus 5:42–44. (in Russian)

Sorokin PA, Soldatova NV, Lukarevskiy VS, Kholodova MV (2011) Genetic diversity and relations of the goitered gazelle (*Gazella subgutturosa*) groups from Uzbekistan, Turkmenistan, and Azerbaijan: analysis of the D-loop of mitochondrial DNA. Biol Bull 38(6):585–590

Stevenson-Hamilton J (1950) Wild life in South Africa. Cassel. London. 206 pp

Sun M, Gao X, Abdukadir A, Shao M (2002) Study trends about Gazella subgutturosa. Arid Zone Res 19(3):75–79. (in Chinese, English abstract)

Svizzero S, Tisdell C (2018) Desert kites: were they used for hunting or for herding? A review of the recent academic literature. J Zool Res 2:7–28

Tancheng D, Hongjun C, Yong C, Hongpan W, Lei H, Yan GE (2016) Genetic diversity and phylogenetic status of Gazella subgutturosa at the mountain Kalamaili ungulate nature reserve, Xinjiang. Acta Theriologica Sinica 36(1):77

Thesiger W (1994) Arabian sands. Motivate Publishing, Dubai, pp 1–306

Thouless CR, Grainger JG, Shobrak M, Habibi K (1991) Conservation status of gazelles in Saudi Arabia. Biol Conserv 58:85–98

Tsalkin VI (1956) Preliminary results of the study of faunal material from the excavations of Dzhebel made by a. P. Okladnikov. Proc South Turkmen Archaeol Expedition 7:220–221. (in Russian)

Underwood R (1982) Vigilance behaviour in grazing African antelopes. Behaviour 79(2–4):81–107

Van Soest PJ (1996) Allometry and ecology of feeding behavior and digestive capacity in herbivores: a review. Zoo Biol 15(5):455–479

Vassart M, Greth A, Dureand V, Cribiu EP (1993) Chromosomal polymorphism in sand gazelles (*Gazella subgutturosa marica*). J Hered 84:478–481

Vekua A, Lordkipanidze D (2008) The history of vertebrate fauna in eastern Georgia. Bull Georg Natl Acad Sci 2(3):149–158

Vereshchagin NK (1939) Goitered gazelle in Azerbaijan. Proc Zoological Institute Azerbaijan Branch USSR Acad Sci 10:89–90. (in Russian)

Vereshchagin NK (1947) Hunting and game animals of Caucasus. Institute of Zoology in Baku, Baku, pp 1–144. (in Russian)

Vereshchagin NK (1959) Mammals of Caucasus. History of fauna evolution. Publication of the Academy of Sciences USSR, Leningrad

Vereshchagin NK (1967) The mammals of the Caucasus. A history of the evolution of the fauna. Israel Program for Scientific Translations, Jerusalem, 816 pp. (translation from Russian)

Vesey-Fitzgerald D (1952) Wild life in Arabia. Oryx 1(5):232–235
Von-Wehrden H, Wesche K, Tungalag R (2006) Plant communities of the Great Gobi B Strictly Protected Area. Mongolian Journal of Biological Sciences 4, 3–17
Vrba E (1985) African Bovidae: evolutionary events since the Miocene. S Afr J Sci 81:263–266
Wacher T, Wronski T, Hammond RL, Winney B, Blacket MJ, Hundermark KJ, Bleidorn C (2011) Phylogenetic analysis of mitochondrial DNA sequences reveals polyphyly in the goitered gazelle (*Gazella subgutturosa*). Conserv Genet 12(3):827–831
Walther FR (1984) Communication and Expression in Hoofed Animals. Indiana University Press, Bloomington
Wang X, Schaller GB (1996) Status of large mammals in western Inner Mongolia, China. J East China Norm Univ Nat Sci 12:93–104
Wang XM, Sheng HL, Bi JH, Li M (1997) Recent history and status of the Mongolian gazelle in Inner Mongolia, China. Oryx 31:120–126
Williamson DT, Delima E (1991) Water intake of Arabian gazelles. J Arid Environ 21:371–378
Wronski T, Sandouka MA, Plath M, Cunningham P (2010) Differences in sexual dimorphism among four gazelle taxa (gazelle spp.) in the Middle East. Anim Biol 60:395–412
Xu W, Xia C, Lin J, Yang W, Blank DA, Qiao J, Liu W (2012) Diet of *Gazella subgutturosa* (Güldenstaedt, 1780) and food overlap with domestic sheep in Xinjiang, China. Folia Zool 61(1):54–60
Yanushevich AI, Aizin BM, Kydyraliev AK, Umrichina GS, Fedianina TF, Shukurov ED, Grebenuk RV, Tokobaev MM (1972) Mammals of Kirgizia. Ilim Publication House, Frunze. (in Russian)
Yılmaz B, Demircioğlu İ, Bozkaya F, İnce NG (2020) Three-dimensional tomographic reconstruction and morphometric analysis of skull in gazelles (*Gazella subgutturosa*). Ankara Üni Vet Fak Derg 67:161–168
Zarudniy NA (1903) The birds of Eastern Persia. Notes of Imperial Russian geographical Society, vol. 36 (2), Sant-Peterburg, p. 124. (in Russian)
Zehzad B, Kiabi BH, Madjnoonian H (2002) The natural areas and landscape of Iran: an overview. Zoology in Middle East 26:7–10
Zeng Z, Song Y (2001) Daily activity rhythm and time budget of golden takin in spring and summer. Acta Theriologica Sinica 21(1):7–13. (in Chinese, with English Abstract)
Zhevnerov VV (1975) The ethology and ecology of goitered gazelle during the rutting period in the Barsa-Kelmes Island (Aral Sea). Proc Acad Sci Kazakh SSR 1:26–33. (in Russian)
Zhevnerov VV (1984) Goitered gazelle of the Barsakelmes Island. Nauka Publishing House of Kazakh SSR, Alma-Ata. (in Russian)
Zhevnerov VV, Bekenov AB, Sludskiy AA (1983) Goitered gazelle—*Gazella subgutturosa* Güldenstaedt, 1780. In: Gvozdev EV, Kapitonov VI (eds) Mammals of Kazakhstan, Bovidae, vol 3 (3). Nauka Publishing House of Kazakh SSR, Alma-Ata, pp 11–54. (in Russian)
Zhiriakov VA (1976) Rare ungulates of the Alma-Ata Nature Reserve and their conservation. In: Rare mammals of the USSR fauna. Nauka Publication: Moscow (in Russian)
Zhirnov LV, Lushchekina AA (1992) The ungulates of the Eurasian arid zone (problkems of conservation and rational use). [In: Proceedings of the International Symposium "Ongulés/ /Ungulates 91", Toulouse, September 2-6, 1991. F. Spitz, G. Janeau, G. Gonzalez and S. Aulagnier, eds]. Société Française pour l'Étude et la Protection des Mammifères, Paris, Institut de Recherche sur les Grands Mammifères. Paris, Toulouse: 339–342
Zohary M (1973) Geobotanical foundations of the Middle East, volume1,2. Gustav Fischer Verlag, Stuttgart
Zverev MD (1948) To the question of running speed of some animals. Proc Alma-Ata Nat Reserve 7:145–147. (in Russian)

Feeding Ecology and Behavior

2

2.1 General Characteristics

The main food of gazelles is herbaceous vegetation that grows in their habitats. According to Heptner et al. (1961), the general list of plants eaten by goitered gazelles contains 35–40 species. Usually, these are species of various grasses and *Salsola* spp., but in some cases, there is a definite selectivity of the species composition of forage plants (Sokolov 1959). Based on the frequency of eating various types of plants, three groups of plants can be identified: main, secondary, and rarely eaten. The main group includes plants eaten by the gazelles in the first place, willingly and in large quantities. On Barsakelmes Island, for example, this main group includes 24 species from six families; 17 of them are used year-round. Plants eaten by gazelles in limited quantities, subject of their availability, were classified as secondary. This group of plants includes 35 species from 11 families on Barsakelmes Island. The group of rarely eaten plants includes those eaten only in some cases, or when there is a lack or inaccessibility of plants of the first two groups. These include 14 species from nine families in the conditions of Barsakelmes Island. It should be noted that such a division is to some extent conditional, since differences in the use of these plants by the gazelles, even of systematically similar plant species, are due to a number of factors and, above all, a change in plant chemistry associated with a change in phenological phases. In certain phenophases, plants can acquire a pungent odor, bitter taste, or an excessive salt content and thereby limit their digestibility (Zhevnerov 1984).

Throughout their habitat, goitered gazelles most commonly eat grasses (Poaceae: *Aeluropus*, *Agropyron*, *Aristida*, *Bromus*, *Cutandia*, *Panicum*, *Pennisetum*, *Poa*, *Sporolobus*), halophytes (Chenopodiaceae: *Anabasis*, *Camphorosma*, *Eurotia*, *Halocnemum*, *Halogeton*, *Halostachys*, *Haloxylon*, *Kochia*, *Kraschennekocina*, *Nanophyton*, *Noaea*, *Salsola*), composites (Asteraceae: *Artemisia*, *Centaurea*, *Helianthemum*, *Jurinea*, *Rhanterium*), legumes (Leguminosae: *Alhagi*, *Astragalus*, *Caragana*, *Halimodendron*, *Hulthemia*), buckwheats (Polygonaceae: *Atraphaxis*,

D. Blank, *The Behavioral Biology and Conservation of Goitered Gazelles (Gazella subgutturosa)*, https://doi.org/10.1007/978-3-031-83504-9_2

Calligonum), caltrops (Zygophyllaceae: *Nitraria*, *Peganum*), ephedras (Ephedraceae: *Ephedra*), borages (Boraginaceae: *Heliotropium*) gourds (Cucurbitaceae: *Citrullus*), leadworts (Plumbaginaceae: *Limonium*), and tamarisks (Tamaricaceae: *Tamarix*) (Blank 1990; Harrison and Bates 1991; Heptner et al. 1961; Mohamed et al. 1991; Mowlavi 1978; Sludskiy 1977; Thouless et al. 1991; Zhevnerov 1975; Zhevnerov et al. 1983). In agricultural areas, goitered gazelles eat alfalfa, fruits or shoots of barley (*Hordeum*), chick peas (*Cicer arietinum*), cotton (*Gossypium*), dates (*Phoenix dactylifera*), maize (*Zea mays*), melons (*Cucumis melo*), onions (*Allium cepa*), sugar cane (*Saccharum officinarum*), and wheat (*Triticum*) (Afanasiev et al. 1953; Blank 1990; Dementyev 1935; Heptner et al. 1961; Pitman 1922; Kingswood and Blank 1996; Farhadinia et al. 2009).

Plants containing a high percentage of water and protein are preferentially chosen and plants that are preferred in some areas may be neglected in others (Gorelov 1972; Heptner et al. 1961; Mohamed et al. 1991). The summer diet in Iran is 85.8% shrubs, 11.0% grasses, and 3.2% forbs. Available vegetation consists of 14.2% preferred plants, 41.3% neglected plants, and 44.5% avoided species (Mowlavi 1978). Goitered gazelles in captivity are fed alfalfa (*Medicago sativa*), oats (*Avena*), pelleted grain enriched with selenium and vitamin E, and provided with a sulfur-free salt block (Reed 1977). For some part of the year, goitered gazelles obtain water from green plants or snow, but they drink regularly if surface water is available (Heptner et al. 1988). During spring and early-summer, surface water is sought, particularly by lactating females; daily water intake is about 2–4 L (Heptner et al. 1961; Mowlavi 1978; Zhevnerov 1984).

The availability of water sources also affects the types of feed that the gazelles feed on. In areas where there is a shortage of reservoirs, they eat food with a high-water content. In habitats where water sources are available all year round, gazelles can eat relatively dry plant food and significantly expand their diet compared to gazelles that do not drink water (Sarukhanova 2021).

The weight of the contents of the stomach of gazelles ranges from 2.7 kg to 3.5 kg, which is 10.5–15.9% of their body weight (Heptner 1956). A goitered gazelle eats ca. 6.0 of forage per day, about 30% of its body weight; stomach contents weigh 2.5–4 kg (Heptner et al. 1961; Zhevnerov 1984). Forage consumed by a 4-month-old female was 362 g dry weight per day; and her daily energy expenditure was 1.657.5 kcal (Mowlavi 1978). Apparent digestibility (in %) of several species of forage plant was: minerals, 53.7; protein, 9.5 to 15.9; fat, 3.7 to 9.7; and crude fiber, 29.6 (Mowlavi 1978; Zhevnerov 1975). Means (and ranges) of nutritional values of forage plants ($n = 8$) were as follows: caloric value, 3.5 kcal/g (2.8–4.4); protein, 5.4 mg/g dry weight (0.6 t 13.7); water content, 64.1% (49.4–72.5); and inorganic content, 5.5% (2.1–9.7—Mohamed et al. 1991). Fecal pellets measure ca. 10 mm in length and 7–8 mm in thickness (Heptner et al. 1961) (by Kingswood and Blank 1996).

To nibble the grass, the fawns begin on the fifth, and more often on the eighth to tenth day, after birth, and after 30 days vegetation food occupies a significant place in the diet. At the age of 3–3.5 months, some lambs completely switch to plant nutrition and can live independently from their mothers (Zhevnerov 1984).

2.2 Barsakelmes Island (45°N 59°E: In Aral Sea; Kazakhstan)

2.2.1 Main Diet

Goitered gazelle food has been studied on Barsakelmes Island more than in any other areas. Goitered gazelles eat 71 plant species in this habitat, including shrubs and semi-shrubs: 18 dwarf shrubs and 12 dwarf semi-shrubs, and 41 herbs (Table 2.1). These plant species belong to 17 families. The first place belongs to the Poaceae (cereals; 18 species), the second place belongs to Chenopodiaceae (13 species), the third to Asteraceae (seven species), and the fourth belongs to Polygonaceae and Fabaceae (four species), followed by a family of Ephedraceae (three species) (Zhevnerov 1984).

In spring, the forage is most variable (85% of all eaten species), especially many herbs and cereals (38 species). The gazelles start the season by eating mainly all green shoots, including those that will not be eaten later, such as *Descurainia sophia* and *Galium verum*. Initially, the entire aboveground part of cereals is consumed, while later during its earing (the appearance of an inflorescence), the gazelles eat mainly inflorescences. The same happens with ephemeras and ephemeroides as *Tragopogon ruber*, *Scorzonera pusilla*, and *Taraxacum bicorne*, which are eaten wholly above ground and later only the inflorescences. In summer, the plant selection is still varied. Gazelles consumes 50 plant species, 24 are herbs and cereals, 12 dwarf semishrubs and dwarf shrubs, and 14 semishrubs and shrubs. The fundamental part of the goitered gazelle diet on the island are species from the family Poaceae (*Elytrigia desertorum*, *E. fragile*, *Eremopyrum orientale*, *E. triticeum*, *Poa bulbosa*), Polygonaceae (*Atraphaxis spinosa*, *Calligonum* spp.), Chenopodiaceae (*Haloxylon aphylla*, *H. persicum*, *Eurotia ceratoides*, *Kochia prostrata*), and Papilionaceae (*Alhagi pseudalhagi*, *Caragana grandiflora*) (Afanasiev et al. 1953; Zhevnerov 1984). The leaves of *Eurotia ceratoides* contain 78 mg/kg of vitamin C (Pankova 1949), and *Calligonum* has a high-water content (60–65%) (Morosova 1972). During summers of dry years, the species important in the diet of goitered gazelles are *Artemisia terrae-albae*, *A. arenaria*, *A. quinqueloba*, and *Anabasis salsa*. *Artemisia* species are highly nutritious, and some are juicy. On June 21, 1971, the content of a gazelle's stomach contained 25% *Alhagi pseudoalhagi*, 20% *Calligonum* ssp., 20% *Atraphaxis spinosa*, 15% *Elytrigia* ssp., 4% *Bromus* ssp., and 2% *Tamarix*. At the end of June into July, gazelles willingly eat the fruits of *Ephedra* ssp. and *Nitraria schoberi*. In July and August, the value of green shoots and leaves of *Calligonum* ssp., *Haloxylon* ssp., *Ceratocarpus arenarius*, *Atriplex tatarica*, *Chenopodium urbicum*, and *Aeluropus littoralis* and young reed shoots increases in their diet (Zhevnerov 1984).

In autumn, goitered gazelles eat herbs and cereals on Barsakelmes Island: 18 species of grasses, 12 dwarf semishrubs and dwarf shrubs, and 15 semishrubs and shrubs, which are species mainly from the families of Chenopodiaceae, Papilionaceae, and Polygonaceae. The portion of cereals starts to increase in October during its secondary vegetation. On October 10, 1974, the stomachs of two goitered gazelles were found to contain: 20% cereals, 15% *Ephedra*, 10% *Artemisia*

Table 2.1 The plant species eaten by goitered gazelle in the Barsakelmes Island

Plant species	Eatability degree												Earten parts
	January	February	March	April	May	June	July	August	September	October	November	December	
Family Ephedraceae													
Ephedra distachya	5	5	4	4	3	3	3	4	4	5	5	5	Aboveground parts, fruits
Ephedra intermedia	5	5	4	4	3	3	3	4	4	5	5	5	
Ephedra strobilacea	5	5	4	4	3	3	3	4	4	5	5	5	
Family Zosteraceae													
Zostera minor	4	4	4	2	2	–	–	2	2	2	4	4	The whole plant
Family Poaceae													
Agropyron desertorum	4	4	5	5	5	3	3	3	4	4	4	4	Leaves, stems, inflorescences (ears)
Agropyron fragile	4	4	5	5	5	3	3	3	4	4	4	4	
Aeluropus littoralis	4	4	3	3	3	3	4	4	4	4	4	4	Aboveground parts
Stipagrostis pennata	3	3	–	–	–	–	–	–	–	–	–	3	Stems, Ears
Bromus tectorum	–	–	5	5	5	4	4	4	3	4	4	4	Aboveground parts
Bromus lanceolatus	–	–	5	5	5	4	4	4	3	4	4	4	
Bromus oxyodon	–	–	5	5	5	4	4	4	3	3	3	3	
Calamagrostis epigejos	–	–	–	2	–	–	–	–	–	–	–	–	Leaves
Heleochloa schoenoides	–	–	–	–	5	5	5	–	–	–	–	–	Aboveground parts
Eremopyrum orientale	4	4	5	5	5	4	4	4	4	4	4	4	
Eremopyrum triticeum	4	4	5	5	5	4	4	4	4	4	4	4	
Elymus racemosus	2	2	–	–	–	–	–	–	–	–	–	–	Leaves
Phragmites communis	4	4	4	5	5	5	5	5	5	5	4	4	
Poa bulbosa	–	–	5	5	5	3	3	3	3	3	3	3	Leaves, panicles
Puccinellia gigantea	2	2	–	–	–	–	–	–	–	–	–	–	Leaves
Secale sylvestre	–	–	4	5	4	3	2	–	–	–	–	–	Aboveground parts

Stipa lessingiana	–	–	–	5	5	3	3	3	–	–	–	–	Leaves, stems, panicales
Stipa caspica	–	–	–	5	5	3	3	3	–	–	–	–	
Family Cyperaceae													
Bolboschoenus maritimus	–	–	–	4	4	3	3	3	3	–	–	–	Leaves, stems
Carex dimorphotheca	–	–	–	4	4	3	3	3	3	3	3	3	
Carex heterolepis	–	–	–	4	4	3	3	3	3	3	3	3	
Family Liliaceae													
Asparagus brachyphyllus	–	–	–	2	2	–	–	–	–	–	–	–	Aboveground parts
Gagea reticulata	–	–	–	4	4	–	–	–	–	–	–	–	
Tulipa buhseana	–	–	–	4	4	–	–	–	–	–	–	–	Leaves, flowers
Family Polygonaceae													
Atraphaxis spinosa	4	4	4	5	5	5	3	3	3	3	3	3	Branches, leaves, flowers
Calligonum acanthopterum	4	4	4	4	5	5	5	5	4	4	4	4	Branches, flowers, fruits
Calligonum aphyllum	4	4	4	4	5	5	5	5	4	4	4	4	
C. platyacanthum	4	4	4	4	5	5	5	5	4	4	4	4	
Family Chenopodiaceae													
Anabasis aphylla	4	4	3	3	2	2	2	3	3	4	4	4	Branches, flowers, fruits
Anabasis salsa	5	5	5	4	4	2	3	3	3	4	5	5	Aboveground parts
Atriplex ornata	5	5	4	3	3	3	3	3	3	4	4	5	Branches, leaves
Atriplex tatarica	5	5	4	3	3	3	3	3	4	4	4	5	

(continued)

Table 2.1 (continued)

Plant species	Eatability degree												Earten parts
	January	February	March	April	May	June	July	August	September	October	November	December	
Ceratocarpus arenarius	4	4	4	4	5	5	5	5	4	4	4	4	Aboveground parts
Chenopodium urbicum	4	4	4	2	2	2	2	2	2	2	2	3	
Eurotia ceratoides	4	4	4	5	5	5	5	4	4	4	4	4	Branches, leaves
Halocnenum strobilaceum	4	4	4	2	–	–	–	–	–	–	–	3	Branches
Haloxylon aphyllum	5	5	5	3	3	4	4	5	5	5	5	5	Branches, fruits, gallica
Haloxylon persicum	5	5	5	3	3	4	4	5	5	5	5	5	
Kochia prostrata	4	4	4	5	5	5	5	5	5	5	5	5	Branches
Salsola arbuscula	3	3	3	4	4	4	4	4	4	4	4	4	Branches, leaves, fruits
Salsola orientalis	3	3	3	–	–	–	–	–	–	–	–	3	Branches
Family Ranunculaceae													
Ceratocephala testiculata	–	–	2	2	2	4	4	–	–	–	–	–	Aboveground parts
Clematis orientalis	2	2	–	–	–	–	–	–	–	–	–	–	Branches
Family Capparaceae													
Capparis spinosa	–	–	5	5	5	5	5	5	5	–	–	–	Branches, leaves
Family Brassicaceae													
Chorispora tenella	–	–	–	–	4	4	–	–	–	–	–	–	Aboveground parts
Descurainia sophia	–	–	–	3	–	3	–	2	2	–	–	–	
Lepidium perfoliatum	–	–	–	2	2	2	–	–	–	–	–	–	
Family Fabaceae													
Alhagi pseudoalhagi	–	–	–	4	4	4	4	4	–	–	–	–	Branches, leaves, flowers
Astragalus transcaspicus	4	–	–	3	3	–	–	–	–	–	–	–	Branches, leaves

Caragana grandiflora	–	–	–	5	5	5	–	–	–	–	–	–	Branches, leaves, flowers
Eremosparton aphyllum	3	3	–	–	–	–	–	–	–	2	2	3	Branches
Family Zygophyllaceae													
Nitraria schoberi	–	–	–	4	4	4	–	–	–	–	–	–	Leaves, fruits
Frankenia hirsuta	5	5	3	3	3	3	3	3	3	3	3	5	Aboveground parts
Family Tamaricaceae													
Tamarix elongata	5	5	4	3	3	3	3	3	3	5	5	5	Branches
Tamarix hispida	5	5	4	3	3	3	3	3	3	5	5	5	
Tamarix laxa	5	5	4	3	3	3	3	3	3	5	5	5	
Family Plumbaginaceae													
Limonium otolepis	3	3	3	3	3	4	4	3	3	3	3	3	Leaves, flowers
Limonium suffruticosum	3	3	3	3	3	4	4	3	3	3	3	3	
Family Orobanchaceae													
Orobanche arenaria	–	–	–	–	2	2	–	–	–	–	–	–	Inflorescence
Family Rubiaceae													
Galium aparine	–	–	–	2	–	–	–	–	–	–	–	–	Aboveground parts
Family Asteraceae													
Artemisia arenaria	4	4	4	2	2	2	3	3	3	4	4	4	Aboveground parts
Artemisia quinqueloba	4	4	4	2	2	2	3	3	3	4	4	4	Aboveground parts
Artemisia terrae-albae	5	5	4	3	2	2	3	3	2	4	4	5	
Artemisia tomentella	4	4	3	3	2	2	3	3	3	4	4	4	
Scorzonera pusilla	–	–	–	4	4	–	–	–	–	–	–	–	
Tragopogon ruber	–	–	–	4	4	–	–	–	–	–	–	–	
Taraxacum bicorne	–	–	–	3	3	–	–	–	–	–	–	–	

5—eat willingly; 4—eat less willingly; 3—eat reluctantly; 2—eaten in isolated cases; 1—not eaten (adapted from Zhevnerov 1984) ssp., 40% shrubs, and 2% *Halocnemum strobilaceum*. In winter, the goitered gazelle ration was found to contain 38 plant species, including two tree species, 23 species of shrubs and semishrubs, and 12 species of herbs and cereals; among them were 11 species of Chenopodiaceae, seven from Poaceae, four species of Tamaricaceae, and four species of Polygonaceae. On the plains in low snow, remaining food is dominated by *Artemisia terrae-albae* and *Anabasis salsa*, and in years with a good autumn vegetation, cereals and various grasses also prevail. The volume of grasses from the total mass of the stomach contents in five gazelles was 10–30% in the winter of 1970–1971. If the snow is loose, then the gazelles will push it apart with their muzzles and eat mainly plants of an *Artemisia-Anabasis* complex. But when the snow is dense (winter 1971–1972), the animals graze in the sand among the bushes, where the snow is looser. There, in the little snow, 30–70% of the remaining foods were *Halocnenum* and *Ephedra*, but when the snow is deep, the role of tall plants (*Artemisia*, *Astragalus*, *Atriplex*, *Calligonum* and *Haloxylon*) sharply increases. In the stomachs of eight gazelles that fell in the harsh, snowy winter of 1971–1972, woody and shrubby plants were 50–70% and herbs were 8–11% of the total content (Zhevnerov 1984).

On Barsakelmes Island and in other areas, plants that are poisonous to livestock are often eaten by goitered gazelles, including *Peganum harmala*, *Zygophyllum turcomanicum*, *Salsola richteri*, *Ephedra distahya*, *E. intermedia*, *E. strobilacea*, *Anabasis aphylla*, *A. cretacea*, *Dodartia orientalis*, etc. For example, the goitered gazelle eats *Anabasis aphylla* in summer and winter, since the plant is rich in protein (on average 19.3%) and is highly nutritious, but this plant is often deadly for livestock (Zhevnerov 1984).

2.2.2 Nutritional Value of the Goitered Gazelle Forage

Among the plants that have the greatest forage value for the goitered gazelle in terms of the number of species, the first place belongs to cereals (Fam. Poaceae—nine species), the second place to the family Chenopodiaceae (six species), the third to Fam. Polygonaceae (four species), and the fourth to Ephedraceae (three species); next come the Fabaceae and Asteraceae families (one species each) (Zhevnerov 1984).

From the cereal family, wheatgrass (*Agropyron desertorum* and *A. fragile*), *Eremopyrum orientale* and *E. triticeum*, *Bromus lanceolatus*, *B. oxyodon*, *B. tectorum*, *Poa bulbosa*, and *Phragmites communis* have the main fodder value for goitered gazelles. Wheatgrass is eaten by the gazelles all year round, but it becomes most important during March or April, depending on the beginning of the growing season. In the early phases of vegetation, cereals contain the maximal amounts of carotene (ca. 100 mg /kg). As the plant develops, the carotene content decreases, and by the end of the growing season, only traces of provitamin A, which is the most important for animals, remain in the feed mass. During the same period, these plants are the richest in proteins, and there is a decrease in nitrogenous substances and an increase in fiber (Table 2.2). By the end of May, wheatgrass becomes coarser and is eaten less readily by gazelles. Its importance in the diet of gazelles increases again

Table 2.2 The chemical composition of some plants in different phases of their development, proportion (%) of absolute weight of dry matter (adapted from Zhevnerov 1984)

Vegetative phase	Ash	Protein	Fat	Fiber
Agropyron desertorum				
Earing	7.3	23.6	1.3	35.3
Fruiting	7.2	14.5	6.9	32.5
Flowering	8.2	15.0	1.8	32.4
Agropyron fragile				
Tiller	7.8	35.6	3.9	24.7
Earing	7.3	22.7	3.0	28.5
Flowering	6.6	19.1	2.8	35.3
Fruiting	6.3	14.4	2.3	28.9
Growing after haymaking	7.7	30.7	3.7	28.7
Seeds	16.1	25.1	2.3	20.1
Artemisia terrae-albae				
Vegetation	7.6	28.0	5.3	27.4
Budding	6.1	16.3	5.8	39.4
Flowering	6.7	21.4	7.7	37.8
Fruiting	4.5	6.3	5.1	38.3
Winter condition	7.0	6.1	3.0	43.1
Phragmites communis				
Before flowering	9.6	19.4	2.5	38.2
Growing after haymaking	13.9	21.1	2.2	32.2
Alhagi pseudoalhagi				
Before budding	8.2	22.2	2.1	22.7
Flowering	6.2	18.9	4.0	28.1
Fruiting	9.5	16.8	5.6	23.5
Poa bulbosa				
Vegetation start	11.7	51.2	3.8	21.8
Bulb forming	11.8	45.0	2.8	21.6
Drying	5.1	16.6	2.9	30.0
Summer dry-stand	3.1	6.0	1.3	38.4
Winter dry-stand	3.2	3.0	-	42.6
Bromus tectorum				
Stemming	11.2	49.4	4.8	17.1
Earing	7.8	32.1	2.0	27.7
Flowering	4.6	20.4	2.3	32.8
Dry-standing	7.1	13.6	1.7	30.6
Ephedra spp.				
Flowering	14.7	16.5	3.0	30.4
Winter standing	8.9	10.1	3.0	37.6
Haloxylon persicum				
Before flowering	12.9	26.3	2.0	18.0
Flowering	19.7	26.9	1.8	17.3
Winter standing	11.9	7.7	1.9	30.6

after autumn rains during its secondary vegetation. In years with good autumn vegetation, part of the green generative shoots goes under the snow, creating a certain vitamin supply for gazelles in winter (Zhevnerov 1984).

Mortuks (*Eremopyrum* sp.) are eaten by goitered gazelles all year round, but they play the greatest role from March to May. During this period, they have high nutritional value and at the beginning of flowering contain 22.3% protein, 17.9% fat, and 11.1% digestible protein. In June, the mortuks dry up and their fodder value drops somewhat. The stems and leaves of the *Eremopyrum* are quite well preserved in dry form, and the gazelles willingly eat them in autumn and winter (Kurochkina and Osmanova 1973).

Brome (*Bromus* spp.) acquires fodder value for the gazelle from the moment of their vegetation (March and April): in the initial phases of development, the *Bromus tectorum* contains 24.37 mg of carotene per 100 g of dry matter. In summer, gazelles eat the seeds and vegetative mass in dry form. The dry-standing *Bromus* is maintained until snowfall, but during the snowy period, it loses its fodder value (Morosova 1972).

Poa bulbosa acquires greatest fodder value in April and May. At the beginning of the growing season, it contains up to 30% protein and at the end 16%. Even in a dry state, the nutritional value of the *Poa* remains quite high. Hundred kilogram of this plant during the growing season contains about 2–3 kg of protein and contains the same amount during the drying period. At the beginning of earing (cereal flowering), the plant contains 35.5 mg of carotene per 100 g of dry matter or 10.2 mg per 100 g of matter with initial humidity. In summer, after drying, animals willingly eat *Poa bulbosa* until autumn (Morosova 1972).

Cane (*Phragmites communis*) is used by goitered gazelles throughout the year, but its consumption varies slightly by seasons; the green parts are most readily eaten at the beginning of the growing season. Cane contains a significant amount of sugary substances, including 6.4% of soluble carbohydrates. Green grass contains from 33.1 to 51.5 mg/g of carotene (Larin et al. 1929).

Anabasis salsa plays a significant role in the diet of the gazelle and is readily eaten in all seasons of the year. The nature of its consumption has a well-pronounced cyclicity associated with the phenophases of the plant as well as a number of external factors. This plant is eaten in the autumn and winter period and in the early stages of its vegetation (April–May). But during the flowering period and before the end of fruiting (June into September), they are only rarely used by gazelles, in spite of the fact that this plant still has good nutritional value during this time. In the flowering phase, this plant contains 7.3% proteins and 11.4% fiber (Kubanskaya 1956). During budding and before the end of fruiting, its fodder value drops. Poor use of *Anabasis salsa* in the summer is determined by the chemistry of the plant, due to the change of phenological phases. It is noted that the accumulation of alkaloids in this plant occurs from the beginning of vegetation until its end (from early spring to autumn). There is a certain relationship between the amounts of alkaloids and the phenological phases of the plant. In noticeable amounts, alkaloids begin to accumulate after the flowering of plants, especially by the time the seeds ripen. After the seeds ripen, the amount of alkaloids decrease. The nutritional value of

Anabasis salsa is high and contains from 6.56% to 10.18% protein, 4.66% to 7.36% carbohydrates, and 1.55% to 2.21% fat (Chaplina 1959; Zhevnerov 1984).

Kohia prostrata is eaten by goitered gazelles throughout the year, most willingly from April to October. The fruits and seeds of this plant contain 34.9% protein and 8.3% fat (Larin 1951).

Ceratocarpus arenarius in the diet of the gazelle acquires significant importance during its growing season. In its fruiting phase, this plant contains 11.7% protein and 1.3% fat. The amount of carotene ranges from 7.0 to 36.7 mg/kg. In winter, *Ceratocarpus arenarius* does not have a noticeable forage value. After its fruit ripening in August, the plant breaks at the root neck and is carried away by the wind. These dried plants partially linger in shrubby thickets and are eaten by gazelles along with their feeding on shrub branches (Kurochkina and Osmanova 1973).

Gray teresken (*Eurotia ceratoides*) is used by gazelles all year round. Leaves and twigs are especially well eaten from the beginning of the growing season to flowering. The green leaves of teresken contain 78 mg /kg of vitamin C, 12.9–53.2% of protein, and 0.8–4.8% of fat (Zhevnerov 1984).

Saxauls (*Haloxylon* spp.) are eaten by goitered gazelles throughout all seasons of the year and is especially important in the snowy period. In spring, during the reign of ephemeral vegetation, the forage value of the saxaul is noticeably reduced; in June it increases again. When seeds appear in the saxaul, the gazelles primarily eat those. During the autumn and winter period, seeds and galls are completely eaten at a height accessible to gazelles and collected from the ground. One hundred kilogram of absolutely dry saxaul feed contains 3.7 kg of digestible protein (Morosova 1972).

Wormwood (*Artemisia terrae-albae*) in its nutritional value is considered one of the best nutritious foods. On the island, wormwood begins to develop in March and ends its growing season at the end of October. In spring there is an intensive growth of twigs and leaves. In early summer, budding occurs. Flowering begins in mid-September, and fruiting ends at the end of October. This plant is eaten well by the gazelles in late autumn and winter. During the spring growing season and during the budding and flowering period in summer, wormwood consumption is very limited by goitered gazelles, even though it has the greatest nutritional value during these periods. The phenophases at these times create the presence of a sharp smell and bitterness which causes the gazelles to avoid this plant; after the frosts in autumn, the smell becomes less sharp and the bitterness decreases. In dry years, gazelles readily eat wormwood in the summer as well, because the generative phase of the plants is depressed, and they do not go through a normal development cycle. Wormwood contains 0.975% calcium and 0.079% phosphorus. The copper content in wormwood depends on the characteristics of the plant biology. Wormwood has two distinct maxima in this respect, one in spring and autumn, and two minima in summer and winter. The condition of the animal's coat depends on the presence of copper, since copper participates in the processes of pigmentation and synthesis of wool carotene (Zhevnerov 1984).

Ephedra shrubs are eaten by gazelles all year round, except when flowering period. It acquires its greatest dietary importance in the winter and early spring. The nutritional value of *Ephedra* species in winter is exceptionally high. One hundred kilogram of absolutely dry food contains 10.8 kg of digestible protein. It is especially valuable that the plant in winter contains many vitamins and a whole complex of plant substances that favorably affect the gazelle body (Morosova 1972).

The goitered gazelle also readily consumes *Calligonum* spp. all year round. In spring and summer, gazelles willingly eat the green twigs and fruits, and in winter, they consume the annual shoots. In the flowering phase, *Calligonum* spp. contains 37.9% protein and 3.6% fat (Kurochkina and Osmanova 1973). Juiciness and high moisture content in *Calligonum* spp. make this food especially valuable during the hot period; in the green mass of this shrub at the beginning of the growing season, moisture is 90–95%, and in July moisture is 55–65%. Green twigs have high nutritional value: 100 kg of absolutely dry food contains 3 kg of digestible protein (Morosova 1972).

This shows that the same plant species in a certain phenophase can change its nutritional value and switch from a main to secondary forage and vice versa (Table 2.2; Zhevnerov 1984).

2.2.3 Features of the Seasonal Changes in Goitered Gazelle Feeding

The change in the forage diet of goitered gazelles is primarily determined by the seasonal development of individual plant species and, consequently, phytocenosis, which in turn is subject to the annual course of meteorological conditions. In this regard, the diet of the gazelle should be assessed against the background of seasonal dynamics of plant communities and individual species, considering pheno-climatic seasons. Pheno-climatic season refers to a period of time during the year that is characterized by a certain regime of temperature, precipitation, and light and is accompanied by corresponding changes in the phenological state of flora and fauna. For Barsakelmes Island, seven pheno-climatic seasons are differentiated within the three seasons of spring, summer, and fall: early spring, the height of spring, the beginning of summer, sultry summer, the second half of summer, the beginning of autumn, and the height of autumn (Zhevnerov 1984).

Spring Period During this season, two periods are well distinguished—early spring and the height of spring. Early spring (late March to early April) is associated with a steady transition of the average daily temperature through 0 °C, and its character is determined by the rate of temperature rise and depends on the snow melting rate. Usually this happens already from the end of February, but in some years with a snowy winter (1964, 1972), snow melts only in the middle of March, which leads to a delay in vegetation growth. In other years (e.g., 1958 and 1970), after a low-snow winter and early melting of snow cover, vegetative development in plants starts much earlier than usual. So, in 1958, in the middle of February, *Eremopyrum* spp. and *Poa* spp. began to vegetate, and on March 4, the leaves of *Rheum tataricum* and *Artemisia terrae-albae* appeared (Zhevnerov 1984).

The spring season is characterized by the beginning of the ephemera and ephemeroids' vegetation. The monotonous grayish-brown background of *Anabasis-Artemisia* complexes begins to revive with an emerald brush of *Eremopyrum* spp. and *Carex* spp. A grayish-green veil covers the bushes of *Artemisia terrae-albae*. Sand complexes continue to retain gray winter shades and greenery appears only in the lower tier. In early spring, cold weather is common, and snow can return, delaying the development of plants; in some years, a sharp cold snap can happen even later. For example, in 1970, the vegetation of cereals began in mid-March. Due to good soil moisture content and positive temperatures, the plants developed favorably until mid-April. Frosts struck from April 19 to 23, which adversely affected the further course of the growing season. In short, the period between the snow cover melting and the mass vegetation start of ephemera and ephemeroids, determined by the degree of temperature increase and soil moisture content, may have significant differences in different years (Table 2.3; Zhevnerov 1984).

In the early spring, the diet of the gazelle has minor differences in composition. Basically, the volume ratio of eaten plants changes. With the melting of the snow cover, the gazelles spread out more evenly throughout the island. Due to this, the vegetation of lowland pastures, consisting of *Artemisia terrae-albae*, *Anabasis salsa*, and dry small grasses, becomes essential in their forage. In sand ridges, shrubs, and semishrubs, as well as dry parts of large-stemmed plants, continue to be used by the gazelles as their feed. The stomach contents of two gazelles, examined in early March 1972, consisted of shrubs (*Tamarix*, *Calligonum*, *Haloxylon*) and semishrubs (*Ephedra*, *Artemisia*, *Anabasis salsa*), as well as dry grasses and herbs. *Ephedra* was predominant in the volume ratio in the stomachs. At this stage of the early spring period, the gazelle is typified by a small set of main foods (10–12 plant species), among which shrubs and semishrubs are predominant. The volume of the eaten mass of various grasses during early spring increases by 15–20% compared to the snowy period (Zhevnerov 1984).

In years with good vegetation development in the preceding autumn and immediately after the snow cover melts, gazelles begin to feed on green vegetative shoots of *Artemisia*, *Agropyron*, *Stipa*, and *Eremopyrum* along with other foods, which are

Table 2.3 The melting of the snow cover and the beginning of vegetation of some plant species on the Barsakelmes Island in some years (adapted from Zhevnerov 1984)

	Snow cover melting		
	17.02.1970	20.03.1973	12.03.1974
Plant species	Vegetative development start		
Bromus tectorum	22 March	4 April	27 March
Stipa lessingiana	30 March	2 May	15 March
Eremopyrum orientale	17 March	8 April	21 March
Eremopyrum triticeum	21 March	10 April	23 March
Tulipa buhseana	18 March	2 April	21 March
Agropyron desertorum	19 March	5 April	21 March
Artemisia terrae-albae	20 March	7 April	21 March
Anabasis salsa	24 March	15 April	28 March

of great importance in replenishing the lack of vitamins in the gazelle body after winter. As the result of this situation, the nature of gazelle distribution on pastures is changing by concentrating on the southern slopes of hills, ravines, and inter-row spaces of fixed sands. The precipitation amounts in August and September of the previous year determine the course of the phenological development of plants. Lack of moisture leads to depression, and many plants have a loss of secondary autumn vegetation completely. So, in 1963, the absence of precipitation in autumn delayed the beginning of secondary vegetation until the end of the autumn season. *Agropyron desertorum* begins to vegetate in early October, and by early winter, it has reached only 4–8 cm *Bromus tectorum* 2–4 cm, while *Artemisia terrae-albae* and *Stipa* spp. do not vegetate at all (Zhevnerov 1984).

Autumn vegetation is not typical for ephemera and ephemeroids on Barsakelmes Island. In most cases, these plants behave like spring plants which differs from southern deserts, where the same species behave like winter crops. However, having a high ecological plasticity, in some years, they behave like winter crops on the island. In 1957, after a heavy downpour on August 4, the vegetation of *Bromus tectorum*, *Eremopyrum*, *Poa bulbosa*, and others started their vegetation that continued until snowfall in December. Representatives of other life forms have no less plasticity. Depending on the timing of precipitation, secondary vegetation begins with *Agropyron desertorum*, *Stipa lessingiana*, *Astragalus* spp., *Artemisia terrae-albae*, and *Anabasis salsa* (Kuznetzov 1976). With the beginning of the ephemera mass vegetation, which is usually observed at the end of March, the main diet of the gazelle consists of the vegetative parts of ephemera. Gazelles give special preference to cereals, of which the entire aboveground part is eaten. At the beginning of the growing season, the protein in plants is at its highest content of ephemera approaching and even exceeding 30% (Zhevnerov 1984).

In April, the group of herbaceous plants in the gazelle diet occupies first place (36 species) and accounts for 58.2% of all plant species used during this period (Table 2.4). Along with herbaceous plants, semishrubs and shrubs remain a constant component in the diet, with some plant species used at the level of the winter period, but with the volume of eaten mass reduced by almost three times. In the stomach contents of gazelles that fell in winter, shrubby vegetation accounted for 50% to 70%, and perennial and annual grasses only 8–11% of the total. The stomach contents of the gazelles, examined in April, consisted of 70% of the green mass of ephemera and about 20% of the remains of shrub species, mainly *Calligonum* and *Haloxylon*; the rest of the volume consisted of the perennial parts of *Halocnemum strobilaceum* and *Anabasis aphylla* (Zhevnerov 1984).

Considering the fact that with the abrupt transition of gazelles to green food, they often have diarrhea; therefore, the use of *Anabasis aphylla* and *Calligonum* spp., along with ephemera, is probably not by accident. Most likely, during this period, these plants play a preventive role in gastrointestinal disorders, supplying tannins and salts to the animal's body. This is confirmed by observations of tamed gazelles, which, with digestive disorders, ate twigs of leafless *Anabasis aphylla* in bulk. *Calligonum* spp. contain up to 5–7% of tannins, and *Halocnemum strobilaceum* up to 32.7% of table salt (Keller 1940; Pavlov 1947).

Table 2.4 The number of plant species which are eaten by goitered gazelles during a year (in parenthesis proportion (%) from total number of eaten species) (adapted from Zhevnerov 1984)

Months	Herb grasses	Dwarf shrub and dwarf semishrub	Semishrubs and shrubs	Total number of eaten species
January	12 (30.0)	12 (30.0)	16 (40.0)	40
February	12 (30.0)	12 (30.0)	16 (40.0)	40
March	17 (38.6)	11 (25.0)	16 (36.4)	44
April	36 (58.2)	10 (16.0)	16 (25.0)	62
May	38 (61.2)	8 (13.0)	16 (25.8)	62
June	24 (49.1)	11 (22.4)	14 (28.5)	49
July	24 (51.0)	11 (23.4)	12 (25.5)	47
August	20 (46.5)	12 (27.9)	11 (25.6)	43
September	19 (45.3)	12 (28.5)	11 (26.1)	42
October	18 (40.0)	12 (26.7)	15 (33.3)	45
November	17 (38.6)	12 (27.3)	15 (34.1)	44
December	15 (35.7)	12 (28.6)	15 (35.7)	42

The second period of a spring—the height of spring (mid-late-April-early-May)—is characterized by the transition of the average daily temperature through +5 °C and usually an increase in precipitation. Mass vegetation of dwarf shrubs, semishrubs, shrubs, herbaceous perennials, and annuals begins. Ephemera and ephemeroids bloom, and *Artemisia* associations acquire a uniform bluish-green shade. In the ravines and small dry riverbeds, the *Caragana grandiflora* is covered with bright yellow flowers. In the sands, the bushes of *Tamarix*, *Atraphaxis* and *Calligonum* are bright green. By the end of spring, a lack of moisture begins, and yellow spots of drying vegetation appear. However, the typical plant activity during the second period of spring is not always repeated annually. In very dry years, in many associations of *Anabasis salsa* and *Artemisia terrae-albae*, these aspects are not expressed at all, since the generative phase of most ephemera and ephemeroids is suppressed. For example, such phenomena were observed in 1965 and 1969, when the absence of these characteristics were due to the complete absence of precipitation and strong drying winds (a combination of soil and atmospheric drought). The height of spring is usually the most favorable time of year for gazelles according to feeding conditions. During this period, gazelles use 62 species of plants for food, which is 85% of the entire list of foods they consume, with the number of herbaceous plant species used increasing to its maximum of 38 (61.2%) (Zhevnerov 1984).

With the increasing role of ephemera in the diet of gazelles, the edibility of some other plants, such as *Artemisia* (*A. terrae-albae*, *A. quinqueloba*, *A. arenaria*) and *Anabasis salsa* decreases. In the use of shrubby and semishrubby vegetation by animals, a characteristic selectivity appears. Among them, in terms of the frequency of eating and the volume of plant mass, the first place is taken by *Calligonum*, *Atraphaxis*, *Caragana*, whose young leaves and inflorescences are bitten off. The role of *Tamarix* and *Haloxylon* is noticeably reduced, and in some years, they generally drop out of the list of the goitered gazelle diet during this period. The gazelles are especially willing to start using the *Calligonum* spp. from the end of May, when the ephemera dry up and the cereals get coarser, thus partially compensating for the need for juicy foods. The moisture content in the green mass of *Calligonum* at the beginning of the growing season is 90–95% (Morosova 1972).

The appearance of green shoots leads to the redistribution of gazelles on the island, where they are concentrated on the slopes of gullies, hollows, and depressions. The height of the vegetation cover in these locations is twice as high as on the plain. Here, there are shoots of *Poa bulbosa*, *Eremopyrum orientale*, and *E. triticeum*, the leaves of which can grow up to 3–5 cm long, and the height of the shoots of *Agropyron desertorum* can reach 10–18 cm, whereas on the plain, the same species are only the size of the early developmental plant growth (Zhevnerov 1984).

With the appearance of the first green shoots, gazelles bite off everything in a row. In the initial stage of vegetation, they eat *Asparagus brachyphyllus*, *Galium aparine*, and *Descurainia sophia*, which subsequently do not eat or only eat after they have dried out. In cereals, the entire aboveground part is initially eaten, and during the earing period, panicles and spikelets are preferred. As the leaves of *Agropyron*, *Bromus*, *Eremopyrum*, and *Poa* start to grow, gazelles begin to prefer the latter, and with the appearance of spikelets in *Eremopyrum*, they primarily eat them. In such ephemera and ephemeroids as *Tragopogon ruber*, *Scorzonera pusilla*, and *Taraxacum bicorne* in the initial stage of vegetation, the entire aboveground part is eaten, and with the appearance of inflorescences, only the latter are bitten off. *Descurainia sophia* is used by goitered gazelles at the beginning of the growing season and after drying. During the flowering and fruiting period, this plant is not eaten at all, though at this stage, this plant contains 0.8–0.9% volatile oils (Zhevnerov 1984). *Ceratocephala orthoceras* is occasionally eaten in the initial stage of vegetation. During the flowering period and the formation of fruits, the goitered gazelles bite off only the leaves. After the fruit ripens and the plant dries, gazelles willingly eat its entire aboveground parts, despite the fact that the seeds have sharp thorns. During the flowering period, the hornbill (*Ceratocephala orthoceras*) contains the slightly toxic substance of protoanemonin, a significant consumption of which can cause poisoning in animals (Morosova 1972).

In the spring period goitered gazelles have the opportunity to choose the best pastures with a high selectivity of food and a specific nature of plants consumption. Pasture vegetation abounds in vitamins and nutrients. This time coincides with the most important biological processes in the life of the gazelle—spring molting, birthing, and suckling of fawns (Zhevnerov 1984).

Summer Period This period on the island covers the time from the end of May to the end of September. According to the nature of development dynamics of plant communities, it is possible to split this period into three parts. The first part is the early summer (end of May to mid-June) when ephemera and ephemeroids bear fruits and dry up and *Tamarix*, *Atraphaxis*, *Ephedra*, and *Calligonum* begin to bloom. The season is associated with the transition of the average daily temperature through +15 °C, a decrease in the amount of precipitation, and the onset of the dry season. The second part is the sultry summer (end of June to July), which is the hottest and driest time of the year with average daily temperatures of more than +22 + 23 °C. The last group of plants is blooming, predominately sand dwellers and plants in habitats with close subsurface waters (*Clematis orientalis*, *Alhagi pseudalhagi*, *Phragmites communis*). Most perennial species enter a summer dormancy period, which is characterized by the absence of an accumulation of organic matter and a delay in the passage of phenological phases (*Artemisia terrae-albae*, *Anabasis salsa*, *Haloxylon* spp., *Ephedra distachya*). Some species at this time lose leaves or green twigs (*Agropyron desertorum*, *Astragalus trascaspicus*, *Calligonum* spp.); shrubs begin to bear fruits (*Atraphaxis*, *Nitraria*, *Calligonum*, *Ephedra*); and most dried stems of ephemeroids break off from the roots and are carried by the wind (Zhevnerov 1984).

At the end of May into early June, the ephemera bear fruit and dry up. Of the cereals in early summer, only some of them remain green, but their nutritional value decreases. Drying of plants leads to the disappearance of vitamins and carotene in their content and the amount of protein drops from 18.37% to 6.70% (Mikheev 1959). Also, the fiber content increases. In the ephemera group, fiber amounts increase from 13.6% at the beginning of the growing season to 46.7% during the dry season. Protein digestibility is dropping sharply (Morosova 1972) and the humidity of plants also changes significantly. For example, during the growing season, the humidity of the *Eremopyrum* is about 40%, but after drying, it is about 3%. Most ephemera and ephemeroids are subject to such change with the differences mainly related to drying plants. So, in 1973, *Eremopyrum* began to dry up on June 20, *Bromus* on June 25, and *Agropyron* on July 25. In 1974, the drying of these species began on June 8, June 4, and July 10, respectively (Zhevnerov 1984).

At different stages of plant vegetation, the nutrient ratio varies both in the plant as a whole and in its various parts. Spikelets and seeds contain the largest amounts of proteins and fats in relation to the vegetative parts of plants. After the plants dry out, these organs become the main reserves of nutrients. At the end of May to June, cereals (*Agropyron*, *Eremopyrum*, *Bromus*) continue to make up a significant part of the goitered gazelle's diet. By this time, the stems of plants are getting coarser, and gazelles bite off mostly spikelets. *Agropyron desertorum* seeds contain more protein than oats (*Avena*). As the ephemera shrinks, while shrubs, semishrubs, and large-stemmed plants become important as a source of vitamins. During this period, gazelles willingly eat the green parts of *Calligonum*, *Atraphaxis*, *Caragana*, *Kochia prostrata*, *Eurotia ceratoides*, and *Alhagi pseudalhagi*. They also eat the flowers of *Caragana* and *Atraphaxis* in large quantities. Shoots and leaves of shrubby vegetation have high nutritional value, and some of them are juicy with a high moisture content. The green mass of *Eurotia ceratoides* contains up to 75.2% water and 78 mg/100 g vitamin C in the leaves (Zhevnerov 1984). The amount of carotene in the *Alhagi pseudalhagi* can reach 16.1 mg/100 g, the moisture content in the green mass of the *Calligonum* is 90% to 95% (Morosova 1972). The contents of a gazelle's stomach, examined on June 21, 1971, contained *Alhagi pseudalhagi* at 25% of the total volume, 20% *Calligonum*, 20% *Atraphaxis*, 15% *Agropyron*, 10% *Eremopyrum*, 4% *Bromus*, 2% *Tamarix*, and 2% *Artemisia* (Zhevnerov 1984).

At the end of June into July, when shrubs and dwarf shrubs begin to bear fruits, gazelles especially readily eat the fruits of *Ephedra* and *Nitraria*. The content of vitamin C in *Ephedra* fruits reaches 160 mg/100 g (Zhevnerov 1984). By this time, most of the dried stems of ephemera and ephemeroids are torn off from the roots and carried by the wind, losing their significant feed value for the gazelle; however, the role of *Artemisia* and *Anabasis* increases. The degree of ingestion of *Artemisia terrae-albae* and *Anabasis salsa* by goitered gazelles is determined by climatic factors and the vegetation stage of these plants. In well-moistened years, the consumption of *Anabasis* in summer months is poor, and during the flowering period (June), its use by gazelles is random. In dry years, its role in the nutrition of the gazelle increases. Eating *Artemisia terrae-albae* has an identical character. During budding (June) and flowering (September), this plant is rarely eaten by the goitered gazelle, and in the rest of the summer, only occasionally. In dry years, the goitered gazelles eat these plants readily during the entire summer period, since the generative phase of them is suppressed and they do not have a full cycle of development (Zhevnerov 1984).

In summer, gazelles try to use everything, even small areas of pastures, with green vegetation. They are constantly found among the thickets of *Caragana*, where *Agropyron* remains green longer, and among the thickets of cappers (*Capparis spinosa*), the leaves, buds, and fruits of which are willingly consumed. Capper fruits contain 18% protein and 36% fat. The content of vitamin C in the capper buds reaches 100 mg to 150 mg/100 g (Zhevnerov 1984).

In July to August, the volume of green mass of shrubs consumed contains *Calligonum* and *Haloxylon*, some plants from the family Chenopodiaceae (*Atriplex tatarica*, *A. ornata*, *Chenopodium urbicum*, *Ceratocarpus arenarius*), and the young shoots of *Aeluropus littoralis* and young shoots of reeds increases. All these plants have good forage qualities. For example, *Aeluropus littoralis* is close to alfalfa in its starch equivalent (Zhevnerov 1984). In the flowering phase (July), *Aeluropus littoralis* contains more protein than the average for cereals (11%) (Larin 1951). Reeds contain a significant amount of sugary substances, including 6.4% soluble carbohydrates (Zhevnerov 1984). In the contents of the stomachs of two gazelles examined in August 1975, woody and shrubby plants (*Haloxylon*, *Calligonum*, *Tamarix*, *Atraphaxis*) were 25–40%, perennial and annual herbs (*Atriplex*, *Aeluropus*, *Limonium*, *Chenopodium*) at 20–40%, dwarf semishrubs and dwarf shrubs (*Artemisia*, *Anabasis salsa*, *A. aphylla*, *Ephedra*) were 7–12% of the volume of plant mass (Zhevnerov 1984).

The second half of summer (August and September) is a hot and dry period. However, a decrease in temperature is already beginning and the amount of precipitation is increasing, which determine the further course of the phenological development of plants. At the end of summer, depending on the timing of precipitation, cereals begin a secondary vegetation, and their role in the nutrition of gazelles increases again. Gazelles begin to appear more often in depressions, along the slopes of gullies and in the coastal strip, where the vegetation of cereals is most intense. Depending on the timing of precipitation, secondary vegetation starts in *Agropyron desertorum*, *Artemisia terrae-albae*, *Anabasis salsa*, and *Limonium suffruticosum*. In summer, the food of gazelles is quite diverse. They eat 50 species of plants, including 24 herbaceous species, 12 dwarf semishrubs and dwarf shrubs, and 14 semishrubs and shrubs (Table 2.4). The main source of the summer diet of the gazelle consists of plants of the family of Poaceae (*Agropyron*, *Eremopyrum*, *Poa*, *Aeluropus littoralis*), Polygonaceae (*Atraphaxis spinosa*, *Calligonum*), Chenopodiaceae (*Haloxylon*, *Eurotia ceratoides*, *Kochia prostrata*), Fabaceae (*Alhagi pseudalhagi*, *Caragana grandiflora*), and Tamaricaceae. The lack of moisture leads to a depression in plant growth and in many plants to the complete absence of secondary vegetation. In dry years during the summer period, plants of the family Asteraceae and mainly *Artemisia* species acquire significant importance in the diet of the gazelle (*Artemisia terrae-albae*, *A. quinqueloba*, *A. arenaria*) as well as *Anabasis salsa* (Zhevnerov 1984).

Autumn Period The beginning of autumn (in October) is characterized by a sharp increase in precipitation and an average daily temperature of 15–10 °C. This is the transition from summer to autumn. Under favorable conditions, *Anabasis-Artemisia* complexes come to life, dominant shrubs begin to grow, and annual species of *Salsola* stand out with dark red fruits. Fresh green leaves of *Agropyron* make their way through the strawy background of cereals. In places, the young growth of *Eremopyrum*, *Poa*, and *Bromus* begins to develop intensively. The height of autumn (late October into November) is accompanied by a drop in the temperature during the day to 0 °C. The vital processes of plants slow down, and the mass fading of leaves and branches begins. The plants of the sands are completely exposed and only gray branches of saxaul and tamarix bushes stand out against their yellowish-brownish background. Ephedra retains its green twigs a little longer, on which only their tops fall off. Semishrubs (*Anabasis salsa*, *Artemisia terra-albae*) begin to go bare and *Limonium*, part of *Frankenia* and *Halocnemum*, stay under the snow cover with their green leaves. In some years, significant cold spells are possible as early as the beginning of the season. However, the average daily temperature steadily passes through 0 °C by the end of November (Zhevnerov 1984).

The autumn diet of the gazelle is a transition from their summer forage to their winter diet. Depending on the climatic year conditions over the year and the vegetative state of plants, autumn may reflect the features of summer or winter nutrition. In the autumn period, goitered gazelles use 18 herbaceous species, 12 dwarf semishrubs and dwarf shrubs, and 15 semishrubs and shrubs (Table 2.4). Plants from the family Chenopodiaceae, Fabaceae, and Polygonaceae still predominate in their diet (Table 2.1). The cereals' portion of the goitered gazelle diet is determined by the conditions of their secondary vegetation. It usually increases in October during the maximum autumn regrowth. At this time, gazelles move widely around the island, noticeably concentrated in areas of pastures with young cereal growth. As in early spring, they bite off all the green shoots in a row. In the stomach contents of two goitered gazelles examined on October 10, 1974, cereals accounted for 30% of the total plant mass, ephedra 15%, wormwood 10%, woody and shrubby plants 40%, and *Halocnenum strobilaceum* 2%. The wormwood group continues to be dominated by *Artemisia quinqueloba* and *A. arenaria*, and among woody and shrubby plants are found *Haloxylon* and *Tamarix*. During the autumn period, gazelles often feed in the growing areas of annual *Salsola*, biting off their upper shoots and fruits. In contrast, during the dry years (such as 1970 and 1971), as early as October, gazelles concentrate in bumpy fixed sands, switching mainly to feeding on shrubs and large-stemmed plants. Ephedra plants are eaten in bulk, and in addition to annual shoots of saxaul, they eat seeds and galls. The role of *Artemisia* spp., *Anabasis salsa*, as well as dry grasses and herbs increase in such years (Zhevnerov 1984).

In the second half of November, the goitered gazelle rutting season begins, and most of the mature individuals, regardless of the climatic conditions of the year, are dispersed across the flat areas of the island. During the rutting season, gazelles feed mainly on the plants of an *Anabasis-Artemisia* complex (Zhevnerov 1984).

Winter Period This period includes December to February. The total range of plants used by goitered gazelles during these months is 37 species belonging to nine families, of these are two woody species, 16 dwarf shrubby and dwarf semishrubby species, seven shrubby species, and 12 perennials, annual grasses, and herbs. The main part of the feed consists of plants from the Chenopodiaceae family (11 species, 29.7%), Poaceae (seven species, 18.9%), and Tamaricaceae and Polygonaceae (four species each, 10.8% total). The remaining 11 plant species belong to the families of Asteraceae, Fabaceae, Frankeniaceae, and Zosteraceae, except *Ephedra distachya* and *E. intermedia*, and *Artemisia terrae-albae* which can be classified as secondary and rare foods. It should be noted that such divisions are to some extent conditional, since the winter diet of the gazelle has a number of features associated with meteorological factors and the forage availability (Zhevnerov 1984).

According to its morphological features, the goitered gazelles are poorly adapted for foraging under the snow with their front hooves like the kulan and saiga living on the island do, and only very rarely use their hooves to shovel snow. In addition, the structure of the hair covering their bodies does not have the density like that of the saiga and poorly resists the blowing wind. Therefore, gazelles are very sensitive to low temperatures and try to hide from the wind more often than the saiga. Consequently, the use of certain pastures by goitered gazelles, as well as the predominance of some foods in their diet, mainly depends on the state of the snow cover and weather conditions (Zhevnerov 1984).

When grazing on flat pastures in periods with a low level of snow, the composition of the diets of the gazelle and saiga is largely similar. *Artemisia terrae-albae* and *Anabasis salsa* predominate among the leftovers of food, and in years with a good autumn vegetation, the role of cereals and various grasses increases markedly, the volume of which from the total mass of stomach contents can be 10–30% (winter 1970–1971, according to five samples). With a snow cover depth of 10–15 cm, especially in years with a summer drought, the main parts of *Artemisia-Anabasis* associations are under the snow. If the snow cover is loose, foraging antelopes can easily push the snow with their muzzles. Therefore, the main plants of this complex still dominate in the diet of goitered gazelles. But on the same pastures, a snow cover with a depth of 8–10 cm and a density of 0.16–0.17 makes it difficult for gazelles to get food. With frequent strong winds, the snow condenses quickly, and, in some years (for example, winter 19711972), in mid-January on the flat areas of the island, the snow reached a density of 0.20–0.27. During these periods, the gazelles are rarely seen on lowland pastures, and instead, they use areas with unevenly distributed snow cover for grazing, as well as places which are dug up by kulans and saigas; that is, they become commensals in relation to ungulates that are able to extract food from under the snow. However, the main mass of goitered gazelles go to the zone of fixed sands, where shrubby vegetation prevails, and the snow is looser (Zhevnerov 1984).

Naturally, the portion of *Anabasis salsa* and *Artemisia terrae-albae* in the diet of the goitered gazelle falls, and these plants become of secondary importance in their food, although they are the best pasture food in terms of their nutritional value. *Anabasis salsa* is characterized by a high protein content, and *Artemisia terrae-albae* is even more nutritious, containing from 9.5% to 15.9% of protein and 3.7% to an especially large 9.7% of fat, which is contained in the leaves of the autumn vegetative development (Kubanskaya 1956). At this time, the main forage content consists of shrubs and semishrubs (*Haloxylon*, *Tamarix*, *Astragalus*, *Atraphaxis*), the nutritional value of which is lower. Therefore, any possibility of grazing on *Anabasis salsa* and *Artemisia terrae-albae* is used by gazelles fully. In snowless areas, under favorable weather conditions, individual groups of animals stay for 2–3 days. When calculating plant remains after grazing in such places, *Artemisia terrae-albae* occupied an average of 58.2%, *Anabasis salsa* 27.3%, and various grasses and herbs 14.5%. The inability of the gazelle to get food from under the snow forces it to feed in limited snowless areas and at the pasture sites of other ungulates, where the accumulation of a large number of animals is not reasonable considering availability. This forces gazelles to use browsing for feeding, which is probably one of the reasons for the low herd size in winter, that has an average of 3.5 (Zhevnerov 1984).

Observations show that in the snowy season, gazelles in large numbers eat annual shoots of tree and shrub species, while during a low snow cover, they continue to partially feed on dwarf shrubs. Among the latter, *Halocnenum strobilaceum* and *Ephedra* spp. are of paramount importance, the food remains of which range from 30% to 70% in the inter-row spaces of the sands. In most cases, the gazelles eat the root parts of *Ephedra* (but not the upper shoots) by pushing away the snow near the bushes apart with its muzzle and then clear the sandy substrate with their hooves. In snowy winters, most of the sites occupied by *Ephedra* are excavated entirely by gazelles. At this time, the role of large-stemmed plants growing in fixed sands sharply increases (*Artemisia arenaria*, *A. quinqueloba*, *Astragalus*, *Atriplex*). When grazing in the saxaul forest (*Haloxylon*), where the height of the snow cover often exceeds 30 cm, goitered gazelles linger near one tree for 2–10 min, biting off annual shoots, and especially willingly eating seeds and galls, which are devoured completely at a height accessible to the gazelles as well as those on the ground. In the contents of the stomachs of gazelles that fell in the harsh, snowy winter of 1971–1972 (according to 8 samples), woody and shrubby plants (*Haloxylon*, *Tamarix*, *Atraphaxis*) accounted for 50–70%, *Ephedra* 20% to 40%, and perennial and annual grasses and herbs 8–11% of the total plant mass (Zhevnerov 1984).

Thus, the importance of certain plants in the diet of the gazelle during the winter season may vary. The general trend is that in the low-snow period, preference is given to dwarf semishrubs, dwarf shrubs, and various grasses, while in the snowy period, larger shrubs and large-stemmed grasses are preferred. With a great depth and density of snow cover, the gazelle is forced to switch almost completely to browsing in their feeding. It should be noted that prolonged feeding with coarse twig foods can causes the goitered gazelle indigestion and bodily exhaustion, which in severe winters is often one of the causes of the death of these ungulates. Among the kulans on the island, eating a large amount of coarse feed also causes digestive disorders. When opening corpses, almost all fallen animals had intestinal inflammation. The degree of exhaustion of gazelles from starvation in severe winters can be judged by the corpses examined in January and February in 1972. The mass of male corpses was 18–23 kg, with a loss from the norm of about 30%. All the fallen gazelles were completely without any fat deposits and the muscle tissues were pale blue. In most cases, the stomachs were filled with vegetable matter, consisting of 90% coarse twig feed (Zhevnerov 1984).

2.2.4 Continental Kazakhstan (*G. s. subgutturosa*)

The food of the gazelle is diverse—lichens, cereals, various grasses, semishrubs, and shrubs (their leaves, shoots, and fruits). In most areas of its range, the gazelle eats about 70 species of plants (Sludskiy 1977).

2.2.4.1 *Western Kazakhstan* (*Ustyurt Plateau, Sam Sands*)

Judging by the rumen contents, the goitered gazelles of the Ustyurt Plateau eat around 40–58 plant species (Mambetzhumaev 1970a, b; Sludskiy 1977). In winter, they feed on mainly semishrubs and shrubs with some admixture of available cereals: *Ephedra distachya*, *Poa bulbosa* and *Bromus tectorum*, *Salsola arbuscula*, *S. rigida*, *S. crassa*, *Halostachys caspica*, *Smirnovia turkestanika*, *Haloxylon persicum*, *Caragana grandiflora*, and *Limonium suffruticosum*. In their spring-summer ration, goitered gazelles prefer new green grasses of *Carex physodes* and *Rheum tataricum* and fresh buds and green shoots and leaves of *Eurotia ceratoides*, *Atraphaxis spinosa*, *Kochia schrenkiana*, *Salsola arbuscular*, *Anabasis salsa*, *Calligonum tortile*, *C. ustjurtense*, *C. spinulosum*, and wormwoods (*Artemisia arenaria*, *A. turanica*, *A. terrae-albae*) (Zhevnerov et al. 1983). During summer in the Ustyurt Plateau, the diet content of the goitered gazelles depends especially on the availability of drinking water. In the southern part of this plateau, for example, when gazelles are deprived of sources for drinking, they absolutely will not eat wormwoods, *Salsola* spp., which in many parts of the area form the basis of the gazelle diet. In summer (June into August), these gazelles stay in the Ustyurt depressions and feeds on the green parts of growing plants that contain a significant amount of moisture: leaves, tops of shoots, and fruits of cappers (*Capparis spinosa*). Although this species is not an abundant plant, the animals' stomachs were found to contain green leaves, tops of shoots, and fruits of cappers with a moisture of about 70% and juicy *Salsola* and *Anabasis*. Apparently, this was enough moisture to satisfy the water needs of the gazelles. To a lesser extent, the gazelles eat the fruits of the camel thorn (*Alhagi camelorum*), tamarisk shoots (*Tamarix* spp.), and very little of the fruits of the salt-bearing shrub *Reaumuria fruticosa*. In summer, the goitered gazelles eat less *Eurotia*, *Ephedra*, and *Haloxylon*. As a source of moisture, gazelles readily eat melon leaves and juicy bulbs of *Ungeria* and leaves of *Phragmites communis* and maize (*Zea mays*) (Dereviagin 1947; Kostin 1955). In contrast to domestic sheep, goitered gazelles do not touch the green shoots of the camel thorn but do eat the cappers that are poisonous for sheep and goats without harm (Minervin 1944). In southern Mangyshlak Peninsula, goitered gazelles eat leaves and fruits of *Frangula alnus*. And in this location, they have the opportunity to drink seawater on the coast of the Caspian and Aral Seas and quench their thirst by eating juicy algae on the seashore. Also goitered gazelles feed on marine plants, such as *Zostera minor* and *Najas marina*, thrown on shore by the waves, and in greater numbers also eat large dried bushy lichen *Teloschistes lacunosus*, resembling *Cladonia* in appearance. Despite the fact that in summer this lichen is very dry and brittle and turns into powder when rubbed between the fingers, 75% of the stomach contents of the hunted gazelles were the remains of this plant. Animals eat it a lot and willingly due to its high nutritional value, and the dryness was compensated for by the presence of drinking water in sufficient quantities. This lichen is also found on the Ustyurt Plateau, but due to the dryness and lack of watering places, it is not eaten here by the goitered gazelles at all (Minervin 1944). In the Sam Sands, goitered gazelles often eat *Aristida pennata* which that is very common in sand deserts (Kostin 1955).

2.2.4.2 *Southern Kazakhstan* (*Kyzylkum Desert*)

In the Kyzylkum Desert (Kazakhstan part), these goitered gazelles eat 42 species of plants, mostly the same as the Ustyurt Plateau (Sludskiy 1977). In the spring, their diet is dominated here by cereals and ephemerals and ephemeroides, including bromes (*Bromus tectorum*), *Eremopyrum orientale*, *Malcolmia karelini*, *Alyssum desertorum*, *Senecio subdentatus*, *Carex physodes*, as well as *Calligonum*. In the summer-autumn period, gazelles feed mainly on succulent plants and shoots and the leaves of shrubs, including *Anabasis aphylla*, *Salsola foliosa*, *S. lanata*, *S. nitraria*, *S. arbuscula*, *Calligonum*, *Haloxylon*, *Halimodendron*, and several species of *Artemisia*. In winter, camel thorn (*Alhagi kirghisorum*) is often eaten, as well as *Artemisia* spp., *Atraphaxis spinosa*, sand acacia (*Ammodendron argenteum*), astragalus (*Astragalus* spp.), and *Tamarix* (*Tamarix elongata*). According to the stomach contents (of 39 stomachs checked), the most preferable plants are *Leptaleum filifolium* (57.8%), *Calligonum* spp. (34.3%), *Convolvulus arvensis* (31.4%), *Peganum harmala* (29%), *Poa bulbosa* (17.6%), and *Ephedra* ssp. (14.7%) (Mambetzhumaev 1970b). From this list of plants, it is clear that goitered gazelles prefer juicy plants in dry deserts, and they eat such plants as *Peganum harmala*, which is also juicy but very poisonous to domestic livestock. The stomachs of hunted goitered gazelles (10 June 1959) were found to contain the stems and fruits of *Delphinium*, leaves and fruits of *Amygdalae amarus*, various cereals, and *Carex* (Ishunin 1961).

In the Betpak-Dala Desert, goitered gazelles eat 47 plant species in their summer diet, with the main portion cereals such as *Poa bulbosa*, *Eremopyrum triticeum*, some feather grasses of *Stipa* spp., as well as several species of sedges (*Carex* spp.), onions (*Allium turkestanicum* and *A. schubertii*), tulips (*Tulipa buhseana*, *T. alberti*), *Kochia prostrata*, leaves of *Salsola arbuscularformis*, *Leontice incerta*, *Descurainia Sophia*, *Rosa* (*Hulthemia*) *persica*, *Astragalus* spp., *Caragana balchaschensis*, *Anabasis salsa*, *Artemisia* spp., and *Glycyrrhiza* spp. As these plants dry out, green twigs of *Spiraeanthus schrnckianus*, *Rosa* spp., *Kochia prostrata*, and camel thorn *Alhagi pseudalhagi* are readily eaten (Sokolov 1959; Zhevnerov et al. 1983). In July and August, *Spiraeanthus* and *Hulthemia persica* and *Alhagi* increase their importance in the diet of goitered gazelles. In September, goitered gazelles eat *Salsola* in great amounts, and in October and November, they eat *Salsola*, *Artemisia*, and *Ephedra* (Heptner et al. 1961).

In the Ili Hollow, these gazelles eat several species of *Artemisia* spp., green shoots of various *Salsola* species, *Calligonum* spp., *Astragalus* spp., *Ephedra* spp., and *Haloxylon* spp. all year round. During spring and summer, they consume *Festuca valesiaca*, *Stipa* spp., *Spiraea* sp., *Gallium verum*, *Allium* spp., and other herbs (Zhevnerov et al. 1983). In the foothills of the Dzungarian Alatau, the main forage plants are yellow bedstraw (*Galium verum*), biyurgun (*Anabasis salsa*), onion heads (*Allium* spp.), târsa (*Stipa capillata*), tipchak (*Festuca valesiaca*), Atraphaxis (*Atraphaxis frutescens*), and even reeds (*Phragmites australis*) (Antipin 1941). There are reports that sometimes gazelles are known to feed on cotton seedlings (Dementyev 1935), and in the Syugatinsky Valley (Kazakhstan), there were cases noted of gazelles eating corn leaves (Afanasiev et al. 1953). During the unfavorable conditions presented by severe winters, in the lower reaches of the Ili River, exhausted in the lower reaches of the Ili River eat a lot of dry beans and the prickly branches of *Halimodendron halodendron* (Zhevnerov et al. 1983).

I believe that goitered gazelles eat almost all kinds of plants in their habitats, including species that are poisonous for livestock, such as *Peganum harmala* and *Anabasis aphylla*, although the basis of nutrition for these plants is only 3–4 dozen species. In the area of the Ili Hollow, I noted the most often plant species most often eaten during spring and summer (Table 2.5). It is cleared up from my data that the most edible plant in the Ili Hollow is *Salsola arbusculiformis*, which belongs to the Chenopodiaceae (now included in *Amaranthaceae*), while the least eaten are trees (*Elaeagnus oxycarpa* and *Populus diversifolia*). However, similar to the study by Zhevnerov (1984), the preference of plants in the goitered gazelle ration can change considerably in autumn and winter, when other plant species are preferable. In the Ili Hollow, drinking water is available all year round; therefore, goitered gazelles can eat plants containing more nutrients and less water, while in conditions of a water deficit during dry summers, the gazelles eat mostly juicy plant species during dry summer (Blank and Li 2022).

Table 2.5 Preferences (%) in plant species in diet of goitered gazelles in the Ili Hollow (adapted from Blank 1990)

Plant species	Observed cases of eating	Proportions (%)
Salsola arbusculiformis	167	47.7
Halimodendron halodendron	50	14.3
Nanophyton erinaceum	40	11.4
Eurotia ceratoides	22	6.3
Atraphaxis frutescens	15	4.3
Caragana balchaschensis	15	4.3
Haloxylon aphyllum	12	3.4
Alhagi kirgisorum	7	2.0
Calligonum junceum	6	1.7
Artemisia terrae-albae	6	1.7
Nitraria sibirica	5	1.4
Elaeagnus oxycarpa	4	1.2
Populus diversifolia	1	0.3

Everywhere in the spring, the goitered gazelles feed on ephemera and ephemeroids, which are rarely found in the Ili Hollow. In the summers in other areas, gazelles also search for the juiciest plants (*Salsola* spp., *Capparis* spp., *Calligonum* spp.), but with a water abundance, these plants are not eaten as often. In the autumn, gazelles eat secondary plants (especially wormwoods, which are bitter in summer); in winter, along with *Salsola* spp. and wormwoods *Artemisia* spp., gazelles eat dry grasses and small branches of shrubs and trees, which are eaten rarely in summer (Zhevnerov et al. 1983; Blank 1990).

From all shrubs, goitered gazelles eat the leaves and shoots. When eating *Alhagi kirghisorum*, with soft thorns, they bite off the entire apical part, but as soon as the thorns harden, they only eat the leaves. With the drying of the plant, they refuse it altogether, but they begin to eat it again in late autumn and winter. With *Nanophyton erinaceum*, only fresh tender shoots are eaten in the spring, but they refuse it later, when the plant becomes prickly. In late autumn and winter, along with other *Salsola* species, this plant again becomes one of the most eaten. Goitered gazelles readily eat wormwood (*Artemisia terrae-alba*) as early as spring, and in late autumn and winter. Spring migrations to the western part of Ili Hollow are related to the early starting of vegetation in that area (Blank 1990).

In autumn and winter, the main food of the goitered gazelles in the Ili Valley are *Kochia prostrata*, *Salsola arbusculiformis*, *Eurotia ceratoides*, *Salsola orientalis*, wormwoods (*Artemisia* spp.), *Haloxylon aphyllum*, *Nanophyton erinaceum*, and *Artrophyton iliensis*. They eat ephemera quite often, but because of the small size of the plants, it is difficult to distinguish specific plants from a distance (Blank 1990).

Goitered gazelles willingly graze around the springs of Mynbulak (Ili Hollow) with typical near-water vegetation. Some of the gazelles stop here on their way to the watering places and back, while other gazelles stay in the area all the time. In June and July of 1985–1986, the same group of females, at first without fawns, and later together with them (three females and six fawns), descended from the foothills of the mountains to the springs. The gazelles grazed here for 1–1.5 h, then climbed back into the foothills for a 2–3-h rest, after which they descended again. This was repeated twice, sometimes three times, a day, and the gazelles came to these springs from all sides. Therefore, the frequency of their meetings at this time was very high at 9.5–10.7 individuals/km along my daily walking route. I believed that not only the proximity of watering holes, but also the rich, much more diverse, longer-growing vegetation around the springs makes these places the most attractive for the gazelles. There is a large set of legumes and cereals located in this area that are valuable in terms of feed. In July 1985, the yield of grasslands was 7.14 c/ha, whereas the yield of the surrounding *Nanophyton-Salsola* associations, the yield was only 0.55 c/ha. However, the distribution of gazelles in the unusually rainy summer of 1987, however, showed that this was not always the case. Having a lot of temporary watering holes throughout the territory, gazelles practically ceased to come across the springs of Mynbulak (in June 1987, the population density here was 0.4 individuals/km). Judging by the distribution, the *Salsola-Nanophyton-Arthrophyton* pastures in the eastern and southern foothills of the Chulak mountains turned out to be the most favorable for them. The average yield here, even in the most favorable years, does not exceed 4 c/ha (Kurochkina and Karibaeva 1986).

In the spring and mid-summer, a small number of gazelles go to the fields of alfalfa, corn, and melons, located mainly in the western part of the Ili Hollow. But more often, they are in this area in the harvested fields in autumn. Sometimes at this time, having crossed the low mountains, gazelles graze on cultural plantings located on the northern exposure slopes of these mountains (Blank 1990).

The goitered gazelles also receive also mineral substances by eating some plants rich in salts (i.e., *Tamarix* spp., *Haloxylon* spp., etc.), the earth on the salt lakes, and by drinking h the water of the seas and salt springs (Heptner 1956).

Thus, the basis of the diet of the gazelle in summer everywhere is the juiciest forages, most often where there are *Salsola*, *Capparis spinosa*, and pistachio (*Pistacia vera*), depending on availability. Selectivity in feed in summer is mainly determined by the moisture content of plants, not their nutritional value. In cases where gazelles stay near open water sources, they can also feed on dry food, such as cereals and lichens. Outside of the period of drought goitered gazelles eat fresh green parts of various plant species.

2.2.5 Caucasus (*G.s. subgutturosa*)

In the Eastern Transcaucasia, goitered gazelles eat mainly 35–40 plant species (Vereshchagin 1939), and about a dozen and a half forage plant of goitered gazelles are known. Especially in spring, the main species belong to cereal ephemeroids and ephemera, with *Poa* spp. playing the greatest role in their diet (Vereshchagin 1939). In the Milsko-Karabakh steppe, xerophytic cereals are the gazelles' favorite food, while in the cultural zone, gazelles feed on agricultural crops; and in winter, they graze in the fields of barley and wheat in Transcaucasia (Vereshchagin 1939).

2.2.6 Turkey (*G.s. marica*)

The composition of rumen ciliate was investigated in goitered gazelles living in Turkey. Three species were found belonging to the genus *Entodinium*: *E. dubardi*, *E. nanellum*, and *E. simplex*. The ciliate density in the rumen contents was 87.5×10^4 mL^{-1}. Goitered gazelles have a ciliate fauna somewhat similar to that previously observed in Mongolian gazelles. Diets also vary between browsing and grazing ruminants with subsequent differences in their fauna. Many of the browsers have all *Entodinium* fauna, while in some cases, only a single species of *Entodinium* has been found. Goitered gazelles have an intermediate type of feeding which includes grazing and browsing. A diet of browse tends to lack fibber, which causes the diet to be rapidly fermented by rumen bacteria with results in the PH value in the rumen to drop. The presence of fauna consisting of entodinia alone has been attributed to a low PH condition in the rumen. The gazelles in this study inhabited an area with small bushes, roots, and grasses. The rumens of intermediate type feeders may have environmental factors more favorable for the growth of entodinia than those of true browsers and true grazers (Gürelli et al. 2012).

2.2.7 Tajikistan (*G.s. subgutturosa*)

In Tajikistan, in the Nature Reserve "Tigrivaya Balka" (Tiger Gorge) and its adjacent desert areas, Dutov (1990) carried out investigations of the goitered gazelle feeding. The short growing season of many plant species in this area determines significant periodic changes in the composition of the goitered gazelle diet. This circumstance entails the enlargement the list of forage plant species. Of the approximately 300 plant species recorded in the desert part of the reserve, 69 plants belonging to 20 families were noted in the goitered gazelle diet. Of these, representatives of five families are of the greatest importance (by the number of species): Poaceae (mainly spring-summer; 23–31% in different seasons), Polygonaceae (mainly early spring and winter; up to 15% during cold seasons), Amarantaceae (Chenopodioideae—all seasons, but especially preferable in winter), Fabaceae (mainly spring-summer), and Asteraceae (mainly spring-summer) (Dutov 1990).

The early spring period (March–April), when it gets warm and a lot of rain falls, is characterized by the rapid development of all desert plants. At this time, the goitered gazelle's diet contains the largest number of plant species (55 species). The gazelles feed mainly on fresh grasses and herbs, among which are plants from families Poaceae, Polygonaceae, Asteraceae, and Fabaceae. At the end of May and all summer, there is hot and dry weather. Goitered gazelles at this time eat the green parts of shrubs and perennial herbaceous plants, grains of cereals, and seeds of some ephemera. The specific weight of Chenopodiaceae species in the list of summer feeds increases due to a decrease in the proportion of spring plants from other families. Cereals still prevail. The specific weight of legumes doubles. The flowering of species from Asteraceae finishes their vegetation, and their specific weight in the diet of the goitered gazelle is significantly reduced. Polygonaceae species in the diet are not marked at all during this period. Gazelles eat a total of 22 plant species at this time (Dutov 1990).

In late autumn and winter, temperatures noticeably decrease, and precipitation increases. Fogs are frequent at this time, and the soil is almost always moist. In November, almost every year, there is a secondary vegetation of some plants (*Carex*, *Poa bulbosa*, *Bromus tectorum*), which are included in the diet of the goitered gazelle along with the juicy green parts of shrubs. With the onset of frost, the aboveground parts of perennial grasses and shrub leaves die off, but the greenery of herbaceous plants of secondary vegetation remains. In terms of the number of plant species eaten by goitered gazelles during this period, the species from family Chenopodiaceae are only slightly inferior to cereals, and Polygonaceae species again appear in the diet of goitered gazelles. The specific weight of legumes is significantly reduced, and there are no species from the family Asteraceae in the diet at all. The total number of plants in the goitered gazelle diet in the autumn-winter period increases to 38 species (Dutov 1990).

The period with snow cover in the study area is short, only 10–20 days, but it is still the most difficult for goitered gazelles. With the onset of frost, shrubs shed their leaves and the aboveground parts of perennial plants die. After snowfall, the growing greenery is also inaccessible to goitered gazelles. At this time, the forage plant list in the diet of the gazelles is reduced to 12 species. The most used from this list is Chenopodiaceae (up to 76.5% cases). Eating of tree buds (*Populus diversifolia*) in winter has been recorded only twice and the branches from which the buds were eaten that were broken off by snow. It was also noted that when kept in captivity, gazelles willingly eat young shoots of *Elaeagnus oxycarpa* and leaves of *Glycyrrhiza glabra*. In the wild, goitered gazelle eating of these plants has not been observed (Dutov 1990).

In the diet of goitered gazelles, plants found in sandy, clay, and salt marsh deserts are of primary importance. Gazelles enter the tugai forests only for recreation, and if they graze, then either in deserted areas or in places where trees are replaced by desert plants. In any case, goitered gazelles avoid dense thickets of reeds and stick to open spaces. Penetrating into the *Populus diversifolia* tugai, goitered gazelles willingly use roads and open areas and walk along the paths of other animals. The gazelles almost never eat tugai plants and visiting the forest is associated with transitions to watering holes, which are located mostly in the tugai area (Dutov 1990).

2.2.8 Turkmenistan (*G. s. subgutturosa*)

In southwestern Turkmenistan, along with wormwood (*Artemisia* spp.), bluegrass (*Poa* spp.), sedge (*Carex* spp.), and wheatgrass (*Elytrigia* spp.) are readily eaten (Sokolov 1959). In the summer in the Badkhyz Nature Reserve, gazelles feed on the most succulent plants (*Salsola*, *Artemisia*) and often their stomachs contain almost exclusively fruits and pistachio leaves. And on hot days in the Murghab Valley, the gazelles will graze on the green cotton fields. Outside of the dry period, *Poa* spp. plays the greatest role in the diet of Badkhyz gazelles (Heptner 1956). Gorelov

(1972) found in the Badkhyz (SE Turkmenistan) that goitered gazelles often dig the ground to get and eat the juicy roots of plants, especially during hot dry summers (July and August). In the Ili Hollow, I noted gazelles digging into the ground very rarely. However, in the semi-captive herd of the Issyk-Kul breeding center, I saw gazelles digging the ground regularly. This is perhaps related to the kind of soil at the different locations: the gazelles were digging in the softer sands in in Badkhyz and Bukhara, but they avoided digging in the hard grounds of the gravel deserts in the Ili Hollow.

2.2.9 Uzbekistan (*G. s. subgutturosa*)

During the spring and autumn vegetation, ephemera and ephemeroids are preferred by goitered gazelles. After May and the drying of ephemera, halophytes and xerophytes, especially *Suaeda* and *Salsola* containing 70–85% water, are eaten in large quantities. In the autumn-winter period, along with *Salsola* and *Artemisia*, they feed on dry herbaceous vegetation and the branches of shrubs and trees. In total, up to 84 plant species are eaten by goitered gazelles during this time (Volozheninov 1983; Mardonov 1993, 1996).

In the southern Ustyurt Plateau, goitered gazelles exist for months during the hottest time of the year with a complete lack of drinking water. They survive by eating juicy green plants. The gazelles of the southern Ustyurt do not undertake any migrations, even small ones, in search of water, but, in turn, are forced to abandon very nutritious resources and eat dry plants, being satisfied with the prickly but juicy capers (Minervin 1944).

In the Kyzylkum Desert (the part located in Uzbekistan), Mambetdzhumaev (Mambetzhumaev 1970b) found 30 plant species that are eaten by goitered gazelles (May 1953–February 1954). The most often eaten are as follows: *Bromus tectorum* (dry stems and leaves), *Poa bulbosa* (dry and green leaves), *Carex physodes* (green leaves), *Calligonum* (shoots), *Kochia prostrata* (shoots and seeds), and *Zygophyllum* (dry shoots and seeds). In less amounts, goitered gazelles also eat the following: *Salsola rigida* (shoots), *Petrosimonia sibirica* (seeds), and *Artemisia* (leaves and seeds). The least chosen are *Haloxylon*, *Ephedra*, *Salsola*, and *Halocnemum strobilaceum* (Mambetdzhumaev 1970b). In spring in this part of the Kyzylkum Desert, cereals and *Carex* are preferable, and in autumn and winter, plants from Chenopodiaceae are dominate, while the gazelles refuse to eat them in other seasons (Ishunin 1961).

In the northern and western Kyzylkum, goitered gazelles live mainly on the takyr-like grounds with *Anabasis salsa* and *Artemisia*. In autumn, goitered gazelles feed mostly on *Halimodendron halodendron* and fallow agricultural lands, where they can find rich pastures with grasses that are green until winter. In November, the gazelles stay in the depressions between sand ridges, where they eat *Salsola* and *Artemisia* and can find water in small springs. In winter, goitered gazelles gather in the saxaul forest (Kim 1953).

In the semi-captive conditions of the Bukhara Breeding Center, goitered gazelles consume 77 plant species from 21 families. Family Chenopodiaceae has the most numerous plant species, which are eaten by goitered gazelles (totally 18 species, including 18 in winter and 11 in summer), Poaceae (total of 11 species, including 11 species in winter and nine species in summer), Asteraceae (total of ten species, including ten species in spring and three species in autumn). The most preferable plants in the diet of goitered gazelles are *Alhagi pseudoalhagi*, *Karelinia caspia*, *Haloxylon aphyllum*, *Aelinia microphylla*, and *Salsola arbuscularformis* (Mardonov 1993).

The study of the nature of plant consumption in semi-voluntary conditions of keeping of the gazelle in the Bukhara nursery showed the proximity of such diet with gazelles that stay in the enclosure. But often, differences in eating the same plant species are characteristic of different biotopes, which is especially noticeable in relation to pasture areas at different distances from watering holes: the closer to the watering hole, the stronger the blend of plants. Thus, significant differences in the alienation of phytomass in areas remote from the watering holes and close to it are 8–30% per plant (Mardonov 1993).

2.2.10 Mongolia (*G. s. hilleriana*)

In Mongolia, out of the 17 species eaten by goitered gazelles, only four provide the main portion of the diet: Gobi grasshopper (*Stipa gobica*), bagalur (*Anabasis brevifolia*), Mongolian onion (*Allium mongolicum*), and halogeton (*Halogeton* sp.). Also available are *Artemisia*, *Dontestomon*, *Peganum harmala*, *Ephedra*, fruits of the *Nitraria*, and various *Salsola*, while wormwood is not eaten at all in summer (Bannikov 1954).

2.2.11 China (*G. s. hilleriana*)

In the eastern Junggar Basin, goitered gazelles prefer two vegetative associations: *Anabasis salsa-Artemisia* sp. and *Stipa glareosa-Kochia prostrata*. The richness of plant species and the vegetative cover are possibly the most important factors determining the foraging habits of goitered gazelles. In the shrub patches, the gazelles select the succulent leaves of *Corydalis glaucescens* in order to get necessary water. In the northern areas of study area, gazelles prefer relatively well-vegetated consolidated and semi-consolidated dunes, which have a greater vegetative cover, plant richness, and vegetative height. On the sand dunes, *Ephedra distachya*, *Kochia prostrata*, and *Ceratoides latens* were browsed by gazelles (Yang et al. 2005).

In the lowlands of Xinjiang Province in China, goitered gazelles concentrated around quagmires with a central water source. The vegetation around these water sources is predominately Poaceae plants, forming azonal steppe-like patches. In the diet of goitered gazelles here, 15 main plant species have been found, mainly cereals and chenopods. The most eaten plants are *Stipa gobica* (Fam. Poaceae) and

Ceratoides latens (Fam. Chenopodiaceae), while *Allium* spp. and *Kochia prostrata* are noted less often in autumn (Hu et al. 1998). Microhistological analysis of the feces demonstrated that in the gravel deserts of Dzungarian Xinjiang (Kalamani Nature Reserve), goitered gazelles eat 47 plant species from 16 families during the year. Among them, 34.2–43.4% are shrubs, 16.8–31.1% graminoids, and 25.6–41.9% forbs. The composition of their diets varies seasonally: 24 plant species in autumn, 17 plant species in winter, 41 plants species in spring, and 30 species in summer. The diversity indices of each season are as follows: 2.87 in autumn, 2.59 in winter, 3.59 in spring, and 2.97 in summer (Xu et al. 2012).

Chenopodiaceae and Poaceae are major foods, comprising 38.8–85.1% of the gazelle diet (Xu et al. 2012). Among plant species, *Stipa glareosa* is eaten readily throughout the year, but most often in spring (7.7%) and in summer (22.1%). *Ceratoides latens* is preferred in spring (9.2%) and summer (10.5%). The portion of *Anabasis*, *Achnatherum splendens*, and *Phragmites australis* in winter are evidently higher than in the other three seasons; the use of *Anabasis brevifolia* was found to be over six times higher in winter than in autumn. Ephemeral plants (mostly from Asteraceae and Brassicaceae) are used mostly during spring (27.0%). Succulent plants, such as *Allium polyrhizum*, *Zygophyllum rosovii* and *Salsola subcrassa* are favored by gazelles during hot and dry seasons (spring and summer). *Ceratocarpus arenarium* and *Salsola subcrassa* are favored by gazelles in autumn and winter (Xu et al. 2008, 2012). Hofmann and Stewart (1972) classified that the mixed feeders' diet contains over 25% browse and no more than 75% fruits, dicotyledonous foliage, and shoots. Based on this information, the goitered gazelle should be classified as a mixed feeder (Xu et al. 2012).

It is widely accepted that the gut microbiota can affect the host's metabolism and help meet the energy and nutritional needs of the host by improving host's digestion efficiency (Nicholson et al. 2012). The gut microbiota can vary across seasons (Zhang et al. 2014) with the bacterial microbiota is commonly dominated by Firmicutes and Bacteroidetes in mammals (Combes et al. 2017). Seasonal variations of diet are the main reason leading to fluctuation in the microbiota community, and the composition and diversity of diets play an important role in shaping the microbiota (Qin et al. 2020). A healthy, diverse diet promotes a more diverse gut microbiota (Zmora et al. 2018). The alpha diversity of the gut microbiota is better in summer, and the seasonal variation in microbiota structures in goitered gazelles between summer and winter can be related to seasonal changes in edible plant species and food quality (Qin et al. 2020). High-fat diets in winter can cause the proportion of Ruminococcaceae to decrease and that of Rikenellaceae to increase, due to changes of the dietary fat content. Indeed, the portion of *Haloxylon ammodendron* increases in the goitered gazelle diet, the crude fat content of which also increases from May to October (Gao and Ji 1996). In addition, the content of crude fiber increases from summer to winter in saxaul, to help the host degrade cellulose, with the relative abundance of Butyrivibrio increasing significantly in winter (Qin et al. 2020).

Although there is a significant difference in gut microbiota composition between summer and winter, the ultimate purpose of the favored microbes appears to be consistent, as all the changes are devoted to maintaining intestinal balance and normal metabolism (Qin et al. 2020). The relative abundance of *Bacteroides* increased significantly in winter, and *B. thetaiotaomicron* was only found in winter. *Bacteroides* enable the host to degrade carbohydrates, especially polysaccharides, as well as proteins, and enhance the nutrient utilization by the host. *Bacteroides* are important members of the gut microbiota, and *Bacteroides* species actively catabolize polysaccharides, and the increase in the relative abundance of *Bacteroides* improves saccharide metabolism in the goitered gazelle. In winter, the enhancement of saccharide metabolism may be a way of compensating for the reduction of nutrients in foods, to ensure that there is no reduction in the total energy intake by the host. The changes in *Gemmatimonadetes* reflect an adaptation to the water shortage problem in winter faced by goitered gazelles. To survive in a water shortage environment, vasopressin-regulated water reabsorption is significantly enhanced (Qin et al. 2020).

2.2.12 Iran

Various factors influence the formation of the spatial structure of the goitered gazelles. At the same time, it is believed that vegetation, namely, the quality and quantity of vegetation cover, is the main determinant of the choice of habitats by the goitered gazelle, as well as other gazelle species. The influence of the quantity and quality of vegetation cover on the spatial distribution of goitered gazelles is shown in the national parks, located on the territory of Iran (Nowzari et al. 2007; Farhadinia et al. 2009; Nezami 2009; Bagherirad et al. 2014).

In the Khosh Yeilagh Wildlife refuge in northern Iran, the goitered gazelles' most important forage is *Artemisia terra-alba* (41%), *Eurotia ceratoides* (26%), *Astragalus* sp. (14%), and *Agropyron pectiniforme* (11%), with forbs comprising only 3% of the diet. These species comprised 93% of the gazelles' total diet. Generally, the average concentrations of mineral elements are significantly higher in fall than in spring. In comparing between species and seasons, the concentrations of macro- and microelements in shrub species are markedly higher in fall, while for grass species, they were higher in spring (Mowlavi 1978). In southern Iran, goitered gazelles eat shrubs, grasses, and forbs in winter and early spring, but browsing increases during the rest of the year. They also feed in farmlands at night and can be trapped by villagers. Analysis of the rumen content of a female in Ghamishloo Wildlife Refuge in October 1992 showed that *Salsola* was the most common component of the diet, followed by grasses (Hemami and Groves 2001).

In the Golistan National Park (GNP), spring supports the greatest diversity of food categories in the GNP study area, and the gazelles eat most of the plants that they encounter during this season. The preference for herbaceous species in spring is probably due to the high quantity and quality of grasses and forbs available in their early growth stages. The protein content of silver feather grass and bulbous bluegrass, the two main perennial grasses, exceed the gazelles' requirements during birthing and feeding fawns after birth (at least 14%). Succulent ephemerals were favored by the gazelles due to their rich nutrients, vitamins, and water content.

Spring is an important season for the gazelles, during which time molting, birthing, and rearing of their young occurs. For these reasons, the gazelles consumed a wider range of plant species to meet their energy and nutritional requirements in spring. Although the content of protein, metabolizable energy, and mineral concentration of *Eremopyrum bonaris*, *Alyssum strigosum*, and other annual forbs are unable to satisfy the food requirements of the gazelles in spring, they are likely to be consumed by the gazelles due to their high palatability, high soluble cell content, and high digestibility. These ephemeral plants appear in the study area most often in spring, when their quality and quantity are at their best. Rapidly decreasing in summer, they are completely absent from the habitat during autumn and winter. Rad et al. (2015) hypothesized that herbaceous species usually provide adequate nutritional forage for fawn growth and good general development during their early growth stages. As spring progresses, the protein levels in grasses decreased more rapidly than in shrubs and forbs to eventually reach below the gazelle's minimum requirements (Rad et al. 2015).

Summer is the driest season in the GNP study area. July and August are the warmest months with a maximum temperature of 45 °C. Water deficiency here can be a critical and determining factor in the survivability of the gazelles. During water stress, the gazelles probably focus more on feeding on succulent plants, such as shrubs, as reliable water sources with more than 50% water, than grasses and forbs. Another study found positive correlations between the pellet group density of the gazelles and the canopy cover of *Aellenia* spp. during summer. Besides high-water content, some shrub species, such as *Aellenia* spp., are also dominant forage, with protein and energy levels that exceed the basic requirements for the goitered gazelle during the hot summer. It seems that the green bushes of *Aellenia* spp., along with other succulent shrubs species, could provide a sufficient food source for the gazelles during summer. A shortage of water in arid areas confirms the use of succulent species by the goitered gazelle as they switch to a diet dominated by forbs with a high-water content, such as *Allium polyrhizum* and *Zygophyllum rosovii* (Rad et al. 2015).

In the Golestan National Park, the cold period begins in October and continues through to April. Also, fluctuations of monthly temperatures showing that January and February are the coldest months, with a minimum temperature of −26 °C and approximately 22 frozen days. The extreme cold weather associated with the reduction in the quantity and quality of forage increases the risk of gazelle mortality and forces them to flee into the surrounding farmlands to find food, which ultimately leads to hunting by villagers (Rad et al. 2015). In Yazd Province, the main food of gazelles contains *Artemisia sieberi*, *Teucrium polium*, *Salsola tomentosa*, and *Stipagrostis plumosa*. *Artemisia sieberi* has the highest rate of protein reaching 11.37% (Dehghani and Parvaneh 2010).

In autumn, as most ephemerals dehydrate, food species diversity in the gazelles' habitat rapidly decreases. Herbaceous plant availability and quality declines with the cold season, while shrubs become the more dominant and nutritional food sources. During autumn, the gazelles need food that has a relatively high crude protein content, because they are starting their rutting behavior. As spring passes, the protein levels of silver feather grass and bulbous bluegrass decrease significantly, reaching below requirement levels for gazelles. In fact, the protein levels of bulbous bluegrass drop to below what is minimally required for gazelle maintenance for the rest of the year. Moreover, the high fiber content of these two perennial grasses increases with maturity more than in shrubs. Instead of perennial grasses, shrubs, mainly white wormwood, Mediterranean saltwort, winter fat (*Krascheninnikovia lanata*), and *Anabasis aphylla*, become the more available forage across autumn and winter. The high levels of crude protein in these plants' metabolizable energy, and their mineral concentrations significantly meet the requirements of the goitered gazelle during the mating and harsh winter seasons. The dominance and distribution pattern of forage frequently has a great deal to do with its preference rating. During the cold seasons, finding enough forage to satisfy the gazelles' basic requirements of the gazelles is difficult; therefore, during food stress periods and the low availability quantity and quality of food, herbivores generally use any edible forage available in their habitat. The energy expended searching for nutritional foods, especially during the harsh seasons, is probably more than the metabolizable energy intake of the forage consumed (Rad et al. 2015). Therefore, as reported by Abaturov et al. (1995), the gazelles and other ungulates spend more time feeding and resting than they do walking. The results of their study indicate that energy expenditure during walking is more than the energy content provided by the available forage (Rad et al. 2015).

Seasonal variations in the mineral-nutrient content of forage species were detected in the gazelles' habitat in my study area. Even though the highest concentrations of mineral elements in herbaceous plants occur in the spring, the levels were below minimum requirements, which can lead to a mineral deficiency and grass tetany for grazing gazelles during calving and lactating. Compared to herbaceous species, Rad et al. (2015) found the highest concentrations of mineral elements in shrubs occurred in fall, when the plants matured and the mineral contents are sufficiently high to meet some macro mineral requirements for browsing gazelles during mating behavior. The highest concentrations of mineral elements for elk forages were also reported at the end of the growing season (Rad et al. 2015).

Most animals require 24 mineral elements for growth and development, health, and survivability, all of which may not be found in any single forage species. Therefore, a variety of plant forms and species, rather than any single plant type, is required by gazelles to meet their nutritional demands for many macro- and microminerals. The most salient findings of their study indicate that, with the exception of Na, Fe, and Mn, most edible plants in the GNP habitat provide limited forage nutrient level and an inadequate mineral content during critical periods required for their survival, growth, and reproduction (Rad et al. 2015). Two shrub species, *Anabasis aphylla* and *Aellenia* spp., have the most protein and energy levels, which exceeded the gazelles' requirements across all seasons. Mineral contents are the limiting nutrients for shrub species in meeting the gazelles' requirements, especially in fall, with white wormwood the only species that can satisfy the gazelles' mineral needs during fall. The potential for maintaining a relatively high density of gazelle populations, given the forage quality levels to fulfil their requirements, declines as spring approaches early summer in the Golestan National Park (Rad et al. 2015).

About 15 herbal species are found in the goitered gazelle diet during winter, with the main preference from the family Poaceae (Jahandizi et al. 2015). In Bamoo National Park in autumn and winter, the density of gazelles is positively associated with three shrub species (*Astragalus* spp., *Ebenus stellate*, and *Achillea eriophora*) and two species of grass (*Poa bulbosa* and *Aegilops umbellulata*); conversely, population density is negatively associated with trichonic *Helichrysum leucocephalum* and thorny *Scariola orientalis* as well as both trichonic and thorny *Centaurea virgata*. Feeding behavior demonstrates that goitered gazelles feed on *Bromus danthoniae* and *Stipa barbata* during autumn and *Astragalus* spp., *Poa bulbosa*, *Aegilops umbellulata*, and *Bromus danthoniae* over winter. That means Poaceae supplies important foods for goitered gazelles (Nowzari et al. 2007).

2.2.13 Arabian Peninsula (Reem or Arabian Sand Gazelles: *Gazella subgutturosa marica*)

Some researchers (Thouless et al. 1991) think that reem are mostly grazers, preferring to feed on ephemeras after rain, but most of the time, they rely to dry grasses, chenopods, and forbs, showing their favor toward *Panicum* and *Stipagrostis* grasses as well as forbs in different areas of Saudi Arabia (Haque 1996; Cunningham 2013). However, everywhere, rainfall has the most noted effect on gazelle diet changes. During the dry season in western Saudi Arabia, reem eat mostly *Salsola spinescens* shrubs and fallen *Acacia tortilis* pods and occasionally *A. tortilis* gum (Cunningham 2013).

In Bahrein, reem prefers mostly *Heliotropium crispum* and *Helianthemum lippii*, which are the most important food items in their diet. These plants are both abundant in the area, and *H. crispum* contains 72% water; however, they both have a low protein content. In contrast, *H. lippii* has 50% of water with considerable amounts of protein. The combination of these plants in the gazelle diet provides a good source of both water and protein and form the major diet of the reem during the nonrainy season, which lasts most of the year (Mohamed et al. 1991).

In Saudi Arabia, *G.s. marica* eats a total of 80 species from 23 families in five protected areas throughout country. Their diet includes 17 species from Poaceae, 11 species from Chenopodiaceae, species from Fabaceae, eight species from Zygophyllaceae, five species from Boraginaceae, four species from each Brassicaceae and Cruciferae, and less plants from other families (Cunningham 2013).

Arabian sand gazelles (reem) are typical intermediate feeders (browsing and grazing) becoming concentrated selectors during the wet season (grazing) with a preference for short grasses and forbs throughout their range. Generally, Arabian sand gazelles consume more grass (58.4%) than browse (41.6%), but during the dry season, gazelles spend more time browsing (51.0%) and less time grazing (49.0%), while in the wet season, browsing is 17.6% and grazing is 82.4% (Schulz et al. 2013). The movement toward areas that had received rain and an appreciable change in diet from browse to graze during the wet season furthermore indicates the importance grazing. Forbs are utilized more than grasses (Cunningham 2013).

These authors (Cunningham and Wacher 2009; Schulz et al. 2013) indicate that in Mahazat As-Sayd, grasses (including seedlings) that are most heavily used are *Ochthochloa compressa* (leaves and occasionally inflorescence) followed by, but not necessarily in order of utilization, *Lasiurus scindicus* (leaves), *Panicum turgidum* (leaves), and *Stipagrostis* sp. (leaves). Dwarf shrubs are also important browse (Harrison and Bates 1991) with Wacher (1995) indicating their occasional use of trees (i.e., browse in southern Saudi Arabia). Mowlavi (1978) noted that browsing forms the largest part of the diet of *G. s. subgutturosa* especially during the drier months, in Iran. In Mahazat as-Sayd, males use *A. tortilis* and *Maerua crassifolia* more, while females consume *S. spinescens* more while browsing (Cunningham 2009). According to Gillet and Launay (1990), new growth of *S. spinescens* from Mahazat as-Sayd had a crude protein content of 17.2% and a mineral content of 19%, while grasses had a lower crude protein content (e.g., *Lasiurus scindicus* 12.2% and *Panicum turgidum* 4.8–6.8%) probably indicating the reason why females selected for *S. spinescens*.

Despite the fact that the gazelles are well adapted to extremely hot conditions, in the southern parts of its range, prolonged droughts can cause a catastrophic decline in the population, as happened on the protected territories in the central part of Saudi Arabia. During the period from 1999 to 2008, drought became a factor that caused the mass death of the Arabian sand gazelle (*G. s. marica*). At the same time, the high mortality of animals was not caused by stress and heat stroke but occurred because of hunger due to the lack of a sufficient number of high-quality forage plants That is, the climatic factors themselves do not directly affect the state of the gazelle populations but instead have an indirect affect through the availability of food (Islam et al. 2010).

2.2.14 Competition for Food Between Goitered Gazelles and Domestic Sheep

It should be added that the distribution of high-quality vegetation and its quantity over an area can be influenced by livestock, acting as another factor influencing the spatial distribution of goitered gazelle. In some cases, there is a strong competition between livestock (sheep) and goitered gazelles for forage resources, especially in winter, which may even lead to a reduction in the number of gazelles, as shown in China (Xu et al. 2012), and to a narrowing of the ecological niche for the gazelle (Karami et al. 2017). In other cases, livestock displaces Arabian sand gazelles from the best pastures, forcing them to use pastures not quite suitable for domestic ungulates (Cunningham and Wacher 2009).

In Kalamali Nature Reserve (Xinjiang, China), during winter, domestic sheep eat 15 species from five families ($H = 2.51$), and goitered gazelles eat 17 species from six families ($H = 2.59$). Similar to the winter diet of gazelles, Chenopodiaceae and

Poaceae plants are also the staple food of domestic sheep. The diet overlaps between goitered gazelles and domestic sheep reaches 0.77 in winter (Xu et al. 2012).

Goitered gazelle habitat loss to livestock during winter is around 48.5%, with the relatively good vegetation areas occupied by domestic sheep; this, in turn, forces the gazelles to live in the open clay desert, where they are seldom seen in other seasons (Li et al. 2010). The quality of vegetation is poor in the open clay desert, with the dominant plant species being *Anabasis* and *Reaumuria songarica* (Li et al. 2009). *Anabasis* is scarcely used by gazelles during its growing season due to its poison content and low nutritional value; the toxicity level falls after it dries up during winter (Jia 1987). Domestic sheep utilized significantly more *Stipa glareosa* than goitered gazelles, while gazelles significantly consume more *Anabasis brevifolia* than sheep. It is suspected that the winter diet of gazelles is a result of selection due to the adaptation to the competition with livestock during the harsh winters (Xu et al. 2012).

The study of Xu et al. (2012) shows that the high degree of dietary overlap between the goitered gazelle and domestic livestock poses a threat to the survival of the gazelle during the cold and food-scarce winters. Competition is considered to be a major selective force causing the differential use of resources between species (Schoener 1974, 1985). Livestock competition forces gazelles to consume secondary foods that are less digestible and have a lower nutritional value (Xu et al. 2012). In Tajikistan, goitered gazelles are displaced by domestic sheep and goats from desert areas and forced to stay in the closed habitats of tugai, which are atypical for them and are avoided in other seasons. The role of livestock is very significant and restricts goitered gazelle access to desert pastures and in the most difficult feeding seasons (Dutov 1990). When competition between gazelles and sheep is recognized, decreasing the domestic livestock numbers could serve as a solution. Xu et al. (2012) suggest that the number of domestic sheep in the reserve should be restricted to prevent competition, especially during the harsh winters, as well as initiating a strict protected area exclusively for gazelles.

In conclusion, my review of the goitered gazelle feeding presented above shows that goitered gazelles can eat various plants in different parts of their natural habitat. But food selectivity of gazelles is associated with the solution of two important goals: supply the body with food of high quality; therefore, they select the most nutritious plant species across seasons when vegetation has its greatest nutritional value and provide the body with sufficient moisture in the hottest and driest seasons of the year. Thus, gazelles eat not only the most nutritious plants and their parts with the lowest fiber content, which is available in spring and early summer, but they also switch to the most succulent plants in the middle of the hot summer. Goitered gazelles also have the ability to dig the juicy roots of plants in loose, sandy grounds, though not in stony-gravel deserts. With the availability of watering holes, gazelles will eat absolutely dry, but nutritious plants, which they refuse when there is a lack of watering holes lack. To support the two important goals, gazelles will change their dietary preferences depending on the season and local environmental conditions.

2.2.15 Feeding Behavior and Daily Movements

In addition to seasonal migrations, gazelles make daily movements with distances of 6–7 km. In the mornings, goitered gazelles move from their night pastures to the places of the daytime rest, and in the evenings, they go back to their night pastures and watering places (Blank 1986). In the Ili Hollow, daily rest sites generally are located in the foothills of Chulak and Kalkan, while in the Saryishikotrau Desert and in the area of Ayakkalkan (Chilik district in the Alma-Ata region), they are in the dunes until nighttime treks to takyrs or sandy plains. In the western part of the Ili Hollow, the gazelles pass from the shore of the Kapchagay reservoir to the foothills of Chulak in the morning and return to pastures that are closer to the reservoir in the evening. Closer to the shore, there are similar movements: to the east, up in the morning; to the west, down in the evening. Such movements in other areas of the hollow look somewhat different, but everywhere, there is the same pattern: in the morning, goitered gazelles move from the plains to the foothills, and in the evening, they go back to the plains (Blank 1990).

Tracking goitered gazelles in Azerbaijan with the help of GPS collars demonstrated that goitered gazelles cover usually 2–5 km per day during grazing from pastures to watering places and back. In some cases, during foraging, goitered gazelles can cover 10–15 km per day, reaching a watering place at the end. In dry summer periods, the gazelles move closer to water sources to shorten their daily trek, but they continue their movements during feeding (Askerov et al. 2021). If quite a lot of puddles are formed after rare summer rains in the Ili Hollow, goitered gazelles stop going to watering places during the day, but they continue to do their movements during their morning and evening grazing times (Blank 1990).

Only very rarely will gazelles stay for more than a few minutes in one place, and they never graze in one place continuously. In limited territories, such as the mountain plateaus, gazelles go in circles or shuttle back and forth while grazing. They also change their pastures even if there is an abundance of watering places. This grazing system allows for the most complete use of the pastures by the gazelles with a uniform use in all of its parts. The constant movement of the animals while grazing, combined with the minor damage that the gazelle causes to plants, allows desert pastures to recover quickly. And even though there is an abundance of watering places in many gazelle habitats, like exist in the Ili Hollow, they do not exist everywhere. Sometimes, the gazelles are forced to wander to water sources daily or even several times a day, changing pastures at the same time (Blank 1990). In captive conditions, where goitered gazelles have restricted movements, many shrubs and grasses are overgrazed (Fig. 2.1).

During their daily movements in the Ili Hollow, goitered gazelles move slowly while constantly grazing. Along the way, they stop to rest at least once, sometimes twice for 20 min to an hour. While resting, they chew their cud, then get up, and move on until they reach their daytime resting site, where they stay until the evening. Sometimes, the gazelles will interrupt their rest by short-term grazing. In the evening, they go back down the slope to their night-time pastures, grazing constantly as they go. The rest stop on the way back in the evening is not as long

Fig. 2.1 In captive conditions, where goitered gazelles are restricted in their movements, many shrubs and grasses become overgrazed. The Issyk-Kul goitered gazelle nursery for reacclimatization, 24.11.2022. Photo by D. Blank

(15–20 min), and as a rule, it happens before dusk. Gazelles do not all move at the same time: some gazelles arrive early to their place of their day rest, while others are just starting to climb, some are resting halfway, and still others climb at a brisk pace without grazing or rest. But the maximum number of moving animals is noted close to dawn and sunset. In addition, the direction of movement can be different. Moreover, the same gazelle can repeatedly change its direction and sometimes unexpectedly turn 180° and go back. However, the general trend of movement of most ungulates to the foothills of mountains from the plain pastures is obvious. In bad weather and before precipitation, daily movements almost stop. Goitered gazelles stay for a day in the mountains, going into saxaul thickets or deep dry riverbeds (Blank 1990).

A female with fawns behaves somewhat differently. The mother does not go far from her young for the first 2 days after giving birth, remaining constantly in the area where her fawn is hidden. While grazing, she walks in circles or back and forth. On the 2nd or 3rd day, she leaves for the first time to drink, but not for long (maximum absence 1.5 h); she often runs to the watering site and back. And only on the 3rd or 4th day will she begins to leave for watering and night pastures and return to her fawns only in the morning. During the day, the movements and activities of a female with a hidden fawn differ markedly from those of other adults and subadults. Then, on approximately the tenth day after birth, the fawn, left to itself, walks in random directions for a short while, before returning to the bed where its mother left it. The movements of the fawns become directional, such as in adults, only at the age of 2–2.5 months (Blank 1990). Territorial males, which are forced to stay only within their own individual territories during the rut, stop daily movements that are typical for most of year and show a different behavior by moving during feeding from one territorial border side to another, checking and refreshing scent marks along the way (Blank 1998).

In gravelly deserts, goitered gazelles graze along dry riverbeds, where they spend more of their time than between the riverbeds, especially during the flora growing season. Almost all the dwarf shrubs and shrubs are concentrated in riverbeds, where green shoots appear early. In late autumn and winter, they are more often located between the reverbeds, where *Arthrophyton* and *Nanophyton* grow, which are the basis of nutrition at this time of year (Blank 1990).

Eating shrubs, gazelles usually linger around them from a few seconds to a minute, rarely longer. I have noted 17 cases when gazelles ate leaves from one bush for as long as 1 h an 8 min; the average for all cases noted is 13.3 min. Often gazelles try to reach leaves as high as possible by stretching their muzzles upward. They also stand with their front legs on the mounds near the shrub in order to reach higher leaves (Fig. 2.2). Sometimes they manage to get the lower, young branches or leaves of trees (*Elaeagnus oxycarpa*, less often *Populus diversifolia*). Eating a shrub, the gazelles pinch the leaves off one by one (this is especially true of such prickly shrubs as *Halimodendron*, *Caragana*, and *Berberis*) or immediately tear off a mouthful (*Eurotia ceratoides*, *Atraphaxis*). In the latter case, the gazelle grabs a

Fig. 2.2 In order to reach higher leaves, goitered gazelles can stand with their front legs on the mounds near the shrub. Issyk-Kul goitered gazelle nursery for re-acclimatization, 26.06.2022. Photo by D. Blank

lower branch and twists its head to the side so that the branch slides between the teeth and lips, scraping off and leaving all the leaves are in the mouth (Blank 1990).

With the beginning of the vegetation of shrubs, gazelles bite mainly the apical green shoots. But in early spring, before the green parts of the plants appear, they eat the leaves and twigs at the base of the shrub or try to squeeze their muzzle into the middle while sometimes scratching the bush with their front foot. Apparently, only dry leaves and thin twigs remain there from winter. During grazing, gazelles will often lick their muzzle and nose with their tongue (Fig. 2.3), but this is more often observed after eating snow in winter (Fig. 2.4) (Blank 1990).

Several times, I noted how gazelles scrape the ground with 1 foot 3–4 times, then eat something in that place. According to Gorelov (1972), digging up juicy plant roots is a fairly common phenomenon with these gazelles in the Badkhyz Reserve in summer. This is how they quench their thirst (Minerin 1955). In the Ili Valley, over my 6-year study (more than 1000 h of observations), I noted soil scraping for the purpose of nutrition only six times. And only once was it possible to say with certainty that the gazelles ate rhizomes. In other cases, the gazelles were scratched with their front feet in a place completely devoid of vegetation or in the track of a dirt road. Sometimes they seemed to eat something found on the road without scratching. The rarity of such behavior in the Ili Valley is probably due to the relative abundance of available watering holes. In contrast, in the captive herd in Issyk-Kul, I saw goitered gazelles, which scratched ground by their front legs quite frequently. This difference in the rate of ground scratching is likely explained by

Fig. 2.3 During grazing, gazelles often lick their muzzles and noses with their tongues. Issyk-Kul goitered gazelle nursery for re-acclimatization, 25.11.2022. Photo by D. Blank

Fig. 2.4 Goitered gazelles often lick their muzzle and nose with tongue after snow eating. Kalamali Ungulate Nature Reserve, Xinjiang, China, 15. 12. 2022. Photo by F. Xu with kind permission

softer sandy ground in the Issyk-Kul enclosure. These captive gazelles have access to water to drink freely, so scratching the ground likely was related to digging up shrub branches or scattering shrubs around or crushing vegetative pads for eating (Fig. 2.5). In addition, the gazelles often collected broken shrub branches from the soft sandy ground without scratching (Fig. 2.6). It is interesting that in captivity, goitered gazelles will gnaw the bark of coniferous trees being used as support posts inside the enclosure, though these trees do not exist in goitered gazelle habitats (Fig. 2.7) (Blank 2022, unpublished report).

Like any ruminant, the goitered gazelle chews the cud after grazing while resting and usually lying down. However, they also chew their cud by interrupting grazing or while resting on their feet. So, of the 64 cases of chewing cud noted, 40 (62.5%) occurred when the gazelle was lying down and 24 when standing (37.5%), including five that interrupted grazing (20.8%) and 19 when standing at rest (79.2%). The duration of chewing varies from 4 min to 1 h 14 min, on average 17.7 min ($n = 18$), and longer while lying down (on average 19.8 min, $n = 12$) than standing (13.5 min, $n = 6$) (Blank 1990). I also observed a wide opening of the mouth for a few seconds during grazing. This movement seemed very similar to yawning and has been observed in both the wild and in captive groups. Is it related to digestion or a real yawn?—a question that needs an answer (Fig. 2.8).

Fawns begin to show interest in vegetation by as early as their 2nd or 3rd day after birth. At this time, they occasionally sniff plants, especially those that their mothers eat. On the 5th to 7th day, they bite off and try to chew bits of the plant, imitating their mothers. By the 10th day, they are already eating plant material, albeit only small amounts and only for a short time. During this time, the fawns try to learn exactly what plants and only what parts of the plant to eat by observing and

Fig. 2.5 Scratching the ground by goitered gazelle likely is related to crashing vegetative parts for eating. Issyk-Kul goitered gazelle nursery for reacclimatization, 25.11.2022. Photo by D. Blank

Fig. 2.6 Goitered gazelles often collect broken shrub branches from the soft sandy ground without scratching. Issyk-Kul goitered gazelle nursery for re-acclimatization, 26.06.2022. Photo by D. Blank

Fig. 2.7 In captivity goitered gazelles can gnaw the bark of coniferous trees, which are absent from the habitats of this species. The Issyk-Kul goitered gazelle nursery for re-acclimatization, 25.11.2022. Photo by D. Blank

Fig. 2.8 This behavior is very similar to yawning and is observed during grazing both in the wild and in captivity. Issyk-Kul goitered gazelle nursery for reacclimatization, 25.11.2022. Photo by D. Blank

trying what the female is eating: the fawn pushes their mother's muzzle away with their head and body, but as soon as the female moves away to another bush, the fawn immediately rushes to it. On their 15th to 17th day, the fawns graze on a par with adults, despite the fact that their mother's milk still occupies a significant place in their nutritional intake (Blank 1985).

References

Abaturov BD, Kassaye F, Kuznetsov GV, Magomedov MRD, Petelin DA (1995) Nutritional Estimate of Populations of Some Wild Free-Ranging African Ungulates in Grassland (Nechisar National Park, Ethiopia) in Dry Season. Ecography 18(2):164–172

Afanasiev AV, Bazhanov VS, Karelov MN, Sludskiy AA, Strautman EN (1953) Goitered gazelle. In: Animal of Kazakhstan. Publication of Kazakh Academy of Sciences, Alma-Ata, pp 484–488. (in Russian)

Antipin VM (1941) Mammals of Kazakhstan. In: Ungulates, vol 3. Publication OGIZ, Alma-Ata, pp 47–54. (in Russian)

Askerov E, Mallon DP, Zazanashvili N, Kochiashvili V, Sarukhanova S, Muradov A (2021) An attempt to restore the population of the Goitered Gazelle (Gazella subgutturosa) in its indigenous range in the Caucasus (Mammalia: Bovidae). Zoology in the Middle East, 67(3):189–197

Bagherirad E, Ahmad N, Amirkhani M, Abdullah M, Mesdaghi M, Kabudi A (2014) Seasonal habitat use of Persian gazelles (*Gazella subgutturosa subgutturosa*) based on vegetation parameters at Golestan National Park, Iran. Arid Land Res Manag 28(4):464–484

Bannikov AG (1954) Persian gazelle. In: Heptner VG (ed) Mammals of Mongolian People Republic. Publication of USSR Academy of Sciences, Moscow, pp 250–258. (in Russian)

Blank DA (1985) Peculiarities of social and reproductive behaviour of *Gazella subgutturosa* in the Ili Valley. Zoological Zhurnal 64:1059–1070. (in Russian, English abstract)

Blank DA (1986) On ecology goitered gazelle in the Ili Valley. In: Rarely animals of Kazakhstan. Nauka of KazSSR Publication, Alma-Ata, pp. 14–19 (in Russian)

Blank DA (1990) Persian gazelle. In: Kovshar AF (ed) Rare animals of desert regions Nauka Press of Kazakh SSR. USSR, Alma-Ata, pp 56–80. (in Russian)

Blank DA (1998) Mating behavior of the Persian gazelle *Gazella subgutturosa* Güldenstaedt, 1780. Mammalia 62(4):409–419

Blank DA (2022) Activity behavior of goitered gazelles in an arid environment. Behav. Proc., p.104804

Blank DA, Li Y (2022) Antelope adaptations to counteract overheating and water deficit in arid environments. J Arid Land 14(10):1069–1085

Chaplina ZP (1959) The chemical content of forage plants of deserts and foothill semi-deserts of middle Asia. Proc Institute Sheep Karakul-Breeding 8:148–158. (in Russian)

Combes S, Massip K, Martin O, Furbeyre H, Cauquil L, Pascal G, Bouchez O, Floch NL, Zemb O, Oswald IP, Gidenne T (2017) Impact of feed restriction and housing hygiene conditions on specific and inflammatory immune response, the cecal bacterial community and the survival of young rabbits. Animal Int J Anim Biosci 11:854–863

Cunningham PL (2009) Observations of the seasonal dietary preference of male *Gazella subgutturosa maraca* Thomas, 1897 (Cetartiodactyla: Bovidae) along foraging trails of Central Saudi Arabia. J Threatened Taxa:445–449

Cunningham PL (2013) Plants included in the diet of Arabian sand gazelle (Reem) from Saudi Arabia. J King Saud Univ Sci 25:167–173

Cunningham PL, Wacher T (2009) Changes in the distribution, abundance and status of Arabian sand gazelle (*Gazella subgutturosa marica*) in Saudi Arabia: a review. Mammalia 73:203–210

Dehghani TM, Parvaneh AE (2010) Dietetics of *Gazella subgutturosa* (Goitered gazelle) and Quality of Consuming Forages in Kalmand–Bahadoran Protected Area in Yazd Province. J Animal Biol 2(2):19–25

Dementyev GP (1935) Mongolian gazelle, goitered gazelle, saiga antelope. KOIZ Publication, Moscow, pp 15–30. (in Russian)

Dereviagin PYA (1947) On seasonal migrations of goitered gazelle from southern pre-Balkhash deserts into inter-mountain valley Kalkan-Matay- Mynbulack. Bull Acad Sci Kazakh SSR 4(25):33–38. (in Russian)

Dutov VYU (1990) About feeding of goitered gazelles in the "Tigrovaya Balka" nature reserve. Proc Acad Sci Tajik SSR Biol Depart Biol Sci 4(121):41–46. (in Russian)

Farhadinia MS, Esfandabad BS, Karami M, Hosseini-Zavarei F, Absalan H, Nezami B (2009) Goitered gazelle, *Gazella subgutturosa*: its habitat preference and conservation needs in Miandasht wildlife refuge, North-Eastern Iran (Mammalia: Artiodactyla). Zool Middle East 46(1):9–18

Gao W, Ji L (1996) An analysis of the nutrient composition of five species of sandy plants growing in west inner Mongolia. J Inner Mongolia Inst Argric Anim Husband 17:23–28

Gillet H, Launay C (1990) Flora analysis of the Mahazat as-Sayd vegetation.—unpublished report. National Wildlife Research Centre, Taif

Gorelov UK (1972) Breeding of goitered gazelle (*Gazella subgutturosa* Güldenstaedt) and problem of restoring of its number in the Badkhyz (SE Turkmenistan). In: Theriology, vol 1. Nauka of Siberian Department of Academy of Sciences of USSR, Novosibirsk, pp 420–424. (in Russian)

Gürelli G, Göçmen B, Yıldız MZ (2012) Rumen ciliates from the Goitered gazelle (*Gazella subgutturosa*) living in the Şanlıurfa Province of Turkey. Biharean Biologist 6(1):42–44

Haque MN (1996) Reintroducing Arabian sand gazelle at Mahazat as-Sayd. Unpublished report. National Commission for Wildlife Conservation and Development, Riyadh, pp 1–31

Harrison GL, Bates PJJ (1991) The mammals of Arabia. Harrison Zoological Museum, Sevenoaks

Hemami MR, Groves CP (2001) Chapter 21. Iran. In: Mallon DP, Kingswood SC (eds) Antelopes. Part 4. North Africa, the Middle East, and Asia. Global survey and regional plans. The IUCN/SSC Antelope Specialist Group, IUCN, Glands, pp 114–118

Heptner VG (1956) Fauna of vertebrate animals of the Badkhyz (S Turkmenistan). Academy of Sciences of Turkmenistan, Ashkhabad

Heptner VG, Nasimovich AA, Bannikov AG (1961) Mammals of Soviet Union. In: Artiodactyla and Perissodactyla, vol 1. Vysshaya Shkola Press, Moscow

Heptner VG, Nasimovich AA, Bannikov AG (1988) Mammals of the Soviet Union. Artiodactyla and Perissodactyla. Smithsonian Institution Libraries and The National Science Foundation, 1, 1147 pp (translation from Russian)

Hofmann RR, Stewart DRM (1972) Grazer or browser: a classification based on the stomach-structure and feeding habits of East African ruminants. Mammalia 11:226–240

Hu D, Zhang D, Lan C (1998) Ecology habitat of *Equus hemionus* and *Gazella subgutturosa* in Kalamaili ungulate Reserve in autumn. J Forestry Res 9(2):130–132. (In Chinese, English abstract)

Ishunin GI (1961) Mammals (carnivores and ungulates). In: Fauna of Uzbek SSR, vol 3. Publication of the Academy of Sciences of Uzbek SSR, Tashkent, pp 165–174. (in Russian)

Islam MZ, Ismail K, Boug A (2010) Catastrophic die-off of globally threatened Arabian Oryx and sand gazelle in the fenced protected area of the arid Central Saudi Arabia. J Threatened Taxa:677–684

Jahandizi EK, Kaboli M, Karami M, Soufi M (2015) The assessment of carrying-capacity of Persian gazelle in Sorkh-E-Hesar National Park, Iran. J Earth Environ Health Sci 1(2):66

Jia SX (1987) Fodder plants in China. Agriculture Press, Beijing

Kalenov GS (1959) The content of carotene in the pasture vegetation of Badkhyz. Bulletin of the Scientific Research Institute of Animal Husbandry and Veterinary Medicine. Ashgabat: 40–47 (in Russian)

Karami P, Esmaeilpour Y, Sharifi M (2017) Investigating the effects of livestock entry on ecological niche changes of the Persian gazelle (*Gazella subgutturosa subgutturosa*) (case study: Qaraviz no hunting area, Kermanshah). J Environ Sci Technol 19(5):507–521

Keller BA (1940) Plant and environment. Publication of Academy of Sciences, Moscow (in Russian)

Kim TA (1953) Mammals of northern and western Kyzylkum Desert. Dissertation for the degree of candidate of biological sciences. Moscow State Pedagogic Institute, Moscow (in Russian)
Kingswood SC, Blank DA (1996) Gazella subgutturosa. Mamm Species 518:1–10
Kostin VP (1955) Materials on ecology and commercial value of goitered gazelle. Proc Institute Zool Acad Sci Kazakh SSR 4:67–82. (in Russian)
Kubanskaya ZV (1956) Vegetation and forage resources of the Betpak-Dala Desert. Academy of Sciences of KazSSR Press, Alma-Ata. 254 pp. (in Russian)
Kurochkina LYA, Karibaeva KN (1986) Forage plants of Kazakhstan deserts. Publication of Kainar, Alma-Ata. 206 pp. (in Russian)
Kurochkina LYA, Osmanova LT (1973) Pastures of sand deserts in Kazakhstan. Academy of Sciences KazSSR, Alma-Ata. 204 pp. (in Russian)
Kuznetzov LA (1976) Observation on seasonal development of desert vegetation of the Barsakelmes Island. In: Systematics, anatomy and ecology of plants of the Asiatic part of the USSR. Leningrad, 171–188 pp. (in Russian)
Larin IV (1951) Forage plants of haymaking areas and pastures in USSR. Vol. 2. USSR Academy of Sciences, Moscow-Leningrad
Larin IV, Tshelkov BK, Kazbekov IS, Ischenko MM (1929) Natural forage South-Western Kazakhstan. Chemistry and digestibility. USSR Academy of Sciences, Moscow. 193 pp. (in Russian)
Li Y, Xu W, Qiao J, Yang W (2009) The spatio-temporal distribution and habitat selection of Gazella subgutturosa in Kalamaili Mountain Nature Reserve in four seasons. Arid Land Geogr 32:261–267. (in Chinese with English abstract)
Li Y, Xu W, Yang W, Qiao J, Liu W, Xia C (2010) Habitat suitability assessment of *Gazella subgutturosa* in Kalamaili Mountain nature reserve. Acta Theriol Sin 30:11–20. (in Chinese with English abstract)
Mambetzhumaev AM (1970a) Goitered gazelle. FAN publication, Tashkent. 196 pp. (in Russian)
Mambetzhumaev AM (1970b) On the number and distribution of goitered gazelles in the Ustyurt plateau and Kyzylkum Desert. Zoological Zhurnal 49(3):431–434. (in Russian)
Mardonov BK (1993) Impact of goitered gazelle on the vegetative cover of desert. PhD thesis, Moscow, 26 pp (in Russian)
Mardonov BK (1996) The nature of impact of goitered gazelle on the vegetative cover in the semi- captive conditions. In: Proceedings of nature reserves of Uzbekistan. Issue 1. FAN Press, Tashkent, pp 12–23. (in Russian)
Mikheev GD (1959) Seasonal changes of nutritional value of forage in the cereal-Carex-physodes pastures. Publication of the Turkmen Research Institute of Crop Production and Veterinary, Ashkhabad, pp 24–32. (in Russian)
Minerin VN (1955) About pasturing and watering regime of sheep in desert. Turkmen Branch of the Academy of Sciences USSR, Ashkhabad. 192 pp
Minervin VN (1944) On the watering-feeding regime of the goitered gazelle and domestic sheep in desert. Proc Turkmenistan filial USSR Acad Sci 5:227–241. (in Russian)
Mohamed SA, Abbas J, Saleh M (1991) Natural diet of the Arabian Rheem gazelle, *Gazella subgutturosa marica*. J Arid Environ 20:371–374
Morosova OI (1972) Desert and semidesert pastures. Publication house Kolos, Moscow. 168 pp. (in Russian)
Mowlavi M (1978) Ecological studies of the goitered gazelle (*Gazella subgutturosa*) in Khosh Yeilagh wildlife refuge, Iran. M.S. thesis. Michigan State University, East Lansing. 40 pp
Nezami B (2009) Goitered gazelle, *Gazella subgutturosa*: its habitat preference and conservation needs in Miandasht wildlife refuge, North-Eastern Iran (Mammalia: Artiodactyla). Zool Middle East 46(1):9–18
Nicholson JK, Holmes E, Kinross J, Burcelin R, Gibson G, Jia W, Pettersson S (2012) Host-gut microbiota metabolic interactions. Science 336:1262–1267
Nowzari H, Hemami M, Behrouzi Rad B (2007) Habitat use by Persian gazelle (*Gazella subgutturosa subgutturosa*) in Bamoo national park during autumn and winter. Acta zoológica mexicana 23(1):109–121
Pankova IA (1949) Herbaceous containing C-vitamin. Proc VIN USSR 2:56–62. (in Russian)

Pavlov NV (1947) Plant resources of southern Kazakhstan. Publication of Moscow Society of Naturalists, Moscow, pp 199 (in Russian)

Pitman CRS (1922) Notes on Mesopotamian mammals. J Bombay Nat Hist Soc 28:317–323

Qin W, Huang Y, Wang L, Lin G, Yang J, Song P, Gao H, Zhang J, Zhang T (2020) Gut microbiota enabled Goitered gazelle (*Gazella subgutturosa*) to adapt to seasonal changes. Pakistan J Zool 52(5):1637–1646

Rad EB, Mesdaghi M, Ahmad N, Abdullah M (2015) Nutritional quality and quantity of available forages relative to demand: a case study of the goitered gazelles of the Golestan National Park, Iran. Rangelands 37(2):68–80

Reed MC (1977) A history of three rare desert antelope at the San Antonio zoo. Am Assoc Zool Parks Aquariums Regional Workshop Proc:198–210

Sarukhanova S (2021) The study level of the goitered gazelle (Artiodactyla, Bovidae) in Azerbaijan. Bull Sci Pract 7(9):42–58. (in Russian)

Schoener TW (1974) Competition and the form of habitat shift. Theor Popul Biol 6:265–307

Schoener TW (1985) Some comments on Connell's and my reviews of field experiments on inter-specific competition. Am Nat 125:730–740

Schulz E, Fraas S, Kaiser TM, Cunningham PL, Ismail K, Wronski T (2013) Food preferences and tooth wear in the sand gazelle (*Gazella marica*). Mamm Biol 78(1):55–62

Sludskiy AA (1977) Goitered gazelle. In: Rare animals of USSR. Publication of Lesnaya Promyshlennost, Moscow, pp 28–61. (in Russian)

Sokolov II (1959) *Gazella subgutturosa* Güldenstaedt—Goitered gazelle. In: Pavlovskiy EN (ed) Fauna USSR, mammals: ungulates (orders Perissodactyla and Artiodactyla). Publication of the Academy of Sciences USSR, Moscow, pp 308–331. (in Russian)

Thouless CR, Grainger JG, Shobrak M, Habibi K (1991) Conservation status of gazelles in Saudi Arabia. Biol Conserv 58:85–98

Vereshchagin NK (1939) Goitered gazelle in Azerbaijan. Proceedings of the Zoological Institute of the Azerbaijan Branch of the USSR Academy of Sciences 10:89–90. (in Russian)

Volozheninov NN (1983) Feeding and feeding activity of goitered gazelle in Bukhara nursery. In: Ecology, № 1: 46-49 (in Russian) (citation by Marmazinskaya and Mardonov 2004)

Wacher T (1995) Sand gazelle monitoring at Uruq Bani Ma'Arid.—unpublished report. King Khalid Wildlife Research Centre, Thumamah, pp 1–15

Xu W, Qiao J, Liu W, Yang W (2008) Food habits of goitered gazelle (*Gazella subgutturosa sairensis*) in northern Xinjiang. Acta Theriologica Sinica 28(3):280–286. (in Chinese, English abstract)

Xu W, Xia C, Lin J, Yang W, Blank DA, Qiao J, Liu W (2012) Diet of *Gazella subgutturosa* (Güldenstaedt, 1780) and food overlap with domestic sheep in Xinjiang, China. Folia Zool 61(1):54–60

Yang W, Qiao J, Yao J, Gao X (2005) Characteristics of foraging habitat of goitered gazelles (*Gazella subgutturosa sairensis*) in eastern Junggar Basin, Xinjiang. Acta Theriologica Sinica 25(4):355–360. (in Chinese, English abstract)

Zhang J, Guo Z, Lim AAQ, Zheng Y, Koh EY, Ho D, Qiao J, Huo D, Hou Q, Huang W, Wang L, Javzandulam C, Narangerel C, Menghebilige J, Lee YK, Heping Z (2014) Mongolians core gut microbiota and its correlation with seasonal dietary changes. Sci Rep 4:1–11

Zhevnerov VV (1975) The feeding of goitered gazelle in the Barsakelmes Island in winter. Ecology 6:95–97. (in Russian)

Zhevnerov VV (1984) Goitered gazelle of the Barsakelmes Island. Nauka Publishing House of Kazakh SSR, Alma-Ata. 144 pp. (in Russian)

Zhevnerov VV, Bekenov AB, Sludskiy AA (1983) Goitered gazelle—*Gazella subgutturosa* Güldenstaedt, 1780. In: Gvozdev EV, Kapitonov VI (eds) Mammals of Kazakhstan, Bovidae, vol 3 (3). Nauka Publishing House of Kazakh SSR, Alma-Ata. 11–54 pp. (in Russian)

Zmora N, Zilberman-Schapira G, Suez J, Mor U, Dori-Bachash M, Bashiardes S, Kotler E, Zur M, Regev-Lehavi D, Brik RBZ, Federici S (2018) Personalized gut mucosal colonization resistance to empiric probiotics is associated with unique host and microbiome features. Cell 174(6):1388–1405

Rutting Season: Morphological and Physiological Changes

3

3.1 Testicle Sizes and Weights: Age and Seasonal Changes

The testes diameters for three young males were 4 mm at 2 weeks of age, 10 mm at 18 months, and 18 mm at 12 months-age; the testes diameters of three adult males (>3 years old) were 24–27 mm (Bland 1985). Male-yearlings have undeveloped testicles of 35 × 19 mm in size with a weight of 4.4 g, while testicles of adult males have a weight of 21.6 g on average (13 May, 1956) (Betpak-Dala Desert, Kazakhstan—Sludskiy 1956). In Turkmenistan, the total weight of testicles with appendages of a young male (1.5 years old) was only 18.4 g on November 16, 1962, while an adult male at the same time had testicles by weighing 47 g (Gorelov 1972). Apart from the autumn rut in November, there is a second period of increasing of testicles sizes, in April–May (Table 3.1) as a result of the spring peak of sexual activity in goitered gazelles. The spring peak is usually nonproductive (see details below).

Testicular weight does not depend on a male's body weight or size: males with a body length of 121 of 125 cm and a weight between 28 and 32 kg had testicles that weighted from 18 to 44 g, while males with a body length of 107–119 cm and a weight of 24–32 kg had testicles that weighed 22–55 g (Zhevnerov 1984). During the rut, the sizes of testicles and their weights increase significantly, reaching their maximum in November (Tables 3.1 and 3.2). During a period of the functional rest of testicular in July and August, they weights of 13.6 to 42.4 g for an average of 27.6 g ($n = 5$), while in November, the weight range was from 30 to 57.5 g with an average of 50 g ($n = 7$) with an increasing weight of 55%, where the appendage of the testicles has a weight of 2–5 g (Sludskiy 1956). After the main rut in January, the testicle weight decreases drastically to 21–32 g with an average of 25.8 g (Zhevnerov 1984). The testicle sizes also increase during the breeding period, with maximal sizes observed during the rut: the length is 42–50 mm with an average of 46 mm and a width of 28–36 mm for an average of 32.3 mm. Testicle sizes during the summer

D. Blank, *The Behavioral Biology and Conservation of Goitered Gazelles (Gazella subgutturosa)*, https://doi.org/10.1007/978-3-031-83504-9_3

Table 3.1 Changing of testicle sizes and weights in adult male of goitered gazelles over months (adapted from Sludskiy 1956)

Date	Testicle size (mm)		Testicle weight with appendage (g)		Number of individuals
	Length, range	Width, range	Range	Average	
January	33–44	24–32	18.9–25.1	19.4	2
March	35–38	25–28	10.0–12.0	11.0	1
April	38–39	25–28	14.2–17.3	15.7	1
May	34–54	25–32	20.5–23.5	21.6	7
June	38–42	25–29	17.1–17.2	17.1	2
July	27–31	19–28	8.2–9.9	9.0	2
August	35–44	24–32	12.0–22.2	17.5	3
November (1–15)	45–50	32–34	20.0–29.0	24.7	2
November (16–30)	42–47	29–36	19.3–28.6	23.7	4
December	37–44	28–32	19.5–22.0	20.0	2
Total					26

Table 3.2 Changing of testicle sizes and weights in adult goitered gazelle males on the Barsakelmes Island over months (adapted from Ismagilov 1954 and Zhevnerov 1984*)

Date	Testicle size (mm)		Testicle weight with appendage (g)	
	Right side (Length × width)	Left side (Length × width)	Right side	Left side
March 05*	41 × 32	41 × 32	18.5	18.0
March 20	35 × 25	38 × 28	10.0	12.0
April 27	38 × 25	39 × 28	14.2	17.3
01 May	39 × 26	39 × 25	10.0	8.5
23 May	34 × 25	35 × 25	8.0	8.6
07 June	31 × 21	27 × 19	8.2	5.4
25 June	42 × 27	40 × 29	17.1	17.2
21 July	30 × 21	30 × 28	9.4	9.9
01 August	31 × 32	41 × 32	19.3	19.3
04 August	35 × 27	35 × 24	12.5	12.0
13 August	44 × 30	43 × 30	22.2	20.0
17 August*	47 × 32	45 × 32	22.5	22.0
15 September*	44 × 31	43 × 30	24.1	23.5
09 November	50 × 34	50 × 33	29.0	28.5
09 November	46 × 34	45 × 32	22.0	20.0
23 November	47 × 36	45 × 35	26.6	25.1
27 November	47 × 33	46 × 31	28.6	25.4
30 November	42 × 28	41 × 26	20.0	19.3
08 December*	45 × 32	43 × 30	28.5	27.5
20 December*	40 × 35	40 × 35	26.0	26.0

at their smallest are as follows: a length between 30 and 44 mm, for an average of 36.2, and a width between 21 and 32 mm, for an average of 25.8 mm.

3.2 Puberty, Spermatogenesis, and Ovulation

Males In the juveniles at the age of 2.5 months (26 July of 1962), one to six spermatogonia were found in the tubules of testicles in the different stages of development and on the periphery were situated follicular cells. Tubules of the epididymis were covered by a line of low prismatic epithelial cells with a narrow strip of cilia at the beginning of their development. Basal cells were absent. In males at the age of 5–6 months, which were taken in October, the number of spermatogonia increased up to ten (two individuals). Tubules of the epididymis were formed completely and were covered by two rows of ciliary epithelium, containing prismatic cells equipped with cilia and cubic basal cells. Spermicides of the first order in a stage of the prophase can be observed, and Sertoli cells can be found (Tsapluk 1972).

In male yearlings (April 2, 1957), most seminiferous tubules had a row of the spermatogonia cells, 1–2 rows of spermicides of the first order, perhaps spermicides of the second order, and 1–3 rows of spermatids which were located closer to the center. Some tubules had spermatozoids in the secretions of the gap, and some degenerative elements were found in other tubules. The spermatogonia cells were dispersed in the secretions of the epididymis tubules and spermatozoids were rarely found. On May 9, a male was caught with advanced involution of his testicles. In the periphery, spermatogonia and spermicides of the first order were distinctive, and then, the cells were resorbed. The resorbed outputs of the spermatogenesis were seen in the epididymis tubules. Thus, the processes of the spermatogenesis for all its stages were observed first in yearlings, though their intensity was very low. The testicles' involution followed after the spring peak of gonad activity (Tsapluk 1972).

On June 28, 1962, a 14-month-old male was observed to have the first stages of spermatogenesis, with spermatogonia found in the periphery of the seminiferous tubules, and after the spermatogonium, there were a few spermatocytes of the first order, and the first spermatids had appeared. The products of resorption were found sporadically in the center of the tubules. Only secretion was observed in the epididymis tubules. On October 3, 1957, spermatogenesis was in progress. Spermatozoids covered the most space of the gap, with their density fluctuating from mid to high density. In November, the testicles of goitered gazelles had signs of a decline of the spermatogenesis, because the conversion process of spermatids into spermatozoids decelerated and spermatids existed in the most seminiferous tubules, some of which had degenerated with fewer spermatozoids. And on January 10, 1965, the seminiferous tubule contained mostly degenerated spermatogonia (Tsapluk 1972).

Females In females, sexual maturity comes earlier in life. According to observations by Sludskiy (1956), some young females become fertile at the age of 7–8 months and can give birth as yearlings. In the Betpak-Dala Desert, Central Kazakhstan, the ovaries of young females had reached their follicle diameter of 4 mm in October and early November and 7 mm in December. In female yearlings of 12 months, the embryo weight was between 0.8 and 2.1 kg. Only in one case, on May 5, 1959, was the embryo weight as high as 2.8 kg, while in adult females, embryo weights fluctuated between 2.1 kg up to 3.3 kg. Hence, young females mate later than adult females, which is apparently related to their time of puberty, and the extended rutting season in goitered gazelles is determined by later fertilization of young females. Thus, only subadult females can participate in the rut during their first-year life, while male yearlings have a reduced intensity of spermatogenesis and the subsequent stage of involution of the gonads; they only experience complete puberty at the age of 17–18 months (Tsapluk 1972).

Although yearling males might effectively sire offspring in the absence of older males, they usually do not mate with females before 1.5–2.5 years of age (Heptner et al. 1961; Tsaplyuk 1972). In the Issyk-Kul nursery, in the absence of adult males, yearling males demonstrated numerous mounts on yearling females in May 2022 (during the spring peak of sexual activity), but no one female got pregnant (Blank 2022, unpublished report) because of the very low intensity of spermatogenesis process in the young males at the age of 11 to 12 months (Tsapluk 1972). Some young females participate in the rut at the age of 6–8 months, giving birth as yearlings. The proportion of young females bearing young depends on environmental conditions, increasing greatly in number during favorable years. Among nine young female yearlings observed in the Betpak-Dala Desert in May 1955, five females were pregnant and four were not, and in 1956, 20 out of 22 female-yearlings were pregnant, while only two, the smallest by size and weight (16.2 kg and 22.4 kg), were not (Sludskiy 1956). Goitered gazelles of both sexes can remain reproductively mature past the age of 10 years; a fawn was sired by a male of about 10 years and 9 months, and a female gave birth at 13 years and 10 months (Carter 1991).

3.3 Seasonal Dynamics of Testicular Activity

Throughout the year, the mass of testes in gazelles is subject to significant changes (Table 3.1). In January, in the somniferous tubules of goitered gazelles, there were mostly degenerative spermatogenic cells, primarily spermicides and spermatogonium (partly new). Some epididymis tubules had spermatozoids and degenerative reproductive cells, others had no spermatozoids, and the third tubules had only secretions (three individuals). On February 6, 1963, the reproductive cells were in different stages of regeneration. The tubules were filled with by-products of decomposition, while some had degenerative spermatogonia, and the testicles were complete and deeply involuted. On April 27, 1957, males had intensive spermatogenesis, and numerous reproductive cells of all stages were found in the testicles, and most tubules had spermatids and numerous spermatozoids. The resorption products were found rarely. The epididymis tubules were filled by dense spermatozoids and rarely were degenerative germ cells observed. A distinctive decrease of spermatogenesis with involuted testicles was found in May, with spermatozoids found only in a few tubules, and mainly products of resorption of the spermatogenic cells were observed. The epididymis tubules were filled by spermatozoids with middle and high-level density. In October, the spermatogenesis was at its height, when spermatids and spermatozoids were predominant. The epididymis tubules were filled by spermatozoids with middle up to high density, while no spermatogenetic cells were found. The nuclei of the high-prismatic epithelium occurred in 4–5 layers. In late November, the initial processes of degeneration were found in epididymis tubules: spermatozoids were observed only occasionally, and in most tubules, products of the spermatogenesis were presented primarily by one row of spermatogonia, one row of spermatocytes of the first order, and 2–3 rows of spermatocytes of the second order and spermatids. However, many of these cells had signs of degeneration. The epididymis tubules were filled by spermatozoids to a high degree with high density. In early December, the spermatogenesis intensity significantly decreased in four males out of six. Many epididymis tubules had a few spermatogenic cells, and only separate spermatozoids were seen, while concentrations of the resorption products were high. In contrast, two other males at that time had active spermatogenesis with a predominance of spermatids. In the epididymis tubules, the spermatozoid density was from middle to high, though with a considerable admixture of products of the last stages of spermatogenesis, which were usually degenerative. In the high-prismatic epithelium, the nucleus vertical layers reached six. At the end of December, most epididymis tubules had spermatogonia in the periphery, part of which were degenerative stages and spermatocytes of the first order having the pyknosis of the nucleus, and then the products of resorption followed. Some tubules had all stages of the spermatogenesis. Another male had only spermatogonia (partly degenerative) in most epididymis tubules. The considerable space of tubules was filled by the products of resorption of the spermatogenic cells. The epididymis tubules were filled by spermatozoids from one third to full space with middle density. In January, the involution of testicles was increasing (Tsapluk 1972).

Thus, the seasonal changes of the histological structure of the testicles and epididymis demonstrated that male goitered gazelles have two periods of increasing sexual activity within a year: the spring peak is weak and usually nonproductive, and the autumn peak is more vigorous and very productive. The changes in the histological structure of ovaries of adult females over seasons also showed two peaks of sexual activity in spring and autumn (Tsapluk 1972).

The histological structure of gonads in adult females, which were observed in mid-May, demonstrated a decline in high sexual activity. The ovaries were still abundantly supplied by blood vessels, and somewhere lacunae were filled with uniform blood elements. However, the largest antral follicle at 6 mm in diameter had marks of deep atresia. There were 30 cavernous follicles out of 39 in the stage of deep atresia, three at the stage of developing atresia, and five at the stage of initial atresia, and only one follicle of 3 mm in diameter had no signs of degeneration. The occurrence of only a single atresia body per ovary demonstrated that high sexual activity had occurred recently, probably in April into early May. Interestingly, that females will sometimes get pregnant in spring (Antipin 1941). Tsapluk's (1972) histological data suggests that female fertilization can happen in the spring, and a June embryo could corroborate of an atavistic phenomenon. After May, ovary activity was in a more passive period based on the absence of atresia bodies. During mating, females can have a second estrus cycle if the first did not result in pregnancy.

At the end of September, sexual activity was quite intensive, which was indicated by a great number of the cavernous follicles. There were three follicles without signs of atresia and seven follicles had begun degeneration. However, 20 cavernous follicles were in a stage of deep atresia, which demonstrated that the period of the intensive sexual activity already had been ongoing for a significant amount of time. The diameter of the largest follicle at this time was equal 4 mm. In October, the ovaries of two females had follicles with a diameter of 6 mm, but they did not develop until ovulation and the process of atresia process. During this time, the atresia bodies formed. One female had eight such bodies, while another had 12. In November, there were numerous cavernous follicles developing. The diameter of largest follicle was 8 mm, which was found in two individuals. Obviously, fertilization occurred at this time, because on December 8, a female had the yellow body or corpus luteum in the most advanced stage of hormone development. Apart from the corpus luteum, this female had a follicle with an 8.5 mm diameter which indicated the existence of two estrus cycles in goitered gazelles, similar to the saiga antelope (Tsapluk 1972).

In Arabian sand gazelles (*G.s. marica*) in Saudi Arabia, a second postpartum estrus season was found in two females in May and even June, which is possible when a female's body is in good condition. In semi-captive conditions in Saudi Arabia, photoperiods, which control prolactin, as well as food and water conditions are favorable for a second peak of breeding in Arabian sand gazelles, as well (Sempéré et al. 2001).

In summary, investigations of the histological structure of goitered gazelle gonads demonstrated that:

(a) Reproductive puberty of females occurs at the age of 7–8 months.
(b) During the breeding period, goitered gazelle females may have a second estrus cycle, if the first mating was unsuccessful.
(c) The first spermatogenesis consisting of all its stages occurs first at the age of 1 year but with lowered intensity. A male's reproductive puberty and potency come at the age of 17–18 months.
(d) During the year, adult goitered gazelles experience two peaks of sexual activity—spring and autumn—with only the latter being highly and predictably productive. Female fertilization can happen during the spring peak, though in the extremely rarely cases. During the spring in males of the Betpak-Dala Desert population in Kazakhstan, spermatogenesis reaches its highest intensity in April and again in October and November; between these peaks, there is a decline of sexual activities.

(e) The spring, nonreproductive peak of goitered gazelle sexual activity, is considered as a fading link to the biorhythms of a previous, more favorable epochs when goitered gazelles had two breeding periods per year. Saiga antelope also have two peaks of sexual activity in a year, but the intensity of the spring peak in goitered gazelles is much higher than in the saiga antelope (Tsapluk 1972).

References

Antipin VM (1941) Mammals of Kazakhstan. In: Ungulates, vol 3. Publication OGIZ, Alma-Ata, pp 47–54. (in Russian)

Bland KP (1985) Some observations based on the testicular histology of Dorcas and Arabian goitered gazelles. Chinkara 1:24–25

Blank DA (2022) Behavior of goitered gazelles in the Issyk-Kul Breeding Center, Kyrgyzstan. Unpublished Report: May, 2022

Carter S (1991) Goitered gazelle north American regional studbook, 1st edn. Sedgwick County Zoo and Botanical Garden, Wichita, p 87

Gorelov UK (1972) Breeding of goitered gazelle (Gazella subgutturosa Güldenstaedt) and problem of restoring of its number in the Badkhyz (SE Turkmenistan). In: Theriology, vol 1. Nauka of Siberian Department of Academy of Sciences of USSR, Novosibirsk, pp 420–424. (in Russian)

Heptner G, Nasimovich AA, Bannikov AG, (1961) Mammals of Soviet Union, Vol.1. Artiodactyla and Perissodactyla. Vysshaya Shkola: Moscow

Ismagilov MI (1954) The goitered gazelle ecology on the Barsakelmes Island. The Bulletin of the Moscow Society of Naturalists, Biol Department 59(6):3-9

Sempéré AJ, Brown N, Pereladova OB, Bahloul K, Lacroix A, Soldatova N (2001) Comparative analysis of reproductive cycles in female Persian gazelle (*Gazella subgutturosa subgutturosa*) (Central Asia) and sand gazelle (*Gazella subgutturosa marica*) (Arabian Peninsula). Gen Comp Endocrinol 121:57–65

Sludskiy AA (1956) Breeding of goitered gazelle. Proceedings of institute of zoology, Kazakh Academy of Sciences, vol 6. Nauka Press of Kazakh SSR, Alma-Ata, pp 78–108. (in Russian)

Tsapluk OE (1972) Age and seasonal dynamics of sexual activity of goitered gazelle. Proc Acad Sci Kazakh SSR Biol Series 3:39–46. (in Russian)

Zhevnerov VV (1984) Goitered gazelle of the Barsakelmes Island. Nauka Publishing House of Kazakh SSR, Alma-Ata. (in Russian)

Rutting Behavior

4

4.1 Introduction

In spite of the most intensive studies of goitered gazelle ecology and behavior conducted in the former Soviet Central Asia (Sokolov 1959; Heptner et al. 1961), there have been disagreements for a long time about the social structure of goitered gazelles during the rut. The most common viewpoint claimed that male goitered gazelles form harems during rutting season (Dereviagin 1947; Afanasiev et al. 1953; Sludskiy 1956, 1977; Zhevnerov et al. 1983). During the Soviet period of gathering and documenting zoological data, the harem structure was attributed to many ungulate species inhabiting Eurasia (Heptner et al. 1961), and it was considered an established fact that goitered gazelles also have harems during the rut (Zhevnerov 1984). This opinion was prevalent until the mid-1980s (Blank 1985). Another group of researchers did not observe harems in goitered gazelles and instead claimed that, during the rutting period, the gazelle males stayed in mixed, unstable male-female groups, where promiscuous mating took place (Ishunin 1961; Gorelov 1972). At the same time, investigations of breeding behavior of antelopes and many gazelles in southern Asia and Africa documented that most of these species do have a territorial social structure during the rut (Walther and Mungall 1983; Estes 1991). Although sexual behavior among antelopes and gazelles was not studied for all species, including the social structure of goitered gazelles, Walther (1984) had no doubt that territorial behavior during the rut is typical for all gazelle species, including goitered gazelles. To study the breeding behavior of goitered gazelles in the wild, I conducted observations on the sexual behavior of these gazelles in the Ili Hollow in southeastern Kazakhstan during the 1980s. I was able to prove that Walther and Mungall (1983) was right and goitered gazelle males establish individual territories and court females inside of their own defined territories similar to other gazelle species (Blank 1985, 1998). In this chapter, I want to present the results of my investigations of goitered gazelle rutting behavior in detail and compare them with points

D. Blank, *The Behavioral Biology and Conservation of Goitered Gazelles (Gazella subgutturosa)*, https://doi.org/10.1007/978-3-031-83504-9_4

Table 4.1 Behavioral patterns (ethogram) observed in goitered gazelle males during courtship (by Walther 1984; Walther and Mungall 1983; Blank 1985, 1998)

Behavioral patterns	Detail description of posture displays
Low-stretch (neck-stretch)	A male stretches his head and neck forward horizontally. His ears are directed backward and usually pressed to his head in a position either parallel to the ground or slightly downward with the foramen inward. Rarely do his ears lay back pressed together or held apart at an angle of 45° (Fig. 4.1)
How-stretch (nose-lift)	A male stretches his head and neck forward and upward, with the top of his muzzle lifted slightly, and his ears pressed back to his head (Fig. 4.2)
Tail position	A male raises his tail to varying degrees, up to a vertical position, depending on his level of excitement (Fig. 4.3)
Bellowing or roaring	When bellowing, a male raises his head and neck slightly and keeps them stretched forward and upward; bellowing is something like a low, rough gurgling, wheezing and snorting with hoarseness and hissing sounds. To bellow, a male inhales, opens his mouth widely, his tongue is raised, and his mouth closed; an air stream is passed through his nostrils under pressure, and the bellow is herd with his mouth closed. During the bellowing process the male makes jerks his head and his entire body moves forward and upward, and his goiter moves up and down. When running, the male's mouth is open, and his bellowing is irregular. These sounds are produced when the mouth is closed or semi-closed and can be heard at a distance of 150 to 400 m. The bellowing of subadult males is not heard at the same distance (Fig. 4.4)
Flehmen	Typically, after sniffing a female's urine spot on the ground, the male raises his head and neck forward and upward, opens his lips and increases the flow of air to the vomerasal organ at the base of the upper lip, and slowly moves his head and neck from one side the other (Fig. 4.5)
Mounting	A male approaches closely to a selected female from behind, keeping his head above her back (Fig. 4.6). His muzzle is raised up and forward, with his tail in a vertical position. He is bellowing constantly at this moment. The male stands on his hind legs in a bipedal posture, with his forelegs wide apart and bent at the knee, hanging down; in this posture, the male moves very closely to the female, touching her only with his pelvis and not with his chest or forelegs. During each mount, the female typically walks two or three steps (five steps noted once) forward and the male steps with his hind legs in the bipedal position, keeping his pelvis close to the female's hindquarters (Fig. 4.7)

of view of other researchers. Behavioral patterns (ethogram) observed in goitered gazelle males during the rutting period are presented in Table 4.1.

4.2 The Rutting Behavior

4.2.1 The Rutting Time Period

In contrast to most other African and Asian gazelle species (Estes 1991; Habibi et al. 1993), the goitered gazelle is a strong seasonal breeder that gives birth once a year (Zhevnerov et al. 1983). All over their range, the rutting season of goitered

Fig. 4.1 Similar to many gazelle species and other antelopes, a low-neck-stretch display is the main pattern of courting behavior of a male after female in goitered gazelles. Issyk-Kul goitered gazelle nursery for reacclimatizing, 25.11.2022. Photo by D. Blank

Fig. 4.2 A high-stretch display is the second in importance pattern of male courting behavior in goitered gazelles. Issyk-Kul goitered gazelle nursery for reacclimatizing, 24.11.2022. Photo by D. Blank

Fig. 4.3 In the state of greatest sexual excitement, the goitered gazelle male raises his tail vertically. Issyk-Kul goitered gazelle nursery for reacclimatizing, 24.11.2022. Photo by D. Blank

Fig. 4.4 Courting goitered gazelle males are roaring frequently, opening the mouth and sticking his tongue out. Issyk-Kul goitered gazelle nursery for reacclimatizing, 24.11.2022. Photo by D. Blank

Fig. 4.5 Male gives a flehmen response (to bare the upper teeth as in a grimace) after sniffing the urine left by a female. Issyk-Kul goitered gazelle nursery for reacclimatizing, 24.11.2022. Photo by D. Blank

Fig. 4.6 A male keeps his head above female's back some time prior to mounting. Issyk-Kul goitered gazelle nursery for reacclimatizing, 24.11.2022. Photo by D. Blank

Fig. 4.7 During each mount, the female typically walks 2-3 steps forward and the male steps with his hind legs in the bipedal position, keeping his pelvis close to the female's hindquarters. Issyk-Kul goitered gazelle nursery for reacclimatizing, 24.11.2022. Photo by D. Blank

gazelles is observed during the months of September through January, though this behavior does not appear simultaneously in all parts of the gazelles' range. The intensive rutting activities of males start earliest in the southwestern populations of the Arabian Peninsula (from late September into October) and then include more and more populations, moving gradually through Sothern and Central Asia to the extreme northeastern populations in Mongolia (December through January) (Allen 1940; Kingswood and Blank 1996; Heptner et al. 1988). Even though the rutting peak is limited to 7–10 days (rarely as long as 1 month), copulation in goitered gazelles has a similar space-time sequence, usually observed in mid-October in Saudi Arabia (Habibi et al. 1993); October through November in Iraq, Iran, Turkmenistan, and Tajikistan (Flerov 1935; Gorelov 1972; Jamsheed 1976; Heptner et al. 1988; Shahooth et al. 2015), the end of November in western Kazakhstan (Zhevnerov 1975; Zhevnerov et al. 1983); early December in eastern Kazakhstan (Blank 1998); and December through January in Mongolia (Bannikov 1954).

In the Transcaucasian region (also known as the South Caucasus), the goitered gazelle rut is observed in November through December (Vereshchagin 1939). Within the territory of Kazakhstan, the rut takes place at essentially the same time, lingering slightly only in the southeast (Sludskiy 1956): in western populations (Ustyurt Plateau, Barsakelmes Island, Telekul Lake area), the rutting season starts earliest and lasts from the end of October through mid-December, while in the eastern populations (Ili Hollow location), the rut starts after mid-November and continues into mid-December (Zhevnerov et al. 1983; Blank 1998). Most authors indicate that the rutting period for goitered gazelles in Kazakhstan occurs from October to December, while Sludskiy (1956), alone, believes that the first signs of an imminent onset of rutting, judging by their behavior, can be noticed as early as September, when there is a significant decrease in daytime temperatures. He observed in the Betpak-Dala Desert that with the arrival of cooler weather in September, the gazelles become more mobile, resting a little during the day and grazing more. There are often games between males that turn into fights, and young gazelles frequently fight with adult males. The games/fights observed among goitered gazelles in September are far inferior in strength and ferocity to the struggles of males during the rut, though at times they will turn into serious fights. On the Barsakelmes Island, for example, a fight between two adult males was observed at night, when the noise from the blows was clearly audible at 100 m and was similar to the crack of a breaking saxaul tree. After a 5-minute fight, the gazelles separated with a groan but then started to fight again. In September, adult goitered gazelle males began to chase females more often, trying to separate them from their groups, as they do during the rut (Sludskiy 1956).

The estrus season in goitered gazelle females is shorter in northern populations and longer in southern populations (Sempéré et al. 2001). In *G.s. marica*, the luteal phase, follicular phase, and estrous cycle in adult female gazelles last from 6 to 19 days (Shahooth et al. 2015). In a captured herd of.

G. s. subgutturosa, a relationship was found between the timing of the onset of estrus and the social status of the female: in the captive group, the first to come into estrus were the dominant adult females, then the middle-ranked, and last, in late

December and early January, females of lower ranks, including adolescents as young as 7 to 8 months. Also, within this semi-captive population, the reproductive success of female gazelles depended on the density of the population: with a high density, adolescent females and a large percentage of adults were not fertilized, and most of the females who gave birth had only one fawn instead of their usual two (Marmazinskaya 1999).

Goitered gazelle males apparently go through unseasonal sperm production (Bland 1985), with a smaller peak of testicular activity in spring (pseudo rut), in addition to the main rut in fall (Tsaplyuk 1972). Sexual activity during spring is less prominent than during fall and is typically nonproductive (Blank and Fedosenko 1983). Nevertheless, though rare, mating can happen in spring, because there are documented cases when females became pregnant and gave birth in fall and even as late as winter (Antipin 1941; Gorelov 1972; Soldatova 1983; Sempéré et al. 2001). The timing of the main rutting season is obviously related to the time of parturition, which corresponds to the most favorable season for rearing young (Essaghaier 1981; Baharav 1983; Habibi et al. 1993). For goitered gazelles, warm spring weather with abundant vegetation first appears in the southwestern arid regions of their habitat in Saudi Arabia before Central Asia populations and even more so the northeastern populations in Mongolia (Kingswood and Blank 1996). In addition to the general geographical differences, the rut depends on local weather conditions, which can change from year to year and can shift somewhat one way or the other (Gorelov 1972; Sludskiy 1977; Zhevnerov et al. 1983), with some studies showing the shift may be limited to within 5 days (Blank 1985) or possibly 7 days (Zhevnerov 1984).

It took as late as 1985 to definitively recognize that goitered gazelle males can perform courting behavior all year round, but during the fall rut, adult males older than 3.5 years establish individual territories and the rate of courting behaviors increases considerably (Blank 1998). Previous authors overlooked the fact that males establish individual territories during rutting season and, instead, identified the start of the rut strictly through rutting activities; therefore, these authors had a distorted view of the start time and length of the actual rut, which led to documentation inconsistences in time and duration of the rutting season in different populations. For example, it is considered that in the Turkmenistan population, the rut lasts from early October to the end of November (Gorelov 1972), and in Uzbekistan, the goitered gazelles participate in reproduction activity from mid-November to the end of December (Mambetzhumaev 1970a, b). Marmazinskaya (1996a), during her study of a semi-captive population of goitered gazelles in the Bukhara Breeding Center (Uzbekistan), reported that the rutting season in goitered gazelles lasted 3 months, because she observed courting activities of males from mid-October until mid-January. Sludskiy (1956) argued that right after hot air temperatures dropped in September, males demonstrated their first signs of rutting behavior by starting to chase females. These courting activities continued until the end of December, while the most violent fights between males occurred in September. It seems likely that these behaviors by the males observed in September and January were being interpreted as an early start and late end of rutting activities (Sludskiy 1956;

Marmazinskaya 1996a), when, in fact, the short duration and relatively rare courting activities are possible to see at any encounter of males with females all year round (Blank 1998). The report by Sludskiy (1956) on the most violent male-male fights in western Kazakhstan mainly occurring in September needs further investigation to confirm if this was truly an early start to the rut.

My observations in the Ili Hollow in southeastern Kazakhstan demonstrated that the goitered gazelle rut has two stages: preliminary or pre-rut and the basic or principal phase. The preliminary phase occurs from mid-October through mid-November, when some adult males establish individual territories, mark their borders intensely, and chase the females that pass through their territories. These males, however, also leave their territories at night, accompanying the last daily migrating female groups, returning only in the mornings. The principal stage lasts from mid-November through mid-December, when territories are already established. This stage usually starts after the first snowfall, when territorial males do not leave their ranges at night but stay there constantly, satisfying their needs for water by eating snow.

The most intensive chases are observed during the second, basic phase, when males stay in their territories around the clock and mark and maintain their borders constantly. Females' cross territorial borders, moving from their night pastures and watering places to their midday resting sites in the mornings and back again in the evenings. Every territorial male courts females when they are in his territory but stop when the females cross into a male neighbor's territory (Blank 1985). Mating occurs within a very limited time of 5–7 days in early December, and the rut fades out after mid-December (more details below). Marmazinskaya (1996a) also observed in her study that the most active marking and rutting activities occurred during November and December. Since the rut has different phases with different intensities of courting behaviors, previous authors may have had different ideas on the time of the rut. However, the general geographical trend of the southwestern populations starting their rut the earliest and the northeastern populations the latest seems to be valid (Blank 1998).

4.2.2 The Spring Peak of Sexual Activity (Pseudo-Rut)

According to my study, the population of goitered gazelles in the Ili Hollow (SE Kazakhstan), apart from the real rut in autumn (October–December), displays a spring peak of sexual activity in April and May (false or pseudo rut), when some adult males establish territories and increase their courting rate compared to preceding and following months, though at a frequency still significantly lower than during the real rut in the fall. Mounting and copulation, as well as morphological changes in the size and color of the goiter, which are typical during the actual rut, are not observed during the spring pseudo-rut (Blank and Fedosenko 1983; Blank 1998). Zhevnerov (1984) also observed spring courting activities (end of May to early June) of male goitered gazelles in the Barsakelmes population. In the Bukhara Breeding Center in Uzbekistan, Marmazinskaya (1996a) similarly observed a higher rate of marking and aggressive behaviors between males in spring, and males courted pregnant and suckling females within large mixed-sex groups, although she did not see any males establishing individual territories or copulations (Marmazinskaya 1999). In Uzbekistan, the spring peak of sexual activity occurred from March through mid-June and was 2.5 months longer than the 1 month in Kazakhstan that runs from mid-April through mid-May (Blank 1998). Histological investigations of testes and ovaries show that, physiologically, goitered gazelles do have two peaks of sexual activity in a year—in spring (April) and fall-winter (October–December)—but only the fall rut is considered reproductive (Betpak-Dala Desert population, Kazakhstan—Tsapluk 1972).

Gorelov (1972) also observed a pseudo-rut in the Badkhyz Nature Reserve (Turkmenistan) among several males, and Marmazinskaya (1999) observed an increase in social and rutting activities of male-yearlings of 11–12 months of age during the spring peak of sexual activity (Uzbekistan). I also observed a pseudo-rut in the Issyk-Kul group of yearlings (Kyrgyzstan), when young males courted and tried to mate with young females, but without any subsequent pregnancies (Blank 2022b, unpublished report). Yearlings actively participate in fighting and horning each other and adult males, in addition to courting females. This increase in aggression and sexual activity of yearlings might be explained by the start of their first spermatogenesis during the spring peak, which goes through all developmental stages, but at a low intensity. Reproductive puberty in male gazelles occurs at the age of 17 to 18 months (Tsapluk 1972).

According to histological investigations, Tsapluk (1972) found that not only do males go through spring spermatogenesis, but females also undergo a spring oogenesis and gonad maturation in April and May. However, spring copulations are usually not observed in the wild (though maybe observed in captive and semi-captive conditions—Blank 2022b, unpublished report); therefore, the spring peak of rutting activities is not typically productive, except in very rare cases of spring fertilization and December births (Gorelov 1972): a 3-month-old embryo was found in a female hunted on June 22 (Antipin 1941), and a 2- to 4-week-old fawn was found in December of 1960 in Badkhyz Nature Reserve (Gorelov 1972).

4.2.3 Social Structure During Rutting Season: Harem Versus Individual Territory

In the past, a widely held belief was that similar to many equid species during rutting season, male goitered gazelles formed harems of several females and then courted only their own females, moving with them constantly and protecting them from rivals by maintaining a distance of 50–200 m (Dereviagin 1947; Afanasiev et al. 1953; Sludskiy 1956, 1977; Zhevnerov et al. 1983). Gorelov (1972) and Ishunin (1961), however, did not observe harems but stated that goitered gazelles formed mostly mixed-sex groups during rutting season with several adult males that courted females within their group; these groups were unstable and easily disintegrated into smaller groups and singles after only a short time together (Gorelov

1972). Since this group composition and accompanying behaviors did not match a harem social structure, these two authors concluded that harems are not typical for this species during the rut in the conditions found in Turkmenistan and Uzbekistan, although they also did not suggest another social structure apart from harems.

It was not until studies of goitered gazelles in Kazakhstan's Ili Hollow (Blank 1985) that it became clear that male goitered gazelles were known to have a territorial social structure during their rutting season (mid-October through mid-December), similar to many other gazelle species (Walther and Mungall 1983; Estes 1991; Habibi et al. 1993). Outside of the rutting period, males more often move freely in small, purely male groups across their habitat, shifting between pastures and watering sites, just following female herds. But during the rut, males establish and stay within their individual territories, which are arranged in areas where females pass through on their daily movements from their night pastures to their midday resting sites in the morning and back in the evenings (Blank 1998). Therefore, the male territories are covered with a network of well-trodden paths, along which females and non-territorial males regularly pass (Marmazinskaya 1996b).

Some adult males, however, remain non-territorial. They usually stay in male groups or more rarely alone and continue to move freely all over their home range; these males are called bachelors, because they do not participate in breeding, though they will still chase females on any given occasion should the opportunity arise. According to Zhevnerov (1984), in the Barsakelmes Island population, bachelors exist during the rut even if the number of males is equal to or more than the number of females. In years with a disproportionate sexual ratio with a bias toward females, a significant number of rutting males increase the number of females within their individual territories, but bachelors still exist. Walther and Mungall (1983) found that the number of territorial males increases with population density in Thomson's gazelles, but adult male bachelors are still always present during the rut. This phenomenon is not related to a lack of females or limited areas to establish territories, but rather a male's physical or psychological status, when bachelor males are too young (0.5–1.5 years old) or too old (more than 6 or 7 years old) to have the stamina for the rigorous fighting involved.

Apart from adult individuals (older than 2 years old), bachelor groups contain several young males, both subadults (5–7 months old) and yearlings (17–18 months old). During their daily migrations, they join female groups and will court females, though this courting is less intensive than the rutting activities of territorial males. Similar to territorial males, bachelors try to block females like rutting males even though they do not have their own territories; but in contrast to territorial males, they do not protect females from other males and 2–3 males can court the same female simultaneously without aggression toward each other. Female groups with bachelors, however, are ephemeral and exist not longer than 1–1.5 h and then disintegrate, with the bachelors abandoning their temporary groups (Blank 1997).

In contrast to territorial males, females do not have territoriality at all but instead visit one individual territory of rutting males after another, changing from one male territory owner to the next as they pass through the network of territories. The males stay continuously in their ranges, marking their borders and protecting them from neighboring owners and wandering bachelors, courting only the females that cross into their defined territories (Blank 1997). Territorial males become quite intolerant to young males (6–7 months old), most of whom continue to follow their mothers, driving them away from their mother's groups (Zhevnerov 1984). Territorial males, however, are not as aggressive to young males as they are to other adult males, and only periodically harass them, after which they stop paying attention to them (Marmazinskaya 1996b). During this time, many 6-month-old males roam between individual male territories (Smirin and Smirin 1991). Young males either form their own groups or, more often, join to bachelor male groups (Blank 1998).

Establishment of a network of individual territories during the rut also is described for the Bukhara goitered gazelle population (Marmazinskaya 1996b) and sand gazelles from Saudi Arabia (*G. s. marica*—Habibi et al. 1993). Marmazinskaya (1999) proposed the original explanation for inconsistences in the various descriptions of goitered gazelles' social structure during the rutting period. Her hypothesis says that goitered gazelles have the ability to change their mating strategy depending on population density, where males strictly protect their "harems" in low-density populations, but shift more attention to protection of their individual territorial borders with an increase in population density (Marmazinskaya 1996b). Based on her interpretation: (a) at the lowest densities (less than 3 individuals/km^2), males form "harems" with several females, protect them from other males, and do not stay in a specific location but instead move their harems widely throughout their habitat; (b) at medium density (3–5 individuals/km^2), males occupy individual territories, mark and protect their borders, and constantly hold their "harems" within their territories; (c) at highest density (22–24 individuals/km^2), males keep their own individual territories, protect them from penetration by other males but have no permanent harems, and instead court females from different groups during their brief presence as the female groups cross one territory after another during their daily migrations (Marmazinskaya 1996b).

In general, this phenomenon of alternating breeding tactics in goitered gazelles, depending on different population densities, is theoretically possible. Rubenstein (1989), for example, proposed that when the habitat productivity diminishes, the population density declines and subsequently the social organization also changes. Sinclair (1983) stated that in areas where female densities are relatively high, males establish territories, but in areas where population densities are low, males follow small groups of females. In other words, population density dictates the size of a group and the degree of inter-male competition for females (Sinclair 1983). The same idea was proposed by Habibi et al. (1993) for sand gazelles, which are strongly territorial during breeding season in captivity, but in the wild, at very low densities, single males are herding females into compact groups, preferring harem polygyny.

The reality of a mating strategy dependent on population density in goitered gazelles, however, needs further and more specific investigation, because in the Ili Hollow, I observed males' territorial structures without permanent harems but with an average population density of 2–3 individuals/km^2 (Blank 1990), which coincides with the lowest density according to Marmazinskaya (1996b). Furthermore, I did most of my observations on rutting behavior in an area where the population density was even lower than the average for the study area (less than 0.5 individuals/km^2) (Blank and Kovshar 1986), which directly contradicts the hypothesis presented by Marmazinskaya (1999). In addition, before the 1980s, researchers in the former Soviet Union widely believed that goitered gazelles had an exclusively harem mating strategy (Sludskiy 1956; Mambetzhumaev 1970a, b; Zhevnerov et al. 1983). It is possible this conviction was a consequence of a low-density populations under observation, which sustains Marmazinskaya's (1996b) the harem mating strategy in low density populations. Indeed, during the 1950s and 1960s, most goitered gazelle populations in Central Asia had a low density: in January of 1960, for example, between 0.01 and 0.82 individuals/km^2 were found in the deserts of Kazakhstan (Sludskiy and Shubin 1963). However, there were still separate populations that sustained a high density: the Sam Desert population density was greater than 25 gazelles/km^2 (Kostin 1955), and the Ustyurt Plateau population density was greater than 27 individuals/km^2 (Sludskiy 1977). In the Ili Hollow, the population density of goitered gazelles was higher during the 1940s (Dereviagin 1947) than during my study in the 1980s (Blank 1990). Unfortunately, even though the previous researchers observed high density populations, none identified the territorial mating strategies of the goitered gazelles at those times. All these examples suggest that these earlier authors did not describe the actual, observed behavior of goitered gazelles but rather fell back on the generally accepted hypothesis that rutting males had exclusively a "harem" mating strategy. Marmazinskaya (1996b) also observed only territory-protection behavior of rutting males while describing "harem" and territorial-harem structures based on other authors' writings (Zhevnerov 1984; Smirin and Smirin 1991), which most likely were also held captive by conventional wisdom of the time. Therefore, the hypothesis of a changing social structure during the rut based on population density in goitered gazelles demands additional specific investigations.

The generally accepted understanding of a harem is a stable group of females with one male held together long enough that the male can service most, if not all, of the females (Walther and Mungall 1983). Estes (1974) stated that all antelope species have not harems as typically seen in wild and domestic horses or plains zebras (Klingel 1972). And Walther and Mungall (1983) argued that males of all gazelle species establish individual territories during the rut and all gazelle species have a permanent (all year round) or temporary (only during the limited rutting period) territorial social structure and "harems" are impossible for them with no relationship to exclusive territories.

4.2.4 General Characteristics of the Rutting Behavior

The onset of the rut is preceded by the formation of groups. In August and September, females begin to gather into herds, which usually number from 10 to 30 individuals (though much larger herds of 100 to 200 gazelles have been observed); in the

Betpak-Dala Desert, the herd index in September on average increased to 17.6 individuals in a group (Sludskiy 1956). The autumn migration starts during this period, when goitered gazelles migrate from relatively closed environments into more open areas with smooth terrain (Zhevnerov 1975). Female herds stay in the open habitats of the plains, large dry riverbeds, and gentle foothills, sometimes with sparse bushes or saxaul trees. Adult males gather in relatively small groups of 2–10 individuals and either walk along with the females or separately and only occasionally joining the female groups. Males begin to divide the female herds into smaller groups by running around a favored group of females and separating off groups consisting of 2–5 females. Each male then walks with his female group, fiercely guarding them from other males. At this time, the young animals also begin to separate from their mothers. All these events were observed by Sludskiy (1956) in the Betpak-Dala Desert and by me in my study in the Ili Hollow (Blank 1997). Sludskiy, on other hand, believed that the goitered gazelles were forming harems, which completely contradicts my long-term observations of rutting behavior in the Ili Hollow population. He was right, though, that during the rutting season, goitered gazelle males do prefer aligned spaces, and only later it became clear that aligned and opened spaces are good for establishing their territories for visual control over as large an area as possible to detect everything that happens around their territorial borders (Blank 1997). This same phenomenon is also observed in the goitered gazelle populations of the Barsakelmes Island and Bukhara Breeding Center (Zhevnerov 1984; Marmazinskaya 1996a).

It is possible to observe goitered gazelle males chasing females and displaying courtship behaviors all year round (Blank 1985). But only during the rutting season from mid-October through December do adult males establish individual territories and greatly increase their rate of courtship displays. For example, in October courtship displays increase by 3.5 times, in November by 13 times, and in December by 20 to 21 times more often compared to July and August. In addition, only during the reproductive rut in fall do rutting males undergo morphological changes. First, the goiter, noticeable in any season, increases in size by several times, which is likely connected to the intensification of bellowing during the rut, when loud sounds are heard constantly (Blank et al. 2014). Second, the coloration of the male's head and neck get lighter, and in some territorial males, the color turns a true pure white. And third, mounting and copulation typically are seen typically during the real rut (Kingswood and Blank 1996).

From mid-October, most mature males establish their own individual territories that coincide with the routes of the female daily migrations. Territorial males mark the boundaries of their ranges and defend them against invading bachelors and neighbors. Rutting males tend to pursue females just within their own borders, although from time to time, they can break the rules and, in the heat of pursuit, cross the border into the territory of a neighbor. In defense, the neighbor will then run toward the intruder, making aggressive displays or directly attacking with horns lowered, forcing the intruder back into his own range. Females do not have territories and wander widely with no territorial boundaries. During their travels, females cross through several male territories, with each male owner chasing them to speed up mating readiness. Every territorial male tries to keep the female groups within his territory for as long as possible, but females always succeed in escaping and crossing the border into a neighboring territory. Once in the foothills, the female groups ascend the slopes to a position slightly above the area of the male territories, where they stay the entire day. In the evenings, the females return to their night pastures on the plains and again are herded and courted by the male-owners as they travel back through the network of male territories (Blank 1998).

During the peak of the rutting season, when males demonstrate courting displays most intensively, the female groups change their travels significantly. In the mornings, females return to the same male territories as previously visited, with some female groups staying in a selected territory during an entire day, leaving only for the night. This period lasts from 7 to 10 days, and most copulations occur during this short time span. The size and composition of female groups changes to some extent over these days, but overall, they remain relatively stable. Distribution of females over the various territories can be extremely irregular, with one male maybe gathering up to 16 females while another remains alone (Blank 1998).

During the rut, territorial males are very aggressive to other adult males that intrude on their exclusive ranges, even if there are no females present. When bachelors arrive in a territory, territorial males stop all their activities, even if they were courting females, and run toward the intruders to expel them. Bachelors retreat without resistance but usually do not run away. Instead, they tend to continue their advances while deviating aside from their initial trajectory to try to run around the male, who, in turn, blocks them in a threat posture. In this situation, bachelors end up galloping in a semicircle in the general direction they were trying to go but with a territorial male chasing behind them. Bachelors cross several territories before reaching their daytime resting places located in the foothills, which they share with the females during the day. Occasionally bachelors do not wait for an aggressive territorial male to approach and chase them but begin running immediately once they have crossed a territorial boundary. These same behaviors can be observed in the evenings, when the bachelors return to their night pastures (Blank 1997).

During the principle phase of the rut, the territorial males remain within their territories continuously, both day and night, using the snow as a source of drinking water so they do not have to leave their areas for water (Blank 1985). Using snow as a water source is typical for most goitered gazelle populations in Kazakhstan. For populations in the more southern regions of goitered gazelle habitat (Uzbekistan, Turkmenistan), where the snowfalls are rare or totally absent, making snow unavailable as a water source most winter seasons, male territories most likely are either established near watering places or territorial males are forced to temporally leave their territories for water. These southern populations have not been studied as thoroughly as the Kazakhstan population, so further investigations into this phenomenon would provide valuable data to better understanding goitered gazelle rutting behavior in relation to the different conditions for these populations. After the principal phase of the rut in mid-December, courting activities decrease considerably, with most males abandoning their individual territories, which in Kazakhstan disappear entirely by the end of December. In contrast, Uzbekistan gazelles continue their rutting activities until the end of January according to Marmazinskaya (1996a).

4.2.5 Rutting Behavior of Territorial Male

In the evening, close to dusk, most female and non-rutting male groups leave the area of individual territories for the night until morning, while territorial males remain in place. The next morning, the female groups return to the male territories rather late (around 9:00 to 10:00). Until the females emerge, males are alone in their areas. At this time, they graze, walk around their entire territory, and mark plants with their preorbital glands while horning saxaul trees and shrubs to refresh their border markings. It is not uncommon for a male to mark the boundaries of his territory, often spending more of his time on the side along the boundary from which the female groups usually come. He spends long periods of time following the movements of a group of females if they appear in the distance. He also attentively observes the courtship of other males, if females go to the neighbors' territories earlier than his. As the intensity of courtship increases during the rut, males become more interested in the courtship of their neighbors, which they observe closely with increasing frequency. The frequency of such behavior increases by several times in November (1.02 cases per hour, $n = 187$) compared to October (0.33 cases per hour, $n = 11$) until it reaches its maximum at the time of the rutting peak in December (1.51 cases per hour, $n = 211$). A male watches neighbor-female activity at a rate of 30.3 ± 3.9 s (range 2–395 s, $n = 205$).

If females finally approach the boundaries of a male's territory, the male runs to meet the group, crossing his neighbor's boundary, if the neighbor is not nearby, and driving the females inside his territory so he can begin his intense courtship behavior. In herding the females, the male tries to drive the females away from the boundary of his territory toward its center, blocking their way toward his boundaries to try to keep them from escaping to a neighboring territory. At the beginning of this process, the male assumes a low-stretch pose (Table 4.1) and stands in this pose behind a female. Then, he walks around the female from the side (Figs. 4.8–4.11) and stands in this pose in front of her, showing the pose to the female from a better perspective. The male freezes with his neck extended, ears pressed back, and tail raised upward. If the female does not respond, he circles the female again and freezes again from the front. From time to time, the male produces bellows with a protruding tongue and raises his tail up to a vertical position, and his muzzle ends up and forward at the same time. According to Marmazinskaya's study (Marmazinskaya 1996a), this behavior is accompanied by the male bobbing his head up and down (common) and wagging his tongue (rarely) (Fig. 4.12), but I did not see these displays in the goitered gazelles of the Ili Hollow (Kazakhstan). When the females begin to run away with their heads down and tails up, the male chases them in at a gallop in a high-stretch pose (Table 4.1), with an open mouth and ears pressed back to his head (Fig. 4.9). During this chasing, the courting male's preorbital glands are

Fig. 4.8 During courting. a male puts his ears back, pressed to each other with earholes directed to the sides as if folded in half. Issyk-Kul goitered gazelle nursery for reacclimatizing, 24.11.2022. Photo by D. Blank

Fig. 4.9 Sometimes, a male demonstrates his tail is turned up over his back during courting. Issyk-Kul goitered gazelle nursery for reacclimatizing, 25.11.2022. Photo by D. Blank

Fig. 4.10 During courting by walk, a male often keeps his tail in a horizontal position or lower position. Issyk-Kul goitered gazelle nursery for reacclimatizing, 24.11.2022. Photo by D. Blank

Fig. 4.11 Initially, a male approaches a female from the front, passes by her and then comes up to her hindquarters. Issyk-Kul goitered gazelle nursery for reacclimatizing, 24.11.2022. Photo by D. Blank

Fig. 4.12 Pre-copulatory behaviors includes the freezing in the low-stretching posture that maybe alternated by the head bobbing and tongue wagging. Drawing by U. Kenzhetaev with kind permission

opened widely (Blank 2022b, unpublished report), and it is possible that this produces a distinctive odor that spreads from the open glands of the courting male during the chasing of females (Walther and Mungall 1983; Fig. 4.13).

Thus, low- and high-stretch poses are the two most commonly displayed postures during male courtship. The frequency of demonstrations of these postures increases as the rut progresses, and the maximum frequency of these expressions can be observed in December. Sand gazelle males also perform front-leg kicking (Laufschlag) when courting females in Saudi Arabia populations, though this is done in a relaxed, languid manner with the leg at about a 45° angle without touching the female (Habibi et al. 1993). Marmazinskaya (1996a) also observed a similar behavior by courting males in a dominant posture with a bent knee in the Bukhara population in Uzbekistan; however, I did not see this presentation during courting displays of goitered gazelles in the Ili Hollow wild population (Blank 1998), but I did observe performing front-leg kicking in a semi-captive Issuk-Kul herd (Fig. 4.14), where males also performed this behavior in a relaxed manner.

There is also a neutral pose in the repertoire of rutting behavior, when the male chases females without any special postures. This courtship behavior is quite rare during the intense phases of the rut, but quite often, the absence of postures can be seen in males chasing females at the beginning of the rut in October. To make the female run, the male freezes and then makes a sudden jerk toward her or assumes a dominant posture. It is very rare for a male to bump a female with his horns. As a rule, these postures force females to start running, while the male then willingly follows her and continues his courtship (Figs. 4.15 and 4.16).

Fig. 4.13 Usually, a goitered gazelle male opens his preorbital glands widely during courting displays. Issyk-Kul goitered gazelle nursery for reacclimatizing, 24.11.2022. Photo by D. Blank

Fig. 4.14 Display of typical front-leg kicking pattern is not observed in goitered gazelles, though this is done in a relaxed languid manner with the leg at about 45° angle without or slight touching the female. Issyk-Kul goitered gazelle nursery for reacclimatizing, 24.11.2022. Photo by D. Blank

Fig. 4.14 (continued)

Fig. 4.15 A goitered gazelle male sometimes court females without any special behavioral displays, just directs his ears back. Issyk-Kul goitered gazelle nursery for reacclimatizing, 24.11.2022. Photo by D. Blank

Fig. 4.16 In the case, if a female does not react to male courting, he assumes a horn threat posture, which is demonstrated usually in male-male conflicts. Issyk-Kul goitered gazelle nursery for reacclimatizing, 25.11.2022. Photo by D. Blank

The male pursues the female with a peculiar series of different, alternating postures. The gaits also change when the male goes from a slow run to a fast gallop, followed by a slowdown and then another increases. This continuous pursuit can last from tens of seconds to a quarter of an hour. After several minutes of pursuit, the male stops and rests standing without movements or less often he lies down. This rest can last from a few minutes to 20 minutes (Blank 1985). The male tries to stay close to a female in estrus and often runs up to sniff her hindquarters or her urine spots on the ground (Zhevnerov 1975).

During the rutting period, it is possible to observe several kinds of gaits, including walking, trotting, galloping, and careering. At the beginning of the rut, slow gaits are most common. The share of fast gaits gradually increases, and at the height of the rut, galloping prevails, sometimes turning into carriering. During the courtship of females, males change their speed of pursuit and gaits so quickly that sometimes it is difficult to separate one gait from another. The longest gait is the gallop (over half a minute); all other gaits are noticeably shorter. The male may pursue a whole group of females at once, but more often, he follows one female after another from the herd. Sometimes the male chases the female to the group of females to keep the entire herd within his territory. Similar behavior is observed in sand gazelles in Saudi Arabia, when male gazelles constantly round up and herd females back into their rutting territories (Habibi et al. 1993). But more often, males separate a female from the group to court only one. As soon as a female in estrus appears, the male courts her until he succeeds in mating with her. Then, he moves on to other females.

A rutting male is very noisy when chasing females and starts to bellow loudly when females appear in his territory. When roaring, the male's muzzle is directed up and forward, and his sounds can be heard at a distance of several hundred meters. Goitered gazelle males can produce roaring sounds out of rutting season, as well, but these sounds are rare and not as loud as during the rutting period. In other goitered gazelle populations, males also accompany their courting with loud bellows (Turkmenistan—Gorelov 1972; western Kazakhstan—Zhevnerov 1984; Saudi Arabia—Habibi et al. 1993; and Uzbekistan—Marmazinskaya 1996a). When a female hears the roar of a male, she immediately lowers her head to the ground and raises her tail vertically and often starts running. Therefore, the male often uses the roar to make the female run away, and the faster she runs away, the more vigorously the male chases her.

The territorial male not only chases after particular females but also runs around the entire group of females to keep them in his territory as long as possible. Females, though, are usually trying to just cross the male's territory to continue their daily movements to their day or night locations. Despite the male's very active efforts, females eventually manage to cross the male's boundary and move to the neighboring territory. Many times, the male continues to run after them, trying to chase them back to his territory, despite threats from his neighbor. Sometimes he succeeds and the female group returns and stays in his territory longer, but eventually the females leave again.

A territorial male will court not only a standing or grazing female but may also attempt to lift up females that are lying down. The male comes close to a lying female, freezes in a pose, and lets out a roar, and if she does not get up, he touches her with his muzzle or horns, forcing her to get up. In Saudi Arabia, male sand gazelles touch laying females' hips with the anterior or lateral surface of their horns or tap the females with a foreleg, forcing them to stand up (Habibi et al. 1993). Occasionally a male will attempt to lift the entire group of females lying down, but he usually has no appreciable success, for while he is getting one female up, the other lifted females lie down again as soon as he moves away from them.

4.2.6 Female Behavior

In contrast to the rutting behaviors of territorial males, females do not define their own territories but regularly visit the territories of rutting males during their daily migrations, often stopping in an individual territory to graze and rest, and then continue on their way. Only Gorelov (1972) is the only author to observe something like marking behavior by a female, as she scratched a place where a male had previously marked. This behavior, however, was thought to be more playful than any real participation in marking behavior since it was so unique. During most of the rutting period, females just cross through male territories, unless they are in estrus with a swollen vulva; then, they will stay in a males' territory for up to several hours, when copulation occurs (Zhevnerov 1984; Blank 1998; Marmazinskaya 1999).

Fig. 4.17 A goitered gazelle female chased by male typically lower their head and raise their tails in a vertical position. Issyk-Kul goitered gazelle nursery for reacclimatizing, 25.11.2022. Photo by D. Blank

When a courting male approaches any female, she raises her tail vertically (Fig. 4.17) and lowers her head to the ground. Sometimes the female will even put her tail on her back. During the first stages of the rut in late October and early November, the female rarely raises her tail (8% in October and 6% in November, $n = 301$) and generally reacts very sluggishly to the male's courting. But as the intensity of the rut increases, however, females raise their tails more often at the approach of a courting male, and at the height of the rut in late November–December, tail raising occurs in up to 20% of the cases ($n = 142$). With her head down and tail raised vertically (Fig. 4.17–6218), sometimes with her tail on her back, the female runs away from the male most often at a gallop. As soon as the female stops running, she immediately raises her head and lowers her tail. Whenever the male approaches, the female tries to turn so her hindquarters are toward the male and her head is pointing in the opposite direction. As the male shifts, the female also changes her body position to keep her hindquarters pointed toward the male. Sometimes a female will raise her tail when just passing the male, although often he pays no attention to her; however, just hearing a male roar is enough for a female to react and lower her head to the ground. In addition, young animals of either sex show these same behaviors all year round, but especially during the rut when young females respond to males more often than adult females, indicating absolute subordination (Ili Hollow, Kazakhstan—Blank 1997). Goitered gazelle females in Uzbekistan (Marmazinskaya 1996a) and Turkmenistan populations (Gorelov 1972) demonstrate a similar behavior when they are chased by males during the rut. In rare cases, a female has been known to run after a male to flirt with him (Gorelov 1972).

If a female wants to go back to her group or seeks to leave the male's territory, she does not go in the direction of the male who is trying to block her path but instead moves a little to the side at an angle to him, striving to go around him, sometimes making quite a large circle around him, aiming to leave him behind so she can escape. If she fails to do so and the male manages to block her exit from his area, the female runs away from the male in a straight line at a gallop and then suddenly makes sharp, acute changes in her trajectory. This maneuver is usually unexpected, and the male does not turn fast enough, making him lose his running speed to continue the pursuit. This same method of escaping pursuit is also used by many antelopes to escape from a predator. If the male does catch up with the female, she runs in a half circle with the pursuing male behind her and flees in the direction she was intending to go originally.

Another way for a female to get rid of a chasing male is to show almost total disregard for his courtship behaviors. Although the male shows various courtship poses, approaches the female from different directions, and roars near her, the female does not react in any way. She just continues to graze and does not even raise her head and instead turns away from him. Sometimes she may just lie down, not paying any attention to his presence. After repeated and unsuccessful attempts to make the female respond to his demonstrations, he leaves her and runs to another female, with the first female getting up almost immediately and going where she wanted to go.

The behavior of females changes during the rut. At the beginning of the rut (in October–November), females rarely respond to male courtship of the males and instead just walk away (31%, $n = 36$) or more often run away immediately (69%) with tail raised (92% cases, $n = 36$). In order to get rid of the male's advances, females try to cross a male's territory as quickly as possible. At the height of the rut (late November–December), females usually respond to courtship by trying to run away from the male, who pursues the females with a great vigor. Such galloping pursuits increase only twice during the rutting peak (0.75 cases per hour in October and up to 1.26 cases per hour in November), while walking displays change insignificantly in these months (0.33 up to 0.39 cases per hour). When escaping from a male's sexual advances, females mostly gallop in a shuttle-like or circular path (76%, $n = 301$). Responding to a male's intense pursuit, a female will urinate, allowing the male to sniff her urine while making a flehmen response (to bare the upper teeth as in a grimace) to check the female's readiness to mate. In the beginning of the rut, these cases are relatively few (0.06 cases per hour in October), but at the height of the rut, the number of female urinations increases fivefold (0.32 cases per hour in November). During the pre-copulation stage in early December, many females are in estrus and, in response to the male courting behaviors, often urinate with a rate increase of more than twice (0.69 cases per hour). At the same time, the females have a weak reaction to the males' displays, running almost half as often compared to November (0.69 cases per hour) and only walking away reluctantly (32% cases, $n = 301$), with lowered heads 1.6 times less often and frequently without vertically raised tails (up to 19%, $n = 142$); they even have been seen to lie down in the male's territory (Blank 1997). During a second estrus, if the first estrus was unsuccessful and did not end with fertilization, a female tends to approach the male as close as possible, following him; sniffing his marking places; rubbing her head against his body; sniffing the male's muzzle, hindquarters, and penis; and even trying to mount the male (Marmazinskaya 1996a). In the Badkhyz gazelle population in Turkmenistan, Gorelov (1972) observed several cases when a female approached and followed a male during the rut.

During most of the rutting period, females usually cross individual male plots without staying in any one place for long periods of time. Only at the very end of the rut, in early December, do females begin to stay in territories of males they have selected for as long as 1–3 days. In general, males mate with females during this time, followed by the females leaving the male's territories. Typically, females are distributed unevenly among male territories. However, it is an interesting observation that after a snowfall, females have been seen to be distributed more evenly between individual territories. The reason is unknown.

4.2.7 Flehmen

As stated above, being continuously chased by a territorial male during the peak of the rut results in the urination of the female. The male usually waits for the female to finish urinating, seldom (1.8% cases, n = 114) approaches her to put his muzzle under her urine stream (1.5% cases, n = 114) (Fig. 4.18), and never engages in

Fig. 4.18 The goitered gazelle male usually waits for the female to finish urinating and seldom cases, approaches her to put his muzzle under her urine stream. Issyk-Kul goitered gazelle nursery for reacclimatizing, 25.11.2022. Photo by D. Blank

rear-sniffing or nuzzling of the female's genitals. He usually approaches to sniff the urine spot on the ground by walking slowly in a low-stretch (29% cases, $n = 114$) or neutral posture (71% cases), with the tops of his ears laid backward and his earholes pointing to the sides. A rutting male rarely ignores a female's urination (22% of cases in November and 13% in December, $n = 46$ and 84, respectively) and, in most cases, sniffs a female's urine spot (78% and 87% cases, respectively). After sniffing her urine (Blank 1998) or even wetting his lips with the female's urine on the ground (Habibi et al. 1993), the male usually shows a flehmen response (Fig. 4.5). Less often, the male gives a flehmen response after sniffing the urine left by a female some time earlier (9.8% cases) or when seeing a urinating female from a distance. Males have also been observed to sniff a female's urine but without having a flehmen reaction (27.5% cases, $n = 40$ in November, and only 4% cases, $n = 79$ in December). A male will pass the urine location while continuing to court the female (15% cases, $n = 114$) but will usually return to search for the urine spot, showing a flehmen response if he finds it or walking back and forth if he cannot.

The male sniffs the urine spot thoroughly for usually 1–4 s (mean 2.9 s, $n = 7$) and very rarely (1 case out of 114) licks it. Habibi et al. (1993) observed a male wetting his lips from the female's urine on the ground, though this particular behavior was not seen in Kazakhstan (Blank 1998) or Uzbekistan (Marmazinskaya 1996a). In performing the flehmen response, the male's muzzle is stretched slightly upward and forward with his mouth opened somewhat as he slowly turns his head from side to side, though not rhythmically (5% cases, $n = 76$). Only in exceptional cases does the male not raise his muzzle during flehmen or turn his head (0.9% cases). Quite often the male will sniff the spot once more (31% cases, $n = 114$) after the flehmen display, and there have been a few cases documented cases when he sniffed the urine spot more than twice: out of 114 cases, a male sniffed 3 times (0.9%); in 2 cases, he sniffed 4 times; and once (3.11.1981), he sniffed 8 times consecutively (0.9% cases, $n = 114$), having a flehmen response after each sniff. A male can also have a flehmen response twice (5.3% cases, n = 114) or even 3 times consecutively (2.6% cases). After having a flehmen reaction, the male either chews (4.4%, $n = 114$) or licks his upper lip (4.4% cases), then resumes chasing females, begins to graze, or sometimes rests for 10 to15 sec while either standing or lying down.

It is possible to observe flehmen presentation all year round, though this display is noted rarely out of the rutting season. Flehmen is noted most often at the time of copulation in early December (0.54 cases/h, $n = 76$), less frequently during the peak of the rut in mid-November (0.16 cases/h, $n = 29$), and ten times less often during the summer months (0.02–0.05 cases/h, $n = 63$). It is also possible that occasionally a male does not react to a female's urination at all. The lack of a flehmen response is most often notable during the pre-rut period from mid-October into mid-November, while it is most often seen during the peak of the rut in mid-November through December (Blank 1985). The duration of flehmen displays is at its highest during the spring peak of sexual activity in April (18.4 s) and the rut in December (17.3 s) compared to 8.3 s in June. In November the duration is significantly shorter. Its shortest duration is in June, when the duration difference between June and both the autumn main rut and the spring pseudo-rut is highly significant. There have been some unusual cases where a male, after sniffing of the urine spot on the ground, urinated and defecated at the same place, though such cases were very rarely observed (Blank 1997).

4.2.8 Pre-copulation and Copulation Behavior

After a long courting session, the male chases a female in estrus that is the readiest for copulation (Blank 1998). Females have a slight swelling of the vulva during estrus, which lasts only about 12 h (Roberts 1977). The male's head and neck are stretched parallel to the ground, his ears are pressed to his head, and his tail is raised vertically, which he throws on his back at his maximum level of arousal (Blank 1998). Based on a study by Marmazinskaya (1996a), this pose is accompanied by a nodding of the head (Fig. 4.12), a roaring (snorting) sound, moving the Adam's apple up and down, and occasionally chattering with an outstretched tongue. In my study, I did not observe any males nodding their heads or flicking their tongues, though the tongue sticking out of the mouth was observed occasionally. Unlike ordinary courtship behavior, during the pre-copulatory period, the male performs courtship poses more carefully and intensely and usually does them when walking, often freezing in a place, while running and especially galloping is rarely observed. A female that is ready for mating no longer runs to escape the male, though she still may take a few short mincing steps away from him (Marmazinskaya 1996a; Blank 1998).

The female usually holds her head raised, or more rarely lowered to the ground, with her chin bent to her neck and her forehead turned forward. Her ears are directed up and backward, with her earholes turned back. Her tail is raised in varying positions up to vertical and is flicking from side to side or up and down. She sometimes will drop her tail down only to raise it again, when the male closely approaches her from behind. The male follows her in a low-stretch posture, often freezing when near the female, finally staying behind her, keeping his muzzle close to her tail. He repeatedly touches and sniffs at her hindquarters, then approaches her with his head raised higher than her back and touches her hindquarters with his chest; his tail is in a vertical position and the male is ready for mounting. Before mating takes place, the male bellows constantly. Marmazinskaya (1996a) asserted that during pre-mating displays, a male demonstrates a dominant posture to the female with his head raised high, his neck stretched out in an arc with his chin pressed against his neck, his tail is away from his body, and his ears are directed to the back of his head. This dominance pose is often accompanied by the flexion of the front leg at the wrist joint and a nodding head. In my study, I did not observe this dominance posture or the raising of the front leg very often during pre-copulatory behaviors, though this display was common during the main courting behavior before the pre-copulatory stage, with the exception of the head nodding, which I did not see at all (Blank 1998). But in the semi-captive group in the Issyk-Kul breeding center (Kyrgyzstan), however, I did observe a goitered gazelle male butt a female from behind with his horns from time to time during pre-copulatory chasing (Blank 2022b, unpublished report).

During copulation preparation, the male and female walk back and forth together from 100 to 150 m for up to 10 min. The female raises her tail to a vertical position when the male approaches her but keeps walking even though the male tries to keep her still, as he bellows and walks around her in a low-stretch posture. The male's intent is to stop the female completely and not let her step away at his approach. If the female is standing still with her head raised, the male touches her from behind and she pulls her tail to the side, and he rests his chest against her rear, holding his head over her back, and he assumes the initial position for the mounting (Blank 1998).

Males demonstrate four forms for mounting in a bipedal vertical posture: 1. the posture without touching the female with his pelvis region and without an erect penis (Fig. 4.19); 2. a similar display, but his erect penis is visible (Fig. 4.20; 3. the male touches the female with his pelvis region, without intromission (Fig. 4.21; and 4. physical copulation, when the male is in a bipedal vertical posture, touching the female with his pelvis only and with intromission of his erect penis (Fig. 4.22. In the Bukhara semi-captive population in Uzbekistan, the number of mounts during one pre-copulation interaction was from 17 to 61 times, while in the natural wild conditions of Kazakhstan, I observed mounts from 22 to 185 times (Marmazinskaya 1996a; Blank 1998; details below).

The male usually attempts to mount the female from behind, but occasionally he mounts her from the side (head above her back) (Fig. 4.23–5926). In doing so, the

Fig. 4.19 Mounting in goitered gazelle: bipedal vertical posture of a male without touching the female with his pelvis region and without an erect penis. Issyk-Kul goitered gazelle nursery for reacclimatizing, 24.11.2022. Photo by D. Blank

Fig. 4.20 Mounting in goitered gazelles: bipedal posture in a male with erected penis. Issyk-Kul goitered gazelle nursery for reacclimatizing, 24.11.2022. Photo by D. Blank

Fig. 4.21 Mounting in goitered gazelles: touching the female with male's pelvis region, without intromission. Issyk-Kul goitered gazelle nursery for reacclimatizing, 24.11.2022. Photo by D. Blank

Fig. 4.22 Physical copulation with erect penis intromission of the goitered gazelle male. Issyk-Kul goitered gazelle nursery for reacclimatizing, 24.11.2022. Photo by D. Blank

Fig. 4.23 Occasionally, a goitered gazelle male mounts a female wrongly from the side. Issyk-Kul goitered gazelle nursery for reacclimatizing, 24.11.2022. Photo by D. Blank

Fig. 4.24 If a female moves forward, a goitered gazelle male walks behind her in bipedal posture, trying to keep a close contact with her hindquarter. Issyk-Kul goitered gazelle nursery for reacclimatizing, 24.11.2022. Photo by D. Blank

male stands or walks bipedally on his hind legs, forelegs wide apart and bent at the knees; he approaches the female on his hind legs very closely and touches her only with his pelvic region, but not with his chest or forelegs. The last mounts are accompanied by a strong push of the pelvis, when a short-term intromission and ejaculation occurs. Sometimes his pelvis push is so powerful that the female is moved away with quick steps. If during mounting the female walks away from him, he walks behind her on his hind legs, trying to maintain contact with the rear of the female (Fig. 4.24). Usually, a mounting male will try to copulate quickly making rapid, short steps on his hind legs behind the departing female and a quick push with his pelvis at the end of mating before the female has time to move away. During the series of mountings, an erection of the penis occurs, with actual mating with a protruding penis usually occurring at the end of the mounting sequence (Marmazinskaya 1996a; Blank 1998).

The male then continues to follow the female, keeping close to her and holding his head over the female's back; repeating his vertical, bipedal posture; and walking or running behind her many times. The female, in turn, stands or moves 2–3 steps (rarely as many as five steps) forward during copulation. After copulation, her tail is erect and deflected to the side, as if making access to her genitals open; then, she shakes her tail from side to side several times (Blank 1998). Copulation lasts only a few seconds (usually not more than 2 s) but is repeated many times (Zhevnerov 1984).

Immediately after copulation, the female walks several meters away from the male and stands motionless for a few minutes and then starts to graze or often lies down, while the male licks his penis and rests, as well (Marmazinskaya 1996a). Sand gazelle females after copulation stay with their group, where they may be repeatedly mounted by other females and juvenile males. During such interactions, the female walks in circle as she is being mounted incessantly, creating a frenzy of activity within the group, while the territorial male remained indifferent to this activity (Habibi et al. 1993). I did not see this behavior in Kazakhstan (Blank 1998) nor was it noted in Uzbekistan (Marmazinskaya 1996a).

A male will mount a female multiple times successively (22 to 103 times—in December 8, 1984). Each mount lasts 1 to 2 sec. On December 14, 1985, in the Ili Hollow, I witnessed a male mounting a female 29 times during a 3-min period, with copulation occurring twice (for 1 sec). Afterward, the male started grazing and the female laid down to rest. Two hours later, the male resumed his courtship, chasing the same female for 40 min and then mounting her again. During a 25-min period, I counted 103 mounts from behind without intromission, 74 improperly oriented mounts (from the side or passage from the side to the hindquarters), and eight copulations for a total of 185 mounts. Even after all this, the male continued to court the same female (Blank 1998). Zhevnerov (1984) observed eight mounts with intromissions, with each lasting 2–3 s.

Pre-copulative courting and actual copulation have been documented to last from 1 min up to 1 h 40 min (Table 4.2). A male can copulate twice with some females with an interval of 2.5 h between the first and second series of pre-copulation behaviors; in one noted case, the second mating was shorter in duration. In the case of an unsuccessful fertilization, the same female can enter the estrus again 18 days later after her first estrus cycle. It was also observed two copulations with the same female. Copulations can be observed at any time of a day and even after sunset. A noted case of female behavior in the second estrus cycle was of an active reunion

Table 4.2 Pre-copulatory behavior of goitered gazelles in captive conditions of the Bukhara nursery (adapted from Marmazinskaya 1996a)

Date	Observation duration, min	Female age	Behavior starting	Behavior ending	Behavior duration
1987					
November 20	304	3.5	11:19	11:36	17
November 24	195	3.5	20:38	20:44	8
November 26	60	2.5	17:30	–	–
December 02	194	1.5	12:30	12:50	20
2nd copulation	–	–	15:05	15:12	7
December 07	641	2.5	13:48	14:04	15.5
2nd copulation	–	–	16:55	18:05	70
1988					
December 02	450	1.5	10:40	11:15	35
2nd copulation	–	–	14:00	14:18	18
December 23, one copulation		7 months	–	–	–
December 25	420	2.5	9:58	11:38	100

with a male as she closely approached and followed him, mounting him and rubbing her head against his back and sniffing his muzzle, tail, and penis, as well as any branches against which the male had rubbed his horns. The male also demonstrated pre-copulatory behavior and mating and then repeated it all again within 1 day (Marmazinskaya 1996a).

The behavior of a 6-month-old male in the presence of an estrus female was interesting: immediately after the female arrived, the subadult male immediately started following her. He demonstrated elements of sexual behavior by relentlessly following her while demonstrating the stretching posture, nodding his head, genital sniffing, and mounts with thrusts but without an erect penis. In addition, when the female stopped, the young male butted her side and rubbed his head against her back, forcing her to walk. In total, the young male followed the female for 83 min, during which time he made 58 mounts. The dominant adult male, standing at a distance of 3–5 m, did not pay much attention to the behavior except to watch the young male and female; then, he followed the two and periodically ran up to the female, demonstrating the stretching pose and touching her croup with his muzzle. The young male was very excited and did not react in any way to the approach of the adult male while continuing to follow the female. In the semi-captive population, adolescent males were also observed to graze on the plots of territorial males. The owners of the plots tolerated their presence, although they periodically followed them, trying to drive them out of the territory (Marmazinskaya 1996a).

Despite the long duration of the goitered gazelle's rut, which lasts around 2 months, most mating takes place within a 1-week period at the beginning of December (in Central Asia), with a marked decrease in sexual activity during the second half of the month. Other researchers also pointed out from their studies that the peak of mating can also be observed from as early as late November into early December (Sludskiy 1956; Zhevnerov 1984). As a consequence, most females give birth to their offspring in the spring within a short period of a few days in mid-May. This is called a mass birthing period (Blank 1985). In the Bukhara semi-captive population, copulation lasts for almost a month from November 18 to December 23, with mass copulations observed over a 20-day period from November 20 to December 10. In late November, dominant adult females are the first to come into estrus and copulate with males which is followed by subdominant females and then the lowest ranked females, including subadults as young as 7 months old (Marmazinskaya 1996a). Sludskiy (1956) and Zhevnerov et al. (1983) also noted that estrus was observed first in adult females that had previously given birth and latest in young females of 7–8 months and 18 to 19 months, which were mating for the first time. All tamed females were fertilized at the age of 7–8 months and gave birth to single fawns, while a year later, they gave birth to twins (Zhevnerov 1984).

In mixed groups, subadult males at the age of 18–19 months try to court females but are chased away by adult males. Subadult males have the ability of courting and copulating at this age because they have active spermatogenesis and the ability for fertilization (Tsapluk 1972) but only in the absence of adult males (Marmazinskaya 1996a).

4.2.9 Masturbation

Adult males will masturbate in any season, while young males demonstrate their first masturbation at the age of about 1 year. Male yearlings can masturbate quite frequently, up to a few times a day, in spring (semi-captive herd in the Issyk-Kul nursery, Blank 2022b, unpublished report). To masturbate, the male first stands still in a tense posture, raises his arched tail slightly and switches it up and down several times (30% cases, $n = 118$). He then turns his muzzle back and licks his erected penis. This process may be more prolonged as the male may stand for several seconds, even letting his tail down and continuing with other activities (four cases out of 118) for some time; but then he raises his tail again, erects his penis, directs it up and forward, and moves his forelegs close to his hind legs, orienting his pelvis upward. In order to facilitate masturbation, the male quite often lifts his hind leg (18%, $n = 118$). He then turns his muzzle back, takes his penis into his mouth, and makes suckling movements in different directions (Fig. 4.25). Simultaneously, he periodically opens and closes his mouth as if he is sucking or licking his penis (Fig. 4.26); although more precisely, released ejaculate is licked rather than suckled with the male's tongue.

At the end of masturbating, the male often opens his mouth widely, as if yawning, and then chews the ejaculate for a few seconds (up to 7 s). There are cases when the male bites off a bunch of grass or regurgitates a portion of the gum and swallows

Fig. 4.25 Masturbation in goitered gazelle: the male makes milking movements with his head, either up and down or from side to side. Issyk-Kul goitered gazelle nursery for reacclimatizing, 26.06.2022. Photo by D. Blank

Fig. 4.26 Masturbation: the male also periodically opens and closes his mouth as if he is licking his penis. Issyk-Kul goitered gazelle nursery for reacclimatizing, 26.06.2022. Photo by D. Blank

this mixture. After that, the male can stand for quite a long time without moving (from 2 to 25 s).

Even though males masturbate year-round, their frequency is very low outside of the rut, for example, in summer (June–July is equal to 0.02 cases/h, $n = 15$), whereas most often this behavior can be observed at the peak of the rut in December (0.27 cases/h, $n = 40$) and less often in November (0.12 cases/h, $n = 25$). The duration of masturbation can vary from a few seconds to 15 s, with an average of 10 s ($n = 105$).

In half of all cases, masturbation is observed while a male is grazing or moving (48%, $n = 110$) and in a quarter of cases (24%) immediately after a long rest; this behavior is observed noticeably less often when scratching (15%) and only occasionally during courtship of females (7%), with minor cases during other behaviors (3–4%, $n = 110$) (Blank 1998).

4.2.10 Snow Eating and Mating Behavior

Sludskiy (1956) stated that goitered gazelle males were especially thirsty during the rut as they often gallop while chasing females or driving off rivals. After establishing their individual territories, male goitered gazelles prefer not leave their territories during the main part of the breeding season. However, these territories are found often in the submontane dry plains with a lack of water, and since males need a high level of energy and water during the rut, the only source for water in this region is snow. Females also eat snow from time to time (Fig. 1.3.22–215-7), though they can visit watering places, as well. Unfortunately, the grounds where goitered gazelles prefer to spend the winters, including their rutting grounds, usually have a thin snow cover that can be unstable and typically melts or is blown away within several days (Blank 1990). Territorial males can survive without water for a few days until the next snowfall, but in some warmer and snowless years, the breeding area gets no snow for long periods (more than a week), so even though they prefer not to, territorial males have to leave their territories for 1–2 h to find drinking water, often running all the way there and back. The snow cover is so important for rutting males that they can increase the intensity of their rutting activities when snow is available as a water source. Territorial males eat snow more often and longer after chasing females and during warm sunny days than during cold and cloudy days. Based on these observations, it is clear that the snow cover is a necessary part of the rutting season and plays a unique role in the reproduction success of goitered gazelles (Blank 1998).

Goitered gazelles experience great difficulty during the early, snowless frosts, when water sources freeze and there is no snow cover. The gazelles graze more intensively and for a longer time in the mornings after a frosty night, when the frost falls on the plants and the grasses remain moist for a longer time. During periods of low snow, the occurrence of the gazelles significantly increases in pastures with sharply crossed terrain, where snow accumulates under the influence of the wind and is available longer. For goitered gazelles, the autumn migration is often caused not by a lack of food but by the freezing of available water sources in the absence of a snow cover. In snowless, frosty winters, gazelles are quickly exhausted because of water drought, sometimes resulting in death (Zhevnerov et al. 1983).

A territorial male consumes snow often, especially after intensive rutting activities. A male will usually eat snow for less than half a minute at a time and for only very short breaks of a few seconds during intensive chasing, but in a relaxed environment, a male will eat snow for a longer time of up to 4.5 min. Snow eating takes from 4 to 30 min (1–6%) of the male's total daily time budget. A dependence rate of snow eating in relation to the intensity of courtship displays is absent, but males chase females more intensively when snow is available. In the Ili Hollow in 1982 and 1984, November was a snowless month with the snow cover not appearing until December, as a result the rate of males chasing females during the rut was almost three times more often in December than in November, while in other years (1985, 1986), the snow cover was available for both months and the courting rate was identical between months.

Rutting males eat snow more often during days when there is a positive temperature (higher than zero) and sunny weather with clear skies. The highest intensity of snow eating in a territorial male is observed after chasing females (1.23 cases/h, n = 121), while when grazing, snow is eaten almost half as often (0.68 cases/h, n = 85). In addition, the intensity of snow eating is very low after territorial conflicts or when males make border rounds of their territories (0.07–0.18 cases/h, n = 20). Territorial males typically eat snow much more often in the middle of the day, compared to morning and evening, and they have been observed to eat snow the longest after galloping after females (115.4 ± 8.3 s, $n = 28$), a shorter time after trotting (72.1 ± 1.8 s, $n = 34$), and the least often when grazing and walking (45.2 ± 1.5 s, $n = 81$) (Blank 1998).

4.2.11 Daily Activity of Territorial Males During the Main Rut

The first females appear on the male plots quite late, usually between 9:00 to 10:00 in the morning. As soon as they arrive, the males begin demonstrating endless courtship displays. Closer to noon, the intensity of the rut increases, but soon the females in the group leave the territory and go to their days resting places, higher up along the slopes. When the males no longer have females to court, they make a tour of their territory and soon go to rest, during which they will even sleep. Closer to 15:00, female groups start to descend the slopes and again cross the network of individual male plots. Males do not miss the opportunity and actively chase the females until they have left the male territories again. Usually, the courtship by males continues until almost dark, and sexual activity subsides by night, when all females have left for their nighttime pastures (Blank 2022a).

As one would expect, the intensity of male courtship displays varies during the rutting period, depending on internal changes of a male's testicular function. Courtship behaviors and intensity are also related to weather conditions. In months with the same weather conditions, the intensity of rutting is consistent, while under different environmental conditions, the months are characterized by different frequencies of courtship displays. In addition, in cold years, the rut can start up to an hour later due to very low morning temperatures. During the rut, the duration of grazing decreases for territorial rutting males, but the most time of these males is spent not on courting females but on protecting and marking territorial boundaries from intrusion by male neighbors and vagrant bachelors (Zhevnerov 1984).

Activities change little in the lives of bachelor males during the rut, as they move from their night pastures to their daytime resting places just as before the rut, spending a long-time grazing. Only if the opportunity presents itself within their daily movements will these males court females, though these courtships are not so long-lasting, since most of the males' time males is taken up with grazing and resting (Blank 2022a).

4.2.12 Starting Age for Participation in Breeding

Some adolescent females have their first estrus cycle in December at the age of 6–8 months, and after engaging in the rut, birth their first offspring in May at the age of 1 year. The rest of the young females have their first estrus at the age of 18–19 months. During a hunt in May 1955 in the central part of the Betpak-Dala Desert, five out of nine young females aged 12–13 months killed during a hunt had recently given birth and four more were infertile; in 1956, only two young females out of 22 females (noted by their very small sizes; their weights were 16.2 kg and

22.4 kg) did not participate in breeding. Since adult females who had fawns in previous years are the first to participate in the rut followed by the younger females whose estrus occurs 1–2 months later, the birthing period for goitered gazelles stretches for 2.5–3 months in the spring (Sludskiy 1956).

Studying the physiological rhythms in goitered gazelles, Tsapluk (1972) found that the reproductive maturity in females can appear as early as 7–8 months, with some having newborns by 1 year of age. In males, the first spermatogenesis occurs at the age of 1 year, while reproductive maturity occurs at the age of 17–18 months (Tsapluk 1972). In a study on Barsakelmes Island, juvenile females (6–7 months of age) stayed mostly in female groups during the rut and composed 35% on average of all females observed (Table 4.3). When chasing female groups, males courted not only adult females, but also adolescent females that also had visible signs of estrus, with some of them mating. In the winter of 1971–1972, pregnancy rates among eight adult females and 14 female yearlings were found to be as follows: among adults, five out of the eight adults (63.3%) were pregnant; four with twins; and three out of 14 yearlings (21.4%) were pregnant, all with just one embryo. According to the observations of Sludskiy (1956) and Zhevnerov (1984), a few young females had their first estrus at the age of 6–8 months in December and gave birth for the first time at the age of 1 year the following May.

There are also differences in embryo sizes and development rates between adult and subadult females. On the 3rd of March, two embryos of an adult female weighted 180 g and 175 g with a body length of 230 mm, while the single embryo of a 10-month-old female weighted 104 g with a body length of 132 mm (also on the third of March). Two embryos found in an adult female on the 20th of March weighted 400 g each with a body length of 300 mm. On the 24th of March, the embryo found in a subadult female of 10 months weighed 320 g with a body length of 217 mm (Zhevnerov 1984). Thus, demonstrating that young females have smaller offspring than adults.

The courtship of female adolescents by territorial males is not observed until the end of November. As a rule, the signs of estrus in young females become visibly noticeably visible in December, during the period of mass mating. The differences in the timing of mating between adult males and young females is also confirmed by the distribution of births. At the beginning of the delivery period in spring, most adult females give birth to two fawns in early May, while newborns of adolescent females appear in the second half of May (Zhevnerov 1984). In the Betpak-Dala

Table 4.3 The proportion of juvenile and adult females in groups of goitered gazelles in the Barsakelmes Island during November and December (adapted from Zhevnerov 1984)

Year	Observed females in groups, total number	Juvenile females in female groups	
		Number	Proportion, %
1969	48	17	35.4
	89	32	35.9
1970	36	11	35.0
	54	21	37.0
Totally	227	81	35.7

Desert, adolescent female gazelles also mated later than adult females (Sludskiy 1956). All tamed females also were fertilized also at the age of 7 or 8 months and gave birth to singletons, while a year later, they birthed twins. Two males that were born from tamed females reached sexual maturity at the age of 18 months, when they participated in the rut and fertilized two females (Zhevnerov 1984).

4.2.13 The Physical Condition of Territorial Males During and After the Rut

Some goitered gazelle males receive a number of injuries in fights during the rut. On the Barsakelmes Island, rutting males are often be seen with bloody sides, and occasionally, fights between males end with the death of one of the combatants. On the island in 1942–1943, where 800 to 900 gazelles were living with no large predators, 14–15 males annually were found to have died from wounds they received during fights. Combat wounds found among the dead males included a gouged eye, large lacerations on the neck, lameness, and other injuries. One old male found dead in Betpak-Dala had one horn broken almost to the base, and his ear was badly torn (Sludskiy 1956). I also observed an adult male in the Kalamali Nature Reserve (Xinjiang, China) with a broken horn almost to the base likely as a result of rutting combat (Fig. 4.27).

Fig. 4.27 The male with a broken horn almost at the base. Kalamali Ungulate Nature Reserve, Xinjiang, China. 06.07.2011. Photo by D. Blank

At the beginning of autumn, before the rut, both male and female gazelles are the most well-fed of the year. In animals harvested in October, the fat layer on the back often reached a thickness of 1–1.5 cm (Sludskiy 1956). But after the start of the rut, males become significantly weakened as early as the first 5 days, and during the rutting period, territorial males lose a lot of weight and almost all of their fat deposits. If in October the average weight of a male is 36 kg ($n = 5$), then in January it is 24 kg ($n = 8$), which is a weight loss of up to 34%. Since females are much calmer and less active and better able to graze regularly during the rutting season, there is no noticeable decrease in their fat layer. The mortality among males after the rut is very high (out of 22 dead bodies found after a rut, 21 gazelles were adult males and only one female) (Gorelov 1972). During the harsh winter of 1971–1972 on Barsakelmes Island, 86 dead bodies of goitered gazelles were found, of which 50 individuals were adult males (58.1%), eight were adult females (9.3%), 14 subadult males (16.3%) and 14—subadult females (16.3%) (Zhevnerov 1984). Goitered gazelles can remain reproductive to around 10 years of age, with documented records in captivity of 10 years 9 months for males and 13 years 10 months for females (Carter 1991).

After the rut, weakened males are relatively easy to be overtaken in large numbers by dogs and wolves, especially in areas where there are no saigas as alternative prey; during jutes, deaths are mainly among adult males. Males that do not die, stay alone or in groups of 2–3 other males. Sometimes they are joined by male adolescents and young females, forming groups of up to ten individuals. Such groups stay in quiet areas where they rest and graze often (Zhevnerov 1984).

4.3 Discussion

Courting displays in goitered gazelles can be observed all year round, but only during the main rut does the rate of courting displays increase significantly (Blank 1985). The rutting season in goitered gazelles is observed during from October into January with some time offsets in different parts of their natural habitat, with the earliest occurring in the southwestern populations (Saudi Arabia) and the latest in the northeastern populations (Mongolia) (Kingswood and Blank 1996). The rutting season typically lasts 2 months, but most copulations occur within a 5-to-7-day window in December (Blank 1998). As a result, most females give birth within a very short period in spring (Blank 1985). The timing of births and the requirements of parental care are a function of prevailing environmental conditions, which coincide with the most favorable season for both the neonates and their mothers (Jarman 1974; Anderson 1979). Food quality and quantity at the time of birthing determines the survival of young and the ultimate reproductive success in ungulates (Jarman 1974; Leuthold and Leuthold 1975).

In goitered gazelle populations in Central Asia, the synchronization of mating and resulting births are an adaptation to their environment, where cold windy springs with an unstable climate and limited vegetation in May can quickly change to hot, dry weather when most fresh food can disappear over a 1-to-2-week period in June (Blank 1990). Since births have to coincide with the most favorable time of year for rearing young and this period is quite short in goitered gazelle habitats, this species has evolved a highly synchronized mating cycle with a short birthing season. The reproductive strategies in ungulates are strongly dependent upon environmental factors (Baharav 1983). The photoperiod regulates and adjusts the sexual cycle in animals that exhibit estrous and anestrous seasons, and the influence of the photoperiod on the sexual cycles is stronger when animals live above a latitude of 30°N (Bubenik 1986; Loudon and Brinklow 1991). As a result, the more southern populations (*G. s. marica* in Saudi Arabia—24°N) have a more extended estrous period during the rut and consequently longer birthing periods compared to northeastern populations (*G.s. subgutturosa* in Central Asia—up to 40–43°N) that are living in harsher conditions (Sempéré et al. 2001).

During the spring peak of sexual activity from April into May, some males behave similarly to the real rut, but with less intensity and without mating (Blank 1998). The nature of this phenomenon is poorly understood, though it has been suggested that it is an artifact from what was once natural during previous, more favorable epochs (Tsapluk 1972). Indeed, many African antelopes breed all year round in favorable climates but start to be seasonal breeders when the environment does not allow continuously successful breeding (Estes 1974). Moreover, some gazelle species regularly exhibit two or even continuous calving seasons a year, with females having a new estrous cycle (postpartum estrus) in early spring occurring some days after parturition that is not inhibited by lactation (Loggers 1992; Baharav 1983). Sand gazelles (*G.s. marica*), the Arabian subspecies of goitered gazelles, have two peaks of birthing per year, February–April (62 offspring) and September–December (46 offspring) (Hague and Smith 1996). Also, mountain (*Gazella gazella*) and Dorcas gazelles (*G. dorcas*) in Israel have been found to be able to breed throughout the year when food and water are abundant. Mountain gazelles in arid environments of Upper Galilee give birth only in spring, while the same species in Lower Galilee (distance of 80 km) with water and food available all year round, birth young all year-round (Baharav 1983). Dorcas gazelles in western Libya have two peaks of rutting behavior in April and November (Essaghaier 1981), though the most important breeding season in spring occurs when food is abundant. These data suggest that female gazelles can have estrus cycles year-round given the availability of resources, with the length of the estrus season and female fertility affected by food and water (Sempéré et al. 2001).

It is interesting that in tropical and subtropical ungulates living between 20°N and 20°S, breeding occurs all year round and there is no distinctive calving season (Loudon and Curlewis 1988). This phenomenon is also found in the reproduction of tropical gazelle species that are not bound to a fixed season for breeding, birthing or territoriality. In species with two seasonal peaks, one peak usually has a higher number of territorial males, while the second has a comparatively lower number of territorial males. The blackbuck in India has reproductive activity from January to April with a second peak from mid-August to mid-October. In the Serengeti, the first reproductive peak of Thomson's and Grant's gazelles lasts from mid-December to May, with another peak in September. In all these species, the lowest level of reproductive behavior is observed from early June to mid-August (Walther and Mungall 1983). Some mountain gazelles (*G. gazella gazella*) and wild populations of Farasan gazelle (*G.g. farasani*) and mountain gazelles in semi-captive conditions of Saudi Arabia, exhibit continuous breeding all year, also with two peaks of births in spring and autumn (Shalmon 1987; Dunham 1997). This means that similar to tropical Antilopinae species, goitered gazelles in the past, when their habitat was suitable, could have had two reproductive peaks a year with territoriality. Environmental conditions, though, have changed adversely so that the current climate is unable to support year-round breeding; however, the goitered gazelles' second peak of sexual activity in April still lingers, though it has become nonreproductive.

The social structure of goitered gazelles is very similar to the behaviors of many antelopes and especially other gazelle species (Walther 1979; Estes 1991). According to Walther and Mungall (1983), breeding males of most gazelle species (Indian blackbuck, springbok, gerenuk, Grant's gazelle, mountain gazelle, Thomson's gazelle, and goitered gazelles, as well as other gazelle species) establish individual territories during the rut. In all gazelle species, it is possible to find same common features in breeding behavior: (a) only adult males become territorial, but not all; (b) usually territorial periods alternate with non-territorial periods during the life of the same individuals; (c) the territory owner aggressively expels other males from his territory, or at least those that are more dominate; and (d) females usually visit the males only temporarily in their territories. The territoriality aspect of breeding is of greatest importance to reproductive success since, in all gazelle species, reproduction is almost exclusively with the territorial males. Non-territorial males herd and court females in the same way as territorial males, even mounting females, though these sexual displays are observed much less often and for shorter periods than in territorial bucks. Grazing rituals and vocalizations are only rarely demonstrated by non-territorial males, while territorial males display these behaviors very often. Generally speaking, these behaviors can be considered as an indication of territoriality in male gazelles. In addition, cases of non-territorial males mating with females, are very rarely observed (Walther and Mungall 1983).

"It has long been thought that all individuals of one sex of a species display species-specific behaviours" (Isvaran 2005). However, the actual situation is now known to be more complex than a simple species-specific strategy, when dominant males aggressively guard females from subdominants (Clutton-Brock et al. 1982; Hogg 1987) or territorial males confront non-territorial males during the breeding season (Blank 1998; Corlatti et al. 2013). For a various animal taxa from arthropods to mammals, a broad intraspecific variation has been found in mating behaviors—or so called alternative mating tactics—within a population, when not only dominant or territorial males have access to mating females, but in some conditions, subdominant satellite males can also participate in population breeding, though yielding low reproductive benefits (Taborsky et al. 2008). Among ungulates, alternative mating tactics have been found in separate taxonomic groups of caprids, cervids, antelopes, as well as perissodactyls (horses, rhinos) (Isvaran 2005; Willisch and Neuhaus 2009), and since many species have yet to be investigated for these behaviors, some authors believe that this phenomenon is much more widespread among ungulates, especially in bovids and cervids, than we know at present (Cassinello and Pieters 2000; Isvaran 2005; Oliveira et al. 2008). Alternative mating tactics have not yet been reported for goitered gazelles, or other gazelle species (Walther and Mungall 1983; Estes 1991), but according to the definitions of Isvaran (2005), Bro-JØrgensen (2011) and Bowyer et al. (2020), it is possible to suppose that the observed agonistic behavior of goitered gazelles reflects the occurrence of alternative mating tactics in this species. Moreover, the simultaneous or alternating coexistence of different mating tactics are known for other ungulate species. For example, in topi (*Damaliscus lunatus*—Bro-Jørgensen and Durant 2003) and fallow deer (*Dama dama*—Apollonio et al. 1992), some males establish clustered leks in the center of breeding activities, while in the periphery of these activities, they defend dispersed individual territories. However, since this phenomenon was not a specific subject of my or other researchers' goitered gazelle studies, I lack information for consideration on this phenomenon. In this book, I do not discuss alternative mating tactics and instead consider only differences in agonistic behaviors between territorial and non-territorial males within a population of goitered gazelles, the aggressive behavior of which has not been the subject of special investigation until now (Blank and Yang 2014; Blank et al. 2015).

The generally accepted understanding of a harem is a stable group of females with one male holding it together long enough that the male can service most, if not all, of the females. Males of all gazelle species with the exception of one establish individual territories during the rut (Walther and Mungall 1983). The only known exception among gazelle species is the Grant's gazelle, but even for the Grant's gazelle, it is only under certain conditions (Walther 1972). A female group in Grant's gazelles sometimes stays in the individual territory of a male for several weeks or months—possibly because the home range of the female group coincides with the territory of the male. In this case, it is possible to call such a group a "harem," which is combined with territorial behavior. Harem formation without territorial behavior of the male, as found in the plain zebras (Klingel 1972), does not exist in Antilopinae (Estes 1974; Walther and Mungall 1983). The home ranges of female groups in mountain gazelles also coincide with male territories, but these females occasionally leave the territory changing the composition of the group from time to time (Walther and Mungall 1983); this is similar to the gerenuk (Leuthold 1978). Since these female groups are not necessarily stable units themselves and they do not stay exclusively with one male long enough for servicing of the entire female group, the term "harem" cannot be applied to these gazelles in the generally accepted sense (Walther and Mungall 1983). Since goitered gazelles have very similar rutting behaviors as many other gazelle species, it is difficult to confidently assert that this species has a special harem social structure, which has not been observed among any other gazelle species.

During the copulation stage in early December, goitered gazelle females will stay within the territory of their chosen male throughout the day and even overnight and into a second day. The number of chases can increase noticeably, but their speed decreases significantly, and a walk is the preferred gait during courting; females run away reluctantly, especially those in estrus. During the testing phase of female readiness for the rut, males chase one female after another, forcing them to urinate so they can sniff the urine spot on the ground or more rarely even put their muzzles under the female's urine stream. In most Antilopinae species, males do not touch the genitals or other regions of the females' bodies with their noses (except mountain gazelles) (Walther and Mungall 1983). Males concentrate their courting activities mainly toward females in estrus and estrus females urinate frequently followed by males displaying a flehmen response almost as if mandatory. Ungulate males check a female's sexual state during their flehmen displays when the vomeronasal organ is activated (Estes 1972); therefore, this reaction is longest and most often noted at the peak of the rut, especially frequently when courting of a female in estrus that is ready for copulation. Most mounts and copulations in goitered gazelles occur at the end of the second stage of the rut. Goitered gazelle males demonstrate flehmen responses, copulation, and masturbation postures that are typical for all other gazelle species, though the quantitative parameters of these behaviors in most gazelle species are not available (Walther 1979).

For actual copulation, most males of gazelle species make a series of mounts prior to copulation. Mounting four times before copulation seems to be a minimum number, while 10–20 mounts are quite common and 30 to 40 mounts before copulation are not rare. Occasionally, the number of mounts in Thomson's gazelles has been observed to reach 164. In my study of goitered gazelles, the number of mounts may be even greater. Females stand or walk during mounting by a male. In all Antilopinae species, the female starts walking at the moment the male mounts her, even if she has been standing still before the mount. In most gazelles, the male follows bipedally behind the female with the mount and copulation taking place while walking. The shortest mating ritual in Thomson's gazelle lasted 3 min, with 22 rituals completed within 5–15 min, 13 rituals within 15–30 min, and two within 45–60 min; only one ritual lasted about 90 min. A similar distribution was found for Grant's gazelles. So, most of the Antilopinae mating rituals are completed within 5–45 min, with copulation itself taking only a few seconds (Walther and Mungall 1983). After copulation, courting activities in goitered gazelle males decrease drastically, and within 7–10 days, the individual territories have disappeared (Blank 1998).

Three factors affect the frequency of snow eating by territorial males: wind, clouds, and the availability of snow. When there is no wind but plenty of snow and sunshine, the intensity of snow eating is at its maximum. As a result, the frequency of male courtship behavior also increases in these conditions. According to my data from Ili Hollow, any changes in temperature results in an increase of sexual activity in males. Thus, in contrast to snow, temperature is not a main factor determining the intensity of courtship, although it affects quantitative parameters of the rut. In addition, the frequency of courtship displays changes during the day, usually with a morning and evening peak of courting activity and a lull from noon to 14:00. The decrease in the grazing activity of a territorial male is not due to intense courtship, but rather to territoriality, since the male has to defend his territory against intruders and spends much of his time on maintenance of his territorial borders (Blank 1998).

Some goitered gazelle males will return to their territories the next year (Marmazinskaya 1999), but not all of them establish the same territory from year to year after a period of absence. For example, only seven out 44 Thomson's gazelle males (or 16%) returned to their former territories or the immediate vicinity after periods of absence. One of them even returned twice (Walther and Mungall 1983).

The goitered gazelle lives further north than other gazelle species, in the sharply continental climate of the Central Asian deserts, having hot, dry, and calm summers and cold, snowy, and windy winters. These conditions differ from the natural habitats of other gazelles, yet the courting behaviors of all gazelles are generally alike (Walther 1979; Walther and Mungall 1983). The neck-stretch display, for instance, is a common posture for all gazelle species, as well as the majority of Bovinae species; though this pose is pronounced demonstrated in only some species and not very clearly in another gazelle species (Walther 1979). The nose-lift posture is less common and mainly a characteristic of gazelles (genus *Gazella*) and gazelle-like species (genera *Antilope*, *Antidorcas*, *Litocranius*). The social structure during the rut, described here for goitered gazelles, when several females stay temporarily with a male in his territory, is the same in all the gazelle species and also for all subfamilies Antilopinae, Aepycerotinae, Alcelaphinae, and most probably for Saiginae (Walther and Mungall 1983).

A more detailed consideration and comparisons of mating activity demonstrates that the goitered gazelle has some marked behavioral differences from other gazelle species. First, there is a very high level of seasonality in rutting, and individual territories are established and marked during November and December (the productive, main autumn rut) and to a lesser extent in April and May (the unproductive, pseudo spring rut), and copulations are observed most often within a several-day period at the beginning of December, resulting in the overwhelming majority of births occurring within a 4–5-day period in mid-May. In contrast, strict seasonality in breeding is the exception in tropical Antilopini, although there are one or two birth peaks in the year (Estes 1991). There is also one obvious reproductive peak in the North African dorcas gazelle (*Gazella dorcas*), but it lasts for two to even 3 months, which makes it possible to see young of different ages and pregnant females simultaneously. Second, the role of snow as a source of drinking water is so important for goitered gazelle males during the rut that the time and intensity of courtship activities can change depending on the snow cover. Third, vocalizations during the rut are apparently more important for goitered gazelles than other species, as males call out almost constantly while chasing females. Males even have a morphological adaptation that is connected to an intensification of the bellow during this period—the considerable growth of the male's goiter. Male courtship in other gazelle species (*G. dorcas*, *G. gazella*) also is accompanied by vocalizations, but these sounds are usually soft and impossible to hear at a distance of more than 150–200 m. And fourth, there are also a number of details in courtship postures and the manner of their displays which differ considerably between gazelles (Blank 1998).

References

Afanasiev AV, Bazhanov VS, Karelov MN, Sludskiy AA, Strautman EN (1953) Goitered gazelle. In: Afanasiev AV (ed) Animal of Kazakhstan. Publication of Kazakh Academy of Sciences, Alma-Ata, pp 484–488. (in Russian)

Allen GM (1940) The mammals of China and Mongolia. Am Museum Nat History 2:621–1350

Anderson JL (1979) Reproductive seasonality of the Nyala *Tragelaphus angasi*; the interaction of light, vegetation phenology, feeding style and reproductive physiology. Mamm Rev 9:33–46

Antipin VM (1941) Mammals of Kazakhstan. In: Ungulates, vol 3. Publication OGIZ, Alma-Ata, pp 47–54. (in Russian)

Apollonio M, Festa-Bianchet M, Mattioti MF, Sarno B (1992) To lek or not to lek: mating strategies of male fallow deer. Behav Ecol 3(1):25–31

Baharav D (1983) Reproductive strategies in female mountain and Dorcas gazelles (*Gazella gazella* and *Gazella dorcas*). J Zool 200:445–453

Bannikov AG (1954) Persian gazelle. In: Heptner VG (ed) Mammals of Mongolian People Republic. Publication of USSR Academy of Sciences, Moscow-Leningrad, pp 250–258. (in Russian)

Bland KP (1985) Some observations based on the testicular histology of Dorcas and Arabian goitered gazelles. Chinkara 1:24–25

Blank DA (1985) Peculiarities of social and reproductive behaviour of *Gazella subgutturosa* in the Ili Valley. Zool Zhurnal 64:1059–1070. (in Russian, English abstract)

Blank DA (1990) Persian gazelle. In: Kovshar AF (ed) Rare animals of desert regions. Nauka Press of Kazakh SSR, Alma-Ata, pp 56–80. (in Russian)

Blank DA (1997) Social and reproductive behavior in the Persian gazelle. Ph.D. thesis. The University of Tel-Aviv, Tel Aviv-Jaffa. 189 pp

Blank DA (1998) Mating behavior of the Persian gazelle *Gazella subgutturosa* Güldenstaedt, 1780. Mammalia 62(4):409–419

Blank DA (2022a) Activity behavior of goitered gazelles in an arid environment. Behav Proc 204:104804

Blank DA (2022b) Behavior of goitered gazelles in the Issyk-Kul Breeding Center, Kyrgyzstan. Unpublished Report: May, 2022

Blank DA, Fedosenko AK (1983) The pseudo-rut in goitered gazelle. In: The behavior of animal in society. Materials of third all-union conference for animal behavior, vol 2. Nauka Press, Moscow, pp 258–259. (in Russian)

Blank DA, Kovshar AF (1986) The number of goitered gazelle populations in deserts of southern and South-Eastern Kazakhstan. In: Flint VE, Prisyazhnuk VE (eds) Conservation and perspectives of the number restoration of goitered gazelle in USSR. All-Union Scientific Institute of Nature Conservation and Nature Reserves, Moscow, pp 8–12. (in Russian)

Blank D, Yang W (2014) Does social status of males change their aggressive behavior repertoire in goitered gazelle (*Gazella subgutturosa* Güld., 1780)? Behav Proc 108:20–26

Blank DA, Ruckstuhl K, Yang W (2014) Roaring function in male goitered gazelles. Behav Process 106:152–159

Blank DA, Ruckstuhl K, Yang W (2015) Seasonal dynamics of agonistic displays in territorial and non-territorial males of goitered gazelle. Zoology 118(1):63–68

Bowyer RT, McCullough DR, Rachlow JL, Ciuti S, Whiting JC. (2020) Evolution of ungulate mating systems: integrating social and environmental factors. Ecology and Evolution. Jun;10(11):5160–78

Bro-Jørgensen J, Durant SM (2003) Mating strategies of topi bulls: getting in the center of attention. Anim Behav 65:585–594

Bro-JØrgensen, J. (2011). Intra-and intersexual conflicts and cooperation in the evolution of mating strategies: lessons learnt from ungulates. Evolutionary Biology, 38, 28–41

Bubenik GA (1986) Regulation of seasonal endocrine rhythms in make boreal cervids. In: Asenmacher I, Boissin J (eds) Endocrine regulations as adaptive mechanisms to the environment. CNRS, Paris, pp 461–472
Carter S (1991) Goitered gazelle north American regional studbook, 1st edn. Sedgwick County Zoo an Botanical Garden, Wichita, p 87
Cassinello J, Pieters I (2000) Multi-male captive groups of endangered dama gazelle: social rank, aggression, and enclosure effect. Zoo Biol 19:121–129
Clutton-Brock TH, Guiness FE, Albon SD (1982) Red Deer. Behavior and ecology of two sexes. University of Chicago Press, Chicago
Corlatti L, Caroli M, Pietrocini V, Lovari S (2013) Rutting behaviour of territorial and non-territorial male chamois: is there a home advantage? Behav Process 92:118–124
Dereviagin PY (1947) On seasonal migrations of goitered gazelle from southern pre-Balkhash deserts into inter-mountain valley Kalkan-Matay- Mynbulack. Bull Acad Sci Kazakh SSR, Alma-Ata 4(25):33–38. (in Russian)
Dunham KM (1997) Population growth of mountain gazelles reintroduced in Central Arabia. Biol Conserv 81:205–214
Essaghaier MFA (1981) Ecology and behavior of Dorcas gazelle. Thesis. University of Idaho, Moscow
Estes RD (1972) The role of the vomeronasal organ in mammalian reproduction. Mammalia 36:35–341
Estes RD (1974) Social organization of the African Bovidae. In: Geist V, Walther F (eds) The behaviour of ungulates and its relation to management, IUCN Publ. No. 24. International Union for the Conservation of Nature and Natural Resources, Morges, pp 166–205
Estes RD (1991) The behavior guide to African mammals. University of California Press, Berkeley-Los Angeles
Flerov KK (1935) Persian gazelle. In: Pavlovskiy EN (ed) Wild animals of Tajikistan in life and significance for man, vol 1. Academy of Sciences of the USSR, Moscow, pp 109–112. (in Russian)
Gorelov UK (1972) Breeding of goitered gazelle (*Gazella subgutturosa* Güldenstaedt) and problem of restoring of its number in the Badkhyz (SE Turkmenistan). In: Theriology, vol 1. Nauka of Siberian Department of Academy of Sciences of USSR, Novosibirsk, pp 420–424. (in Russian)
Habibi K, Thouless CR, Lindsay N (1993) Comparative behavior of sand and mountain gazelles. J Zool 229:41–53
Hague MN, Smith TR (1996) Reintroduction of Arabian sand gazelle *Gazella subgutturosa marica* in Saudi Arabia. Biol Conserv 76:203–207
Heptner VG, Nasimovich AA, Bannikov AG (1961) Mammals of Soviet Union. In: Artiodactyla and Perissodactyla, vol 1. Vysshaya Shkola Publishing House, Moscow (in Russian)
Heptner VG, Nasimovich AA, Bannikov AG (1988) Mammals of the Soviet Union. Artiodactyla and Perissodactyla. Smithsonian Institution Libraries Natl Sci Foundation 1:1–1147. (translation from Russian)
Hogg JT (1987) Intrasexual competition and male choice in Rocky Mountain bighorn sheep. Ethology 75:119–144
Ishunin GI (1961) Mammals (carnivores and ungulates). In: Fauna of Uzbek SSR, vol 3. Publication of the Academy of Sciences of Uzbek SSR, Tahkent, pp 165–174. (in Russian)
Isvaran K (2005) Variation in male mating behaviour within ungulate populations: patterns and processes. Curr Sci 89:1192–1199
Jamsheed R (1976) Big game animals of Iran (Persia). Rashid Jamsheed, 122 pp
Jarman RJ (1974) The social organization of antelope in relation to their ecology. Behaviour 48:215–267
Kingswood SC, Blank DA (1996) Gazella subgutturosa. Mamm Species 518:1–10
Klingel H (1972) Social behaviour of African equidae. Zool Afr 7(1):175–185
Kostin VP (1955) Materials on ecology and commercial value of goitered gazelle. Proc Institute Zoology Acad Sci Kazakh SSR 4:67–82. (in Russian)

Leuthold W (1978) On social organization of the gerenuk *Litocranius walleri* (Brooke, 1878). J Anim Ecol 47:471–490

Leuthold W, Leuthold BM (1975) Temporal patterns of reproduction in ungulates of Tsavo east National Park, Kenya. East Afr Wildl J 13:46–74

Loggers CO (1992) Population characteristics of Dorcas gazelles in Morocco. Afr J Ecol 30:301–308

Loudon ASI, Brinklow BR (1991) Reproduction in deer: adaptation for life in seasonal environments. In: Brown RD (ed) 2nd international symposium on the biology of deer. Springer Verlag, New York, pp 261–278

Loudon ASI, Curlewis JD (1988) Cycles of antler and testicular growth in an aseasonal tropical deer (*Axis axis*). J Reprod Fertility 83:729–738

Mambetzhumaev AM (1970a) Goitered gazelle. FAN Publication House, Tashkent. 196 pp. (in Russian)

Mambetzhumaev AM (1970b) On the number and distribution of goitered gazelles in the Ustyurt plateau and Kyzylkum Desert. Zool Zhurnal 49(3):431–434. (in Russian)

Marmazinskaya NV (1996a) Breeding of goitered gazelle *Gazella subgutturosa* (Artiodactyla Bovidae) in the ecological center "Dzheiran". Proceedings of the Uzbekistan nature reserves, 1. FAN Publication House, Tashkent: 53–61. (in Russian)

Marmazinskaya NV (1996b) Territorial and marking behavior of goitered gazelle *Gazella subgutturosa* (Artiodactyla, Bovidae) in the Bukhara ecological center "Dzheiran". Zool Zhurnal 75(11):1737–1751. (in Russian, English summary)

Marmazinskaya NV (1999) Social behavior of goitered gazelle (*Gazella subgutturosa* /Güld., 1780/) during breeding period in the conditions ecological center "Goitered gazelle". Ph.D. thesis in zoology. Institute of zoology of Uzbek Academy of Sciences, Tashkent, pp 1–20. (in Russian)

Oliveira RF, Taborsky M, Brockmann HJ (2008) Alternative reproductive tactics: an integrative approach. Cambridge University Press, Cambridge

Roberts TJ (1977) The mammals of Pakistan. Ernest Benn, London, p 361

Rubenstein DI (1989) Life history and social organization in arid adapted ungulates. J Arid Environ 17:145–156

Sempéré AJ, Brown N, Pereladova OB, Bahloul K, Lacroix A, Soldatova N (2001) Comparative analysis of reproductive cycles in female Persian gazelle (*Gazella subgutturosa subgutturosa*) (Central Asia) and sand gazelle (*Gazella subgutturosa marica*) (Arabian peninsula). Gen Comp Endocrinol 121:57–65

Shahooth MA, Zalzala SJ, Hatif SA (2015) Study of reproduction in the local Reem gazelle (*Gazella subgutturosa marica*). World J Pharm Pharm Sci:810–814

Shalmon B (1987) The Arava gazelle. Israel Land Nat 13:15–18

Sinclair ARE (1983) The adaptation of African ungulates and their effects and community function. In: Bourliere F (ed) Tropical Savannas. Elsevier, Oxford, pp 401–426

Sludskiy AA (1956) Breeding of goitered gazelle. In: Proceedings of institute of zoology, Kazakh Academy of Sciences, vol 6. Nauka Press of Kazakh SSR, Alma-Ata, pp 78–108. (in Russian)

Sludskiy AA (1977) Goitered gazelle. In: Rare animals of USSR. Publication of Lesnaya Promyshlennost, Moscow, pp 28–61. (in Russian)

Sludskiy AA, Shubin IG (1963) Airplane survey of saiga, goitered gazelle and these ungulate resources in the Kazakhstan deserts. In: The fauna resources of game animals in USSR and their accounting. Publication of the Academy of Sciences, Moscow. USSR: 84-91 (in Russian)

Smirin VM, Smirin YM (1991) Animals in nature. Publication of Moscow University, Moscow. 225 pp. (in Russian)

Sokolov II (1959) *Gazella subgutturosa* Güldenstaedt—Goitered gazelle. In: Pavlovskiy EN (ed) Fauna USSR, mammals: ungulates (orders Perissodactyla and Artiodactyla). Publication of the Academy of Sciences USSR, Moscow, pp 308–331. (in Russian)

Soldatova NV (1983) To behavior of goitered gazelles Bukhara breeding center. In: Flint VE (ed) Ecology of plants and animals of nature reserves of Uzbekistan. FAN Publication House, Tashkent, pp 86–90. (in Russian)

Taborsky M, Oliveira RF, Brockmann HJ (2008) The evolution of alternative reproductive tactics: concepts and questions. In: Oliveira RF, Taborsky M, Brockmann HJ (eds) Alternative reproductive tactics: an integrative approach. Cambridge University Press, Cambridge, pp 1–22

Tsapluk OE (1972) Age and seasonal dynamics of sexual activity of goitered gazelle. Proc Acad Sci Kazakh SSR, Biological series 3:39–46. (in Russian)

Vereshchagin NK (1939) Goitered gazelle in Azerbaijan. Proc Zool Institute Azerbaijan Branch USSR Acad Sci, Baku 10:89–90. (in Russian)

Walther FR (1972) Social grouping in Grant's gazelle (*Gazella granti* Brooke, 1872) in the Serengeti National Park. Z Tierpsychol 31:348–403

Walther FR (1979) Das Verhalten der Hornträger (Bovidae). Hand Zool 8, 10(30) 177 pp

Walther FR (1984) Communication and expression in hoofed animals. Indiana University Press, Bloomington

Walther FR, Mungall EC, Grau G.A (1983) Gazelles and their relatives. A study in territorial behaviorNoyes Publication, Park Ridge., 239 pp.

Willisch CS, Neuhaus P (2009) Alternative mating tactics and their impact on survival in adult male alpine ibex (*Capra ibex ibex*). J Mammal 90:1421–1430

Zhevnerov VV (1975) The ethology and ecology of goitered gazelle during the rutting period in the Barsa-Kelmes Island (Aral Sea). Proc Acad Sci Kazakh SSR 1:26–33. (in Russian)

Zhevnerov VV (1984) Goitered gazelle of the Barsakelmes Island. Nauka Publishing House of Kazakh SSR, Alma-Ata. 144 pp. (in Russian)

Zhevnerov VV, Bekenov AB, Sludskiy AA (1983) Goitered gazelle—*Gazella subgutturosa* Güldenstaedt, 1780. In: Gvozdev EV, Kapitonov VI (eds) Mammals of Kazakhstan, Bovidae, vol 3 (3). Nauka Publishing House of Kazakh SSR, Alma-Ata, pp 11–54. (in Russian)

Home Range, Territoriality, and Marking Behavior

5

5.1 General Remarks

Most mammal species have a well-developed olfactory sense which is used in social communication (Ralls 1971). Communication and transmission of certain information between individuals is vital in any social system, including animal societies (Gillam 2012), and animals send visual, acoustic, and olfactory signals to each other (Davies et al. 2012). Visual and acoustic signals have undergone significant changes in the process of evolution (Blank 2021), whereas scent signals are a relatively ancient way of communication in the animal world (Lévy et al. 2004). Despite the relative antiquity, scent signals have not lost their importance and play an exceptional role in recognizing their own young among strangers and in distinguishing familiar conspecifics and strangers within their home range; individual distinction is also important for determining the social status of an individual in their group hierarchy. During the rut, scent cues are also of particular importance in protecting a male's individual territory from his territorial neighbors or from male bachelors (Gosling 1985). To this end, many territorial species use deposits of urine and feces in certain locations as territorial boundary markers (Müller-Schwarze 1999). The scent marks of a territorial male not only have a deterrent effect on other intruding males, but also have confidence-giving effect. Males only inside their territory, where their own smell prevails, can behave freely and participate in courting activities. Scent markings are characterized by efficiency over long distances, in absolute darkness, preservation of odor signals over time, allowing to perform their function of protecting the territory for several days even in the absence of the animal who left the mark (Mykytowycz 1972; Gosling and Roberts 2001), whereas acoustic and visual signals require the constant presence of the signaler (Alberts 1992).

Walther (1984) defined scent marking as an action for depositing odoriferous substances contained in an animal's urine and feces or in the secretions of the

D. Blank, *The Behavioral Biology and Conservation of Goitered Gazelles (Gazella subgutturosa)*, https://doi.org/10.1007/978-3-031-83504-9_5

specialized cutaneous glands at specific sites or on objects in the environment. Studies of quantitative data on the frequency and duration of urination-defecation acts and their seasonal fluctuations in various species are quite rare. In fact, only three studies for three species are known, Thomson's gazelle, the Chinese water deer (*Hydropotes inermis*) and the oribi (*Ourebia ourebi*) (Walther 1978a; Sun et al. 1994; Brashares and Arcese 1999b). The frequency of marking seems to vary for different antelope species, and some species mark often (goitered gazelles), while others rarely (Dorcas gazelle—*Gazella dorcas*), and some leave small dung marks (gerenuk—*Litocranius walleri*), and others leave large marks (springbok—*Antidorcas marsupialis*) (Walther et al. 1983). Walther (1978a) was the first to investigate the frequency of urination-defecation acts in Thomson gazelles. Gosling (1981), as well as Roberts and Lowen (1997), found the optimal placement of urination/defecation marks for protection of individual territories in males of some African antelope species.

Goitered gazelle males are known to have individual territories during the main rut in November–December and to a far lesser degree during the spring peak of sexual activity (pseudo rut) in April–May. During these months, territorial males protect their territories against intruders and intensively mark their territorial borders, while throughout the rest of the year, they move freely within their home range for distances of 6–10 km (Blank 1985). In goitered gazelles, marking activities are demonstrated only by adult males, while females do not mark at all since females have no individual territories (Walther et al. 1983), though Gorelov (1972) observed once a female which scratched a male's marking site.

Having a home range is a much different concept from an individual territory. A home range is an area in which a gazelle or group of gazelles stay long enough to establish a subjective space-time system, but without real boundaries. The home range of neighboring herds may overlap, and individuals may readily switch from one herd in one home range to another herd in another home range (Mechkour et al. 2008). The average home range of goitered gazelles females is estimated at 3.61 ± 0.47 km^2 in winter, 3.96 ± 0.44 km^2 in spring, 4.55 ± 1.35 km^2 for the first half of summer, 2.26 ± 0.20 km^2 for the second half of summer, 3.38 ± 0.44 km^2 in autumn, 1.37 ± 0.50 km^2 during the mating season, and 1.66 ± 0.50 km^2 during the birthing season (Durmuş 2010). However, different home range sizes are observed for different groups at different sites, as in Azerbaijan, where a female's home range fluctuates between 100 and 120 km^2 over seasons. Moreover, the difference between home range sizes of goitered gazelle males and females is insignificant, though between seasons their home range sizes have big differences. This is true especially in autumn, when goitered gazelles gather into large groups for rutting and do not move far from the male territories, decreasing their home range sizes by almost twice compared to summer (40–50 km^2) (Askerov et al. 2021). It is possible that this phenomenon is observed in Azerbaijan due to the fact that goitered gazelles acclimating to a new place have large, free territories and a low population density, and with an increase in the number and density of the population, the sizes of their home range would decrease. However, some authors (Gorelov 1972) believe that gazelles in Turkmenistan are very loosely tied to the territory in which they live. This is the exact opposite of Gorbunov's (1983) opinion from observing the herds of a small population of goitered gazelles in the Krasnovodsk Peninsula (Turkmenistan), who concluded that goitered gazelles have a fairly strong attachment to their home range, especially to its protective elements. He supposes that gazelles know their area very well and, if necessary, move around it quickly and covertly.

For comparison, the home range of springboks (*Antidorcas marsupialis*) is about 0.15 km^2 in females and 0.07–0.1km^2 in males (Mechkour et al. 2008), which is about 10 times smaller than that of the goitered gazelle. It is possible that such a big difference is due to the semi-captive conditions for the observed springbok population, though in the wild, the springbok home range fluctuates between 0.1 and 0.7-km^2 (Williamson 1985), which is still much less than that of the goitered gazelles.

Individual territories, unlike a home range, have obvious boundaries or a zone that is marked by the male-owner with urine and feces, by secretions of his preorbital glands, and by butting vegetation, mainly shrubs and saxaul. This territorial marking provides boundary protection by the owner and allows him to have dominance over any male who enters (territorial neighbor or bachelor) his individual territory (Gosling 1982).

Generally, big animals have large individual territories and small animals have small territories, but this is not always true. Primarily, the size of male territories is a species-specific matter. In addition, territorial density has an impact on its size, when territories are larger in areas with a low population density and smaller in areas with a high density. In addition, a territorial male cannot keep an area larger than he is able to fulfill the activities necessary to maintain his territorial status. This limit varies with species-specific territorial activities. The range between the maximum and minimum territory size is not very large within a species and its shrinkage and growth outside an average size of a territory have narrow limits (Walther et al. 1983).

Marking behaviors of goitered gazelles have been described since the early 1940s (Antipin 1941; Sludskiy 1956), but the territoriality nature of this behavior was rejected, attributing harem behavior to the goitered gazelle during the rut without any consideration of its relationship to territory (Mambetzhumaev 1970; Zhevnerov 1984). Though, in connection with the mention of rutting marks of goitered gazelles, Sludskiy (1956) did suggest the existence of individual territories in males, or at least their possible existence in the past. Building a network of individual territories by rutting males was first described in 1985 in the Ili Valley, SE Kazakhstan (Blank 1985). Territorial behavior of goitered gazelle males also was found in the semi-captive populations of the Bukhara Breeding Center (Marmazinskaya 1996).

Males guard their individual territories by patrolling and marking their borders, courting females that pass within these boundaries. Individual territories of males are located in places teeming with dry sandy riverbeds with sparse thickets of saxaul (*Haloxylon persicum*), teresken (*Eurotia ceratoides*), and other shrubs like *Athraphaxis spinosa* running along them. Often the borders of these territories run along these channels. In the Ili Hollow (SE Kazakhstan), observations were made of six territorial males. Each male was a territory owner, the size of which could not be accurately determined since their boundaries were often distinguishable only by gazelles. However, judging by the movements of the males, the size of these territories ranged from 50 to 80 ha (Blank 1985). In the semi-captive population of the Bukhara Breeding Center, the territorial plots were smaller at 20–25 ha. It is possible that the size of territorial areas in different populations depends on the biotope, feeding conditions, and population density. Adult males (3.5 years and older) establish a network of individual territories with each rutting male staying within his area during the entire rutting season (mid-October through December) (Blank 1998). It is possible that some males occupy the same areas from year to year though this has been documented in wild population only for some individuals of Thomson's gazelles (Walther et al. 1983). In at least one case, semi-captive goitered gazelle male in the Bukhara Breeding Center occupied the same plot for 2 years in a row (Marmazinskaya 1990). The maximal sizes of a territory in other gazelle species vary between 10 and 20 ha (Thomson's gazelle and Indian blackbucks), 60–70 ha (Grant's gazelles and springbok), and up to 130–340 ha in gerenuk (Walther et al. 1983). When a territory is enlarged beyond an optimal size, the owner cannot fulfill his territorial duties to safeguard the entire area; conversely, a territory also has a lower limit, as it must be large enough to guarantee the owner a sufficient food supply (Walther et al. 1983).

Territorial males mark the borders of their plots in various ways: marking with urine and feces, marking shrubs with secretions from the preorbital glands, and marking by horning shrubs and butting the ground, and some kind of marking through vocalizations. There are also interdigital and metacarpal glands, but it is not clear how these glands work and what role, if any, they serve in marking behavior of goitered gazelles.

5.2 Marking by Urination-Defecation

5.2.1 Marking Urination-Defecation Ritual

Blank et al. (2014b)—Licence number—547541506089 (full text).

Male goitered gazelles mark their territories as most other gazelle and gazelle-like species (Dubost and Feer 1981; Habibi et al. 1993; Bärmann et al. 2021). Most gazelle species leave a mark on the bare ground, devoid of vegetation, apparently in order for the mark to be immediately noticed by other males or, possibly, so the old mark is visible to the marking male which makes it easier for him to move from one old mark to another to refresh them (Estes 1991). Moreover, most ungulates prefer to deposit droppings in the places having already urine/feces marking points (Walther 1984). Approaching an old feces mark, the male first sniffs it, then he scratches it, alternating the front legs one after the other several times, throwing out the old feces regardless of whether it was his own fecal mark or from another male. The scratching leaves clearly visible longitudinal stripes (from 5 to 12 lines), always oriented from the narrowed end of the stroke to the wider end. Then in a stretching posture, the male separates his hind legs extremely far backward from his front legs, lowers his belly closer to the ground as is possible, and urinates. After that he moves his hind legs forward with one jump and defecates in a crouched position sinking anus as close to the ground as possible (Table 5.1). During defecation, the male's

Table 5.1 Description of the recorded behavioral patterns (ethogram) during urination and defecation of goitered gazelle (according to Walther et al. 1983; Walther 1984; Blank 1985—License Number—5475920987776)

Behavior	Description
Adult males: full succession of urination–defecation acts following one after another without pauses	
Sniffing	Lowering the head and sniffing the ground or pellet pile (Fig. 5.1)
Scratching or pawing	Scratching the ground or pellet pile 2–4 times, alternating both forelegs; several strokes with one foreleg and then several strokes with the other in turn; the head is commonly lowered during scratching. Scratching or pawing is not obligatory element of this marking ritual (Fig. 5.2)
Urination	Stepping forward and stretching the body, lowering the belly toward the ground, the hind legs sloping backward; urination during 30–40 s with a lifted tail in this posture (Fig. 5.3)
Defecation	Abrupt bringing of hind legs forward so that hooves are placed lateral to the stationary forelegs, squatting closer to the ground and taking a crouched posture with his anus only a few cm above the ground and defecating for 15–20 s. In this way, urine and feces are deposited on the same spot. The hair of the white rump patch is fluffed, and the tail is lifted to a horizontal position. After defecation, the male resumes a normal posture, wags his tail, and walks off (Fig. 5.4)
Sub-adult males have incomplete succession: Sniffing and scratching usually are absent, and urination and defecation acts are often not linked	
Adult females: Urination and defecation acts	
Urination	Raising the tail, squatting close to the ground, and taking a crouched posture with the vulva only a few centimeters above the ground and urinating (Fig. 5.5)
Defecation	Raising the tail and defecating while walking or grazing without a special posture (Fig. 5.6)

Fig. 5.1 A goitered gazelle male lowers the head and sniffs the urine or pellet pile on the ground. Issyk-Kul goitered gazelle nursery for reacclimatizing, 25.11.2022. Photo by D. Blank

Fig. 5.2 Male scratches the ground or pellet pile 2-4 times, alternating both forelegs. Issyk-Kul goitered gazelle nursery for reacclimatizing, 25.11.2022. Photo by D. Blank

Fig. 5.3 Male urination posture in goitered gazelles: male sets the hind legs back and stretches the body as long as possible. Issyk-Kul goitered gazelle nursery for reacclimatizing, 25.11.2022. Photo by D. Blank

Fig. 5.4 Male defecation posture in goitered gazelles: male takes crouched posture with his anus only a few cm above the ground. Issyk-Kul goitered gazelle nursery for reacclimatizing, 25.11.2022. Photo by D. Blank

Fig. 5.5 Female urination posture in goitered gazelles: female takes a crouched posture with the vulva only a few centimeters above the ground. Issyk-Kul goitered gazelle nursery for reacclimatizing, 25.11.2022. Photo by D. Blank

Fig. 5.6 Female defecation posture in goitered gazelles: female defecates while walking or grazing without a special posture. Issyk-Kul goitered gazelle nursery for reacclimatizing, 25.11.2022. Photo by D. Blank

fecal pellets fall exactly on the urine spot. This sequence of marking actions manifests itself throughout the entire period of rutting (Blank et al. 2014b). For example, in November and December 1971, 312 rutting urine-feces marks were examined, and in all cases, they contained a pile of feces with the characteristic scratching furrows from hooves at the bottom of the making site (Zhevnerov 1984). During the ritual marking, after sniffing and scratching the male shows the urination posture, followed by defecation in obligatory order (Blank 1985). A detailed ethogram of postures related to urination and defecation is presented in Table 5.1. According to my observations, the whole urination/defecation marking process takes 45–60 s. In the Bukhara Breeding Center, similar results were noted for semi-captive population of goitered gazelles (30.1 ± 0.21 s (n = 29; range is 18–60 s). In the Bukhara Breeding Center, rutting marks were located either directly on the trails or adjacent to them, or removed at a distance of 40 cm to 2.5–3 m; 2–5 marks also were located at distances of 2–15 m from each other (Marmazinskaya 1996).

Scraping before urination-defecation is described for males of many antelope species (Leuthold 1977; Estes 1991) and deer species (Sun et al. 1994; Johansson and Liberg 1996; Zaitsev 2006). In steenbok, both males and females scrape the ground and they do so before, in between and after deposition of urine and feces marks (Walther 1984). Like in other antelope species, scraping in goitered gazelle is observed primarily in males, especially often in territorial owners, while bachelors only occasionally scratch the ground before urination/defecation. In contrast, goitered gazelle females do not scratch during excretions at all. This is logical since they do not have individual territories to mark, though they are known to scratch the ground when preparing resting places. Females of antelope species that maintain their own permanent individual territories (e.g., steenbok—*Raphicerus campestris* and dik-dik *Madoqua kirkii*—Hendrichs 1975) do scratch the ground in their ranges. Musk deer females (*Moschus moschiferus*) also have their own territories and therefore they scratch the ground and mark the ground with their fecal pellets (Zaitsev 2006). In order to attract the attention of rivals, territorial males not only leave urine and fecal marks, but they also make noticeable changes to the area around these marks (Roberts et al. 2001), but quite actively and purposefully causing damage to the ground and vegetation around the marked points, such as the visible spoors caused by scratching the ground. Among such damage is clearing shrubs of leaves and small twigs, biting the tops of large grasses and small twigs or scraping or gnawing trunks with teeth or horns—this behavior is typical for deer and antelopes (Gosling 1985; Alberts 1992; Sun et al. 1994; Johansson and Liberg 1996).

Undoubtedly, this is a completely logical explanation for the scraping of the ground before urination/defecation acts. However, there is another point of view on this behavior. Walther (1984) believes that scrapping is closely related to agonistic behavior. In his opinion, aggression and scraping are essentially two sides of the same behavior. And, indeed, marking with urine and feces is often associated with territorial conflicts during rutting and, to one degree or another, it is part of aggression directed toward a territorial neighbor in goitered gazelles. And it is no coincidence that only territorial males scratch almost without fail, whereas outside the rut, scrapes before urination-defecation do not always occur; i.e., scraping is not a mandatory procedure. Moreover, the bovids had hornless ancestors, so they fought and defended with their front legs, as modern deer do when they shed their horns. Therefore, according to Walther, the use of the front legs in fights evolved into scraping the ground with the front legs. Therefore, the scraping of the ground is a part of aggressive behavior by its origin, as well.

Outside of periods with increased sexual activity, urination and defecation in adult males are often not ritualized: no sniffing or scratching the ground before urination/defecation, and defecation often occurs on the move away from the place of urination, with fecal pellets widely scattered on the ground. Although summer agonistic interactions between males occur frequently, they are much less often accompanied by a demonstration of urination and defecation postures, as happens during periods of increased sexual activity during the rut. Urination/defecation occurs mainly when grazing and during transitions between pastures and watering holes, as well as after resting (Marmazinskaya 1996).

For the most part, goitered gazelle males have insignificant differences in their urination/defecation marking rituals compared to other Antilopinae species. Goitered gazelle males keep their tails in a horizontal or vertical position during their marking ritual (Figs. 5.3 and 5.4), as other antelope species do. Blackbuck males (*Antilope cervicapra*), however, have been known to sometimes circle in this situation and exhibit a flehmen response after sniffing the dung, as well as keeping their tails curled over their backs, and dibatag males (*Ammodorcas clarkei*) are the only species that have been identified to sometimes beat their long tails between

Fig. 5.7 Sometimes, goitered gazelle males may bed down at the dung pile. Issyk-Kul goitered gazelle nursery for reacclimatizing, 25.11.2022. Photo by D. Blank

their hind legs when urinating, wetting their tails with urine in the process. Occasionally, blackbuck males will also bed down at the site of dung pile (Walther et al. 1983), a behavior similarly seen in goitered gazelle males, though only rarely (Blank 2022, unpublished report—Fig. 5.7). Young goitered gazelle males will sometimes urinate and then, after walking a few steps, defecate. Grau (1974) asserts that linking urination and defecation acts is rarely observed in non-territorial mountain gazelle *(Gazella gazella)* males during the rut, while in other gazelle species, the sequence is almost obligatory in adult males. Even when a male occasionally defecates without urination, he almost always displays the urination posture first, thus fulfilling the vacuum activity in the ritual (Walther et al. 1983).

In goitered gazelle territorial males, connection of urination/defecation acts is almost strictly required, but only during the rutting period, while bachelors during the rut and adult males outside of breeding only occasionally demonstrate this relationship. In contrast, females, fawns, and sub-adult males usually have no such linkage and often urinate and defecate in different places (Blank 1997). In general, acts of urination and defecation can be observed at any time, but goitered gazelles typically urinate and defecate after a long rest, and in relation to territorial conflicts between males (Habibi et al. 1993; Alberts 1992). In captivity dibatag males, female's urination and/or defecation has been thought to possibly trigger urination-defecation in captive dibatag males: a male sniffed at a female's urine or pellets, performed the flehmen response, and then pushed the female's rump with his horns and forehead to move her away. He scraped the ground and then urinated and defecated on top of hers. This was usually followed by the female returning and sniffing the male's urine and also performing flehmen response (Walther 1963). This specific ritual is unknown for any other gazelle species. It is possible that this behavior is possible to observe only in captivity due to its rarity. It is also possible that this activity is related to behavioral contagion, when one gazelle urinates-defecates, and another close individual also urinates-defecates. The most frequent case of contagious behavior possible to observe is between two territorial neighbors involved in a territorial conflict. The striking male's urination-defecation postures may be used to emphasize his sexual potency and possibly these postures maybe suppressed in the presence of dominant male. Otherwise, the dominant male will move toward subdominant displaying aggression, forcing him to abandon his stereotypical postures (Walther et al. 1983).

5.2.2 Function of Urine/Feces Marking

Among scent marking, urination and defecation occupy a special place. Urine and feces are the physiological by-products of the body's use of foods as the body excretes undigested and unneeded remnants. At the same time, urine and feces contain information on the hormonal and social status of the individual, as well as genetic information and the quality of food available (Hurst and Rich 1999; Čupić et al. 2023). This marking procedure, out of all others, utilizes materials that are already naturally excreted from the body, so this type of marking is basically energetically cheap (Gosling 1981, 1985). During the autumn rut, urination and defecation can be considered an act of marking territorial boundaries by adult territorial males, while outside the rut, the excretion of urine and excrement can be considered as a purely physiological act of excretion. Urination and defecation in females, adolescents, and male bachelors, on the other hand, play an exclusively excretory function all year round, though these products of excretion also carry some information that is not connected to territoriality (Hurst and Rich 1999).

Territorial marking by leaving urine stains and fecal pellets is very widely used among all species of gazelles (Arabian gazelles *Gazella arabica*; Mountain gazelles *Gazella gazella*; Grant's gazelles *Nanger granti*; Thomson's gazelle *Eudorcas thomsonii*—Walther et al. 1983; Habibi et al. 1993; Wronski et al. 2013) and many antelopes (beira *Dorcatragus megalotis*; dik-dik *Madoqua kirkii*; gerenuk *Litocranius walleri*; springbok *Antidorcas marsupialis*—Hendrichs 1975; Gosling 1981; Mechkour et al. 2008; Giotto and Gerard 2010).

What are territorial marks for? In order to advertise the identity, borders, and range of a male's exclusive territory, and deter intruders from entering his range, it is necessary for the male to have his own unique smell throughout his protected area (Johnson 1973; Gosling 1982). Scent marking is also related to dominance in many animal species. Gosling (1982) proposed his own hypothesis to explain territory marking. All aggressive interactions have a high energetic cost, especially if one of the opponents is seriously injured in a fight. Territorial marking determines that one specific male (and potential adversary) is owner of the territory and is therefore dominant within his area ("dominance fixed in space"), which makes it is easier for him to hold onto a territory than for a new male to take it over. When a male enters that territory and encounters multiple marks, he receives information about the owner and his potential dominance. Thus, an intruding male who enters the territory will try to avoid an encounter with the dominant, eliminating the need to find out who is stronger; a male that has "illegally" entered the territory prefers not to engage in an open battle with the owner and tries to escape from the territory as quickly as possible. The owner of the territory then avoids energy-consuming fights with invading males.

Goitered gazelle males have territories only during rutting season in November and December; therefore during this time, all urination-defecation acts of breeding males are related mostly to territorial border conflicts, as well as when patrolling of their own ranges. Beyond that, rutting males use these markings more frequently in the presence of females, both before and after courting females. It is also very possible that urination and defecation by males have some kind of chemical impact on females, possibly accelerating and synchronizing their estrous cycles (Coblentz 1976; Owens 2014). The relatively short female estrous cycles are particularly important in goitered gazelles since they have a very narrow window of time for birthing in the spring (Blank et al. 2012). Outside the rut, urination/defecation of goitered gazelle males has predominantly a physiological function and occurs mostly when feeding. In goitered gazelle females, fawns, and sub-adults these acts are always observed mainly during grazing (Blank et al. 2014b). During a study of marking activity of klipspringers (*Oreotragus oreotragus*), males were also found to mark more often in the presence of a female, especially if the female is in estrous (Roberts and Dunbar 2000).

Since ungulates live in a certain territory, they must know their habitat well. To do this, their home range is marked for orientation in space and for contact with their conspecifics. Such marks are especially important to protect their individual territory from the invasion of other males from outside (Grau 1976). There is a very strong link between marking frequency of an individual and his high social status or dominance (Ralls 1971). However, there are certain limitations in using of urine and feces for marking, specifically that animals cannot produce the quantity necessary for constant and intensive marking of an individual territory at their normal production rate. Such markings are necessary to try to prevent another male from violating the boundaries of an established territory, which would require the owner to exert all his energies to expel the intruder from his range. Since border conflicts between territorial males occur quite frequently during the rut, territorial males of many antelope species usually increase the frequency of urine-fecal pellet marking. However, the amounts of urine and fecal pellets produced are reduced per mark, because of the physiological constraints in producing excrement amounts per time unit. Research of marking behavior in some ungulate species has shown that males are able to regulate and decrease the amounts of urine and fecal pellets per mark when intensifying their marking of their territories. Males of Chinese water deer (*Hydropotes inermis*) from cervids, for example, have the ability to leave marks with very limited amounts of urine (literally drops) and smaller size fecal piles per mark during the rut compared to other seasons (Sun et al. 1994). The males of oribi antelopes (*Ourebia ourebi*) also can manage producing limited amounts of pellets per mark, leaving scent marks along the boundaries of their individual territories (Brashares and Arcese (1999b).

5.2.3 Frequency and Duration of Urination/Defecation Acts

I decided to check if goitered gazelle males increase the frequency of urination/ defecation acts during the rutting period compared to other seasons and weather they also have the ability to regulate the amounts of urine and feces during the rut, decreasing the sizes of their marks. And I found that goitered gazelle males indeed increase significantly (1.5–2 times) the rate of urination and defecation during the rutting season compared to other seasons, when marking rate is low and fluctuates insignificantly (Fig. 5.8). In addition, the urination/defecation acts in males occur more frequently during the rut than in adult females and sub-adults in the same

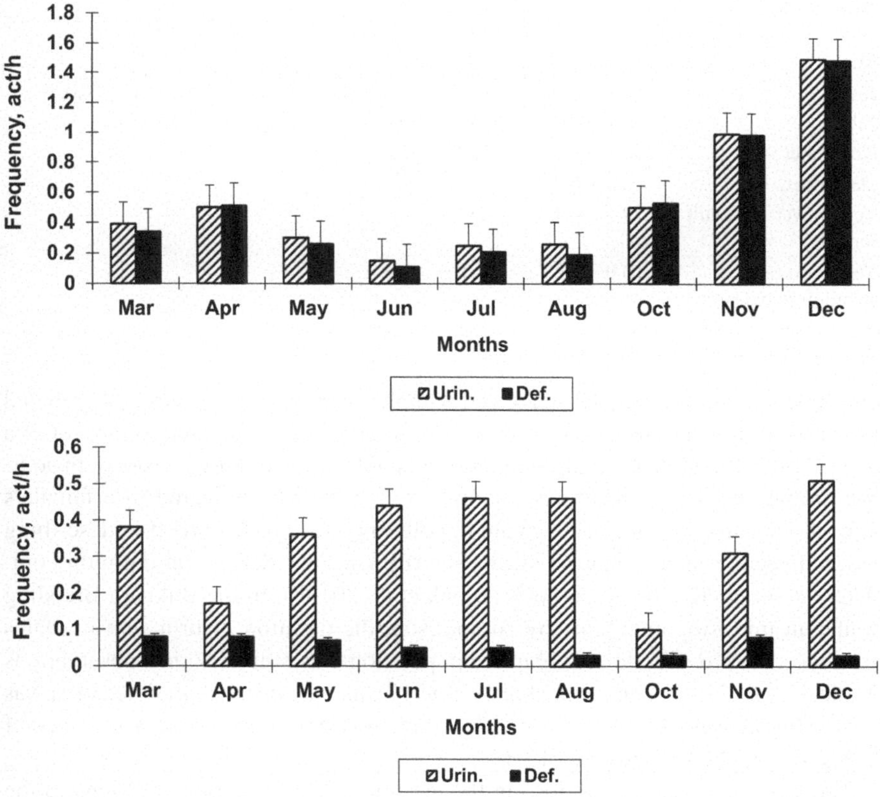

Fig. 5.8 Changes of urination and defecation frequency in adult males (**a**) and females (**b**) of goitered gazelle over different calendar months (by Blank et al. 2014b—License number—5475920461589)

Table 5.2 Fluctuation of the mean duration of urination (seconds) in adult males and adult females over various months and yearly average of sub-adult males and females (seconds) by Blank et al. (2014b). (License number—5475920987776)

Month	Mean duration, X ± SE	N	SD	Kruskal–Wallis test			
				H_3 April–July	P	H_5 April–December	P
Adult males							
April	14.1 ± 0.9	48	5.9				
May	17.5 ± 1.4	79	12.8				
June	15.6 ± 2.0	18	8.5	1.212	0.750	75.104	0.0005
July	13.5 ± 1.6	18	6.6				
November	9.4 ± 0.5	124	5.2				
December	10.1 ± 0.5	135	5.6				
Yearly average, adult males	12.1 ± 0.4	423					
Yearly average, subadult males	16.1 ± 1.6	21					
Adult females							
April	7.9 ± 1.0	13	3.7				
May	11.3 ± 0.6	117	6.9				
June	10.1 ± 0.6	123	6.7	9.854	0.020	12.763	0.026
July	8.1 ± 0.8	37	4.6				
November	10.9 ± 2.2	14	8.2				
December	6.6 ± 0.8	8	2.2				
Yearly average, adult females	10.1 ± 0.4	288					
Yearly average, subadult females	10.0 ± 1.2	26					

period. Unlike males, females defecated with the same frequency during the rut and outside of it. Rutting males also urinate for a shorter time compared to the rest of a year (Table 5.2), while defecation duration is stable in males all year round. In addition, a male's act of defecation is shorter during the rut compared to a female's defecation. Thus, the "material saving hypothesis" was partly correct in describing goitered gazelle male urination during the rut, but their defecation duration completely contradicted this hypothesis (Blank et al. 2014b). In the Bukhara Breeding Center in mid-May and June, by comparison, the duration of urination in males from the goitered gazelle semi-captive population was similar to my observations (Table 5.2): 6–29 s, for an average of 18 s ($n = 7$), and the duration of defecation was 3–24 s, for an average of 9 s ($n = 6$). The whole marking process took an average of 25.8 s ($n = 7$) (Marmazinskaya 1996).

The same seasonal fluctuations in the urination/defecation rate are found in the semi-captive population of the Bukhara Breeding Center, when the least number (9.7–10.9%) of marked sites are found at the end of January to February, after the end of the autumn rut, and before the beginning of the spring sexual activity. During the spring period of activity (March–April), the frequency of marked points increases and reaches a maximum in May (81% of marked points). In June, the proportion of marked points decreases almost to the level of April (60%). Ritualized marking with urine and excrement is rarely observed in July, while in August, the number of marked latrines increased. Based on a survey conducted on the 25th of September, all urination/defecation marks were found in an area with a high concentration of goitered gazelles. Thus, after a month's break in July, in the pre-rutting period (August to October), the frequency of urinations and defecations increases sharply, and the proportion of ritualized markings also increases. At the height of the autumn rut (November), the maximum number of marked points (79.9%) is observed for this period (Marmazinskaya 1999).

Interestingly, a captive goitered gazelle male that had strongly imprinted to humans established a whole chain of dung piles with short intervals between them along the fence which separated him from human visitors (Walther et al. 1983). Surprisingly, though, the average duration of defecation in the goitered gazelle males is the same in November as in other months, and even longer in December. In contrast to adult males, females have insignificant fluctuations of urination-defecation rates over biological seasons (Blank et al. 2014b).

5.2.4 Seasonal Stability of the Duration of Defecation in Goitered Gazelles, Possible Causes

Why is such an ambiguous result obtained? The fact is that the phenomenon of saving excrement has been found for two species, the oribi and the Chinese water deer. The oribi is observed in the African tropics, and the Chinese water deer is found in Eastern China. In the climate conditions of the African tropics, seasonal changes in vegetation, although occurring during the rainy and dry seasons, are not as significant as in the northern deserts of Kazakhstan. The Chinese water deer is found on the border of the tropics and temperate zone, in eastern China. There the seasonal change of vegetation occurs, as well, but to a rather limited extent. In the goitered gazelles' environment in Kazakhstan, there is a huge difference between warm and cold seasons. In late fall and winter, vegetation loses most of its moisture and nutrition and the goitered gazelles have to feed on dry vegetation with a high fiber content, which is only partially digested and most of it is excreted through great amounts of feces. As a result, the gazelles likely have more excrement during the cold period than in spring and summer, so they do not have to save material to use it during intensive marking of rutting territories. Therefore, I detected a stability in the duration of the defecation act.

The high fiber content of the forage probably has an effect on the increase in the amount of feces. But, in my opinion, a more likely reason for the absence of seasonal differences in the duration of defecation is in the methodical approach. In reality, male gazelles probably do reduce the portion of pellets in each mark during the rut, but I determined the duration of defecation according to the length of time the male was in the defecation posture. If I was counting the number of pellets in each mark, I most likely would have found fewer pellets during the rut compared to other seasons. As for measuring the duration of urination and defecation according to the posture in which the male is positioned, there are some problems. The fact is that the urination and defecation postures are not always related to the actual amount of urine or feces excreted. The postures themselves have a strong symbolic meaning and are performed for an opponent in an emphasized demonstrative manner, where the poses are prolonged to have a maximum effect on the opponent and advertise the territorial male's sex, territorial, and dominant status and take time to display (Müller-Schwarze 1999). As a result, the pose itself may be demonstrated for much longer than the actual act of urination or defecation required (Eilam et al. 2006). During the rutting period, adult males display the complete set of urination/defecation postures and movements, including sniffing, scratching the ground, and the stretching posture of urination and squatting posture of defecation (90–98%, $n = 480$) more often than in other seasons (60–75%, $n = 350$). Outside of the rut, there are quite often cases (up to 20%) when males skip some successive elements or urinate and defecate separately. Walther et al. (1983) noted that at times a territorial male performs only a pose, and nothing is actually excreted.

Urination was noticeably shorter during the rut than in other seasons and is probably influenced not only by the need to save urine due to the large number of marks, but also by the deficit of moisture. Goitered gazelles eat very dry food in winter with minimal moisture, and rutting males get very thirsty due to sexual excitement and their almost constant chasing of females, often at a very fastest gait (gallop, career) (Sludskiy 1956). Territorial males try not to leave their territories and generally do not go to watering holes; instead they attempt to satisfy their thirst by continually eating snow. However, snow cover fluctuates and can be unavailable for long periods of time, resulting in males remaining without drinking water and experiencing a great deficit of moisture against the background of frequent use of urine for territorial marking.

Thus, it is likely that during the rut, male gazelles do reduce the amount of urine and feces to establish as many territorial marks as possible, but my methodical approach of determining the duration of urination and defecation through the demonstrated postures yielded results only for urination but not for defecation.

5.2.5 Optimal Placement of Marks in a Territory

In addition to the freshness, density, and chemical properties of the scent signal, its location is very important for performing the main function of transmitting the signal to the recipient; it is necessary to advertise the scent mark in the right way and increase its detectability, placing it in the location where it will be detected as soon as possible (Roberts et al. 2001). Similar to males in other antelope species (e.g., klipspringer *Oreotragus oreotragus*—Roberts and Lowen 1997; oribi *Ourebia ourebi*—Brashares and Arcese 1999b; gerenuk *Litocranius walleri*—Gosling 1981;

hartebeest *Alcelaphus buselaphus*—Gosling 1974), goitered gazelle territorial males are not able to leave their marks everywhere within their territories, or even the entire periphery of their individual ranges, because of limited urine and feces; instead they leave their marks in the most significant points for protecting their territories from rivals. Studies of the placement of territorial marks of a number of species have shown that even here ungulates save urine and feces, placing their marks more often along the borders of their individual ranges and on the sides from which a rival is most likely to violate the border (Gosling and Roberts 2001; Zaitsev 2006; Mechkour et al. 2008). Inside their own borders, males mark much less often. I have observed this same phenomenon in goitered gazelles. Territorial males mark their territories most often during territorial conflicts, when a male crosses or even just comes close to the border of a neighbor. Then both males begin to rigorously mark their boundaries in front of their neighbor (mainly urination/defection and horning vegetation, less preorbital marking), as if emphasizing with their marks where exactly the territorial border was passed. As a result, most of the marks are concentrated along the borders. And as seen for other species, inside their ranges, territorial males rarely leave marks, with the number determined at only a few (Blank et al. 2014). Other authors (Marmazinskaya and Mardonov 1996) also found that goitered gazelles mark mostly along their borders, leaving a lot of urine-feces marks there. In contrast, bachelor males urinate and defecate in a random way (Walther 1984).

The Grant's gazelle and Dorcas gazelle males have been found to use landmarks (termite heaps, rocks, single trees) for their few urination/defecation marks (Walther 1972; Soultan et al. 2021). Goitered gazelles also can use landmarks and prefer to leave their marks in soft sandy soil of dry river beds, which are often borders between neighboring individual territories of males (Blank 1998). Ungulates can mark places marked by other animals with their own marks, using them for greater detectability (Gosling 1980). In the Chinese Xinjiang, I have observed that goitered gazelle males use the kulan's (*Equus hemionus*) marks for depositing urination-defecation markings in. The kulan marks represented areas cleared of vegetation by scraping with their forelegs and large piles of dung. The marks of the goitered gazelles were located next to or on top of such marks in areas without vegetation and dung. An unusual preference of goitered gazelle males for their urine-feces marking was found on the central Iranian plateau (Kolah-Qazi National Park). About 84% of the gazelle pellet aggregations were observed on ant nests (*Messor* sp.), even though nest sites covered only about 0.29% of the sampled transects. That means gazelles actively selected ant nest sites, where soil content and annual herb vegetation differed significantly from areas without ant nests (Esmaeili and Hemami 2013). Gazelles usually use natural landmarks for demarcating their territories (Walther et al. 1983; Attum and Mahmoud 2012). In that study area, the landscape lacked conspicuous features that could be used for marking territories, and apparently ant nests sites were the only landmarks available (Esmaeili and Hemami 2013).

5.2.6 Rutting Marking vs Latrines

Rutting marks are represented by a pile of pellets on top of a urinary spot with fecal pellets from previous marks scattered around the new mark. These are called the rutting marks that can be found during the rut (November–December) in large numbers in the areas of individual territories. Fewer marks are found during the spring peak of sexual activity in April–May (Blank 1985). These territorial rutting marks are usually small in size and consist of the pellets of one territorial male, or rarely, also the pellets of the male neighbor. In November–December 1971, 312 rutting urine-feces marks were examined, and in all cases, they contained only one pile of feces (Zhevnerov 1984).

Other kinds of urine and feces sites are observed in goitered gazelles outside the rut. These are the so-called latrines. These are latrines, which are also supplied only by males, but not by just one territorial male, as in the case of territorial markings; all males of the home range leave their pellets. Outwardly, the latrines look like bare plots of land filled with a large amount of gazelle droppings, surrounded by a border of herbaceous plants. A latrine can be seen from a distance of several tens of meters. The attractiveness of latrine use by goitered gazelles is confirmed by the multilayered accumulation of excrement of different sizes and different stages of decomposition, meaning it was left by different individuals at different times. Latrines are located on trails or near trails, which in most cases lead from the forest to the desert, where plants start to vegetate earlier than in the riparian forests. All together a patch of soil cleared by gazelle hooves, the accumulation of excrement, a border of bright dense greenery, and the smell of droppings left by many individuals at different times make the gazelle latrines very visible, stable attractors in the signal field (Pavlov et al. 2013). And it has been found that reintroductions of goitered gazelles are more successful in areas where tracks and latrines of the local gazelles are available (Pereladova 2022).

During a survey of gazelle habitat in the Tigrovaya Balka Nature Reserve (Tajikistan), 42 latrines were documented along a route of 28 km. The area occupied by a latrine had an oval shape with a length ranging from 20 to 170 cm (average length of 76.0 ± 34.5 cm) and a width from 20 to 100 cm ($x = 50.0 \pm 20.2$ cm). The male gazelles cleared the latrine area with their feet, scattering the droppings all around it and deepening the pit (sometimes up to 10 cm). Latrines are relatively numerous in the areas of the reserve inhabited by gazelles. In the forest zone, latrines are located in open areas with a width of 10–30 m and a distance between adjacent latrines from 5 to 182 m ($x = 70.0 \pm 59.1$ m; $n = 19$). In the ecotone (transition zone from forest to desert), the adjacency distance is from 7 to 67 m ($x = 24.0 \pm 18.6$ m; $n = 15$), and in the desert from 40 to 846 m ($x = 314.0 \pm 297.8$ m; $n = 8$). The border of plants can be asymmetrical due to the uneven distribution of pellets in and around the toilet, but more often a complex of plants forms a neat border around the entire perimeter of the latrine. The width of the plant strip varies between 10 and 100 cm and, according to preliminary observations, depends on the age of the latrine (Pavlov et al. 2013).

Due to the fact that all males rake over old excrement and then leave their own pellets, over time cone-shaped depressions with a diameter of up to 1 m and with numerous excrement deposits in the center and around the edges form in certain places. These are not rutting marks, but rather registration points for information where all males in the area note their presence, playing a "visit card" type role having certain information for orientation of goitered gazelles in the space. In the Ili Hollow, "visit card" latrines were rarely found (Blank 1985), while in the Bukhara Breeding Center they were common (Marmazinskaya 1999). Perhaps the number of latrines increases with an increased population density, or in the Bukhara Breeding Center, maybe there were no distinctions between "rutting marks" and "registration latrines" (Marmazinskaya 1996). Also, Zhevnerov (1984) described all defecation marks as one kind (see below), though these two kinds of sites have different functions. Before the start of the rut, common latrines serve as a kind of "guest book" where gazelles can find out what other gazelles are in a given territory; this may contribute to the uniform distribution of latrines in an area. Latrines undoubtedly help animals living alone or in small groups to communicate with each other (Zhevnerov et al. 1983).

In the Barsakelmes Island population, goitered gazelle males deposit feces throughout the year (Zhevnerov 1984). On trails, abandoned roads, and just areas with soft soil, they make ellipse-shaped depressions or simply scratch the ground with their front hooves and immediately leave their droppings. These latrines are called "strokes" by hunters. In areas with a lot of gazelles, "strokes" are sometimes found every 20–70 m, though more often every 200–400 m. Encountering a "stroke" with excrement, the gazelle sniffs it, rakes it completely with his front hooves, and then always deposits his excrement and urine in its center. The furrows made by the hooves are clearly visible at the bottom of the "puddles". Outside of the rut period, goitered gazelle males make "public latrines" that occur in areas of constant concentration and in places convenient for resting and protecting gazelles from adverse weather conditions. Outside of the rut period, several males use one "stroke." By the end of summer, the number of latrines increases, as well as their sizes, up to 45 cm long and 15 cm deep. "Public latrines" are somewhat similar to depressions of resting beds, but more often are cone shaped. The depth, depending on the duration and frequency of use, can vary from 3 to 15 cm. Near them, on the edges of the expanded area, old feces are scattered, and in the recess itself, there is always one fresh pile (Zhevnerov 1984).

The leaving of urine/feces marks by dominant males in captivity generally occurs in the following situations: (1) before the pursuit of a female; (2) after the pursuit of females (in the intervals between the courting of the same female or different females); (3) after the flehmen ritual; (4) after the rise of one or more females from their beds; (5) after the appearance of a male in a neighboring enclosure; (6) after the pursuit of a male juvenile; (7) after a conflict with an subdominant male; (8) when groups were kept in adjacent enclosures with a common wall between adjacent dominant males—prolonged agonistic interactions and butting through a grid, leaving a series of synchronous marks by both males in the same way as territorial males marking their territorial borders; and (9) after resting in a calm situation (Marmazinskaya 1996).

Marking behavior in enclosure conditions has its own characteristics, presumably due to the limited area, the presence of a mesh fence between groups, and the constant presence of the same gazelles in the group. While moving around the enclosure, males periodically sniff their excrement in the latrines, the ground, and urinary spots and piles of female droppings for 10–20 s. The females do not defecate in the latrines, but they often sniff the marks of males at these sites (Marmazinskaya and Mardonov 1996). In the wild, goitered gazelle females also do not put their excrement into either rutting marks or registration or registration latrines and generally do not sniff or show any interest in them (Blank 1998).

5.2.7 Development of Urination-Defecation Marking Behavior

In a semi-captive group, half-year-old males (6–8 months) during the autumn rut and 1-year-old males (10–12 months) during the spring peak of sexual activity do not demonstrate marking behaviors. Urination and defecation very rarely manifested themselves into what might be considered marking and demonstrative behavior and instead simply performed an essentially physiological function. At the age of 12 months, a male is already actively marking latrines, mainly with feces (1–10 piles of feces in one toilet). At the age of 14 months, the number of toilets with urinary points and the number of points in one toilet increase (from 1 to 7). When the male is 16 months old, the number of latrines with urinary marks even exceeds the number of latrines marked with pellets. At this age, he also begins to leave scrapes in the latrines, secretion markings from the preorbital gland, and head-butting plants. Thus, in this individual development, marking of latrines with urine/feces forms first; then almost immediately came marking with the preorbital gland, followed by head-butting of plants (Marmazinskaya 1996). In the absence of adult males in a captive group, a 1.5-year-old male demonstrates all types of marking activities during the entire rutting period, although the poses of young males are poorly expressed. Thus, all three types of marking activity in male gazelles seem to be fully formatted by the onset of puberty, but in nature they are more or less suppressed by the presence of adult males in groups. In the wild, males 2–2.5 years and older display all types of marking activity (Marmazinskaya 1996).

5.3 Marking with the Preorbital Gland

5.3.1 Preorbital Gland Functions and Marking Ritual

Blank et al. (2014a)—Open excess.

Olfactory communication is especially important to ungulates and many species have a number of skin scent glands that produce secretions (Wyatt 2003). The odorous signals may be used by territorial holder as a deterrent factor for intruding males and supply information about age, competitive ability, aggression, physical condition, physiological state, and spatial orientation. The skin gland secretion can include odors of dominant territorial or subdominant non-territorial males (Rajagopal et al. 2011). There are a number of scent glands shaped as specific skin organs and can be found in various parts of the ungulate body from the muzzle, neck, and chest to front and hind legs, and even the genital preputium area depending on the lifestyle of species (Mykytowycz 1972). In addition, there may be other smaller and non-shaped glands not in the form of clear organs, but rather scattered glands in the form of glandular fields, and therefore, some of them maybe still unknown to researchers. The widespread distribution of skin glands among ungulates indicates their importance in the life of mammals. In all gazelle species, including the goitered gazelles, the specific preorbital scent glands are located in the area of the anterior corner of the eye (Estes 1991). The preorbital glands are paired dermal exocrine sac-like formations equipped with connective tissue and sometimes a muscular membrane and are located in front of the eye. These glands are groove-shaped slits from dark blue to black, having an almost naked skin, and extend from the medial canthus of each eye. They are lined with a combination of sebaceous and apocrine (sweat) glands and produce secretions containing pheromones and other semi-chemical compounds (Sokolov and Chernova 2001).

A number of gazelle species emit a sticky secretion of a dark, almost black color from the preorbital glands, which males rub on the tips of grasses and small shrubs (Albone 1984) to actively mark their territories especially during the rut. Not all gazelle species use their preorbital glands, however. Large gazelle species (*Nanger dama*, *N. granti*, *N. soemmerringii*) and some small gazelles (*G. gazella*, *G. dorcas*), for example, do not use these glands in marking vegetation and do not excrete any secretions, even though these glands are present similar to other gazelle species (Walther et al. 1983), while other gazelle species, including goitered gazelles, actively use their glands to mark territories. In goitered gazelles both males and females possess preorbital glands, but these glands are active and especially large in males, looking like prominent bulges during the rut. Since goitered gazelle females do not have territories and not need to protect borders, they do not use their preorbital glands for marking and the glands remain inactive. Only once did I ever observe a female move her head near a shrub similar to a male's marking behavior; however, it is hard to know if it was real or simulated marking or just some odd movement. Marmazinskaya (1996) had only two documented cases from her observations of females in the semi-captive population in the Bukhara Breeding Center (Uzbekistan) marking with their preorbital glands. In contrast, female marking is a typical phenomenon in small African monogamous antelopes (klipspringer *Oreotragus oreotragus* or Kirki's dik-dik *Madoqua kirkii*), because they protect territorial borders along with males and thus have active preorbital glands that they use for marking (Gosling 1982; Roberts and Dunbar 2000).

These preorbital glands are present in many species of artiodactyls, and since many ungulate species are territorial and protect clearly defined boundaries, these glands' secretions most often serve as territorial marks to signal the quality and competitive ability of the territorial owner (Roberts et al. 2001). Secretions of preorbital glands produce a complex olfactory signal for other conspecifics. The chemical composition of these secretions may vary individually, carrying information about a specific identity including physical characteristics (Lawson et al. 2000; Brennan and Kendrick 2006). During the rut, a goitered gazelle male's preorbital glands swell, becoming so prominent that they are visible from a distance. The surplus of the pitch-like secretions from these glands flows down the male's muzzle leaving a strip of dark, wet fur that is especially noticeable, while outside the rut these secretion marks are not visible at all (Blank 1985; Habibi et al. 1993). A similar phenomenon is found in territorial males of suni antelope (*Nesotragus moschatus*), where their preorbital glands are very active during the rut and emit surplus secretions, making the suni males' muzzles wet below the preorbital glands, while the muzzle is dry in subdominant and yearling males (Somers et al. 1990). Territorial males use preorbital gland marking especially intensely when courting receptive females more in December than November, when most females are not yet in estrous. Marmazinskaya (1996) also observed in the captive herd enclosure intensive marking by rutting males when they detected the receptive female. Many other antelope species (klipspringer, dik-dik, grysbok *Raphicerus melanotis*) similarly demonstrate their highest rates of preorbital gland marking rate during the rut, especially when the receptive females appear (Dunbar and Dunbar 1974; Novellie et al. 1984; Kranz 1991; Roberts and Dunbar 2000). It is also possible that adrenaline arising from agonistic interactions can stimulate preorbital gland secretions (Gray et al. 1989).

In addition to territorial marking, these glands also have other functions (Gosling Blank; Sokolov et al. 1986; Rajagopal and Archunan 2011). A gerenuk male marks a female by rubbing his preorbital gland on her chest, on her croup, or on her rump, likely as a part of courting behavior (Walther et al. 1983). Similarly, larger and lesser mouse deer male (*Tragulus napu* and *T. javanicus*) mark their females with their skin glands during mating ritual (Ralls et al. 1975). In duikers (Cephalophus ssp.) males and females mark each other with preorbital glands (Ralls 1969), and a mother marks her young with secretion of this gland (Frädrich 1964). In cervids preorbital glands not only have an olfactory role, but also visual significance (Ceacero et al. 2014). Red deer (*Cervus elaphus*) open the preorbital gland widely during aggressive interactions with other males (Bartoš 1983) and when roaring (Butzler 1974). Muskox males (*Ovibos moschatus*) are nonterritorial and use their preorbital gland secretion for marking firm objects or their own front legs in most intense aggressive conflicts with other males as a part of threat displays and display for readiness to fight, while in stress situation all herd members (males, females, and sub-adults) do such marking (Gray et al. 1989). And the preorbital gland also plays a significant role in mother-fawn interactions in red deer (Hatlapa 1977) and in goitered gazelles (Marmazinskaya 2008), when a fawn produces a signal to its mother with the help of this gland when he is hungry (Bartoš et al. 2005) or when is full (Bartošová et al. 2012); the preorbital glands in young also allow them to show fear, pain, stress, and excitement (Ceacero et al. 2014).

Marking an individual territory with the secretions from the preorbital glands in male goitered gazelles has been known for a long time (Sludskiy 1956). Other functions of this gland in goitered gazelles are still unknown, although the behavior of males shows that the preorbital glands certainly have additional purposes. The preorbital glands play some role during male courtship, since when chasing females, the preorbital glands of territorial males are widely open. It is possible that the visual effect of the open entrance of the gland shows the high arousal of the male and somehow stimulates the female to respond and accelerate her readiness to mate. It is also very possible that the preorbital glands exude some kind of pheromones that somehow affect the acceleration of synchronization of the sexual cycles of females and males (Blank 2022, unpublished report). The wide opening of the preorbital glands also occurs during aggressive contacts of the territorial males with other males, mainly territorial neighbors (Habibi et al. 1993). An intensive emission of odor from the preorbital glands might work as some kind of self-stimulation for the producer or affect the opponent in a frightening way (Walther et al. 1984). All these additional functions are just hypotheses since no special research has been conducted on these, or even other, potential functions of these glands.

All Antilopinae species mark vegetation with the preorbital gland in the same manner (Walther et al. 1983). In the Bukhara Breeding Center, the goitered gazelle male usually starts his marking by carefully sniffing the tip of the grass or twig of a shrub, and only rarely he marks without first sniffing (6% cases, $n = 33$). Sometimes the male sniffs the potential mark site several times (7 times in one case) and then does not leave a mark (Marmazinskaya 1996). In contrast, rutting goitered gazelle males in the Ili Hollow rarely sniff (less than 4% cases), most often leaving a preorbital marking without any initial sniffing. In some antelope species (klipspringer and Thomson's gazelles *Eudorcas thomsonii*), marking males lick old marks or even slightly bite them to release the scent for greater stimulus during the marking procedure (Walther 1978b; Roberts 1998), though this behavior is not observed for goitered gazelles. To apply the secretions, the rutting male typically turns his head, slanting one side toward the ground (87.5–94.7% observed cases, $n = 410$ in goitered gazelles); rarely (5.3–12.5%) does a male keep his head parallel to the ground with only a slight tilt toward the bush. The male then opens his mouth widely, apparently trying to widen the entrance to the gland as much as possible; this mouth opening during marking is observed rarely most of months (1–4%) and only in April the frequency of this behavior increased up to 12.5% ($n = 65$).

The male lowers the preorbital region of his head toward the plant stem, bringing the stem's tip or twig into the widely opened gland. He then gently rubs the outer and inner surfaces of the gland against the tip of the thin dry twig (Fig. 5.9) on which he deposits a few drops of the secretion that harden in the open air. To position himself for scent marking, the male typically stretches his head upward to get his preorbital gland as high as possible, although he never raises his body to stand on just his hind legs for this purpose. During the marking procedure, the male's ears stayed slightly apart with the ear holes oriented forward. Sometimes a goitered gazelle male will sniff the branch after applying his preorbital gland secretions to it, or when he marks a twig several times with the same gland or when he changes to the other side, marking it with both glands (up to 16% observed cases). Most often, the secretions of the preorbital gland are applied once on a branch (from 66% to

Fig. 5.9 A goitered gazelle male rubs the preorbital gland against the thin tip of bush branch. Issyk-Kul goitered gazelle nursery for reacclimatizing, 25.06.2022. Photo by D. Blank

91% in different months and years), much less often twice (6–24%), and very rarely (2–11%, $n = 435$) several times in a row (up to 5 times) (Blank et al. 2014a).

5.3.2 Fluctuations of Marking Behavior Rate and Duration over Seasons

The preorbital glands have a number of functions, but one of the most important is defining a rutting male's territory. In goitered gazelles, individual territories appear during the autumn rut in November and December and to a lesser extent during the spring peak of sexual activity in April and May. Therefore, it would be logical to assume that the highest frequency of marking behavior would be found during the autumn rut and less often during the false rut in spring, when all elements of sexual behavior are noticeably lower than during the autumn rut, and very rarely is this marking behavior observed during other months. The analysis of my data from the goitered gazelle population in the Ili Hollow completely supports this hypothesis that preorbital gland marking of rutting males occurs most often during the autumn rut (2.22–3.56 acts/h in November–December), less frequently during the spring peak (0.80 acts/h in April), and least often in other months (0–0.45 acts/h) (Table 5.3). In addition, intervals between marking acts are shorter during the rut (33–36 min, $n = 215$) and longer out of rut (44–60 min, $n = 36$). These marks are periodically renewed and thoroughly examined by each trespassing male (Blank

Table 5.3 The monthly rate changing of preorbital gland marking behavior of males over months

Month	Observed marking cases	Observation duration (hours)	Rate (acts per hour)
March	17	38.1	0.45
April	146	181.7	0.80
May	166	469.8	0.35
June	30	374	0.08
July	46	173	0.27
August	0	58.8	0
September	0	15.0	0
October	9	60.2	0.15
November	497	224.1	2.22
December	526	147.7	3.56

et al. 2014a). The efficacy of these marks decreases with time deposition, though the gland secretions can retain their effect (the persistence time) for several days, up to a maximum of 7 days in klipspringer (*Oreotragus oreotragus*—Roberts 1998), 2.8 days in Thomson's gazelles (*Eudorcas thomsonii*—Gosling 1985), 4 days in suni antelope (*Nesotragus moschatus*—Somers et al. 1990), and 4.5 days in gerenuk (Gosling 1981). Moreover, males can distinguish their own marks from those of other males and determine the age of their mark, renewing only the old ones and leaving the fresher ones untouched (Roberts 1998).

In goitered gazelles, the preorbital gland marking can last from 1 s up to 28 s (6.4 ± 0.21 s on average, $n = 370$), but the most often observed durations of such markings fluctuate between 2 and 7 s (Blank et al. 2014a). In comparison, the duration of preorbital markings in suni antelope males has almost the same duration as in goitered gazelles 7.5 sec (range 1.5–28 sec) (Somers et al. 1990). In the Bukhara Breeding Center, the marking duration in semi-captive males lasts longer (12.4 ± 0.12 sec in average, $n = 27$), fluctuating from 4 s up to 20 s (Marmazinskaya 1996). Goitered gazelle males demonstrate preorbital marking on average longer during the real rut in November (6.0 ± 0.3 s, $n = 160$) and December (7.3 ± 0.4 s, $n = 132$) and the pseudo-rut in April (7.4 ± 0.9 s, $n = 31$), and for a significantly shorter time out of the rut (4.8–5.0 s, $n = 46$ in May and June). It is interesting that preorbital marking are seen slightly more frequently at the end of pseudo-rut in May (Fig. 5.10). The development and secretion activity of skin glands is under the control of the steroidal sex hormones (mainly androgens), and castration leads to the inhibition of skin gland activity (Mykytowycz 1972). Preorbital glands in goitered gazelle males function more intensely and with greater secretions (increase in size and density of sebaceous and apocrine cells of the gland) during the rutting period compared to other seasons of the year, and may be also associated with the increase in sex hormones (testosterone), which are especially active in mature territorial males. These glands have insufficient potency in bachelor males and little activity in young males (Marmazinskaya 1996; Rajagopal and Archunan 2011).

The intensity of male marking of territorial boundaries with preorbital gland secretions is influenced by the number of surrounding territorial male neighbors.

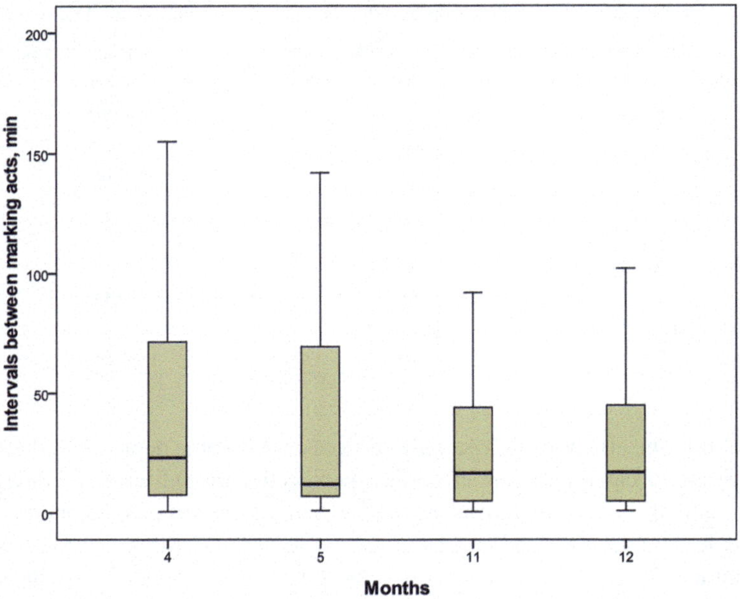

Fig. 5.10 Intervals between marking act by preorbital glands in male goitered gazelle in April–May (4–5) and November–December (11–12) (by Blank et al. 2014a—open access)

Territorial males mark more intensely along the border of a neighboring territory that is occupied by several males jointly (dominant male with subdominants is observed for oribi and klipspringers) compared to territories with a single male (Brashares and Arcese 1999a; Kruger et al. 2002). The marking rate also increases when the territorial male finds the mark of another male (Roberts 1998). In goitered gazelles, territorial males do not allow any other males within their boundaries but mark the borders with other active territorial neighbors more intensely. In addition, the intensity of the preorbital marking is different for males in different group sizes (Fig. 5.11), and there is an inverse relationship between the size of the male group and the intensity of making; the smaller the group, the more males use the preorbital gland for marking. It is possible that in large groups, males interfere with each other's markings which may drive others away since markings are, to a certain extent, a demonstration of dominance over others. Therefore, male marking activity negatively correlates with group size. The highest intensity of marking is observed in single males. This is not surprising, since territorial males mark their territories with the preorbital gland when rutting, demonstrating dominance within their own territories. These males always keep apart from other males, and expel any male that enters their territories, which is why single males tend to have the highest marking frequency (Blank et al. 2014a). This is true also for Thomson's gazelles (*Eudorcas thomsonii*), where single territorial males mark much more intensely than migrating and non-territorial males (Walther 1978b).

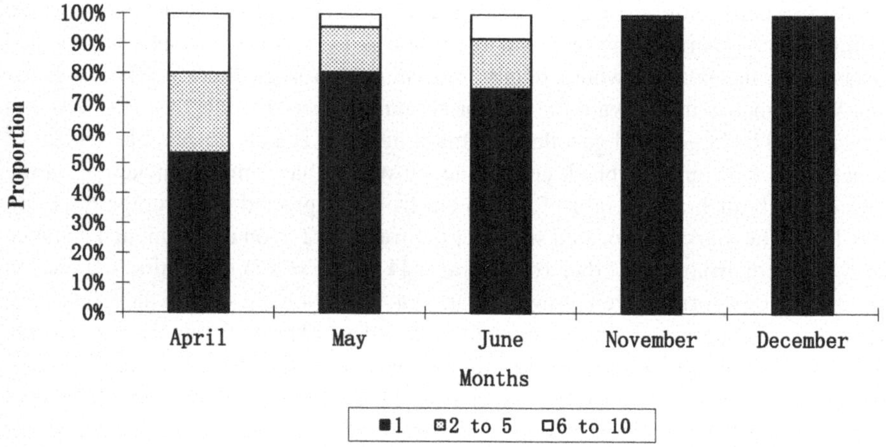

Fig. 5.11 Scent marking activity of single males (1 gazelle) and bucks in groups of intermediate (2–5 individuals) and large sizes (6–10 individuals) in different months. Totally, 501 cases were analyzed. According to Blank et al. (2012) the proportion of single males fluctuated around 40–50% for the whole year, increasing up to 71% only in November and decreasing up to 45% in December (by Blank et al. 2014a—open access)

5.3.3 The Marking Points and Preferable Objects for Secretion Markings

After entering a rival's territory, the trespasser actively searches for territorial markings in order to avoid energy-costly fights with the territorial owner. This may be why there are certain "rules" for marking locations, so they can be quickly found (Alberts 1992; Roberts 1998). Preorbital scent marks usually have a narrow, vertical distribution, though they can be located above or below some preferable height (Roberts et al. 2001). Gerenuk (*Litocranius walleri*) and oribi (*Ourebia ourebi*) deposit their preorbital gland secretions on branches at head height of a walking intruder so the marks can be detected immediately (Gosling 1981). Oribi lives in places where the grass often is taller than its height, so this little antelope first bites off the plant and makes it lower, at the level of its head, and then marks the stump tip with its preorbital gland (Gosling 1972). Goitered gazelle males also mark objects at body height; however, they are known to mark twigs that are as high as they can reach stretching up on all four legs, though never rising on their hind legs like other Antilopinae species (Blank 1997). Goitered gazelle males hold their heads in a position so a suitable branch can be inserted into the widely opened preorbital gland for rubbing and depositing secretions; consequently, shrubs close to a male's head height are marked more often than lower grasses or higher saxaul trees. The Thomson's gazelle (Walther et al. 1983), gerenuk (Gosling 1981), blackbuck (*Antilope cervicapra*—Dubost and Feer 1981), and klipspringer antelope (Roberts 1997) also typically scent-mark at the height of their heads, though Thomson's gazelle males are known to mark lower grasses and twigs down to a very few centimeters above the ground (Walther et al. 1983).

Additionally, unlike marks made by urination-defecation, which are located along the periphery of the territory, scent marks by preorbital glands are located away from the borders where plants important for foraging are marked, but the marked branches face toward the territorial boundaries of the territory (Roberts and Lowen 1997). In goitered gazelles, the marking point is a clearly visible deposit of a hardened dark gray or black crystalline substance that forms a thickening at the end of the branch at a height of 40–86 cm from the ground (on average 63.4 cm, $n = 101$). The marks are located singly or in groups of 2–6 on one plant at a distance of 3–183 cm from each other (on average 44 cm, $n = 27$). Sometimes secretion accumulations form frozen growths hanging from branches, in which hairs are included. The marking points shine in the sun and are clearly visible from a distance as well as having a characteristic smell. The more often the ends of the branches are marked, the thicker the secretion layer is and the darker its color. Long-term (not renewed) marks are light gray, dusty, odorless, and poorly visible (Marmazinskaya and Mardonov 1996).

The scent marks are usually located in places with local topographical features or some prominent characteristics that attract attention: a free-standing or tall tree or bush, damaged or dry vegetation. In addition, males try to create noticeable changes around the mark to attract the attention of a potential scent signal receiver: they peel off the bark or break off branches of trees or shrubs, causing visible damage to vegetation, and scrape the ground with their front hooves (Roberts et al. 2001. Thus, the scent marks have also a visual element, which contributes to a faster discovery of the marks. So, in addition to marking with the preorbital glands, quite often males leave urination-defecation and vegetation horning marks at the same point (more information in the next chapter). Goitered gazelle males also likely reject the common grasses and choose the more conspicuous branches (dry, broken, or highest) instead (Marmazinskaya 1996). The gerenuk (Gosling 1981) and klipspringer (Roberts 1997) prefer to mark distinctive dead trees or dry branches, as well.

In goitered gazelles, marking points are concentrated near the trails, at a distance of 0.3–7 m. Other antelope species (pronghorn antelope *Antilocapra americana*—Gilbert 1973; gerenuk—Gosling 1981; dik-dik *Madoqua kirkii*—Gosling 1985) also often place marks near to regularly used pathways, while suni antelopes deposit secretions in open areas away from pathways. This difference is related to nutrition, since the first group of species are browsers or grazers, while suni antelopes live in woodland and with a diet related mostly to the leaves of trees. Obviously, there is a dependence of the distribution of marks on the distribution of plant species that is important for ungulate foraging (Somers et al. 1990).

Certainly, the distribution of marking points in goitered gazelles is also influenced by vegetation and the density of concentration of the gazelles in areas with a particular plant association. For example, males do not mark plants with a height less than 40 cm, plants with very thin and unstable branches, or plants with many small sprouts at the end of the branch. A high number of marking points have been noted in plant associations in which *Tamarix*, saxaul, *Ephedra*, *Astragalus*, *Salsola arbuscula* grow, as well as where there are dried reeds (*Phragmites communis*) and

Karelinia caspica. In plant communities where *Salsola arbuscula* grows alongside *Astragalus*, males often prefer to mark *Astragalus* bushes. In the same areas where *Ephedra* grows with these species, there are more marks on the Ephedra. In areas of the Bukhara Breeding Center where the seasonal concentration of animals is high and vegetation is not suitable for marking, marked plants are rare or completely absent. This is true for extensive, purely *Halocnemum strobilaceum* and *Salsola gemmascens* or *S. gemmascens-Artemisia* associations without the presence of other plant species, since smooth, stable, and well-standing branches are very rarely found on *Halocnemum strobilaceum* and *S. gemmascens* bushes. *S. gemmascens* bushes are also low to the ground and not convenient for marking. A low density of marking points is noted for salt marshes as well (Marmazinskaya and Mardonov 1996). The use of habitat areas in different seasons of the year can be judged by the appearance of the marking points (degree of use, the color, thickness of the secretions). In the areas of territorial males, marking points are distributed throughout the area near the trails on plants convenient for marking (Marmazinskaya 1996). Other antelope species (suni antelopes, klipspringer) also prefer firm, sharp twigs and sprouts of vegetation for marking (Roberts 1997), but they have the ability to mark plants lower than body height and as far down as ground level (Somers et al. 1990).

Male goitered gazelles usually mark shrubs that gazelles commonly eat, such as *Eurotia ceratoides*, *Salsola arbusculaeformes*, *Athraphaxis*, and *Calligonum*; among trees, they usually mark saxaul (*Haloxylon aphyllum*); and among grasses, *Phragmites communis*. They most often use the twigs of dry shrubs and semi-shrubs for secretion markings (91–98% cases, $n = 489$), while marking the tops of tall grasses and branch tops of saxaul trees is observed considerably less often (1–6% in different months). As previously stated, this tendency is demonstrated especially clearly during the rutting period (96–98% in November and December, $n = 379$) and slightly less often (albeit not significantly) in other months (91–94%, $n = 120$); this ratio shows an insignificant change (Blank et al. 2014a). In the semi-captive conditions of the Bukhara Breeding Center, the same phenomenon was observed when males marked with the secretions of their preorbital glands the ends of strong, stable, well-standing, and often the highest dry branches of *Tamarix hispida*, saxauls (*Haloxylon aphyllum*), *Karelinia caspica*, *Phragmites communis*, *Calligonum microcarpum*, *Aellenia subaphylla*, *Artemisia* sp., *Halocnemum strobilaceum*, *Halostachis caspica*, *Limonium* sp., *Astragalus unifoliatus*, *A. villosissimus*, *Ammothamnus Lehmannii*, *Salsola arbuscula*, *Ephedra strobilacea*, *Cousinia* sp., *S. gemnascens*, *Zygophyllum eichwaldii*, *Smirnowia turkestana*, and *Iris songarica* (Marmazinskaya and Mardonov 1996). In areas with few apposite plants available for marking, marked plants are rarely found, if at all, even when the population density of goitered gazelles is high (Marmazinskaya 1996). However, in the semi-captive conditions of the Issyk-Kul nursery (Kyrgyzstan), males often mark the tops of thin and unstable stems of the grass *Achnatherum* (*Lasiagrostis*) *splendens* with their preorbital glands (Fig. 5.12–5775) possibly because gazelles had a limited number of shrubs suitable for such markings (Blank 2022, unpublished report).

Fig. 5.12 Goitered gazelle males often mark tops of thin and unstable stems of grass *Achnatherum (Lasiagrostis) splendens* by preorbital glands. Issyk-Kul goitered gazelle nursery for reacclimatizing, 24.06.2022. Photo by D. Blank

5.3.4 The Interaction of Preorbital Gland Markings with Other Kind of Markings

The marking of plants with preorbital gland secretions occurs both during the day and after dark (Zhevnerov 1984). Generally, marking with preorbital glands is performed as a single, separate act (62–91%, $n = 528$), but occasionally, it is linked to other types of marking, such as alternating preorbital gland markings with urination-defecation acts (4–28% cases) and/or acts of object-horning (4–10% of cases). In addition, the proportion of combined markings fluctuated over seasons from a minimum of 8.7% in the summer months (June through August) up to a maximum of 30.9% and 37.5% in November and December, respectively (Blank et al. 2014a).

5.3.5 Activities when Preorbital Gland Marking Is Observed Most Often

During the first half of the autumn rut in November, males used preorbital gland marking mainly when patrolling their territorial boundaries (Table 5.4). Even more, territorial males prefer to re-mark (or refresh) previously marked sticks than establish new mark points (Walther et al. 1983). In general, the lifespan of a scent mark is limited so that the mark needs to be periodically refreshed (Alberts 1992). This is understandable, since in November (first half of the rut), territorial males are more

Table 5.4 Different behavioral patterns of goitered gazelles, which triggered scent marking with preorbital glands during the rutting period (November–December) and "false" rut (April–May). Numerator is absolute frequency value and denominator is proportion of this frequency within a month (%) (by Blank et al. 2014a—open access)

Months	Feeding	Territory patrol	Territorial conflict	Chase males	Chase females	Walk, run
April	43/82.7	3/5.8	–	2/3.8	4/7.7	–
May	28/62.2	11/24.5	–	–	1/ 2.2	5/11.1
November	45/20.8	97/44.9	12/5.6	15/6.9	42/19.5	5/2.3
December	35/21.4	43/26.2	2/1.2	3/1.8	81/49.4	–
Log-linear model test	Z = 3.153, P = 0.002	Z = 3.422, P = 0.001	–	–	Z = 4.219, P = 0.0005	–

concerned with establishing and maintaining their territorial dominance status against rivals than courting of females. But already in December, males are busier chasing females, with preorbital gland marking occurring mostly during the chase. However, though the entire rutting season, in both November and December, goitered gazelle males pay attention to patrolling and remarking their territories each day, which can take up to 23.9% of their daytime activity budgets (Blank 1990). During the spring peak of sexual activity (or "false" rut) in April, rutting males devote only 5.5% of the daily time for patrolling and they mark mostly when grazing (Table 5.4). Contrary to widely held belief, territorial males rarely engage in preorbital gland marking during territorial conflict situations (primary threat displays) with neighbors during the rut, while the rate of horning (against plants) and urination–defecation marking usually increase during border conflicts between neighbors. Outside of the rut, males do establish territories; therefore, preorbital gland marking is quite rarely done and mainly when grazing or sometimes when changing locations (Blank et al. 2014a).

5.3.6 The Development of Preorbital Marking Behavior in Young Males

In the wild, young goitered gazelle males start secretion marking with the preorbital gland at the age of 1 year (in May), but only rarely, with just 9 such cases recorded (in May–July) from all preorbital gland marking acts observed during my entire study (1.7%, $n = 528$). In the semi-captive group in the Bukhara Breeding Center, half-year-old males (6–8 months) during the autumn rut and 1-year-old males (10–12 months) during the spring peak of sexual activity did not demonstrate this marking behavior. The earliest observation of preorbital gland marking was noted on July 19 for a young male at the age of 14 months. A male at the age of 15 months on August 11 marked a plant alternately with both preorbital glands and performed a complete ritual of urine/feces marking. At the age of 18 months, sexual maturity of males begins, which coincides with the beginning of their second real, autumn rut. In the semi-captive grouping, marking of plants with the secretions of the preorbital gland by young males was periodically observed. Thus, similar to all types of marking activities, preorbital marking is fully formatted in male gazelles by the onset of puberty, but often this behavior is more or less suppressed by the presence of adult males in the groups, while adult males 2–2.5 years and older demonstrate all types of marking activity in natural conditions (Marmazinskaya 1996).

For other antelope species, blackbuck young males (*Antilope cervicapra*) start preorbital marking at the age of 2 years (Dubost and Feer 1981), though Benz (1973) noted marking behavior at the age of 6 months. In Thomson's gazelles, the preorbital glands start to excrete their secretions at the age of 1 year, though yearlings rarely demonstrate marking behavior. In some cases of preorbital marking, the young male is more likely to be mimicking the marking movements of an adult male than being motivated to actually mark the plant (Walther et al. 1983).

5.4 Marking by Object-Horning

Blank and Yang 2014—License Number—5477550687665 (full text).

5.4.1 Object-Horning Ritual

Object-horning is often observed in cervids and bovids. It is essentially a special form of fighting behavior with inanimate objects, when a male attacks vegetation, mostly shrubs and trees, less often grass, and horn-butts the ground and rocks, as well as fences and feeders in captivity (Walther et al. 1983; Estes 1991; Adams et al. 2001). During object-horning the male uses his horns to strike at vegetation and/or the ground as if he is fighting an imaginary opponent, performing all the same movements he usually uses when fighting with a living opponent (Walther 1984).

Goitered gazelles perform object-horning in the same way as other Antilopinae species. The male approaches a bush or tree in a threatening posture with his forehead and horns exposed forward and upward and the tips of his ears directed apart horizontally and pointing a little backward with his earholes turned to the ground; his tail is raised. Sometimes he sniffs the previously worn point of the trunk. Then he pushes his head toward the trunk or bush until it is between his horns as far as to their base and makes up and down movements with his head. With great strength, he pushes at the plant shoving it from side to side as far as possible. Males not only rub their horns against plants, but also the area between the horns and almost the entire front part of the muzzle, sometimes muzzle sides of the muzzle. During this display, the male breaks off small branches and pieces of bark from the trunks. He stops periodically from horn-butt, touching the worn area with his nose and sniffs and then begins to butt his horns-forehead against the trunk again. Often, object-horning is replaced by marking the same bush or tree with his preorbital gland, followed by either continued object-horning or leaving.

Horning of grass and ground is performed somewhat differently. The goitered gazelle male approaches his chosen marking point and moves his horns from side to side rather than a direct forward horn-butting. During this long display (up to 1 min), he makes rather sharp, side to side shaking movements with his head and occasionally picks up grass with one horn and then the other. Then similar to shrubs and trees, the male touches the worn place with the front end of his muzzle and either he sniffs the horning place or maybe he applies the secretion of the preorbital gland, flowing down to the mouth, and then continues rubbing with the horns again. A male can rub not only his horns, but also forehead and muzzle's sides on the grass. During the rut, two territorial males-neighbors can perform this behavior simultaneously, with each staying on his own side of the boundary area (Blank and Yang 2014). Object-horning is especially expressive in Grant's gazelles (*Nanger granti*) with the horns moving in undulating lateral movements from side to side for a long time (up to 15 min) with only small interruptions (Walther 1984).

Similar to other gazelle species (Walther 1978a, c), goitered gazelle males display demonstrative horning of vegetation in front of other males and is very often observed during border territorial conflicts between neighbors. Male gazelles also create visual anomalies within their territories like breaking twigs of bushes and/or branches of saxaul trees and peeling the bark off trunks. Whether males deposit secretions on these visual spots with some kind of skin gland remains a mystery. Therefore, in this chapter I will try to answer the question of what the primary intention of object-horning is for goitered gazelle males, aggression or territorial marking.

5.4.2 Frequency and Duration of Object-Horning over Months

According to my research in Kazakhstan, goitered gazelles perform object-horning of vegetation with various frequencies over different months. Most often, male butting of plants is observed during the autumn rut in November (0.81 cases per hour) and December (0.93 cases per hour). This behavior is observed noticeably less often during the spring peak of sexual activity in April (0.27 cases per hour), and very rarely do male goitered gazelles horn vegetation in other seasons (0.04 cases per hour in June–August) (Blank and Yang 2014). In the Bukhara Breeding Center (Uzbekistan), the semi-captive goitered gazelles also show their highest increase in the rate of vegetation horning during the autumn rut in November–December, with a smaller increase during the spring peak of sexual activity in April. In young goitered gazelle males, the first attempt at horning vegetation occurs at the age of 1.5 months. Regular horning of bushes and trees is observed in adolescents at the age of 1 year, though it is still quite rarely. Goitered gazelle sub-adult males butt vegetation very rarely in any season (Marmazinskaya 1996), which is also true for sub-adult blackbucks (*Antilope cervicapra*) and bushbucks (*Tragelaphus scriptus*) with object-horning observed extremely rarely or not at all (Dubost and Feer 1981; Wronski et al. 2006) (Fig. 5.13).

Goitered gazelle males exhibit the longest duration of object-horning during the rutting season in December, when they experience their most intense courtships of females; in the spring peak of sexual activity in April and May, object-horning has the second longest duration (42–46 s), with the shorter horning procedure during the

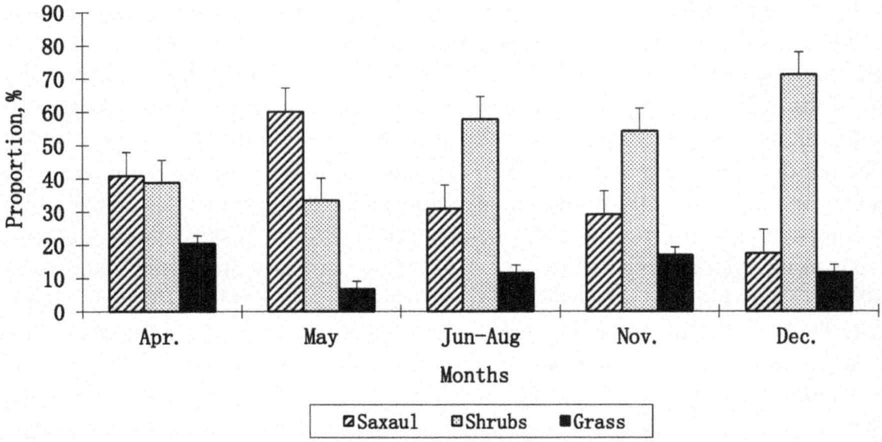

Fig. 5.13 Plant preference for object-horning in goitered gazelle over months (by Blank and Yang 2014—License number—5476290436207)

Table 5.5 Duration of object-horning acts (sec) in males of goitered gazelle over months (by Blank and Yang 2014—License Number—5476291061876)

				Mann-Whitney U test	
Months	Mean duration, X ± SE	N	SD	Z	P
March	22.7 ± 4.9	6	12.1	−0.887	0.375 (December)
April	45.8 ± 6.8	35	40.1	−2.732	0.006 (November)
May	42.6 ± 6.7	33	38.8	−2.096	0.036 (November)
June–August	16.2 ± 3.4	10	10.9	−2.719	0.005 (April)
				−2.316	0.021 (may)
November	31.0 ± 2.9	136	33.4	−1.167	0.243 (June–august)
December	45.5 ± 5.2	94	50.3	−2.452	0.014 (November)
				−2.076	0.038 (June–august)

first half of the main rut in November (31 s); outside the rut, the duration is the briefest, if object-horning is seen at all (Table 5.5; Fig. 5.14) (Blank and Yang 2014). The duration of butting vegetation in captive conditions (relatively small cages) of the Bukhara Breeding Center last from 12 to 80 s (on average 47.1 ± 0.42 s, $n = 10$), and in semi-captive conditions (large enclosures) lasts from 15 to 140 s (on average 65 ± 0.51 s, $n = 20$) (Marmazinskaya 1996). I never saw the object-horning behavior in females.

Object-horning is a typical behavior mostly for territorial owners; therefore, during the rut in November and December, only single males perform object-horning (98–100% of observed cases); while outside of the rut both single males and males in the groups display object-horning in approximately the same ratio. Bachelors during the rut also butt vegetation only not as frequently. During the autumn rut, horning of vegetation is observed mostly during territorial conflicts in the presence of another male, less often when chasing females, and rarely when males are alone

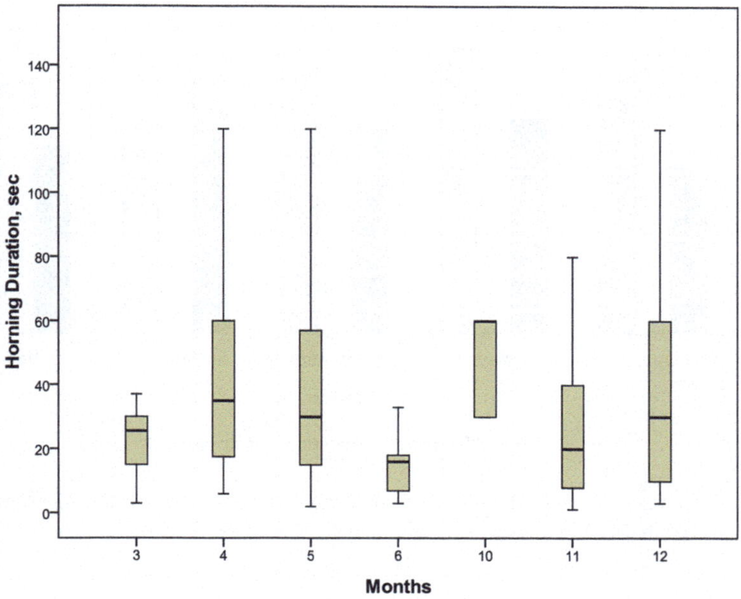

Fig. 5.14 Changes of object-horning duration in goitered gazelle over months (Blank and Yang 2014—License number—5476290436207)

grazing; in the spring peak of sexual activity in April and outside of the rut, object-horning is observed mostly during grazing (Table 5.6 - Blank and Yang 2014). In captive conditions, the frequency of object-horning increases sharply if a female in estrus appears in the group, or an adult male arrives in a neighboring enclosure. In the semi-captive herds, the frequency of vegetation butting increases if a territorial male finds females or males entering his individual territory (Table 5.7). When the population density of goitered gazelles is decreasing significantly, vegetation butting in males is almost never observed (Marmazinskaya 1996).

Walther (1978b, 1984) found that for Thomson's gazelle males, the presence of other males in the vicinity of an owner's territory incites intense object aggression on the part of the territory owner, even if the other males do not cross his territorial borders. In this case, object-horning has a visual display function and the male owner advertises not only his presence and position but also his very aggressive state, so that these acts can be considered as threats to the potential intruder.

Compared to the rate of scent marks (with preorbital glands and deposits of urine/feces), object-horning of vegetation is demonstrated relatively less often by males during the autumn rut in December (26%) and the spring peak of sexual activity in April (23%). While in other months, butting of vegetation is observed noticeably more often, almost equal to the frequency of marking with the preorbital glands in May (31%) and in the summer months (30%), reaching the rate of the urination–defecation acts in November (31%). However, this difference is not very large, since the portion of preorbital markings fluctuates between 28% and 39% and urination/

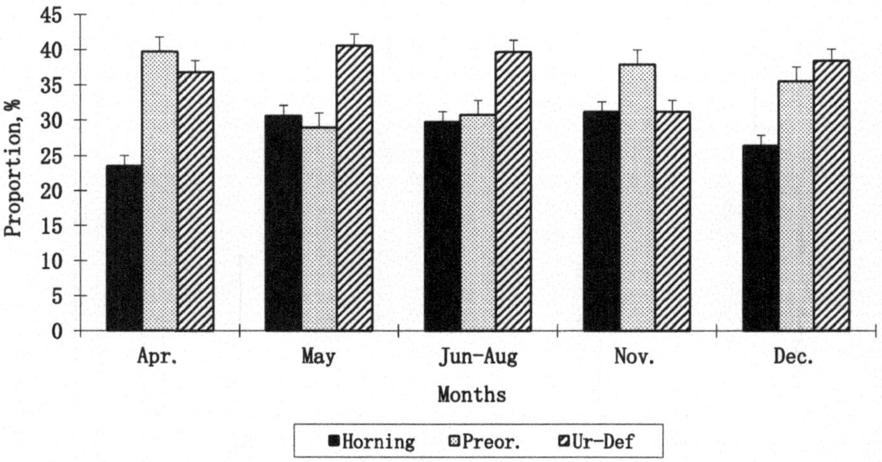

Fig. 5.15 Proportions of object-horning, preorbital scent marking, and urination–defecation acts over months (Blank and Yang 2014—License Number—5476290436207)

defecation acts vary between 31% and 40% (Fig. 5.15—Blank and Yang 2014). Marmazinskaya (1996) also found that adult males display object-horning least often, marking more often with urination-defecation and most often with preorbital glands. Moreover, object-horning is always demonstrated least often, even in semi-captive conditions and young males.

In the Bukhara Breeding Center, when calculating the average density of various types of marks per 100 m^2, the following results were obtained. In areas with a low concentration of goitered gazelles (0.07–0.13 individuals/ha), the density of urination/defecation marks was 0.1; preorbital marking points 1.2; and broken bushes 0.1. In areas with an average concentration (0.14–0.25 individuals/ha), the density of markings was equal to 0.3, 0.5, and 2.4 respectively, and with a high concentration (0.36–0.91 individuals/ha), density was 0.8, 3.0, 2.4, respectively. The calculation of the average density of marks per 100 m of the route yielded the following results: in areas with a low concentration of individuals, the density of latrines was 0.6, marking points 0.3, and broken bushes 0.2; in areas with an average concentration of individuals the density was 1.0, 0.6, and 0.6, respectively; and with a high concentration of individuals the density was 2.9, 2.1, and 2.7 (Marmazinskaya and Mardonov 1996).

5.4.3 Selection of Plants and Location of Object-Horning Sites

Many antelopes performing object-horning do it within a narrow height position, though physically they have the ability to change widely the vertical location of a mark (Gosling 1981; Roberts 1997). In goitered gazelle males, object-horning of vegetation, similar to preorbital marks, occurs at the approximate height of the shoulders (higher than 70 cm) up to the head level. Sometimes the male stretches his body up, trying to get to higher branches. Similar to other antelopes, goitered gazelle males do not rise on their hind legs, while some cervid species do, for example, barasingha (*Rucervus duvaucelii*) and axis deer (*Axis axis*) stand on their hind legs to rub the branches as high as possible (Walther 1984).

Broken shrubs from object-horning, as well as preorbital marking points and urination/ defecation deposits, are concentrated near trails (at a distance of 0.3–7 m) and often located at the end of branches hanging down to the trail. Very often, all three types of marks are located between two parallel trails, as well as at the intersection of several trails stretching in different directions. Since males periodically produce a series of several types of markings, their habitats can contain a large number of marking complexes (consisting of one or more urination/defecation points, broken bushes, plants marked with the secretions of the preorbital glands) located near each other (Marmazinskaya and Mardonov 1996).

In cervids, a male chooses trees and shrubs for object-horning in accordance with the size of the trunks and other characteristics of trees and shrubs; it is also believed that the aromatic properties of a plant play a role in the degree of its attractiveness for object-horning (Bowyer et al. 1994; Oehler et al. 1995; Massei and Bowyer 1999). Goitered gazelle males have only saxaul (*Haloxylon aphylla*) trees in their desert habitats, with most trunks having small diameters that pass freely between the males' horns, allowing the males to easily horn them and peel off bark. Therefore, it is often possible to find saxaul in the habitats of goitered gazelles with broken branches, located on trees from a meter high down to the ground. I found nine such trees along a 3 km section of a dry riverbed. Males usually select trees that are different in some way from other trees: the largest, tallest, or standing aside from others or with broken branches. Such objects attract special attention of goitered gazelle males and they are selected many times for repeated acts of object-horning (Marmazinskaya 1996). Repeated attacks of a particular tree can lead to its drying up and dying. As for odor, it is possible that the smell from worn trunks of a saxaul somehow *affects males, but this specific aspect has not yet been studied (Blank and Yang* 2014*).*

Among the other plants that are chosen for object-horning, shrubs from the family Chenopodiaceae and Polygonaceae are most often noted, including Krascheninnikovia (or *Eurocea) ceratoides*, *Salsola arbusculaeformes*, and *Atraphaxis frutescens*, and sometimes *Caragana* (Fig. 5.16–6285). This choice by males is primarily explained by the fact that these bushes are the most common plants in the area of my study. Additionally, these plant species are eaten by goitered gazelles and are their preferable food most of the year (Zhevnerov 1984; Blank 1990; Kingswood and Blank 1996). Walther et al. (1983) suggest that gazelles and other antelopes do not so much choose a suitable plant for object-horning, but rather use objects that they come across along their path. Other researchers (Benner and Bowyer 1988; Massei and Bowyer 1999; Wronski et al. 2008), however, claim the opposite is true: that ungulates are very selective in their choices of butting objects, which are located at strategically important points for a particular male. In addition, the choice of plant object-horning does not necessarily depend on their palatability.

Fig. 5.16 Object-horning in goitered gazelles: male butts *Caragana* by horns. Issyk-Kul goitered gazelle nursery for reacclimatizing, 26.06.2022. Photo by D. Blank

For the goitered gazelles in my study, the most available shrubs, which also happen to be the plant species most readily eaten, were those chosen for object-horning. So, my observations rather confirm Walter's point of view.

In general, goitered gazelles prefer to do horning of various shrubs, followed by saxaul trees; herbaceous plants are the least preferable. However, the preference for object-horning between these three groups of plants changes over months. During the spring peak of sexual activity, butting males tend to choose saxaul trees most often (in April 41% and May 60% of cases), with shrubs less attractive (33–38%). During the real rut in November and December and during summer, males do horning of mostly shrubs (54–71% of cases). Vertical trunks and branches are most preferred by males for horning (57% of observed cases), angled branches are less interesting (26%), and horizontally branches are least attractive (17%) (Blank and Yang 2014).

When demonstrating "ground" horning, goitered gazelle males sniff the low grasses and scratch the ground with their horns, goring the plants and bare land (Fig. 5.17) and as if trying to shift this ground. Males tend to choose places to do this that have low vegetation (mainly dwarf-shrub *Nanophyton erinaceum* and herbaceous plants) and bare ground, often with soft soil. Butting grass and soil have an even more aggressive character than butting bushes; therefore, it is observed exclusively during territorial conflicts along the boundaries of individual territories. These borders often run along sandy dry riverbeds, where soil butting occurs most often. Obviously, butting of low vegetation has a demonstrative character and is

Fig. 5.17 A goitered gazelle male sniffs at low grasses and dwarf shrubs and starts to horn, scratch and gore the plants and naked ground. Issyk-Kul goitered gazelle nursery for reacclimatizing, 26.06.2022. Photo by D. Blank

often performed in full view of the male's opponent, and it is not uncommon for two territorial males to simultaneously butt the ground, each on his own side of the boundary area. Compared to object-horning, goitered gazelle males butt low vegetation and ground relatively infrequently, with the highest rate in April (20%) and May and in less than 7% of cases in other months (Blank 1997).

In the Bukhara Breeding Center, males butt saxaul trees, shrubs and large-sized herbs. Among broken shrubs are observed: *Halostachys belangeriana*, *Kalidium* sp., *Salsola arbusculiformis*, *Haloxylon aphyllum* (Fem. Chenopodiaceae or Amaranthaceae), *Ephedra* spp. (Fem. Ephedraceae), *Convolvulus* spp. (Fem. Convolvulaceae), and *Tamarix* spp. (Fem. Tamaricaceae). The average level of damage to a plant (the proportion of broken shoots) for an individual species is 10–60%. The accounting of damaged plants shows an average of up to 443 individual plants of *Halostachys belangeriana*, 50%—*Kalidium* sp., 30%—*Tamarix* sp., 15%—*Halocnemum strobilaceum*, and 10%—*Salsola gemmascens* (Mardonov 1993).

The species composition of plants these gazelles eat that are butted by males is as follows: *Haloxylon aphyllum*, *Tamarix hispida*, *Halostachys caspica*, *Karelinia caspia*, *Kalidium capsicum*, *Ephedra strobilacea*, *Astragalus* spp., *Salsola arbuscula*, *Artemisia* spp., *Salsola gemmascens*, *Convolvulus hammala*, and *C. divaricatus*. The species composition of plants broken down by males varies in different seasons of the year, as they depend on the biotopic timing of seasonal concentrations of animals and plant associations characteristic of the season. During the spring peak of sexual activity (false rut), the main concentration of goitered gazelles shifts to the northern part of the territory of the Bukhara Breeding Center where *Ephedra*, *Calligonum*, *Haloxylon*, and *Astragalus* are most exposed to horning there. In the second half of spring, the concentration of gazelles shifts to the middle part of the territory with *Zygophyllum eichwaldii*, *Tamarix* spp., *Halocnemum strobilaceum*, and *Haloxylon aphyllum*. In summer and during the autumn rut, the concentration of goitered gazelles is in the southern part of the Bukhara Breeding Center and the species composition of the butted plants changes to basically *Kalidium caspicum*, *Tamarix* spp., *Karelinia caspia*, *Halocnemum strobilaceum*, and *Halostachys caspica* (Marmazinskaya and Mardonov 1996).

Broken shrubs are found mainly along the trails. Sometimes there are areas with several bushes (up to 10) damaged by goitered gazelles located at a distance of several meters from each other. In the areas of territorial males, the broken bushes are distributed throughout the territory, and not concentrated along the borders. Once a broken plant is established as an olfactory and optical signal, it is repeatedly damaged by territorial and other males. Plants broken by goitered gazelles' object-horning, along with other marking types, are markers of territorial areas. The usage of habitats in different seasons can be judged based on the density and distribution of broken bushes and other markings found in a specific area (Marmazinskaya 1996; Marmazinskaya and Mardonov 1996).

5.4.4 Is Object-Horning Marking or Aggression

There are several interpretations of object-horning behavior; however, most reports interpret this behavior either as a part of aggressive or marking behavior or consisting of both elements (Walther et al. 1983). Only Adams et al. (2001) suggest that horning of vegetation has some impact on the synchronization, or priming, of the estrus cycle in females. Several studies of African antelopes (*Antidorcas marsupialis*, *Connochaetes gnou*, *Kobus ellimsiorymnus*, *Redunca arundinum*) have showed that horning of vegetation is related mostly to aggression or a biased sexual pattern (Estes 1972; Jungius and Claussen 1975; Spinage 1969). When a lone male displays object-horning of vegetation, in the absence of any other conspecifics, this activity looks like a playful behavior or possibly just a release of excess aggression. When object-horning alone, males use movements with their heads and horns similar to real fights with rivals (Estes 1991; Wronski et al. 2008). Horning of vegetation often occurs in the presence of another male and done as a demonstrative threat behavior. In this case, it has a visual aspect of communication through postures and horning movements (Geist 1965), since ungulates of open habitats use many postures and movements for this purpose (Kiley-Worthington 1978). Object-horning is purely an external expression of prominence and, in fact, is related to and likely comes from agonistic behavior (Walther 1984). Dominant males are especially motivated to vigorously demonstrate marking behaviors when strongly aroused by the presence of other males, advertising a readiness to attack (Ralls 1971).

Goitered gazelle males often use object-horning of vegetation as a redirected threat toward bachelors or territorial male-neighbors which come close to the territorial borders, even without entering. The male-owner shows his intention to drive away an approaching male, but in order to do so, he needs to leave his own territory, where he is dominant, and enter the neighboring territory, where he cannot demonstrate his dominance. This situation of uncertainty forces the male to stay in his own territory and reduce his aggressive tension through his displaced behavior and butting plants. By doing this, males demonstrate not only their presence and status, but also their aggressive intentions. It is possible to observe this same behavior in some cervids (*Dama dama*, *Capreolus capreolus*) (Johansson and Liberg 1996) and antelopes (*Aepyceros melampus*, *Antelope cervicapra*, *Antilocapra americana*, *Litocranius walleri*, *Taurotragus oryx*, *Tragelaphus angasii*) (Walther 1984).

Vegetation horning is observed not only in territorial species. Many deer species, for example red deer (*Cervus elaphus*), white-tailed deer (*Odocoileus virginianus*), and axis deer (*Axis axis*), do not have territories, but almost all of them exhibit object-horning, which indicates to others the presence of the male, past or present, and potentially his dominant status, but with no territorial marking involved (Walther 1984). Horning vegetation is not only about aggression, but also has an aspect of marking behavior, since in some ungulate species the forehead has glandular fields. This provides the potential for leaving aromatic secretions, while butting grasses and goring the ground seems to be almost entirely associated with aggression. Ungulates usually do not use a straight-on butting movement in a forward direction but use a more sideways movement that has the capability to leave secretions for scent marking (Walther 1984).

If object-horning is a making behavior in addition to a display of aggression, then it is a special way for males to identify their own presence and advertise of their dominant territorial status. During object-horning, the male positions the forehead area between his horns or antlers against the vegetation and then rubs the entire front of his head against the shrub or tree, most likely depositing secretions (scent marking) from non-specific glandular skin of the head; this is a behavior observed in many deer species (Marchinton and Hirth 1984). Through this process, the ungulate also changes the environment (scrapes on the ground, destroyed ant and termite hills, broken branches, and striped patches of bark and scarring of the bark) which is a way of leaving visual markings (Walther 1978a, c). Object-horning is reported for many ungulate species for both territorial species (waterbuck *Kobus ellipsiprymnus*, Uganda kob *Kobus kob*, springbok *Antidorcas marsupialis*, and most other gazelles, fallow deer *Dama dama*, roe deer *Capreolus capreolus*) and non-territorial ungulates (moose *Alces alces*, red deer *Cervus elaphus*, caribou *Rangifer tarandus*, white-tailed deer *Odocoileus virginianus*, and mule deer *O. hemionus*) (Walther 1984; Adams et al. 2001).

Marking with skin scent glands, and especially markings with the scent secretions of the preorbital glands, are widespread and common among Antilopinae. So, why then do antelopes also have horning plants? The fact is that just marking with scent glands has certain disadvantages. The most important is that scent marks lose their effectiveness quite quickly, since the secretions are made of highly volatile substances in order to detect the mark from a distance. Scent marks are most often used on stationary objects, for example on branches of shrubs and trees, and less lethal chemical compounds are used for individual specificity and greater information from the mark. But with the limitation that such a mark can only be detected at a close distance. In this regard, object-horning vegetation and its consequences (broken branches, worn trunks) and raking the soil around a horned bush or tree are clearly visible at a distance and remain for a long time. These marks, however, are reduced or completely devoid of individual specificity, because they are not always accompanied by scent marking. Therefore, the best solution to accommodate marking limitations may be a combination of horning vegetation, visible at great distances, and scent marks with high specificity in spite of a transitory nature (Carranza and Mateos-Quesada 2001).

Goitered gazelle males have a rich array of scent marking behaviors, including urine/feces deposits and preorbital gland markings. In addition, they perform ground scraping, shrub horning, tree barking, and ground goring, making highly visible changes in the environment. These are all forms of marking and may also possess secretions from skin glands, though this has not yet been confirmed by research, although it cannot be completely ruled out (Blank 1985). In goitered gazelles, object-horning seems to be quite strongly associated with marking by the preorbital glands and urine/feces deposits, since butting of vegetation and scent marking can be observed at any time of year. The relationship of these behaviors, however, varies by month. Males show their highest rate of linked cases of object-horning with scent marking and acts of urination/defecation during rutting in December (46%), while during the spring peak of sexual activity in April, it does not exceed 24%; during summer, only separate acts of object-horning are noted without any connection with other types of scent marking.

Other antelope species also show a close relationship between object-horning and scent marking. When blackbuck males horn vegetation, the orifice of the preorbital gland is opened wide and the secretions drip onto a patch of grass that is the focus of the butting male. Impalas (*Aepyceros melampus*) and bushbucks (*Tragelaphus scriptus*) horn vegetation frequently and the front parts of their heads (forehead, cheeks, crowns, necks, and ears) are richly equipped with skin scent glands, oily secretions of which are excreted when butting vegetation. Bushbucks usually display object-horning when alone and do not perform this ritual even during territorial conflicts with neighbors, which may mean that, for this species, butting vegetation is not intended as a visual function and is more likely connected to marking than to aggression (Wronski et al. 2008).

My observations of the goitered gazelle males during the rutting season have shown that they most often show object-horning of vegetation in the presence of another male and very often during territorial conflicts. Butting of vegetation has a strong visual threat component when it is addressed to another male. Thus, object-horning in goitered gazelles is clearly closer to agonistic behavior, but only during the rutting period. Outside of the rut, goitered gazelle males perform object-horning most often when alone and mainly when grazing, though horning vegetation can be observed as threat displays all year round. Sometimes males display object-horning to each other even in the absence of individual territories and without borders to protect. Goitered gazelles may also apply some kind of secretion to the vegetation from a facial glandular field scattered over the scalp, though the presence of such a glandular field is not yet a proven fact. So, significance of object-horning in goitered gazelles can be defined as either "biased aggression" or marking behavior or both. In any case, although butting vegetation looks like a purely external show of aggression, it is in fact a multifactorial behavior that, in addition to obvious aggression, has elements of scent marking and communication between males (Blank and Yang 2014). Walther (1984) holds the same opinion.

5.5 Rutting Vocalizations

Blank et al. (2014c)—License Number—5477550976506 (full text).

5.5.1 General Remarks

The life of a society is impossible without interactions and the exchange of signals between its members. Acoustic signals are especially important and occupy a special place in mammalian communication. Vocal signals between males and females are vitally important during the rut as males of many ungulate species usually use their vocalizations for courting females (Andersson 1994). In this regard, bovids and cervid males have a particularly rich repertoire of vocal signals during the rutting season that are interesting to consider (Vannoni et al. 2005; Frey et al. 2007, 2008a, 2011). Different species of deer, for example, produce a very diverse vocalization during the rut from whistling (*Hydropotes*) and cackling (*Muntiacus*) to grunts (*Alces*, *Capreolus*, *Rangifer*), groans (*Dama*), and loud roaring (*Cervus*) (Cap et al. 2008). Rutting vocalizations of antelopes are also widely varied, but these sounds are especially diverse in loudness in different species, from a loud roar with a widely opened mouth to a quiet grunt with the mouth closed to an even quieter clicking of the tongue (Walther 1984). Gazelle species other than goitered gazelles use soft nasal sounds for their courting rituals to achieve mating success, while only goitered gazelles use loud oral sounds in their rutting vocalization behaviors (Frey et al. 2011).

Large animals have relatively long vocal tracts and produce lower sounds compared to smaller species. A number of studies have demonstrated that the lower-pitch sounds of male roaring are more attractive to females and scare off other males-rivals (Reby et al. 2005; Charlton et al. 2007). A relatively large body size is one of the most important indicators of male quality among ungulates, and larger males are preferred by females more often than smaller ones (Fitch and Hauser 2003). As a result, large males have greater breeding success than their smaller competitors. Thus, lower-formant sounds are a sign of a large male, which means that he will be chosen as a high-quality breeding partner. Generally, there is a natural selection for a larger body size, but through evolving changes, males can find ways to continuously increase their body size to be more and more attractive to females. However, this is not always possible in all environmental conditions, where forage of high quality and quantity is only available for a very limited time. In this case, evolutionary shifts and special adaptations arise in males so that they can produce lower sounds to attract females, while at the same time retaining their relatively small body size; in other words, the male is able to acoustically create the impression of a larger body size without the stature to go with it (Fitch and Reby 2001; McElligott et al. 2006; Frey et al. 2007).

The production of low sounds in ungulates is achieved in various ways: increasing the mass of the vocal cords (*Budorcas taxicolor*); a special adaptation in the ventrorostral ventricle, which increases the volume of the laryngeal air spaces (*Ovibos moschatus*); retraction of the larynx (*Cervus elaphus*, *Procapra gutturosa*, *Gazella subgutturosa*); and nose elongation and the appearance of a trunk-like form of nose (*Saiga tatarica* and *Rhynchotragus guentheri*) (Frey and Hofmann 1996; Gebler and Frey 2005; Volodin et al. 2009). The goitered gazelle has evolved an enlarged mobile larynx that lengthens the vocal tract significantly, allowing the males to produce low pitch rutting roars (Frey et al. 2011). The peculiarities of the vocal apparatus structure indicate that the vocal behavior of males is closely related to the rut. Roaring characteristics play an important role in the courtship of females and project the males' quality and likelihood of reproductive success (Reby and McComb 2003; McElligott et al. 2006; Frey et al. 2008a; Vannoni and McElligott 2008). A direct relationship has been found between sexual success of ungulates and the time males spend roaring during the rut (McElligott et al. 1999), and a male's ability to produce rutting grunts of lower formants (resonances) determines his better breeding success by attracting estrus females and restraining opponents (Reby and McComb 2003; Briefer et al. 2010). For red deer, McComb (1991) did not find that females have a preference between red deer males producing low-pitch and high-pitch roaring, but she did find that females prefer males with a higher roaring rate, which correlates to their fighting ability.

5.5.2 Goitered Gazelle Vocalizations of Rutting Adult Males

Both males and female goitered gazelles have a larynx that clearly protrudes from the upper third part of the neck close to the head. The females' small larynx is clearly visible in summer when the body is covered by short hair, but it is almost completely hidden in winter when covered by long hair. Compared to adult females, the adult males' larynx is almost 2 times larger, which is very noticeable all year round, especially during the rut, when it increases by almost twice in size. The larynx slightly sways during male's movements or shifts from side to side when shaking of the head and neck. In addition, the larynx moves freely upward when swallowing food and water and downward when vocalizing rutting roars. The larynx of bachelors also increases during the rutting season, but not as significantly as in territorial rutting males (Kingswood and Blank 1996).

The male larynx, which is sensitive to steroidal hormones under the influence of androgens, begins to increase significantly before the rut. The larger larynx of males has longer vocal cords which determine the lower sounds of their roars by vibrating at a lower frequency compared to the sounds that females make. In addition, the male's larynx is very mobile and, during roaring, can move to an extremely low position along the vocal tract to the neck-thorax junction, significantly elongating it by up to 150%. Thus, the lower tones of a male's roar are amplified even more by lower formants, producing sounds of unusually low frequencies for animals of their sizes (Frey et al. 2011). Males produce mainly oral rutting roars with an open mouth, making these sounds quite loud (Marmazinskaya 1996), while other conspecifics (females and fawns) have the ability to make both oral and nasal sounds (Volodin et al. 2009). Besides goitered gazelle males, an enlarged and retracting, low-resting larynx is also found in Mongolian gazelles (*Procapra gutturosa*), fallow deer (*Dama dama*), and red deer (*Cervus elaphus*), as well as in human males. The larynx in these males also has the ability to significantly elongate the vocal tract during rutting calls although to varying degrees in different species; in *Cervus elaphus*, the vocal tract is known double in size (Fitch and Reby 2001; McElligott et al. 2006; Frey et al. 2008a, b).

To roar, goitered gazelle males lift their muzzles quite high and forward, stretching their necks and heads in a straight line (Table 5.8; Blank 1998). According to Frey et al. (2011), this position allows the males to lengthen their vocal tracts more, further lowering the diapason when roaring. Walther et al. (1983), though, argue that raising the head during a roar is typical for any ungulate individual with a dominant status, regardless of any oral or nasal vocalization, and that an elevated head may play two functions simultaneously (lengthen vocal tract and dominant position display). In contrast, roaring goitered gazelle females raise their muzzles up only slightly, if at all. Goitered gazelle males become especially noisy only during the autumn rut in November and December, with their roars heard from a distance of 400 m, at least; while at other times of the year, they usually do not make loud sounds at all, even when courting females during the spring peak of sexual activity in April, with their occasional roars inaudible from an observation distance of 100–150 m (Blank 1997).

Frey et al. (2011) assert that the rutting vocalization of goitered gazelle males consists of three kinds of sounds: roaring, growling, and grunting, which are produced all together in combination or singly. Each sequence lasts about 1.3 s, and four calls last 2.7 s. All these sounds are emitted when the mouth is wide open, and the larynx is retracted maximally down to the neck-thorax junction, staying in the lowest position until the end of the roar and then rising back up to its normal resting position. The tongue stays low inside of mouth cavity during the roar. My non-instrumental hearing impression when observing the production process of a roar by a rutting male is slightly different. First, I heard the males' roar as a single but complex sound consisting of sonorous, hissing, and wheezing components. I usually did not hear each of these components separately, although I did hear short staccato calls similar to grunts when males chased females at a fast gallop. In addition, growls and grunts are not very loud sounds, so perhaps this made it difficult to hear these sounds from my observation point at 50–150 m. Second, before roaring, the male took a deep breath, opened his mouth quite widely, lifted his tongue, then closed his mouth, and blew air through his nostrils at the same time. When roaring, the larynx usually made limited movements up and down.

When observing a male's roaring, I had the impression that the male's mouth was partly closed, but the loudness of the sounds in the roar indicates that the vocal cords participate in the formation of the roar and that these sounds come out through the mouth. At the same time, there are also gravelly, hoarse sounds, most likely created from the vibration of the nasal cavities. Therefore, it seems to me that both vocal cords (oral sounds) and nasal sounds are involved in creating rutting vocalizations in goitered gazelle males.

To the observer's ear, the rutting vocalization can be described in different ways. These sounds are similar to something like the snorting of a dog blowing through its nostrils, low hoarse sounds similar to blowing air through a plastic tube, sneezing sounds or even snorting with hoarseness, hoarse gurgling, or rough wheezing mixed with hissing and sneezing. One thing is clear that the roar consists of both vocal and hoarse nasal sounds. My description of the vocalizations of goitered gazelle males during rutting differs from the descriptions of professional researchers of the vocal behavior of ungulates (Frey et al. 2011). Such contradictions indicate the need for additional research on the vocal sounds and behaviors of goitered gazelle males.

The males' roaring can be observed actually at any time of the year, but for most of months, males are relatively silent, since their roaring, such as it is, is quite soft and not heard from a distance of more than 50 meters, while they usually do not emit the loud sounds. Certainly, they can use the soft nasal sounds that are also not audible at a distance of observations (50–100 meters). Efremova (2011) confirms that the ability to make loud oral sounds is typical mainly for young individuals and the proportion of these sounds decreases with age, while adult gazelles produce mostly soft nasal sounds most of the year. The situation changes dramatically during the autumn rut, in November and December, when rutting adult males become the noisiest among conspecifics, almost continuously or periodically emitting loud oral roars while intensely courting females. The oral roaring sounds are formed by the vibration of the vocal cords (Frey et al. 2011) and loud enough to be audible at a distance of several hundred meters. The roar of a breeding adult male is perceived as a low gurgling, wheezing sound that usually produced repeatedly 2–3 times in a row. Therefore, the roaring of multiple rutting males is heard as a loud constant noise which is heard almost continuously from a distance of hundred meters around the network of individual male territories. The frequency of males' roaring during rutting season increases dramatically by almost 40 times in November and December compared to July, when roaring is quite rarely observed (Fig. 5.18). Although the frequency of male roaring also increases during the spring peak of male sexual activity in April, the vocalizations increase only by 6 times compared to July, and these roars are not always audible from a distance.

Adult males use loud roaring not only to court females, but also to expel other males from their individual territories. The frequency of such aggressive contacts increases during rutting, but this increase is insignificant. For a yearly average

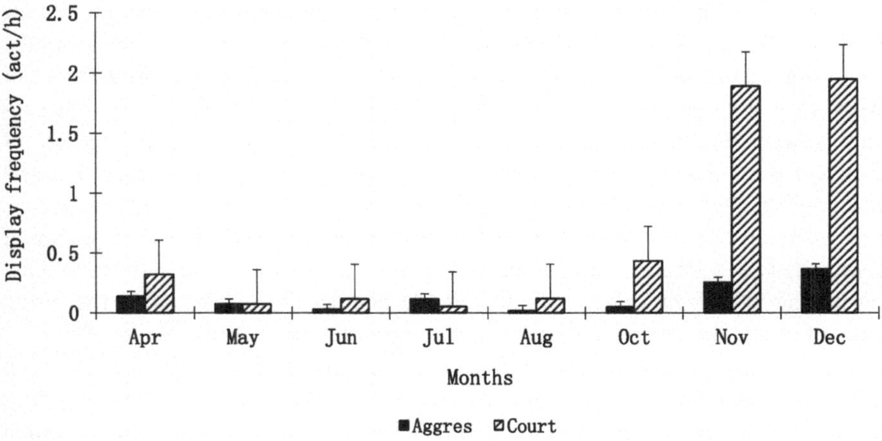

Fig. 5.18 Frequency (act/h) fluctuations of vocalized courting (Court) and aggressive (Agress) displays of adult males over months. Totally, it was observed 204 cases of vocalized aggressive and 955 cases of vocalized courting behaviors during 1690 h of focal observations (Blank et al. 2014c—License Number—5476360815518)

frequency, males are 2.4 times more likely to use a roar during courtship of females than using a roar in aggressive interactions. From this it is clear that the roar of males is more likely to be associated with sexual behavior than with agonistic behavior. Outside the rut, males use a roar in courtship and aggression in almost equal proportions, although the total number of such cases is insignificant compared to the rut (Fig. 5.18). Most often, males roar during courtship of females, especially during rutting season in November and December (98% of all cases, $n = 911$) and the spring peak of sexual activity (94% of all cases), although the total number of observed cases in April is small (Fig. 5.18). During the rut, not only does the number of males chasing of females increase, but also the proportion of courtship activities accompanied by vocalizations, which grows rapidly and, in November, at least half of all courting displays are accompanied by male roaring. Males usually roar while courting females but rarely before the start of the pursuit, with rare exceptions. The usual posture in which a male roars is either a low-stretch or high-stretch, when the head and neck are aligned, and the vocal tract is lengthened which makes the roar sound lower (Frey et al. 2011). Roaring is produced exclusively in these poses during the rut (99% of cases, $n = 913$), while outside of the rut, the use of these postures is still quite high, though the proportion of roars in these poses decreases (88%). When roaring, the tips of the male's auricles are usually pressed back (70% up to 99%, $n = 753$) and his tail is raised to a horizontal position (95%–98% cases). During the rut, adult males most often roar during a fast gallop (41% cases, $n = 697$), somewhat less often during a quiet walk (31–39%), and significantly least often during a trot (24–27%). Rarely do males roar in a standing position (1.5%) or when jerking the whole body forward (2.6%). Outside the rut, males prefer to roar when pursued at a calm pace, trotting or walking (30–55%), and in July and August generally when standing still (11–38%).

While courting, males most often use their roars when chasing a single female (26% $n = 903$), roaring a little less often when chasing two or three females at the same time (21% and 22%), and very rarely when chasing a larger group (7%) or when they are alone in their individual territories (1%). There is a strong relationship between the frequency of a male's roaring and the number of female groups of different sizes that visit his territory. During the rutting season, single territorial males most often roar when courting females, whereas bachelors in male groups roar very rarely when chasing females. This difference between the frequency of roaring between single males and bachelors in male groups is especially significant during the rutting season in November and December, much less during the spring peak of sexual activity in April, with almost no difference the rest of the year.

As for agonistic interactions during the rut, territorial male roars more often when he encounters a single opponent and noticeably less often when an entire group of males enters his territory. This behavior is reversed outside the rut, when single males roar more often in groups than when they are dealing with a single male. During agonistic exchanges, territorial males use roaring mostly when chasing bachelors-intruders, and rarely in confrontations with territorial neighbors. Most of the year, when chasing other males, roaring males are most often walking (75%), while they roar less often when trotting or galloping (up to 30%), During the rut, rutting males roar more often when running (trotting or galloping—up to 45%), although the proportion of quiet walking remains high (up to 50%) (Blank et al. 2014).

During the courtship of females, a rutting male usually roars intermittently, although some distinct series of roars are quite long. Continuous roaring for a long time without any break, though, does not occur very often; during the most intense courtship in November, the proportion of continuous roaring can increase up to 28%. When in agonistic interactions, territorial males use single, relatively long roars (2–3 s) infrequently when expelling trespassing bachelors from their territories, while outside of the rut, males pursue each other with short (less than 1 s) calls repeated many times.

In yearlings, their roaring when courting females and vocalized agonistic contacts with peers are first observed in April with some increased frequency in May into July and then stops. Yearling males vocalize much less frequently (3.4%, $n = 58$ in April; 15%, $n = 98$ in May; 40%, $n = 44$ in June and 12%, $n = 8$ in July) compared to adult males, and agonistic interactions between males, which are supported by roaring, are observed even less often (4%, $n = 59$ in April and May and 10%, $n = 10$ in June) (Blank et al. 2014c).

From this analysis, it can be seen that the roar of males plays a very important role in the rutting behavior of goitered gazelles. And this loud vocalization is so important for the courtship of rutting males for improvement of reproductive success through female-male preferring and male-male competition that, especially for rut, the larynx, which is the main source of sound, significantly increases in size due to an increase in the mass of the muscles of the vocal cords up to 50%. Goitered gazelles have a much larger larynx than many domestic and wild ungulate species of comparable size. In addition to increasing the size, the larynx moves freely along the neck to its lowest point and significantly increases the vocal tract; as a result, males produce a roar of unusual low frequency and resonance for a hoofed animal of such size as goitered gazelle (Frey et al. 2011; Volodin et al. 2011). Low roar sounds are a sign of large body size, and females always prefer larger males as a sign of highly competitive ability. As a result, males with a lower voice have a great advantage in reproduction, which led to the rapid evolution of high laryngeal mobility and maximal lengthening of the vocal tract (Fitch and Reby 2001; Reby and McComb 2003). Male goitered gazelles have the ability to lengthen their vocal tract by 47% (Frey et al. 2011).

Although the roar of goitered gazelle males is used both for courtship of females and agonistic interactions with other males, the rutting vocalizations likely play a more important role in the reproduction success of these gazelles. It has been found that a more intense roar in red deer provides a greater advantage to the male in the choices that females make (Clutton-Brock and Albon 1979). In addition, male roaring has been noted to intensify ovulation in females (McComb 1987), and apparently performs the same important role for goitered gazelle in the intensification and synchronization of sexual cycles and a faster onset of estrus in females.

5.5.3 Rutting Roaring as an Advertisement of Territorial Dominant Status

In addition to a major role in courting females, roaring during the rut advertises the special arousal physiological state and territorial dominance of the goitered gazelle male in a territory where there can be no other dominants. In the Issyk-Kul Breeding Center, I observed 5 adult males with several females in a small enclosure. During the rut, only one dominant male roared, while the others did not, though they still tried to court females occasionally. Later, when this herd was moved to larger enclosure, two adult males started roaring after establishing their own territories (Blank 2022, unpublished report). The same phenomenon was found by Walther (1984), who observed four red deer stags and some females in one very large enclosure. During the rutting season, only one dominant male roared and other three males did not, though they started roaring every evening when the dominant male was moved to another enclosure. Advertisement of territorial status and dominance through vocalizations is possible to find in many antelope species: strophic laugh in pronghorn (*Antilocapra americana*), shrill and strophic blares in black wildebeest (*Connochaetes gnou*), and whistling of Uganda kob (*Kobus kob*) (Walther 1984).

In addition, roaring can play the role of a static-optical presentation and the most obvious way of sound marking the rutting male's territory. In order for the territorial male to be clearly visible, he periodically stands in an open area or even rises to a hill. He does this when he is alone in his territory making the circuit around his territory. Then the male stands almost motionless and examines his surrounding territory as if supporting the principle of "seeing and being seen." At the same time, he occasionally roars. The roar probably sends an additional signal to attract attention to the male in self-presentation and making himself recognizable to any potential opponents. Such behavior can be interpreted as a visual self-exposure of the male's presence, position, and territorial status, or another way of marking his territory. Even if the goitered gazelle male is among the females, his dominant territorial status can be determined by holding his head higher than the females and male bachelors. Other males probably determine his dominant territorial status by this behavior. Similar patterns are observed in other antelopes of open environments, such as hartebeest (*Alcelaphus buselaphus*) and topi (*Damaliscus lunatus*) that stand on termite mounds in self-presentation quite frequently. Wildebeest males (*Connochaetes taurinus*) will stay apart from conspecifics and self-present themselves or run around producing low-pitched calls as another way of self-exposure of his position and status (Walther 1984).

Since rutting vocalizations are a means for the male to present his dominant status and territory ownership, the roar sound should be loud enough to be not only directed at the closest sexual partners or expelled opponents, but also at all potential rivals from neighboring territories. Though most of the sounds that are produced by antelope species are quite soft, there are also those that can be heard at a great distance, especially sounds made during the rut. One of these examples is the goitered gazelle male, which has unique morphological adaptations for the production of very loud sounds (enlarged larynx and elongation of the vocal tract due to a retracted larynx). The male of the Grant's gazelle (*Nanger granti*), for example, rarely emits a loud roar with its mouth wide open when pursuing females. The Thomson's gazelle male, by contrast, quite often makes loud noises during courtship of females or during the pursuit of bachelors or rivals. Like most antelope species, Thomson's gazelles make these vibrating, repetitive sounds through their noses. Louder oral sounds are repeated many times by the blackbuck (*Antilope cervicapra*) during courtship, while he abruptly raises his head and sharply exhales a portion of air through his open mouth. This behavior is typical for any ungulate with a dominant status. Unexpectedly loud vocalizations for nasal sounds are observed in springbok (*Antidorcas marsupialis*). These males are among the noisiest of antelope species, since they almost constantly emit a rutting roar, clearly audible from a distance, while courting females. When courting females or expelling the bachelors or rivals, a male greatly intensifies his roaring (Walther et al. 1983).

5.5.4 Loud Rutting Vocalizations as an Adaptation to an Unstable Environment

Walther et al. (1983) reported that the vast majority of antelopes have soft rutting vocalizations which are not audible at a distance. Most medium- and large-sized antelopes are inhabitants of open environments, and in order not to attract too much attention to themselves from predators, they have very softvocalizations even during the breeding season. Why do only a few species use loud oral vocalizations for rutting activities? Goitered gazelles live in the really unstable environments of the deserts of Central and Southern Asia, where hot and dry summers replace snowy and very cold winters in the northern habitats or very hot and dry summers change to only a short rainy season in the southern habitats. In both cases, the favorable time of year for raising young in this gazelle is very short, occurring between cold, windy days in spring and the onset of summer heat, when all vegetation quickly burns out. Under these conditions, it is necessary that the majority of females have a high degree of breeding synchronization and give birth in a very short time window when forage is available and abundant enough to feed mothers and fawns.

Most goitered gazelle females give births within a 4- to 5-day time window in mid-May which usually coincides with the most favorable month of the near, when fresh, high-quality green vegetation is available to females in abundance. A plentiful diet of quality food allows mothers to provide their offspring of one, two or, rarely, even three fawns with a very intense milk diet in their first month of nursing. This, in turn, enables their young to grow very quickly in order to be ready on their own and successfully escape predators as soon as possible. In addition, offspring need to develop fast enough to switch to a mainly plant-based diet, because during the fawn's second month (after mid-June), the amount of milk available through nursing decreases sharply (Blank and Yang 2015a, b). Since the best time of year for foraging usually lasts only several weeks, the synchronization of most births needs to be extraordinarily high to achieve maximum offspring survival.

To achieve this high synchronization, it is necessary that mating also reaches the same high synchronization during rutting, since the gestation periods remain fairly constant. In order for most females to come into estrus at the same time, the rut itself must be very intense and most females come into estrus as quickly as possible. And indeed, the rut of goitered gazelles is very intense and most mating takes place within a very short time of a few days. In addition, one of the most important elements of the sexual behavior in ungulates are rutting vocalizations, which accelerates ovulation and synchronizes the sexual processes of both sexes faster (McComb 1987). The more intense the loud rutting vocalizations of males, the faster females become ready for fertilization (McElligott et al. 1999). Also, the increase in the volume and frequency of the roar and lowering the sound pitch increases the attractiveness of males to females and accelerates their readiness to mate (McComb 1991). From my perceptive, the emergence of a loud low-pitch roaring in goitered gazelles is not accidental; it is instead associated with the realization of high synchronization in mating and birthing that is necessary for successful reproduction and population survival in the unstable desert environments of Central and Southern Asia.

Another important adaptation of goitered gazelles to their environmental instability is the birth of twins by most females. Unfortunately, favorable conditions to support the birth of twins do not happen every year, and when severe conditions occur, most twins die in their first weeks from lack of milk. In favorable years, though, most twins survive and quickly replenish the population. This feature of giving birth to twins is further related to the challenges of nursing multiple fawns and hitting a very narrow window of opportunity for raising fawns in an optimal environment.

As for other gazelle species, most live in much more favorable conditions of subtropical and tropical savannas and southern steppes, where rainy periods are replaced by just drier ones, and abundant green vegetation is replaced by drier conditions. Therefore, specific breeding periods are more an exception and gazelle populations breed almost all year-round. There are usually two breeding peaks—one during the rainy season and the other 6 months later. Twins are practically the exception in African gazelles. The favorable seasons last for months, and male rutting periods last several months; some species have individual territories continuously during the entire year, with females giving birth twice a year or even three births in 2 years (Estes 1991). Under these conditions, male courtship of females is not as intense as in goitered gazelles and vocalizations are not as loud or frequent so to attract the attention of predators. Thomson's gazelles (*Eudorcas thomsonii*) in East Africa's savannas and grasslands, for example, have two birthing periods per year and males produce sputtering calls when chasing females. In Eastern Africa, Grant's gazelles (*Nanger granti*) breed throughout the year and males vocalize quite rarely when rutting and only with soft snorting (Blank 2021). And Acacia gazelle males (*Gazella gazella acacia*) in Israel have continuous territories and breed all year round with more than two peaks of birthing per year. Their rutting vocalizations are soft nasal calls.

Another example of a gazelle-like species with an enlarged, retractable larynx is the Mongolian gazelle (*Procapra gutturosa*—Frey et al. 2008a, b). Similar to goitered gazelles, Mongolian gazelles have strict seasonal breeding. Males establish individual territories for only 2–3 weeks in November and most males mate synchronously within several days in December. Females give births once a year within a few days of each other in mid-June, and twinning is found quite rarely (up to 8%). Young grow very quickly, and within 5–12 days after birth, mothers with their offspring start to slowly migrate to other localities. Mongolian gazelles inhabit grass-onion and grass-feather steppes and migrate to the north or south according to the snow cover. The lifestyle of the Mongolian gazelle is in many ways similar to the goitered gazelle: the rut is intense with constant loud, oral low-pitched roaring, and mating and birthing are observed to take place within a very short time of just a few days for most individuals. Mongolia's climate has frosty winters with little snow and warm, humid summers with monsoon rains. Thus, Mongolian gazelles are usually provided with fodder for the entire summer period (June through September). Like goitered gazelles, the Mongolian gazelles' loud roaring, vigorous rutting period, and high synchronization of parturition are associated with a severe winter climate and a quite short warm period that is favorable for the birthing and rearing of young, as well as the need for seasonal migrations related to the winter snow cover (Sokolov and Lushchekina 1997).

Another example of loud rutting vocalizations is in saiga antelope (*Saiga tatarica*) (Frey et al. 2007, 2008a, b). This species is found in almost the same natural environment as the goitered gazelles—the semi-deserts and deserts of Central Asia (Frey et al. 2007). There is a striking similarity in the lifestyles of saiga antelopes and goitered gazelles. In breeding, saiga also have intense rutting, which is accompanied by loud, low-formant vocalizations, although not oral like goitered gazelle males, but rather a loud, nasal roaring generated by a trunk-like nose and increasing nasal cavities—a special morphological adaptation (Frey et al. 2007). Similarly, saiga antelopes have a condensed time window for mating and an equally short 5–8 days for birthing, with most females producing twins. Saiga behaviors and their high synchronization of birthing and raising young are associated with an unstable climate and a brief period favorable for fawn survival, necessitating the need for constant migrations (Fadeev and Sludskiy 1982; Bekenov et al. 1998).

As for the ungulates living on the high-altitude plateaus of Tibet, such as the Tibetan antelope or chiru (*Pantholops hodgsonii*) and the Tibetan gazelle (*Procapra picticaudata*), they also live in conditions of constant negative temperatures for prolonged periods of time, while relatively warm daytime temperatures occur for only a very short time—quite literally just a few weeks. Both of these antelopes also demonstrate quite loud, intense vocalizations during the rut: deep-throated roars in chiru and grunting roars in Tibetan gazelles. Antelopes also have a very short summer birthing period of around 1 week when most females demonstrate a high rate of synchronized birthing. The Tibetan antelope does not have such a trunk-like nosed as the saiga, but their nasal cavities are greatly enlarged in the form of walnut-sized bulges of their nostrils (Schaller 1998). The Tibetan gazelle has an elongated nose, like the saiga, but smaller, and the males inflate them when making rutting vocalizations (Pohle 1974). Chiru produces both oral and nasal vocalizations (Schaller 1998), though the vocal behaviors of this species need detailed investigations.

From all these examples, it becomes clear that the primary reason for loud, low-formant vocalizations when rutting is related to the instability of environmental conditions that create a narrow time window for birthing and raising offspring. Male vocalizations, especially loud ones, have been found to stimulate and synchronize female estrus cycles for a short, intense mating time window, which, in turn, leads to a very high synchronization of birthing in spring. Morphological adaptations for extracting low-formant roaring in different species are quite diverse: increasing the length of vocal cords and larynx mobility, increasing the size of nasal cavities, and elongating the nose to be more trunk-like. But no matter the variation, the goal is the same—for rutting males to increase the volume and lowering the tone of their roars to make themselves the most attractive to females as possible.

As for the antelope species living in the sharply continental climate of temperate climate zones and high-elevated cold deserts, my explanations seem quite plausible. But there are species of antelopes with loud vocalizations in a much different zone of subtropical Africa, such as the springbok (*Antidorcas marsupialis*) and impala antelope (*Aepyceros melampus*). The springbok is an antelope that is taxonomically close to gazelles and lives in the arid southern part of the African continent and is common in the Kalahari Desert. Springbok can breed throughout the year or be a strictly seasonal breeder, depending on their habitats (Bigalke 1972; Smithers 1983). This antelope makes loud nasal sounds during rutting, and this species is one of the noisiest among the antelopes (Walther 1981). Despite the fact that the springbok lives in subtropical Africa, this species also demonstrates a high synchronization in mating and birthing young. Springboks usually give birth at the beginning of the rainy season, but this season may begin in different months or be completely absent in some years. Therefore, this species will reproduce in different months and may not breed at all in dry years. The unpredictability of the arid environment of southern Africa leads to the situation when springboks begin to breed at any appropriate moment, as soon as the favorable conditions arrive (Skinner and Louw 1996). The duration of the rainy period can also vary, so female springboks are at times forced to give birth with high synchronization with the arrival of the rain. How do springbok males conduct intensive courtship 6 months before females give birth, as if knowing when the rainy season will begin? That needs investigation (Tables 5.6, 5.7, and 5.8).

As for the impala antelope, it was discovered recently that this antelope also emits a loud roar during the rut (Frey et al. 2020). Different impala populations can breed all year round or have strictly seasonal breeding as in the more arid south of Namibia (Volodina et al. 2022). The vocalizations of impala males in a population with seasonal reproduction were studied, and males were found to have the ability to lengthen their vocal tract due to the movement of the mobile larynx downward, emitting a low-pitched roar of high volume, similar to the roars of lions (Frey et al.

Table 5.6 Impact of different behavioral patterns, which triggered object-horning reaction in goitered gazelle (number of observed cases—upper and proportions lower—by Blank and Yang 2014—License Number—5476291061876)

Patterns-triggers	April	May	June–August	November	December	Log-linear model test
Female chasing	5 7.9%	2 5.3%	2 10.0%	20 11.3%	17 16.0%	Z = 2.761 P = 0.006
Marking territory	1 1.6%	5 13.2%	1 5.0%	28 15.8%	14 13.2%	Z = 2.466 P = 0.014
Grazing	33 52.4%	26 68.4%	9 45.0%	35 19.8%	10 9.4%	Z = 1.953 P = 0.051
Male chasing	9 14.3%	1 2.6%	1 5.0%	15 8.5%	11 10.4%	Z = 2.093 P = 0.036
Territorial conflict	6 9.5%	–	–	68 38.4%	48 45.3%	Z = 4.421 P = 0.0005
Threat to another male	9 14.3%	4 10.5%	7 35.0%	11 6.2%	6 5.7%	Z = 1.200 P = 0.230

Table 5.7 The marking frequency in territorial males during the rut (number marks per 1 h of observations) (numerator is a range; average (M ± SE) is in the denominator) (adapted from Marmazinskaya 1996)

	Marking type		
Males type	Urine/feces	Preorbital gland marks	Object-horning
Male (3.5 years age)	0–2.2	0–10.9	0–2.2
	1.4 ± 0.07	2.8 ± 0.2	0.6 ± 0.06
Male (1.5 years old)	0.3–1.6	0–6.4	0–1.9
	0.8 ± 0.07	1.6 ± 0.2	0.6 ± 0.05
Male from semi-captive population (3.5 years old)	0.9–2.7	0–0.4	0.3–2.6
	1.4 ± 0.03	0.1 ± 0.01	1.2 ± 0.04

Table 5.8 Description of the recorded courting displays and roaring (ethogram) in adult males goitered gazelle during the rut according to Walther et al. (1983), Blank (1998), and Blank et al. (2014c—License Number—5476360815518)

Behavioral displays	Description
Courting postures	
Low-stretch or neck-stretch	Stretching the head and neck forward horizontally with the nose pointing horizontally forward. Auricles are directed backward and usually pressed to the head in a position either parallel to the ground or slightly downward with the aperture inward. Rarely auricles lay back, pressed to each other, or held one apart of the other at a 45° angle. Tail is lifted in different degrees, up to vertical position (Fig. 4.1)
High-stretch or nose-lift or head-up	Head and neck are stretched forward and upward up to 45° above the horizontal level, the top of the muzzle lifted forward and upward, and auricles pressed to the head (Fig. 4.2)
Roaring	Roaring consists of gruff, wheezing, or hissing sounds resembling gurgling or sneezing. In producing these sounds, the male opens his mouth widely, lifts his tongue slightly, and takes a deep breath and then passes air through his partly closed mouth and possibly as well as his nostrils. During a long and fast gallop, the male's mouth is open and his tongue out. Posture: Muzzle is raised to an angle of up to 45° from a horizontal position. Auricles are directed backward and tightly pressed, more or less, to the head by apertures inside. Abruptly, the male jerks his head and whole body forward and upward, and his protruding thyroid cartilage moves down and up. During the run, the roaring is intermittent (Fig. 5.19; Fig. 5.20)
Aggressive postures	
Threat	Horns are directed forward. Ears are directed forward-sideward; tail is normal position (Figs. 6.2 and 6.3)
Horn-pressing	The rivals approach each other with lowered heads to the ground without preliminary displays of threat and lean the frontal surfaces of their horns against each other. They fight while standing on all four legs with the forelegs widely spread, and hind legs in normal position or relatively close together. Each of them tries to press the horns of the opponent to the ground and move him back. Then everyone hits the rival's head with maximum power and presses the head to the ground and back again, marking jerks. The horns almost invariably interlock (horn of one between horns of another). This fight can transfer to fight-circling, when rivals do circles with locked horns. In this situation, they pivot with their hindquarters while keeping their heads in the circle center. During fighting, the ears are directed sidewise by top horizontally or backward sometimes and tail is pressed. This fight is common for non-territorial males (Fig. 6.2)

2020). The males also conduct their courtship very vigorously and in the relatively short period of time of 2–3 weeks. In Botswana, impala birthing is highly synchronized, with 90% of females giving birth within a 2-week time period at the beginning of the rainy season, when food quality is highest (Moe et al. 2007). As for impala populations in more favorable conditions with an even climate, impala also synchronize their birthing, though with a lower proportion of birthing females (78%). Some authors have explained this difference as an antipredator strategy and reducing predation risk (Moe et al. 2007). The simultaneous birth of most of the

Fig. 5.19 Roaring male in goitered gazelles: A male stands on the elevated points, demonstrating its ownership for the territory. Drawing by U. Kenzhetaev with kind permission

Fig. 5.20 Male roaring is observed often in goitered gazelles during courting of females. Issyk-Kul goitered gazelle nursery for reacclimatizing, 24.11.2022. Photo by D. Blank

young includes the dilution effect (safety in numbers), when even when attacked by predators, some of the young remain alive in any case (Delm 1990).

Some cervid species during rutting have the ability to lengthen their vocal tracts by lowering the movable larynx and expressing a low-pitched roar, while others lack this ability (Frey et al. 2021). However, I think that my explanation about the connection of a loud roar with the synchronization of birthing in females due to habitat conditions may be applied to cervid species, as well. Intense roaring is seen more often in the northern species with limited favorable time for breeding and less often in deer species living in an even climate of the subtropics and tropics. I will not discuss this issue in detail here, however, since deer are less related to gazelles and the interspecific and intraspecific taxonomy of cervids is still insufficiently studied.

Thus, from all these examples, an obvious conclusion arises: the morphological and anatomical adaptations of the goitered gazelles (enlargement of the larynx and its retraction) to produce a loud roar of low frequency when rutting is primarily associated with their adaptation to a sharply continental climate and a short time window favorable for rearing offspring. To support the survival of the population, goitered gazelles have developed the ability to maintain a high level of synchronization in mating and the subsequent birthing of young.

References

Adams CA, Bowyer RT, Rowell JE, Hauer WE, Jenks JA (2001) Scent marking by male caribou: an experimental test of rubbing behaviour. Rangifer 21(1):21–27

Alberts AC (1992) Constraints on the design of chemical communication systems in terrestrial vertebrates. Am Nat 139:62–89

Albone ES (1984) Mammalian semiochemistry: the investigation of chemical signals between animals. The John Wiley and Sons Press, New York

Andersson M (1994) Sexual selection. Princeton University Press, Princeton

Antipin VM (1941) Mammals of Kazakhstan, Ungulates, vol 3. Publication OGIZ, Alma-Ata, pp 47–54. (in Russian)

Askerov E, Mallon DP, Zazanashvili N, Kochiashvili V, Sarukhanova S, Muradov A (2021) An attempt to restore the population of the Goitered Gazelle (*Gazella subgutturosa*) in its indigenous range in the Caucasus (Mammalia: Bovidae). Zool Middle East 67(3):189–197

Attum O, Mahmoud T (2012) Dorcas gazelle and livestock use of trees according to size in a hyper-arid landscape. J Arid Environ 76:49–53

Bärmann EV, Klappert AN, Chen A (2021) *Litocranius walleri* (Artiodactyla: Bovidae). Mamm Species 53(1005):65–77

Bartoš L (1983) Some observations on the relationship between preorbital gland opening and social interactions in red deer. Aggress Behav 9:59–67

Bartoš L, Víchová J, Lancingerová J (2005) Pre-orbital gland opening in red deer (*Cervus elaphus*) calves: signals of hunger? J Anim Sci 83:124–129

Bartošová J, Ceacero F, Bartoš L (2012) Pre-orbital gland opening: part of sucking behavior in red deer (*Cervus elaphus*) calves. J Anim Sci 90(9):3207–3212

Bekenov AB, Grachev YA, Milner-Gulland EJ (1998) The ecology and management of the Saiga antelope in Kazakhstan. Mamm Rev 28(1):1–52

Benner JM, Bowyer RT (1988) Selection of trees for rubs by white-tailed deer in Maine. J Mamm 69(3):624–627

Benz H (1973) Zum Sozialverhalten des Sasin (Hirschziegenantilope, *Antilope cervicapra* L. 1758). Zool Beitr 19:403–466

Bigalke RC (1972) Observations on the behavior and feeding habits of the springbok, *Antidorcas marsupialis*. Zool Afr 7:333–359

Blank DA (1985) Peculiarities of social and reproductive behaviour of *Gazella subgutturosa* in the Ili Valley. Zool Zhurnal 64:1059–1070. (in Russian, English abstract)

Blank DA (1990) Persian gazelle. In: Kovshar AF (ed) Rare animals of desert regions. Nauka Press of Kazakh SSR, Alma-Ata, pp 56–80. (in Russian)

Blank DA (1997) Social and reproductive behavior in the Persian gazelle. Ph.D. dissert., The University of Tel-Aviv, p 189

Blank DA (1998) Mating behavior of the Persian gazelle *Gazella subgutturosa* Güldenstaedt, 1780. Mammalia 62(4):409–419

Blank DA (2021) Artiodactyl vocalization. In: Rosenfeld CS (ed) Neuroendocrine regulation of animal vocalization. Elsevier, pp 159–188

Blank DA (2022) Behavior of goitered gazelles in the Issyk-Kul breeding center. Unpublished report. Kyrgyzstan

Blank DA, Yang W (2014) Object-horning in goitered gazelle: agonistic or marking behaviour? Behav Proc 103:165–172

Blank D, Yang W (2015a) Strategy of goitered gazelle suckling behavior and its adaption to the environment. Mamm. Res., 60(4):393–401

Blank DA, Yang W (2015b) Suckling behavior in goitered gazelle: do females invest more in twins or singletons? Zoology, 118:348–356

Blank DA, Ruckstuhl KE, Yang W (2012) Social organization in goitered gazelle (*Gazella subgutturosa* Güldenstaedt, 1780). Ethol Ecol Evol 24:306–321

Blank DA, Ruckstuhl K, Yang W (2014a) Secretion marking with preorbital glands in goitered gazelle, *Gazella subgutturosa* (Artiodactyla: Boidae). Folia Zool 63(2):127–136

Blank DA, Ruckstuhl KE, Yang W (2014b) The economics of scent marking with urine and feces in goitered gazelle (*Gazella subgutturosa*). Mamm Res 60(1):51–60

Blank DA, Ruckstuhl K, Yang W (2014c) Roaring function in male goitered gazelles. Behav Proc 106:152–159

Bowyer RT, Van Ballenberghe V, Rock KR (1994) Scent marking by Alaskan moose: characteristics and spatial distribution of rubbed trees. Can J Zool 72:2186–2192

Brashares JS, Arcese P (1999a) Scent marking in a territorial African antelope: I. The maintenance of borders between male oribi. Anim Behav 7:1–10

Brashares JS, Arcese P (1999b) Scent marking in a territorial African antelope: II. The economics of marking with faeces. Anim Behav 57:11–17

Brennan PA, Kendrick KM (2006) Mammalian social odours: attraction and individual recognition. Philos Trans R Soc B Biol Sci 361(1476):2061–2078

Briefer E, Vannoni E, McElligott AG (2010) Quality prevails over identity in the sexually selected vocalizations of an ageing mammal. BMC Biol 8:1–15

Butzler W (1974) Kampf und Paarungsverhalten, sozial Rangordung and Aktivitatsperiodik beim Rothirsch, Beiheft. Z Tierpsychol 16. Paul Parey Verlag, Hamburg, Berlin

Cap H, Deleporte P, Joachim J, Reby D (2008) Male vocal behavior and phylogeny in deer. Cladistics 24(6):917–931

Carranza J, Mateos-Quesada P (2001) Habitat modification when scent marking: shrub clearance by roe deer bucks. Oecologia 126:231–238

Ceacero F, Landete-Castillejos T, Bartošová J, García AJ, Bartoš L, Komárková M, Gallego L (2014) Habituating to handling: factors affecting preorbital gland opening in red deer calves. J Anim Sci 92(9):4130–4136

Charlton BD, Reby D, McComb K (2007) Female red deer prefer the roars of larger males. Biol Lett 3:382–385

Clutton-Brock TH, Albon SD (1979) The roaring of red deer and the evolution of honest advertisement. Behaviour 69(3–4):145–170

Coblentz BE (1976) Functions of scent-urination in ungulates with special reference to feral goats (*Capra hircus* L.). Am Nat 110(974):549–557

Čupić S, Cassinello J, Kušta T, Ceacero F (2023) Differences in faecal nutrition components in three species of Saharan gazelles on standard diets in relation to species, age and sex. MDPI Anim 13:3408

Davies NB, Krebs JR, West SA (2012) An introduction to behavioral ecology, 4th edn. Wiley-Blackwell, Oxford

Delm MM (1990) Vigilance for predators: detection and dilution effects. Behav Ecol Sociobiol 26:337–342

Dubost G, Feer F (1981) The behaviour of the male Antilope cervicapra L., its development according to age and social rank. Behaviour 76:62–127

Dunbar RIM, Dunbar EP (1974) Social organization and ecology of the klipspringer (*Oreotragus oreotragus*) in Ethiopia. Z Tierpsychol 35:481–493

Durmuş M (2010) Determination of home range size and habitat selection of gazelles (*Gazella subgutturosa*) by GPS Telemetry in Şanliurfa. Master's thesis, Middle East Technical University, Turkey

Efremova KO, Volodin IA, Volodina EV, Frey R, Lapshina EN, Soldatova NV (2011) Developmental changes of nasal and oral calls in the goitred gazelle Gazella subgutturosa, a nonhuman mammal with a sexually dimorphic and descended larynx. Naturwissenschaften 98:919–931

Eilam D, Zor R, Szechtman H, Hermesh H (2006) Rituals, stereotypy and compulsive behavior in animals and humans. Neurosci Biobehav Rev 30:456–471

Esmaeili S, Hemami MR (2013) Utilization of harvester ant nest sites by Persian goitered gazelle in steppes of Central Iran. Basic Appl Ecol 14(8):702–711

Estes RD (1972) The role of the vomeronasal organ in mammalian reproduction. Mammalia 36:315–341

Estes RD (1991) The behavior guide to African mammals. University of California Press, Berkeley-Los Angeles

Fadeev VA, Sludskiy AA (1982) The Saiga in Kazakhstan. Kazakh Academy of Sciences Press, Alma-Ata. (in Russian)

Fitch WT, Hauser MD (2003) Unpacking "honesty": vertebrate vocal production and the evolution of acoustic signals. In: Acoustic communication. Springer, New York, pp 65–137

Fitch WT, Reby D (2001) The descended larynx is not uniquely human. Proc R Soc Lond B 268:1669–1675

Frädrich H (1964) Beobachtungen zur Krezung zwischen Schwarzrucken-ducker, *Cephalopus dorsalis*, Gray 1846, und Zebraducker, *Cephalophus zebra* (Gray 1838). Z Saugetierk 29:46–51

Frey R, Hofmann RR (1996) Evolutionary morphology of the proboscideal nose of Guenther's dik-dik (*Rhynchotragus guentheri*—Thomas, 1894) (Mammalia, Bovidae). Zool Anz 235:31–51

Frey R, Volodin I, Volodina E (2007) A nose that roars: anatomical specialization sand behavioural features of rutting male saiga. J Anat 211:717–736

Frey R, Gebler A, Olson KA, Odonkhuu D, Fritsch G, Batsaikhan N, Stuermer IW (2008a) Mobile larynx in Mongolian gazelle: retraction of the larynx during rutting barks in male Mongolian gazelle (*Procapra gutturosa* Pallas, 1777). J Morphol 269:1223–1237

Frey R, Gebler A, Olson KA, Odonkhuu D, Fritsch G, Batsaikhan N, Stuermer IW (2008b) Head anatomy of male and female Mongolian gazelle—striking example of sexual dimorphism. In: Endo H, Frey R (eds) Anatomical imaging- towards a new morphology. Springer, Tokyo, pp 1–13

Frey R, Volodin I, Volodina E, Soldatova NV, Juldaschev ET (2011) Descended and mobile larynx, vocal tract elongation and rutting roars in male goitered gazelles (*Gazella subgutturosa* Guldenstaedt, 1780). J Anat 218(5):566–585

Frey R, Volodin IA, Volodina EV, Efremova KO, Menges V, Portas R, Melzheimer J, Fritsch G, Gerlach C, von Dörnberg K (2020) Savannah roars: the vocal anatomy and the impressive rutting calls of male impala (*Aepyceros melampus*)–highlighting the acoustic correlates of a mobile larynx. J Anat 236(3):398–424

Frey R, Wyman MT, Johnston M, Schofield M, Locatelli Y, Reby D (2021) Roars, groans and moans: anatomical correlates of vocal diversity in polygynous deer. J Anat 239(6):1336–1369
Gebler A, Frey R (2005) Anatomical structures involved in non-human vocalization. ZAS Pap Linguistics 40:33–43
Geist V (1965) On the rutting behaviour of the mountain goat. J Mamm 45:551–568
Gilbert BK (1973) Scent marking and territoriality in pronghorn (*Antilocapra americana*) in Yellowstone National Park. Mammalia 37:25–33
Gillam E (2012) An introduction to animal communication. Nat Educ Knowl 3(10):70
Giotto N, Gerard JF (2010) The social and spatial organization of the Beira antelope (*Dorcatragus megalotis*): a relic from the past? Eur J Wildl Res 56:481–491
Gorbunov AV (1983) About ecology and behavior of goitered gazelles in the Krasnovodsk Peninsula. Proc Turkmen Acad Sci Ser Biol Sci 3:66–68. (in Russian)
Gorelov YK (1972) Breeding of goitered gazelle (*Gazella subgutturosa* Güldenstaedt) and problem of restoring of its number in the Badkhyz (SE Turkmenistan). In: Theriology, vol 1. Nauka of Siberian Department of Academy of Sciences of USSR, Novosibirsk, pp 420–424. (in Russian)
Gosling LM (1972) The construction of antorbital gland marking sites by male oribi (*Ourebia ourebia*, Zimmerman, 1783). Z Tierpsychol 30:271–276
Gosling LM (1974) The social behaviour of Coke's hartebeest (*Alcelaphus busephalus cokei*). In: Geist V, Walther FR (eds) The behavior of ungulates and its relation to management, vol no. 24. IUCN Publ. New series, Morges, pp 488–511
Gosling LM (1980) Defense gilds of savannah ungulates as a context for scent communication. Symp Zool Soc Lond 45:195–212
Gosling LM (1981) Demarcation in a gerenuk territory: an economic approach. Z Tierpsychol 56:305–322
Gosling LM (1982) A reassessment of the function of scent marking in territories. Z Tierpsychol 60:89–118
Gosling LM (1985) The even-toed ungulates: order Artiodactyla. In: Brown RE, MacDonald D (eds) Social odours in mammals. Oxford University Press, Oxford, pp 550–618
Gosling LM, Roberts SC (2001) Testing ideas about the function of scent marks in territories from spatial patterns. Anim Behav 62(3):F7–F10
Grau GA (1974) Behavior of mountain gazelle in Israel. Ph.D. dissertation, Texas A&M University, College Station
Grau GA (1976) Olfaction and reproduction in ungulates. In: Mammalian olfaction, reproductive processes, and behavior, vol 219, p 242
Gray DR, Flood PF, Rowell JF (1989) The structure and function of muskox preobital glands. Can J Zool 67:1134–1142
Habibi K, Thouless CR, Lindsay N (1993) Comparative behavior of sand and mountain gazelles. J Zool 229:41–53
Hatlapa HH (1977) Zur biologischen Bedeutung des präorbital Organs beim Rotwild, Pragung, Individualgeruch, Orientierung. Tierartzl Wochenschr Berl Munch 90:100–104
Hendrichs H (1975) Changes in a population of dikdik, *Madoqua kirkii*. Z Tierpsychol 38:55–69
Hurst JL, Rich TJ (1999) Scent marks as competitive signals of mate quality. In: Johnston RE, Muller-Schwarze D, Sorensen P (eds) Advances in chemical communication in vertebrates. Plenum, New York, pp 209–226
Johansson A, Liberg O (1996) Functional aspects of marking behaviour by male roe deer (*Capreolus capreolus*). J Mamm 77:558–567
Johnson RP (1973) Scent marking in mammals. Anim Behav 21:521–535
Jungius H, Claussen CP (1975) The horn bases of the reedbuck *Redunca arundinum*. Koedoe 18:131–137
Kiley-Worthington M (1978) The causation, evolution and function of the visual displays of the eland (*Taurotragus oryx*). Behaviour 66(3/4):179–222
Kingswood SC, Blank DA (1996) Gazella subgutturosa. Mamm Spec 518:1–10
Kranz KR (1991) Monogamy in the dikdik. Appl Anim Behav Sci 29:87–105

Kruger M, du Bothma P, Kruger JM (2002) The effect of neighbouring klipspringer on the scent-marking behaviour of a group of klipspringer in the Kruger National Park. Koedae 45(1):87–92

Lawson R, Putnam R, Fielding A (2000) Individual signatures in scent gland secretions of Eurasian deer. J Zool 251(3):399–410

Leuthold W (1977) African ungulates. A comparative review of their ethology and behavioral ecology. Springer, Berlin

Lévy F, Keller M, Poindron P (2004) Olfactory regulation of maternal behavior in mammals. Horm Behav 46(3):284–302

Mambetzhumaev AM (1970) Goitered gazelle. FAN Publication House, Tashkent, p 196. (in Russian)

Marchinton RL, Hirth DH (1984) Behavior. In: Howells LK (ed) White-tailed deer: ecology and management. Stackpole Books, Harrisburg Pennsylvania, pp 129–168

Mardonov BK (1993) Impact of goitered gazelle on the vegetative cover of desert. PhD thesis, Moscow, p 26 (in Russian)

Marmasinskaya NV (2008) Mother-offspring interactions in the goitred gazelle *Gazella subgutturosa* Guldenstaedt, 1780. In: Azimov DA, Vashetko EV (eds) Proceeding in the natural reserves of Uzbekistan. State Committee of Nature of Uzbekistan, Tashkent, pp 84–102. (in Russian)

Marmazinskaya NV (1990) Social behaviour of goitered gazelles during the reproductive period. In: V congress of all-union theriological society, abstracts, vol 3, pp 32–33. (in Russian)

Marmazinskaya NV (1996) Territorial and marking behavior of goitered gazelle *Gazella subgutturosa* (Artiodactyla, Bovidae) in the Bukhara ecological center. Zool Zhurnal 75(11):1737–1751. (in Russian, English summary)

Marmazinskaya NV (1999) The social behavior of goitered gazelles (Gazella subgutturosa Guld 1780) during breeding period in the conditions of the Dzheiran Ecocenter. PhD thesis, VNIIPriroda, Tashkent

Marmazinskaya NV, Mardonov BK (1996) Density and distribution characteristics of different mark types in goitered gazelle. In: Proceedings of nature reserves of Uzbekistan issue 1. Fan Publication House, Tashkent, pp 62–68. (in Russian)

Massei G, Bowyer RT (1999) Scent marking in fallow deer: effects of lekking behavior on rubbing and wallowing. J Mamm 80(2):633–638

McComb K (1987) Roaring by red deer stags advances the date of oestrus in hinds. Nature 330:648–649

McComb KE (1991) Female choice for high roaring rate in red deer, *Cervus elaphus*. Anim Behav 41:79–88

McElligott AG, O'Nell KP, Hayden TJ (1999) Cumulative long-term investment in vocalization of fallow bucks, *Dama dama*. Anim Behav 57:1159–1167

McElligott AG, Birrer M, Vannoni E (2006) Retraction of the mobile descended larynx during groaning enables fallow bucks (*Dama dama*) to lower their formants frequencies. J Zool 270:340–345

Mechkour F, Maublanc ML, Bideau E, Gerard JF, Pépin D (2008) Spatial organization and spatial distribution of activities within home ranges in a springbok (*Antidorcas marsupialis*) captive population. Zoo Biol 27(1):19–35

Moe SR, Rutina LP, Du Toit JT (2007) Trade-off between resource seasonality and predation risk explains reproductive chronology in impala. J Zool 273(3):237–243

Müller-Schwarze D (1999) Signals specialization and evolution in mammals. In: Johnston RE et al (eds) Advances in chemical signals in vertebrates. Springer Science+Business Media, LLC, New York, pp 1–14

Mykytowycz R (1972) The behavioural role of the mammalian skin glands. Naturwissenschaften 59:133–139

Novellie PA, Manson J, Bigalke RC (1984) Behavioural ecology and communication in the cape grysbok. South Afr J Zool 19:22–30

Oehler MW, Jenks JA, Bowyer RT (1995) Antler rubs by white-tailed deer: the importance of trees in a prairie environment. Can J Zool 73:1383–1386

Owens B (2014) Male scent stimulates female goats' fertility. Nature 10:1038. https://doi.org/10.1038/nature.2014.14788
Pavlov VN, Vanisova EA, Nikolskii AA (2013) The vegetation on the signaling points of goitered gazelle *Gazella subgutturosa*. Bull Moscow Soc Natural 118(3):44–48. (in Russian, English abstract)
Pereladova OB (2022) The acoustic and optical signaling field as an indicator of the state of populations of rare species (on the example of some ungulate species of arid ecosystems). In: Biological signaling field of mammals, p 150
Pohle C (1974) Haltung und Zucht der saiga-Antelope (Saiga tatarica) im Tierpark Berlin. Zool Garten 61:263–266
Rajagopal T, Archunan G (2011) Histomorphology of preorbital gland in territorial and non-territorial male blackbuck Antelope cervicapra, a critically endangered species. Biologia 66:370–378
Rajagopal T, Manimozhi A, Archunan G (2011) Diurnal variation in preorbital gland scent marking behaviour of captive male Indian Blackbuck (*Antelope cervicapra* L.) and its territorial significance. Biol Rhythm Res 42(1):27–38
Ralls K (1969) Scent-marking in Maxwell's duiker, Cephalophus maxwelli. Am Zool 9:1071
Ralls K (1971) Mammalian scent marking. Science 171:443–449
Ralls K, Barash C, Minkowski K (1975) Behavior of captive mouse deer, *Tragulus napu*. Z Tierpsychol 37:356–378
Reby D, McComb K (2003) Anatomical constraints generate honesty: acoustic cues to age and weight in the roars of red deer stags. Anim Behav 65:519–530
Reby D, McComb K, Cargnelutti B, Darwin CJ, Fitch WT, Clutton-Brock TH (2005) Red deer stags use formants as assessment cues during intrasexual agonistic interactions. Proc R Soc Lond B 272:941–947
Roberts SC (1997) Selection of scent-marking sites by klipspringers (*Oreotragus oreotragus*). J Zool (Lond) 243:555–564
Roberts SC (1998) Behavioural responses to scent marks of increasing age in klipspringer *Oreotragus oreotragus*. Ethology 104:585–592
Roberts SC, Dunbar RIM (2000) Female territoriality and the function of scent-marking in a monogamous antelope (*Oreotragus oreotragus*). Behav Ecol Soc 47:417–423
Roberts SC, Lowen C (1997) Optimal patterns of scent marks in klipspringer (*Oreotragus oreotragus*) territories. J Zool 243:565–578
Roberts SC, Gosling LM et al (2001) The economic consequences of advertising scent mark location on territories. In: Marchkewska-Koj N (ed) Chemical signals in vertebrates 9. Kluwer Academic/Plenum, New York
Schaller GB (1998) Wildlife of the Tibetan steppe. University of Chicago Press, Chicago, IL
Skinner JD, Louw GN (1996) The Springbok (*Antidorcas marsupialis*—Zimmermann, 1780). Transvaal Museum, Pretoria
Sludskiy AA (1956) Breeding of goitered gazelle. In: Proceedings of Institute of Zoology, Kazakh Academy of Sciences, vol 6. Nauka Press of Kazakh SSR, Alma-Ata, pp 78–108. (in Russian)
Smithers RHN (1983) The mammals of the Southern African subregion. University of Pretoria, Pretoria
Sokolov VE, Chernova OF (2001) The skin glands of mammals. GEOS Press, Moscow
Sokolov VE, Lushchekina AA (1997) Procapra gutturosa. Mamm Species 571:1–5
Sokolov VE, Gromov VS, Chernova OF (1986) Skin glands of preorbital area of the roe deer: morphology and functional significance. In: Sokolov VE (ed) Chemical communication in animals. Theory and practice. Nauka, Moscow, pp 220–230. (in Russian)
Somers M, Rasa OAE, Apps PJ (1990) Marking behaviour and dominance in suni antelope (*Neotragus moschatus*). Z Säugetierkd 55:340–352
Soultan A, Nagy A, Attum O (2021) Midden site selection in Dorcas gazelle: larger is not always better. Ecol Evol 11(20):13661–13667
Spinage CA (1969) Territoriality and social organization of the Uganda defassa waterbuck *Kobus defassa ugandae*. J Zool 159:329–361

Sun L, Xiao B, Dai N (1994) Scent marking behaviour in the male Chinese water deer. Acta Theriol 39:177–184

Vannoni E, McElligott AG (2008) Low frequency groans indicate larger and more dominant fallow deer (Dama dama) males. PloS one 3(9), p.e3113

Vannoni E, Torriani MVG, McElligott AG (2005) Acoustic signaling in cervids: a methodological approach for measuring vocal communication in fallow deer. Cogn Brain Behav 9(3):551–565

Volodin IA, Volodina EV, Efremova KO (2009) Antelope, calling through the nose: structure of sounds and effect of sexual selection on the vocal behavior of the saiga (*Saiga tatarica*). Zool Zhurnal 88(1):113–124. (in Russian, English abstract)

Volodin IA, Lapshina EN, Volodina EV, Frey R, Soldatova NV (2011) Nasal and oral calls in juvenile goitred gazelles (*Gazella subgutturosa*) and their potential to encode sex and identity. Ethology 117(4):294–308

Volodina EV, Volodin IA, Frey R (2022) Male impala (Aepyceros melampus) vocal activity throughout the rutting period in Namibia: daily and hourly patterns. Afr J Ecol 60(1):95–99

Walther FR (1963) Einige Kampfvehalten des Gerenuk (*Ammodorcas clarkei* Thomas, 1891). D Zool Garten 27:233–261

Walther FR (1972) Social grouping in Grant's gazelle (*Gazella granti* Brooke 1827) in the Serengeti National Park. Z Tierpsychol 31:348–403

Walther FR (1978a) Quantitative and functional variations of certain behaviour patterns in male Thomson's gazelle of different social status. Behaviour 65(3/4):212–240

Walther FR (1978b) Mapping the structure and the marking system of a territory of the Thomson's gazelle. East Afr Wild J 16:167–176

Walther FR (1978c) Forms of aggression in Thomson's gazelle; their situational motivation and their relative frequency in different sex, age, and social classes. Z Tierpsychol 47(2):113–172

Walther FR (1981) Remarks on behavior of springbok, Antidorcas marsupialis Zimermann 1790. D Zool Garten 51:81–103

Walther FR (1984) Communication and expression in hoofed animals. Indiana University Press, Bloomington

Walther FR, Mungall EC, Grau GA (1983) Gazelles and their relatives. A study in territorial behavior. Noyes Publication, Park Ridge, 239 pp.

Williamson JE (1985) Aspects of the behavioural ecology of springbok (*Antidorcas marsupialis*, Zimmermann, 1780) in the Central Kalahari Game Reserve, Botswana. PhD Thesis, University of Witwatersrand, pp 1–171

Wronski T, Apio A, Baranga J, Plath M (2006) Scent marking and territorial defense in male bushbuck (*Tragelaphus scriptus*). J Zool 270:49–56

Wronski T, Kabasa JD, Plath M, Apio A (2008) Object-horning as advertising and marking behaviour in male bushbuck (*Tragelaphus scriptus*)? J Ethol 26:165–173

Wronski T, Apio A, Plath M, Ziege M (2013) Sex-difference in the communicatory significance of localized defecation sites in Arabian gazelles (*Gazella arabica*). J Ethol 31(2):129–140

Wyatt TD (2003) Pheromones and animal behaviour: communication by smell and taste. Cambridge University Press, Cambridge

Zaitsev VA (2006) The musk deer: ecology, population changes, and conservation prospects. Publicatin of Biodiversity Conservation Center, Moscow. (in Russian)

Zhevnerov VV (1984) Goitered gazelle of the Barsakelmes Island. Nauka Publishing House of Kazakh SSR, Alma-Ata. (in Russian)

Zhevnerov VV, Bekenov AB, Sludskiy AA (1983) Goitered gazelle—*Gazella subgutturosa* Güldenstaedt, 1780. In: Gvozdev EV, Kapitonov VI (eds) Mammals of Kazakhstan, Bovidae, vol 3(3). Nauka Publishing House of Kazakh SSR, Alma-Ata, pp 11–54. (in Russian)

Agonistic Behavior of Adult and Subadult Males

6

6.1 General Remarks

Most ungulates live in specific social relationships with each other and gather into groups of different sizes. Social behavior is very closely related to agonistic behavior, and any social community is based on aggressive interactions of its members in order to constantly maintain its hierarchical system and coordinate the herd's activities and movements (Walther 1974). Social processes in a herd are based on signals exchanged between individuals. In antelopes, agonistic contacts are observed in all age and sex groups (Walther 1974). Agonistic behavior can be divided into two main groups: aggressive behavior during physical contacts and threat displays at a distance without direct contact. Distant threat displays arise from contradictory tendencies, intention to fight or running away (Eibl-Eibesfeldt 1961).

Antlers and horns play a distinctive role in the aggressive and fighting behaviors of bovids and cervids. Hornlike formations appeared several times in the evolution of different taxonomical groups of mammals, and they developed on different parts of the head and from different tissues. Horns were evolving, appearing, and disappearing repeatedly in systematically close groups, especially in ruminants and rhinoceroses. Horns, however, always have an adaptive value, evolving with a certain shape, size, and location that brought an evident advantage to the horns' owners. According to the different sizes and shapes of the hornlike organs in different species, their fighting techniques vary (Walther 1974). The main function of horns is to guide a rival's head when butting and bind their horns during a tournament. The function of spirals, bumps, and ridges on the horns of different antelopes is to hold the horns of the two opponents together during the fight, allowing each to catch

Blank and Yang (2014)—License Number—5477510619124.
Blank et al. (2015)—License Number—5477510968766.

D. Blank, *The Behavioral Biology and Conservation of Goitered Gazelles (Gazella subgutturosa)*, https://doi.org/10.1007/978-3-031-83504-9_6

most of the attack with their horns and use their full strength when head butting, while at the same time being relatively harmless to the combatants. Sometimes in the midst of a vicious fight, however, opponents begin to strike almost any part of the body, which indicates that the body remains a principled target during truly serious fights. Horns have appeared as a protective weapon with additional functions as a guarding and binding instrument or as a means of aggressive display. The loss of one or both horns can lead to the disappearance of a male's dominant status (Walther 1978; Geist 1966).

Antilopinae neither rise on their hindlegs when fighting nor do they drop down on their "knees" (carpal joint) like wildebeest (*Connochaetes gnou*), topi (*Damaliscus lunatus topi*), and kongoni (*Alcelaphus buselaphus*) usually do. Instead, they fight while standing on all four legs with the forelegs widely spread and their hind legs in a normal position or relatively close together. Horizontal forward and backward leaps with all four legs are more common in some the species than in others. In fights of great intensity, movements of the torso are also observed, mainly humping the back, stretching the torso, deep ducking the chest region, and a vehement throwing of the whole body to the side while horns are firmly interlocked. The leg and torso movements are used to support the actions of the head and neck which are the most important fighting movements for gazelles (Walther et al. 1983).

It is possible to distinguish several degrees of intensity in fights. Low intensity fights are horning type and carried out by head movements, sometimes involving the neck. These fights hardly ever occur in territorial males. Fights of medium intensity involve considerable movement of the head and neck, with added leaps. In high intensity fights, forceful aggressive movements of the head and the neck, intensive "legwork," and often a strong involvement of the torso in the actions are common. Clash-fighting, push-fighting, twist-fighting, and front-pressing are typical for high intensity fights. The fighters move forward and backward according to their offensive and defensive maneuvers, but they usually do not move each other over long distances, because both opponents try to remain in place. Territorial males prefer fights of high intensity. In fights of highest intensity, attackers one after another in turn can displace the opponent over distances of 10–20 m by push-fighting and front-pressing. This highest intensity is reached in long and very severe fights, but even then, such intensity lasts just a short time (Walther et al. 1983). Evolution is favorable for individuals who experience very rapid changes or create contrasting actions around an opponent, namely, rapid sharp movements, loud harsh sounds, or penetrating pungent odors (Geist 1966).

The duration of fights may range from of a second up to half a day, though fights of one or more hours are very rarely observed. The majority of fights last only from a few seconds to 5 min. A fight usually consists of several bouts interrupted by pauses during which the opponents stand in front each other, display threats, mark objects, urinate-defecate, or graze. Usually, the rivals remain at the same place during pauses, but sometimes they move while grazing parallel to each other (parallel march). In fights among territorial males, an end with an explicit loser is extremely rare. The less successful fighter will walk away, while the winner remains standing behind, continuing to make threats or following him in a pursuit march. Usually, however, there is no obvious winner and loser in fights between territorial males in Antilopinae (Walther et al. 1983).

Agonistic behavior is a powerful mechanism of intra-sexual selection among males leading either to sexual morphological dimorphism, such as size and shape of horns and/or body size and weight (Geist 1966; Janis 1982; Pérez-Barberia and Gordon 2000; Bro-Jørgensen 2007), or to behavioral dimorphism, such as agility and aggressiveness (Linklater 2000; Cappellini 2007). Both forms of dimorphism give significant advantages in male-male competition (Corlatti et al. 2012). The main purpose of agonistic behavior is obtaining or retaining resources (Senar 1990), including mating opportunities, over which males compete during breeding season (Clutton-Brock et al. 1979). Competition for receptive females is a typical male behavior in many ungulates, especially polygynous species when a few males defend females from rivals and monopolize access to most matings (Andersson 1994; Corlatti et al. 2012). Males of many antelope species establish individual territories and maintain and protect their borders to prevent the intrusion of other males; this is a common way to obtain exclusive mating opportunities (Walther et al. 1983).

Some ungulate species among separate taxonomic groups (antelopes, cervids, caprids, perissodactyls—Isvaran 2005; Willisch and Neuhaus 2009) demonstrate alternative breeding tactics depending on external and/or internal stimuli within a population. Such alternative breeding tactics have not yet been observed among gazelle species, and I also did not observe this phenomenon in goitered gazelles, which have only a known territorial structure during the breeding season. Marmazinskaya (1996, 1999), though, believes that goitered gazelles change their breeding tactics depending on the population density. In my opinion, alternative breeding tactics are possible for goitered gazelles, but this is rather unlikely. In any case, I will not touch on this issue, in this chapter, but will look into the main question, of whether there are any differences in the agonistic behavior of goitered gazelle males with different social statuses, including territorial males, bachelors and males outside the rut, as well as immatures.

6.2 Agonistic Behavior in Goitered Gazelle Males

A detailed description (ethogram) of the main elements of the agonistic behavior of goitered gazelle males is presented in Table 6.1.

6.2.1 Seasonal Fluctuations in the Rate of the Aggressive Displays

Like other antelope species, goitered gazelle adult males establish individual territories during the preparatory stage of the rut in October and protect the borders of these territories by expelling territorial neighbors and other adult males-intruders throughout the entire rutting season in November and December (Blank et al. 2015).

Table 6.1 Description of the recorded aggressive displays (ethogram) in adult goitered gazelle males during observations based on Walther et al. (1983), Blank (1998), (Blank and Yang 2014—License Number 5476401078241)

Posture	Description
Clash-fighting of territorial males	After preliminary threat displays, rivals run toward each other and clash together, which can be heard at the distance of 700–800 m. When the horns are almost invariably interlocked, each rival tries to press the horns and head of the opponent to the ground and push him back. They then separate again, moving backward and repeat the same clashing actions. At the end of the fight, both rivals return to their own territories from opposite directions. During a long, intensive fight, the combatant's heads and necks are twisted aside, and they fight with their necks, when the horns are interlocked; the rival tries to twist the neck of his opponent upward-sideward, when the neck is twisted back. Sometimes, the muzzle, neck, or front parts of the breast are hooked by the rival's horns, which hit these parts. Opponents open their mouths widely and roar constantly. Their tails are pressed to their hindquarters (Fig. 6.1)
Horn-pressing fights of adult males	Typically, without preliminary threat displays, rivals approach each other with heads lowered to the ground and lean the front surfaces of their horns against each other (Fig. 6.2). They fight while standing on all four legs with the forelegs widely spread and hind legs in a normal position or relatively close together. Each tries to press the horns of the opponent to the ground and move him backward. Then, every rival hits the opponent's head with maximum power and presses his head to the ground and back, making jerks. The horns almost invariably interlock (horn of one between horns of another). This fight can transfer to fight-circling, when rivals with locked horns move in circles. In this case, they pivot with their hindquarters while keeping their heads in the center of the circle. During fighting, the ears are apart horizontally and directed sidewise at the top or sometimes backward, and the tail is pressed to the rump. This fight is common for non-territorial males (Figs. 6.3 and 6.4 and 6.5)
Frontal or aside threat posture	Horns are directed toward or aside opponent. Ears are directed backward-sideward; tail is in normal position (Figs. 6.6 and 6.7)
Dominance	Rivals stand in lateral T-position, parallel, or, most commonly, a reverse-parallel orientation. The posture is erect, with the neck vertically erected upward "proudly." The entire body is as high as possible due to all four legs stretched up. The muzzle top is pressed to a throat, and the horns are directed forward with the head turned in a more or less pronounced fashion away from the opponent. Tail is uplifted, making waving movements from one side to the other; ears are apart horizontally and directed sidewise at the tops, with the ear holes directed forward or downward. All walking movements are unusually slow (Fig. 9.1)
Shrub horning	In the case of shrub or saxaul (*Haloxylon aphyllum*) horning, the male approaches the tree in a dominance posture (head level with the body, horns directed forward, ears apart), sniffs the bare trunk, positions the trunk between the horns, and thrashes with the forehead, goring the trunk with the power of his whole body; he cleans small branches from bigger ones and breaks some shrub's branches, with pauses to smell the scratched vegetation, which he touches with his muzzle top after every 2–3 thrashing acts before continuing to horn. The ears stay apart horizontally and slightly backward (45°) at the tops and downward by the ear holes. Head movements are up-down along the branch and from side to side, sometimes round the head axis, and he pushes the shrub with the entire bod, trying to scratch as high as possible (Fig. 6.8)

(continued)

Table 6.1 (continued)

Posture	Description
Ground horning	A buck sniffs points of low grass and starts to horn, scratching and goring the plants and naked ground, making sharp horning movements from one side to another and setting his horns against the ground, as if trying to shift the ground like he does with his real rival (Fig. 6.9)
Grazing ritual	Two territorial males meet at the boundary of neighboring territories, threatening and fighting each other. After several clashes, they simultaneously begin to graze and step backward. If they do not renew the fight, permanent grazing leads to enlargement of the distance. The longer the grazing ritual lasts, the greater the distance becomes. Sooner or later, they turn into a reverse (hindquarters to hindquarters) position and move farther and farther away from each other, each of them grazing back toward the center of their own territory (Fig. 6.10)
Parallel walk	A lateral display of two males engaging in a slow, tense, regular walk in parallel same or opposite directions to each other; horns are held high, and often their hair is raised (Fig. 6.11)
Looking to opponent	One male looks attentively for any movements of the other male
Display freezing	One or both opponents freeze in an agonistic posture and hold it for several seconds, then continue their displays (Fig. 6.12)
Urination	Stepping forward and stretching the body, lowering the belly toward the ground, with the hind legs sloping backward; urination with a lifted tail (Fig. 5.3)
Defecation	Abrupt bringing hind legs forward so that the hooves are placed lateral to the stationary forelegs, squatting closer to the ground and taking a crouched posture and defecating (Fig. 5.4)
Chasing	One male follows the other, often roaring toward his opponent and demonstrating threat or dominant postures. This is a typical behavior of territorial males that expel intruders from their territories
Low-stretch	Stretching the head and neck forward horizontally with the nose pointing horizontally forward. Ears are directed backward and usually pressed to the head in a position either parallel to the ground or slightly downward with the foramen inward. Rarely ears are laid back, pressed to each other, or held one apart of the other at a 45° angle. The tail is lifted to varying degrees, up to a vertical position (Fig. 6.13)
High-stretch	Head and neck are stretched forward and upward up to 45° above the horizontal, the top of the muzzle is lifted forward and upward, and the ears are pressed to the head (Fig. 6.14)
Preorbital marking	The ritualized act of marking with the male's preorbital glands rarely begins with sniffing the plant stem or twig. Instead, the goitered gazelle male opens both preorbital glands, sometimes opening his mouth at the same time, and carefully lowers the preorbital region toward the plant stem, bringing the stem's tip or twig into the widely opened gland. With quivering movements of the gland's margin, he deposits some secretions, which soon hardened in the open air. The male sometimes marks the same plant stem a second time using either the same or his other preorbital gland. To position himself for scent marking, the male commonly stretches his head upward to get his preorbital gland as high as possible, although he never raises his body to stand on just his hind legs for this purpose (Fig. 6.15)

(continued)

Table 6.1 (continued)

Posture	Description
Roaring	Roaring consists of gruff, wheezing or hissing sounds resembling gurgling or sneezing. In producing these sounds, the male opens his mouth wide, lifts his tongue slightly, and takes a deep breath, then closes his mouth and passes air through his nostrils or possibly through his partly closed mouth Posture: Muzzle is raised to an angle of up to 45° from a horizontal position. Ears are directed backward and tightly pressed to the head with the ear holes inside. Abruptly, the male jerks his head and whole body forward and upward, and his protruding thyroid cartilage moves up and down. When running, roaring is intermittent (Fig. 6.16)
Close approaching	One male approaches another to a very close distance, breaking an individual distance and forcing the opponent to retreat or start battling (Fig. 6.17)
Muzzle touching	One male approaches another very closely and touches his opponent's body (usually hindquarters) with his muzzle (Fig. 6.18)
Mating posture	In a mating bipedal posture, one male follows the hindquarters of another male keeping in a bipedal position for several seconds
Sniffing opponent	One male approaches the other to a close distance, trying to sniff any body part (often hindquarters) of his opponent

Fig. 6.1 Clash-fighting in goitered gazelles: Males run toward each other and clash together, which can be heard at the distance of 700–800 m; each rival tries to press the horns and head of the opponent to the ground and push him back, with locked horns. Drawing by U. Kenzhetaev with kind permission

These individual territories are important to rutting males for obtaining exclusive rights to mating females (Walther et al. 1983). Some of adult males do not occupy territories, and these bachelors stay non-territorial throughout the rut, though they are also fertile (Skinner and Louw 1996). Similar to females, bachelors continue their daily migrations during the rut as they do outside of rutting season. During most of the rest of the year, all adult males do not establish or maintain territories, moving freely and widely without showing any signs of territoriality. The only exception is the spring peak of sexual activity, and though many males remain non-territorial at this time, some adult males demonstrate signs of territorial behavior in

Fig. 6.2 Horn pressing fight in goitered gazelle males: males approach each other with heads lowered to the ground and lean the front surfaces of their horns against each other. Issyk-Kul goitered gazelle nursery for reacclimatizing, 25.06.2022. Photo by D. Blank

Fig. 6.3 Horn pressing fight (upper position) in goitered gazelle males: Each male tries to press the horns of the opponent to the ground and move him backward, making jerks. Issyk-Kul goitered gazelle nursery for reacclimatizing, 25.06.2022. Photo by D. Blank

Fig. 6.4 Horn pressing fight (lower position) in goitered gazelle males: Males fight trying to move opponent backward. Issyk-Kul goitered gazelle nursery for reacclimatizing, 25.06.2022. Photo by D. Blank

Fig. 6.5 Horn pressing fight (side by side position) in goitered gazelle males: Males fight trying press horns of the opponent to the ground. Issyk-Kul goitered gazelle nursery for reacclimatizing, 25.06.2022. Photo by D. Blank

Fig. 6.6 Frontal threat posture in goitered gazelle males: horns are directed toward opponent. Ears are directed backward-sideward; tail is in normal position. Issyk-Kul goitered gazelle nursery for reacclimatizing, 25.06.2022. Photo by D. Blank

Fig. 6.7 Aside threat posture in goitered gazele males: Horns are directed aside opponent. Ears are directed backward-upward; tail is in normal position. Issyk-Kul goitered gazelle nursery for reacclimatizing, 25.11.2022. Photo by D. Blank

Fig. 6.8 Shrub horning in goitered gazelle males: Head movements are up-down along the branch and from side to side, sometimes round the head axis. Issyk-Kul goitered gazelle nursery for reacclimatizing, 24.11.2022. Photo by D. Blank

Fig. 6.9 Ground horning in goitered gazelle males: A buck scratches and gores the plants and naked ground, making sharp horning movements from one side to another and setting his horns against the ground. Issyk-Kul goitered gazelle nursery for reacclimatizing, 25.11.2022. Photo by D. Blank

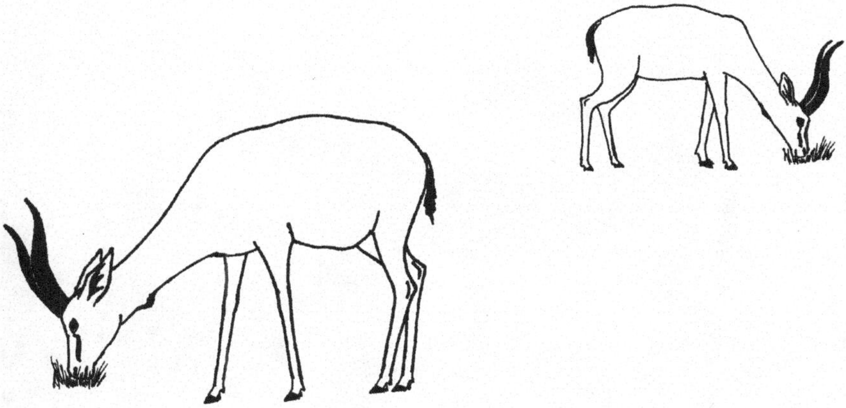

Fig. 6.10 Grazing ritual in goitered gazele males: After fighting two territorial males simultaneously begin to graze and step backwards. Drawing by U. Kenzhetaev with kind permission

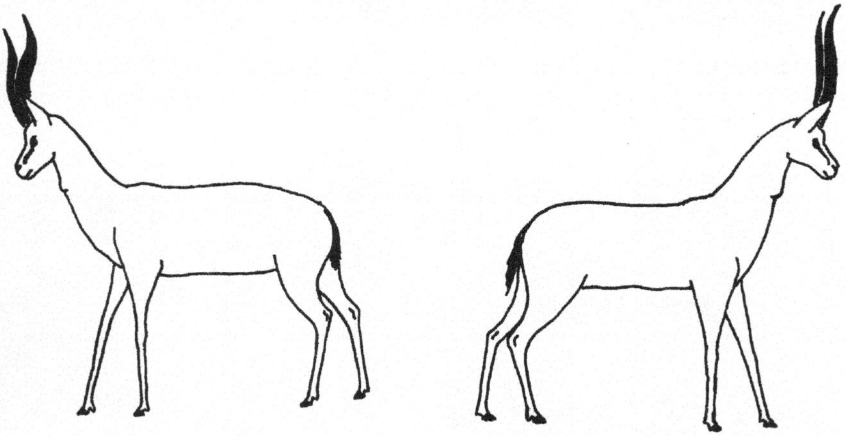

Fig. 6.11 Parallel walk display in goitered gazelle males: A lateral display of two males engaging in a slow, tense, regular walk in parallel same or opposite directions to each other; horns are held high. Drawing by U. Kenzhetaev with kind permission

mid-April to early May (false or pseudo rut) and show increased courting and agonistic activities (Blank et al. 2015).

The highest frequency of aggressive interactions in goitered gazelles is observed during the main rutting season in November and especially in December. Many fewer (half or even less) conflicts arise between males during the spring peak of sexual activity (pseudo rut) in April and May and very rarely in other months of the year (Fig. 6.19). This is not surprising, since histological studies of the testes of males and ovaries of females have shown that the highest activity of the genitals is observed during the autumn rut in November and December, while during the spring

Fig. 6.12 Display freezing in goitered gazelle males: An opponent freeze in an agonistic posture and hold it for several seconds. Issyk-Kul goitered gazelle nursery for reacclimatizing, 25.11.2022. Photo by D. Blank

Fig. 6.13 Low-stretch posture of a dominant male toward another male in goitered gazelles: Stretching the head and neck forward horizontally with the nose pointing horizontally forward. Issyk-Kul goitered gazelle nursery for reacclimatizing, 25.11.2022. Photo by D. Blank

Fig. 6.14 High stretch posture of a dominant male toward another male in goitered gazelles: Head and neck are stretched forward and upward up to 45° above the horizontal. Issyk-Kul goitered gazelle nursery for reacclimatizing, 25.11.2022. Photo by D. Blank

Fig. 6.15 Preorbital marking during conflict between two males: male lowers the preorbital region toward the plant stem, bringing the stem's tip or twig into the widely opened gland. Issyk-Kul goitered gazelle nursery for reacclimatizing, 25.11.2022. Photo by D. Blank

Fig. 6.16 Roaring in goitered gazelle males: Male roars often during courting of a female. Issyk-Kul goitered gazelle nursery for reacclimatizing, 24.11.2022. Photo by D. Blank

Fig. 6.17 Close approaching in goitered gazelle males: a male approaches another to a very close distance, breaking an individual distance and forcing the opponent to retreat or start battling. Issyk-Kul goitered gazelle nursery for reacclimatizing, 25.06.2022. Photo by D. Blank

Fig. 6.18 Muzzle touching in goitered gazelle males: a male touches his opponent's body with his muzzle. Issyk-Kul goitered gazelle nursery for reacclimatizing, 26.06.2022. Photo of D. Blank

pseudo-rut in April and May (Blank and Fedosenko 1983), the activity of the genitals is noticeably lower compared to the real autumn rut (Tsaplyuk 1972). A similar phenomenon is observed in chamois (*Rupicapra rupicapra*), when males show reduced aggressiveness during the second spring peak of territoriality compared with the main autumn-winter rutting season. Male chamois, too, showed very high levels of testosterone and cortisol during the autumn rut and much lower levels of hormones during the spring pseudo-rut (Creel 2001; Corlatti et al. 2014). The higher level of androgens (testosterone) is related to stronger dominance and greater physiological capabilities of a male expressed in territorial aggressiveness and courtship displays (Miles et al. 2007). Aggressiveness of male gerenuks (*Litocranius walleri*), Dorcas gazelles (*Gazella dorcas*), and scimitar horned oryx (*Oryx dammah*) is also related to higher levels of testosterone (Penfold et al. 2002).

Moreover, males of different ages and social statuses behave in different ways. Territorial males exhibit their most aggressive behaviors during the rutting period in November and December. These high intensity agonistic activities are much greater than in non-territorial bachelors and are likely related to the stress of domination (Creel 2001; Corlatti et al. 2014). During the pseudo-rut in April and May, however, aggression of territorial males towards other males is quite rarely observed (Fig. 6.19a). In other months, territorial males are completely absent. In contrast, non-territorial males, staying in groups, show their aggression to each other most often in April and May, whereas during the main rut, agonistic contacts with other bachelors are much less frequent (Fig. 6.19b). Apparently, the stress of domination of bachelors in April and May is weaker due to lower testicular activity (Tsaplyuk

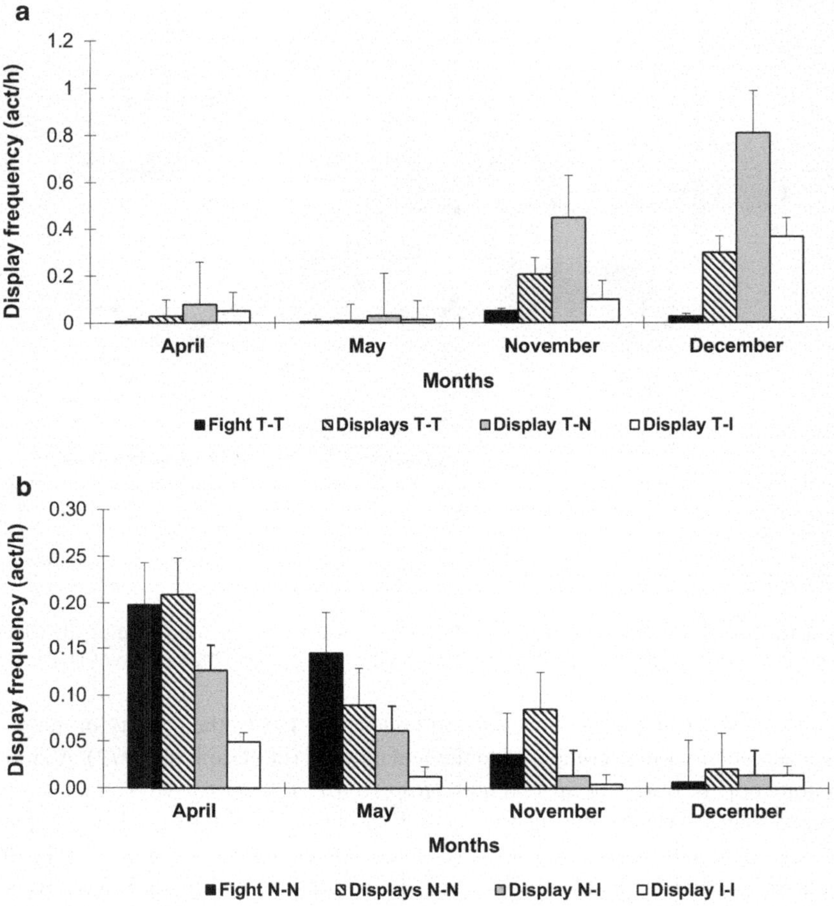

Fig. 6.19 Frequency of fights and noncontact agonistic displays (number of events observed per hour of observation) observed in territorial males (**a**) and non-territorial males (**b**) during spring onset of territoriality (April–May) and rutting period (November–December). Abbreviations: Fight T-T, fights observed in territorial males; display T-T, agonistic displays between territorial males; display T-N, dis-plays between territorial and non-territorial males; display T-I, displays between territorial males and immature; fight N-N, fights observed in non-territorial males; display N-N, agonistic displays between non-territorial males; display N-I, displays between non-territorial males and immature; display I-I, displays between immature males (Blank et al. 2015—License Number—5476400246045)

1972) and, thus, aggressiveness of territorial males has a lower level in the spring peak of sexual activity. In addition, non-territorial males in spring display low intensity aggressive behavior as fights often turn into easy playful interactions. This agonistic behavior of bachelors is related more to establishing hierarchical positions in the group and likely for improving fighting skills, rather than a struggle for the right to reproduce. By game theory models, fighting intensity closely depends on resource value (Parker and Rubenstein 1981; Enquist and Leimar 1987); therefore, there is a great difference in agonistic behaviors between territorial and non-territorial males.

During the rut, territorial males are engaged in fighting mostly with bachelors and much less often with territorial neighbors, but this happens infrequently compared to the noncontact aggressive displays. During the spring peak of sexual activity, fights of territorial males occur even less often compared to the autumn rut. Bachelors, in contrast, engage in contact fighting (mainly horn-pressing) even more often with other males during the rut compared to the fight rate of territorial males. During the spring rut, non-territorial males horn butt much more often compared to territorial males and compared to the autumn rut (Fig. 6.19a, b) (Blank et al. 2015).

6.2.2 Distant Noncontact Displays Vs Direct-Contact Fights

Goitered gazelle territorial males, which are common during the autumn rut and relatively rarely observed during the spring pseudo rut, mainly use demonstrations of threatening noncontact displays in their agonistic interactions, and the proportion of such contacts reaches 94–98% of all noted cases of agonistic behavior, whereas direct fights of territorial males with physical contact with any other males irrespective of their social status (territorial or bachelor) are surprisingly rare and constitute only 2–6% of all aggressive interactions (Figs. 6.19 and 6.20). It is especially rare for territorial males to engage in direct contact fights with equal rivals, such as other territorial neighbors, and instead use mostly noncontact threat displays from a distance (Ili Hollow—Blank and Yang 2014). Zhevnerov (1984) observed a similar phenomenon in the Barsakelmes Island population (Aral Sea), where territorial males avoided direct contact interactions and mainly performed threat displays at a distance during aggressive displays. Direct contact fighting is unadaptable for most

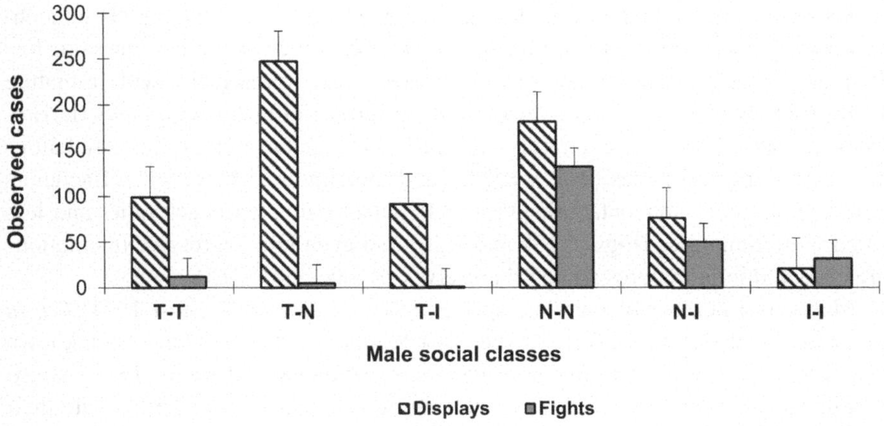

Fig. 6.20 Frequency of aggressive displays and fights (number of events observed per hour of observation) between males of different social and age classes. Abbreviations: T-T, territorial male vs. territorial male; T-N, territorial male vs. non-territorial male; T-I, territorial male vs. immature; N-N, non-territorial male vs. non-territorial male; N-I, non-territorial male vs. immature; I-I, subadult male vs. subadult male (Blank et al. 2015—License Number—5476400246045)

ungulate species. Overt fighting is more typical for species having good defensive technics and/or weapons of low effectiveness that do not lead to serious life-threatening injuries (Geist 1966). That's why territorial males of most antelope species demonstrate mainly noncontact displays in their agonistic interactions, with only some species (springboks) having a very limited repertoire of over fighting displays and others (Grant's gazelles) with a very intricate and striking set of these behaviors (Walther 1965, 1981).

Similar to territorial males, non-territorial males also have fewer direct contact fights compared to noncontact displays. However, direct contact fights in bachelors are noticeably more common compared to territorial males but with a less significant difference between the frequency of fights and demonstrations at a distance in non-territorial males compared to territorial males (Fig. 6.20). Immature males, on other hand, do not show preferences and demonstrate butting contacts and distant aggressive displays in approximately equal proportions (Blank et al. 2015).

6.2.3 Frequency of Noncontact Displays and Fights Territorial Males with Peers, Non-territorial Males, and Immature Males

It is known that a high content of the steroid hormone androgen is closely associated with high activity of territorial and aggressive behaviors in males (Miles et al. 2007). During the rut, territorial males possess a higher social status than other males and demonstrate aggressive displays most often due to the higher content of testosterone in their blood (Li et al. 2004; Shargal et al. 2008). And indeed, during the autumn rut in November and December, goitered gazelle territorial males most often show their aggressive intentions (noncontact displays and chases) toward adult bachelor males who wander into their established territories, while conflicts (mainly noncontact displays) with other territorial males occur 2–2.5 times less often. The same low frequency of aggressive interactions is observed in territorial males with immature males (mainly chasing). Real contact fights of territorial males with peers (territorial neighbors) happens quite rarely (10%, $n = 111$), and even more rare are battles with non-territorial males (2%, $n = 251$) and immatures (1%, $n = 92$). Bachelors demonstrate more noncontact displays and contact fights with each other and less often with immatures (Figs. 6.19b and 6.20), but even less aggressive interactions are observed in immatures among themselves (Blank et al. 2015).

Males of a number of ungulate species (*Ovis canadensis*—Geist 1971; *Capra pyrenaica*—Alados 1986; *Gazella* species—Walther et al. 1983) have been known to interact most often with males of the same age and social status. In my study, non-territorial bachelors definitely display aggressive patterns most often with their peers and significantly less often with immature males. However, territorial males most often come into contact with non-territorial males and demonstrate most of their aggressive actions to them as representatives of a lower social class. The longest duration of the agonistic interactions is observed when rivals are both territorial males, while aggressive contacts between territorial males and non-territorial and immature males are of a much shorter duration. Thus, territorial males have less contact with their peers than with males of another lower social class (non-territorial), and non-territorial males interact with their peers more often than with immatures, but these contacts are considerably shorter than in territorial males. According to the energetic "war-of-attrition" and cumulative assessment models (Maynard Smith and Price 1973; Payne and Pagel 1996), the duration of aggressive contacts is the most important sign of fighting ability and an indicator of the quality of a male. The territorial male should show his strength and endurance only to his peers which are equal to him in social status, whereas to non-territorial and immature males, the territorial male is a priori dominant within his territory. Short aggressive interactions between non-territorial males can be explained by the possibility that fights are a means to resolve their conflicts, so there is no need for lengthy noncontact displays (Blank et al. 2015).

The dependence of the intensity of agonistic behavior on duration, when the aggressive element is repeated many times, is determined by the persistence and success of displays, a phenomenon found in many species (Clutton-Brock and Albon 1979). Moreover, territorial males have higher levels of hormones (testosterone and cortisol), greater dominance, and physical capabilities compared to non-territorial males (Miles et al. 2007; Corlatti et al. 2012). A male's possession of a territory has a high energy cost, because it requires constant monitoring and maintenance to send signals about its ownership and the male's dominance (Müller-Schwarze 1999). Goitered gazelle males are almost constantly engaged in marking their territories and demonstrating aggressive displays during the rut, leaving very limited time left for grazing (Blank 1990). As a result, rutting males can lose a third of their body weight during the rut and become so weakened after rutting that many die during the winter, a mortality many times greater than that of females (Zhevnerov 1984). Thus, only males of high quality and physiological resistance and endurance can become territorial males (Payne and Pagel 1997).

6.2.4 Duration of Agonistic Interactions

The most diverse repertoire of agonistic activities and the duration of these distant displays between territorial males are the greatest among males ($X \pm SE = 534.61 \pm 54.26$ s, $n = 69$), while aggressive demonstrations of a territorial male toward bachelors are much less diverse and noticeably shorter ($X \pm SE = 190.36 \pm 26.22$ s, $n = 187$). To immature males, territorial males show the most limited variety of aggressive displays, and these interactions are the shortest ($X \pm SE = 63.71 \pm 26.22$ s, $n = 76$). Non-territorial males make threatening noncontact displays to each other for a much shorter time ($X \pm SE = 173.22 \pm 28.63$ s, $n = 60$) than territorial males do to each other. In addition, the duration of these demonstrations in non-territorial males among themselves and with immature males, as well as immature males among themselves, is almost equal and does not change seasonally. Non-territorial males and immatures show a wider variety repertoire of distant displays during aggressive contacts with each other. The duration of contact fights is much shorter than distant displays, and fights in territorial males last less than half as long (X ± SE = 20.67 ± 8.42 s, $n = 9$) as similar fights in non-territorial males (X ± SE = 53.41 ± 7.89 s, $n = 125$) and only a little longer compared to the fights of immature males among themselves (X ± SE = 15.52 ± 3.62 s, $n = 27$). I observed a very prolonged fight between two territorial males, which lasted 32 min, but this was the only time I noted such a prolonged fight during my entire project observing of goitered gazelles (Blank et al. 2015).

6.2.5 Agonistic Behavior of Territorial Males

Territorial males, by establishing and protecting individual territories, create exclusive access to mating females and have a high dominance rank among other males. Thus, in territorial species, males are quite similar in appearance and have approximately the same size of horns and body and maintain the same size of territory, with their signal of their dominance coming mainly from the possession of territory (Geist 1966). Territoriality demands significant changes in male behavior, with the main part of aggressive behavior devoted to guarding territorial borders from penetration by other males; intensive marking of boundaries by territorial males plays the role of an intimidating mechanism against other males (Walther et al. 1983). Males do not strictly respect the boundaries of neighboring territories and periodically run into a neighbor's range, especially when pursuing a female group trying to exit the previous individual's area or when the other male runs toward an approaching female group. Therefore, border threat displays at a distance between male neighbors occur quite frequently, though these noncontact displays very rarely turn into direct contact fights quite rarely. Aggressive male displays can be divided into two broad categories: reciprocal, when both interacting males perform threat postures, and one-sided displays, when only one male makes threat displays to another male, which does not return the threat but instead retreat.

Territorial males have an equally high dominant status, so they demonstrate reciprocal poses of threat and dominance, each in their own territory (86% cases, $n = 78$), while one-sided demonstrations are quite rarely found (14%). But as soon as one crosses his marked boundary, he loses this dominant status and retreats when the owner of the neighboring territory appears. As soon as the retreating male returns to his territory, he immediately regains his dominant status, stops retreating and begins to demonstrate his threat poses and dominance again. Similar behavior is described for territorial blackbucks (*Antelope cervicapra*) and Thomson's gazelles (Walther 1978).

Since territorial males always occupy a dominant position in relation to other males who enter their territories, males of lower social status behave differently when meeting a territorial male and rarely (adult non-territorial males) or never (immature males) respond with the same threatening poses. When a non-territorial bachelor, or even more so an immature, crosses into an individual male's territory, its owner immediately rushes (fast trot or gallop) toward him ready to demonstrate horn threat and poses of aggressiveness and dominance. Quite often, however, the territorial male is able to expel the intruder without any threat displays, except perhaps a roar; the intruder usually does not resist or even demonstrate any poses in response but immediately seeks to escape from the dominant male. Thus, one-sided displays of threat are observed only from the territorial male (Fig. 6.21). Such behavior is typical for most antelope species (Walther 1984). Territorial males of Thomson's gazelles, for example, also display reciprocal displays to their peers, while one-side displays are observed between territorial males and non-territorial males (Walther 1978).

Despite the fact that goitered gazelle territorial males rarely engage in direct fights with their peers, when they do, they use clash-fighting (80%, $n = 15$, Fig. 6.22), where rivals forcefully jump or run toward each other. Clash-fighting is also a typical kind of fighting between territorial males in Thomson's gazelles (Walther et al. 1983) and sand gazelles (*G. s. marica*—Habibi et al. 1993). Horn-pressing is mainly seen in struggles between a territorial male and a non-territorial (80%, $n = 20$) and rarely seen between territorial peers. Fights between territorial males and immatures are very rarely observed, but if they happen, then horn-pressing is only used. Springboks (*Antidorcas marsupialis*) have similar behavior, and territorial males usually avoid fights with their neighbors, but if they do, then it lasts for only a short time, while they fight longer with bachelor males (Skinner and Louw 1996). Impala territorial males also use mainly clash-fighting during territorial conflicts (Schenkel 1966).

At the end of a border conflict, the territory owner does not chase his neighbor but instead follows him making threatening displays in a walking pursuit march. Faster gaits (trotting and galloping) are used much less often when expelling a territorial male-intruder. When chasing non-territorial bachelors and immatures from their sites, territorial males often use fast gaits and pursue non-equal males at either

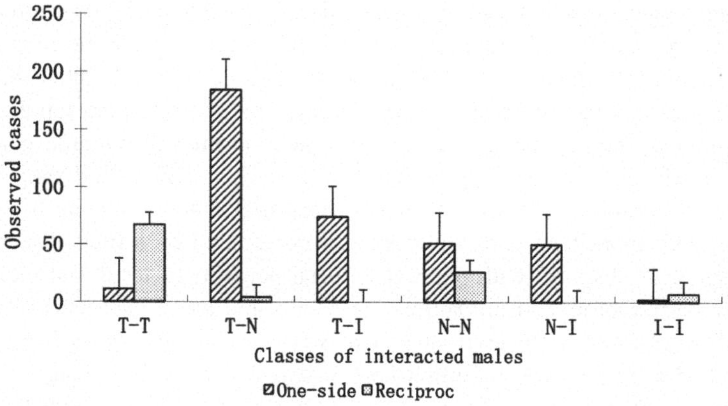

Fig. 6.21 Proportion of reciprocal (Reciproc) and one-side displays (one-side) in encounters of males with different social status: T-T, territorial male vs. territorial male; T-N, territorial male vs. non-territorial males; T-I, territorial male vs. immature; N-N, non-territorial male vs. non-territorial male; N-I, non-territorial male vs. immature; I-I, immature vs. immature (Blank and Yang 2014—License Number—5476401078241)

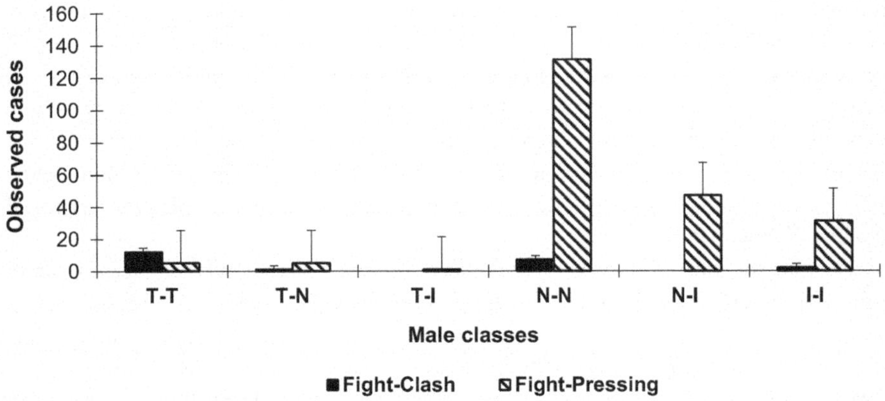

Fig. 6.22 Proportion of fighting type (fighting-clash vs. fighting-pressing) in encounters of males with different age and social status: T-T, territorial male vs. territorial male; T-N, territorial male vs. non-territorial males; T-I, territorial male vs. immature; N-N, non-territorial male vs. non-territorial male; N-I, non-territorial male vs. immature; I-I, immature vs. immature (Blank and Yang 2014—License Number—5476401078241)

a trot or gallop. During the rut, territorial males expel their equal rivals (territorial neighbors) mainly by walking (60–76%, $n = 59$), although fast gaits (gallops) are also becoming more frequent (up to 33%, $n = 48$ in December) due to a decrease in the number of males trotting. During the spring peak of sexual activity April and May, territorial males pursue their territorial opponents most often at a walk (78–80%, $n = 20$). Territorial males pursue both non-territorial males and immatures by walking or running in comparable proportions (40% to 46% and 24% to 39%, $n = 164$ and 150, respectively), with a decrease in the proportion of trotting (Blank and Yang 2014).

Among the postures that territorial males demonstrate to each other, threats with horns forward in a frontal orientation to each other, bush/ground horning and the grazing ritual (53% in sum, $n = 308$) are most often noted, while other threat displays are observed less often (Table 6.2). Ground and bush horning represent both aggression and marking behaviors, though object-horning in goitered gazelles is likely more closely related to aggression than to marking (Blank and Yang 2014).

The least common response is for one territorial male to pursue a territorial neighbor, with more than half occurrences accompanied by a roar (8%, $n = 49$). Unlike agonistic displays with peers (territorial neighbors), territorial males very often pursue bachelors and immatures (58%, $n = 439$ and 77%, $n = 113$, respectively) after only short-term aggressive displays (42% and 24%) or no displays at all (16% and 53%, respectively). Furthermore, chasing with roaring (a sign of unconditional dominance) is most often addressed to non-territorial bachelors, three times less often to peers (dominance only within his own territory) and immature males (unconditional dominance). Since the roar enhances the intimidating effect of the pursuit, it is primarily intended for non-territorial males, who in general are the most likely potential contenders for territory competition. A rare roar addressed to

Table 6.2 Frequency of various aggressive patterns in encounters of males with different age and social status (Blank and Yang 2014—License Number—5476401078241)

Behavioral patterns	Territorial versus			Non-territorial versus		Immature versus
	Territorial	Non-territorial	Immature	Non-territorial	immature	immature
Frontal threat	123	89	19	60	24	13
Bush horning	90	40	2	11	3	4
Grazing ritual	95	3	0	12	1	1
Parallel walking	49	5	0	19	0	0
Looking to opponent	41	50	7	10	1	0
Ground horning	43	2	1	9	0	0
Display freezing	35	1	0	8	1	2
Urination-defecation	36	19	1	3	0	0
Chasing	21	254	87	116	89	21
Dominance	0	27	2	37	17	6
Low-high-stretch	3	25	32	1	19	0
Preorbital marking	5	7	0	1	0	0
Chasing roaring	28	111	30	83	24	14
Close approaching	0	0	0	30	20	9
Muzzle touching	0	0	0	2	12	0
Mating posture	0	0	0	9	11	0
Sniffing opponent	0	0	0	6	4	2

another territorial male is only effective within the roaring male's territory, since the two males are of the same social status; the dominance of one male over another is possible only for a male within his own territory and ends immediately at his boundary (though it may still give him a slight advantage). When a territorial male encounters an immature, he represents unconditional dominance, and any threatening demonstrations, roaring even more so, are simply unnecessary, because the young male always retreats from a male of the highest social status.

Territorial males show the longest and most diverse distant displays to their neighbors entering their territory, but at the end of the conflict, they very rarely pursue the trespasser. In relation to non-territorial bachelors, territorial males show a smaller set of aggressive displays but chase them more often than other categories of males. The least number of threatening postures is addressed to immatures, but they often chase them at a run (Table 6.2). The variety of agonistic displays of goitered gazelle territorial males is similar to the aggressive behaviors seen in territorial males of other gazelle species (Walther 1984; Blank and Yang 2014).

Seeing a bachelor who has violated his border, the territorial male will rush to meet him. The trespasser, as a rule, trying to avoid a fight, begins to run in a circle, making attempts to continue his path. The owner of the territory does not necessarily catch up with his opponent, since as soon as the pursued male slows down, the owner does the same. Such a chase can continue until the rivals are completely exhausted, after which both, resting, walk side by side, without showing aggression toward each other. Gradually, the owner of the territory slows down and returns to his females. This can happen several times until the trespasser leaves (Zhevnerov 1984).

During pauses in aggressive encounters, some male displays are not clearly agonistic in intent but more like marking behaviors, though, in fact, they are still threat displays and related to aggressive actions because they are often performed in the presence of an opponent. Called "space-claim" displays, these behaviors are not direct aggression toward a rival, since they are directed toward the ground and vegetation in the near vicinity (Walther et al. 1983). Space-claim behaviors include object-horning, marking with urine/feces and preorbital glands, looking directly at an opponent, parallel walking, performing the grazing ritual, bush horning, and frontal threat (patterns are ordered by increasing the performing frequency). Other elements of space-claim behaviors are observed quite rarely except for chasing. These displays are demonstrated by goitered gazelle territorial males during territorial conflicts between neighbors but are not typical behaviors directed toward non-territorial males (Blank and Yang 2014).

One of the most important elements of the agonistic behavior of ungulates is the orientation of the male's body relative to his rival. The frontal threat posture demonstrates a readiness to fight, and the male makes the same movements in the air that he would do when head butting with a real opponent but without touching him; the defending male will occasionally freeze in this posture for up to several seconds or more (Walther 1984). Thomson's gazelle territorial males show threat postures to other territorial males, and, in most cases, these postures lead to actual fighting (Walther 1978). Springboks (*Antidorcas marsupialis*) rarely use frontal threat displays, while Grant's gazelles use them often (Estes 1991). The threat from the high presentation of horns is typical and pronounced for goitered gazelles, as well as for Dorcas gazelles, while rarely observed in Grant's and mountain gazelles and absent completely gerenuk (*Litocranius walleri*) and blackbuck (*Antelope cervicapra*) (Walther et al. 1983).

Another typical agonistic pattern is a parallel walking, when the movement of opponents may be in the same direction or opposite directions and with the horns of both males directed forward and down in a threat position. Parallel walking is typical for territorial males but rarely observed in bachelors and never in immatures. This behavior is noted in many animal species, including most antelopes and cervids (Clutton-Brock et al. 1979; Walther 1984; Estes 1991). Lateral threat displays and lateral fighting are relatively antiquated behaviors used mostly by ancient bovid and cervid species with short horns and often canine tusks, since they have no other way to demonstrate their dominance and threats. Lateral displays are often the final element of aggressive displays before the end of a conflict (Jennings et al. 2003). Though evolution, antelopes began to use more frontal threat displays and as a result developed relatively long horns, while the parallel orientation of their bodies during the threat presentations began to play a much lesser role for them. Modern deer, sheep, and bulls also have abandoned lateral displays and prefer a frontal threat body position, which allows them to meet and repel their opponent's blows with their horns and thus avoid serious injury (Geist 1966; Barrette 1977).

The grazing ritual is one of the most frequently used elements among space-claim behaviors, is typical for territorial males during conflicts (95%, $n = 91$), and is addressed to another territorial male. During the grazing ritual, both opponents lower their heads closely to the ground and create the impression of calm grazing. In fact, neither opponent grazes but instead closely monitors the movements of their opponent until the conflict is over and the two leave the battleground space. This grazing ritual is rarely used by territorial males to non-territorial males (1.4% cases, $n = 219$) and never to immatures. Non-territorial males rarely use the grazing ritual rarely and immatures do not use it at all. Zhevnerov (1984) also observed the grazing ritual used by the mature males in the Barsakelmes Island population. This ritual of horned ungulates apparently originates from the ancestral behavior of displaced aggression. The grazing ritual is typical for territorial males of all gazelle species and number of antelopes (Estes 1969; David 1973; Gosling 1974; Walther 1984), and in Thomson's gazelles territorial males, almost 40% of border conflicts end with this ritual (Walther et al. 1983).

Bush and ground horning of shrubs and the soil are also demonstrated quite often, but less often compared to the grazing ritual. These actions are directed by territorial males toward neighbors who have violated their boundaries (67–74%, $n = 52$ and 74), especially during the autumn rut in November and December, but still less often compared to the grazing ritual and threat displays. In contrast, non-territorial males use bush and ground horning rarely in their agonistic repertoire. Thomson's gazelle males have similar behaviors (Walther 1978) and gerenuk males (*Litocranius walleri*) use ground horning also for solving territorial conflicts, especially as a particularly strong threat toward young males (Estes 1991). Furthermore, territorial males often show a "direct look" at their rival with freezing as a part of their threat displays. Usually, males stop during a contest and freeze for several seconds while at the same time closely monitoring each other's movements. This behavior is very typical for all gazelle species (Walther et al. 1983).

Urination and defecation, usually marking activities by goitered gazelles, can also be part of aggressive behaviors and used as space-claim displays, though significantly less often than the grazing ritual (22–29%). Using preorbital gland marking as a demonstration of aggression is not used very often by territorial males and rarely by non-territorial males. In contrast, some antelope species use preorbital marking and urination-defecation acts in their agonistic demonstrations much more frequently (Walther 1984). Goitered gazelle territorial males demonstrate space-claim behaviors to non-territorial bachelors but much less often than to other territorial males. In addition to distant frontal horn threats, territorial males prefer head butting bushes among other spatial varieties of threats (urination-defecation, preorbital markings) (Table 6.2). Territorial males do not make space-claim displays toward immatures at all, and non-territorial males, and especially immature males, occasionally but rarely use these behaviors (Blank and Yang 2014).

6.2.6 Agonistic Behavior of Non-territorial Males and Immatures

Non-territorial males demonstrate their aggressive behaviors in all seasons of the year with almost no changes seasonally. During the autumn rut and during the spring peak of sexual activity, they become bachelors and do not engage in rutting activities. Most conflicts between non-territorial males are related to hierarchical relationships within the group. Compared to territorial males, non-territorial males rarely demonstrate reciprocal (mutual) threats to each other and are much more likely to show one-sided (unilateral) displays during conflicts with other non-territorial males or with immatures, although reciprocal threats are observed noticeably more often in non-territorial males in conflicts with each other than during encounters with immatures or immatures among themselves (Fig. 6.22). Immature males demonstrate mainly reciprocal displays with their peers. Unlike territorial males, when non-territorial bachelors fight, they rarely use clash-fighting and much more often use horn-pressing in conflicts with peers and immatures, though bachelors rarely fight during rutting season. Outside of the rut, all adult males become non-territorial males and fight with each other more often than during the rut, most often using horn-pressing during encounters. Immature males fight with peers preferring horn-pressing contests, with clash-fighting very rarely observed (6%, $n = 33$) (Blank and Yang 2014). Horn-pressing and head butting are observed in all antelope species, though some species display it frequently (Grant's gazelle), while others rarely (Thomson's gazelle) (Walther et al. 1983).

At the end of an aggressive conflict, non-territorial males drive away other bachelors, chasing them at a walk (85%, $n = 40$) and less often chasing them at a trot (7–23%) or gallop (4–21%); however, they prefer faster gaits when chasing away immatures. In preference to slower gaits when chasing other males, non-territorial males are similar to territorial males. Also, like territorial males, bachelors avoid direct fights, when possible, and instead prefer aggressive demonstrations from a distance in conflicts with rivals; these conflicts of aggressive behavior in non-territorial males, though, occur much less frequently than in territorial males (Table 6.2). Non-territorial bachelors only pursue other males more often than perform the distant threat displays. Moreover, non-territorial males demonstrate chasing and roaring while chasing just their peers, while territorial males do the opposite and address their chasing mainly to males of other social classes (non-territorial males and immatures). Some aggressive displays, including closely approaching, sniffing, touching, and mating postures, are displayed by non-territorial males and immatures but are not noted during conflicts between territorial males (Table 6.2). Space-claim activities are observed quite rarely in non-territorial males. For example, the grazing ritual is very rarely observed in non-territorial males in conflicts with peers (9.2%, $n = 130$) and even less often with immatures, whereas in territorial males, it is one of the most frequent patterns of aggressive demonstrations. Immature males, when interacting with peers, only use the grazing ritual in exceptional cases (Blank and Yang 2014).

6.2.7 Injury of Goitered Gazelle Males From Fighting

From all the above, it becomes clear that goitered gazelle males try to avoid direct interactions and fights, preferring noncontact threats at a distance from their rivals, but fights still happen on occasion and can be quite fierce. It is possible that fights between males occur more often in populations where the sex and age composition is disturbed and there is an increased number of adult males. Sludskiy (1956), observing the artificially created goitered gazelle population on the island of Barsakelmes, where predators were completely absent, thought that during competition for females, males have numerous violent fights during the rutting season, as adult males with bloody bodies were commonly observed and some fights between males ending in the death one combatant: in 1942 and 1943 on Barsakelmes Island, between 14 and 15 wounded and dead males were found per year. Among the damages were a gouged-out eye; large, lacerated wounds on the neck; broken and damaged legs; and other injuries. In the Betpak-Dala Desert in Central Kazakhstan, an old male was observed with a horn broken almost to the base and a ripped ear (Sludskiy 1956). It should be noted that most wounds were inflicted on the horns and head (forehead, eyes, ears), which is due to the fact that antelope males almost never strike their horns into an opponent's body but instead prefer to direct their horns towards the head and horns of their rival. The opponent also directs his head toward the attacking male and meets the blow with his horns. Thus, antelopes primarily use only their horns and heads for fighting and no other parts of their bodies (Geist 1966).

In the Ili Hollow in SE Kazakhstan, serious battles are rarely observed. In most cases, territorial males have horn-to-horn fights, and only very rarely does one male hit his opponent's body with his horns while in pursuit. The power of such blows is very limited and cannot seriously injure the opponent. Extremely rarely, though, one male may accidently inflict serious injuries on the head and legs of the other male. However, I did not see any seriously injured males during or after the rut in the Ili Hollow population, and no males were found dead as a direct result of male-male fights in the Barsakelmes population in western Kazakhstan. On February 14, 1970, however, a dead male body was discovered on the island with a large abscess on his forehead, a scar near the right elbow joint, and damaged digits on his right hind leg. This male died of sepsis and emaciation (his weight was 22 kg), presumably as a result of bodily damage from the rut. Other injuries noted during the rut: on November 26, 1970, one male limped on his left leg, and he had dried blood in the area of his right eye; and on December 19, 1970, a male was seen with a limply hung ear with a large bloody spot at his base (Zhevnerov 1984).

Goitered gazelle males in captivity will not only strike with their horns into the horns or the head of the rivals but also make direct blows with their horns into the body, causing quite serious wounds to their opponents. In a small group of gazelles artificially raised with bottles and transported to the enclosures of the Issyk Kul Breeding and Acclimatization Center in Kyrgyzstan, the most active male (at the age of 3.5 years) inflicted serious bodily wounds on a young male (2.5 years old) during the rut, after which the young male died. Moreover, this male killed two females (2.5 and 3.5 years old) in the same way, inflicting large wounds on their bodies with his horns (Dauletbakov A., oral report). Cases of males killing females during the rut are practically impossible under natural conditions, or at least none have ever been observed by anyone in the wild. Under conditions of captivity in the Bukhara Breeding Center, though, these cases have been noted during rutting season (Soldatov V., oral report).

References

Alados CL (1986) Aggressive behaviour, sexual strategies and their relation to age in male Spanish ibex (*Capra pyrenaica*). Behav Process 12:145–158

Andersson M (1994) Sexual selection. Princeton University Press, Princeton

Barrette C (1977) Fighting behaviour of muntjac and the evolution of antlers. Evolution 31(1):169–176

Blank DA (1990) Persian gazelle. In: Kovshar AF (ed) Rare animals of desert regions. Nauka Press of Kazakh SSR, Alma-Ata, pp 56–80. (in Russian)

Blank DA, Fedosenko AK (1983) False rut in Persian gazelle. In: Zakharov AA (ed) Proceedings of the third all-union conference on animal behaviour, vol 2. Nauka of the Academy of Sciences of the USSR, Moscow, pp 258–259. (in Russian)

Blank D, Yang W (2014) Does social status of males change their aggressive behavior repertoire in goitered gazelle (*Gazella subgutturosa* Guld., 1780)? Behav Proc 108:20–26

Blank DA, Ruckstuhl K, Yang W (2015) Seasonal dynamics of agonistic displays in territorial and non-territorial males of goitered gazelle. Zoology 118(1):63–68

Blank DA (1998) Mating behavior of the Persian gazelle Gazella subgutturosa Guldenstaedt, 1780. Mammalia 62(4):409–419

Bro-Jørgensen J (2007) The intensity of sexual selection predicts weapon size in male bovids. Evolution 61(6):1316–1326

Cappellini I (2007) Dimorphism in the hartebeest. In: Fairbairn DJ, Blanckenhorn WU, Szekely T (eds) Sex, size and gender roles. Oxford University Press, Oxford, pp 124–132

Clutton-Brock TH, Albon SD (1979) Roaring of red deer and the evolution of honest advertisement. Behaviour 69:145–170

Clutton-Brock TH, Albon SD, Gibson RM, Guinness FE (1979) The logical stag: adaptive of fighting in red deer (*Cervus elaphus* L.). Anim Behav 27:221–225

Corlatti L, Bethaz S, Hardenberg A, Bassano B, Palme R, Lovari S (2012) Hormones, parasites and male tactics in Alpine chamois: identifying the mechanisms of life history trade-offs. Anim Behav 84:1061–1070

Corlatti L, Palme R, Lovari S (2014) Physiological response to etho-ecological stressors in male Alpine chamois: timescale matters! Naturwissenschaften 101:577–586

Creel S (2001) Social dominance and stress hormones. Trends Ecol Evol 16:491–497

David JHM (1973) The behaviour of the bontebok, *Damaliscus dorcas dorcas* (Pallas1766), with special reference to territorial behaviour. Z Tierpsychol 33:38–107

Eibl-Eibesfeldt I (1961) The fighting behavior of animals. Scient Am 205:112–122

Enquist M, Leimar O (1987) Evolution of fighting behaviour: the effect of variation in resource value. J Theor Biol 127:187–205
Estes RD (1969) Territorial behaviour of the wildebeest (*Connochaetus taurinus* Burchell, 1823). Z Tierpsychol 26:284–370
Estes RD (1991) The behaviour guide to African Mammals. University of California Press, Los Angeles
Geist V (1966) The evolution of horn-like organs. Behaviour 27:175–214
Geist V (1971) Mountain sheep. University of Chicago Press, Chicago
Gosling LM (1974) The social behaviour of Coke's hartebeest *Alcelaphus buselaphus cokei*. In: Geist V, Walther FR (eds) The behaviour of ungulates and its relation to management, IUCN Publ. No. 24. IUCN, Morges, pp 488–511
Habibi K, Thouless CR, Lindsay N (1993) Comparative behaviour of sand and mountain gazelles. J Zool 229:41–53
Isvaran K (2005) Variation in male mating behaviour within ungulate populations: patterns and processes. Curr Sci 89:1192–1199
Janis C (1982) Evolution of horns in ungulates: ecology and paleoecology. Biol Rev 57:261–318
Jennings DJ, Gammell MP, Carlin CM, Hayden TJ (2003) Is the parallel walk between competing male fallow deer, *Dama dama*, a lateral display of individual quality? Anim Behav 65:1005–1012
Li C, Zhigang J, Yan Z, Coie Y (2004) Relationship between serum testosterone, dominance and mating success Pere Davids deer stags. Ethology 110:681–691
Linklater WL (2000) Adaptive explanation in socio-ecology: lessons from the Equidae. Biol Rev 75:1–20
Marmazinskaya NV (1996) Territorial and marking behavior of goitered gazelle *Gazella subgutturosa* (Artiodactyla, Bovidae) in the Bukhara ecological center. Zool J 75(11):1737–1751. (in Russian, English summary)
Marmazinskaya NV (1999) Social behavior of goitered gazelle (*Gazella subgutturosa*)/Güld., 1780/during breeding period in the conditions ecocenter "Goitered gazelle". Ph.D. thesis in zoology, Institute of Zoology of Uzbek Academy of Sciences, Tashkent, pp 1–20. (in Russian)
Maynard Smith JM, Price GR (1973) The logic of animal conflict. Nature 246:15–18
Miles DB, Sinervo B, Hazard LC, Svensson EI, Costa D (2007) Relating endocrinology, physiology and behaviour using species with alternative mating strategies. Funct Ecol 21:653–665
Müller-Schwarze D (1999) Signals specialization and evolution in mammals. In: Johnston RE, Müller-Schwarze D, Sorensen P (eds) Advances in chemical signals in vertebrates. Plenum Press, New York, pp 1–14
Parker GA, Rubenstein DI (1981) Role assessment, reserve strategy, and acquisition of information in asymmetric animal conflicts. Anim Behav 29:221–240
Payne RJH, Pagel M (1996) Escalation and time costs in displays of endurance. J Theor Biol 183:185–193
Payne RJH, Pagel M (1997) Why do animals repeat displays? Anim Behav 54:109–119
Penfold LM, Ball R, Burden I, Jöchle W, Citino SB, Monfort SL, Wielebnowski N (2002) Case studies in antelope aggression control using a GnRH agonist. Zoo Biol 21(5):435–448
Pérez-Barberia FJ, Gordon IJ (2000) Difference in body mass and oral morphology between the sexes in the Artiodactyla: evolutionary relationships with sexual segregation. Evol Ecol Res 2:667–684
Schenkel R (1966) On sociology and behaviour in impala *Aepyceros melampus suara* Matschie. Z Süugetierkunde 31:177–205
Senar JC (1990) Agonistic communication in social species: what is communicated? Behaviour 112:270–283
Shargal D, Shore L, Roteri N, Terkel A, Zorovsky Y, Shemesh M, Steinberger Y (2008) Fecal testosterone is elevated in high ranking female ibexes (*Capra nubiana*) and associated with increased aggression and a preponderance of male offspring. Theriogenology 69(6):673–680
Skinner JD, Louw GN (1996) The Springbok (*Antidorcas marsupialis* Zimmermann, 1780). Transvaal Museum, Pretoria

Sludskiy AA (1956) The breeding in goitered gazelle. Proc Inst Zool Acad Sci KazSSR 6:78–108. (in Russian)

Tsaplyuk OE (1972) Age and seasonal dynamics of sexual activity in *Gazella subgutturosa* Gueld. Proc Acad Sci KazSSR Biol Ser 3:39–46. (in Russian)

Walther FR (1965) Verhaltensstudien an der Grantgazelle (Gazella granti Brooke, 1872) im Ngorongoro-Krater 1, 2. Z Tierpsychol 22(2):167–208

Walther FR (1974) Some reflections on expressive behavior in combats and courtship of certain horned ungulates. In: Gest V, Walther F (eds) The behaviour of ungulates and its relation to management. IUCN. Publications new series, No 24, Morges, pp 56–106

Walther FR (1978) Forms of aggression in Thomson's gazelle: their situational motivation and their relative frequency in different sex, age, and social classes. Z Tierpsychol 47:113–172

Walther FR (1981) Remarks on behavior of springbok Antidorcas marsupialis Zimmerman, 1780. Zool Garten 51:81–103

Walther FR (1984) Communication and expression in hoofed animals. Indiana University Press, Bloomington

Walther FR, Mungall EC, Grau GH (1983) Gazelles and their relatives. Noyes Publications, Park Ridge

Willisch CS, Neuhaus P (2009) Alternative mating tactics and their impact on survival in adult male Alpine ibex (*Capra ibex ibex*). J Mammal 90:1421–1430

Zhevnerov VV (1984) Goitered Gazelle of the Barsa-Kelmes Island. Nauka Press of the Kazakh SSR, Alma-Ata. (in Russian)

Birthing and Related Behavior

7

7.1 Introduction

7.1.1 Birthing

Behavior associated with birthing and rearing fawns in some African and partly Asian gazelles and gazelle-like species, including gerenuks *Litocranius walleri*—Kirchshofer 1963; Reif and Klingel 1991; Thomson's and Grant's gazelles—Walther 1965, 1968; Dorcas gazelles (*Gazella dorcas*) and mountain gazelles (*G. gazella*) (Baharav 1983), was investigated with varying degrees of detail and reported in English publications several decades ago. In contrast, birthing and suckling behaviors of goitered gazelles generally remain unknown for most English readers, since previous data on this Central Asian species have primarily been published in Russian (Zhevnerov 1984; Blank 1985; Pereladova and Pereladov 1986; Marmazinskaya 1999). Only in recent years has the birthing behavior of this species started to be published in English (Blank 2017; Blank and Yang 2015a, b), while most basic information on the behaviors of birthing and rearing goitered gazelle young has remained untranslated until now. In this chapter, I will translate and present these behaviors, along with my own, more recent observations.

7.2 Birthing. Results and Discussion

7.2.1 Pregnancy and Preparation for Birth

For a long time, an approximate duration of the goitered gazelle pregnancies was estimated at 5.5–6 months (Sludskiy 1956; Heptner et al. 1961). Zhevnerov (1984), however, determined a much more precise duration of pregnancy in his tame females: the first female was fertilized between December 28 and 30 and the second one was fertilized between January 22 and 25; the first female gave birth on May 28

D. Blank, *The Behavioral Biology and Conservation of Goitered Gazelles (Gazella subgutturosa)*, https://doi.org/10.1007/978-3-031-83504-9_7

and the second on June 23. The significant difference in the timing of the females' estrus cycle was due to differences in their physical development: the first female was stronger and had a better fat layer. Thus, these females' pregnancies lasted 149–152 days and 148–151 days, respectively, which equals an average of 5 months. On the Barsakelmes Island (Aral Sea), mass mating is observed at the end of November into early December with mass birthing beginning after May 10, making the pregnancy duration of this population slightly longer than 5 months (Zhevnerov 1984). Judging from periods of mass copulation and parturition in mainland populations, gestation also lasts slightly more than 5 months (Heptner et al. 1961). Other authors have pointed out similar pregnancy durations from other parts of the goitered gazelles' natural habitat, with this duration also over 5 months: 148–159 days in Pakistan (Roberts 1977; Frazier and Hunt 1994), 160–164 days in Turkmenistan (Pereladova and Pereladov 1986), and 154–158 days in Uzbekistan (Marmazinskaya 1996). In Iraq, pregnancies of *G.s. marica* pregnancy last 152–165 days (Shahooth et al. 2015), and in Saudi Arabia, the duration is 148–160 days (Al-Eissa et al. 2011). Considering the influence of various external factors, conditions of detention, and individual differences in the timing of pregnancy, I can safely say that the duration of pregnancy in goitered gazelles lasts within 148–165 days (or 5–5.5 months) in most cases.

Throughout the winter and early spring, females usually stay in large female or mixed groups with males, freely moving from one group to another (Zhevnerov et al. 1983). Single adult females, with rare exceptions, cannot be found from January to April (Zhevnerov 1984). Though some pregnant females leave their herds as early as the end of March (Blank and Yang 2013), most separate from their groups and relocate to places suitable for birthing at the end of April into May. At this time the number of single females begins to increase sharply, especially in May (Tables 7.1 and 7.2). This process of each female relocating from their groups proceeds gradually throughout the entire pre-birthing period of all pregnant females. A comparison of the dates for the first appearance of a single female at her chosen birthing sites and the date of her first newborn suggests that pregnant females leave their herds about 1–4 days before giving birth. This timing was also observed in tamed females on Barsakelmes Island (Zhevnerov 1984). Some females in the wild, however, will stay alone much earlier (up to 1.5 months) than most females (Blank and Yang 2013). In addition, pregnant tamed females usually separate and stay apart from other gazelles as much as possible as early as a month before giving birth,

Table 7.1 Occurrence of single females observed on the Barsakelmes Island during May of 1970–1975 (adapted from Zhevnerov 1984)

Month of May in different decades	Total number of observed females	Single females from them	
		Number	%
First	590	155	26.3
Second	620	335	54.0
Third	540	380	70.4

Table 7.2 Time of goitered gazelle birthing in the Central Betpak-Dala area, May 1955 and 1956 (adapted from Sludskiy 1956)

	Observed females					
	Pregnant		Afterbirth		Total	
Date	number	%	number	%	number	%
1955						
6–15 May	15	83.3	3	16.7	18	100
16–26 May	5	38.4	8	61.6	13	100
1956						
1–15 May	5	45.5	6	54.5	11	100
16–30 May	2	22.3	7	77.7	9	100

using any barrier available in the enclosure to isolate themselves from conspecifics (Pereladova and Pereladov 1986).

Pregnant females move to their birthing sites where they graze and rest, waiting until they give birth. These birthing sites have a well-defined relationship with the nature of the terrain (Zhevnerov 1984). During their isolation, females noticeably change their behavior toward being more aggressive toward other gazelles. As a result, the concentration of pregnant females in one particular "maternity" area, which is typical for saiga antelopes (Fadeev and Sludskiy 1982), is not seen in goitered gazelles. Some authors (Heptner et al. 1961), however, have previously described seeing a concentration of pregnant goitered gazelle females in "birthing places" on the Badkhyz Plateau (southeastern Turkmenistan), but at this time, these concentrations of birthing females are absent in that area, as well (Gorelov 1972).

7.2.2 Birthing Time

Goitered gazelle fawns are commonly born during March and April in Saudi Arabia with a 3-week peak during March (Habibi et al. 1993) and in Iraq and southern Central Asia (Turkmenistan and Tajikistan) during mid-April and May (Flerov 1935; Gorelov 1972; Heptner et al. 1988). In the Transcaucasus region, newborns appeared at the mid and end of May (Vereshchagin 1951). In Uzbekistan (NW Kyzylkum Desert) birthing occurs from May 10 to 20 (Mambetzhumaev 1970a, b), or according to another source, from April 22 to 25 to May 10 to 13 in the Bukhara Breeding Center (Marmazinskaya 1996). And after reintroduction at the Issyk-Kul Breeding Center, Kyrgyzstan, goitered gazelle females originally from Uzbekistan and Turkmenistan give birth between April 20, 2023, and May 4, 2024 (Dauletbakov in litt.).

In almost the entire area of goitered gazelle distribution in Kazakhstan, the birth of young occurs almost everywhere within a 2.5-month time window between the end of April (first birth) to the first days of July (last birth) (Kostin 1955; Sludskiy 1956; Zhevnerov et al. 1983), while mass birthing is very synchronized in most females with the appearance of the largest number of fawns born between 15 and 25 May (Blank 1985). In the Betpak-Dala Desert (Central Kazakhstan), the first female to give birth was found on May 5 and the last one on May 30 (Sludskiy 1956—Table 7.1). In the Telekul Lakes area of the Betpak-Dala Desert, mass birthing of gazelles occurs between May 10 and 20 (Afanasief and Sludskiy 1947). According to Kostin (1955), birthing takes place on the Ustyurt Plateau starting in April run and into the first half of May; on the Barsakelmes Island, birthing begins in early May with mass birthing from May 10 to 20. In southeast Kazakhstan, in the Ili Valley and adjacent areas, birthing of gazelles also begins in the second half of May (Antipin 1941), where the first newborn fawns were found on May 17 and 23. Birthing takes place for most females during a very short period and pregnant females are almost never seen after May 25 (Antipin 1941).

Some cases of delayed pregnancy have been noted. On the Ustyurt Plateau (W Kazakhstan), Kostin (1955) observed pregnant females and newborn fawns on June 29. In the Kyzykum Desert, some females give birth at the end of June or early July (Fig. 7.1—N8405), though fawns born in May can reach close to adult sizes by this time and, therefore, they have a better chance of successful survival during the winter period (Fig. 7.2—M5410). Some females birth later in the Barsakelmes population, with the latest births noted in 1946 on July 4 and 5 (Ismagilov 1954); on June 7 of 1971, a 2- to 3-day-old gazelle fawn (estimated birth June 4 or 5) was seen to run up to a human observer and poke its muzzle into the top of the observer's boot. In 1972 and 1973, 2 newborns were found on June 4 and 12 (Zhevnerov 1984). On June 22 in the Ili Valley, a pregnant female was hunted and killed with embryos at the age of 3–4 months (Antipin 1941). On June 5 of 1955, a fawn was caught in the

Fig. 7.1 In goitered gazelles, some females give birth unusually late. A female with fawn of 2-3-week-old (mid-July) in the Kyzylkum Desert, Southern Kazakhstan. 18.07.2017. Photo of Zhaskairat Nurmukhambetov with kind permission

Fig. 7.2 In July, fawns at the age of almost 3 months can reach the sizes which are close of sizes of their mother. A female (in the front) and 3 fawns near watering place. Southern Ustyurt Chink (Kaplankyr), SW Kazakhstan. 05.08.2019. Photo of the camera-trap of the Ustyurt State Nature Reserve team with kind permission

Ili Valley at an estimated age of about 3 days (estimated birth June 2), since the umbilical cord had not yet fallen off (Sludskiy 1956). A significant delay in birthing in some females can be explained in two ways: by the fact that females who are not fertilized during the rutting period can come into estrus a second time after the end of the primary mating period (Tsapluk 1972) and by the fact that late-birthing females are mostly young animals participating in reproduction for the first time. As a rule, in all late births, there was just one fawn, which is common for young, first-time mothers (Zhevnerov 1984).

In Mongolia and China, the breeding period of gazelles is from June to early July, a delay of a whole month or even more compared to Central Asia (Allen 1940; June 30 and July 4—Bannikov 1954). This timing difference can be explained by a prolonged, cold spring, which is typical of these countries with a severe continental climate. In addition, spring in Mongolia and China is extremely dry, with no ephemeral vegetation; meanwhile gazelles during and for the first days after birthing need a lot of water to drink. In Mongolia, the maximum precipitation falls in June through August and perhaps the late spring rains are directly related to the late onset of birth (Bannikov 1954).

The seasonality of births in Saudi Arabia and Central Asia coincides with the maximum food availability that follows winter and spring rains (Habibi et al. 1993; Kingswood and Blank 1996). It should be noted that in the northern parts of the goitered gazelles' natural habitat, mass birthing coincides with the onset of warm weather and the cessation of spring frosts, which favors the survival of newborn fawns. The average minimum air temperatures in the Betpak-Dala Desert for May are +9.1 °C in the first week, +9.9 °C in the second, and + 12.6 °C in the third. The average maximum temperatures are +24.1 °C, +26.7 °C, and +30.6 °C, respectively, though the absolute maximum of the air temperatures has reached +32.9 °C, while on the surface of the soil, the temperature can rise even as high as +49.8 °C. On Barsakelmes Island, the birth of gazelles also occurs during the most favorable time of year, which is characterized by a transition of +5 °C in the average daily air temperature and the end of the spring frosts. The average minimum air temperatures in May on Barsakelmes Island over a 10-year period of observations were equal to +5.4 °C in the first week, +6.5 °C in the second, and +10.6 °C in the third. Average maximum temperatures in the first week were +18.6 °C, in the second +20.3 °C, and in the third +22.9 °C. During this period, there is an increase in precipitation and the mass growth of vegetation with shrubs, herbaceous perennials, and annuals; small puddles of freshwater are available in the depressions (Zhevnerov 1984). Snowfalls or sudden drops in temperature can delay parturition by several days (Blank 1997).

Despite such high temperatures in May, there are still frosts in the first half of the month. In 1940, frost was observed on May 17 (−2.5 °C), and in 1941, the last frost was recorded on May 5 (−3.2 °C), while the amount of precipitation for May was negligible (in 1941 it was only 5.6 mm). In May 1955, 15.4 mm of precipitation fell, of which 13.6 mm fell in the first week; in the second week, precipitation was completely absent and in the third 1.8 mm fell. In the second half of May of this year, there was a sharp transition to a sultry summer in the Betpak-Dala Desert. Usually at this time of year, the gradual yellowing and dying of the leaves of *Salsola arbuscula*, *Anabasis salsa*, and *Artemisia terrae-albae* begins. Ephemeroids are drying up. The drying of *Rheum* leaves is especially noticeable with its large yellow and light brown leaves dotting all around among the green wormwood and *Salsola arbuscula*, giving a mottled color to the vegetation of the desert. This time of drying *Rheum* leaves coincides with the mass birthing of goitered gazelles and especially saigas (Sludskiy 1956).

G. subgutturosa marica (maybe *G. marica*) in Syria also had a highly synchronized birthing period with most births taking place after a migration from Saudi Arabia, Jordan, and southern Syria to the banks of the Euphrates River. These migrations were related to the need for water during lactation and rearing young (Legge and Rowley-Conwy 1987). In Saudi Arabia, *G.s. marica* females can give birth all year round with two peaks: in March (most females) and in April and May (3–10 times less often compared to March, respectively) and in September and October (least often) (Riesch et al. 2013).

Meanwhile, during birthing in May and June and in the first month after it, female gazelles have a great need for water due to lactation, and they can almost always satisfy it. During this period, in all northern deserts, not only clay, but also sandy deserts, there is a lot of water in puddles and other temporary reservoirs located on takyrs (a type of relief similar to salt flats that are typical for many deserts in Central Asia); there are often temporary streams and rivers and just gullies with good freshwater. In addition to temporal water sources, there are still many green, juicy plants despite the partial drying of ephemeroids during years with very dry Mays. Parturition of all species of desert gazelles everywhere coincides with the period of precipitation and intense plant growth (Zhevnerov et al. 1983). The frequency of reproduction, which has developed in the wild homeland of gazelles, is so fixed hereditarily in their behavior that it persists even when they are kept in captivity, despite the uniform supply of food and water throughout the year. Captive births in North America have also been seasonal: 70% ($n = 299$) of *G. s. marica* births were during March and April and 84% ($n = 729$) of *G.s. subgutturosa* births were during April and May (Carter 1991). Births in the London Zoo peaked during May (Roberts 1977).

7.2.3 Birthplace

Female gazelles everywhere stay alone or in small groups when giving birth and mothers generally occupy a large area. For example, in the Betpak-Dala Desert, the birthing area for goitered gazelles covers almost the whole southern half of this vast desert (Fig. 1.42). Female goitered gazelles do not gather in large herds during the birthing period, most likely due to the shortage of feed and the lack of water in the desert (Sludskiy 1962). Published information is contradictory regarding the birthplaces of gazelles, but most authors point to semi-closed areas of the landscape, thus concentrating the bulk of mothers in certain areas (Ismagilov 1954; Sludskiy 1956; Sokolov 1959). The attraction for females before giving birth to areas where there are shelters such as bushes, hollows, cliffs, and chinks can be explained by the fact that they instinctively seek to shelter their newborns from strong cold winds, often blowing in the desert in May, as well as from possible predators. Severtsov (1873) had a rather dated view that female gazelles went into the thickets of the saxaul forests to birth their young and came out after the young grew up; though in reality, goitered gazelle females avoid dense thickets and birthing takes place on the outskirts of saxaul forests or in highly sparse plantings (Zhevnerov et al. 1983). In March on the Ustyurt Plateau, female goitered gazelles migrate to the internal parts of the plateau to the saxaul thickets, where they give births (Kostin 1955). And in the western part of their range, gazelles gather in significant numbers in the fixed sands of Sam that are overgrown with rare saxaul and semi-shrubs (W Kazakhstan) (Sludskiy 1956). Females also birth their young in the Barsakelmes Basin (W Uzbekistan), which is an area that is warmer and drier than the surrounding landscape. This basin is a huge depression with a large salty lake close to the chinks (cliffs) of the Ustyurt Plateau. In the southwestern parts of the Barsakelmes Basin, females with fawns can be found in spring (March–April), where they drink rainwater that accumulates in depressions and on takyrs. With the onset of the summer season, the gazelles most likely migrate to places of permanent watering holes

(Lake Sarykamysh), and where the food supply is adequate (Marmazinskaya et al. 2012).

In the Kyzylkum Desert, females give births on the sands of barchan ridges that have a protective effect and early spring vegetation (Ishunin 1961). Female gazelles also gather for birthing young in the chinks of the Ustyurt Plateau, where they stay and hide their fawns among the sparse bushes growing on the outskirts of the saxaul forests. In addition, there are good watering holes with freshwater in all these places I have just described. When the fawns get stronger, the females along with their fawns begin to move further north of the Plateau (Kostin 1955). Less often, females choose to give birth in the narrow valleys occupied by swollen salt marshes, which are sparsely overgrown with large shrubs of *Achnatherum splendens*, where females seek out open areas. A small number of females give birth near trails and slopes of low desert hills and in the saxaul forest along the Chui River, choosing areas where thickets alternate with large, flat open areas or a hollow that is sheltered from the cold winds. They will also quite often choose to give birth to their young on the outskirts of bumpy sands along the slopes of foothills (Sludskiy 1977).

On Barsakelmes Island, female gazelles disperse throughout the area to give birth, where it is not typical to find closely located hiding young or the presence of two or more females, even in years with a high population density (the population noted in one season reached 800–900 gazelles) (Ismagilov 1954). Before giving birth, female gazelles choose a relatively protected environment, tending to hide in areas with bushes, hollows, or cliffs, which are better protected from the cold, strong winds of spring (April–May) and predators. So, on the Barsakelmes Island, births usually occur on the slopes of gullies, on ravines, or in depressions with thickets of *Caragana* spp. About 70% of newborns were once found near the upper part of the slopes of the ravines but rarely seen at the wet bottom with long-lasting springs. Some females give birth in a sparse saxaul forest, in the glades of the forest, or on the outskirts.

To a lesser extent, birth sites on the island are confined to solid shrubby thickets and ridge sands. In the depths of the dune strip, fawns were discovered only twice: on May 17, 1970, one helpless fawn was found near the ridge of the dune, and on May 21, 1971, a second newborn was found in the depths of the dunes lying near the second dune ridge from the sea among ephedra bushes (*Ephedra*) (Zhevnerov 1984).

In dry years, the number of births in areas directly adjacent to shrubby thickets and fixed sands increases markedly. This was observed in the spring of 1970. The previous winter was snowless, so the grass cover of the main part of the island developed very poorly and dried up early. Temporary freshwater reservoirs formed from rains, and in the ridge sands and strip of land near them, the quality of the food was better. In addition, in the absence of freshwater, the sea remained the only water source for gazelles, which probably was an important contributing factor for the relocation of the birthing location to the near-barkhan (sand dunes) zone. It is significant that in 1970, saigas also changed their choice of birthing sites from the open plain to the border of the fixed sands and closer to the water sources (Zhevnerov 1984). With the appearance of newborns, the need for water increases in females, as indicated by the intensity of their visits to water sites during lactation. From the second week of May, they can be regularly found on the coast throughout the daylight hours. Tame females, before giving birth, drank water 1–2 times a day. With the beginning of lactation, their need for water increased to 3–4 times a day (Zhevnerov 1984).

In the Betpak-Dala Desert (Central Kazakhstan), birthing of gazelles takes place in more or less defined areas which lie between the 45°N and 46°N parallels near the northern edge of the desert zone. In the central and eastern Betpak-Dala Desert, births occur in the low-elevated mountains and foothills, more often among Spirea bushes in small hollows of chinks with a southern exposure or on the extensive, slightly undulating valleys, where thickets of *Spiraeanthus schrenkianus* and *Caragana* alternate with large glades overgrown with sagebrush. Near the mountains, the females birth their young in dry channels and gullies with bushes, but also in open areas (Zhevnerov et al. 1983). Observations in the Badhyz State Nature Reserve in Turkmenistan demonstrated that birthing females use vast grounds to disperse widely enough so as to preserve the lives of their newborns. If females are too concentrated in an area, fawns become vulnerable to predation and the effects of human economic activities (Kuznetzov 1986).

In the southeast of Kazakhstan, a small number of gazelles give birth on the western edge of the sands of Sary-Ishikotrau in a rare saxaul forest with large open areas in the vicinity of the Bakanas dry river-bed. Also, many birthing females gather in the area of the Malai-Sary Ridge and its foothills, where they stay in the fixed sands overgrown with bushes of *Eurotia ceratoides*. In the Ili Valley, a significant number of pregnant females go to the southwestern spurs of the Dzungarian Alatau and climb to heights of 1500 m ASL. Birthing of gazelles occurs there at the bottom of the channels of dried mudflows overgrown with brushwood. Gazelles also go to the Boguty Mountains for birthing (Ili Valley) (Sludskiy 1956). In Turkmenistan, pregnant gazelles also climb into the foothills and enter the gorges that open onto the plains. When the fawns get stronger, they re-enter the open areas (Vinogradov 1952).

As a rule, in all the locations described above, there are good watering sites in the form of springs and small wells, in which the water rises almost to the very surface. In addition, from April into June, there are a lot of puddles with freshwater after frequent spring rains. My long-term observations in the Ili Valley show that in May and June and later, gazelles are concentrated from 1 to 10 km around watering sources. Other authors (Sludskiy 1956; Heptner et al. 1961; Sokolov 1959) also observed that pregnant females do not go more than 10 km from watering places. In contrast, Gorelov (1972) argued that pregnant females can stay several tens of kilometers from water when birthing their young and satisfy their water demands in the spring from consuming succulent plants. Drinking water is also very important for females of saiga antelopes; therefore, they choose birthing places not far from watering sites, because they need drinking water regularly when nursing their young (Fadeev and Sludskiy 1982). The birthing area was located especially close to watering places for the Mongolian subspecies of saiga antelope (Buuveibaatar et al. 2014).

7.2.4 Birthing

The first noticeable signs of udder enlargement in pregnant females appear 17–21 days before delivery (Zhevnerov 1984; Pereladova and Pereladov 1986). Enlarging of the back part of the belly occurs 7–10 days before birth. Based on observations of a captive population, the pregnant female is constantly alert at this time; she avoids other members of the group, goes to lie in the corner of the enclosure farthest from the group, and moves away from the feed when other members of the group approach. The udder's further development is insignificant until a day or two before delivery, when it sharply increases in volume. The maximum growth of the udder occurs 10–12 h before birth (Pereladova and Pereladov 1986). Simultaneously with the onset of the udder swelling, secretory activity of the inguinal glands increases. Their secretions, a yellowish liquid, begin to accumulate in the pockets of the glands and, as the birth approaches, are released onto the udder (Zhevnerov 1984). One to two days before giving birth, the female begins to lie down often. The last day she spends 30–40% more time lying down than usual and rarely, and only when alone, does she go for food. During the last day before giving birth, the female no longer leaves when other members of the group approach, but aggressively drives them away, including adult males, and retains her preferred resting position. Two to three hours before delivery, a change in the shape of the belly can be seen, and labor pains begin. The female at this time moves along the grid of the enclosure pacing 10–15 m back and forth in a "shuttle movement." Such movements in all gazelles were recorded in an excited state (Pereladova and Pereladov 1986).

In general, a female leaves her group and chooses a new site a few days (sometimes more) before giving birth, which she uses mostly for grazing and resting. For 2–3 h before giving birth, she begins to show uneasiness: she often lies down and gets up about every 5–10 min, walks and sniffs the ground, and often urinates. Discharge from the vulva is observed 30–60 min before birth. The female finally chooses the place for birthing her young, which is an area of soil without vegetation. The female slightly loosens the ground with a few blows from her front legs though not very carefully. At the beginning of the birth, the female lays usually on her side, somewhat closer to her back, bending her front legs and stretching out her hind legs (Zhevnerov 1984). The birth is accompanied by intense labor pains and attempts to get rid of the fetus, getting up and lying down several times and licking her anogenital area which lasts 15–20 min (per Pereladova and Pereladov 1986) or 26–40 min (per Zhevnerov 1984). When more than half of the fetus comes out, the mother rises abruptly and takes 2–3 steps, after which the newborn falls to the ground, breaking the umbilical cord. The afterbirth usually comes out 1.5–3 h after delivery (Zhevnerov 1984).

In other cases, the female does not stand up, but lies down until the end of the birth and then eats the umbilical cord. With the onset of labor and the appearance of the newborn's forehead, the female stopped licking her body, lowered her leg, and began to touch her belly with her nose. The time from the beginning of the appearance of the head to the completion of the birth of the first newborn was 3–4 min: at

1.5 min, the head appeared, at 0.5 min the neck appeared, at 1.5 min, the shoulder girdle appeared, and the eyes opened, and at 0.5 min the pelvis and hind legs appeared. Immediately after birth, the female began to lick her newborn. When other gazelles approached, the female turned her back, moving so as to be between the intruder and her fawn. Occasionally (one time noted is 2–3 min), the female interrupted her licking, raised her head high, and looked around warily. After 14–15 min after the birth of her first young, the head of the second baby appeared, and the female laid down again. The delivery of the second young took 8 min, and after the appearance of the head and neck, the female got up, walked for 2.5 min along the grid until the appearance of the shoulder girdle, then laid down again, and intensively licked the second newborn after the birth. For 5 min, the female continued to lick the newborns from a lying position, then got up, and alternately licked the babies and herself, looking around anxiously at times. About 40 min after the end of labor, the female fed her fawns for the first time, and after suckling, she licked the anal areas of both babies for 3–7 min (Pereladova and Pereladov 1986).

For another hour, the female moved very little after birth, often lying down next to her newborns and licking them when she got up to feed. After 2 h, the female left her newborns in another corner of the enclosure, and then immediately ate the afterbirth. After that, the activity of the female sharply increased, running around the entire enclosure while being chased by a male. The female's abdominal muscles began to contract sharply and within 3 h after giving birth, her body was practically back to the proportions of other adult female gazelles (Pereladova and Pereladov 1986).

In the natural conditions of the Ili Hollow in southeastern Kazakhstan, I observed the first birth of offspring at 19:55 in the evening on May 15, 1983. Three hours before giving birth, the female behaved as usual, although she did not separate from her group before giving birth, as most pregnant females do, and continued to be with other gazelles. She was lying with her group of 3 pregnant females, a female 1-year-old, and a male 1-year-old. At 18:00, this female began to descend with the rest of the herd from the foothills to the plain. At 19:00, part of her amniotic sac came out. The female continued to graze actively, but separated from her group. She squatted several times as if urinating, trying to get rid of her fetus. At 19:44, she crouched down to the ground and sat in this position for 30–40 s, then got up, and left my field of view. At 19:55, when I was able to observe this female again, the baby was already born. At 19:56, the female walked a little, and the baby began to move, making even small attempts to run, but crouching to the ground, on half-bent hind legs. At 19:57, I observed the first suckling which lasted 44 s, during which the mother carefully licked the anal area of the fawn. After the second suckling at 19:59, she licked her whole body. Then the female moved away, and the fawn raced to catch up with her. Suckling followed for 11 s, and the mother again licked and sniffed the anal area of the baby; he began to suck from the second nipple, which lasted for 18 s. The first and second suckling was stopped by the fawn itself, and the third by the mother. At 20:15, the female left her offspring, and she began to climb to the mountains while grazing (Blank 1985).

In observations of tame gazelles by Zhevnerov (1984), with the beginning of the separation of the afterbirth, the female left her fawn at a distance of 150–300 m 1 h after delivery. This was obviously caused by the need to not leave traces of the birth near her newborn by removing the afterbirth which had begun to come out after about 2 h and 40 min. At that time, she was resting at a distance of 50 m from the newborn. Having risen, she moved to the side, occasionally laid down, and got up again. The separation of the afterbirth lasted 8 min. During this time, the female shifted from her initial lying position by 150 m. All females have similar behaviors, which explains why there is often no afterbirth at the birth sites of goitered gazelles, even near newborns that have not yet dried off. Although some mothers eat the afterbirth, others do not. Some authors (Sludskiy 1956; Sokolov 1959) claim that female gazelles do eat the afterbirth, and according to Sludskiy (1956), in the Betpak-Dala Desert, eating the afterbirth is a common phenomenon. Zhevnerov (1984), on the other hand, believes that eating the afterbirth is the exception rather than the rule. In May 1955 in the Betpak-Dala Desert region, a placenta was found in the stomachs of 5 out of 11 females which had recently given births. Females eating the placenta after birthing their young were also observed by Antipin (1941) in southeastern Kazakhstan. This phenomenon is usually explained by a protein and mineral deficiency in females after birthing. In addition, if the afterbirth remains at the birth site, it can attract predators and flies to the fawns. Dried afterbirth is rarely found near 1- to 2-day-old fawns (Zhevnerov et al. 1983).

A mother sniffs her newborn and immediately begins to lick it, starting with the head and neck and then the body and legs. The duration of licking is around 12–20 min. The mucus covering the body of the newborn is eaten by the mother, and sometimes the mucus remains are also eaten from the place of birth. The newborn learns movements very quickly. Attempts by a fawn to stand are observed from the first minutes of its life and follow one after another, becoming successful after 17–25 min. Having risen, the newborn immediately tries to find its mother's udder. At first, it randomly pokes its muzzle into its mother's neck, chest, and belly of the mother, making suckling movements with its lips. At the same time, it often falls, but after resting for 15–20 s, it crawls under the mother again, who continues to lick it. The first suckling is usually noted 45–60 min after birth, although sometimes it happens earlier (Zhevnerov 1984).

After finding the nipples, an important factor for nursing is the stimulation of the anal area of the newborn by the mother's tongue. Licking the neonate's anal area in all cases is accompanied by an increase in the activity of the offspring. During artificial feeding, the newborn immediately got up and found the nipple of the bottle when stroked in the anal area with a human hand. This stimulation in most cases also caused the effect of urination. At the first feeding, the newborn satisfied its needs for milk usually after a 2–3 short bouts (7–10 s) of suckling, which followed at intervals of 3–10 min. At this time, the fawn tended to follow its mother and thus moved away with her from the place of birth. When the fawn was sated, it laid down and hid at the site of the last suckling. Thus, the first experiences in the life of a fawn consist of being licked by its mother, developing initial movements, feeding, and leaving its place of birth, all of which takes 60–80 min (Zhevnerov 1984). In Grant's gazelles, the baby gets up 20 min after birth and the female leaves the place of birth after 40 min (Estes 1991).

In captivity, 5–6 h after giving birth, the female, in the process of the next feeding, transferred her fawns to another corner of the enclosure and left them there as she rejoined the group. A group of adult gazelles restored their normal use of the enclosure, with a bed in their preferred eastern corner where the birth took place, and the fawns laid in the southwestern corner of the enclosure, in the grass, under a bush, or near reed mats by the fence that were little used by adults (Pereladova and Pereladov 1986).

Intensive lactation in goitered gazelles lasts about 3 months until mid-August. Some females, whose fawns appear in June or even in early July, feed their young with milk until October (Sludskiy 1956). In the Ustyurt, according to Kostin (1955), lactating females were found from June 16 to August 3. On the Barsakelmes Island, females' lactation lasts 65–80 days and suckling fawns were found most often from May to July, with the latest seen on September 12 (Zhevnerov 1984). Detailed research data on suckling behavior in goitered gazelles are presented in Chap. 8: Suckling Behavior.

7.2.5 Number of Young

During years with favorable environmental conditions lasting through breeding and birthing seasons, goitered gazelle fertility can support an increase in population of 70–80% by autumn (Heptner et al. 1988). Goitered gazelles (*G. subgutturosa subgutturosa*) usually have one (mainly in young females) or two fawns (mainly in adult females), but occasionally a female will give birth to three (Fig. 7.3) and even four (Sokolov 1959; Heptner et al. 1961; Carter 1991). Between 3 and 7 years of age, females generally give birth to twins, but younger and older females usually produce single fawns (Kingswood and Blank 1996). Some adult females will birth just one fawn as a result of poor individual body health and/or environmental conditions (Marmazinskaya 1996). In the Bukhara Breeding Center before September 1989, 72.5% of pregnant females had one fawn and 27.5% had twins ($n = 49$); in 1991, 66.3% of pregnant females had one fawn and 33.7% birthed two ($n = 98$). Triplets are rarely observed among goitered gazelles in the wild but seen occasionally in captivity (Fig. 7.4). In 1989, the number of barren females was 43.4%, and on average, there were 1.38 fawns per female. In 1991, the percentage of barren females in Bukhara Center was 19.8–29.8%, and there were 1.04 fawns born per female on average. The decrease in the number of fawns was due to a high percentage of young females who did not give birth which resulted in establishing of the maximum allowable density of the population in the semi-captive conditions of the Bukhara Breeding Center (Marmazinskaya 1996). In Kazakhstan, the proportion of barren females is usually 8.7–11.7% (up to 24.6%), and the average number of fawns per single female is normally 1.68 (Sludskiy 1956; Zhevnerov 1984).

Fig. 7.3 Some goitered gazelle females have 3 fawns in wild environmental conditions of the Southern Ustyurt Chink, SW Kazakhstan. A female and 3 fawns (2 males and a female) in the watering place. Southern Ustyurt Chink (Kaplankyr), W Kazakhstan. 07.08.2019. Photo of the camera-trap of the Ustyurt Nature Reserve team with kind permission

Fig. 7.4 In goitered gazelle, triplets are rather rarely observed in the wild, though may be more common in captivity. A goitered gazelle female with three fawns in the enclosure of the Bukhara Breeding center. Bukhara, Uzbekistan. 22.05.2014. Photo by M. Pestov with kind permission

In the natural conditions of Kazakhstan, 75% of female gazelles give birth to 2 young and about 14% have single fawns; triplets are rarely found and this phenomenon is probably less than 1% ($n = 164$) of the time (Sludskiy 1956). In unfavorable years, females decrease their productivity when the proportion of mothers giving births to twins may decrease by up to 2.6% (Kingswood and Blank 1996). In Turkmenistan, goitered gazelle females usually birth two fawns, less often singles, and least often three (Dinnik 1910; Gorelov 1972). The same numbers are also indicated for Tajikistan (Flerov 1935). In Uzbekistan, in the northwestern part of the Kyzylkum Desert, the number of fawns is 1 or 2 for each female (Mambetzhumaev 1970a, b). In Turkey, *G. s. marica* has 71% of pregnancies resulting in twins (61 twinning females from 84 females giving birth) (Gürler et al. 2015). In Transcaucasus, females have more singles than twins, and in Mongolia, most females usually have only one fawn each (Bannikov 1954; Heptner et al. 1961). There are exceptions noted that gazelles have triplets and even 4 fawns each: a female with 3 embryos was shot in the Badkhyz Plateau (SE Turkmenistan) (Gorelov 1972). An employee at the weather station in the Guryev region of Kazakhstan found 4 embryos in a pregnant female, extracted on April 29, 1965 (Mambetzhumaev 1970a, b), and Zhiryakov (1976) also reported on the meeting of a female with 3–4 fawns.

In general, the weight of a pregnant adult female in May ranged between 21 kg and 33 kg for an average of 27 kg, and the weight of yearling females varied between 16.2 kg and 22.9 kg for an average of 21.4 kg. Females with 1 fawn were also mainly yearling females (Sludskiy 1956). In the spring and summer months in the Betpak-Dala Desert, only 24.0–61.2% of females were with 2 fawns; the rest had one. In 1955, 75% of females were with two embryos, with a smaller number of twins found later due to the death of one of the fawns within their first weeks after birth. On August 15, 1951, Sludskiy (1956) found a female with 3 young of the same age near the Atasu River in the NE Betpak-Dala Desert. Out of 234 females, she was the only one with 3 fawns (0.4%), and she stayed apart from the other gazelles. In Ustyurt and Mangyshlak (W Kazakhstan) in May and June 1964, out of 29 pregnant females and those seen with fawns, 21 (72.5%) females had twins and 8 (27.5%) had a single fawn each (Sludskiy 1956).

Ismagilov (1954), who studied the breeding of gazelles on the Barsakelmes Island during the 1940s, noted that females with one fawn were less common in gazelles than those with two, and 3 fawns were found in exceptional cases. Also, on the island, from 1970 through 1975, 47-gazelle offspring were found, of which 33 (70.2%) were twins and 14 cases (29.5%) were singletons (Table 7.3). From observations of young over 5 breeding seasons, an average of 50.2% were twins and 30.6% were singles. A decrease in the percentage of twins through this period is explained by a significant death rate of young animals in their first months (Sludskiy 1956; Table 7.4). Of the 6 females in Saryishikotrau Desert (Ili Basin) on November 1–4, 1980, 4 females were with twins (66.6%) and two females had singletons (33.4%). According to Zhiryakov (1976), in the Alma-Ata Nature Reserve, out of 85 encounters of female gazelles, 44 (51.8%) had twins, 39 (45.8%) had singletons, one female had 3 fawns (1.2%), and one mother had 4 fawns (1.2%). Antipin (1941)

Table 7.3 Number of newborns per family and proportion of barren females in goitered gazelles of the Barsakelmes Island in June–July (adapted from Zhevnerov 1984)

Year	Number of observed females	Including					
		Female and 2 juveniles		Female and 1 juvenile		Barren females	
		Number	%	Number	%	Number	%
1970	104	46	44.2	40	38.5	18	17.3
1971	168	81	48.2	69	41.1	18	10.7
1972	154	76	49.3	40	26.0	38	24.7
1973	176	92	52.3	77	43.7	7	4.0
1974	123	70	56.9	44	35.8	9	7.3
1975	149	78	52.3	60	40.3	11	7.4

Table 7.4 The number of fawns born to goitered gazelle mothers in the Betpak-Dala Desert in 1951, 1954, and 1955 (adapted from Sludskiy 1956)

Month	Females observed						
	Total	With 1 young		With 2 young		With 3 young	
		Number	%	Number	%	Number	%
May	10	4	40	6	60	–	–
June	40	20	50	20	50	–	–
July	87	66	76	21	24	–	–
August	67	25	37.3	41	61.2	1	1.5
September	30	27	90	3	10	–	–

also found that most females have twins, and among two hundred females, 75% gave birth to twins. It seems that everywhere in Kazakhstan, the number of fawns born to females within a population more often is two rather than one. The same is true in Uzbekistan, Tajikistan, and Turkmenistan, where most females have twins (Sludskiy 1956).

Previously in other subspecies of goitered gazelles, such as the Arabian sand gazelle (*G. s. marica*), a high proportion of twins were noted among newborns (Vesey-Fitzherald 1952). Current studies, however, show that the twinning rate has dropped to 27% from 75 females by Hague and Smith (1996), or the ratio of singletons to twin births has become remarkably even 50% from 192 families by Riesch et al. (2013). No evidence has been found of any influence on a mother's birth history (whether a mother was a single or twin at birth or her birth weight) on her propensity to give birth to singletons or twins. Among females that gave birth at least three times, some females had only singletons and others only twins. However, many females had both singleton and twin births. More experienced females of *G.s. marica* are more likely to give birth to twins than singletons. In addition, twinning was especially high in March (53%, $n = 434$), while mostly singletons were observed during births in September (87%, $n = 23$) and October (100%, $n = 24$), and the

Table 7.5 Ratio of fawn sexes in goitered gazelle births on the Barsakelmes Island (adapted from Zhevnerov 1984)

Ratio of sexes	Number of cases	%
Male and female	10	21.3
2 females	12	25.5
2 males	11	23.4
1 male	6	12.8
1 female	8	17.0
Totally	47	100.0

Table 7.6 Ratio of sexes among newborns in goitered gazelles in Betpak-Dala Desert (adapted from Sludskiy 1956)

Family combination	Male + female	2 males	2 females	Male	Female	Total
Number of cases	15	1	8	8	4	36
Proportion (%)	41.7	2.6	22.3	22.3	11.1	100

proportion of twin births was positively correlated with the mean monthly precipitation and negatively correlated with the mean monthly temperature. According to the sex ratio, the proportion of twins was different at birth: female–female: male–male: mixed sex = 1: 0.83: 0.76 (Riesch et al. 2013). In the population on the Barsakelmes Island (Aral Sea), the sex ratio of goitered gazelle twins was comparable (Table 7.5), while in the Betpak-Dala Desert, proportions were different (Table 7.6).

It was once estimated that in Mongolia only 2.6% of females have twins, while most females have only one fawn (Bannikov 1954). Later, however, it was determined that this statement was not so correct and that, under favorable conditions, most goitered gazelle females in Mongolia give birth to twins, similar to other populations in Central Asia (Kaczensky in litt.). In Saudi Arabia, Rietkerk et al. (1992) proposed that the high reproductive capacity of goitered gazelles actually may be an adaptive response to the unpredictability of favorable conditions (rainfall and food availability) in a hyper-arid environment, where it would be best for the population's survival to maximize reproduction whenever possible because the next suitable opportunity is so unpredictable. This is also true for all other areas of goitered gazelle habitat, where favorable conditions can vary from 1 year to the next (Zhevnerov 1984; Kingswood and Blank 1996).

7.2.6 Barren Female Proportion

About 5–12% of females usually remain barren each year (Heptner et al. 1988; Reed 1977). In the Barsakelmes Island, the proportion of barren females during favorable years does not exceed 7.4%, while in unfavorable years barrenness can reach 24.6% (Zhevnerov 1984). For example, in May of 1955, after a favorable winter in the Betpak-Dala Desert, only 4 out of 36 adult and yearling females that were checked were found to be barren (11.1%): 5 females had one fawn (13,9%) and 27 mothers had 2 fawns (75.0%), supporting an average yearly female to birth ratio of 1.64 up to 1.84 newborns per year (Sludskiy 1956). In the same location in May 1956, after a winter with heavy snow and ice cover, there were 2 barren females (8.7%), 8 females (34.7%) had one embryo or fawn, and 13 mothers (56.6%) had 2 fawns out of a total of 23 females. Thus, in favorable years, the proportion of barren females is low at 8.7–11.1%, and mainly young females aged at about a year turned out to be barren (Sludskiy 1956).

In the semi-captive Bukhara Breeding Center, the proportion of barren females may be higher: in 1989, the number of barren females was 43.4%, and in 1991, the barren females were 19.8–29.8% of the female population (Marmazinskaya 1996). In Mongolia, 67.8% of females with fawns were counted at the end of July into August, which indicates a quite high proportion of barren females in this country (Bannikov 1954).

7.2.7 Sex Ratio

The sex ratio of gazelles at birth is close to 1:1. Of 59 newborns examined in 1955 in the Betpak-Dala Desert, 28 were found to be male (47.4%) and 31 were females (52.6%). On the Barsakelmes Island, out of 80 newborns examined, 42 (52.5%) were females and 38 (47.5%) were male. In Turkey, from 47 newborns 22 (46.8%) were females and 25 (53.2%) were males (Gürler et al. 2015). The distribution of sexes within populations is shown in Tables 7.6 and 7.7. Among adult gazelles, the sex ratio has been shown to change significantly toward a predominance of females. Of 1023 captive births of known sex, 506 were males and 517 were females. About 27–34% of twin births were both females, 22–33% were both males, and 33–50% were male and female (Carter 1991). In Saudi Arabia, sand gazelle newborns (*G. s. marica*) have a male to female sex ratio of 0.8:1.0 (Hague and Smith 1996).

In summer, males make up only 5.4–21.1% of adult gazelles, while females range from 78.9% to 94.6% (Table 7.7—Sludskiy 1956). A small percentage of adult males in the population can be explained by the fact that every year a lot of them die after the rut. During the rutting season, territorial males are busy courting

Table 7.7 The sexes ratio in adult goitered gazelles in the Betpak-Dala Desert (adapted from Sludskiy 1956)

		Including				
		males		females		
Date	Total gazelles	number	%	number	%	Female number per male
August 1951	93	7	7.5	86	92.5	12.3
May 1954	109	24	21.1	85	78.9	3.5
May 1955	370	20	5.4	350	94.6	17.5
May 1956	377	27	7.2	350	92.8	12.9
Total	949	78	8.2	871	91.8	11.2

females and fighting with other males and end up eating less than in other seasons of the year. The large expenditure of energy during the rut, starvation, and injuries severely deplete the health of these males, so after the rut, they become easy prey for wolves and other predators. During dzhuts (mass death), males weakened after rutting die in mass, while females, having retained their strength and health, tolerate winter deprivation much better than males (Sludskiy 1953).

In summary, in the very north of the gazelles' natural habitat (Kazakhstan), the birth of young is recorded from the end of April to the beginning of July (2.5 months); mass birthing takes place in the second half of May. In the south of their range (Saudi Arabia), fawns appear almost a month earlier than in Kazakhstan, while in the east (in Mongolia) their appearance is a month later than in Central Asia. The duration of pregnancy is quite stable for every species. Everywhere, mass birthing of offspring occurs during the transition period when the weather switches from spring to summer, when abundant vegetation is available, and when frosts end. Though also influenced by the state of the health of the gazelles (e.g., fatness) and the weather, the timing of mass birthing is the primary factor determining the timing of the rut, and not vice versa.

7.3 Female Aggression to Conspecifics During Prepartum and Postpartum Periods

7.3.1 General Remarks

It is widely known that aggression between females can occur during basically all four reasons: First and foremost, it may be competition for limited food resources and grazing areas (Espmark 1964; Zhevnerov et al. 1983); second, it can be related to the struggle for one female to establish or maintain her own hierarchical rank among females (Clutton-Brock et al. 1982; Côte 2000); third, and one of the most common causes of aggression is competition for comfortable resting sites (Prins 1989; Thouless 1990; Hass 1991; Alados and Escós 1992; Lovari and Locati 1993); and fourth is the defense of a chosen birthing site to obtain resources for the purpose of raising offspring (Clutton-Brock 2009). Usually, female aggression in ungulates is most often observed in the last weeks of pregnancy and during the first month postpartum, when females expend their greatest amount of energy (Hanwell and Peaker 1977; Oftedal 1985). Self-isolation is an essential stage of the birthing process for hiding species, when a pregnant female decreases her home range significantly just before giving birth. Prime-aged females (more than 3 years old) reduce their area to a greater extent than females who give birth for the first time. As a result, the survival rate of fawns born to experienced adult females is much higher and with a correspondingly low mortality rate (less than 9%) compared to primiparous females that have a very high mortality in young (up to 63%). Since adult females establish smaller birthing ranges, they remain much closer to their fawns, which gives them the ability to protect their offspring quickly. In addition, multiparous females show much more aggressiveness toward their conspecifics, which minimizes postpartum losses (Ozoga et al. 1982). Female–female aggression in goitered gazelles is observed most often in May and less in June, and the object of aggression is mainly intrusive sub-adult females. Although this aggression is related to the protection of the fawn's hiding places, it is still not entirely related to the protection of resources for raising offspring as there are also other reasons why the fawns' locations are strictly protected from the invasion of other females. To understand this issue and using a goitered gazelle as an example, I have created this separate sub-chapter on female–female aggression in this species.

Goitered gazelle females usually stay in small groups with other females and fawns and single females are not typical for goitered gazelles most of year. At the end of April into early May, before the start of labor, pregnant females begin to separate from their groups and retreat to places they find convenient to begin the birthing process. According to my Ili Hollow data, the first females can leave their groups as early as the end of March, to begin leading a solitary lifestyle, but most females leave their groups in the second half of April. As this happens, the number of single females is noticeably increasing. The process of females leaving their groups proceeds gradually throughout the entire birthing period. Concentration of pregnant females in one specific maternity area, as seen in saigas, is not observed in goitered gazelles, though goitered gazelle birth sites do have a connection with the rough terrain (Zhevnerov 1984; Blank 1985). Zhevnerov et al. (1983) suggest that pregnant females begin to prepare for birthing a few days early, since his tamed females showed signs 2–4 days before giving birth. Based on my observations, however, most females in the wild begin to lead a solitary lifestyle at least 2–3 weeks before giving birth (Blank and Yang 2013). In captivity, females also avoid other conspecifics 2–3 weeks before birthing their young (Pereladova and Pereladov 1986). As soon as females choose a site for birthing, their behavior changes noticeably and they become aggressive toward other individuals, retreating only in front of adult males (Zhevnerov 1984).

A description (ethogram) of the recorded displays of aggression during female–female contacts in goitered gazelles within my study period in Ili Hollow is presented in Table 7.8.

7.3.2 Prepartum and Postpartum Aggression of Goitered Gazelle Females

Aggressive interactions between females are most often observed in May (57.0%, $n = 151$; Fig. 7.6), less frequently in June (24.2%), and least often in other months of the year (19%, $n = 149$). Most often, the dominant adult female expels the subordinate younger female, chasing her in a completely neutral position, with only her tail lifted (in 39%, $n = 56$ in May and 54%, $n = 35$ in June); a neutral posture and threat posture (forehead directed forward) are most common (40.2% and 49.8%,

Table 7.8 Description of the recorded displays of aggression (ethogram) during female–female contacts in goitered gazelles within the study periods according to Walther (1984), Blank (1998), Blank and Yang (2015b), and Blank (2019—License Number—5476431236455)

Behavioral patterns	Description
Threat posture	Female approaches another female with her chin pressed to her neck and the sinciput (part of horn placement in males) directed toward her opponent or lowered toward the ground; ears are directed forward-sideward or backward and pressed to the neck; tail is in a normal position or raised in an arc (Fig. 7.5—FemaleAgression)
Dominance posture	A female slowly walks, approaching another in a dominance posture, with her chin pressed to her neck and forehead directed forward; ears are set apart with the foramen directed downward
Roaring	Roaring consists of gruff, wheezing, and hissing sounds. The female often raises her head, opens her mouth wide, takes a deep breath, then closes her mouth, and passes air through her nostrils; roaring sometimes consists of a series of short sounds
Butting	Rivals approach each other with heads lowered to the ground and lean their foreheads against each other. Each tries to push her opponent backward. When butting, the ears are apart horizontally and with the tops directed sideways or sometimes backward; tail is pressed to the rump
Low-stretch (neck-stretch) posture	The head and neck are stretched forward horizontally; ears directed backward and pressed to the head in a position either parallel to the ground or slightly downward with the foramen inward; sometimes the ears are held with one apart from the other at a 45° angle; tail is raised to a horizontal position
Walk approach, neutral posture	A female approaches closely another in a slow walk in a neutral posture and drives the other away; sometimes she makes a down-up movement with her muzzle as if this motion is a signal to compel her opponent to leave
Sharp jerk	Unexpected fast running of one female after another female with her ears pressed back to her head
Playing run	Female's muzzle is lowered to the ground; she looks toward the other female through the grass, waves her tail, and presses the front body part to the ground like a dog playing

n = 107, respectively) in aggressive interactions; a dominance posture is displayed less often (10%). Females display some of their threat postures with roaring (34.7% of all cases, n = 49), though it is impossible to hear these sounds from a distance, because they are gentle and soft unlike male roars. With almost the same frequency, a forehead threat in females shifts into forehead butting (32.6%). The pursuit of one female by another occurs most often by walking (68.9%, n = 151; Fig. 7.6), less often by running (24.5%), and very rarely by a sharp jerking lunge (6.6%, n = 151) (Blank 2019).

An adult female usually threatens a sub-adult female (57.6%), and a little less often another adult female (42.4%, n = 139). The agonistic behavior of one female toward another is observed most often when a sub-adult wanders into the area chosen by the adult female for her hiding fawn (61.7%, n = 120). Less often (24.2%), a female reacts aggressively to the approach of another female that violates the established individual distance between them; the rarest aggressive interactions are

Fig. 7.5 Goitered gazelle females butt heads in aggressive interactions. Drawing by U. Kenzhetaev with kind permission

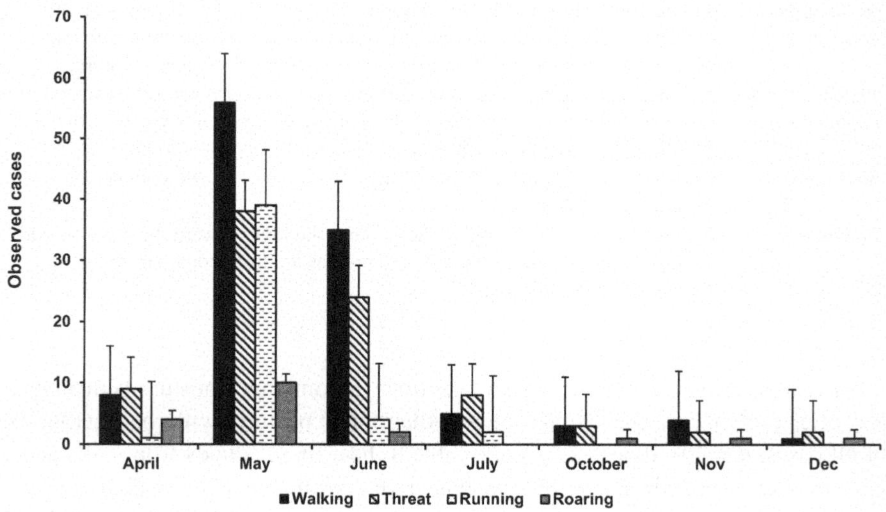

Fig. 7.6 Number of aggressive female–female interactions over months (Blank 2019—License Number—5476431236455)

related to the displacement of a female from her resting place (14.1%). The first reason for aggressive demonstrations by one female toward another and the expulsion of other females from the area of hiding the fawn is most often observed in May (58.8%, $n = 120$), when other reasons are quite rarely noted. While in July, competition for a resting spot becomes one of the main causes of aggressive reactions between females (38.1%).

One of the most frequent aggressive actions observed in goitered gazelles is approaching too close to a lying or standing conspecific, forcing her or him to stand up or retreat. This aggressive behavior is typical for both females and males of any age. In females, this type of threat is observed most often in May (62.2% cases, $n = 37$; Fig. 7.6), less in June (21.6%, $n = 37$; Fig. 7.6), and least in other months (10.8%, $n = 37$ and 5.4% in April and in July, respectively). A dominant female usually just approaches another resting female (77.1%), though sometimes she also touches her body with her muzzle (22.9% cases, $n = 48$) or, much more rarely, sniffs her hindquarters (9.8%, $n = 41$). Subdominant females most often simply leave by walking away from the dominant female (61.3% cases, $n = 80$); more rarely, the sub-adult will run away (38.7%). Attacking and retreating of females are observed most often in May (57.6% cases, $n = 139$), less often in June (24.5%) and April (13.7%), and least often in July (4.3%). Retreating is a typical response of subdominant females (81.6%), and rarely is another behavior displayed (18.4%), such as just staying and turning her muzzle toward the attacking female or threatening the aggressor with her forehead exposed forward. When threatening, the dominant female often holds the ends of her ears back (75%, $n = 32$), pressing them to her head, and less often with the ends of her ears pointed to the sides (25%) (Blank 2019).

Thus, my data indicate that goitered gazelle females leave their groups before birthing and begin to lead a solitary lifestyle, which is totally atypical in other seasons when they are only seen in groups. Different authors give different dates, from several days to several weeks and even a couple of months, when females isolate themselves from their conspecifics before giving birth (Zhevnerov 1984; Marmazinskaya 1999; Blank and Yang 2013). It is possible that different females self-isolate at different times or older females begin their solitary life before giving birth much earlier than young primiparous females. It is also possible that the females separate from their herds and then rejoin them for a while before leaving again. In short, behavior variations are possible, but one thing is obvious that this process is gradual and different females begin isolation at different times. Studying the sex ratio in the Ili population, I found that as early as the end of March, i.e., 1.5–2 months before birthing begins, the proportion of females in the population drops quite sharply, shifting the ratio toward males up to parturition in May.

Abandoning their groups before birthing is an essential stage in the process and is known for pregnant females of many ungulate species. One of the main reasons why females separate and stay apart from conspecifics is protection of their newborns from predators. However, the most important reason for isolation is the need to create a maternal–neonate bond between mother and young within the first hours and days after birth. Learning and knowing an offspring's smell, sounds, and appearance is very important for a mother to unmistakably distinguish her own fawn from others (Ciuti et al. 2006). In follower species, recognition of young takes a relatively short time, because mother and young are together all the time (Lent 1974). In hider species, however, a mother has limited time to learn her fawn's cues, because they are separated from each other most of time and uniting for only for short periods for suckling (Blank and Yang 2017). Therefore, mother–young bonds may take weeks to form in some species (Ozoga et al. 1982). Any disturbance during the first 2 days after birth may lead to a breakdown in the mother's fragile bond and lead to the death of the fawn (Cook et al. 1971).

During my research I found that pregnant goitered gazelle females change their behavior after leaving their group and do not tolerate the company of other females, driving them away with threat demonstrations, tilting their heads so their foreheads are forward, using the same threatening posture as males. Most maternal aggression is observed in May with the highest level toward any female approaching her hiding fawn occurring for at least during the first several weeks postpartum. For the first few weeks, a mother recognizes her fawn only by its smell and the location of its hiding place. Over time, smell continues to be the main form of recognition factor of her fawn, but gradually the mother gets to know her fawn better by its appearance and voice at an increasing distance. Until then, the mother does not allow other females to approach her fawn or let other fawns nearby. As a result, the females giving birth are scattered over a large area with each female in her own isolated plot with a hiding fawn. This all looks like territorial behavior in females similar to the territorial structure of males during the rut. However, these sites do not have distinct clear boundaries that are constantly defended like the territories of rutting males; instead, they are limited margin areas around a resting fawn that the mother protects from intruders (Marmazinskaya 1999).

An interesting question is why females who have given birth expel sub-adult females more often than adult females, especially in the first week or two after giving birth. The fact is that almost all adult females have their own newborns, and for the first 3–4 days, they do not move far from their hiding fawns, not even to go for water. Therefore, adult females almost never encroach on other females' plots, although after a few days postpartum, they begin to leave their fawns and increasingly cross into neighboring plots of other females. Most sub-adult females, on the other hand, do not give birth and continue to follow their mothers in the second year and sometimes the third year of their lives, forming family mother–daughter groups (Marmazinskaya 1999). During their mothers' next birthing cycle, daughters still try to follow, violating the mother's isolation of her neonate. And even though they are related, their mothers behave aggressively and try to expel them from their sites (Blank 2019). The daughter retreats, but does not go far from the mother's location. Staying close, the daughter approaches the fawn as soon as her mother goes away and even tries to sniff the newborn, forcing it to stand. Mothers almost constantly monitor their hiding fawn and immediately notice the appearance of a sub-adult female and rush in to expel her. Thus, mature females who have given birth are initially much more likely to expel their sub-adult females, especially their own daughters. If a sub-adult female approaches a fawn, the fawn immediately rushes to try to suckle from her. Greatly frightened by such an unexpected behavior, the sub-adult female runs away and tries to avoid approaching the hiding fawn by bypassing that location. Thus, over time, the number of cases of aggressive interactions between mothers and other adult females increases as contacts with sub-adult females decrease. Both adult and sub-adult females showing interest in a hiding fawn try to sniff it or even force it to stand, breaking the isolation and secrecy of the hiding place. Perhaps, this is another reason why mothers tend to drive all other females from the site.

As the fawn grows, it spends more and more time with its mother, giving the mother extra time to get to know her fawn by voice and appearance. Moreover, the fawn increasingly leaves its bed site and makes long, independent movements from its hiding place and back, improving its physical capabilities. During June, the mother is less likely to monitor the fawn's resting places, because they are gradually losing their importance. As a result, the female ceases to expel other females from her site and sub-adult females resume following their mothers (Blank et al. 2012). The mothers' intolerance and aggressiveness toward other females decrease quite quickly in July, and thus the females' site protective behavior disappears for the rest of the hiding period. In the absence of protection of the fawn's hiding place, female–female aggressiveness drops to a minimal level and is associated only with competition for favorable resting places, violation of the acceptable individual distance between individuals, and approaching to another group (Marmazinskaya 1999). Adult females are especially impatient with the approach of young females to the group, chasing the sub-adult at every opportunity and making head-butting movements in the air. The young female takes a submissive pose, lowering her head to the ground and raising her tail and quickly retreats, trying to stay closer to her mother (Gorbunov 1983).

There are many hiding species among antelopes (Grant's gazelles *Nanger granti* and Thomson's gazelles *Eudorcas thomsonii*) and deer species (roe deer *Capreolus capreolus*, reindeer *Rangifer tarandus*, moose *Alces alces*), the females of which leave their groups, isolating themselves from their conspecifics (Estes 1967; Ozoga et al. 1982). Similar to goitered gazelles, isolated females behave quite aggressively toward all other females entering their fawn's hiding site and expelling them as soon as they appear close to the hidden young (Lent 1974; Autenrieth and Fichter 1975; Danilkin and Hewison 1996). White-tailed deer (*Odocoileus virginianus*) mothers, for example, aggressively behave (threats, pursuit, or physical strikes) against conspecifics for about 4–6 weeks postpartum, and yearling females are harassed most often. Conspecifics soon avoid the female's area. In white-tailed deer, after selecting a certain area, the mother uses the same area annually. If a mother loses her young, she will stop her aggressive behavior toward other females. After losing her fawn, a young, 2-year-old female returned to the behavior of a yearling and tried to find company with other non-mother females (Ozoga et al. 1982). Mule deer females (*Odocoileus hemionus*) show very similar behavior (Riley and Dood 1984). In Thomson's gazelle, the females follow two different strategies: some traditionally leave their group and stay apart from all others for the first few weeks after giving birth but give birth in sites with a high grassy cover; other females stay in the group, or give birth not far from the group, and leave their young in open areas with a low grass cover. According to Roberts (1996), both strategies are successful against predators. But nothing is said about the need to isolate a female with her fawn in order to create a reliable mother–young bond between them for subsequent recognition. Argali females also leave their herd before giving birth and aggressively expel all other females from the area where they hide their young, while the hiding period of wild sheep is short and lasts only a few days (Fedosenko and Blank 2005).

Ungulate females do not always leave their groups. In some species, females give births not far from the herd and stay with conspecifics, visiting their offspring from time to time only for suckling (Bromley 1967). Despite the fact that Mongolian gazelles (*Procapra gutturosa*) are typically a hiding species, they behave differently from most of others. Before giving birth, the females of this species separate from their herd for only 3–8 days, remaining in their group before and after giving birth. Mongolian gazelle females visit their offspring periodically but spend most of their time with other females without showing specific aggression toward them. A few of the females, however, especially in small groups, behave like a typical hiding species and separate from their group along with their fawns for several weeks. In typical cases, by the age of 7–10 days, the fawns with their mothers join the group and from that time on they stay constantly in the herd, which starts its usual migration. Such a rapid postpartum development of the fawn and a very short period of isolation in Mongolian gazelles and rapid joining to the herd are associated with a high socialization of the species and a need for migration; because of such a high density, these gazelles cannot stay in one place for too long time (Kiriluk and Zvychainaya 2011). Though females of Saiga antelope (*Saiga tatarica*) leave their group before birthing, this antelope has similar behavior to Mongolian gazelles. They gather in special birthing areas in great numbers (up to 100 thousand), where offspring of many females stay close to each other. Mothers do not leave their groups for long periods of time, staying in female herds and only occasionally visiting their offspring for suckling. The hiding period lasts 8–10 days, when offspring from different females stay together in a "children's garden" (or creche) and mothers do not show their aggressiveness to other females. This concentration of birthing females and newborns in one place seems to be associated with a decrease in losses from predation. The main maternity areas of saigas do not geographically coincide with the territories of wolf families, with only the peripheral groups partially coinciding with the wolf brood sites (Fadeev and Sludskiy 1982; Bekenov et al. 1998). It is interesting to note that the Mongolian subspecies of saiga antelope (*Saiga tatarica mongolica*) that is less numerous uses a dispersion strategy during the hiding period where mothers are secluded from each other during the first weeks after birth. That means, the Mongolian subspecies of saiga antelope uses the same antipredator strategy as goitered gazelle females, because of low population density (Buuveibaatar et al. 2014).

In spite of variations in female birthing behavior, self-isolation before giving birth and for a few weeks postpartum is apparently the most predominant behavioral tendency in most hiding ungulate species (Lent 1974; Ozoga et al. 1982; Danilkin and Hewison 1996).

Thus, goitered gazelles belong to the classification of typical hiding species where females separate from their groups before giving birth and establish an isolated site for their newborns, from which all other females are aggressively expelled. The self-isolation and aggressiveness of goitered gazelle females in relation to all other females is not related to competition for food resources in the area of their hidden fawns, but with the ability to gradually learn to recognize the individual characteristics of their fawns and create mother–young bonds, so as the fawns grow, they can unmistakably recognize their own fawn among others by smell, voice, and appearance.

References

Afanasief AV, Sludskiy AA (1947) Materials on mammals and birds of Central Kazakhstan. In: Proceedings of Academy of Sciences of KazSSR, vol 36. Publication of Academia of Sciences of Kaz SSR, Alma-Ata, pp 48–64. (in Russian)

Alados CL, Escós JM (1992) The determinants of social status and the effect of female rank on success in Dama and Cuvier's gazelles. Ethol Ecol Evol 4:151–164

Al-Eissa MS, Alhamidi AR, Mohamed AEKS, Ibraheem AS, Hamad AY, Al-Farraj S, Saad A, Al-Dahmash B (2011) Scrotal, testicular and semen characteristics of the mountain gazelle (*Gazella gazella*) males. Afr J Microbiol Res 5(28):5137–5141

Allen GM (1940) The mammals of China and Mongolia, vol 2. The American Museum of Natural History, New York, pp 621–1350

Antipin VM (1941) Mammals of Kazakhstan, Ungulates, vol 3. Publication OGIZ, Alma-Ata, pp 47–54. (in Russian)

Autenrieth RE, Fichter E (1975) On the behavior and socialization of pronghorn fawns. Wildl Monogr 42:3–111

Baharav D (1983) Reproductive strategies in female mountain and Dorcas gazelles (*Gazella gazella* and *Gazella dorcas*). J Zool 200:445–453

Bannikov AG (1954) Persian gazelle. In: Heptner VG (ed) Mammals of Mongolian People Republic. Publication of USSR Academy of Sciences, Moscow-Leningrad, pp 250–258. (in Russian)

Bekenov AB, Grachev IA, Milner-Gulland EJ (1998) The ecology and management of the saiga antelope in Kazakhstan. Mammal Rev 28:1–52

Blank DA (1985) Peculiarities of social and reproductive behaviour of *Gazella subgutturosa* in the Ili Valley. Zool Zhurnal 64:1059–1070. (in Russian, English abstract)

Blank DA (1997) Social and reproductive behavior in the Persian gazelle. Ph.D. thesis, The University of Tel-Aviv, 189 pp

Blank DA (1998) Mating behavior of the Persian gazelle Gazella subgutturosa Guldenstaedt, 1780. Mammalia 62(4):409–419

Blank DA (2017) Antipredator tactics are largely maternally controlled in goitered gazelle, a hider ungulate. Behav Proc 136:28–35

Blank DA (2019) Female-female aggression in goitered gazelles: the desire for isolation. Behav Proc 164:186–192

Blank DA, Yang W (2013) Sex ratio in goitered gazelles (*Gazella subgutturosa* Güldenstaed, 1780). Acta Theriol 58:73–78

Blank DA, Yang W (2015a) Strategy of goitered gazelle suckling behavior and its adaption to the environment. Mamm Res 60(4):393–401

Blank DA, Yang W (2015b) Suckling behavior in goitered gazelle: do females invest more in twins or singletons? Zoology 118:348–356

Blank DA, Yang W (2017) Mother-young recognition in goitered gazelle during hiding period. Behav Proc 142:21–28

Blank D, Yang W, Xia C, Xu W (2012) Grouping pattern of the goitered gazelle, *Gazella subgutturosa* (Cetartiodactyla: Bovidae) in Kazakhstan. Mammalia 76(2):149–155

Bromley PT (1967) Pregnancy birth behavioral development of the fawn and territoriality in the pronghorn (*Antilocapra americana* Ord) on the National Bison Range Moiese Montana. Master Degree, University of Montana

Buuveibaatar B, Fuller TK, Young JK, Berger J (2014) Calving location selection patterns of saiga antelope in Mongolia. J Zool 294(4):241–247

Carter S (1991) Goitered gazelle North American regional studbook, 1st edn. Sedgwick County Zoo an Botanical Garden, Wichita, p 87

Ciuti S, Bongi P, Vassale S, Appollonio M (2006) Influence of fawning on the spatial behaviour and habitat selection of female fallow deer (*Dama dama*) during late pregnancy and early lactation. J Zool 268:97–107

Clutton-Brock T (2009) Sexual selection in females. Anim Behav 77:3–11
Clutton-Brock T, Albon SD, Guinness FE (1982) Competition between female relatives in a matrilocal mammal. Nature 300:178–180
Cook RS, White M, Trainer DO, Glazener WC (1971) Mortality of young white-tailed deer fawns in South Texas. J Wildl Manag 35:47–56
Côte SD (2000) Dominance hierarchies in female mountain goats: stability, aggressiveness and determinants of rank. Behaviour 137(11):1541–1566
Danilkin A, Hewison AJM (1996) Behavioural ecology of Siberian and European roe deer. Chapman and Hall Publication, London
Dinnik NY (1910) Animals of Caucasus, part 1, Cetaceans and ungulates. West Caucasus Dep Imper Russ Geogr Soc 27(1):215–220. (in Russian)
Espmark Y (1964) Studies in dominance-subordination relationship in a group of semidomestic reindeer (*Rangifer tarandus* L.). Anim Behav 12:420–426
Estes RD (1967) The comparative behavior of Grants and Thomson's gazelles. J Mamm 48(2):189–209
Estes RD (1991) The behavior guide to African mammals. University of California Press, Berkeley
Fadeev VA, Sludskiy AA (1982) The Saiga in Kazakhstan. Kazakh Academy of Sciences Press, Alma-Ata. (in Russian)
Fedosenko AK, Blank D (2005) Ovis ammon. Mamm Species 773:1–15
Flerov KK (1935) Persian gazelle. In: Pavlovskiy EN (ed) Wild animals of Tajikistan, their life and significance for man. Academy of Sciences of the USSR, Moscow, pp 109–112. (in Russian)
Frazier H, Hunt K (eds) (1994) Zoo infant development notebook, vol 2. American Association Zoo Keepers, Topeka, pp 921–1755
Gorbunov AV (1983) About ecology and behavior of goitered gazelles in the Krasnovodsk Peninsula. Proc Turkmen Acad Sci Ser Biol Sci 3:66–68. (in Russian)
Gorelov UK (1972) Breeding of goitered gazelle (*Gazella subgutturosa* Güldenstaedt) and problem of restoring of its number in the Badkhyz (SE Turkmenistan). In: Theriology, vol 1. Nauka of Siberian Department of Academy of Sciences of USSR, Novosibirsk, pp 420–424. (in Russian)
Gürler Ş, Bozkaya F, Özüt D, Durmuş M (2015) Some morphological characteristics and neonatal weights of reintroduced gazelle (*Gazella subgutturosa*) in Turkey. Turkish J Zool 39(3):458–466
Habibi K, Thouless CR, Lindsay N (1993) Comparative behavior of sand and mountain gazelles. J Zool 229:41–53
Hague MN, Smith TR (1996) Reintroduction of Arabian sand gazelle *Gazella subgutturosa marica* in Saudi Arabia. Biol Conserv 76:203–207
Hanwell A, Peaker M (1977) Physiological effects of lactation on the mother. Symp Zool Soc Lond 41:297–312
Hass CC (1991) Social status in female bighorn sheep (*Ovis canadensis*): expression, development and reproductive correlates. J Zool 225:509–523
Heptner VG, Nasimovich AA, Bannikov AG (1961) Mammals of Soviet Union, Artiodactyla and Perissodactyla, vol 1. Vysshaya Shkola Publishing House, Moscow. (in Russian)
Heptner VG, Nasimovich AA, Bannikov AG (1988) Mammals of the Soviet Union. In: Artiodactyla and Perissodactyla, vol 1. Smithsonian Institution Libraries and The National Science Foundation, 1147 pp. (translation from Russian)
Ishunin GI (1961) Mammals (carnivores and ungulates). In: Fauna of Uzbek SSR, vol 3. Publication of the Academy of Sciences of Uzbek SSR, Tashkent, pp 165–174. (in Russian)
Ismagilov MI (1954) On ecology of goitered gazelle in the Barsakelmes Island. Bull Moscow Nat Biol Dep 59(6):3–9. (in Russian)
Kingswood SC, Blank DA (1996) Gazella subgutturosa. Mamm Species 518:1–10
Kirchshofer R (1963) Das Verhalten der Giraffengazelle, Elenantilope und des Flachlandtapirs bei der Geburt; einige Bemerkungen zur Vermehrungsrate und Generationsfolge dieser Arten im Frankfurter Zoo. Z Tierpsychol 20:143–159
Kiriluk VE, Zvychainaya EY (2011) Mother-young behavior of Mongolian gazelle (*Procapra gutturosa*). Sci Notes Trans-Baikal State Univ 1(36):74–83. (in Russian)

Kostin VP (1955) Materials on ecology and commercial value of goitered gazelle. Proc Inst Zool Acad Sci Kazakh SSR 4:67–82. (in Russian)

Kuznetzov VI (1986) Goitered gazelles in Badkhyz. In: Flint VE, Prisyazhnuk VE (eds) Conservation and perspectives of the number restoration of goitered gazelle in USSR. All-Union Scientific Institute of Nature Conservation and Nature Reserves, Moscow, pp 31–37. (in Russian)

Legge AJ, Rowley-Conwy PA (1987) Gazelle killing in Stone Age Syria. Sci Am 257(2):76–83

Lent PC (1974) Mother—infant relationships in ungulates. In: Geist V, Walther F (eds) The behaviour of ungulates and its relation to management. IUCN, Morges, pp 14–55

Lovari S, Locati M (1993) Intrasexual social behaviour of female Apennine chamois *Rupicapra pyrenaica ornate* (Neumann, 1899). Ethol Ecol Evol 5:347–356

Mambetzhumaev AM (1970a) Goitered gazelle. FAN Publishing House, Tashkent, 196 pp. (in Russian)

Mambetzhumaev AM (1970b) On the number and distribution of goitered gazelles in the Ustyurt Plateau and Kyzylkum Desert. Zool Zhurnal 49(3):431–434. (in Russian)

Marmazinskaya NV (1996) Breeding of goitered gazelle *Gazella subgutturosa* (Artiodactyla Bovidae) in the ecological center "Dzheiran". In: Proceedings of the Uzbekistan nature reserves, issue 1. FAN Publication House, Tashkent, pp 53–61. (in Russian)

Marmazinskaya NV (1999) Social behavior of goitered gazelle (*Gazella subgutturosa*/Güld.), 1780/ during breeding period in the conditions ecocenter "Goitered gazelle". Ph.D. thesis in zoology, Institute of Zoology of Uzbek Academy of Sciences, Tashkent, p 1–20 (in Russian)

Marmazinskaya NV, Gritsina MA, Mitropolskiy MG (2012) Terrestrial vertebrates of arid ecosystems. In: Materials of the international conference dedicated to the memory of Zarudny, Tashkent, pp 204–211. (in Russian)

Oftedal OT (1985) Pregnancy and lactation. In: Hudson RJ, White RG (eds) The bioenergetics of herbivores. CRC Press, Boca Raton, pp 215–238

Ozoga JJ, Verme LJ, Bienz CS (1982) Parturition behavior and territoriality in white-tailed deer: impact on neonatal mortality. J Wildl Manag:1–11

Pereladova OV, Pereladov SV (1986) The goitered gazelle breeding in nursery enclosures of the Syunt-Khasardak Nature Reserve. In: Flint VE, Prisyazhnuk VE (eds) Conservation and perspectives of the number restoration of goitered gazelle in USSR. All-Union Scientific Institute of Nature Conservation and Nature Reserves, Moscow, pp 49–59. (in Russian)

Prins HHT (1989) Buffalo herd structure and its repercussions for condition of individual African buffalo cows. Ethology 81:47–81

Reed MC (1977) A history of three rare desert antelope at the San Antonio Zoo. In: American Association of Zoological Parks and Aquariums Regional Workshop Proceedings, pp 198–210

Reif U, Klingel H (1991) Hiding behavior in wild Gerenuk (*Litocranius walleri*) fawns. Z Saugetierkunde 56:159–168

Riesch R, Martin RA, Lerp H, Plath M, Wronski T (2013) Size and sex matter: reproductive biology and determinants of offspring survival in *Gazella marica*. Biol J Linn Soc 110(1):116–127

Rietkerk FE, Lindsay N, Tatwany H, Mubarak S, Badri O, Williamson DT (1992) Population dynamics of a captive herd of Arabian sand gazelles. Ongulés/Ungulates 91:379–382

Riley SJ, Dood AR (1984) Summer movements, home range, habitat use, and behavior of mule deer fawns. J Wildl Manage:1302–1310

Roberts TJ (1977) The mammals of Pakistan. Ernest Benn, London

Roberts G (1996) Why individual vigilance declines as group size increases. Anim Behav 51:1077–1086

Severtsov NA (1873) Vertical and horizontal distribution of Turkestan animals. Proc Soc Nat History Anthropol Ethnogr Lovers 8(2):1–157. (in Russian)

Shahooth MA, Zalzala SJ, Hatif SA (2015) Study of reproduction in the local reem gazelle (*Gazella subgutturosa marica*). World J Pharm Sci:810–814

Sludskiy AA (1953) The mass death of wild ungulates in the Kazakhstan deserts and its impact on the animal numbers. In: Proceedings of the Institute of zoology, vol 2. Publication of Academy of Sciences KazSSR, Alma-Ata, pp 3–30. (in Russian)

Sludskiy AA (1956) Breeding of goitered gazelle. In: Proceedings of Institute of zoology, Kazakh Academy of Sciences, vol 6. Nauka Press of Kazakh SSR, Alma-Ata, pp 78–108. (in Russian)
Sludskiy AA (1962) Interrelation between predators and prey. In: Proceedings of the Institute of zoology. Publication of Academy of Sciences KazSSR, vol 17, pp 24–143. (in Russian)
Sludskiy AA (1977) Goitered gazelle. In: Rare animals of USSR. Publication of Lesnaya Promyshlennost, Moscow, pp 28–61. (in Russian)
Sokolov II (1959) *Gazella subgutturosa* Güldenstaedt—Goitered gazelle. In: Pavlovskiy EN (ed) Fauna USSR, mammals: ungulates (orders Perissodactyla and Artiodactyla). Publication of the Academy of Sciences USSR, Moscow, pp 308–331. (in Russian)
Thouless CR (1990) Feeding competition between grazing red deer hinds. Anim Behav 40:105–111
Tsapluk OE (1972) Age and seasonal dynamics of sexual activity of goitered gazelle. Proc Acad Sci Kazakh SSR Biol Ser 3:39–46. (in Russian)
Vereshchagin NK (1951) Mammals. Wildlife of Azerbaijan. Publication of Azerbaijan Academy of Sciences, Baku, pp 84–251. (in Russian)
Vesey-Fitzherald LDEF (1952) Wildlife in Arabia. Oryx 1(5):232–235
Vinogradov BS (1952) Mammals of the Krasnovodsk district of the Western Turkmenia. In: Proceedings of the Zoological Institute of Academy of Sciences USSR, vol 10. Publication of the Academy of Sciences USSR, Moscow, pp 7–44. (in Russian)
Walther FR (1965) Verhaltensstudien an der Grantsgazelle (*Gazella granti* Brooke, 1872) im Ngorongoro-Krater. Z Tierpsychol 22:167–208. (in German)
Walther FR (1968) Verhalten der Gazellen. A. Ziemsen Verlag, Wittenberg Lutherstadt. (in German)
Walther FR (1984) Communication and Expression in Hoofed Animals. Indiana University Press, Bloomington
Zhevnerov VV (1984) Goitered gazelle of the Barsakelmes Island. Nauka Publishing House of Kazakh SSR, Alma-Ata, p 144. (in Russian)
Zhevnerov VV, Bekenov AB, Sludskiy AA (1983) Goitered gazelle—*Gazella subgutturosa* Güldenstaedt, 1780. In: Gvozdev EV, Kapitonov VI (eds) Mammals of Kazakhstan, Bovidae, vol 3 (3). Nauka Publishing House of Kazakh SSR, Alma-Ata, pp 11–54. (in Russian)
Zhiryakov VA (1976) Rare ungulates of the Alma-Atinskiy Nature Reserve and their conservation. Rare mammals of the fauna of USSR. Nauka Publishing House, Moscow, pp 141–154. (in Russian)

Suckling Behavior

8

8.1 General Remarks

Ungulates are one of the most precocial and fastest growing mammals, with the rapid development of both motor and sensory abilities (Lent 1974). Maternal care of the offspring, and especially nursing behavior, is of great importance for the rapid growth of the young and its survival. In addition, lactation is a big burden on the energy balance of the females themselves, when 75–80% of additional body energy is necessary to compensate for the demands of nursing (Oftedal 2000). The sucking process itself is not only for nutrition, but also has a social significance. Milk is the only source of energy for offspring during their first weeks after birth. Gradually, though, milk loses its significance as a source of nutrition when the young switches to plant food, but suckles continue to play a role more in a social context, strengthening the ties of the new generation with their mother and with other conspecifics (Scornavacca et al. 2018). In a temperate climate, birth and lactation occur during the most favorable time of the year, when the weather is warm enough so there is a large abundance of high-quality green forage. The abundance of high-quality food allows females to quickly recover after giving birth and overcome the difficult period of providing their young with milk for a fairly long period, until the offspring's physiology has developed to allow for vegetative forage. The growth of offspring is influenced by both the environmental factors and the effectiveness of the mother's nursing, with the positive or negative dynamics of the entire population dependent on the survival rate of the young (Cassinello 2001; Scornavacca et al. 2018).

Blank and Yang (2015b)—License Number—5477490647947 (full text).
Blank and Yang (2015a)—License Number—5477490461356 (full text).

D. Blank, *The Behavioral Biology and Conservation of Goitered Gazelles (Gazella subgutturosa)*, https://doi.org/10.1007/978-3-031-83504-9_8

It has been established that multiparous females, due to their experience, provide much more effective care of their young, with a higher success rate in their survival compared to primiparous females (Ozoga and Verme 1986; Cameron et al. 2000). In addition, ungulates, and especially the young, are very susceptible to death from predator attacks, so a very important part of the mother's care for her offspring is to balance her need to provide intensive nursing to satisfy the high nutritional requirements of her young, and her strategy to protect against predators, with vigilance and a high level of caution so as not to attract the attention of predators (Toïgo 1999; Hamel and Côté 2008). Generally speaking, there are two main anti-predator strategies in caring for newborns—following and hiding, but at the same time, there is a wide variety of behaviors in each species in order to successfully protect their offspring from different predators in various conditions (Lent 1974). Each of these strategies has its advantages and disadvantages, and neither strategy provides absolute protection for the young, but statistically both strategies work and have allowed the species to exist for thousands of years. Like most antelopes, goitered gazelles have a hiding strategy, and the anti-predator behavior characteristics of this species will be discussed in more detail in the chapters devoted to the mother/young interaction and defensive behavior. In this chapter, I will consider mainly suckling indices for investigating goitered gazelle nursing behavior and its changes with fawn growth, as well as the features of nursing in females with twins and singletons (Blank and Yang 2015b).

8.2 Suckling Posture and Nursing Procedure in Goitered Gazelles

Newborns can perform suckling movements without any contact with the mother's nipples (Kirchshofer 1963). Like most ungulates, goitered gazelle fawns usually suckle standing under their mother's belly or side by side with her (reverse-parallel orientation), as well as with their hindquarters to the female's muzzle (reverse-parallel position), giving her the opportunity to lick and sniff the anus while nursing. In effect, the mother's behavior induces her young to take this position. Some fawns can suckle standing behind their mothers and reaching up to the nipples between her hind legs; a few females allow them to be successful in this position, though most of these suckling attempts end with the mother's refusal (Blank 1985). Young of a few species can suckle from between the mother's hind legs and this suckling position is observed frequently in bison (Marjoribanks-Egerton 1962), caribou (*Rangifer tarandus*—Lent 1966), and chamois (*Rupicapra rupicapra*—Kramer 1969), and occasionally in red deer (*Cervus elaphus*—Burckhardt 1958), pudu (*Pudu pudu*—Hick 1969), and mountain sheep (*Ovis canadensis*—Geist 1971). Suckling a young from the rear position is more typical for the follower species due to their greater activity. For example, caribou fawns can generally suckle from this rear position while moving (Lent 1974).

During suckling, the goitered gazelle fawn lifts its tail vertically and shakes it from side to side, and also steps with its hind feet (from the female's side) or sharply raises its legs, alternating them toward its mother's belly and down again. Perhaps, the fawn does all these movements out of pleasure, when it gets the opportunity to suckle milk and satisfy its hunger. It is also believed that ungulate offspring, while wagging their tails, disperse the smell from the anus, which is necessary for individual recognition of young, and the vertical position of the tail gives the mother free access to sniffing and licking the fawn's anus (Blank 1997). In addition, fawns poke and push their mother's udder with their muzzles when nursing, which increases in frequency and vigor as the fawn ages. The result of this behavior is to apply stimulation of the udder and nipples to release more milk (Hafez et al. 1969). This pushing into the udder is apparently painful enough for the gazelle mother to quickly pivot or move away or respond with aggression, threatening her young with her forehead or even head butting her fawn (Walther 1965). Some species (fallow deer *Dama dama*) are even known to bite their young (Lent 1974).

Making its first suckling attempts within its first 1–2 days (rarely longer), the offspring does not immediately find the udder and nipples. As soon as the fawn gets to its feet, it immediately makes attempts to find the nipples, paying special attention to the angles between the ventral parts of the mother's body and her limbs. Most researchers believe that success in finding the mother's udder the first time is based on trial and error and scent does not play much of a role if any (Lent 1974). In its initial search, a goitered gazelle fawn will walk around its mother and from the front stick its muzzle in between her front legs, poke its muzzle into her neck, then try to suckle the udder from behind between her hind legs, try to find nipples on the sides of its mother's body sides, or at the top of her front legs stay under her belly. When the fawn cannot find nipples near the front legs, it will make several circles around the mother's legs, looking for her nipples from the front and behind. This can be repeated up to three times. Finally, the fawn turns around under the mother's belly, and starts to look for nipples between her hind legs and then bumps into the udder. Some goitered gazelle females direct their fawns with their muzzles, and through licking of the anal region, move them toward the hindquarters and nipples so that they can find the udder faster (Blank and Yang 2015a). These behaviors of goitered gazelle mothers are typical for many ungulate species (Walther 1984).

During the fawn's first suckles, it often loses the nipple, sometimes several times in a row, and then looks for it with its muzzle all over the mother's belly from below, even going around her front and hind legs again, taking a long time before reconnecting with the nipple. Like roe deer (Sokolov et al. 1985), a single fawn suckles from both nipples alternately, changing from one to the other sometimes very quickly (every 1.5–2 s) (for *G.s. subgutturosa* in Kazakhstan—Blank and Yang 2015a). Uztemur et al. (2024) also found that singles of sand gazelle fawns suckle both nipples in turns (for *G.s. marica* in Turkey). For the first 5–7 days (up to 14 days in some cases) after birth, the mother of twins puts her offspring in different places, nursing and transferring her twins alternately, with each twin getting used to having access to both nipples, as a single fawn does. But when the mother brings them together in one place (Fig. 8.1), she tries to nurse them simultaneously. Particularly at the beginning of co-nursing, both twins will sometimes run up to the same side of the mother, pushing each other away until each finds its own nipple.

Fig. 8.1 After 7–10 days, a goitered gazelle mother brings twins together and since then they have been together. South Gobi Region, Mongolia. 08.07.2017. Photo by P. Kaczensky with kind permission

Such a struggle for the same nipple can be repeated one or two more times during subsequent nursing bouts. But in such a position, it is very uncomfortable for one of the twin to suckle, and very quickly (after several suckling attempts) the fawns learn to approach their mother each from their own side (Blank and Yang 2015a). In Turkey, in sand gazelle fawns each twin also suckles from only one own nipple (Uztemur et al. 2024).

As they suckle, the mother sniffs and licks their anal areas in turn. By the age of 2.5 months, the muzzle of the fawn has become higher than the level of the udder making it uncomfortable for nursing, so the fawn spreads its front legs wider and bends over them to reach a nipple; as the fawn grows even more, it kneels down on its front legs to suckle (Fig. 8.2) (Blank 1985). Suckling while leaning on the carpal joint for larger young is reported in a number of ungulates, including caribou and lechwe (*Kobus lechwe*—Lent 1966, 1969), wildebeest (*Connochaetes taurinus*—Bubenik 1965), nilgai (*Boselaphus tragocamelus*—Hediger 1950), red deer (*Cervus elaphus*—Hafez et al. 1969), and mountain sheep (*Ovis canadensis*—Geist 1971).

Usually, goitered gazelle mothers are the initiator of nursing, especially in the first 2 weeks after birth. They come closer to the hiding fawn, making an inviting quiet nasal call (86–94% cases, $n = 50$ and 17, in May and early June, respectively). The fawn jumps up and runs to suckle from its mother immediately, making bleating sounds while it is running. Without the invitational sounds from its mother, the fawn does not get up from its hiding site, even if its mother passes as close as 1 m. In the first 2 weeks, the female makes this call almost every time it is necessary for

Fig. 8.2 Suckling in goitered gazelles: As the fawns grow older, it becomes more and more uncomfortable for them to suckle while standing on four legs under the mother's belly. Southern Ustyurt Chink (Kaplankyr), SW Kazakhstan. 31.07.2019. Photo of the camera-trap of the Ustyurt Nature Reserve team with kind permission

nursing (Blank 1985). In many species of ungulates, mothers very often begin feeding by calling out to their young. This behavior has been observed in several species of gazelles (*Gazella*), American pronghorn (*Antilocapra americana*), great kudu (*Tragelaphus strepsiceros*), white-tailed deer (*Odocoileus virginianus*), caribou (*Rangifer tarandus*), and others (Walther 1984). During the nursing itself, the goitered gazelle mothers usually do not make any sounds, which is true for most ungulate species, even though the fawns can make soft sounds while running to their mothers to suckle. The meaning of this audio communication is not yet clear (Walther 1984). Sometimes a goitered gazelle fawn will not respond to its mother's call, so she approaches close to her lying fawn and touches it with her muzzle. This touch makes the fawn stand up and start suckling. Such touching is quite rarely observed behavior and noted in only 14% of cases in May and 6% in early June, and thereafter it is not observed at all (Blank and Yang 2015a). Such tactile stimulation of an offspring by touching it with the muzzle, horns, or even a foot is a rarely observed phenomenon among ungulates, though it is noted in roe deer (*Capreolus capreolus*—Kurt 1968), reindeer (*Rangifer tarandus*—Espmark 1971), and musk oxen (*Ovibos moschatus*—Lent 1974).

A female makes sounds not only when she approaches her lying fawn, but also if it runs away from her. These are soft sounds which are difficult to hear from a distance. However, when a female does not find her fawn in the hiding place where she left it, she starts walking around and producing quite loud sounds (low roaring), which are clearly heard at a distance of approximately 100 m (Blank 1985). The fawn also makes sounds when it runs up to its mother, sounding like the bleating of a goat kid (Zhevnerov 1984). When a fawn is hungry and the mother is late for her visit, the fawn will get up periodically and walk around its hiding place, producing distress calls, which are unusually low for its small size, sounding as bass-like bleating similar to the bleating sounds of an adult domestic sheep (Blank 2023).

As fawns grow, mothers no longer make sounds to call their young to nurse, and the fawns, themselves, increasingly become the initiators of suckling bouts. Initially, the fawns do not recognize their mothers and tend to try to suckle from any passing gazelle, both females and males, adults and sub-adults. Alien gazelles immediately run away from the fawns, which tend to follow them for some distance; the non-parental gazelle then stretches its neck out to look back toward the fawn. Sometimes a female will return and try to sniff the fawn, but after the fawn's next attempt to suckle, the female will run away again with her tail raised vertically (Blank and Yang 2017). Once (May 20, 1983) an adult male accidentally came across a hiding fawn while grazing. He stopped and looked at the fawn for a few seconds and then stretched his muzzle toward it. The offspring jumped up, ran up to the male, and began to poke its muzzle in the usual place between his hind legs. The male immediately made a jump to the side, and the offspring ran away in the opposite direction and laid down. After 5–6 s, the male started grazing again but at a distance from where the offspring laid down (Blank 1985). When the fawns are older and start walking alongside their mothers, they use every opportunity to try to suckle: if the female just stops for vigilance or freezes in place or squats down to urinate, her fawns immediately rush to her to suckle. If the female just turns her head to look toward her fawns, then this is definitely perceived as an invitation for nursing, and the fawn immediately rushes to her and makes another attempt to get some milk. As a rule, though, this is usually met with a refusal from the mother (Blank and Yang 2017).

The female directs her muzzle toward the approaching fawn in order to have time to sniff its muzzle and the front part of its body before the young runs up to her udder and starts to suckle. Then, when nursing, she licks or sniffs the fawn's anal area, and the fawn shakes its tail quickly from side to side and pokes its muzzle and entire body into the udder while nursing. In addition, the fawn lifts up and lowers its one or alternately both hind legs quite quickly during suckling, as if from impatience. For the first 7–10 days after birth, the mother allows her fawn to suckle as much as it wants, and the female allows him to do it. But by the second or third week, the mother begins to limit suckling more and more often, by just moving away, jerking her hind leg, pulling her nipples out of the fawn's mouth, or stepping over or jumping over her suckling fawn. There have been observations of goitered gazelle fawns, in order to suckle more, standing up in front of their mother's front legs to stop them from walking away. And there also have been cases when a fawn has approached its lying mother, apparently having rested too long, and prodded her with its muzzles and/or scratched her with a front leg, usually making the female stand up so the fawn can immediately try to suckle. Similar behaviors are observed in chamois (Kramer 1969), bighorn sheep (*Ovis canadensis*—Geist 1971), Marco Polo sheep (*Ovis ammon polii*—Walther 1961), greater kudu (*Tragelaphus strepsiceros*), and sitatunga (*Tragelaphus spekii*—Walther 1964).

Immediately after giving birth, the mothers carefully lick their newborns to release them from the mucus covering their entire bodies. Mothers continue licking mostly the anus region for another 2–3 weeks (Fig. 8.3), although sometimes they also lick the fawn's muzzle, neck, and even the back. This procedure lasts during an entire suckling bout (sometimes with short breaks) and even between suckling sessions when females approach their fawn from behind and lick the anus area. Licking the anus not only promotes urination and defecation in the fawn, but also stimulates it to look for its mother's nipples and to suckle (Blank 2023). After a few weeks the licking duration gradually decreases until it mostly disappears by the time the fawn is 1 month old (Blank and Yang 2015b).

As the fawn grows up and the mother licks the anus area less and less often, she instead sniffs mostly its anus, or less often its muzzle and least often the chest, shoulders, body side, and/or back before nursing. Anus sniffing is a mandatory procedure for mothers before nursing to recognize their own fawn. A female sniffs the fawn's anus even if it comes to her every 2–3 min. Such thoroughness in sniffing the anus continues for almost the entire hiding period, but by the end of hiding in July, the female does not smell her fawn every time they meet. This is because fawns spend most of the daytime with their mothers, and the mothers are able to recognize

Fig. 8.3 Suckling in goitered gazelles: A female sniffs her fawn's anal area and licks it during the whole suckling bout. Issyk-Kul Breeding Center for goitered gazelle reacclimatizing, 06.05.2023. D. Blank

their fawns by their appearance, eliminating the need for scent checking. However, mothers do continue to sniff the anus or at least other body parts of their fawns from time to time, despite the fact that the fawns are with their mothers all the time (Blank and Yang 2015b).

Some goitered gazelle mothers lift their hind leg sideways so that it is more convenient for the fawn to suckle. This behavior is atypical for most ungulate species and seen quire rarely in most mothers while nursing their young. It is noted in species of more ancestral species, such as muntjacs (*Muntiacus*) (Dubost 1971) and all mouse deer species (*c*) (Davis 1965; Ralls et al. 1975; Walther 1984).

8.3 Behavior of Females and Fawns During the Nursing Period

As a hiding species, the first 2 months of a goitered gazelle fawn's life is spent in hiding and motherless after nursing until it's time for the next feeding. Between nursing sessions, females initially walk away from their hiding young to a distance of several hundred meters up to 2–3 km to graze or rest (Fig. 8.4—Zhevnerov 1984; Blank 1985). The maximum distance noted was 10 km when a female went to a watering site. After 5–7 days, the mother will usually decrease the distance from her hiding fawn to 50–100 m. When leaving her offspring, the mother looks back several times as she walks away. For the first 2–3 days after birth, the offspring lies down quietly, stretches its head and neck forward while pressing its whole body to the ground, and does not move in the absence of its mother (Fig. 8.5). Within 4–5 days, the fawn starts to raise its head, examines the area around it, watches passing gazelles (Blank 1985), and rises only occasionally to stretch and perhaps cool off in the wind (Zhevnerov 1984). And at the age of 7–10 days, the young fawn is already independent enough in the absence of its mother to be able to change bed sites on its own, looking for a more comfortable place to rest. Before this, the bed sites are determined only by the mother (Blank 2017) as she stretches her muzzle to the ground, pointing to her young where to lie down; more often than not, the fawn lies down by literally falling like a stone, as the mother abruptly runs away, even in the absence of any danger (Blank 1985).

When a mother returns to her offspring, she walks quickly, sometimes even running until the last few hundred meters when she slows down and approaches very carefully (Zhevnerov 1984). If she senses no danger, the mother finally moves closer to the fawn and calls to it (Zhevnerov et al. 1983). The fawn immediately jumps up and runs to its mother to suckle. If it is asleep and does not see or hear her, she approaches closer and touches its body with her muzzle to get it up; the offspring immediately runs to its mother, and after the mother sniffs the baby's muzzle, suckling begins. If a female does not produce summoning sounds, fawns do not stand up or go to the mother, even if she passes closely to the resting place. In captivity, a bottle-suckling fawn walked away when the person with the bottle approached it, because the fawn did not receive the summoning sounds of a mother. Goitered gazelle fawns need these sounds only during their first week after birth,

Fig. 8.4 A goitered gazelle mother leaves her fawns alone after suckling, walking away to graze or rest from her hiding offspring at a distance. Drawing by U. Kenzhetaev with kind permission

Fig. 8.5 Hiding in goitered gazelle fawns: First days the offspring lies down quietly and does move, stretching its head and neck forward while pressing its whole body to the ground. Great Gobi B, Mongolia. (May 2003) Photo by P. Kaczensky with kind permission

when they do not move from their resting places without their mothers. Older fawns, as previously noted, will run to any female (or male or sub-adult) that happens to walk nearby without any motivational calls.

Fig. 8.6 A few weeks after birth, goitered gazelle fawns move for hundred meters in the absence their mother. Great Gobi A, Mongolia. 28.06.2005. Photo by P. Kaczensky with kind permission

Motor activity of fawns in the first days of life is closely related to their nutrition. Stimulation from the consumption of their mothers' milk determines their rhythm of activity (Zhevnerov 1984). During each visit, fawns receive a series of suckles before being left alone again for several more hours. During each suckling period, the female leads her fawns to a new hiding site (Blank 1985). During the first 1–3 days, the displacement of fawns from one bed site to the next is relatively insignificant at 10–15 m; after 3–4 days; however, they relocate 50–100 m between sites, a distance that can take 30–40 min to traverse (Zhevnerov 1984; Blank 1985), and after 10–15 days, they move as far as 400–500 m (Fig. 8.6—Blank 1985). Moreover, during the mother-escorted transitions, the mother increasingly runs with her fawn for training, first in one direction and then in the other. Concurrently, the time a female spends with her fawn increases, while at the same time, the number of times the fawn suckles in a bout, but the bout duration decreases (Tables 8.2 and 8.3). So, by the end of June, a sucking bout lasts an average of 10 min, but the female spends 1.5–2 h with her fawn; this is the time when the fawn's milk (food) dependency from its mother is replaced by a mother–young social relationship and the fawns increased consumption of plant food (Blank and Yang 2015a).

It has been reported that when approaching her twins for feeding, the mother will position herself exactly in the middle between them before the suckling session (Zhevnerov 1984); however, I did not observe this behavior in my study. When it's time for nursing, the fawns, as if on command, jump up from their beds at almost the same time and run to their mother from opposite sides and simultaneously begin to suckle; after which the mother, stepping over them, leaves her fawns alone until the next feeding. If for some reason one of her babies does not react to her voice, the mother begins to feed the first but does not allow it to suckle for very long before interrupting the feeding to search for the second. The first fawn follows its mother until the second one appears; then the feeding of the first restarts. Sometimes, the female will search for her missing fawn for hours, not letting the first fawn to suckle. The point of this before seems to be to not give one twin more milk than its sibling. So, it is not by chance that twin fawns usually develop more or less at the same rate, a phenomenon called "synchronous feeding" of young (Zhevnerov 1984); I did not observe this behavior, either. An interesting note: while feeding fawns in captivity with a bottle, the fawns always exuded a peculiar, strong odor (smelling "like mice"), which, perhaps, contributes to an increase in the flow of milk from the mother (Zhevnerov et al. 1983).

My observations of the behavior of goitered gazelle mothers in Kazakhstan have demonstrated that they nurse their own fawns exclusively. When a fawn runs up to her, the mother sniffs the fawn and only allows her own fawns to suckle. If the fawn is not hers, the females almost immediately refuse to let it suckle and run away from them. During my study in the Ili Hollow, I observed about 4000 cases of suckling young (Table 8.1) and not once was an alien fawn successful in stealing milk. Allosuckling has not been documented in goitered gazelles in other studies in the wild, either (Zhevnerov 1984; Marmazinskaya 1996, 1999), though unexpectedly, I observed this behavior in the Issyk-Kul Breeding Center (Blank 2023. Unpublished report): while one female was feeding a fawn, another female approached the suckling fawn and began to lick his anal area; then the same fawn moved to another female and began to suckle from her without any resistance. Moreover, this second female began to intensively lick the anal area of another fawn who had just suckled from a different female. This phenomenon was observed twice during the day. This may be just some kind of unique behavior among these females or more likely the result of these adult females being together in the same enclosure in their initial life in the Bukhara Breeding Center and their bottle-feeding together from their first days after birth. It is difficult to say. But this phenomenon warrants further study. During the following year (3–5 May of 2024), the same female likely nursed as many as 3–4 fawns that approached her to suckle. I observed this phenomenon 2–3 times a day (Blank 2024. Unpublished report).

Regarding other gazelle species, allosuckling is also observed extremely rarely, even in captivity (Walther 1984); Dittrich (1968) noted this phenomenon in captive Soemmerring's gazelles (*Nanger soemmerringii*) but only once. In contrast, observations of fallow deer (*Dama dama*) females during nursing season have shown that suckling of alien fawns is a common behavior for this species and about half (43%—Ekvall 1998; or even 73%—Pélabon et al. 1998) of all suckling cases noted nursing from alien mothers. In addition, some reindeer females demonstrate tolerance to alien fawns suckling (up to 15% of cases) (Espmark 1971; Engelhardt et al. 2016), and the Apennine chamois (*Rupicapra pyrenaica ornata*—Scornavacca et al. 2018) is also willing to suckle alien young. In general, however, allosuckling is noted in a limited number of ungulate species in the wild, though it is common for captive greater kudu (*Tragelaphus strepsiceros*), waterbucks (*Kobus ellipsiprymnus*), suni

Table 8.1 Changes in suckling duration (per half month) during May–December in females with single vs twin juveniles (by Blank and Yang 2015a—License Number—5476440889606)

Number of juveniles	Mean ± SE (s)	Range (s)	N	GLM, post hoc Bonferroni test, P
15–31 May	21.8 ± 1.6	1–151	200	
Single	28.7 ± 0.8	2–275	1896	
Twin				
01–15 June	17.7 ± 1.4	2–133	186	0.012
Single	27.1 ± 1.1	2–287	1066	
Twin				
16–30 June	14.0 ± 0.9	1–92	232	0.0005
Single	18.0 ± 1.8	1–190	178	
Twin				
01–15 July	9.8 ± 0.6	1–25	76	0.0005
Single	5.7 ± 0.3	1–14	87	
Twin				
16–31 July	6.5 ± 2.1	2–16	6	
Single	4.3 ± 0.4	3–6	7	
Twin				
01–15 August	4.0 ± 0.6	2–6	6	
Single	6.5 ± 0.7	5–9	6	
Twin				
16–31 August	3.6 ± 0.7	1–6	7	
Single	2.0	2	1	
Twin				
01–15 September	1.5	1–2	2	
Single	3.0	3	1	
Twin				
15–30 October	2.0	2	1	
Single				
Twin				
01–15 December	2.0	2	2	
Single				
Twin				

(*Nesotragus moschatus*), bushbucks (*Tragelaphus sylvaticus*), dik-dik (*Madoqua kirkii*), sable antelope (*Hippotragus niger*—Lent 1974; Hnida 1985), and the common hippopotamus (*Hippopotamus amphibius*—Pluhácek and Bartošová 2011).

When a mother feeds her twins, they often start or finish suckling at different times, but the female stops their suckling and jumps away from them at the same time. Thus, the duration of suckling in twins can be different. In reality, the situation is even more complicated, since one of the twins can stop its suckling in the middle of feeding and then starts suckling again, repeating the stops again and again. At the same time, the other twin can continue to suckle without interruption or take breaks, regardless of what the first fawn does. In this case, the total duration of suckling in twins can be very different (Blank and Yang 2015a).

Males chase females even during the birthing period and rearing of young. In the case of a prolonged courtship, the female will lead the male away from the place where her fawn is hiding, sometimes running toward him to drive him away (Blank 1985).

Thus, goitered gazelles have a long hiding period for their fawns, which lasts about 2 months, though it is important to note that this hiding period has several stages. The first stage lasts for about 5–7 days, when the fawns remain almost motionless on their beds, sprawled on the ground and pressing their head and neck to the ground (end of May). The second stage lasts for another couple of weeks (early June), when the fawns lie with their heads up, watching what is happening around them and periodically getting up to change their position, though a prone response to danger is still preferable for fawns during the first 14–16 days after birth. At the third stage (end of June), the fawns get up, move from place to place in the absence of their mothers, change bed sites on their own, graze on their own, and gather in groups with other fawns to play and walk together, though they periodically return to where they last parted from their mothers for the last time. And the fourth stage is when the fawns walk with their mothers almost all day, but stay alone at night, when their mothers go to their night pastures (early July). By the middle of July, most of fawns begin to follow their mothers both day and night (Blank 2017; Blank and Yang 2015a, b). Other gazelle and antelope species (Thomson's gazelle, Dorcas gazelle, Indian blackbuck) have relatively shorter hiding periods, lasting 2–6 weeks (Walther 1968). But given the fact that these species also likely have different stages of hiding, it is very possible that these antelopes have nearly the same duration of hiding as goitered gazelles. Many antelopes have comparable durations in their hiding periods, fluctuating between 2 and 4 weeks (klipspringer—*Oreotragus oreotragus*, pronghorn—*Antilocapra americana*, kudu—*Tragelaphus strepsiceros*). The shortest hiding period is in wild goats and sheep, as well as red deer, and lasts 3–5 days; the longest hiding period lasts 2–4 months (Uganda kob—*Kobus kob thomasi* and reedbuck—*Redunca redunca*) (Lent 1974).

8.4 Suckling Modes at Different as the Fawn Ages

According to my observations of goitered gazelles in the wild, fawns stand and walk within 5–7 min after birth and the mothers are able to nurse them for the first time within 12–15 min after giving birth. In captivity, Pereladova and Pereladov (1986) observed that the first suckling occurred about 40 min after delivery. Compared to goitered gazelles, the young of other ungulate species begin to stand and walk after a longer period of time after birth: the saiga antelopes' (*Saiga tatarica*) first suckle is 20 min after birth (Kokshunova 2007), which is about the same as the Mongolian gazelle (*Procapra gutturosa*) (Kiriluk and Zvychainaya 2011); offspring of the Grant's gazelle (*Nanger granti*) and impala offspring (*Aepyceros melampus*) start to stand and walk and suckle after 19–21 min, pronghorns 26–31 min, dorcas gazelles (*Gazella dorcas*) 29 min, and gerenuk 60 min. Maybe only newborn wildebeest begin to stand and walk slightly faster compared to goitered gazelles at 3–5 min after birth (Lent 1974).

Table 8.2 Changing of interval duration between suckles and suckling bouts during May–July in females with single and twin juveniles (Blank and Yang 2015a—License Number—5476440889606)

Juvenile number	Mean ± SE, min	Range, min	N	Mann-Whitney test Z	Mann-Whitney test P
Intervals between suckles					
15–31 May					
Single	2.18 ± 0.25	0.05–25.77	145		
Twin	1.04 ± 0.04	0.02–26.50	1325		
01–15 June					
Single	4.46 ± 0.43	0.02–32.78	149	−4.865	0.0005
Twin	1.47 ± 0.06	0.02–26.78	994	−8.291	0.0005
16–30 June					
Single	23.95 ± 3.68	0.02–203.00	101	−7.463	0.0005
Twin	28.96 ± 5.54	3.00–118.00	24	−6.886	0.0005
01–15 July					
Single	64.00 ± 14.45	5.00–192.00	14	−3.662	0.0005
Twin	16.36 ± 3.78	2.00–45.00	14	−1.484	0.144
Intervals between suckling bouts					
15–31 May					
Single	182.8 ± 52.3	35–478	8		
Twin	335.7 ± 21.4	49–707	61		
01–15 June					
Single	202.5 ± 43.6	61–406	8	−0.420	0.721
Twin	303.7 ± 37.8	38–555	15	−0.646	0.518
16–30 June					
Single	55.9 ± 7.6	3–192	44	−3.526	0.0005
Twin	33.7 ± 8.8	2–120	17	−4.589	0.0005

During the first 2 weeks after the mass birthing of young in mid-May, fawns have the most frequent and longest suckling bouts (up to 30 suckles per bout) with small 2- to 3-min intervals between suckles. The vast majority of suckling (73.8%, $n = 1530$ in twins; 80.1%, $n = 181$ in singletons) does not last more than 30 s, with suckling of 10 s in twins and 10–15 s in singletons observed most often. After mid-June, short suckles increase in number, reaching 92% ($n = 221$) in twins and 92.7% ($n = 113$) in singletons, with suckles of 5–15 s most numerous and suckling durations of less than 5 s and longer than 15 s being observed less and less often. In contrast, intervals between nursing bouts, being minimal during a fawn's first weeks in May, increase continuously with age: in twins by 1.4 times in early June, 27.8 times in late June, and 15.7 times in early July; in single young by 2 times, 11 times, and 19.4 times, respectively, compared to late May (Table 8.2). It should be noted that these intervals are always shorter for twins compared to singles during all periods (Blank and Yang 2015a). Interestingly, female sand gazelles (*G.s. marica*) living in semi-free conditions in Turkey have identical nursing patterns for fawns, when twins suckle longer and more often than singles, especially in the first weeks after birth (Uztemur et al. 2024). Later, this difference between twins and singles is leveled, although in the sandy gazelle it occurs already in early June (Uztemur et al. 2024), whereas in the Kazakh gazelles only by the end of June (Blank and Yang 2015a). In some cases, the duration of suckling in wild fawns from Kazakhstan coincides so precisely that it leads to the idea of the taxonomic proximity of the goitered gazelle and the sand gazelle, and likely these are more subspecies of the same species than different species.

Goitered gazelle mothers visit their hiding offspring to suckle them usually 2–3 times a day, though some mothers have been observed to feed their fawns 4 times a day (Blank 1985). This was also seen in the Issyk-Kul Breeding Center, where, in some cases, suckling occurred more often than 4 times a day (Blank 2023. Unpublished report). Generally, suckling is observed early in the morning, at noon, and in the evening around 20:00 to 21:00 or sometimes later (Ishunin 1961), with intervals between suckling bouts many times longer (3.1–5.5 h or even up to 8 h) compared to intervals between suckles within a bout (Table 8.2). Therefore, during May and June, suckling is distributed very unevenly over the day, as the number of suckles in a suckling bout decreases with the age of the fawn. The longest suckles last more than 2 min, with an observed maximum duration of 4.5 min, though the shortest suckles lasting between 5 and 20 s are noted most frequently (Blank and Yang 2015a). Females of other gazelle species, such as Thomson's and Grant's gazelles, visit their fawns 3–5 times a day (Walther 1968). The white-tailed deer (Huegel et al. 1985), mule deer (Taylor 2013), and roe deer (Aanes and Andersen 1996) also nurse their fawns from 2 to 4 times a day. The number of times a female nurses her young per day apparently depends on the individual preferences of the mothers in all these species (Blank 2017). Similar patterns have been found in female sand gazelles in Turkey (Uztemur et al. 2024). Although sand gazelles reduce the total suckling duration faster, when already in early June this value decreases by almost 10 times, whereas in Kazakh gazelles it occurs later, in the second half of June (Blank and Yang 2015a). This difference is obviously related to the earlier drying of natural vegetation in Turkey compared to Kazakhstan.

For the first 1.5 months (mid-May and June), the female nurses her fawn from 3 to 5 up and to 30 times in a row during a suckling bout, making small intervals of 1–3 min between separate suckles. In May, during its first 2 weeks after birth, a fawn suckles the longest with the duration shortening gradually in early June, and abruptly in late June and especially in early July (Table 8.1; Blank and Yang 2015a, b). An analysis of goitered gazelle milk showed that it has a high fat content of 39%, a Turner acidity of 20°, and a specific gravity of 1.032 (milk on the fourth day after delivery). A newborn receives only about 4–5 g of milk at the first suckling, about 12 g at the second (after 5 min), and more than 30 g 3 days after birth (Ishunin 1961).

At the age of 2–3 days in captivity, fawns drink 50–100 cm^3 or 300 g, sometimes up to 400 g of cow's milk during four feedings a day, and finish it during 2–2.5 min feedings. At a week old, the milk was diluted with boiled water to avoid digestive disorders. At the end of May, Zhevnerov (1984) collected up to 400 g of milk from a nursing female at a single milking. At the age of 2 weeks, the fawn drank between 300 and 400 g of the cow's milk through a bottle in one feeding over a period of 30–50 s. Further, the amount of milk increased every 5 to 7 days, and by the age of

1 month, the fawns drank up to 750 to 1000 g of milk at a time. Starting from the age of 1 month, the rate of milk to young gazelles was reduced, giving them only 500 g per day in two feeding sessions. Most likely, fawns in their natural habitat receive about the same amount of milk as in captivity. In one case in the central part of the Betpak-Dala Desert, a female with one 10-day-old fawn was revealed to have given her fawn 400 g of milk at a time. On June 25, 1954, at 4 o'clock in the morning, a female fawn at the estimated age of about 1 month (weight of 7.1 kg), taken in the Betpak-Dala Desert, had about 100 g of curdled milk in her stomach, along with *Anabasis salsa* mixed with a lot of her own wool, all of which was tinted brown from water drunk from a puddle located on the red clay. This suggests that in addition to milk and moisture from plants, nursing fawns need additional fluids and even as early as 1 month of age go to watering places. Fawns in captivity drink 2–3 l of water a day (Sludskiy 1956; Zhevnerov et al. 1983).

As time passes, goitered gazelle mothers stay with their fawns for longer periods of time, and as a result, the nursing bouts expand from half an hour in May up to more than 1 h in late June for twins, and from 19 min up to 40 min for singletons. As a consequence of bouts expanding, the intervals between them shorten considerably during June, and especially drastically at the end of June when they become almost equal intervals between separate suckles within a bout. In other words, nursing bouts essentially disappear at the end of June and early July, turning into separate instances of single sucking (Blank and Yang 2015a). At the same time, the duration of each suckle gradually decreases over time (Table 8.1). Single suckling cases are observed mainly into late August, so the suckling period lasts around 3.5–4 months in goitered gazelle fawns. Even as late as September into December (and April), I noted separate, very short (2–3 s) suckles, but these were observed very rarely (1–2 times a month) and looked more like unsuccessful suckling attempts (quickly terminated by the mothers) than real suckling (Blank and Yang 2015a). For bighorn sheep (*Ovis canadensis*), Shackleton and Haywood (1985) considered suckles shorter than 5 s as unsuccessful suckling attempts, but for goitered gazelles, it is worth to define shorter duration suckles as unsuccessful (<2 s), since suckles as long as 2 s may also be nutritive (Haley et al. 1998). For the first 2 weeks, most suckling is stopped by the fawn itself, followed in time by some suckling being stopped by the mother, which increases over time. By the second half of June, females limit almost all their fawns' suckling due to the transition of the fawns to a predominantly vegetarian diet, though the mother's milk supply still continues until almost early October.

The number of unsuccessful suckling attempts, which stopped early in the attempt, is minimal in May (7.2%, $n = 138$ in single young; 16.6%, $n = 66$ in twins). After mid-June, the number of unsuccessful attempts increases to its maximum (51.4% in single young and 34.8% in twins in early July) and then drops drastically in August to 1–6% as the fawns are mostly weaned.

Thus, goitered gazelle females make their most intensive maternal investment (supplying the largest amount of milk) during a fawn's first 4 weeks (mid-May to mid-June) after giving birth, especially within the first 2 weeks when in most cases the mothers allow their fawns to suckle as much as possible without limitation in most cases (84.5%, n = 1508 in twins; 72.4%, n = 181 in singletons), and when considerably less suckling is terminated by their mothers (15.5% and 27.6%, respectively). This is when the maximum amount of milk is transferred from mother to young, which contributes to the fawn's fastest possible growth. Later, females gradually reduce milk amounts for the fawns older than 1 month. After mid-June, mothers limit the majority of suckling (91.9%, n = 222 in single young; 98.2%, n = 114 in twins), though females with single young restrict suckling more often than females with twins. By early July, all suckling is stopped by the mothers, and the number of suckles and their duration sharply drop down to the point where the fawn's suckling bouts disappear by the age of 2 months, indicating the start of the weaning process. As a result, milk for a fawn loses its principal significance as an energy source and instead becomes supplemental nutrition for a mainly vegetative diet (Blank and Yang 2015a). After 2 months (late July and August), suckling has a very short duration and is observed only rarely (Table 8.1); by this age, milk has no appreciable contribution to the offspring's energy supply and has more of a social rather than a nutritive significance. By the end of August into mid-September, goitered gazelle females generally stop supplying milk to their offspring altogether (Blank and Yang 2015a). So, for goitered gazelles, the suckling period lasts a total of 3–4 months, although the milk is only of significant nutritive importance during the first 2 months after birth (from late May to early July). This is a relatively short suckling period for rearing young.

According to Ismagilov (1954), the lactation period of goitered gazelles is 2 months on average, although young can survive without milk even as early as the age of 1 month due to the very high, rich milk fat (39%). Other authors (Ishunin 1961; Zhevnerov et al. 1983) assert that this is a mistake and gazelle milk is no thicker than that of a cow (4%). Sludskiy (1956) and Sokolov (1959) indicated that the period of mass lactation lasts about 3 months into mid-August. Observations of tamed females during three breeding seasons show that the lactation period of goitered gazelles lasts 65–80 days. In general, the duration of lactation is somewhat dependent on the individual characteristics of the fawns and their mothers, the timing of the birth, and the state of the environment and vegetation. The vital need for milk nutrition of fawns persists until the age of 1 month, which coincides with the beginning of their consuming significant amounts of vegetation and acquiring watering skills (Zhevnerov 1984).

Goitered gazelle fawns begin to show an interest in vegetation first by imitating their mothers and sniffing and nibbling at grass as early as their second or third day of life. After 5–7 days, the fawn bites and chews some leaves, and after 15–17 days, the fawn is already grazing, though for short intervals (4–5 min), which are interrupted periodically by suckling and play. The fawn begins by eating leaves from the same shrub branch as its mother, pushing her muzzle away, and later, he begins to recognize and choose edible plants on its own. At the age of 1 month, young consume significant amounts of solid forage (plants) (Blank 1985, 1990). Obviously, 1 month is a crucial age for goitered gazelle young, when their mothers begin to force them to alternate between vegetation and milk. From this moment on, milk is no longer the young's only nourishment and becomes a supplementary food source, the portion of which decreases quickly over the following weeks (Blank and Yang 2015a). At the age of 3–3.5 months, most fawns are weaned and completely adjust to plant nutrition, making them able to live independently of their mothers (Sludskiy 1956).

Family groups consisting of females and fawns exist for about 4–4.5 months—from the end of April to the end of August. In August, September, and October, the separation of young from their mothers and the disintegration of family groups begin. At this time, single females and sub-adult males begin to come together, as well as groups consisting exclusively of just several sub-adults. On October 20, 1998, a group of 4 sub-adult females were observed, and in August into October, single sub-adults were seen, as well as groups consisting of adults and sub-adults—a total of 12 encounters in the Bukhara semi-captive population (Marmazinskaya 1996).

8.5 Features of Female Behavior During Suckling of Twins and Singles

As the fawn grows, both twins and single fawns decrease their suckling duration, but these processes take place in different ways. During their first 4 weeks, twins have considerably longer suckling bouts than singletons and twins get greater amounts of milk compared to singletons. However, this difference disappears after mid-June and in July singletons start to suckle longer than twins (Table 8.1). In May, females with singletons restrict suckling of their fawns more often than females with twins. During the next 2-week period before mid-June, females restrict suckling more and more as compared to May, limiting single young more often (35.4%, $n = 198$) than twins (20.5%, $n = 1079$). From August on, the difference in the suckling duration in twins and singletons is absent (Blank and Yang 2015a).

As the fawns grow, the total suckling time of a bout (sum of all suckles) shifts differently for twins and singletons. In twins, this total sum increases at the beginning (in early June compared to May) and then decreases sharply after mid-June. In contrast, singletons demonstrate a continuous decrease from May to July (Table 8.3). In mid-May to early June, the total suckling time is significantly longer in twins compared to singletons, while after mid-June, singletons have a longer total sum of suckles than twins. Intervals between suckling bouts are longer in twins compared to singletons, but the opposite is observed after mid-June, with a general trend of shortened intervals during June, especially quickly after mid-June (Blank and Yang 2015a; Table 8.3).

The ratio of successful and unsuccessful suckles in twins and singletons changes in a similar way in May and early June, but after mid-June singletons are rejected by their mothers 3.7 times more often than twins, and the total number of unsuccessful suckles by singletons is 2.1 times more numerous than in twins. The portion of unsuccessful suckles was at its maximum in August; one unsuccessful suckle is accompanied by successful one (Blank and Yang 2015b).

Table 8.3 Suckling bouts in females with single versus twin juveniles: sum of suckles into bouts, duration of bouts, and intervals between them during May–July (Blank and Yang 2015a—License Number 5476440889606)

Juvenile number	Mean ± SE	Range	N	Mann-Whitney test	
				Z	P
Total time suckling into the bout (sec)					
15–31 May					
Single	169.8 ± 26.5	19–466	24		
Twin	354.4 ± 22.9	35–1091	125		
01–15 June					
Single	83.4 ± 12.2	12–324	42	−3.113	0.002
Twin	438.1 ± 36.1	41–1269	69	−2.280	0.023
16–30 June					
Single	32.1 ± 4.5	3–242	78	−4.962	0.0005
Twin	18.6 ± 2.0	4–57	49	−9.147	0.0005
Bouts' duration (min)					
15–31 May					
Single	19.2 ± 3.2	3–54	21		
Twin	30.0 ± 2.2	2–88	72		
01–15 June					
Single	24.7 ± 3.4	1–98	41	−0.633	0.527
Twin	32.5 ± 3.0	3–93	41	−0.860	0.390
16–30 June					
Single	40.1 ± 4.7	3–143	47	−2.514	0.012
Twin	65.4 ± 15.1	11–171	13	−1.285	0.199

My observations show that twins suckle more frequently and have longer suckles with shorter intervals between suckles than singletons. In addition, singletons have more frequent unsuccessful suckles than twins. Successful suckles of singletons are terminated by mothers more frequently than successful suckles of twins. The suckle frequency and total suckling time are greater in twins than in singletons, which indicates that twins get significantly larger amounts of milk than singletons and mothers make more of an investment in twins, than do females with singletons, especially during the first month when young are almost completely dependent on milk. There are many examples (domestic sheep and cattle, moose *Alces alces*, eland antelopes *Taurotragus oryx*) that indicate that females with twins are able to produce much more milk than single mothers (Gifford 1953; Yazan and Knorre 1964; Treus and Kravchenko 1968; Peart 1982). The twins of white-tailed deer (*Odocoileus virginianus*) and black-tailed deer (*O. hemionus*) can get almost twice the milk amounts compared to singletons (Sadleir 1980; Gauthier and Barrette 1985). It follows that females have the ability to increase milk production, depending on the number of fawns in their families. Not all females have this ability, however. It seems only adult multiparous females in good nutritional condition are able to give birth to twins and produce more milk, while young primiparous mothers or old or sick females in bad body condition tend to have only singletons requiring less milk production (Blank and Yang 2015b). Investigation into suckling behavior of other ungulate species shows that twins usually nurse more frequently compared to singletons (Lent 1974).

Thus, the main reproduction strategy of goitered gazelle females is to prefer the birth of relatively small size twins and then provide them with significant nutrition to promote intensive growth and successful survival. This preference is observed especially during favorably warm, rainy years with an abundance of high-quality vegetation, which supports the highest percentage of females with good body conditions bearing twins (up to 75%—Kingswood and Blank 1996). Because of these positive factors, the population grows rapidly in favorable years. In cold, dry years, when there are limitations to high-quality food, the percentage of females with poorer body conditions increases significantly, resulting in a sharp increase in the number of singletons, many of relatively larger size to possibly compensate for the poorer maternal investment and the inability to provide them with abundant nutrition (Blank and Yang 2015b).

8.6 Bottle-Suckling of Goitered Gazelle Fawns

Methods for artificially feeding and keeping goitered gazelles in enclosures have been developed in the Bukhara Breeding Center since 1978. Over a 7-year period, 44 gazelles were raised using manual bottle-feeding methods. In captivity, four kinds of milk for feeding newborn fawns were tested, which differed from each other in the composition of the diet: cow milk, goat milk, a sheep milk substitute, and a cow milk substitute. As a result of this research, a method of artificial feeding of goitered gazelles based on the cow milk substitute was developed. In total, 34 fawns were fed using this food alternative. The death rate was about 3%, and the deaths occurred by causes unrelated to feeding (Flint and Soldatova 1986).

Groups for artificial feeding were formed from newborn fawns caught in the open areas of the Breeding Center. Newborns were caught by various methods, but the most productive was catching them at night with a special net after being detected from the headlights of a car. The duration for the formation of manual feeding groups was usually limited to a short time period: April 23–30 to May 5–6. The age difference of the captured fawns did not exceed 3–4 days, with the exception of two fawns aged 12–15 days. It was difficult to accustom these gazelles to the artificial nipple, which made them more susceptible to stress during feeding, and resulted in poor weight gain during the initial period. The captured fawns initially underwent a preventive examination and weighing and spent their first day indoors under constant human supervision (Flint and Soldatova 1986).

The first feeding was done very responsibly, since a wrong approach could cause the fawn to develop a negative reflex not only to the bottle, but also to the person. The first feeding occurred after the fawn calmed down from the excitement caused by trapping and transporting. Usually after delivering the captured fawns to the base, they were not disturbed for 2 h before the first attempt at feeding was made. During the first feeding, the fawn was initially lying down; then the person gently stroked and massaged the anogenital area, supporting and lifting to help it stand on its legs. Feeding needed to be carried out with the fawn in a standing position and it was necessary to make the fawn suckle the nipple itself, which was put into its mouth (Fig. 8.7). Gentle pressing on the corners of the mouth caused swallowing movements. Most of the time the first attempts to feed the fawn ended in failure. It either refused to suckle or, after swallowing several times, quickly released the nipple, and after several unsuccessful attempts to suckle, laid down. At this point the fawn was left alone until another attempt at feeding occurred after 1–1.5 h. Usually during the first day, after several attempts, the fawn managed to get used to the nipple and developed the reflex for suckling the bottle. First feedings were abundant and took place in a calm environment, using soft and attentive methods to teach the baby to suckle so as not to instill a defensive reaction of freezing or suckling while lying down (Flint and Soldatova 1986).

During the first feeding, it was necessary to feed the fawn one raw egg stirred in water in order to improve the functioning of the gastrointestinal tract. The total amount of the mixture did not exceed 100 g. This mixture was easily digested and promoted the formation of gastrointestinal flora which helped to move the fawns from one type of food to another without complications. The first maternal feeding is very important for the health of the baby and the formation of its physiology. Fawns deprived of this are distressed and grow poorly. Therefore, only goitered gazelle fawns that have already been suckled by their mothers should be caught from nature (Flint and Soldatova 1986).

Fig. 8.7 Bottle milk feeding for goitered gazelle fawns: Feeding needed to be carried out with the standing fawn, massaging of the anogenital area, put nipple into fawn's mouth. Issyk-Kul Breeding Center for goitered gazelle reacclimatizing, 06.05.2023. Photo by D. Blank

It was established through experience that the daily nutrients needed for growing goitered gazelle fawns can be fully satisfied with 150 g of cow milk substitute. This standard remained nearly stable during the entire artificial feeding period, starting from the age of 1 week. On the first day of feeding, the baby got 1/4 of the standard norm, from a milk mixture with a concentration of 1 part of milk substitute to 5 parts of water. In the following days, the total volume of the milk mixture and the concentration of the milk substitute were gradually increased so that by the end of the week the baby was given three suckling sessions with a concentration of 150 g of the mixture per 1 l of water. Subsequently, while maintaining the daily amount of the mixture, its concentration was reduced by increasing the volume of the milk mixture to 1.5 l (0.5 l per feeding). Before feeding, the milk mixture was heated to 35–38 °C and it could not be shaken, since that would lead to the separation of fat. For the entire period of feeding, 18 kg of the milk mixture was consumed per baby. The total duration of manual suckling was 4 months. In the case of intestinal disorders, one feeding was skipped, and in the next suckling, the fawn was given a weaker concentration of the milk mixture and raw egg and water. If the intestinal function was not restored, the fawn was given medicine with a good therapeutic effect (Flint and Soldatova 1986).

Upon reaching the age of 5 days, the captive fawns began to exhibit more complex feeding on green alfalfa, water, and free grazing within the enclosure. At the age of 8–10 days, the fawns were actively interested in roughage and begin eating it in small quantities. By the end of their need to suckle, they completely switched to self-feeding on vegetation. The total duration of manual feedings was determined by the health and development of the fawn rather than in calendar terms. The milk mixture was not removed immediately from the diet, but gradually, reducing the daily amounts. First, the fawns were transferred to a two-time feeding schedule, followed by a short period of one-time milk feedings. The duration of the final transition period from a dairy diet to self-nutrition was from 2 to 3 weeks. Fawns raised in a captive environment on a dry milk mixture did not differ in terms of appearance, health, or disposition from young gazelles raised in a natural environment. An important indicator of the animal's condition is its weight. It has been empirically established that by the time they stop suckling milk, female fawns should gain at least 15 kg, and male fawns should gain 16–17 kg. All tamed half-year-old females that by autumn reached a weight of 20 kg participated in reproduction, and by the age of 1 year, they had already produced valuable, viable offspring (Flint and Soldatova 1986).

For the successful growth and development of goitered gazelle fawns in captivity, the shape and size of the enclosure in which they are kept are important. The most convenient area for working with goitered gazelles is a round enclosure, or one with rounded corners, and a fence made of wire mesh. The wire mesh is especially important, because it eliminates the opportunity for the fawns to stick their heads through the fence which can cause injury. The area of the enclosure should be sufficient for playing and running. In the Bukhara Breeding Center, a herd of up to 20 animals was kept throughout the year in an enclosure measuring 60 × 100 m. Adjacent to this enclosure, blocks of different configurations were built, in which females were kept during birthing and rutting groups and young animals were also isolated. The enclosure complex was equipped with sun and windproof canopies and barriers. Special attention during the period of bottle-feeding was paid to the cleanliness of the enclosure, since the fawns showed great interest in any objects, which after sniffing were often swallowed. And trapped garbage in the digestive tract can cause poisoning with a fatal outcome (Flint and Soldatova 1986).

In the process of manual bottle-feeding, the fawns developed positive reactions to humans. Habituation to people was supported by positive contact when activities, like daily weighing, were accompanied by a positive food response (the young received crackers and other appealing food). Growing gazelles that go through repeated weighing and measuring procedures during lactation can easily tolerate other handling activities in the future: inspections, capture, and transportation. Systematic work with goitered gazelle fawns, suckled from a bottle, allowed the creation of a more cooperative captive herd in the Bukhara Breeding Center. Tame gazelles were used when catching wild gazelles that lived freely in the main part of the Breeding Center. Usually, the tamed animals were allowed to approach wild gazelles that had entered a stationary trap from the open territory of the park, after which the wild animals moved into an enclosure complex under the guidance of the tame gazelles. Further adaptation of wild gazelles to the conditions of involuntary detention was facilitated when there were tame gazelles among them, demonstrating a positive reaction to the presence of human beings (Flint and Soldatova 1986).

References

Aanes R, Andersen R (1996) The effects of sex, time of birth, and habitat on the vulnerability of roe deer fawns to reed fox predation. Can J Zool 74:1857–1865

Blank DA (1985) Peculiarities of social and reproductive behaviour of *Gazella subgutturosa* in the Iliisky Valley. Zool Zhurnal 64:1059–1070. (in Russian with English abstract)

Blank DA (1990) Persian gazelle. In: Kovshar AF (ed) Rare animals of desert regions, Desert animals. Nauka Press of Kazakh SSR, Alma-Ata, pp 56–80. (in Russian)

Blank DA (1997) Social and reproductive behaviour of the Persian gazelle. PhD Thesis, University of Tel Aviv, Tel Aviv

Blank DA (2017) Antipredator tactics are largely maternally controlled in goitered gazelle, a hider ungulate. Behav Proc 136:28–35

Blank DA (2023) Birthing season in goitered gazelles in conditions of the Issyk-Kul Breeding Center. Unpublished report for the Institute of Biology, Kyrgyz Academy of Sciences

Blank DA (2024) Goitered gazelle females' behavior during rearing fawns in semi-captive conditions of the Issyk-Kul Breeding Center. Unpublished report for the Institute of Biology, Kyrgyz Academy of Sciences

Blank DA, Yang W (2015a) Strategy of goitered gazelle suckling behavior and its adaption to the environment. Mamm Res 60(4):393–401

Blank DA, Yang W (2015b) Suckling behavior in goitered gazelle: do females invest more in twins or singletons? Zoology 118:348–356

Blank DA, Yang W (2017) Mother-young recognition in goitered gazelle during hiding period. Behav Proc 142:21–28

Bubenik AB (1965) Beitrag zur Geburtskunde und den Mutter-Kind-Beziehungen des Reh- und Rotwildes. Z Saugetierk 30:65–228

Burckhardt D (1958) Observations sur la vie sociale du Cerf au Parc National Suisse. Mammalia 22:226–244
Cameron EZ, Linklater WL, Stafford KJ, Minot EO (2000) Aging and improving reproductive success in horses: declining residual reproductive value or just older and wiser? Behav Ecol Sociobiol 13:243–249
Cassinello J (2001) Offspring grazing and suckling rates in a sexually dimorphic ungulate with biased maternal investment (*Ammotragus lervia*). Ethology 107(2):173–182
Davis JA (1965) A preliminary report of the reproductive behavior of the small Malayan chevrotan *Tragulus javanicus* at the New York Zoo. Int Zoo Yearb 5:42–44
Dittrich L (1968) Keeping and breeding gazelles at Hanover Zoo. Int Zoo Yearb 8:139–143
Dubost G (1971) Observations éthologiques sur le Muntjak (*Muntiacus muntjak* Zimmermann, 1780 et *M. reevesi* Cgilby, 1829) en captivité et semi-liberté. Z Tierpsychol 38:387–427
Ekvall K (1998) Effects of social organization, age and aggressive behaviour on allosuckling in wild fallow deer. Anim Behav 56(3):695–703
Engelhardt SC, Weladji RB, Holand Ø, Røed KH, Nieminen M (2016) Allonursing in reindeer, *Rangifer tarandus*: a test of the kin-selection hypothesis. J Mamm 97(3):689–700
Espmark Y (1971) Mother-young relationship and ontogeny of behaviour in reindeer. Z Tierpsychol 29:42–81
Flint VE, Soldatova NV (1986) Bukhara nursery. Report 6. Experience of bottle feeding of goitered gazelles. In: Flint VE, Prisyazhnuk VE (eds) Conservation and perspectives of the number restoration of goitered gazelle in USSR. All-Union Scientific Institute of Nature Conservation and Nature Reserves, Moscow, pp 70–74. (in Russian)
Gauthier D, Barrette C (1985) Suckling and weaning in captive white-tailed and fallow deer. Behaviour 94:128–149
Geist V (1971) Mountain sheep: a study in behavior and evolution. University of Chicago Press, Chicago
Gifford W (1953) Records-of-performance tests for beef cattle in breeding herds. Milk production of dams and growth of calves. Ark Agric Exp Sta Bull:531
Hafez ESE, Cairns RB, Hulet CV, Scott JP (1969) The behaviour of sheep and goats. In: Hafez ESE (ed) The behavior of domestic animals, 2nd edn. Williams and Wilkins, Baltimore, pp 296–348
Haley D, Rushen J, Duncan I, Widowski T, de Passile AMB (1998) Effects of resistance to milk flow and the provision of hay on nonnutritive suckling by dairy calves. J Dairy Sci 81:2165–2172
Hamel S, Côté SD (2008) Trade-offs in activity budget in an alpine ungulate: contrasting lactating and nonlactating females. Anim Behav 75(1):217–227
Hediger H (1950) Beiträge zur Säugetier-Soziologie. In: Grasse P (ed) Structure et physiologie des sociétés animates. CNRS, Paris, pp 297–323
Hick U (1969) Successful raising of pudu, *Pudu pudu,* at Cologne Zoo. Int Zoo Yearb 9:110–112
Hnida JA (1985) Mother-infant and infant-infant interactions in captive sable antelope: evidence for behavioral plasticity in a hider species. Zoo Biol 4(4):339–349
Huegel CN, Dahlgren RB, Gladfelter HL (1985) Mortality of white-tailed deer fawns in south-central Iowa. J Wildl Manag 49(2):377–380
Ishunin GI (1961) Mammals (carnivores and ungulates). In: Zakhidov TZ (ed) Fauna of Uzbek SSR, vol 3. Publication of the Academy of Sciences of Uzbek SSR, Tashkent, pp 165–174. (in Russian)
Ismagilov MI (1954) On ecology of goitered gazelle in the Barsakelmes Island. Bull Moscow Nat Biol Dep 59(6):3–9. (in Russian)
Kingswood SC, Blank DA (1996) Gazella subgutturosa. Mamm Species 518:1–10
Kirchshofer R (1963) The birth of a dwarf chimpanzee: *Pan paniscus* Schwarz 1929 at Frankfurt zoo. Int Zoo Yearb 4(1):76–78
Kiriluk VE, Zvychainaya EY (2011) Behavior of mother and young in Mongolian gazelle (*Procapra gutturosa*). Sci Notes Trans-Baikal State Univ 1(36):74–83. (in Russian)
Kokshunova LE (2007) Nursing behavior of Saiga antelope fawns (*Saiga tatarica*) in captive conditions. News of Univ North-Caucasus Region 5:62–64. (in Russian)

Kramer A (1969) Soziale Organisation und Sozialverhalten einer Gemspopulation der Alpen. Z Tierpsychol 26:889–964

Kurt F (1968) Das Socialverhalten des Rehes. Eines Feldstudie. P. Parey-Verlag, Berlin, p 102

Lent PC (1966) Calving and related social behavior in the barren-ground caribou. Z Tierpsychol 23:702–756

Lent PC (1969) A preliminary study of the Okavango lechwe: Kobus leche leche gray. Afr J Ecol 7(1):147–157

Lent PC (1974) Mother-infant relationships in ungulates. In: Geist V, Walther F (eds) The behaviour of ungulates and its relation to management. IUCN, Morges, pp 14–55

Marjoribanks-Egerton P (1962) The cow-calf relationship and rutting behavior in the American bison. Master's thesis, University of Alberta, Edmonton, p 155

Marmazinskaya NV (1996) Breeding of goitered gazelle *Gazella subgutturosa* (Artiodactyla Bovidae) in the ecological center "Dzheiran". In: Proceedings of the Uzbekistan nature reserves, issue 1. FAN Publishing House, Tashkent, pp 53–61. (in Russian)

Marmazinskaya NV (1999) Social behavior of goitered gazelle (*Gazella subgutturosa* /Güld.), 1780/during breeding period in the conditions ecocenter "Dzheiran". Ph.D. thesis in zoology, Institute of zoology of Uzbek Academy of Sciences, Tashkent, pp 1–20. (in Russian)

Oftedal OT (2000) Use of maternal reserves as a lactation strategy in large mammals. Proc Nutr Soc 59(1):99–106

Ozoga JJ, Verme LJ (1986) Relation of maternal age to fawn-rearing success in white-tailed deer. J Wildl Manag 50:480–486

Peart JN (1982) Lactation of suckling ewes and does. In: Coop I (ed) Sheep and goat production. World animal science, v. Cl. Elsevier, Amsterdam

Pélabon C, Yoccoz NG, Ropert-Coudert Y, Caron M, Peirera V (1998) Suckling and allosuckling in captive fallow deer (Dama dama, Cervidae). Ethology 104(1):75–86

Pereladova OV, Pereladov SV (1986) The goitered gazelle breeding in nursery enclosures of the Syunt-Khasardak Nature Reserve. In: Flint VE, Prisyazhnuk VE (eds) Conservation and perspectives of the number restoration of goitered gazelle in USSR. All-Union Scientific Institute of Nature Conservation and Nature Reserves, Moscow, pp 49–59. (in Russian)

Pluhácek J, Bartošová J (2011) A case of suckling and allosuckling behaviour in captive common hippopotamus. Mamm Biol 76:380–383

Ralls K, Barasch C, Minkowski K (1975) Behavior of captive mouse deer, *Tragulus napu*. Z Tierpsychol 37(4):356–378

Sadleir RMFS (1980) Energy and protein intake in relation to growth of suckling black-tailed deer fawns. Can. J. Zool., 58:1347–1354

Scornavacca D, Cotza A, Lovari S, Ferretti F (2018) Suckling behaviour and allonursing in the Apennine chamois. Ethol Ecol Evol 30(5):385–398

Shackleton DM, Haywood J (1985) Early mother–young interactions in California bighorn sheep, *Ovis canadensis californiana*. Can J Zool 63:868–875

Sludskiy AA (1956) Breeding of goitered gazelle. In: Proceedings of Institute of Zoology, Kazakh Academy of Sciences, vol 6. Nauka Press of Kazakh SSR, Alma-Ata, pp 78–108. (in Russian)

Sokolov II (1959) Gazella subgutturosa Guldenstaedt – Goitered gazelle. In: Pavlovskiy, E.N. (ed.) Fauna USSR, Mammals: Ungulates (Orders Perissodactyla and Artiodactyla). Publication of the Academy of Sciences USSR: Moscow and Leningrad, 308–331 pp. (in Russian)

Sokolov VE, Gromov VS, Danilkin AA (1985) Ontogenesis of behavior of the Siberian roe deer (*Capreolus capreolus pegarcus*). Zool Zhurnal 64(6):915–926. (in Russian)

Taylor CA (2013) Behavior & cause-specific mortality of Mule Deer (*Odocoileus hemionus*) fawns on the National Rifle Association Whittington Center of North Central New Mexico. Doctoral dissertation

Toïgo C (1999) Vigilance behaviour in lactating female alpine ibex. Can J Zool 77:1060–1063

Treus V, Kravchenko D (1968) Methods of rearing and economic utilization of eland in the Askaniya-Nova Zoological Park. Symp Zool Soc Lond 21:395–411

Uztemur A, Orman A, Demir M (2024) Feature of suckling behaviour of Gazella marica fawns in semi-captive conditions. J Anim Plant Sci 34(5). https://doi.org/10.36899/JAPS.2024.5.0802

Walther F (1961) Einige Verhaltensbeobachtungen am Bergwild des Georg von Opel-Freigeheges. Jahrberichte Georg von Opel-Freigeheje 3:53–89

Walther F (1964) Verhaltensstudien an der Gattung Tragelaphus De Blainville, 1816 in Gefangenschaft, unter besonderer Berücksichtigung des Sozialverhaltens. Z Tierpsychol 21(4):393–467

Walther F (1965) Verhaltensstudien an der Grant-gazelle (Gazella granti Brooke, 1982) im Ngorongoro-Krater. Z Tierpsychol 22:167–208

Walther F (1968) Verhalten der Gazellen. In: Die Neue Brehm Bücherei, no. 373. A. Ziemsen Verlag, Wittenberg-Lutherstadt, p 144

Walther FR (1984) Communication and expression in hoofed animals. Indiana University Press, Bloomington

Yazan Y, Knorre Y (1964) Domesticating elk in a Russian National Park. Oryx 7:301–304

Zhevnerov VV (1984) Goitered gazelle of the Barsa-Kelmes Island. Nauka Press of the Kazakh SSR, Alma-Ata. (in Russian)

Zhevnerov VV, Bekenov AB, Sludskiy AA (1983) Goitered gazelle—*Gazella subgutturosa* Güldenstaedt, 1780. In: Gvozdev EV, Kapitonov VI (eds) Mammals of Kazakhstan, Bovidae, vol 3 (3). Nauka Publishing House of Kazakh SSR, Alma-Ata, pp 11–54. (in Russian)

Behavior of Young

9

9.1 General Remarks

Research of goitered gazelle biology in Central Asia was carried out especially intensively during the Soviet period, though primary attention was paid to the morphological characteristics and growth intensity of fawns of different ages. The development of behavioral displays and social behavior of fawns began to be studied only in the late 1980s and early 1990s. Zhevnerov (1984) described some behavioral features of fawns in the Barsakelmes Island population of goitered gazelles. And particularly detailed observations of the development of various behavioral patterns in goitered gazelle fawns were made by Marmazinskaya (2000) in the captive conditions of the Bukhara Breeding Center during the 1990s. So far, individual aspects of fawn behavior have been considered by different authors in different articles, so in this book I want to gather in one place all information on this issue and dedicate a separate chapter to the development of behavioral demonstrations and social behavior of goitered gazelle fawns.

The existence and development of any social structure is impossible without building social connections between individuals (Miranda-de-la-Lama and Mattiello 2010). The fawn's first social connections arise with its mother. Mother/young communication is vitally necessary for the survival of young because they need to translate each other's information on their presence and position, status, and alarm signals. And in order to build a certain relationship with her, it is necessary for the fawn to learn to recognize its mother and distinguish her from other females (Walther 1984). In follower ungulate species, mothers and their young develop mutual visual and audio recognition in the first days, and even first hours, after the newborn's birth. This is necessary for the mother, because she must recognize and feed only her fawn(s) among a whole group of fawns, each of which tries to suckle from any female (Poindron et al. 2003). Individual features of the sounds made by young ungulates of follower species were found for caribou (*Rangifer tarandus*—Lent

D. Blank, *The Behavioral Biology and Conservation of Goitered Gazelles (Gazella subgutturosa)*, https://doi.org/10.1007/978-3-031-83504-9_9

1966), musk ox (Lent 1974), bison (*Bison bison*), Dall's sheep (*Ovis dalli*), domestic cattle, and others (Walther 1984). In hiding young, though, individual recognition of the mother takes longer compared to follower species. The goitered gazelle is a typical hider species, and its fawns start to recognize their mothers only at the end of their hiding period. During their entire hiding period, mothers only recognize their young based on olfactory cues and according to the location of their fawn's bed-site (Blank and Yang 2017).

In addition to mother/young bonds, fawns begin to come into contact with conspecifics, especially often with other fawns. By creating whole groups consisting only of fawns, they begin to build certain mutual relationships among their peers and gradually integrate into the social structure of adults. I will consider this process of fawns' forming social behaviors in particular detail, since the social hierarchy of the group is important for the life and survival of not only individuals but of the population as a whole (Lent 1974, Ruckstuhl and Ingold 1999). A dominance hierarchy allows individuals certain access to important resources, including food, water, shelter, and breeding opportunities (Clutton-Brock et al. 1986; Côté 2000). In addition, occupying a certain place in an established hierarchy, an individual does not waste energy on constant fights and avoids potential injuries as a result of possible head butting (Clutton-Brock et al. 1979; Hand 1986). In some species, such as bighorn sheep (*Ovis canadensis*—Festa-Bianchet 1991), red deer (*Cervus elaphus*—Thouless and Guinness 1986), and mountain goats (*Oreamnos americanus*—Gendreau et al. 2005), the hierarchical structure that appears as a result of contacts between young individuals may also persist into their second year of life.

9.2 Young Hiding Behavior

Goitered gazelle fawns are born well-developed. They are born sighted and covered with thick fur, and a few hours after birth, the strongest can already stand on their feet and even run for short distances (up to 15 m). As a rule, however, the majority of newborn fawns have weak limbs for the first 3–4 days and cannot run long distances. They are barely able to stay on their feet and step only on their phalanges with their hooves widely spread. Fawns tremble and fall when they try to move; they almost always lie down with their legs bent under the body and their necks stretched forward, with their ears lowered to the side and down (Zhevnerov 1984; Pereladova and Pereladov 1986). Less often they lie curled up with their heads on their sides. The color of a fawn is very close to the color of the general background of the clay-gravelly desert and provides good camouflage, making it difficult to notice a fawn lying motionless. Sometimes the fawn is betrayed only by its shining eyes, which are usually kept open while hiding. When a human approaches a hiding fawn, the fawn sometimes tries to get to its feet and run; if taken in hand, it screams and breaks free. The instinct to hide in young gazelles sometimes persists for a very long time, even into adulthood. As an example: in the Betpak-Dala in May, single adult females and groups of 2–3 females were occasionally observed to quickly lie down in the thickets of *Salsola arbuscula* and hide after noticing humans at a distance.

Goitered gazelle twin newborn fawns, after drying out and getting a little stronger after birth, leave their birth bed and, with an uneven, wobbly gait, walk several tens of meters, and lie down between 40 and 100 m apart from each other. Gorelov (1972) also observed that twins bed down several tens of meters apart as well. This distance—the "twin distance"—is preserved later and reduces the likelihood of the death of both twins from predators (Zhevnerov et al. 1983).

The most detailed observations of a female and her fawns during the first hours after birth were made in the enclosures of the Sunt-Khardkskiy Nature Reserve (Turkmenistan) (Pereladova and Pereladov 1986). Immediately after birth, the fawn lies on the ground, putting its legs under its body, closing its eyes, and lowering its ears sideways and down. The first attempts of newborns to get up are observed 17 to 18 minutes after birth. After 21 to 35 min, the fawn gets up on all four legs for the first time, stands for 3–5 s, then lies down again, constantly attempting to get up and stand. This continues until several suckling attempts occur every 0.5–2 min. The formation of motor skills of other muscle groups begins at this time; in particular, after 11–29 min, the fawn's ears are already horizontal; 40–50 min later, they can already hold their ears at an angle of 45°; and after an hour and a half, the ears are vertical. After 25–40 min after birth, the fawn can walk 3–4 m already. The fawn's movements gradually improve, with confident standing with a straightened back noted after 60–70 min and confident walking after 1.5–2 h after birth. However, when the fawn runs away from a threat (in this case, being released after weighing), its hind legs can still bend for another 1.5 days. After confidently walking, the fawn stops constantly trying to get up and for the first 3 days lies on a bed-site in a hiding (frozen) position between feedings. It does not rise either at the approach of a person or adult gazelles, including its mother, if she does not give the appropriate sound signal—her invitation to feeding (Pereladova and Pereladov 1986). I also observed this phenomenon of not getting up until summoned in my study in both the wild in Ili Hollow (Blank 1985) and under the semi-captive conditions of the Issyk-Kul Breeding Center (Blank 2023).

At birth, eight incisors of newborns are fully erupted, white, and sharp; two medium teeth are especially large and directed to the sides, creating a "triangular" space between them that is probably an adaptation to allow suckling (Fig. 9.1—MilkTeeth—Sludskiy 1956). In one case of twining, the first suckling, which lasted 3–5 s, was noted after 40–48 min after birth. In 3–7 min after birth, the female licked the anal areas of both newborns. For another hour, the female moved very little, lying down next to her newborn(s), and then licking them when she gets up for suckling. The mother then left her fawns after 2 h to go to another side of the enclosure, where she ate the afterbirth (a few females do not do this. After that, the activity of the mother increased significantly, and a male in the enclosure started to chase her. The muscles of the female's belly started to contract, and within 3 h, the mother's body had returned to its shape from before her pregnancy (Marmazinskaya 2000; Blank and Yang 2015)

Here are several examples of fawn behaviors within a few days after birth: On May 13, 1955, in Betpak-Dala, Sludskiy (1956) found two fawns (a male and a female) still with attached umbilical cords. They were already running, but after

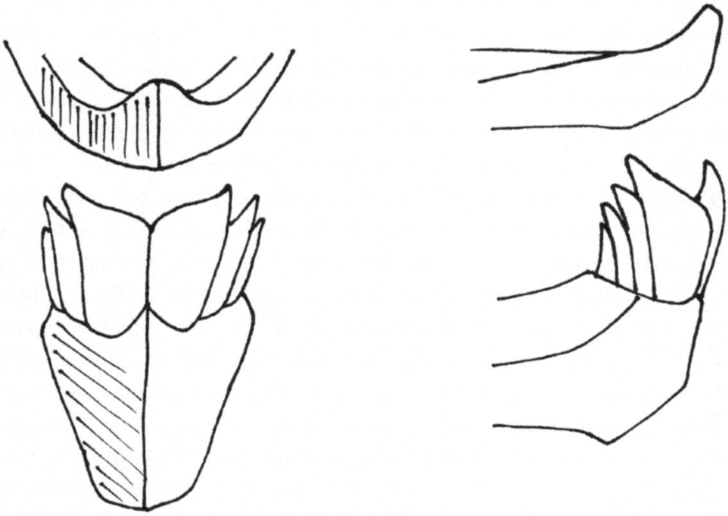

Fig. 9.1 Suckling adaptationn in goitered gazelles: Two medium teeth are especially large and directed to the sides, creating a "triangular" space between them that is probably an adaptation to allow suckling more comfortable without hurting a nursing mother. Drawing by U. Kenzhetaev with kind permission

running about a hundred meters, they laid down and hid. On May 11, 1946, two newborn fawns were found on Barsakelmes Island, lying 3 m apart (Zhevnerov 1984). Other authors (Pereladova and Pereladov 1986) also found that even in captive conditions, twins do not stay more than 2–3 m from each other during the first 4–14 days of their life, though in some cases, it was a distance of 15–20 m. On May 12, newborn fawns moved poorly and did not step on their hooves, and when standing, their hind legs were shaking violently. On the third day after birth, the fawns began to walk and run over short distances. On May 14, 1946, in the Betpak-Dala Desert, two fawns from the same mother were found on the northern slopes among *Elytrigia* and wormwood. When Sludskiy (1956) approached, the male did not get to his feet, but the female jumped up and tried to run away. Fawns, even at the age of only 2–3 days, if they are left for a long time by their mothers and are hungry, will get to their feet and begin to slowly wander around their bed-sites, crying for their mothers (Sludskiy 1956). By the age of 3–4 days, the fawns manage to get strong enough that when approached, they will jump up and run to a distance of several hundred meters (Zhevnerov 1984). And after another day or two, it is difficult to catch up with them even on horseback; fawns can also jump up to 1 m high (Ishunin 1961).

The formation of an attentive reaction in newborns develops very quickly: after 2 days, a wary turn of the head in the direction of an approaching mother; at 4 days, from a lying position, a turn of the head in different directions to check the area around the bed-site; at 7–8 days, a clear pose of an attentive reaction; at 12–15 days, a complex pose of attention + lifting and waving the tail + an audible danger

Fig. 9.2 At the age of one month, goitered gazelle fawns can wander from place to place independently from their mother. Great Gobi A, Mongolia. 28.06.2005. Photo by P. Kaczensky with kind permission

signal + sometimes a blow with the front foot on the ground and turning in a semicircle. Hair fluffing in the rump white patch during an attentive reaction can be observed from the age of 7 days. Noticing a mother's alertness, her fawn approaches her and completely repeats her actions of head turns and other movements from age of 10-15 days. When the fawn gets too close to her, the mother steps aside, wagging her tail and continuing to observe the danger threat. If the potential source of danger is not very strong, the fawn stops its attentive reaction before the female and switches to another activity (grazing, suckling). Trotting or galloping activity in combination with a signal jump forms by the age of 15 days. At the age of about 1 month, the fawn moves freely from one place to another, independent of its mother (Fig. 9.2—0152). Left lying down, the fawn shows a complete adult stereotypical reaction of observing the danger and fleeing by the age of 1+ month (Marmazinskaya 2000) (Fig. 9.3).

In stressful, dangerous situations, urination and/or defecation occurs in adult goitered gazelles as a reaction to tension. Display of urination or defecation by one individual stimulates the same response in other individuals of the group, so apparently these behaviors and distinctive poses are used as signals of danger to conspecifics. It is possible that the release of volatile substances with the smell of anxiety in combination with urine and feces is important, as well as the release of body metabolites before fleeing. In captivity, the stress reaction in fawns occurs as early as the first days after birth as a result of being picked up for weighing. A stress reaction to danger in a female who grew up in complete isolation from other gazelles

Fig. 9.3 The goitered gazelle fawns show a complete adult stereotype reaction of observing danger. South Gobi Region, Mongolia. 08.07.2017. Photo by P. Kaczensky with kind permission

Table 9.1 Encountering females with their fawns and without them (Betpak-Dala Desert)

	Number of females				
		Including			
		Without fawns		With fawns	
Months	Total number	Number	%	Number	%
May	412	402	97.5	10	2.5
June	59	19	32.2	40	67.8
July	127	40	31.5	87	68.5
August	89	22	24.7	67	75.3
September	34	4	11.7	30	88.3

Adapted from Sludskiy (1956)

was observed for the first time at the age of 3.5 months, although in nature this behavior appears much earlier as a result of imitation of a mother's reaction (Marmazinskaya 2000).

For the first 2 weeks, and sometimes longer, hiding fawns rarely go with their mothers but instead lie hidden most of the day (Fig. 8.2.1—GazelleFawn). If they get up, they stay close to their bed-site. Between the 17th and 30th of May 1954, in the Betpak-Dala Desert, when most goitered gazelle females had already birthed their fawns, Sludskiy (1956) found 86 females but only 2 were accompanied by their young. At the same location between the sixth to the 28th of May 1955, he observed 350 females, of which only five were walking with their fawns. Between June 18th and 23rd of 1964, on the Buzachi Peninsula on Kuchak Bay (Caspian Sea), Sludskiy (1956) saw 36 females with 10 escorted by their fawns (Table 9.1).

This prolonged hiding period keeps fawns less visible to predators, making it more likely they will survive (Zhevnerov et al. 1983).

At the age of 3–4 days, the fawns are able to actively avoid danger. Although they have a strong instinct for hiding during this period, they allow the object of danger to be no more than 3–6 m away before jumping up abruptly and running; after running 100–200 m, the fawns suddenly fall among the plants and hide; twin fawns scatter in different directions. On the eighth or ninth day after birth, depending on a newborn's individual characteristics, elements of behavioral independence appear. At this time, activity is already losing its consistency with the periods of feeding by their mothers. The fawns get up and move even in her absence but stay close to their beds. The duration of following their mothers after suckling increases, while the fawns learn to graze. At first, the fawns just chew the twigs without tearing them from the plants but eventually begin to actually consume food, although much smaller plants in much smaller amounts (Zhevnerov 1984).

In a captive group of bottle-fed fawns, the usually short-term behavior of lying frozen in place at the approach of a person has been observed in some fawns up to 32 days of age. In addition, up to the age of 65 days, elements of standing frozen can be observed in the same situation, when the fawn lowers its head and neck down after an orienting reaction. While resting on a bed, a juvenile female at the age of 7 months was observed to press her neck to the ground and raise her tail when a dominant male approached; and several times similar behavior has been observed in adult females. Additionally, at the approach of a person, hiding reactions of bedded gazelle females in the late stage of pregnancy were twice reported (Marmazinskaya 2000).

A fawn's first sampling of plants is seen at the age of 4–5 days, while regular consumption of green food occurs between 25 and 30 days; the first grain feeding is seen at the same time as green food consumption. Fawns try (nibble at) almost all possible plants, as early as 7–10 days, and show a preference for some plants over others: the flowerings of *Barbarea vulgaris* and grape leaves are most preferable; at 15–20 days, alfalfa and other legumes are chosen; and at 30–40 days, *Salsola*, *Alhagi*, and wormwood are eaten. Cereals in the presence of other types of food are consumed very reluctantly. At the autopsy of three dead fawns aged 1–2 days and one a 5-day-old, only the latter fawn had a vegetable mass (47 g), consisting mainly of green leaves of *Agropyron desertorum* and *Lepidium perfoliatum*, in its stomach in addition to milk. This shows that the first attempts of gazelle fawns to nibble vegetative food coincides with the beginning of their free movements apart from their mothers. By the age of 1 month, plant food already occupies a significant place in their diet. In a 1-month-old fawn, the mass of plant food in the stomach contents was 950 g (Zhevnerov 1984).

After 3–4 days after birth, fawns begin to walk easily, and from 10 to 15 days, they are able to keep up with their mothers and run a considerable distance with them (Fig. 9.4). In my study (Blank 1985), newborns were able to walk short distances (several steps) within the first half hour after birth, and observations of the first females with following fawns were seen on May 18—1–2 days after birth. Sludskiy (1956), however, claims that females with young are rarely seen on

Fig. 9.4 Goitered gazelle fawns' ability: Starting from two weeks, fawns do not lag behind their mothers and are able to run a considerable distance behind her. Great Gobi B, Mongolia. 11.07.2005. Photo by P. Kaczensky with kind permission

pastures before June 15 (at the age of 1 month), since mothers hide their babies for a long time even if they can run after them effectively. In my opinion, this is an erroneous statement, probably related to the observation methodology used when route calculations were carried out using a car or on foot and observations were sporadic over a large area. Only focal subject observations made over many hours of surveillance can provide the basis for an accurate vision of hiding behavior. From mid- to end-June, females and their young gather in small groups (Sokolov 1959).

At the end of June, pairs and small groups of goitered gazelle fawns can be seen grazing and even going to watering places without adults (Zhevnerov et al. 1983). In my Ili Hollow study, the fawns' high mobility reaches its maximum activity in late-June when they often travel independently from their mothers (Blank 1985). Fawns start to follow their mothers from their first hours of life, very briefly at first, then for longer periods of time as they age; at the same time, their initially long hiding periods continuously decrease. Only when the hiding period ends in mid-July and the fawns are 2 months old do mothers and fawns start staying together both day and night (Blank 1997). Among other species: offsprings of Grant's gazelles (*Nanger granti*) are known to stand and nurse within 19 min after birth and follow their mothers at 41 min (Walther 1965); newborn saiga antelope can stand and nurse within their first 20 min after birth (Kokshunova 2007); newborn Tibetan antelopes (*Pantholops hodgsoni*) can stand and follow their mothers at 12 min after birth (Schaller et al. 2006), and wildebeest young can follow their mothers within their first 10 min (Estes 1966).

A clear definition of the end of the fawns' hiding period based on observations is difficult for some authors (Ralls et al. 1987) due to the fact that the hiding period has several stages. Based on my study of goitered gazelles, I defined five stages. In the first stage, which lasts for several days, the fawns lay flat against the ground almost constantly and hardly move without their mothers. The second stage lasts for another 5–7 days, when the fawns occasionally raise their heads to watch passing gazelles and sometimes stand up to change sides. The third stage begins at the age of 10–15 days, when, in the absence of their mothers, the fawns' activities increase significantly and they leave their hiding place for as far as hundreds of meters and graze or play in the company of other fawns, periodically returning to their resting site. The fourth stage is observed in the second half of June, when the fawns walk widely away from their bed-site on their own, change their bed-sites on their own, walk, graze and play with other fawns, and follow after alien females. During this period, fawns will run up to and try to suckle from any female that approaches and looks toward them with erected ears. This behavior is not observed in May and rarely during the first half of June (13% cases, $n = 54$), but in late June and early-July, it is common to see (44% and 43%, respectively). The fifth and final stage occurs in the middle of July, when the fawns begin to follow their mothers constantly both day and night. I believe that this is the real end of the hiding period, even though the third and fourth stages can hardly be considered hiding in a literal sense. And it is very likely that other hiding ungulate species have different stages of hiding. Other authors, though, using records made along specific, limited routes or observing ungulates within the closed conditions of captivity, cannot accurately characterize the end of a hiding period (Ralls et al. 1987).

Similar to goitered gazelles, Mongolian gazelles (*Procapra gutturosa*) are also a hiding species (Sokolov and Lushchekina 1997), and their fawns' activities grow with age—threefold by day 25 compared to day 2. However, young of Mongolian gazelles travel much longer distances than goitered gazelle fawns and move up to 6.6 km on average at the age of 5–8 days, covering up to 63 km at the age of 1 month. Unlike sedentary goitered gazelle fawns during the hiding period, migrating Mongolian gazelles have a relatively short hiding period of 7–8 days (Odonkhuu 2009) when they are sedentary; after 2–3 weeks, the fawns join an already large aggregation of more than a thousand individuals and continue the migration with the adults (Olson et al. 2005; Odonkhuu et al. 2009).

9.3 Development of Behaviors in Young

Danger Signals In captivity, fawns begin to sound danger signals from the age of 3–5 days: when captured, when approached by an observer, when approached with an outstretched bottle of milk, or as a reaction to adults in a neighboring enclosure. The frequency of the danger signals issued has an individual characteristic (Marmazinskaya 2000).

Sniffing From the first days after capture (age 3–14 days), a high frequency of sniffing is recorded of various body parts, primarily the head region (ear, forehead, nostrils, eyes), of other fawns in the artificial feeding groups. Three types of reactions are noted: indifference, mutual sniffing, and shaking of the head. Subsequently, naso-nasal sniffing in adulthood is formed from this interaction, which differs from similar behavior in fawns: low frequency, short duration, and sniffing only the end of the muzzle. In enclosure groups with a mixed sex and age composition, fawns often initiate nasal sniffing with adult gazelles. The most frequent initiators of this contact are female adolescents aged 1–8 months and, less often, males under the age of 20 months. The second most important form of behavior in establishing personal connections between individuals is sniffing various parts of the body—flanks, back, neck, legs, hooves, rump patch edges, and croup. Naso-anogenital sniffing between fawns is recorded from 4 to 6 days of age. Often (14.3%) such sniffing occurs during urination or defecation of another fawn, while a high frequency of sniffing of urine spots and urine streams is also noted. The stimulus for naso-anogenital sniffing is the other's raised tail when stretching. Juvenile males at the age of 6–12 months initiate naso-anogenital sniffing with adult and young females. In semi-captive groups, naso-anogenital and naso-nasal sniffing occurs between fawns of the same or different families, in mixed groups during movements and in pastures, as well as between young and adult individuals (naso-nasal sniffing is more often observed with males and usually without aggressive reactions). Sniffing of excrement in fawns is observed from the age of 13 days (n = 37) to the age of 8 months (Marmazinskaya 2000).

Demonstration of a curiosity pose (ears in a vertical position, with a slight forward tilt + neck stretching forward and down + tail lowered or raised into a horizontal position without waving; the pose is demonstrated to the other animal standing or during slow walking). This behavior is formed during the first month of life. The motivational state of gazelles in this case is not intense alertness but a conflict between defensive and inspective behaviors. The behavior of fawns in this situation occurs when approached by a person or when seeing unfamiliar objects or animals or adult gazelles, with an unexpected violation of the minimum distance between members of a group of the same age (Marmazinskaya 2000).

Scraping the Ground with the Front Foot In the individual development of fawns, this behavior is first formed as a result of bedding preparation. The complete formation of this behavior occurs by the age of 8–10 days.

Kicking the Ground with the Front Foot In the first days of life, this behavior is noted as a defensive reaction to annoying flies. From the age of 12–15 days, it is used defensively when coupled with an alarm reaction. At the ages of 2.5–7 months, the kick from the front foot on the ground is observed as a threat from a fawn when it is necessary to lift another fawn from a lying position, when the individual distance during grazing is violated, and as a reaction to naso-anogenital sniffing by another individual. In adult animals, this element has been considered only as a defensive behavior (Marmazinskaya 2000).

Rubbing the Head Against Different Parts of Another's Body At the age of 6–30 days, four cases of this behavior were observed after a violation of the minimum individual distance and after naso-anogenital sniffing prior to the demonstration of a dominance pose. Among fawns, such head rubbing can be initiated by both males and females. Among adult gazelles, the initiators of head rubbing are exclusively females (Marmazinskaya 2000).

Touching an Individual's Forehead This behavior is observed before the age of 14 days and is directed to a lying individual in order to lift him from a prone position (Marmazinskaya 2000).

Threat Poses, Butting, and Skirmishes In captive conditions, the first demonstrations of a threat pose (the head lowered (in males the horns are exposed) and the ears held parallel to each other and directed toward the back of the head or to the side-back at an angle of 45°; during a jump with a threat demonstration, the tail is held out from the body in a horizontal position without waving, which were observed in fawns at the age of 6–8 days; butting movements were seen at the age of 2–7 days; and mutual threats and clashes occurred at the age of 14 days. All these manifestations of aggressive behavior occurred as the result of a violation of the minimum individual distance between conspecifics, often when approaching a resting individual or fighting for a bed-site. From about 1 month of age, alternating demonstrations of threat and dominance, mounting, and sniffing are common. There are also turns of the head toward the other individual and pressing the ears to the back of the head which are considered an incomplete threat. In the artificial feeding group, individuals with different frequencies of aggressive demonstrations can be observed during the first weeks (Marmazinskaya 2000).

The Rise of Another Fawn From a Prone Position From the first days, fawns get up from their beds with a neutral approach to other fawns and attempt to lie down next to them. A purposeful approach to a lying fawn in order to raise it from its bed can be observed from the age of 10–14 days. In these cases, the active partner uses threat demonstrations, danger and threat sound signals, head butting, touching or pushing with its forehead, striking with a front foot (rarely hitting), digging up the ground near the lying fawn, and touching the end of the muzzle. This behavior demonstrates that young gazelles, compared to adults, use a more diverse set of physical acts and short-term interactions to try to raise the other from its bed. In adult animals, the raising of another individual from a lying position is more ritualized and with more, but rare, cases of direct aggression (Marmazinskaya 2000).

Agonistic behavior of fawns in the artificial feeding group was accompanied by a high frequency of two types of sound signals: grumbling with a closed mouth (threat signal) and snorting with a sharp exhalation of air through the nostrils (danger signal) (Pereladova 1988). Threat signals were issued from the age of 6–10 days, often when approaching another's bed-site or trying to lie down next to another fawn, during threat demonstrations, threats during skirmishes, when eating food, following another individual, or as a reaction to sniffing. Situations using both threat signals and danger signals in the context of aggressive behavior practically coincide in fawns. In addition, danger signals were observed as a reaction to naso-anogenital sniffing, an attempt to be touched, to rise from a lying position, as a reaction to an aggressive demonstration by another, before mounting, and in combination with touching the end of the muzzle with another fawn. Threat and danger signals were often emitted one after the other during the same contact or interaction. The use of danger signals by young animals outside of situations of anxiety and aggression were also detected during a game, when moving around in the enclosure, and for no apparent reason. The highest frequency of threat and danger signals was recorded in the first 4 months of life and quite often produced up to the age of 12 months (Marmazinskaya 2000).

Demonstration of Submission Showing submission consists of stretching the neck forward and down with the ears parallel to each other and directed to the back of the head with the tail raised to varying degrees, including throwing it across the back, without waving (Fig. 9.5—SubmissivePosture). From the very first hours of life, fawns demonstrate a vertical position of the tail with or without waving during comfortable behavior, such as suckling, scratching, urination, and defecation. When a little older, a pose of anxiety and flight from danger is formed, in which the tail is raised. The raising of the tail is also observed during agonistic contacts between individuals when one animal leaves or runs away from an active partner. Simultaneously, when lifting the tail, there is a slight lowering of the neck (Fig. 9.6); with age, the ears also are pressed to the back of the head. In the semi-captive population, a submissive posture was formed by the age of 1 month when the fawns were still suckling milk. Submission also is demonstrated by young gazelles as a result of aggressive demonstrations between 1-year-olds and adult males and seen less often with a neutral approach. By the end of October, the frequency of demonstrating a submissive posture by young (age 6 months) increased even with neutral approaches from adult males, which is associated

Fig. 9.5 Fawn submissive posture in goitered gazelles: A fawn stretches its neck forward and downward with approaching of an adult male. Drawing by U. Kenzhetaev with kind permission

Fig. 9.6 Fawn (2.5-3-month-old) lifts its tail throwing on the back and lowers its neck with approaching of an adult male or threatens by horns. Southern Ustyurt (Kaplankyr), SW Kazakhstan. 30.07.2019. Photo of the camera-trap of the Ustyurt Nature Reserve team with kind permission

with the approach of the rut. For the fawns in the artificial feeding group, deprived of communication with adult gazelles, the submissive pose lasted significantly longer. Immediately after joining with a group of adult gazelles, young gazelles at the age of 2.5 months ran away from males chasing them, as they do in the case of danger. They showed a complete submissive pose by the age of 5 months. Thus, normal training for the submissive pose occurs when the fawns have the opportunity to be in contact with adult gazelles. And frequent demonstrations of submission by adult females, including their mothers during the spring peak of sexual activity, are of great importance in teaching this behavior to the fawns (Marmazinskaya 2000).

Demonstration of the Low-Stretch Posture In this pose, the male's ears are parallel to each other and directed to the back of the head + the head and neck are stretched forward parallel to the ground + the tail is raised to varying degrees, up to vertical, + the pose is accompanied by periodic nodding of the head, roaring and up and down movements of the Adam's apple. In young gazelles, this stretching posture looks incomplete. In the youngest cases, a pronounced demonstration of stretching can be observed in males at the age of 7 months during contacts with females in estrus. Young males at the age of 11 to 12 months demonstrate a high frequency of harassment toward females. Demonstration of stretching with a sound signal (snoring) is first observed in a male at the age of 15.5 months (Marmazinskaya 2000).

Demonstration of Dominance In the artificial feeding group, some elements of dominance behavior can be observed in 5- to 10-day-old fawns, though still rare and with poorly coordinated movements. During the first month, dominance demonstrations and mounts are performed mainly by males, much less often by females. By the age of 20–30 days, dominance is seen almost completely in young animals, while the frequency of their demonstrations increases. Unlike adults, the demonstration of dominance in young gazelles is often accompanied by naso-anogenital sniffing, threat demonstrations, touching the horns or forehead to various parts of the other's body, and butting and shaking of the head. Flexing of the front leg as a part of a dominance pose was first recorded in a 3-month-old male. The nodding of the head in young gazelles was observed only 3 times: in males of 53 days and 8 months of age and a female of 35 days (during a dominance pose addressed to a male, she accompanied her head nod with a sound signal); this behavior is not observed in adult females. The dominance pose is observed less often in semi-captive (free range) conditions than in small enclosures. The demonstration of dominance is first formed as a one-sided behavior, with males aged 7–12 months old quite often initiating a one-sided demonstration of the dominance pose with adult males and one-year-olds. Upon reaching the age of one and a half years, they begin to participate in reciprocal demonstrations (Marmazinskaya 2000).

Mounts At first, this behavior is formed partly as an element of dominance and partly as a game. I observed one fawn mount another at the age of 2 weeks (Blank 2024). From the age of 1.5 years, male mounting is the beginning to a complex of pre-copulation behaviors (Marmazinskaya 2000).

Formation of Marking Behavior Marking activity is typical for male goitered gazelles and includes marking with urination-defecation, preorbital gland secretions, and head-butting plants (Blank 1985; Marmazinskaya 1996; Marmazinskaya and Mordonov 1996). In the artificial feeding group, forehead rubbing against various objects (an enclosure pole, a reed mat, a net, *Calligonum* bushes) was observed 9 times in young males aged from 35 days to 8 months. One important element was marking the ground by scratching with a front foot. Long before the formation of this behavior, scraping the ground was noted in fawns as early as the age of 2–3 days to prepare a more comfortable bed, to lift a conspecific from a bed, and digging up food in the compound. Males aged 2–2.5 years and older demonstrate all types of marking activities (Marmazinskaya 1996b).

Forming of Courting Behaviors Most elements, such as naso-anogenital sniffing, urine spot sniffing, mounting a female, demonstration of submission, touching the end of the muzzle of various parts of the other's body in postures of dominance or a low-neck stretch, courtship in the position of a low-stretch and dominance, touching the croup with the breast, and demonstration of urination and defecation poses, are formed during the first year of life (Marmazinskaya 1996a). In the Issyk-Kul Breeding Center, I also observed sniffing a female's urine spot and expressing a flehmen response in a male at the age of 1 year (Figs. 9.7 and 9.8).

Fig. 9.7 Goitered gazelle yearling One-year old male sniffs the female's urine spot. Issyk-Kul goitered gazelle nursery for reacclimatizing, 05.05.2023. Photo by D. Blank

Fig. 9.8 In goitered gazelles, one-year old male shows flehmen after sniffing of female's urine spot. Issyk-Kul goitered gazelle nursery for reacclimatizing, 05.05.2023. Photo by D. Blank

Urination by a fawn in response to the external stimulation of its mother's licking its anogenital area manifests itself immediately after birth. In the artificial feeding group, females from the age of 4 days and males urinate or simply squat in the urination pose in response to touching and sniffing their anogenital areas by other gazelles. In females, provoked urination as a reaction to the approach or pursuit by a male is formed by the age of 5.5–6 months; a behavior that is already sexually motivated, rather than purely physiological. The earliest case of a flehmen response can be observed in a male as young as 10 months, while females reach puberty much earlier than males, usually by the age of 8 months, entering estrus at the end of December. The behavior of young receptive females does not differ from the behavior of adult females, up to and including mating (Marmazinskaya 1996a).

Marmazinskaya (1996a) observed four cases in captive conditions, when goitered gazelle males aged 7–8 months actively courted females in a state of estrus and performed some typical elements of sexual behavior: relentless following of the female in the low-stretch posture, nodding of the head, naso-anogenital sniffing, and mounting with thrusts but without an erect penis. In addition, at the moment the female stopped moving, one of the adolescents butted her in the side of her body and rubbed his head against her back, forcing her to move again. In one case, an adolescent male aged 6–7 months followed a female for a total of 83 min during which he mounted her 58 times (Marmazinskaya 1996a). In the case of a repeated estrus cycle in a female (during a single rut), the behavior of male adolescents is much more intense than during the first estrus. Dominant males react indifferently to the courtship of females by young males (Marmazinskaya 1996a).

To summarize observations of the formation process of various behavioral patterns in young gazelles (Marmazinskaya 2000):

1. During the development of behavior in fawns, a significant number of elements are formed during contacts with other young and mothers, as well as other adult gazelles as early as the first month of life. All these elements are executed in an incomplete form and are not always well expressed. But through further development, these elements are performed better and better until they become identical to the behaviors of adults.
2. In captivity, fawns perform some behavioral patterns more frequently than adults: naso-nasal and naso-anogenital sniffing; smelling of the urine stream; smelling of different body parts; and tactile contacts when eating, resting, and moving.
3. Captive fawns demonstrate postures in different ways compared to adults: threats together with naso-anogenital contacts and smelling of different body parts; threats accompanied by vocalizations (signals of danger and threats to peers, which are used in different contexts); digging ground near a resting conspecific and kicking the other with the front hoof to drive it away from the bed-site.
4. In captive conditions, fawns at the age of 6 months perform behavioral patterns which disappear in adults: licking of another's muzzle; head nodding in young

females (35 days of age)—this behavior is absent in adult females; scratching the ground + threat signals when another approaches; and rubbing the head against various parts of the other's body. This last behavior was demonstrated in both young males and among females with the same frequency, but in adulthood, it was performed only by females.

5. Female fawns provoke more naso-anogenital contacts than male fawns at the age of 1–5 months, while male fawns more often start these contacts compared to female fawns at the age of 5 months and older. The number of aggressive contacts (threat pose, touching body with horns or forehead, butting) are initiated by young males and young females with approximately the same frequency.
6. During observations, aggressive interactions and naso-anogenital sniffing are observed most often.

9.4 Interactions of Fawns with Conspecifics and Social Behavior Development

In my study of the social behavior of goitered gazelle fawns, I sampled several behavioral patterns during the interactions of fawns with conspecifics (nonmaternal females, subadult males and females, and adult males), as well as fawns with each other; the ethogram is presented in Table 9.2.

Immediately after birth, the frequency of interactions between goitered gazelle fawns and other gazelles is very low. After a week, contacts gradually begin to increase until the frequency reaches its maximal level in late June; then, in early July, the rate begins to rapidly decrease. Among conspecifics, fawns have the most frequent contacts with adult females, since fawns do not distinguish their mothers well from other adult females and try to suckle from any who happen to be nearby (92.8%, $n = 153$). These erroneous attempts rarely occur in May due to the limited independence of fawns in the first 2 weeks after birth. In June, the mobility of fawns in the absence of their mothers increases, especially in the second half of June and, accordingly, the number of attempts to suckle from alien females reaches its maximum (Fig. 9.9). In July, the number of fawns' contacts with other adult females decreases, as the fawns spend most of the daytime with their mothers (Blank et al. 2017).

In the first weeks of their life, fawns are still relatively small and weak and many times not persistent enough for suckling, so alien adult females, after sniffing the anal area and determining that the fawn is not their offspring, usually perform a nonaggressive departure: by just walking away or jumping away if the fawn has already grabbed a nipple. As fawns grow, they become more energetic and more persistent in their attempts to suckle other females. Therefore, adult females increasingly demonstrate aggressive behavior toward them, waving them off with their heads, making threatening movements with their foreheads, and even chasing and

Table 9.2 Description of the recorded categories of social interactions (ethogram) between fawns and other conspecifics in goitered gazelles

Behavioral category	Behavior description
Butting	An adult or subadult gazelle approaches with head (and horns in males) lowered toward the fawn, freezes in this posture, and then hits the fawn's head or another body part. While butting, the ears of the aggressor are directed sidewise or sometimes horizontally or backward, and the tail is pressed to the rump
Chasing	An adult or subadult gazelle follows the fawn by walking or running, usually demonstrating neutral postures, but sometimes threat or dominant postures, with the tail pressed to the rump or slightly raised
Forehead threat	An adult or subadult gazelle directs its forehead toward the fawn. Ears are directed forward or sideward and the tail is in its normal position
Jumping away	An alien fawn comes so quickly (usually by running) to an adult or subadult female to suckle that she has no time for a sniff check, but after checking the hindquarters during suckling, the female jumps away
Closely approaching	An adult or subadult gazelle nears a fawn in a neutral posture to a very close proximity, forcing the fawn to retreat
Walking after checking	An adult female comes to a fawn, sniffs, usually the hindquarters, and then moves away
Mounting posture	In a mating bipedal posture, an adult or subadult male (or another fawn) follows the hindquarters of a fawn for several seconds without touching it (Fig. 9.10)
Head movements	A gazelle tosses its head toward the fawn as if it wants to drive the fawn away
Dislodgement	This is an action in which an individual approach a recumbent fawn and induces it to get up. The aggressor walks to the recumbent's rear, lowers its head, and slowly thrusts its nose toward or into the recumbent's hindquarters. This approach, with or without contact, usually causes the recumbent fawn to get up and move away

Blank et al. (2017)—License Number—5476891261102

butting them, though usually displaying this aggression in a gentle manner (Table 9.3), with the proportion of aggressive demonstrations against fawns relatively low (56% from all interactions, $n = 210$). By mid-July, the hiding period ends, and fawns are constantly with their mothers, so cases of fawns attempting to suckle from alien females practically disappear (Fig. 9.9; Blank et al. 2017).

In pronghorn (*Antilocapra americana*), fawns and alien females have contacts with each other from the first days of the fawn's life, and similar to goitered gazelles, the response of pronghorn adult females is avoidance if possible or knocking down persistent fawns (Bromley 1967; Autenrieth and Fichter 1975). In white-tailed deer (*Odocoileus virginianus*—Huegel 1985), as well as pronghorns (Autenrieth and Fichter 1975), contacts of fawns with alien adult females increase with the fawns' age, which, like goitered gazelle fawns, is related to the increase in activity as fawns' gain independence from their mothers. The maximal frequency of these

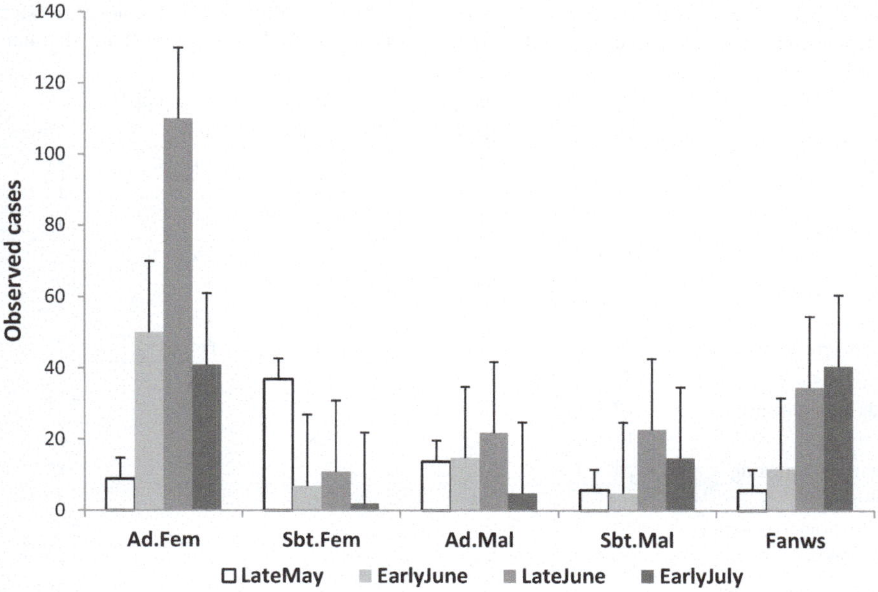

Fig. 9.9 Observed cases of aggressive demonstrations of adult and sub-adult goitered gazelles of both sexes toward fawns during the hiding period (Late-May-Early-July) (Blank et al. 2017)

Fig. 9.10 A goitered gazelle male fawn (4 month-old) shows mounting behavior on a female fawn. Though this is a social behavior demonstrating the dominance one over the other. Southern Ustyurt Chink (Kaplankyr), SW Kazakhstan. 28.07.2019 Photo of the camera-trap of the Ustyurt State Nature Reserve team with kind permission

Table 9.3 Observed cases (upper) and proportions (lower, %) of aggressive and nonaggressive demonstrations of adult goitered gazelles toward fawns during the hiding period (Late May into Early July)

	Adult females	Subadult females	Adult males	Subadult males	Fawns
Butting	55	11	14	14	21
	14.8	18.6	15.2	16.1	11.1
Chasing	54	11	25	34	12
	14.5	18.6	27.2	39.1	6.3
Forehead threat	69	13	13	15	7
	18.5	22.1	14.1	17.2	3.7
Closely approaching	21	1	14	6	23
	5.6	1.7	15.2	6.9	12.1
Jumping away	42	12	1		
	11.3	20.3	9.8	-	-
Grazing, resting together	24	2	7	1	80
	6.4	3.4	7.6	1.2	42.1
Walking away after checking	77	2			
	20.8	3.4	-	-	-
Mount each other			4	12	46
	–	–	4.4	13.8	24.2
Head movements	30	7	6	5	1
	8.1	11.9	6.5	5.7	0.5

Blank et al. (2017)—License Number –5476891261102

contacts occur as the fawn reaches 3–4 weeks of age, followed by a sharp decrease starting at 6 weeks (Bromley 1967). In contrast to goitered gazelle adult females, adult does of white-tailed deer behave very aggressively to alien fawns and will hit them forcefully with their legs and heads (Taylor 2013).

Interactions of fawns with subadult females are also mainly related to the offspring's attempts to suckle (87.5% cases, $n = 40$); however, there is another dynamic of interactions between fawns and subadult females, which is observed most often during the first 2 weeks after birth. Subadult females are usually last year's daughters who continue to follow their mothers. And although they are constantly being expelled by their mothers, they still keep a certain distance and are very interested in their mothers' newborns, approaching them carefully to try to sniff them. Since the subadult female behaves the same way as its mother, the fawn reacts the same way and jumps up to suckle. However, the subadult female does not expect such an abrupt reaction and either jumps away from surprise or, more often, threatens the fawn, chases it, and butts at it. In general, subadult females behave much more aggressively toward fawns than adult females (71.2%, $n = 57$), and after such an unpleasant experience, subadult females begin to carefully and purposefully bypass lying fawns. As a result, after a couple of weeks (early June), the frequency of contacts between fawns and subadult females decreases sharply and then fluctuates at a low level (Fig. 9.9; Blank et al. 2017). Similar behavior is seen in pronghorns as fawns run to any conspecifics, including subadult females, that approach their hiding places; with pronghorns, though, subadult females behave even more aggressively toward fawns than in goitered gazelles. And, as with goitered gazelles, subadult pronghorn females learn to avoid hiding fawns, resulting in a decrease in aggressive displays toward fawns by the age of 10 days (Autenrieth and Fichter 1975).

Adult and subadult males behave even more aggressively toward fawns than subadult females, threatening, chasing, butting, pushing, and shoving them forcefully, sometimes even knocking them to the ground (more than 93% of cases, $n = 48$). Since males of any age also try to avoid meeting fawns after their first encounters, these interactions with fawns are observed significantly less often (Table 9.3). And although fawns also try to suckle from adult males, the proportion of these cases is much lower than in females (at 53.8%, $n = 24$), while other interactions with females are related to competition for resting places (46.2%). Subadult males come into contact with fawns mainly by driving them away from their resting places (70% of cases, $n = 10$), and the rest of the cases are related to suckling attempts (Blank et al. 2017). Contacts between fawns and males are also observed in pronghorns, and males react in a very similar way with quite a bit of aggression toward approaching fawns (Autenrieth and Fichter 1975).

A completely different dynamic of interactions is observed between goitered gazelle fawns, when the frequency of their contacts constantly increases during their entire hiding period. These interactions begin with a minimal rate during their first weeks after birth in May and grow to their maximum in mid-July, despite the fact that the fawns' behaviors are strongly synchronized with the activities of their mothers at the end of hiding period in July (Fig. 9.9). Such dynamics are associated solely with the increase in the mobility of the fawns in the absence of their mothers. As soon as the activity of the fawns increases by mid-June, the fawns begin to gather together and form crèches with several fawns from different mothers walking, running, and playing together. Unlike contacts with adults and subadults, where fawns run to them to suckle, the nature of fawn/fawn contacts is a social attraction toward each other, when they willingly approach and just walk, graze, or play with each other for a long time. Purely aggressive elements of behavior between fawns are minimal (21.6%, $n = 94$), with only mounting and dislodgment observed most often (Table 9.3; Blank et al. 2017). Though mounting is more typical as a dominant behavior of males in many ungulate species (Walther 1984), in pronghorn fawns both males and females will mount each other during play (Autenrieth and Fichter 1975), as do young black-tailed deer (*Odocoileus hemionus*—Müller-Schwarze 1968). Due to mutual social attraction and the lowest level of aggressive reactions, fawns prefer the company of other fawns and avoid being in the same group with males and alien females; fawns even spend less time with their mothers than with their peers, especially after mid-June until the end of the hiding period.

Fawns of many ungulate species form crèches, such as bighorn sheep (*Ovis canadensis*—Geist 1971), barren-ground caribou (*Rangifer tarandus arcticus*—De

Vos 1960), impala antelope (*Aepyceros melampus*—Fraser 1968), red deer (*Cervus elaphus*—Bubenik 1965), and muskox (*Ovibos moschatus*—Lent 1974). In these groups without adults, young stay closer to each other than in the presence of their mothers (Autenrieth and Fichter 1975). These groups ensure the early interaction of young animals, teach them to build their first social connections among themselves, and create the foundation for future social interactions (Collias 1962). Young ungulates of follower species begin their social contacts with peers quite quickly, from their very first days after birth. Fawns of barren-ground caribou start their social interactions and gather in groups with other fawns by day four of their lives (De Vos 1960). Calves of bison (*Bison bison*) begin their social life in herds of young 2–4 weeks after birth (McHugh 1958). As described, young of hiding ungulate species gather in groups, but not immediately after birth. Despite the fact that pronghorn fawns begin to play with each other from the age of 1 week, they form more or less permanent groups at an age of slightly more than 1 month; at the age of 6 weeks, the fawns start to form a kind of hierarchical structure when small-sized fawns, which are born later, are subordinated to a lower position, because their social behavior is less developed (Bromley 1967; Autenrieth and Fichter 1975; Byers 1997). In white-tailed and mule deer fawns (*Odocoileus virginianus*; *O. hemionus*), obvious hierarchical relationships appear at the age of 4–5 months, which can last for a lifetime (Lingle 2003; Holand et al. 2004; Taillon and Côté 2007). In my study, though goitered gazelle fawns start their first contacts with other fawns from the age of 7–10 days, a hierarchical structure is likely not established until the age of 4–5 months (Blank et al. 2017). Goitered gazelles, however, do form loose social connections within the herd where members can join or leave a group at any time (Marmazinskaya 2000).

Thus, all contacts of young fawns with adults and subadults are mainly associated with attempts to get an additional portion of milk, while contacts with their peers are exclusively related to building social relationships and the foundation of a dominant hierarchy, which can extend into adulthood.

References

Autenrieth RE, Fichter E (1975) On the behavior and socialization of pronghorn fawns. Wildl Monogr 42:3–111

Blank DA (1985) Peculiarities of social and reproductive behaviour of *Gazella subgutturosa* in the Ili Valley. Zoological Zhurnal 64:1059–1070. (in Russian, English abstract)

Blank DA (1997) Social and reproductive behaviour of the Persian gazelle. PhD thesis. University of Tel Aviv, Tel Aviv

Blank DA (2023) Birthing season of May 2023 in goitered gazelles in conditions of the Issyk-Kul breeding center. Unpublished report for the Institute of Biology. Kyrgyz Academy of Sciences, Bishkek

Blank DA (2024) Birthing season of May 2024 in goitered gazelles in conditions of the Issyk-Kul breeding center. Unpublished report for the Institute of Biology. Kyrgyz Academy of Sciences, Bishkek

Blank D, Yang W (2015) Strategy of goitered gazelle suckling behavior and its adaption to the environment. Mamm Res 60(4):393–401
Blank DA, Yang W (2017) Mother-young recognition in goitered gazelle during hiding period. Behav Proc 142:21–28
Blank DA, Ruckstuhl K, Yang W (2017) Development of juvenile goitered gazelle social behavior during the hiding period. Behav Proc 144:82–88
Bromley PT (1967) Pregnancy birth behavioral development of the fawn and territoriality in the pronghorn (*Antilocapra americana* Ord) on the National Bison Range, Moiese, Montana. M.D. thesis. The University of Montana, Missoula
Bubenik AB (1965) Beitrag zur Geburtskunde und den Mutter-Kind-Beziehungen des Reh- und Rotwildes. Z Saugetierk 30:65–228
Byers JA (1997) American pronghorn. Social adaptations and the ghosts of predators past. University of Chicago Press, Chicago
Clutton-Brock TH, Albon SD, Gibson RM, Guinness FE (1979) The logical stag: adaptive of fighting in red deer (*Cervus elaphus* L.). Anim Behav 27:221–225
Clutton-Brock TH, Albon SD, Guinness FE (1986) Great expectations—dominance, breeding success and offspring sex-ratios in red deer. Anim Behav 34:460–471
Collias NE (1962) Social development in birds and mammals. In: Bliss EL (ed) Roots of behavior. Harper and Brothers, New York, pp 264–273
Côté SD (2000) Dominance hierarchies in female mountain goats: stability, aggressiveness and determinants of rank. Behavior 137:1541–1566
De Vos A (1960) Behavior of barren-ground caribou on their calving grounds. J Wildl Manag 24:250–258
Estes R (1966) Behavior and life history of the wildebeest (Connochaetes taurinus Burchell). Nature 212(5066):999–1000
Festa-Bianchet M (1991) The social system of bighorn sheep: grouping patterns, kinship and female dominance rank. Anim Behav 42:71–82
Fraser AF (1968) Reproductive behavior in ungulates. Academic Press, New York
Geist V (1971) Mountain sheep: a study in behavior and evolution. Univ. of Chicago Press, Chicago
Gendreau Y, Côté SD, Festa-Bianchet M (2005) Maternal effects on post-weaning physical and social development in juvenile mountain goats (*Oreamnos americanus*). Behav Ecol Sociobiol 58:237–246
Gorelov UK (1972) Breeding of goitered gazelle (*Gazella subgutturosa* Güldenstaedt) and problem of restoring of its number in the Badkhyz (SE Turkmenistan). In: Theriology, vol 1. Nauka of Siberian Department of Academy of Sciences of USSR, Novosibirsk, pp 420–424. (in Russian)
Hand JL (1986) Resolution of social conflicts: dominance, egalitarianism, spheres of dominance, and game theory. Q Rev Biol 61:201–220
Holand O, Gjosein H, Losvar A, Kumpula J, Smith ME, Roed KH, Nieminen M, Weladji RB (2004) Social rank in female reindeer (*Rangifer tarandus*): effects of body mass, antler size and age. J Zool (London) 263:365–372
Huegel CN (1985) Predator-avoidance behaviors of white-tailed deer that favor fawn survival. PhD thesis. Iowa State University, Iowa
Ishunin GI (1961) Goitered gazelle. In: Zakhidov TZ (ed) Fauna of Uzbek SSR, mammals (carnivores and ungulates), vol 3. Publication of the Academy of Sciences of Uzbek SSR, Tashkent, pp 165–174. (in Russian)
Kokshunova LE (2007) Nursing behavior of Saiga antelope fawns (*Saiga tatarica*) in captive conditions. News of universities. North-Caucasus Region 5:62–64. (in Russian)
Lent PC (1966) Calving and related social behavior in the barren-ground caribou. Z. Tierpsychol. 6, 701–756
Lent PC (1974) Mother-infant relationships in ungulates. In: Geist V, Walther F (eds) The behavior of ungulates and its relation to management, vol 1. IUCN Publication New Series 24, Gland, pp 14–55

Lingle S (2003) Group composition and cohesion in sympatric white-tailed deer and mule deer. Can J Zool 81:1119–1130
Marmazinskaya NV (1996a) Breeding of goitered gazelle *Gazella subgutturosa* (Artiodactyla Bovidae) in the ecological center "Dzheiran". Proceedings of the Uzbekistan nature reserves, issue 1. FAN Press, Tashkent, pp 53–61. (in Russian)
Marmazinskaya NV (1996b) Territorial and marking behavior of goitered gazelle *Gazella subgutturosa* (Artiodactyla, Bovidae) in the Bukhara ecological center. Zool Zhurnal (Moscow) 75(11):1737–1751. (in Russian, English summary)
Marmazinskaya NV (2000) Development of social behavior in goitered gazelle (*Gazella subgutturosa subgutturosa* Güeldenstaedt, 1780). Biodiversity conservation in the specially protected areas of Uzbekistan. Chinor ENK, pp 87–94 (in Russian)
Marmazinskaya NV, Mardonov BK (1996) Density and distribution characteristics of different mark types in goitered gazelle. Proceedings of nature reserves of Uzbekistan, issue 1. Fan Press, Tashkent, pp 62–68
McHugh T (1958) Social behavior of the American buffalo (Bison bison bison). Zoologica 43(1):1–40
Miranda-de la Lama GC, Mattiello S (2010) The importance of social behavior for goat welfare in livestock farming. Small Rumin Res 90:1–10
Müller-Schwarze D (1968) Play deprivation in deer. Behaviour 31(1–2):144–162
Odonkhuu D, Olson KA, Schaller GB, Ginsberg JR, Fuller TK (2009) Activity, movements, and sociality of newborn Mongolian gazelle calves in the eastern steppe. Acta Theriol 54:357–362
Olson KA, Fuller TK, Schaller GB, Lhagvasuren B, Odonkhuu D (2005) Reproduction, neonatal weights, and first-year survival of Mongolian gazelles (*Procapra gutturosa*). J Zool 265:227–233
Pereladova OV (1988) Acoustic behavior of goitered gazelle in the nursery. In: Communicative mechanisms of population structure regulation in mammals. All-union conference of VTO, Moscow: 126-129 (in Russian)
Pereladova OV, Pereladov SV (1986) The goitered gazelle breeding in nursery enclosures of the Syunt-Khasardak nature reserve. In: Flint VE, Prisyazhnuk VE (eds) Conservation and perspectives of the number restoration of goitered gazelle in USSR. All-Union Scientific Institute of Nature Conservation and Nature Reserves, Moscow, pp 49–59. (in Russian)
Poindron P, Hernandez H, Serafin N, Terrazas A (2003) Early recognition of newborn goat kids by their mother. I. Nonolfactory discrimination. Dev Psychobiol 43:82–89
Ralls K, Lundrigan B, Kranz K (1987) Mother-young relationships in captive ungulates: behavioral changes over time. Ethology 75(1):1–14
Ruckstuhl KE, Ingold P (1999) Aspects of mother-kid behavior in alpine chamois, *Rupicapra rupicapra rupicapra*. Z Saugetierk 64:76–84
Schaller GB, Aili K, Xinbin CAI, Yanlin L (2006) Migratory and calving behavior of Tibetan antelope population. Acta Theriologica Sinica 26(2):105–113
Sludskiy AA (1956) Breeding of goitered gazelle. Proceedings of institute of zoology, Kazakh Academy of Sciences, vol 6. Nauka Press of Kazakh SSR, Alma-Ata, pp 78–108. (in Russian)
Sokolov II (1959) *Gazella subgutturosa* Güldenstaedt—Goitered gazelle. In: Pavlovskiy EN (ed) Fauna USSR, mammals: ungulates (orders Perissodactyla and Artiodactyla). Publication of the Academy of Sciences USSR, Moscow, pp 308–331. (in Russian)
Sokolov VE, Lushchekina AA (1997) Procapra gutturosa. Mamm Species 571:1–5
Taillon J, Côté SD (2007) Social rank and winter forage quality affect aggressiveness in white-tailed deer fawns. Anim Behav 74:265–275
Taylor CA (2013) Behaviour and cause-specific mortality of mule deer (*Odocoileus hemionus*) fawns on the National Rifle Association Whittington Center of north-Central New Mexico. M.D. thesis. Texas Tech University, Lubbock

Thouless CR, Guinness FE (1986) Conflict between red deer hinds: the winner always wins. Anim Behav 34:1166–1171

Walther FR (1965) Verhaltensstudien an der Grantgazelle (*Gazella granti* Brooke, 1872) im Ngorongoro- Krater 1, 2. Z Tierpsychol 22(2):167–208

Walther FR (1984) Communication and expression in hoofed animals. Indiana University Press, Bloomington

Zhevnerov VV (1984) Goitered gazelle of the Barsakelmes Island. Nauka Publishing House of Kazakh SSR, Alma-Ata. (in Russian)

Zhevnerov VV, Bekenov AB, Sludskiy AA (1983) Goitered gazelle—*Gazella subgutturosa* Güldenstaedt, 1780. In: Gvozdev EV, Kapitonov VI (eds) Mammals of Kazakhstan, Bovidae, vol 3 (3). Nauka Publishing House of Kazakh SSR, Alma-Ata, pp 11–54. (in Russian)

Mother: Young Interactions 10

10.1 Antipredator Female Behavior

Blank et al. (2015b)—License Number—5477460868007—full text.
Blank (2017)—License Number—5477461388644—full text.
Blank and Yang (2017)—License Number—5477470300111—full text.

10.1.1 General Remarks

Maternal behavior is adapted to environmental conditions and depends on food quality and quantity, habitats features, and predation pressures (Geist 1974). In natural conditions, predation is the main factor of a high mortality rate, especially of the young (Jarnemo et al. 2004). The predation risk and mortality is greatest (up to 50%) within the first month of a fawn's life, when offspring are too small, have no ability to run fas, and no experience to avoid a hunting predator (FitzGibbon 1990). Therefore, the antipredator strategy of mothers is of primary importance for successfully rearing offspring, decreasing mortality, and contributing to the growth of the population (Lent 1974; Johnstone-Yellin et al. 2009). Females change their behaviors as early as the last stages of pregnancy due to a significant increase in energy demands, including lactation, as well as increased requirements for antipredator behavior and protection of newborns from potential predator attacks (Grignolio et al. 2007). As a result, experienced multiparous adult females are much more successful in raising offspring—more vigilant and better at avoiding predators with a significantly lower mortality rate (Toïgo et al. 1996)—compared to a young female giving birth for the first time. In addition, larger females produce larger fawns, which have a better chance of survival than smaller individuals. There are also other factors that affect the survival of fawns (Kunkel and Mech 1994).

I mentioned earlier that, in general, all offspring rearing strategies can be divided into two major groups: young following the mother and young hiding. The first

D. Blank, *The Behavioral Biology and Conservation of Goitered Gazelles (Gazella subgutturosa)*, https://doi.org/10.1007/978-3-031-83504-9_10

category includes species whose young stay with their mothers from their first minutes after birth. In hiding species, the mothers leave their young alone and hiding for a period of time (from several days to several months depending on species) and only return for suckling (Lent 1974; Fisher et al. 2002). Most ungulates (up to 80%) are hiding species and all antelopes are hiders (Lent 1974; Estes 1976; Ralls et al. 1986). The wild and domestic sheep and goats have very short hiding periods of 2–5 days (Hewson and Verkaik 1981; Fedosenko and Blank 2001, 2005), while fawns of many antelopes can hide for 1.5–2 months (Estes 1991), with a few having very long hiding times up to several months (*Kobus kob*, Leuthold 1967; *Redunca arundinum*, Jungius 1970). Goitered gazelles' hiding time lasts around 2 months, from mid-May until mid-July (Blank 1985).

Young of hiding species are quite weak, especially in the first few days after birth, and cannot escape from predators by following their mothers. Therefore, in the first days of life, in the absence of its mother, the fawn is safer in hiding, which is assisted by the fawn's cryptic coloring and very limited movements: the fawn presses its body to the ground and does not move, even breathing is very smooth and imperceptible, and the cardiac rhythm slows down (Walther 1968, 1969; Jacobsen 1979). In addition, a fawn's sweat glands practically stop working to reduce the spread of odor, and afterbirth is usually eaten by the mother (Sokolov 1959), as is any urine or excrement of the young during nursing. All these behavioral features are aimed at minimizing the likelihood of a predator finding a fawn (Lent 1974; Autenreith and Fitchter 1975). This hiding strategy can be so effective that a carnivore has been known to pass a hiding fawn at a distance of 2 m (Kruuk 1972; Spinage 1969). The transition from a hiding strategy to following the mother is a gradual process, and fawns, during this transformation, increase the duration of their activities in the absence of the mother. This is the time in their development that fawns most often become victims of predators, but the success of the hiding strategy not only depends on the activities of the fawn, but also on the behaviors of the mother, who must not transmit any information about the location of her fawn to the predator (Byers and Byers 1983).

10.1.2 Features of Antipredator Behavior in the Goitered Gazelle Females

First of all, before delivery, goitered gazelle females leave their groups to remain solitary. When giving birth, they do not gather in great associations like saiga antelope. Instead, the females disperse over a large area at a distance of several hundred meters from each other and each mother has her own area where she can hide her fawn (Antipin 1941; Sludskiy 1956). Mothers usually do not enter any other mother's plot with their fawns. This dispersal strategy is designed so that if a predator finds one fawn, the others are more likely to not be found and survive. Therefore, it is possible to see many single females everywhere just before giving birth, as well after mass birthing, when most have hiding young (Table 10.1). In southern Turkmenistan, during the previous periods of goitered gazelle abundance, large

Table 10.1 Group size in goitered gazelle females in the central parts of the Betpak-Dala in May 1954–1955

	Number of individuals in a group												
	1	2	3	4	5	6	7	8	9	10	11	12	
Deca-des of May	Number of cases												Total cases
1	23	14	13	7	3	–	1	3	–	1	–	1	66
2	104	38	26	8	5	2	2	1	–	–	1	–	187
3	129	24	13	3	–	–	–	–	–	–	–	–	169
Total	256	76	52	18	8	2	3	4	1	1	1	1	422

Adapted from Sludskiy (1956)

clusters of females were described in certain maternity areas, where females gather from great areas (Heptner et al. 1961). Female gazelles fleeing from predators often run for significant distance and do not return to their fawns for a long time, keeping them in hiding (Sludskiy 1962). Zhevnerov et al. (1983) states that mothers with fawns usually prefer to be in the herd during daytime, but leave the herd at night. My observations do not support this statement, though a female can sometimes join to other females during the day, she does it without her fawns and only for a short time. At night, a mother joins other females during the grazing night pastures, while the fawns stay alone in hiding until the next day (Blank 1997, 2017, Blank et al. 2015b).

Although in total, the hiding season in goitered gazelles lasts about 2 months (from mid-May until mid-July), this period has different stages, of which only the first two stages can be called actual hiding in the full sense of the word. The last stages are characterized by quite active independent movements of fawns without their mothers (details in the previous chapter Young behavior). Nevertheless, all stages of this period are conventionally called the hiding period, meaning that a mother initially leaves her fawns for a long time, returning only for nursing, and then as the fawns grow, she gradually increases her time with them, staying in their company longer and longer. Fawns over the age of 1 month begin to show more and more independence in the absence of their mothers. Even in the presence of their mothers, fawns increasingly show great freedom in their movements and can simply leave her at the most unexpected times and in the most unexpected places. Despite this independence, the mothers try to regain control of their fawns; continuing to monitor them, watching their movements, waiting when fawns are too slow, calling or even running after them if they move away for a long distance. At the end of the hiding period, fawn independent activity decreases considerably, and the mother regains her control over the actions of her fawns as they transition from hiding to following their mothers 24-h a day, copying all the mother's movements and behaviors (Blank 1985).

One of the most important factors of a mother's antipredator behavior is to maintain a sufficiently far enough distance to her fawn's hiding place so that, according to the location of the mother, the predator cannot find her fawn in her immediate vicinity. Therefore, after every nursing bout, the mothers leave their fawns and go to graze at a distance of no closer than 200–400 m (at least) from the hiding site unless potential danger appears (Fig. 8.4 from Suckling Behavior chapter). In midday, females go their watering sites, going as far as 10 km away from their hiding fawns. When they return, mothers nurse their fawns and leave them again, while they graze and rest at a distance. The same process is repeated in the evening. After the evening feeding, the females lie down to rest, also at a distance away. Thus, a female leaves her fawn alone after each feeding, and then stays a certain distance from him. Looking for fawns at night in May with the help of a headlight, Sludskiy (1956) several times raised resting females located 80–200 m from their young gazelles. For the first 7–9 days, the female grazes and rests near the site where the fawns are located and actively protects them from small predators (Sludskiy 1962). She leaves her location only in forced cases, such as human intervention or a need for water. For example, a tamed female after giving birth was constantly at her fawn's bed-site, leaving it only for water. In one noted case, a fox appeared at the site and found the fawn. The female was returning at the time and noticed the predator from about 250 to 300 m away. With a loud snort, she rushed toward the fox forcing it to run away, saving her fawn (Zhevnerov 1984). For some females, the instinct to protect their fawns is developed so strongly that they lose their sense of fear of human beings (Gorelov 1972). According to my observations, females stay overnight with their fawns in the first few days after giving birth, but later they usually go to watering holes and graze at night, leaving the fawns alone, and returning to them only at sunrise (Blank 2017; Blank et al. 2015b).

Goitered gazelle females always keep a distance from their hiding fawns, especially in the first 2 weeks after birth, when the fawn is small and can easily become a victim of any predator, even a small one, from which it cannot escape. It is no coincidence, then, that it is in these first weeks of a fawn's life in May that the distance between a mother and her fawn's hiding place is the farthest of the entire period, reaching more than 1 km in some cases (Table 10.2). As the fawns mature, the females reduce this distance dramatically, initially up to three-fold in early-June compared to May and then more than another four-fold in late-June compared to the first half of June. After that, the female continues the gradually reduce the distance she stays away from her fawn, but not as drastically as during June (Table 10.2; Blank et al. 2015b). Pronghorn (*Antilocapra americana*) females behave in a

Table 10.2 Observed distances between mothers and their fawns' hiding period (May–July)

Juvenile age; time of year	Mean ± SE, in meters	Range in meters (min-max)	n	Mann-Whitney test	
				Z	*P*
2 weeks (~15–31 may)	161.2 ± 39.3	3–1200	39		
4 weeks (~01–15 June)	52.0 ± 14.5	1–250	23	−2.755	0.006
6 weeks (~16–30 June)	12.1 ± 3.4	1–40	16	−1.982	0.048
8 weeks (~01–15 July)	6.8 ± 3.3	1–30	9	−1.434	0.152
10 weeks (~16–31 July)	4.0 ± 0.8	1–6	9	−0.137	0.891

Ili depression, Kazakhstan, n = 96. Statistical comparisons are shown between 5 neighboring time periods (2 and 4, 4 and 6 weeks etc.) (by Blank et al. 2015b—License Number—5476940126030)

similar way, when a mother and her hiding young have the greatest distance between them within the first 2 weeks postpartum followed by the female staying closer and closer as the fawn ages (Barrett 1984). This behavior clearly has protective antipredator significance. Predators find young not only during nursing, when they are in an active phase with their mothers, but also by behaviors of the mother. If a mother is too close to her hidden young, the female can give the predator her fawn's proximate location by her behavior, and the predator may begin an intensive search around her to find her offspring. Therefore, the mother must keep a sufficiently large enough distance from her lying fawn so that the predator cannot find it in her immediate vicinity. At best, the female should be at such a distance from her fawn that a predator would not be able to easily find it (FitzGibbon 1993). If a grizzly (*Ursus arctos*) identifies the general area of a hiding moose calf, the calf does not have a high chance of survival (Bowyer et al. 1998).

If the mother is too far from her hidden fawn, she will not be able to see the predator approaching it and will not have possibility to arrive in time to protect it (Lent 1974). In different species and in different circumstances, the reaction of females may be different when protecting their young from a predator, from running away to an overt aggressive attack. Females may have several antipredator strategies, including staying at a distance and attempting to drive or lure the predator away to attacking the predator when it finds her young (Byers and Byers 1983; Bleich 1999). Staying at a distance from the hiding young is a typical strategy in most hiding species, while these other strategies have a fairly high risk of injury and even death for the mother; these strategies are not observed in all species (Andersson et al. 1980). Females of most ungulate species can only defend their young from small-sized predators (notably smaller than the female), which are dangerous to young and practically harmless to adults. Goitered gazelle females will attack foxes (*Vulpes vulpes*—Zhevnerov 1984), Grant's gazelles defend their young from jackals (Walther 1965), white-tailed deer (*Odocoileus virginianus*—Garner and Morrison 1980), mule deer *O. hemionus*—Grovenburg et al. 2009), pronghorn (*Antilocapra americana*—Byers and Byers 1983) and bighorn sheep (Ovis canadensis—Berger 1978) are known to attack coyotes (*Canis lutrans*), elk (*Cervus elaphus*—Altman 1963) and mountain goats (*Oreamnus americanus*—Coté et al. 1997) have the ability to protect their calves from wolves (*Canis lupus*); and moose (*Alces alces*) will attack wolves, American black bears (*Ursus americanus*) and grizzly bears (*Ursus arctos horribilis*—Stephenson and van Ballenberghe 1995; Murie 1981). There have been cases when a predator has already killed a fawn, but the mother still continues to protect it (Byers and Byers 1983).

On the open plains of the Ili Valley, goitered gazelle females are usually at an average distance of 161 m away from their hidden fawns. In the grassy prairies of Alberta, with relatively limited visibility compared to the Ili Valley habitat, and where mostly cayotes are potential predators of young, pronghorn females stay farther away (269 m) from their young compared to goitered gazelle females (Barrett 1984), perhaps because their fawns are better hidden in the grassy habitat. In Montana, where there are not only coyotes, but also golden eagles commonly attack young, females decrease their distance to their hiding young to up to 70 m on average—more than 2 times closer than mother-fawn distances in goitered gazelles possibly to the time it takes for a mother to arrive and protect her young from a deadly raptor (Byers and Byers 1983). The choice of hiding place and the distances a mother keeps away from them mainly depends on the mother's awareness and the specific threats to her young, especially in the first few weeks after birth; she improves her skills with age and experience to choose the most secure places for hiding her fawn to avoid detection and to respond when needed (Ozoga and Verme 1986).

Another parameter important for a mother's antipredator behavior is the frequency of approaches to her fawn and the duration of its nursing, when they actively move together from one place to another. If the fawn lies motionless and merges with the surrounding environment during the hiding period, it is difficult for predators to detect and is relatively safe. Therefore, it is potentially more productive for the predator to wait and watch a mother for when she goes to her fawn to nurse, and the fawn will become visible. For this reason, then, the mother obviously needs to approach her fawn as rarely as possible during the day and nurse it for relatively short periods of time (FitzGibbon 1993). This is why the duration of the hiding periods, which are the intervals between active feeding periods, is longest and suckling bouts are shortest during the first 4 weeks after birth; after that the duration of hiding periods drastically decreases by four times in singletons and ten times in twins (observed for fawns over a month-old) (Table 10.3). At the same time, the periods of activity when the mother feeds her fawn and they move together, making young clearly visible to predators, are the shortest in the first 4 weeks of a fawn's life (May and early June). In the second half of June, when the fawn is able to run along the adults, the duration of active periods is doubled. Despite the fact that active periods increase after mid-June, the total duration of suckling during the bouts declines by the reduction in the frequency and duration of suckling (Table 10.4). This suggests that the lengthening of the active period is not related to nursing, but rather to the need for the gradual increase in the physical connection and communication of the female with her fawn (Blank et al. 2015b). Similar behavior is observed in pronghorn females whose fawns have the longest hiding time (225.5 min in average) within the first several weeks after birth (Byers and Byers 1983), which is comparable to that in goitered gazelles (Blank et al. 2015b).

Another feature of maternal antipredator behavior is that in the first few days after birth, the mother puts each of the twins in different hiding places and feeds and transfers them from one bed to another separately. A week or two later, depending on the choice of the female, the mother brings her twins together, feeding and transferring them at the same time and letting them rest in the same in the same hiding site (Fig. 9.3 from Behavior of Young chapter). Interestingly, the time of twins being reunited was different in different years: within 7 days in 1985, 12 days in 1984, and 14 days in 1983. Perhaps this difference was related to the varying weather conditions across years. Twin fawns can be hidden several hundred meters apart from each other (on average, mean ± SE = 272.4 ± 4.1 m, $n = 14$) in May, while in June, this distance is reduced by 10 times (on average 2.4 ± 0.3 m, $n = 10$) and from that time on they lay together. The separation of fawns is obviously related to the

Table 10.3 Changes in the duration in the lengths of hiding and active (suckling) periods in goitered gazelle females with single and twin juveniles from the Ili Hollow in Kazakhstan

				Mann-Whitney test	
Juvenile number	Mean ± SE, min	Range, min	n	Z	*P*
Length of the hiding periods					
15–31 May					
Single	182.8 ± 52.3	35–478	8		
Twin	335.7 ± 21.4	49–707	61		
01–15 June					
Single	202.5 ± 43.6	61–406	8	−0.420	0.721
Twin	303.7 ± 37.8	38–555	15	−0.646	0.518
16–30 June					
Single	55.9 ± 7.6	3–192	44	−3.526	0.0005
Twin	33.7 ± 8.8	2–120	17	−4.589	0.0005
Length of the active periods					
15–31 May					
Single	19.2 ± 3.2	3–54	21		
Twin	30.0 ± 2.2	2–88	72		
01–15 June					
Single	24.7 ± 3.4	1–98	41	−0.633	0.527
Twin	32.5 ± 3.0	3–93	41	−0.860	0.390
16–30 June					
Single	40.1 ± 4.7	3–143	47	−2.514	0.012
Twin	65.4 ± 15.1	11–171	13	−1.285	0.199

Shown are mean durations, in minutes, Z-test scores and *P*-values (comparing singles for neighboring time periods and separately twins for the same periods) (Blank et al. 2015b—License Number—5476940126030)

Table 10.4 Changing in average suckling bout duration (in seconds) with fawn age; comparing three-time segments from May to July; Ili Hollow, Kazakhstan

				Mann-Whitney test	
Juvenile number	Mean ± SE	Range	n	Z	*P*
15–31 May					
Single	169.8 ± 26.5	19–466	24		
Twin	354.4 ± 22.9	35–1091	125		
01–15 June					
Single	83.4 ± 12.2	12–324	42	−3.113	0.002
Twin	438.1 ± 36.1	41–1269	69	−2.280	0.023
16–30 June					
Single	32.1 ± 4.5	3–242	78	−4.962	0.0005
Twin	18.6 ± 2.0	4–57	49	−9.147	0.0005

Shown are data for singleton or twin offspring. Mean duration of suckling bouts per time segment, sample size (n) and statistics (comparing singletons for neighboring time periods and separately twins for same periods) (Blank et al. 2015b—License Number—5476940126030)

Table 10.5 Changing proportions of mother's leading vs following of her fawn, different paces during movements (walking vs running) and changes of bed sites vs repeated use in May–July

Mother's leading position			
Months	Mother leads her fawn	Mother follows her fawn	Proportion changes of a mother leading vs following her fawn
Late-May	90	6	0.94
Early-June	65	6	0.91
Late-June	49	10	0.83
Early-July	16	5	0.76
Movement time	Walking	Running	Proportion changes of paces: Walking vs running
Late-May	71	6	0.92
Early-June	49	1	0.98
Late-June	23	3	0.89
Early-July	12	1	0.92
Time of changes vs. repeated use of bed site	Bed site changes	Repeated use of same bed site	Proportion changes of bed sites vs repeated use
Late-May	74	4	0.95
Early-June	52	2	0.96
Late-June	27	5	0.84
Early-July	12	1	0.92

Blank (2017)—License Number—5476940463790

possibility of saving at least one of the twins in case one is found by a predator; the second twin, hiding 2–3 hundred meters away, would be much more likely to survive (Blank et al. 2015a). Pronghorn females also hide their twins separately during their first weeks after birth, and only after 7–21 days are the twins reunited and after that stay together (Autenreith and Fitchter 1975; Barrett 1984). White-tailed deer and mule dee females also separate their fawns separated for the first 18–32 days after birth (Ozoga et al. 1982; Hirth 1985: Bishop 2007). Barrett (1984) found that in pronghorn twins laying in different bed sites have a higher chance of at least one surviving over the first 4–15 days after birth, while Bishop (2007) found only a slight difference in the fate of mule deer twins, but in any case, the placement of twins in different places in their first weeks of life is an obviously effective behavior that contributes to the survival of at least one twin (Barrett 1984).

Another feature of the defensive behavior of mothers during the hiding period is that they change their fawns' hiding places during each active period, leaving their fawns in a new place. Changing the hiding place is a typical behavior in goitered gazelles; repeated use of the same resting place happens, but the number of such cases is negligible (Table 10.5, Blank 2017). During every nursing bout, consisting of several separate suckles, females lead their fawns from their previous hiding place to a new location, defining the direction, duration, and speed of travel. While walking, the female changes the direction of her movement constantly, resulting in she and her fawns moving within the same area with a radius of 300–700 m (Zhevnerov 1984). The mother is forced to wait or calls a fawn to hurry if it lags behind or to get up if it lies down halfway. Fawns can lay several times (up to five

Table 10.6 Frequency changes (cases per hour) of different behavioral patterns of mothers, and cases of fawn's wandering off from its mother during the hiding period (May–July)

Months	Observed cases	Frequency, cases/hour	Log-linear model test
Fawn's wandering off from the mother			
May	20	0.071	
Early-June	19	0.170	Z = 5.540; $P < 0.0001$
Late-June	49	0.322	Z = 7.086; $P < 0.0001$
Early-July	14	0.147	Z = 4.821; $P < 0.0001$
Mother's looking after her fawn			
May	4	0.014	
Early-June	8	0.071	Z = 3.200; $P = 0.001$
Late-June	16	0.105	Z = 3.866; $P < 0.001$
Early-July	6	0.059	Z = 3.273, $P = 0.001$
Mother's following her fawn			
May	4	0.014	
Early-June	3	0.027	Z = 1.353; $P > 0.05$
Late-June	9	0.059	Z = 2.199; $P = 0.028$
Early-July	6	0.059	
Mother's waiting			
May	3	0.011	
Early-June	6	0.054	Z = 2.873; $P = 0.004$
Late-June	20	0.132	Z = 3.141; $P = 0.002$
Early-July	2	0.020	Z = 1.347; $P > 0.05$
Mother's calling back vs looking after			
May	9	0.032 vs 0.014	Z = − 3.200, $P = 0.001$
Early-June	2	0.018 vs 0.071	Z = 3.848, $P < 0.0001$
Late-June	4	0.026 vs 0.105	Z = 3.548; $P < 0.0001$
Early-July	–	–	–

Log-linear model tests show significance of changes in frequency of the mother's behavioral patterns with age of her fawn (June–July vs May); calling back was not recorded in early-July (Blank 2017—License Number—5476940463790)

times) per trip. Such calls are most often observed in May, less often in June, when mothers prefer to just keep watch their fawns instead of calling them back, and completely disappear in July (Table 10.6). In the remaining cases, the mother leaves her fawn alone on its resting site and goes away, but very soon returns, picks up her fawn and takes it to another hiding place. This behavior is observed much less frequently than any other (waiting or following). Thus, in May, mothers most often make calling sounds to bring back the departing fawns, whereas in June they mostly watching, following its movements with their eyes, whereas they do waits or pursue the fawn less often (Table 10.6; Blank 2017).

In a calm environment and absence any danger, a female with her fawn walk at a quiet pace, but often stop for vigilance. Sometimes the female starts running ahead, and the fawn tries to keep up with her, running closely after her. When a mother and her fawn travel together, the female is anxious all the time and looking back to see if her fawn is following her. Sometimes the fawn does not follow its mother, but instead runs in another direction, starts playing and running in different directions; the mother then stops and begins to follow the fawn very closely with her eyes. Such disobedience of a fawn and his refusal to follow its mother is rarely observed in the first weeks in May, but by June, it occurs more and more, reaching a maximum by the end of June. In July the mothers again regain control over the movements of their fawns (Table 10.5). If the fawn continues to leave and moves too far away, then the mother tries to catch up in order to bring it back. Such cases are also rare in May, but become common towards the end of the hiding period in June and early July. The mother's pursuit of her fawn is observed almost as often as a simple eye tracking of the fawn's movements (Table 10.5; Blank 2017).

The constant relocation of the hiding place is an obvious adaptation that makes it difficult for predators to track down the location of hiding fawns by watching the females. Different mothers establish different frequencies of nursing visits to fawns, from two to four times a day, so females with twins, keeping them separated, make transitions with her fawns from four to eight times a day. The number of visits and relocations do not change over time for each particular female, although for mothers with twins, the number of transfers is halved from the moment she reunites her fawns in one hiding place, but the duration of the transition increases all the time, and if in May it is on average equal to a few hundred meters (mean ± SE = 311.7 ± 3.5 m, n = 14), then after a couple of weeks in early June, the duration of the transition increases by almost 7 times, up to 2 km (mean ± SE = 2033.3 ± 40.9 m, n = 10). This behavior of goitered gazelle mothers suggests that these transitions are not only directed against predators finding fawns, but also have a training effect, when mothers force their fawns to constantly improve their motor skills and endurance which prepares them to be physically ready to escape from a predator in case of an attack (Blank et al. 2015b). Thus, by transferring the fawn from one hiding place to another, the female deprives the predator of the opportunity to gradually, over time, find a hiding place. In addition, the distance of these transitions increases, increasing the fawn's physical training, as it grows. As a result, fawns become more and more prepared to escape from a predator. The same training and constant lengthening of transitions for fawns are also seen in females of pronghorns (Byers and Byers 1983; Barrett 1984), white-tailed deer (Ozoga and Verme 1986), and mule deer (Riley and Dood 1984).

Changing hiding sites of young with each suckling bout is common behavior among hiding ungulates, including many antelope and cervid species. This behavior is described for pronghorns (Barrett 1984), southern reedbucks (*Redunca arundinum*—Jungius 1970), Grant's gazelles (*Nanger granti*—Walther 1965), and gerenuks (Reif and Klingel 1991), as well as for roe deer (*Capreolus capreolus*—Aanes et al. 1998), mule deer (Riley and Dood 1984) and white-tailed deer (Ozoga and Verme 1986). Moreover, older white-tailed deer mothers make longer transitions to new bed-sites with their fawns than younger ones. And as a result, the older and more experienced mothers better prepare their fawns for predator attacks, and therefore suffer much fewer losses (Ozoga et al. 1982; Mech and Roberts 1990).

Fig. 10.1 Sometimes, a mother nods, lowering her head towards the ground and then raising it, as if inviting a fawn to nurse. Drawing by U. Kenzhetaev with kind permission)

During the approach of the mother to her fawn's hiding place, it is very important that a predator is not watching her, so she, herself, does not indicate the fawn's location. The female can approach the hidden fawn either slowly or quickly, but for the last few hundred meters (or less, as few as 30–40 m have been noted) she stops and looks around for a long time (sometimes moving only 200–300 m over 3 h), and then walks carefully and very slowly to her hiding fawn, with numerous stops, repeatedly freezing in place and frequent prolonged inspections and listening to ambient sounds to detect possible danger in the vicinity. The female is very tense and tries to catch the slightest movement or faintest sound and is ready to run at the slightest sign of danger. Sometimes she goes back, moving away from the hiding place, and then returns again closer or goes around her fawn. If, after all these maneuvers, the mother is convinced that there is no danger, she moves close to her young and makes soft "mooing" sounds. If the fawn is afraid to go closer, the mother sometimes does makes nodding movements, lowering her head to the ground and raising it up as if inviting the fawn to suckle (Fig. 10.1). Females of many hiding species use inviting calls for their young, such as pronghorns (Autenreith and Fitchter 1975), saiga antelope (Kang 2004), and gerenuks (Reif and Klingel 1991). Similar behavior to the slow approach to hidden young, making many stops and showing signs of vigilance is found also in pronghorns (*Antilocapra americana*—Byers 1997) and gerenuks (*Litocranius walleri*—Reif and Klingel 1991) and possibly in red deer (*Cervus elaphus*—Clutton-Brock and Guinness 1975). Though I suppose that this behavior is inherent to many other hiding species, it has not been identified and documented yet.

In goitered gazelle females, the distance for giving this sound to her fawn in the first 2 days after birth is 2–4 m; on the fifth or sixth day, this distance increases to 10–15 m and remains within this limit until the end of the hiding period. It is only at the very end of the hiding period that the females and fawns recognize each other at a much greater distances and run to each other from afar. Even at this time, however, a fawn is probably more likely to be focusing on the behavior of the female looking in its direction and heading toward it than on any real recognition, and only by running up close to each other can they accurately recognize each other, and then probably more by smell than by appearance or sound. The same behavior of a

Table 10.7 Changing proportions of different behaviors when a female leaves her fawn

Behaviors/months	May	Early-June	Late-June	Early-July
Female waits for fawn to lie down and leaves after that	76 69.7	31 62.0	41 54.7	13 61.9
Female does not wait and leaves before fawn lies down	20 18.3	9 18.0	8 10.7	1 4.8
Fawn walks/runs away from mother	9 8.3	5 10.0	10 13.3	5 23.8
Female lies down and waits for fawn to lie down	4 3.7	5 10.0	16 21.3	2 9.5
Total	109 100	50 100	75 100	21 100

(Upper number = number of observed cases; lower number = proportion, %) over May–July (Blank 2017—License Number—5476940463790)

gradual increase in distance for reuniting a mother and fawn as the fawn ages is described for white-tailed deer, mule deer, pronghorns, and Thomson's gazelles (Autenreith and Fitchter 1975; FitzGibbon 1990; Schwede et al. 1994; Byers 1997; Taylor 2013; Costelloe and Rubenstein 2015).

The mother-young separation is not accompanied by prolonged or frequent vigilant acts, since the female tends to leave her fawn after suckling as soon as possible so as not to attract a predators' attention to the hiding place. After completing a suckling bout and approaching the chosen new location, the female stops and waits usually in a standing position to give her fawn the opportunity to find a suitable place to settle and hide at a distance of several meters around her. This maternal behavior is a characteristic for the entire hiding period, though it is observed more often in May and less often in June and July (Table 10.7). Much less often, the mother leaves her fawn before it finds a suitable place to rest and lie down, or when the fawn simply stops following her. The fawn's mobility rises as it grows older, and toward the end of the hiding period (early July), it tries to follow its mother everywhere. If the female needs to leave her fawn, she lies down and waits until the fawn chooses its own bed site and also lies down; then she gets up and leaves. Females behave this way quite rarely in May, while later at the end of June, the number of these cases increases before decreasing again (Table 10.7). Even less often are cases when a fawn simply walks or runs away from its standing mother. Fawns rarely behave this way in May and more often thereafter, gradually increasing their own initiative in activities and independence from their mothers as they age (Table 10.7; Blank 2017).

When the mother stays at a certain distance from her hiding fawn, she knows exactly where its hiding place is and periodically closely monitors in a distance what is happening around it, always ready to run to protect her fawn if a fox (*Vulpes vulpes*) or another small predator suddenly appears in the vicinity (Blank 2017). During the first 2 weeks after birth, when the fawn is strongly attached to its hiding place and almost never gets up in the absence of its mother, the mother often attentively watches the fawn from a distance (0.19 cases/h, *n* = 53). Therefore, she often orients her muzzle and her entire body toward her hiding fawn, with 38% of these cases she does during her bedding. In early-June, the frequency of mother orienting toward her hiding fawn decreases markedly (0.10 cases/h, *n* = 16) and is negligible thereafter (0.03 cases/h, *n* = 5). Similar to goitered gazelle mothers, pronghorn females also spend longer time facing her hiding young. Thomson's gazelle (*Eudorcas thomsonii*) mothers, in contrast, do not spend as much time looking toward their hiding fawns, but instead demonstrate more vigilance (FitzGibbon 1993). There is a misconception that predators can determine the location of a hidden fawn by the direction of the female's muzzle, but observations show that although many predators (cheetah *Acinonyx jubatus*; jackals *Luputella mesomelas* or spotted hyena *Crocuta crocuta*) search for a fawn in the vicinity of Thomson's gazelle females, they do not do this systematically and unlikely use the direction of the muzzle to more accurately determine the location of the fawn (FitzGibbon 1993). In contrast to most other predators, coyotes will search systematically for hiding young, but this does not occur often and only in years when other prey is scarce (Byers and Byers 1983).

Goitered gazelle females show the most anxiety and nervousness in the first weeks after giving birth in May. At this time, they display a higher level of vigilance and make repeated inspections of the area most often (0.86 cases/h, *n* = 235). Within 2 weeks (in early-June), mothers appear to worry about their fawns less often and show vigilance almost half as much (0.48 cases/h, *n* = 75), followed by a further and significant decline in the frequency (0.26 cases/h, *n* = 40 in late-June and 0.05 cases/h, *n* = 5 in early-July). Not only does the frequency of vigilance drop, but there is also a constant drop in its duration, with the maximal vigilance duration observed in May (Mean ± SE = 78.1 ± 15.8 s, *n* = 354), with a three-fold drop in early-June (25.3 ± 10.4 s, *n* = 74) and even less in late-June (15.8 ± 2.9 s, *n* = 47). The mother demonstrates especially frequent and lengthy acts of vigilance when she approaches her hiding fawn. She repeatedly stops and makes long cautious and attentive inspections, ready to instantly run in case of any danger (details below). The first 4 weeks, a significant part of the time she takes to approach her young is dedicated to vigilance (38–48%), while the rest of her time is developed to other activities (standing, walking, sometimes laying or running). Thereafter (late-June), the mother is much less vigilant (only 14%). In May, mothers will also show increasing vigilance when leaving their fawns after a suckling bout, the duration of which is also long (Mean ± SE = 88.2 ± 35.1 s, *n* = 13). Later in June–July, this kind of vigilance disappears almost completely; however it is possible to observe separate cases of vigilance, though very rarely (Blank et al. 2015b). There is a direct correlation between the intensity of acts of vigilance and the survival rate of fawns. A female who makes more frequent and longer acts of vigilance is more likely to see an approaching predator from a greater distance and transmit a danger signal to her fawn, and the sooner the fawn falls to the ground and freezes before the predator approaches, the more likely it will avoid detection and survive (FitzGibbon 1990). Older more experienced females are usually more alert and vigilant, detecting predators more quickly and protecting them, if necessary, resulting in fewer losses

Table 10.8 Changing proportions of vigilance duration

Behaviors/months	Late-May	Early-June	Late-June	Early-July
Changing proportions of vigilance duration (min) in a female, when she stays alone vs when she stays with her fawn				
General mother vigilance when fawn hides	381.4 94.9	46.0 89.5	15.5 90.6	2.8 82.4
General mother vigilance with her fawn	20.3 5.1	5.4 10.5	1,6 9.4	0.6 17.6
Total	401.7 100	51.4 100	17.1 100	3.4 100
Changing proportions of vigilance duration (min) of female during Reunion, separation and when her fawn is hiding				
Reunion process	240.0 62.3	29.9 67.4	50 84.7	1.9 57.6
Vigilance during fawn's hiding	122.1 31.7	12.3 27.8	5 8.5	1.1 33.3
Separation process	23.2 6.0	2.15 4.8	4 6.8	0.3 9.1
Total	385.3 100	44.4 100	59.0 100	3.3 100

Total min spent vigilant by all observed females during the given time period) over May–July, when a female stays alone vs when she stays with her fawn, as well as her vigilance duration (min) during reunion and separation process and during fawn's hiding (upper number = observed vigilance duration (min), lower number = proportion, %. (Blank 2017—License Number—5476940463790)

compared to young mothers (Ozoga and Verme 1982; Ozoga et al. 1982; Byers 1997; Kjellander 2000; Andersen et al. 2000).

Mothers demonstrate much more intensive acts of vigilance when they are at a distance and separated from their fawns; they are less vigilant when they walk in the company of their fawns during their active phases (Table 10.8). At first glance, this contradicts logic, since the female should be more vigilant when she is with her fawn, especially if it is still small and helpless, than when she is alone, and her fawn is in its safe hiding place (FitzGibbon and Lazarus 1995; Hunter and Skinner 1998; Costelloe and Rubenstein 2015). However, mothers behave differently. While she moves with her fawn, she occasionally freezes for a few seconds, but usually does not scan the surrounding areas, even though in practice, she can simultaneously nurse her fawn and scan the environment. It is likely that it is so important for the mother to not attract the attention of predators that she reduces their active time together to a minimum, devoting all her time to feeding milk with no time for other activities, including vigilance, particularly within first weeks after birth. Only after the mother breaks up with her fawn does she become very restless again, demonstrating frequent and long acts of vigilance (Blank et al. 2015b). As the fawn grows, the mother devotes less and less time to vigilance until the end of the hiding period, when she spends almost the whole day with her fawn and the frequency and duration of vigilant acts become negligible, despite the fact that during this period the fawns do not hide, openly walk with or without their mothers and become clearly visible from far away. Clearly at this age (about 2 months), fawns are able to take care of their own safety and escape from a predator. Thomson's gazelle females also spend less time for vigilance as their fawns grow (Costelloe and Rubenstein 2015).

In addition to general vigilance, females spend a long time inspecting their fawn's hiding area from a distance, most often in May (0.60 cases/h), with a decreasing rate thereafter (0.45 cases/h and 0.24 cases/h in early- and late-June, respectively) to a negligible frequency in July (0.05 cases/h, $n = 8$). In comparing the duration of vigilant acts, the longest acts are observed during the reunion of the mother and her fawn, vigilance duration is half as long when she is looking after the fawn's hiding place and the shortest acts of vigilance are noted during the separation of the mother and the fawn after sucking (Table 10.8). Cases when a mother passes near the hiding fawn to check if everything is in order are also rarely seen and only in May and early-June (0.01–0.02 cases/h, respectively) (Blank 2017).

Frequent and prolonged inspections of the hiding place are common behavior for many hiding ungulate species. Among them are Thomson's gazelles (*Eudorcas thomsoni*—FitzGibbon 1990; Costelloe and Rubenstein 2015), Grant's gazelles (*Nanger granti*—Walther 1965), and saiga antelopes (*Saiga tatarica*—Kang 2004), as well as a number of species of deer, such as white-tailed deer (*Odocoileus virginianus*—Ozoga et al. 1982), and moose (*Alces alces*—White and Berger 2001).

When nursing time comes, the female goes to her hiding fawn for their reunion. However, she never immediately approaches the hiding place, and instead repeatedly stops to carefully inspect the surroundings. Most often, the mother makes 4–5 stops for vigilance, noticeably less often 10–15 stops, more stops for acts of vigilance are exceptionally rare (observed up to 67 stops). This prolonged approach to the fawn usually takes from 2 to 10 min, but in some cases, her approach can take much longer (up to 3 h) with frequent and prolonged inspections of the surrounding area. The longest approaches are observed in the first 2 weeks in May (from 2 to 178 min, 21.4 ± 7.6 min, $n = 27$ on average); in June, approaches are much shorter (from 1 min to 59 min, 12.7 ± 3.9 min, $n = 15$ on average), and in July extended approaches are rare (noted only once for 8 min). During a very slow approach, the mother may lie down for quite a long time (from 3 to 45 min) and continue to carefully inspect the surroundings for danger while in a prone position. Then she gets up and continues her slow approach with numerous stops until she comes close to her hidden fawn to nurse. In rare cases (observed only four cases), the mother will advance close to her fawn then suddenly run away to begin her slow approach again with many stops. These long-lasting approaches with numerous stops for vigilance are most often observed in the first 2 weeks after giving birth in May (0.08 cases/h), then throughout June, these cautious approaches gradually declining (up to 0.03 cases/h) until July when they are no longer observed. The long approaching behavior with many stops and acts of vigilance acts is also found in pronghorn (Autenreith and Fitcher 1975) and gerenuk (*Litocranius walleri*—Reif and Klingel 1991).

In case of danger, fawns run away in different directions since the herd instinct at this age is still not strong. Even in the first days of following their mothers, frightened fawns often run away in the opposite directions from their mothers. Meeting back up with the mother occurs at their place of separation, or during the hiding period, at the place of the last feeding, as repeatedly seen when observing tamed gazelles. The ability to memorize the place of the last feeding (or hiding site) is established in gazelles from 3 to 4 days of age (Zhevnerov 1984). Caught when feeding, the fawns, having run away after their mothers for several tens of meters, then hide. The female begins to circle at a distance, moves away, then approaches the human, who in this case was the threat (Zhevnerov 1984). If the newborn fawns are in danger and the mothers are close, the mothers often try to divert the predator or ward off the threat. Several examples follow. The mother was at a distance of 50 m and ran up to the human being near her young, then moved away, walking around him at a distance of 200–300 m, then re-approached and moved away again; sometimes the female ran toward the cry of her fawn, when it was picked up in the person's hands (Zhevnerov et al. 1983). Once or twice Sokolov (1959) was able to observe females who had fawns somewhere nearby. They noticed the man and let him get to within 200–250 m, but when he tried to get closer, they slowly left, all the time watching him and moving in a circle, somewhere inside which there were newborns. In the central part of the Betpak-Dala Desert, the female took people away from the little fawns on May 20 and 25, 1954. For example, on May 25, while walking along the bottom of a small valley overgrown with various grasses and rare bushes, Sludskiy (1956) startled a female lying near the fawns. Having risen, she went straight at him and came up to 50 m, then jumped aside and started walking around people at a distance of 200–300 m, then approaching and then moving away again (Sludskiy 1956). Kostin (1955) observed a similar behavior in female gazelles on the Ustyurt Plateau in Western Kazakhstan. A female sometimes resorts to making the sound of a fawn's cry, if a man takes her fawn in his hands, but does not come close. Seeing a person, the mother characteristically begins to jump up and down and tries to distract the threat; if necessary, she will risk her life for the sake of her fawn. On May 25, 1959, a fawn was caught running at night under the headlights. The fawn struggled and bleated when caught and held in the man's hands, and his mother first ran away into the darkness, then ran back into the light again at a distance of 20 m from the car (Gorelov 1972).

Female gazelles, protecting their fawns from attack by large birds of prey, ravens or foxes, rise on their hind legs and beat the predator with their front hooves and butts at them with their heads (Habibi et al. 1993; Blank 1997). It is interesting to note that in cases when small planes fly low over the heads of gazelles, they sometimes do not run but instead jump up and try to hit the plane with their heads or horns, suggesting that this behavior is the same instinct as protecting young from the attack of birds of prey (Sludskiy 1956). Even with a mother's protection, foxes can still be successful in hunting hiding young. After checking of eight fox burrows in May of 1962 in Badkhyz (Turkmenistan), fawns' remains were found around two burrows and fresh remains were found near a burrow in May of 1965 (Gorelov 1972).

For one adult female who gave birth in the enclosure, the protection reaction for her newborn was first noted 6 days after birth and only with a direct threat to her fawn. In another young mother, aggression to protect her newborns appeared immediately after birth and was observed very often. At the same time, this young mother was excited by the general restlessness of other gazelles in the enclosure (with rapid and abrupt movements of a person, the entry of strangers), and not by any actual threat to her newborns. When calmly catching babies for weighing by a constant observer, the female did not show aggression, even from a distance of 5–6 m (Pereladova and Pereladov 1986).

In captivity, the level of excitement of a mother postpartum is significantly reduced and she does not show alertness even when a familiar person approaches her fawns. When a man tries to take her newborns in his arms, aggressiveness of the female is noted but not in all cases. In some females, aggressiveness is noted for the first time only after their third birth and selectively toward human individuals, and can be manifested to protect both her own and another female's fawns. In her aggression, the female, with a wheezing growl, repeatedly rushes the person, opening her mouth and slightly sticking out her tongue, trying to push the person away from the newborns. The reaction is stimulated by the active protests of the newborns through screams and active attempts to escape, as a result of being approached or sometimes simply by a rapid, careless movement by a person around the enclosure near the fawns' beds. It is interesting to note that the capture of the fawns, themselves, is not accompanied by stress indicators in the fawns (stable rhythm of breathing and heartbeat) as they are only moved a short distance and then returned to their bed sites; a procedure the fawns constantly experience (Pereladova and Pereladov 1986).

When the fawns are growing up and starting to run well, a mother, noticing the danger approaching their direction, runs to where they are lying and takes the young ones with her, encouraging them by her example and vocalization. If the fawns cannot be saved by running, the mother will sometimes rush straight at the enemy herself; females have been seen to do this during the entire period they nurse their young (Sludskiy 1956). Kostin (1955) once chased a family of gazelles (a female and two fawns) with a car. When the fawns got tired and stopped after 20 min, despite all the attempts of the mother to encourage them, she suddenly turned around and with a short cry rushed to butt the car (Blank et al. 2015b).

From all the above, it can be concluded that mothers control most behaviors and movements of their fawns at least in the first month after birth. During the fawns' second month, their mobility increases in the absence of their mothers, when the fawns show more and more independent activities and longer independent movements and the mother's control noticeably weakens, until the end of the hiding period, when mothers and fawns are in constant contact. Goitered gazelle antipredator behavior is quite similar to other hiding species, including pronghorns, white and black-tailed deer, and fallow and roe deer. This suggests that there are signs of parallel evolution in antipredator behaviors between bovids and cervids, despite the fact that these species have diverse evolutionary history and live in different environments on different continents. Despite the widespread use of the same elements of the antipredator behavior, it is not always effective, and when the density of the predator population is high, this standard behavior does not protect against high losses of fawns from red fox attacks, as shown by the example of interactions between red fox and roe deer in Sweden (Jarnemo et al. 2004).

10.2 Mother-Young Recognition During the Hiding Period

10.2.1 General Remarks

In ungulates, feeding offspring with mother's milk is an energetically high-cost mechanism for raising young. Milk has everything necessary for the survival and fast growth of young: excellent nutritional value with large amounts of proteins, fats and carbohydrates, as well as all required microelements. The production of milk takes a large amount of energy from the mother's body, and when nursing, females experience a great energetical drain. Therefore, it is no coincidence that the birth of ungulate young occurs in the spring season that produces the maximum amounts of high-quality forage. With the start of warm weather and heavy rains, abundant green forage with the highest protein and least fiber content, appears everywhere. However, favorable weather conditions with an abundance of precipitation do not occur every year and it is not uncommon in spring, when high-quality food appears in insufficient quantities. Females who have given birth experience a significant lack of forage for milk production and rearing offspring, though mothers still try to feed them. Under these conditions, distinguishing their own young from others is a vitally important skill of a mother to put all her energetic bodily resources into nursing only her their own offspring, because milk quantity has a grateful impact on the growth and ensure its survival (Torriani et al. 2006).

In general, ungulate mothers establish mutual bonds with their young usually very quickly, within a few hours after giving birth. After that, females easily recognize and find their own young, which often stay among their peers. Such maternal behavior is typical for most ungulate species (Poindron et al. 1993). Although allosuckling is found in some species, mothers of most ungulate species supply milk only to their own young and reject all others (Torriani et al. 2006). To establish selective bonds with her offspring, the female leaves her group to give birth alone. Separating females from the herd is a very important and necessary part of their birthing behavior, since this isolation is necessary to establish selective bonds between the mother and her young during the critical imprinting period immediately after parturition. If the mother stays in her group during birthing, there is a great danger that the female will establish contact with someone else's young and begin to lick and feed another female's fawn (Keyserlingk and Weary 2007).

Mothers can recognize their young among others in three main ways: odor, sound, and visual cues. The most important olfactory contacts between mother and offspring are obviously established first (Searby and Jouventin 2003). Soon after birth, amniotic fluids covering the fawn attract the mother to her newborn, promoting intensive licking, which, in turn, acquaints the mother with the individual olfactory characteristics of her young (Nowak et al. 2007). Within 5–10 min of licking, the mother establishes a selective bond with her young through these odor cues. Generally, this procedure lasts for quite a long time, up to 20–30 min with short breaks, until the newborn's coat becomes almost dry. In most cases, females show stable selectivity for their young after 0.5–2 h postpartum (Keller et al. 2003). Gubernick (1981) suggests that ungulate mothers "label" their young when licking, while other authors (Romeyer et al. 1994; Poindron et al. 2007a) believe, as do I, that females just learn their offspring's features based on olfactory cues. The mother-young recognition process also continues during each suckling session, when the mother licks and sniffs her young, especially its anogenital area.

Bonds based on sounds and visual cues between mother and young appear in ungulates likely during the first 6–12 h after birth (Keller et al. 2003), and in some cases even earlier (2–4 h postpartum) (Poindron et al. 2007a). In this short period of time, mothers establish strong selective connections with their offspring without resorting to the help of olfactory cues. In addition, acoustic signals and visual cues have a certain advantage over odor, since the former can be used both at a long and short distance, while odor is available only when close (Torriani et al. 2006).

It is widely known that ungulates have two strategies for raising offspring: followers and hiders (Lent 1974). In most ungulate species, the young are well developed at birth, and they have the ability to follow the mother in the first hours postpartum (Nowak et al. 2007). The young of follower species constantly follow their mothers everywhere almost from the first hour after birth. In this case, the mother and her young have a constant opportunity to learn each other's calls, so she can remember the individual characteristics of her young's voice and appearance in order to easily distinguish her fawn's voice and visual cues from other offspring. This feature is very important for this group of ungulates, since the young stay constantly near not only their mother, but also often come into contact with other mothers and young within a herd. Without a good knowledge of the voice and appearance of her young, every mother would be doomed to constant mistakes when feeding, as generally each fawn will try to suckle from any female. Such intensive and constant use of call characteristics for recognition young has been investigated in a number of ungulate follower species, such as reindeer (*Rangifer tarandus*—Espmark 1971, 1974), bison (*Bison bonasus*—Daleszszyk 2004), and domestic sheep (*Ovis aries*—Sebe et al. 2010). Some authors assert that many ungulates use mostly acoustical signals for recognition and do not need olfactory cues at all. Domestic goats (*Capra hircus*), for example, begin to accurately distinguish their kids by acoustic signals and visual cues during the first 48 h postpartum (Terrazas et al. 2003), and domestic sheep (*Ovis aries*) within 6–24 h after birth (Keller et al. 2003). In fact, though, mothers of even these follower species use olfactory cues, as well, for confirmation of their young's identity before each suckling bout (Poindron et al. 2007a).

This situation is noticeably more complicated in the case of hider species, since the young of these species stay at a distance from their mothers for a relatively long time postpartum, especially in the first 2–3 weeks. Females approach their young from time to time only for feeding, and then they leave their offspring alone, hidden again. In this case, mothers use mainly olfactory signals when reuniting and nursing after each separation, and acoustic signals would not be helpful in this case. Moreover, contact oral signals can attract predators, which can easily detect where the fawn is hiding and still too young to escape (FitzGibbon 1990). Therefore, acoustic signals are used noticeably later, when the young are older and have the ability to run away from a predator. In addition, as they age, young start to move around widely in the absence of their mothers, spending time with other peers, when oral vocalizations for recognizing their own young becomes more practical. Whereas in the first weeks postpartum, vocal signals play almost no role in the mutual recognition of mother and young, if at all (Poindron et al. 2007a).

This scenario is more typical for species of open habitats. As for the species living in closed habitats, searching for the hidden young in forested areas without the sound signals would be very difficult. Therefore, in most deer species living in closed biotopes, mothers use mainly sound signals for searching, as well as for recognizing and rejoining with their fawns. For example, in the Siberian wapiti (*Cervus elaphus sibiricus*), both mothers and fawns have similar calls with individual characteristics suggesting that they find and recognize each other mostly by acoustical cues (Sibiryakova et al. 2018).

In cervids, there are examples of a one-way use of acoustic signals. Females of fallow deer (*Dama dama*—Torriani 2006) and Iberian red deer (*Cervus elaphus hispanicus*—Sibiryakova et al. 2015) cannot distinguish the calls of their fawns from the calls of others, whereas fawns know their own mothers' voice and respond only to it. A study of acoustical behavior has shown that in these species of deer, mothers and fawns have different contact calls, and the calls of females are characterized by high individuality, unlike the fawns' vocalizations. Therefore, fawns distinguish their mothers from other females by their voice, whereas females cannot distinguish their own fawns by sound alone.

10.2.2 Mother-Fawn Recognition During the Hiding Period in Goitered Gazelle

Despite their reputation as a silent animal, goitered gazelles make quiet sounds regularly and produce loud sounds from time to time (Volodin et al. 2011, 2017). Males roar loudly and very intensely only during the autumn rut, when the roar of males is heard almost constantly at a distance of several hundred meters. Females begin to make louder oral calls more often after birthing offspring, especially in the first few weeks postpartum. All vocalizations of a female are aimed at contacts with her fawn (Zhevnerov 1984; Kingswood and Blank 1996). Mothers produce constant quiet closed-mouth nasal contact grunting calls while grazing with their fawns, which also makes similar sounds in exchange, as if echoing their mothers. Mothers also produce nasal sounds in the case searching of their fawns, especially when fawns are approaching to them (Zhevnerov 1984; Pereladova and Pereladov 1986; Marmazinskaya 2008). When a female cannot find her fawn in its hiding place, she begins to walk back and forth or around the immediate area and producing low, hoarse bellowing mooing-like calls (distress calls) looking for her lost fawn. These sounds are noticeably louder than in any other female vocalizations and can be heard from several tens of meters away. When protecting a fawn from a predator (or a human), the mother makes hoarse growling sounds (Zhevnerov 1984; Pereladova and Pereladov 1986; Marmazinskaya 2008). Fawns make sounds similar to low bleating when they run up to their mothers (Blank 1985). When the mother does not come to suckle the fawn soon enough, then it starts to produce discomfort calls and walk randomly around the area. The fawn makes loud low bleating from hunger or low mooing sounds (milk begging calls) and walks away from its rest site for several tens of meters, periodically returning to the place where it last parted with its mother (Zhevnerov 1984; Pereladova and Pereladov 1986). Fawns produce grumbling noises with a closed mouth (nasal sounds) when in contact with other fawns and snorting when in danger (also nasal sounds) (Marmazinskaya 2000). Nasal sounds are lower in frequency and intensity, and a different distribution of energy in the sound spectrum compared to oral sounds (Efremova et al. 2011a).

Similar to other gazelle species, goitered gazelle females and their fawns have the ability to make both relatively quiet nasal sounds, as well as louder oral calls. Nasal sounds are quieter, as vocal energy is intensely absorbed by the nasal epithelium, whereas the energy of oral sounds is absorbed much less by passing through the oral cavity, making oral sounds louder (Fitch 2000), but for most of a year, adult goitered gazelles prefer to use quiet nasal sounds to avoid attracting the attention of predators, although there have been some observations (Zhevnerov 1984: Lapshina et al. 2012) that a female will produce intensive oral calls for a short time immediately after giving birth, but then switches to soft nasal contact sounds for communication with her fawn. Only in unusually stressful situations will adult females use loud oral sounds. Birthing and the first days and weeks of her offspring's life are stressful for mothers and can produce a high arousal factor that promotes oral calls. At the beginning of a fawn's hiding period, especially the first 2–3 weeks postpartum, females can use oral calls to get their hiding fawns to get up, but later oral calls disappear in mothers. Fawns also can vocalize, producing nasal and oral calls. Within 4 weeks of life, fawns have the ability to produce oral calls (in more than 50% of observed cases). Gradually the proportion of oral calls decreases and by the age of 6 months, young gazelles produce mostly soft nasal calls (Volodin et al. 2011). The same behavior is typical for adult females and young of saiga antelope (*Saiga tatarica*) and many species of deer (Efremova et al. 2011a, b).

Closer to birthing, the goitered gazelle female leaves her group, occupies a certain parcel of area with sparse and low vegetation. During this period, she leads a solitary lifestyle, aggressively driving away all other females from her site, including her daughter, born the previous year, who follows her mother everywhere until the next birth. During the prenatal period, the pregnant female does not tolerate anyone's presence nearby, isolating herself and her neonate from conspecifics. Such a craving for loneliness in females happens only in the prenatal period, whereas in the rest of the year, females are almost always in a group with their offspring and conspecifics, and it is almost impossible to see a single female outside the group during the whole rest of the year (Soldatova 1983; Zhevnerov 1984). Isolation behavior of females during the pre- and postnatal period is interpreted as creating favorable conditions for establishing a selective mother-young bonds, as well as an anti-predator protection behavior for the young (Lent 1974; Ciuti et al. 2006).

Table 10.9 Changing proportions of mothers nursing with and without sniffing fawns and preference in sniffing different parts of the fawn's body

Behaviors/months	May	Early-June	Late-June	Early-July
Sniffing before nursing	602 97.7	242 96.0	174 87.0	111 92.5
Nursing without sniffing	14 2.3	10 4.0	26 13.0	9 7.5
Total	616 100	252 100	200 100	120 100
Sniffing fawn's anogenital parts	529 87.9	218 90.1	140 80.4	70 63.1
Sniffing fawn's muzzle	31 5.1	23 9.5	13 7.5	3 2.7
Sniffing fawn's back	42 7.0	1 0.4	21 12.1	38 34.2
Total	60 100	242 100	174 100	111 100
Licking during nursing	361 17.2	160 12.8	19 4.7	10 0.6
Nursing without licking	1735 82.8	1092 87.2	391 95.3	153 99.4
Total	2096 100	1252 100	410 100	163 100

(Upper number = observed cases; lower number = proportion, %) during May–July (by Blank and Yang 2017—License Number—5476931371190)

My observations of the goitered gazelle wild population in Kazakhstan (Blank 1985) and semi-captive group in Kyrgyzstan (Blank 2023) show that the female almost always sniffs the fawn's anogenital area before letting it suckle, particularly in the first weeks of the fawn's life. Only in rare cases does the mother not sniff the fawn before allowing it to suckle. By the second and third months of a fawn's nursing from its mother, the percentage of sniffing decreases, but not so significantly and the cases of suckling without prior acquaintance with the fawn's smell is noticeably less than 10% (Table 10.9). All this suggests that olfactory cues play an exceptional role for mothers in recognizing their fawns. The female also uses her vision, but only to remember the place where she left her hiding fawn. To do this, after leaving the fawn, she stops several times and looks attentively back at the hiding fawn (Blank and Yang 2017).

Why does this work for goitered gazelles? The fact is that these gazelles avoid gathering into large concentrations during and after birthing and mothers do not keep all their offspring in one place, but instead disperse their young throughout a vast birthing area. Each female has her own plot, where other females with fawns are not allowed. Therefore, mothers not only know the location of their area by local landmarks, but they also memorize their fawn's hiding site precisely. Monitoring the hiding area, the mother periodically watches toward her hiding fawn from a distance and usually goes directly to the hiding place where her fawn is lying to reunite for nursing. After each suckling bout, the mother leads the fawn to another place of hiding, but still located inside her small, defined area (Blank and Yang 2017). A similar situation is typical for other hiding species (Lent 1974; Huegel 1985), where mothers remember the location of their young's hiding site and monitor it from a distance. This phenomenon has been found for Thomson's gazelles (*Eudorcas thomsoni*—FitzGibbon 1990), pronghorns (Byers and Byers 1983), and mule deer (*Odocoileus hemionus*—Lingle et al. 2007).

In natural conditions (Ili Hollow, Kazakhstan), when a mother approaches her hiding fawn, she produces soft, nasal contact calls, prompting her fawn to leap up and immediately start suckling. In semi-captive conditions (Issyk-Kul Breeding Center, Kyrgyzstan), mothers produce contact calls during their entire way to the hiding place (Blank 2024), but before feeding her young and while nursing, the female sniffs and licks the fawn's anogenital area. These observations give the complete impression that the fawn recognizes the voice of its mother and therefore jumps up and runs to suckle (Blank 1985). Moreover, without this call, the fawn remains on its bed, especially in the first days after birth, even if the mother passes silently and very close (2–3 m) to her fawn (Zhevnerov 1984; Pereladova and Pereladov 1986). But in fact, this is not the case. First, a mother does not always call her fawn (25% from all observed cases of approaching), and more often just approaches closely, and he just gets up and tries to suckle. Thus, the vocal cues of the female calls are not necessary for a fawn to rejoin its mother. Moreover, the fawn will run to any conspecifics (other adult females, subadult females and even males) approaching too close to its hiding place and tries to suckle (Blank 1997). In semi-captive conditions, I watched as a fawn suckled on his mother, then ran over to another female and followed alien female, periodically trying to suckle on her. After that, he ran over to the third female, trying to suckle on it too. In addition, four fawns several times tried to suckle from male-yearling, which pushed them with his horns (Blank 2024).

Generally, cases when a fawn mistakenly runs to someone else's mother are relatively rare in May and early June. The fact is that in this first month of life, fawns are very attached to their hiding places and very rarely change locations independently, if they do it even, they return back to the point where they last separated from their mothers. Mothers know the hiding places perfectly well and always return to it precisely to feed their fawns. That's why there are almost no mistakes as the mother always finds her fawn during the first weeks of the hiding period. Starting in mid-June, the fawns begin to walk widely in the absence of their mothers, moving away from their bed sites by 1.5–2 km—changing their bed sites, grazing independently, and following other fawns (Soldatova 1983; Zhevnerov 1984). Therefore, at this point in the development stage of their fawns, mothers only know the approximate location of their fawns and return to the site of their last separation; if mothers cannot find their fawns, they walk around the area looking, emitting a calling roar to try to call them back. The fawn also periodically returns to the place of its last separation and runs to any female (or conspecific) that happens to be at that place. After sniffing the anogenital parts of the fawn, the female feeds it if it is her own or walks or runs away, or even butts it, if it is unrelated. Therefore, in the second half of June

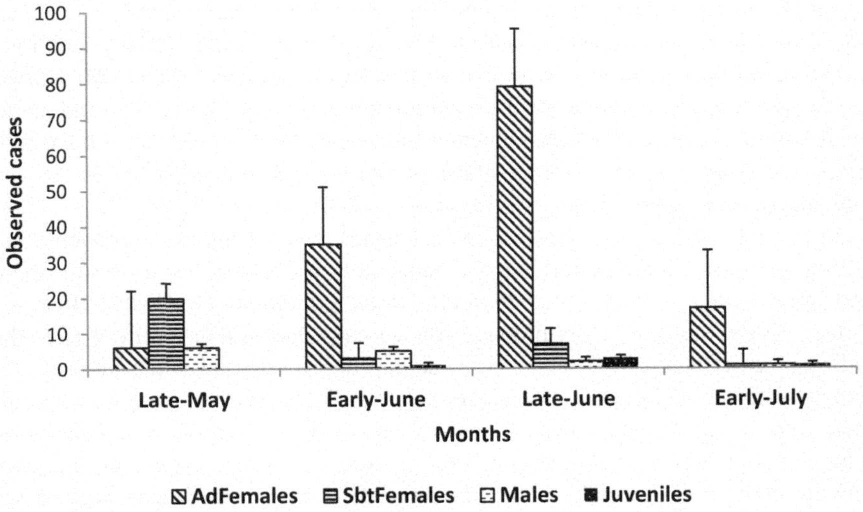

Fig. 10.2 Observed cases of fawns' mistaken attempts to suckle from stranger adult and sub-adult females, males and juveniles during May–July (470 h observations in Late-May, 188 h in Early-June, 186 h in Late-June, and 173 h in Early-July) (Blank and Yang 2017—License Number—5476931371190)

and into July, more mistakes occur in matching fawns with their mothers and quite often (in 11–18% of cases) the fawn runs to someone else's mother. All these observations convincingly show that fawns individually do not distinguish their mother's voice and appearance from other females, as they react to anyone (Blank and Yang 2017). In contrast, Zhevnerov (1984), watching how a mother and her fawn react to each other's voices, is sure that both mother and fawn distinguish voice of each other individually. However, this is more an assumption than a proven fact and females will respond to the plaintive sounds of any fawn, and a fawn will respond to the sounds of any female, and even a crude imitation of these sounds by a human (Walther et al. 1983).

A fawn so unexpectedly runs up to any gazelle and actively and even aggressively tries to suckle, which greatly scares females-yearlings and even adult males. In the beginning, fawns try to suckle mainly from female-yearlings (Fig. 10.2), as they follow in the footsteps of their mothers, despite the fact that the mothers drive them away when new offspring appear the following year. Young females are particularly curious about newborns, so they often approach the fawn in the absence of their mother to sniff it, but it immediately rushes to suckle, which scares off the young female. After this experience, more and more young females try not to come close to the lying fawns, quickly reducing the number of such erroneous attempts to suckle in young females during the first weeks. Adult males usually show no interest in the new generation, but sometimes they accidentally stumble upon a lying fawn, which rushes to suckle. Such cases also teach males not to get too close to lying fawns.

Table 10.10 Observed cases of a fawn running to its own mother compared to mistakenly running to an unrelated gazelle

Behaviors	May	Early-June	Late-June	Early-July
Observed cases of a fawn running to its own mother	2095 98.5	1252 96.0	410 81.3	178 89.0
Cases of a fawn running to an alien gazelle	32 1.5	47 3.6	94 18.7	22 11.0
Total	2127 100	1299 100	504 100	200 100

(Upper number = number of observed cases; lower number = proportion, %) during May–July (Blank and Yang 2017—License Number—5476931371190)

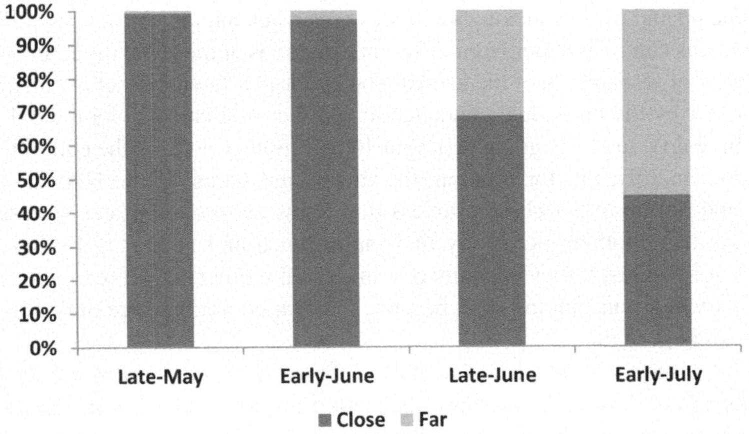

Fig. 10.3 Changing of proportions (%) between different behaviors of goitered gazelle mothers and their fawns over May–July: cases of approaching to fawn's hiding place for close distances (Close) versus cases of mutual mother-young recognition from far distances (Far) (Blank and Yang 2017—License Number—5476931371190)

The situation with adult females is different. Fawns mistakenly run up to them quite rarely in May (Fig. 10.2). This is due to the fact that each female stays on her own plot not far from her fawn and does not enter neighboring plots of other females. Already in June, all females begin to walk more widely and increasingly come across other females' fawns, which rush to suckle them as soon as she is nearby. Therefore, such cases grow in June with an increase the mobility of mothers and fawns in the absence of their own mothers (Table 10.10). This continues until the beginning of July, when the fawns (at the age of 1.5–2 months) improve their recognizing abilities and begin to distinguish their mothers from other adult females in a distance (Fig. 10.3), but rather by visual appearance and behavioral response to a fawn approaching and likely by smell in close distance. Mothers produce oral calls mostly during first 2–3 weeks postpartum (in May) to nurse their fawn, but later

(June–July) they almost do not use oral vocalization, preferring the soft nasal calls (Efremova et al. 2011a, b). So, the audible signals unlikely play a significant role in fawn-mother recognition even in the beginning, since voice signals are produced by females not every case and a fawn does not distinguish mother's calls. When mother-fawn recognizing improves significantly and fawn runs well and does not lag behind the adult gazelles, the mothers take their fawns away from the area of individual plots doing more long-lasting movements with fawns together. At this time, the hiding period of fawns ends and they begin to follow their mothers constantly as during days, as during nights (Blank 1997).

Thus, during first 2–3 weeks of their life, fawns do not distinguish its mother from other gazelles and tends to suckle on any conspecifics that comes up to it, regardless of its age and gender. Herewith, the gazelle only needs to approach the lying fawn even in complete silence, and this is enough for him to run to the one who came up and tries to suckle, but in the cases when a female produces calls, then several fawns can rush to her from different directions at the same time, refuting the assumption of researchers of the acoustic behavior that fawns somehow distinguish their mother by the individual characteristics of her oral calls (Volodin et al. 2011, 2014). In reality, fawns begin to distinguish their mother only by the end of the hiding period, in July, but for reunion, the fawns first focus on the behavior of the female and run up to her closely, unless they show aggression. Later, they already recognize their mothers purely by their appearance and obviously by her smell, whereas vocal signs, if they play any role in mutual mother-young recognition, then a secondary and non-binding role, because females do not produce oral calls at the end of hiding period.

If do the mothers recognize their fawns? Even here the females hardly distinguish their fawns by voice, because they often do not produce oral sounds when meeting her or other gazelles. More than that, in the initial postpartum period, goitered gazelle females respond and come to the calls of any fawn, even to a very imperfect imitation of the fawn's cries by human suggesting that females do not know individual characteristics of their fawns' calls (Soldatova 1983; Zhevnerov 1984). Indeed, fawns do not produce oral calls for reunion with their mothers and they usually do not open their mouth. Instead, mothers come very closely to hiding fawn for reunion and sometimes even touches fawn by their muzzle without resorting to help of calls, because closeness and touching are quite enough for starting of nursing. I have observed the vocalization of fawns only three times during the entire my observation period (6 years). In all cases, it was discomfort calls related to fawn's hungry.

However, mothers have ability to memorize the hiding place of their fawns according local landscape, and for the first few weeks of the hiding period, the females frequently (twice an hour) look at this place, monitoring it from afar (Blank et al. 2015b). If they need to nurse the hiding fawns, females can directly approach this place. They make a call from a close distance and the fawns immediately run to suckle, although they can run up to the females without any calls or run to unrelated female.

Before suckling, the mothers check the smell of fawns' anogenital part and only after this allow to suckle. Obviously, sniffing the fawn's anogenital parts so important for the female to recognize her fawn that the first month (until mid-June) she does this without fail before almost every nursing (Table 10.9). To identify a fawn, sniffing its anogenital area is most important, whereas sniffing the muzzle and back plays an insignificant role (less than 10% of cases) for the first 1.5 months (until end-June), and only at the end of the hiding period (In July), the smell of the whole body (back) begins to play a more significant role (up to a third of the observed cases) (Table 10.9). Next month (end-June-early-July), the mother sometimes feeds without prior sniffing, but there are relatively few such cases (up to 13%), in the vast majority of cases (almost 87–90%) she still sniffs the fawn. At the same period, mothers are increasingly recognizing their fawns purely visually from an increasing distance, in the last weeks of the hiding period they demonstrate the ability to recognize their fawns from 1.5 to 2 hundred meters. Such recognition is most likely based on the appearance of the fawns and knowledge of their approximate location. Fawns can also already recognize their mother by a set of signs: outlook appearance, movement of the female towards the fawns, her behavior in searching for her offspring. Whether oral signals from the mother's side or from the fawns are involved in this, most likely not, because they both don't open their mouths when looking for and meeting each other (Blank and Yang 2017). Nasal sounds may be involved, but they have a low level of individualization (Volodin et al. 2011). Despite a noticeable improvement in the recognition of the mother and fawns by the end of the hiding period (mid-July), the female still sniffs the anal part before nursing in the vast majority of cases to check the correctness of her definition, made by other signs from a distance.

In addition to sniffing the fawns' hindquarters, mothers carefully lick their anogenital area. Such licking has physiological effect and stimulates urination and defecation in newborns, because in the beginning fawns have certain difficulties performing their excretory functions themselves (Levy et al. 2004). But besides that, while licking the anus and genitals, the mother also gets acquainted with the individual smell of the fawn, learning the olfactory cues; this is an important part of the mother-young recognition process (Romeyer et al. 1994; Poindron et al. 2007a). It is no coincidence that immediately after birth, the mother is engaged in licking the fawn's anogenital area very carefully for a quite long period of time (sometimes up to 20 min). Licking anogenital area is observed in most ungulate species, and females usually do it most intensively during 1.5 hours postpartum (Autenrieth and Fichter 1975; Keyserlingk and Weary 2007; Poindron et al. 2007a; Ungerfeld et al. 2008; Kokshunova 2012). Anogenital licking is such an important stage in the process of mother-young recognizing that females deprived of the opportunity to lick the anal area often refused to feed their young (Keyserlingk and Weary 2007; Poindron et al. 2007a).

In the first few weeks after birth, goitered gazelle mothers continue to lick the anus and genitals, although not every nursing, and licking is observed half as often as sniffing. The female licks the anogenital parts less and less often in June and already in the second month (after mid-June) such cases are rarely observed

(Table 10.9). So, it turns out that in an ordinary situation, the female recognizes her fawn only by olfactory cues, fawns produce calls so rarely that their vocal cues are simply absent. It follows from this that the individual distinction of a fawn with the help of contact calls is almost not used, because at the beginning of the hiding period the fawn lies on its hiding site most of the time (more than 90% of the time) and does not move in the absence of the mother. The female always finds her fawn in his place, and since other fawns do not lie nearby, it is impossible to confuse her own among other fawns due to their absence. Although the mother uses oral sounds to raise the fawn, it does not distinguish them individually and reacts to any sound and any approaching gazelle (Blank et al. 2015b).

Thus, there is a unique situation in the case of mother-fawn recognition process in goitered gazelles during the first month of hiding period, when neither the female knows the fawn by voice and outlook appearance (only by smell), nor the fawn knows the mother even by smell. Only the mother visually knows the location of the hiding place and fawn's smell. As far as I know, this is the first description in published literature that both mother and fawn do not recognize each other individually, either by voice or appearance. There are studies of a number of deer species (*Dama dama, Cervus elaphus, Ozotoceros bezoarticus*), where only the fawn recognizes its mother by voice, but the mother likely knows her fawn likely only by smell (Vannoni et al. 2005; Torriani et al. 2006; Ungerfeld et al. 2008). Only after 3 or 4 weeks grazing and moving together do deer females begin to individually recognize the voice of their fawn.

In the first weeks of life, offspring of other antelope species also are not able to distinguish the calls and appearance of their mothers from conspecifics, or even calls from humans. For example, a blesbok young (*Damaliscus pygargus philippsi*) begins to distinguish its mother by her calls at the age of 31 days, and before that it responds to any stranger and any call; and the female distinguishes her young only by smell (Walther 1969). A similar situation is observed in pronghorn (*Antilocapra americana*), where females recognize their fawns mainly by olfactory cues, while mothers and fawns do not know each other by appearance and calls during the first weeks postpartum; and only 3–4 weeks later do mothers and young show a mutual recognition among conspecifics (Autenrieth and Fichter 1975).

Both females and fawns of goitered gazelles emit quite loud oral sounds in stressful situations that can be heard at distances of up to 50–70 m. Oral sounds are produced in more emotional situations compared to nasal calls, and oral sounds have an increased air speed on exhalation. Volodin et al. (2017) divide this vocalization into two categories: stressful when the animal is caught by a predator, and uncomfortable when the fawn is hungry and looking for its mother. I think this division is quite conditional, since a hungry state is also a stressful situation for the fawn. In both cases, this vocalization is the result of intense emotional tension and excitement, which reveals individual characteristics of the caller's voice (Briefer 2012; Sibiryakova et al. 2018). For females of species that can protect their fawns from predators, this distress call is a signal that the young needs protection prompting the mother runs to protect her offspring from any small predator (fox, wildcat, jackal) (Lingle and Riede 2014; Lingle et al. 2012).

I only heard these loud oral calls from fawns three times during my 6 years of observations. In May, the loud bleating came from hungry fawns, when their mothers did not return early enough to feed them. In each of these cases, all the females who happened to be nearby ran to the screaming fawn, and the fawn rushed to suckle from them; though after sniffing the anal area, they all ran away from the alien fawns and watched them from a distance. These observations suggest that none of these goitered gazelle females were able to recognize their fawns by their discomfort calls from a distances, in spite of the fact that bioacoustics investigations show that oral calls encode a caller identity (Volodin et al. 2011) In the case when a mother can identify her fawn using topographic cues, then the highly individualistic calls are no longer required (Torriani et al. 2006) and fawn's discomfort calls likely have fewer identifying characteristics. Females recognize their young only with the help of olfactory cues from close distances. Females of mule deer (*Odocoileus hemionus*) also respond to distress calls from any fawn, and probably also do not distinguish the voice of their own fawn, although perhaps the females respond because they are ready to protect any fawn from predators (Lingle et al. 2007).

Females also can produce stressful calls, which are wheezing-mooing sounds when protecting fawns from predators (or humans) (Pereladova and Pereladov 1986; Marmazinskaya 2008). These vocalizations are relatively loud oral calls (a low bellowing roar) and observed in only rare situations, usually when mothers cannot find their fawns immediately and walk around in different directions, producing contact calls. This occurs such as when fawns are running away from danger following adult conspecifics, however, within about half an hour, the fawns usually return to the bed site and reunification with their mothers. The females return much later after that (after 1 h) and immediately go to the last hiding place of the fawn and usually she will find there without any vocalization. Therefore, cases of loss of fawns after scaring all gazelles by predators or humans, and the producing of loud oral sounds is very rarely observed. I observed two such calling females, but in both cases, I did not observe the presence of the offspring, as the mother usually disappeared from my field of vision. These observations demonstrate that even stressful vocalizations, despite their relatively high identity potential (Volodin et al. 2017), are unlikely to be used by females or fawns for mutual individual recognition.

There are other opinions, however, about the ability of goitered gazelle fawns to distinguish their mothers from non-related females. In the opinion of Zhevnerov (1984), a fawn recognizes its mother, seemingly by her voice (contrary to my study) and especially by her smell. In captivity, baby gazelles quickly get used to the person who regularly feeds them. The fawn will not suckle their bottle from a new face until the new person puts on the clothes of the familiar one. They are also afraid of their regular feeder, if the person wears new clothes (Zhevnerov et al. 1983).

In the first months of a goitered gazelle fawn's life, activation of the inguinal glands emit an oily, dirty yellow secretion with a sharp mouse-like smell that has been associated with the socialization between and fawn by being spread in the air during suckling, with the mother learning the individual smell of her fawn and its scent when afraid (Marmazinskaya 2000). Another idea for individual recognition of fawns is related to an increase in the functioning of the inguinal glands in females before giving birth, which, apparently, is of significant importance in the female's recognition of her offspring. When the udder fills with milk, the nipples approach the pockets of the inguinal glands, and during the act of suckling, the offspring comes into contact with them with its nose. Before feeding in the first 2–3 days, the female sniffs the end of the muzzle of the fawn that runs up to her possibly trying to recognize her own smell. With the cessation of nursing and the functioning of the mammary gland, secretions of the inguinal gland also stop, which indicates an interconnectedness between the secretory activity of the inguinal and milk glands (Zhevnerov 1984).

10.2.3 Mother-Young Recognition in Other Hiding Species

Saiga antelopes (*Saiga tatarica*) inhabit the same arid areas as the goitered gazelles, only saiga make almost long, constant migrations from north to south and back depending on the season (Fadeev and Sludskiy 1982). Unlike goitered gazelles, saiga mothers and young produce sounds before nursing when they reunite during the hiding period in obligatory order (Volodin et al. 2014). Both goitered gazelles and saiga antelopes are hider species and use the same antipredator strategy, only with different durations of hiding period: 2 months in goitered gazelles and 2–3 weeks in saiga antelope. However, these antelope species have a difference in the placement of their offspring. As previously described, goitered gazelle females disperse throughout the breeding grounds and each female has her own area where she hides her fawn, separating him from all others. Saiga females, in contrast, gather in groups for birthing and their newborns are also placed in groups (Fadeev and Sludskiy 1982). Female goitered gazelles remember the location where they hide their fawns and can always find them, then after sniffing the fawn's anal area knows it is really her own offspring. For this case, exchange audible signals between them are completely optional. It is another matter for saiga, where the young are kept in a group of several dozen, and sometimes hundreds of the young, and in order for a mother to find her own offspring, the exchange of calling signals become necessary; otherwise the mother would have to sniff each young one by one. Therefore, oral calls of saiga antelope young have a higher level of individual characteristics compared to goitered gazelle fawns (Volodin et al. 2014).

In goitered gazelles, mothers and fawns rarely produce loud oral calls and only happens in critical situations (discomfort or distress calls—Volodin et al. 2017). This is because the main strategy of goitered gazelles, like any other hiding species, is to not attract the attention of predators. The fawn should not only lie still and have no smell, but not make any sounds, especially in the first 2–3 weeks postpartum, because the young is absolutely helpless in this age for any predator. The female also rarely makes loud oral calls, tries to approach the lying fawn closely and only then she makes only soft calls, possibly using nasal sounds, and mothers make loud oral calls only when looking for lost offspring. Such cautious behavior of the mother is to minimize the danger of attracting predators to her fawn's hiding site.

In follower species, there is another antipredator strategy in which the young follow their mothers constantly from their first day of his life and stay under the protection of their mother and other adults in the herd. In follower species, mothers and young have visual and auditory recognition of each other from the first day, and they continuously produce contact calls between each other. Vocalizations in follower species are the most important channel of communication between mothers and young and oral calls support the survival of their offspring, though the importance of smell recognition does not lose its role (Nowak and Lindsay 1992; Sebe et al. 2010). A number of bovid and cervid species are followers, including bison (*Bison bonasus*—Daleszszyk 2004), African buffalo (*Syncerus caffer*—Ryan et al. 2007), domestic sheep (*Ovis aries*—Terrazas et al. 2003), and reindeer (*Rangifer tarandus*—Espmark 1974).

Laboratory studies of the recognition process in domestic ungulates have shown that the odor distinction of lambs and kids in females of sheep and goats appears within the first hour after birth, and although visual and sound memorization in females also appears during the first day, females still use smell checks to accurately discriminate their offspring (Keller et al. 2003; Poindron et al. 2007a). Anosmic goats, deprived of odor receptors, lose the ability to distinguish their own kids from others and willingly accept any alien young to nurse, despite the fact that these females could distinguish their own offspring from others by voice and appearance (Poindron et al. 2003). Odor discrimination has a big disadvantage compared to visual and vocal recognition, since it is possible only at a very close distance (less than 25 cm). At the same time, the ability to distinguish by smell allows females to distinguish their own from others with a high level of efficiency, even distinguishing between twins. Perhaps odor discrimination remains the main control testing factor, even if vocal and visual differentiation is already well developed (Poindron et al. 2007b).

Even though the saiga antelope is a hider species, saiga mothers and young often make loud oral contact calls, when necessary (Volodin et al. 2017). Saiga use a different antipredator strategy compared to goitered gazelles by gathering their young in one place, expecting that most of the young will remain in the group in case of a predator attack. This is known as the dilution effect of antipredator behavior in that the chances of any one fawn dying are lessened in a group of many (Pulliam 1973; Dehn 1990). Foxes will attack young, but only for the first 2–3 days postpartum, and several mothers can protect their young from these small predators. As for the wolf, it is a very territorial species, which drives all conspecifics away from the pack's territory. Thus, only one wolf family can attack the saiga young, which develop very quickly and after 5–7 days saiga young can escape from a wolf by running away. After another week or two, saiga young begin to follow their mothers and move with the whole herd as they continue their migration (Fadeev and Sludskiy 1982).

Thus, goitered gazelle females remember only the location of the fawn's hiding place and distinguish their own fawn from someone else's only by smell, whereas sound and visual signals do not play a significant role in the mutual mother-fawn recognition until the end of the hiding period, when the female and her fawn begin to distinguish each other visually and acoustically. This behavior is different from follower species, which do mother-young recognition mostly through acoustic and visual cues at once after birth, though olfactory checking is also actual. Females of saiga antelope being a hider species use mostly vocal cues for young recognizing during reunion with young that brings this species closer to behaviors of follower ungulates.

References

Aanes R, Linnell JD, Perzanowski K, Karlsen J, Odden J (1998) Roe deer as a prey. In: Andersen R, Duncan P, Linnell JDC (eds) The European roe deer: the biology of success. Scandinavian University Press, Oslo, pp 139–159

Altman M (1963) Naturalistic studies of maternal care in the moose and elk. In: Rheingold HL (ed) Maternal behavior in mammals. John Wiley and Sons, New York, pp 233–253

Andersen R, Gaillard JM, Linnell JDC, Duncan P (2000) Factors affecting maternal care in an income breeder, the European roe deer. J Anim Ecol 69:672–682

Andersson M, Wiklund CG, Rundgren H (1980) Parental defense of offspring: a model and an example. Anim Behav 28:536–542

Antipin VM (1941) Mammals of Kazakhstan. In: Ungulates, vol 3. Publication OGIZ, Alma-Ata, pp 47–54. (in Russian)

Autenreith RE, Fitchter E (1975) On the behavior and socialization of pronghorn fawn. Wildl Monogr 42:1–111

Barrett MW (1984) Movements, habitat use, and predation on pronghorn fawns in Alberta. J Wildl Manag 48(2):542–550

Berger J (1978) Maternal defensive behavior in bighorn sheep. J Mammal 59:620–621

Bishop CJ (2007) Effect of enhanced nutrition during winter on the uncompahgre plateau mule deer population. Ph.D. dissertation. Colorado State University, Fort Collins

Blank DA (1985) Peculiarities of social and reproductive behaviour of *Gazella subgutturosa* in the Ili Valley. Zool Zhurnal 64:1059–1070. (in Russian, English abstract)

Blank DA (1997) Social and reproductive behavior in the Persian gazelle. Ph.D. thesis. The University of Tel-Aviv, Tel-Aviv. 189 pp

Blank DA (2017) Antipredator tactics are largely maternally controlled in goitered gazelle, a hider ungulate. Behav Proc 136:28–35

Blank DA (2023) Birthing season of may 2023 in goitered gazelles in conditions of the Issyk-Kul breeding center. Unpublished report for the Institute of Biology. Kyrgyz Academy of Sciences, Bishkek

Blank DA (2024) Birthing season of may 2024 in goitered gazelles in conditions of the Issyk-Kul breeding center. Unpublished report for the Institute of Biology. Kyrgyz Academy of Sciences, Bishkek

Blank DA, Yang W (2017) Mother-young recognition in goitered gazelle during hiding period. Behav Proc 142:21–28

Blank DA, Ruckstuhl K, Yang W (2015a) Seasonal dynamics of agonistic displays in territorial and non-territorial males of goitered gazelle. Zoology 118(1):63–68

Blank DA, Ruckstuhl K, Yang W (2015b) Antipredator strategy of female goitered gazelles (*Gazella subgutturosa* Güld., 1780) with hiding fawn. Behav Proc 119:44–49

Bleich VC (1999) Mountain sheep and coyotes: patterns of predator evasion in a mountain ungulate. J Mammal 80(1):283–289

Bowyer RT, van Ballenberghe V, Kie JG (1998) Timing and synchrony of parturition in Alaskan moose: long-term versus proximal effects of climate. J Mammal 79:1332–1344

Briefer EF (2012) Vocal expression of emotions in mammals: mechanisms of production and evidence. J Zool 288(1):1–20

Byers JA (1997) American pronghorn: social adaptation and the ghost of predators past. University of Chicago Press, Chicago

Byers JA, Byers KZ (1983) Do pronghorn mothers reveal the location of their hidden fawns? Behav Ecol Sociobiol 13:147–156

Ciuti S, Bongi P, Vassale S, Apollonio M (2006) Influence of fawning on the spatial behaviour and habitat selection of female fallow deer (*Dama dama*) during late pregnancy and early lactation. J Zool 268:97–107

Clutton-Brock TH, Guinness FE (1975) Behaviour of red deer (*Cervus elaphus*) at calving time. Behaviour 55:287–300

Costelloe BR, Rubenstein DI (2015) Coping with transition: offspring risk and maternal behavioural changes at the end of the hiding phase. Anim Behav 109:217–225

Coté SD, Peracino A, Symard G (1997) Wolf, *Canis lupus*, predation and maternal defensive behavior in mountain goats, *Oreamnos americanus*. Can Field-Nat 111:389–392

Daleszszyk K (2004) Mother-calf relationships and maternal investment in European bison *Bison bonasus*. Acta Theriol 49(4):555–566

Dehn MM (1990) Vigilance for predators: detection Anim. Behav and delution effects. Behav Ecol Soc 26:337–342

Efremova KO, Volodin IA, Volodina EV, Frey R, Soldatova NV, Lapshina EN, Makarov IS, Gorbunov KS (2011a) Effect of sex and age on structural features of nasal calls and body size in fawns of the goitered gazelle, *Gazella subgutturosa* (Artiodactyla, Bovidae). Biol Bull 38(7):736–746

Efremova KO, Volodin IA, Volodina EV, Frey R, Lapshina EN, Soldatova NV (2011b) Developmental changes of nasal and oral calls in the goitred gazelle Gazella subgutturosa, a nonhuman mammal with a sexually dimorphic and descended larynx. Naturwissenschaften 98:919–931

Espmark Y (1971) Individual recognition by voice in reindeer mother-young relationship. Behaviour 40:295–301

Espmark Y (1974) Individual characteristics in the calls of reindeer calves. Behaviour 54:50–59

Estes RD (1976) The significance of breeding synchrony in the wildebeest. Afr J Ecol 14(2):135–152

Estes RD (1991) The behavior guide to African mammals. University of California Press, Berkeley-Los Angeles

Fadeev VA, Sludskiy AA (1982) The Saiga in Kazakhstan. Kazakh Academy of Sciences press, Alma-Ata

Fedosenko AK, Blank DA (2001) Capra sibirica. Mamm Species 675:1–13

Fedosenko AK, Blank DA (2005) Ovis ammon. Mamm Species 773:1–15

Fisher DO, Blomberg SP, Owens IPF (2002) Convergent maternal care strategies in ungulates and macropods. Evolution 56(1):167–176

Fitch WT (2000) The phonetic potential of nonhuman vocal tracts: comparative cineradiographic observations of vocalizing animals. Phonetica 57:205–218

FitzGibbon CD (1990) Anti-predator strategies of immature Thomson's gazelles: hiding and the prone response. Anim Behav 40:846–855

FitzGibbon CD (1993) Antipredator strategies of female Thomson's gazelles with hidden fawns. J Mamm 74(3):758–762

FitzGibbon CD, Lazarus J (1995) Antipredator behavior of Serengeti ungulates:individual differences and population consequences. In: Sinclair ARE, Arcese P (eds) Serengeti II: dynamics, management, and conservation of an ecosystem. The University of Chicago Press, Chicago, pp 274–296

Garner GW, Morrison JA (1980) Observations of interspecific behavior between predators and white-tailed deer in southwestern Oklahoma. J Mammal 61:126–130

Geist V (1974) On the relationship of ecology and behaviour in the evolution of ungulates: theoretical considerations. In: Geist V, Walther F (eds) The behaviour of ungulates and its relation to management. IUCN, Morges, pp 247–273

Gorelov UK (1972) Breeding of goitered gazelle (*Gazella subgutturosa* Güldenstaedt) and problem of restoring of its number in the Badkhyz (SE Turkmenistan). In: Theriology, vol 1. Nauka of Siberian Department of Academy of Sciences of USSR, Novosibirsk, pp 420–424. (in Russian)

Grignolio S, Rossi I, Bassano B, Apollonio M (2007) Predation risk as a factor affecting sexual segregation in alpine ibex. J Mamm 88(6):1488–1497

Grovenburg TW, Jenks JA, Jacques CN, Klaver RW, Swanson CC (2009) Aggressive defensive behavior by free-ranging white-tailed deer. J Mammal 90(5):1218–1223

Gubernick DJ (1981) Parent and infant attachment in mammals. In: Gubernick DJ, Klopfer PH (eds) Parental Care in Mammals. Plenum Press, New York, pp 243–305

Habibi K, Thouless CR, Lindsay N (1993) Comparative behavior of sand and mountain gazelles. J Zool 229:41–53

Heptner G, Nasimovich AA, Bannikov AG (1961) Mammals of Soviet Union, Vol. 1. Artiodactyla and Perissodactyla. Vysshaya Shkola Publishing House, Moscow. (in Russian)

Hewson R, Verkaik AJ (1981) Body condition and ranging behaviour of blackface hill sheep in relation to lamb survival. J Appl Ecol 18:401–415

Hirth DH (1985) Mother-young behavior in white-tailed deer, *Odocoileus virginianus*. South Naturalist 30(2):297–302

Huegel CN (1985) Predator-avoidance behaviors of White-tailed deer that favor fawn survival. PhD thesis. Iowa State University, Iowa

Hunter LTB, Skinner JD (1998) Vigilance behavior in African ungulates: the role of predation pressure. Behaviour 135(2):195–211

Jacobsen NK (1979) Alarm bradycardia in white-tailed deer fawns (*Odocoileus virgianus*). J Mammal 60:343–349

Jarnemo A, Liberg O, Lockowandt S, Olsson A, Wahlstrom K (2004) Predation by red fox on European roe deer fawns in relation to age, sex, and birth date. Can J Zool 82:416–422

Johnstone-Yellin TL, Shipley LA, Myers WL, Robinson HS (2009) To twin or not to twin? Trade-off in litter size and fawn survival in mule deer. J Mammal 90(2):453–460

Jungius H (1970) Studies on the breeding biology of the reedbuck (*Redunca arundinum* Boddaert, 1785) in the Kruger N.P. Z Säugetierkunde 35:129–146

Kang A (2004) Behavioral research of the Saiga antelope, *Saiga tatarica*, in its Main reproductive seasons in a semi-natural condition. PhD thesis. The East China Normal University, Shanghai

Keller M, Meurisse M, Poindron P, Nowak R, Ferreira G, Shait M, Levy F (2003) Maternal experience influences the establishment of visual/auditory, but not olfactory recognition of the newborn lamb by ewes at parturition. Dev Psychobiol 43:167–176

Keyserlingk MAG, Weary DM (2007) Maternal behavior in cattle. Horm Behav 52:106–113

Kingswood SC, Blank DA (1996) Gazella subgutturosa. Mamm Species 518:1–10

Kjellander P (2000) Density dependence in roe deer population dynamics, doctoral thesis. Swedish University of Agricultural Sciences (SLU), Uppsala

Kokshunova L (2012) Ecology and behavior of the Saiga antelope (*Saiga tatarica tatarica* L., 1766) in conditions of extreme anthropogenic press. PhD thesis. Veterinary Academy, Moscow. (in Russian)

Kostin VP (1955) Materials on ecology and commercial value of goitered gazelle. Proceedings of the institute of zoology, vol 4. Academy of Sciences of Kazakh SSR, Alma-Ata, pp 67–82. (in Russian)

Kruuk H (1972) The spotted hyena. A study of predation and social behavior. University of Chicago Press, Chicago

Kunkel KE, Mech LD (1994) Wolf and bear predation on white-tailed deer fawns in northeastern Minnesota. Can J Zool 72(9):1557–1565

Lapshina EN, Volodin IA, Volodina EV, Frey R, Efremova KO, Soldatova NV (2012) The ontogeny of acoustic individuality in the nasal calls of captive goitered gazelles, *Gazella subgutturosa*. Behav Process 90:323–330

Levy F, Keller M, Poindron P (2004) Olfactory regulation of maternal behavior in mammals. Horm Behav 46(3):284–302

Lent PC (1974) Mother–infant relationships in ungulates. In: Geist V, Walther F (eds) The behaviour of ungulates and its relation to management. IUCN Publications, Morges, pp 14–55

Leuthold W (1967) Beobachtungen zum Jugendverhalten der Kob-Antilopen. Z Saugetierk 32:59–62

Lingle S, Riede T (2014) Deer mothers are sensitive to infant distress vocalizations of diverse mammalian species. Amer Naturalist 184(4):510–522
Lingle S, Rendall D, Pellis SM (2007) Altruism and recognition in the antipredator defense of deer: 1. Species and individual variation in fawn distress calls. Anim Behav 73:897–905
Lingle S, Wyman MT, Kotrba R, Teichroeb LJ, Romanow CA (2012) What makes a cry a cry? A review of infant distress vocalizations. Curr Zool 58(5):698–726
Marmazinskaya NV (2000) The development of social behavior of the gazelle (*Gazella subgutturosa subgutturosa* Güldenstaedt 1780). In: Conservation of biodiversity in specially protected areas of Uzbekistan. Chinor ENK, Tashkent, pp 87–94
Marmazinskaya NV (2008) Maternal child relationships in goitered gazelles *Gazella subgutturosa* Güldenstaedt, 1780. In: Proceedings of nature reserves of Uzbekistan. Issue 6. State Committee of Nature of Uzbekistan, Tashkent, pp 84–102. (in Russian)
Mech LD, Roberts RE (1990) Survival of white-tailed deer fawns in relation to maternal age. J Mamm 71(3):465–467
Murie A (1981) The grizzlies of Mount McKinley, 14. United States National Park Service, Fauna of the National Parks Series, pp 1–251
Nowak R, Lindsay DR (1992) Early discrimination of the mother in sheep: important for survival? Appl Anim Behav Sci 34:61–74
Nowak R, Keller M, Val-Laillet D, Levy F (2007) Perinatal visceral events and brain mechanisms involved in the development of mother-young bonding in sheep. Horm Behav 52:92–98
Ozoga JJ, Verme LJ (1982) Physical and reproductive characteristics of a supplementally-fed white-tailed herd. J Wildl Manag 46:281–301
Ozoga JJ, Verme LJ (1986) Relation of maternal age to fawn-rearing success in white-tailed deer. J Wildl Manag 50:480–486
Ozoga JJ, Verme LJ, Bienz CS (1982) Parturition behavior and territoriality in white-tailed deer: impact on neonatal mortality. J Wildl Manag 46(1):1–11
Pereladova OV, Pereladov SV (1986) The goitered gazelle breeding in nursery enclosures of the Syunt-Khasardak nature reserve. In: Flint VE, Prisyazhnuk VE (eds) Conservation and perspectives of the number restoration of goitered gazelle in USSR. All-Union Scientific Institute of Nature Conservation and Nature Reserves, Moscow, pp 49–59. (in Russian)
Poindron P, Nowak R, Levy F, Porter RH, Shaal B (1993) Development of exclusive mother-young bonding in sheep and goats. Oxf Rev Reprod Biol 15:311–364
Poindron P, Hernandez H, Serafin N, Terrazas A (2003) Early recognition of newborn goat kids by their mother. I. Nonolfactory discrimination. Dev Psychobiol 43:82–89
Poindron P, Gillin G, Hernandez H, Serafin N, Terrazas A (2007a) Preference of 12-h-old kids for their mother goat is impaired by pre-partum-induced anosmia in the mother. Anim Behav 1(9):1328–1334
Poindron P, Terrazas A, Montes de Oca MLN, Serafin N, Hernandez H (2007b) Sensory and physiological determinants of maternal behavior in the goat (*Capra hircus*). Horm Behav 52:99–105
Pulliam HR (1973) On the advantages of flocking. J Theoret Biol 38:419–422
Ralls K, Kranz K, Lundrigan B (1986) Mother–young relationships in captive ungulates: variability and clustering. Anim Behav 34:134–145
Reif U, Klingel H (1991) Hiding behavior in wild gerenuk (*Litocranius walleri*) fawns. Z Saugetierkunde 56:159–168
Riley SJ, Dood AR (1984) Summer movements, home range, habitat use and behavior of mule deer fawns. J Wildl Manag 48:1302–1310
Romeyer A, Poindron P, Orgeur P (1994) Olfaction mediates the establishment of selective bonding in goats. Physiol Behav 56:693–700
Ryan SJ, Knechtel CU, Getz WM (2007) Ecological cues, gestation length, and birth timing in African buffalo (*Syncerus caffer*). Behav Ecol 18(4):635–644
Schwede G, Hendrichs H, Wemmer C (1994) Early mother-young relations in white-tailed deer. J Mammal 75(2):438–445
Searby A, Jouventin P (2003) Mother-young acoustic recognition in sheep: a frequency cording. Proc R Soc B, BiolSci 270:1765–1771

Sebe F, Duboscq J, Aubin T, Ligout S, Poindron P (2010) Early vocal recognition of mother by lambs: contribution of low- and high-frequency vocalizations. Anim Behav 79:1055–1066
Sibiryakova OV, Volodin IA, Matrosova VA, Volodina EV, Garcia AJ, Gallego L, Landete-Castillejos T (2015) The power of oral and nasal calls to discriminate individual mothers and offspring in red deer, *Cervus elaphus*. Front Zool 12:2. https://doi.org/10.1186/s12983-014-0094-5
Sibiryakova OV, Volodin IA, Volodina EV (2018) Advertising individual identity by mother and adolescent contact calls in Siberian wapiti *Cervus elaphus sibiricus*. Ethology 124:733–742
Sludskiy AA (1956) Breeding of goitered gazelle. Proceedings of institute of zoology, Kazakh Academy of Sciences, vol 6. Nauka Press of Kazakh SSR, Alma-Ata, pp 78–108. (in Russian)
Sludskiy AA (1962) Interrelation between predators and prey. Proceedings of the institute of zoology, vol 17. Publication of Academy of Sciences KazSSR, Alma-Ata, pp 24–143. (in Russian)
Sokolov II (1959) *Gazella subgutturosa* Güldenstaedt—Goitered gazelle. In: Pavlovskiy EN (ed) Fauna USSR, mammals: ungulates (orders Perissodactyla and Artiodactyla). Publication of the Academy of Sciences USSR, Moscow, pp 308–331. (in Russian)
Soldatova NV (1983) To behavior of goitered gazelles Bukhara breeding center. In: Flint VE (ed) Ecology of plants and animals of nature reserves of Uzbekistan. Fan Publishing House, Tashkent, pp 86–90. (in Russian)
Spinage CA (1969) Territoriality and social organization of the Uganda defassa waterbuck *Kobus defassa ugandae*. J Zool 159:329–361
Stephenson TR, van Ballenberghe V (1995) Defense of one twin calf against wolves, *Canis lupus*, by a female moose, *Alces alces*. Can Field-Nat 109:251–253
Taylor CA (2013) Behaviour and cause-specific mortality of mule deer (*Odocoileus hemionus*) fawns on the National Rifle Association Whittington Center of north-Central New Mexico. MsD thesis. Texas Tech University
Terrazas A, Serafin N, Hernandez H, Nowak R, Poindron P (2003) Early recognition of newborn goat kids by their motherII. Auditory recognition and evidence of an individual acoustic signature in the neonate. Dev Psychobiol 43:311–320
Toïgo C, Gaillard JM, Michallet J (1996) La taille des groupes: un bioindicateur de l ’ effectif des populations de bouquetin des Alpes (*Capra ibex ibex*)? Mammalia 60:463–472
Torriani MVG, Vannoni E, McElligott AG (2006) Mother-young recognition in an ungulate hider species: a unidirectional process. Am Nat 168(3):412–420
Ungerfeld R, Gonzalez-Pensado S, Bielli A, Villagran M, Olazabal D, Perez W (2008) Reproductive biology of the pampas deer (*Ozotoceros bezoarticus*): a review. Acta Vet Scan 50(1):1–16
Vannoni E, Torriani MVG, McElligott AG (2005) Acoustic signaling in Cervids: a methodological approach for measuring vocal communication in fallow deer. Cogn Brain Behav 9:551–566
Volodin IA, Lapshina EN, Volodina EV, Frey R, Soldatova NV (2011) Nasal and oral calls in juvenile goitered gazelles (*Gazella subgutturosa*) and their potential to encode sex and identity. Ethology 117:294–308
Volodin IA, Sibiryakova OV, Kokshunova LE, Frey R, Volodina EV (2014) Nasal and oral calls in mother and young trunk-nosed saiga antelopes, *Saiga tatarica*. Bioacoustics 23(2):79–98
Volodin IA, Efremova KO, Frey R, Soldatova NV, Volodina EV (2017) Vocal changes accompanying the descent of the larynx during ontogeny from neonates to adults in male and female goitred gazelles (*Gazella subgutturosa*). Zoology 120:31–41
Walther FR (1965) Verhaltensstudien an der Grantsgazelle (*Gazella granti* Brooke, 1872) im Ngorongoro-Krater. Z Tierpsychol 22:167–208
Walther FR (1968) Verhalten der Gazellen. A. Ziemsen Verlag, Wittenberg Lutherstadt
Walther FR (1969) Ethologische Beobachtungen bei der kunstlichen Aufzucht eines Blessbockkalbes (*Damaliscus dorcas philippsi* Harper, 1939). Zool Gar 36:191–215
Walther FR, Mungall EC, Grau GA (1983) Gazelles and their relatives. A study in territorial behavior. Noyes Publication, Park Ridge, NJ, USA, pp 239
White KS, Berger J (2001) Antipredator strategies of Alaskan moose: are maternal trade-offs influenced by offspring activity? Can J Zool 79:2055–2062
Zhevnerov VV (1984) Goitered gazelle of the Barsakelmes Island. Nauka Publishing House of Kazakh SSR, Alma-Ata. (in Russian)
Zhevnerov VV, Bekenov AB, Sludskiy AA (1983) Goitered gazelle—*Gazella subgutturosa* Güldenstaedt, 1780. In: Gvozdev EV, Kapitonov VI (eds) Mammals of Kazakhstan, Bovidae, vol 3 (3). Nauka Publishing House of Kazakh SSR, Alma-Ata, pp 11–54. (in Russian)

Group Size and Social Composition 11

11.1 General Remarks

Generally, ungulates are the most social animals among mammals (Groves and Grubb 2011) and most of them have a tendency for living in groups (Krause and Ruxton 2002; Dunbar and Shultz 2010). Social ungulates create and live mainly in groups of the conspecifics, though in some cases, they can stay in groups of taxonomically close heterospecifics (Schmitt et al. 2014). Essentially, group size is the quantitative measure of social units (Reiczigel et al. 2008). The size and compositions of groups are important characteristics of ecological and behavioral processes (Jarman 1974; Ruckstuhl and Neuhaus 2000, 2002; Seebacher and Krause 2017). The group size depends first on habitat structure and food availability and abundance (Isvaran 2007). On the one hand, group behavior allows ungulates to quickly detect a predator and transmit this information to the entire group (the "many eyes" effect), as well as protect the greatest number of the group members from predation (the dilution effect). On the other hand, there is a cost for a larger group size lifestyle, which includes a greater likelihood of encountering and being attacked by predators due to the greater visibility of the herd, increased competition for food within the herd, and a higher and more rapid transmission of parasites and infectious diseases within the group (Kasozi and Montgomery 2020). All these factors affect group size and their social composition and promote a certain balance of factors that create conditions for the enlargement or reduction in the size of the herd. Thus, the size of groups and their dynamics can serve as characteristics of a population's adaptation to their environmental and social conditions (Seebacher and Krause 2017).

A group is a community of individuals who are located relatively close to each other. If they move, they do it in the same direction and most of their activities are synchronized with all or almost all engaged mainly in grazing or resting (Gerard et al. 2002). Two general factors have an impact on the size of groups. First groups grow in size with an increasing population density (Johnson 1983; Wirtz and

D. Blank, *The Behavioral Biology and Conservation of Goitered Gazelles (Gazella subgutturosa)*, https://doi.org/10.1007/978-3-031-83504-9_11

Lorscher 1983), and second, group size is dependent on habitat: groups are smaller in forested environments, where it is easier for small groups to hide from predators, while they are larger in open habitats (Jha and Isvaran 2022), where it is difficult to hide irrespective of group size; larger groups have certain advantages, such as the dilution effect, when there is a high probability that another member of the group will become the victim (Estes 1974; Jarman 1974; Hillman 1987). There is also one more trend, where smaller body-sized species are more vigilant compared larger ones (Berger and Cunningham 1988), resulting in smaller species being able to stay in relatively small groups, compensating for the small group size with a higher vigilance rate (Creel et al. 2014). In addition, the number of predators and human activities have a significant impact on ungulate group size (Worku et al. 2023).

Goitered gazelle group size was the subject of past investigations for a number of scientists. Some of these studies asserted that these gazelles were able to make up large herds of hundreds up to several thousand individuals (Dementiev 1935; Dereviagin 1947; Heptner et al. 1961). These sizes, however, were seen at a time with widespread populations and an overall abundance of these ungulates. In contrast, other scientists (Sludskiy 1956; Sokolov 1959) considered goitered gazelles as a species for which small herds, rather than large herds, were typical. During the 1950s and 1960s, there was a considerable increase in human activity everywhere in the deserts of Central and Eastern Asia (Mongolia, China, and the former Union of Soviet Socialist Republics), which drastically changed the conditions for the goitered gazelles and forced them out of many regions of their natural habitat. By this time, their formerly continuous population was split into numerous and separate micro-populations with low numbers and density (Mallon and Kingswood 2001). Their ranges had been halved in Kazakhstan and reduced even more in Uzbekistan and Turkmenistan, and they inhabited <3% of the area they previously had occupied in the Caucasus and Tajikistan (Flint and Prisyazhnuk 1986).

During the 1970s and 1980s, some conservation measures for goitered gazelles allowed for an increase in their numbers inside the protection of nature reserves. Throughout this time, some investigations of goitered gazelle populations in western Kazakhstan and Uzbekistan, where found that these gazelles prefer small groups, with single gazelles are common (Mambetzhumaev 1970; Zhevnerov 1984).

Over the last 20 years, some information has been reported on the gregarious behavior of the Arabian sand gazelles (*Gazella subgutturosa marica* Thomas 1897) in Saudi Arabia (Habibi et al. 1993) and the goitered gazelles (*Gazella subgutturosa hillieriana* Heude 1894) in China (Gao et al. 1996 for Xinjiang), and in recent years, authors have further investigated the group size and composition in these two subspecies (Qiao et al. 2008; Cunningham and Wronski 2011).

In this chapter, I will describe the preferred group size of the goitered gazelle, and the dynamics of the sizes and composition over seasons of the year, and I will try to explain what factors determine the relatively small group sizes of this species.

11.2 The Preferable Group Sizes in Goitered Gazelles

In the Ili Hollow goitered gazelle population, most individuals were identified according to sex and age. Among 6017 groups, I found approximately the same number of adult males and females (32.8% and 34.8%, respectively) and young individuals, including 13.6% yearlings and 18.8% fawns. Among adult animals, the sex ratio was almost equal between males and females (49:51) during all observed years. This balanced sex ratio indicates a good state of the Ili Hollow goitered gazelle population, while other goitered gazelle populations have often shown a bias to varying degrees in favor of females. This occurs as a result of the sudden onset of an unfavorable period associated with harsh weather or epizootics. Once the adversity has passed, the population gradually regains sexual balance up to an equal ratio between females and males (Zhevnerov 1984).

Goitered gazelles can gather into groups of several tens of individuals, though the common size of most groups is typically less than three individuals. Despite the fact that the size of the groups varies in a very wide range (from 1 to 70 individuals), the majority of goitered gazelles prefer to be in small groups. Singles make up 40% of all groups ($n = 8870$), groups of two individuals make up 23%, groups of three gazelles are likely to occur at around 16%, and herds of four are quite rare at only 7% (Fig. 11.1), so all these small groups, from singles to groups of four, reach a total of 86%, indicating a greater preference for small groups. Thus, the distribution of groups of different sizes has a very strong deviation towards small groups and the frequency of observations of groups of different sizes decreases sharply with the increase in their sizes (Blank et al. 2012a). A similar distribution of group sizes is

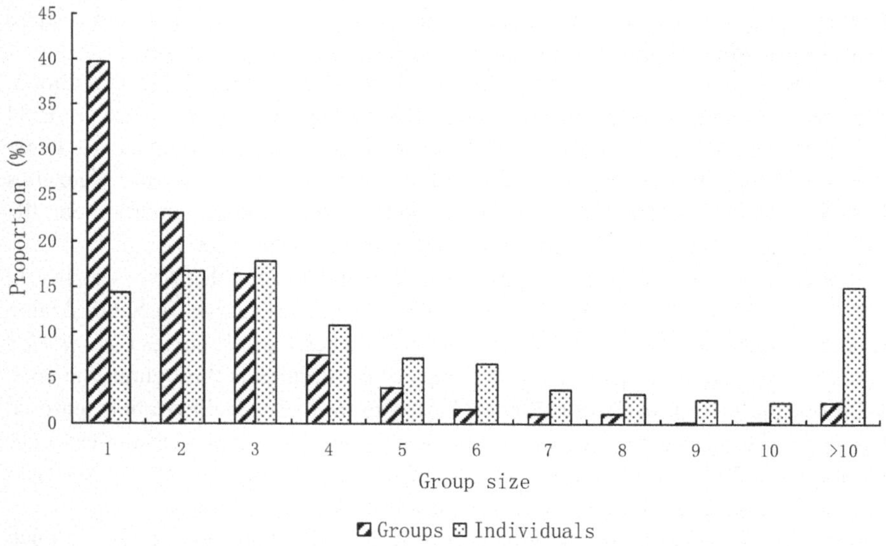

Fig. 11.1 Proportion of groups and individuals sighted in the different group sizes in the Ili Hollow goitered gazelle population (Blank et al. 2012b—License Number—5471380418423)

observed in the Barsakelmes Island population (Aral Sea, W Kazakhstan) (Zhevnerov 1984), and sand gazelles (*G.s. marica*) in Saudi Arabia (Cunningham and Wronski 2011). In the Xinjiang Province (NW China), it was found that goitered gazelle groups of three individuals were the most common (20%), followed by groups of two (15%) and then singles (12%) (Qiao et al. 2008).

The difference with other populations, such as in Xinjiang, may be related to a higher population density (Chu et al. 2009) compared to the Ili Hollow population (Blank et al. 2012a), since it was found that group size increases with population density in wild ungulate species (Lawes and Nanni 1993; Toïgo et al. 1996). That said, the group size distributions of all goitered gazelle populations are still very similar. First, singles and groups of two to three gazelles are the most common. Second, solitary individuals can be observed often in populations in any season, particularly during rutting and birthing/rearing of young (Zhevnerov et al. 1983). Third, the portion of large herds is very low compared to the small groups and occur mainly during migrations (Blank et al. 2012a).

The difference between the number of gazelles that are in small groups (up to four individuals) compared with those individuals that gather in larger groups is less skewed when comparing the number of groups of different sizes, but still the number of gazelles in small groups is noticeably higher compared to those staying in large groups (Fig. 11.1) (Blank et al. 2012b). Most gazelles still prefer to stay in groups of 2 and 3 individuals (totally 33%, $n = 24{,}577$), and the number of single individuals is so large that it exceeds the number of gazelles in aggregations of more than 10 individuals (14% and 15%, respectively) (Blank et al. 2012a).

My study of goitered gazelles in the Ili Hollow (SE Kazakhstan) demonstrated that the mean group size for the entire observation period was equal to $M \pm SE = 2.77 \pm 0.03$ ind/gr ($n = 8870$) (Blank et al. 2012b). In chinkara (*Gazella bennetti*), also living in arid areas, group size is similar at 2.55 ± 1.61 ind/gr (Gajera et al. 2014), while sand gazelles (*G.s. marica*) in Saudi Arabia have relatively larger groups on average at 4.79 ± 3.80 ind/gr ($n = 116$) (Habibi et al. 1993). In addition, the mean group size changed over years fluctuating from 2.28 ± 0.06 in 1984 ($n = 1007$) to 3.42 ± 0.14 ind/gr in 1981 ($n = 438$), and the general impact of a year was significant. In the savanna areas of Eastern Africa, Soemmerring's gazelles have a mean groups size of 4.4. ind/gr, and in the most favorable environment, the size can reach as high as 16.8 ind/gr (Gebremedhin and Yirga 2005).

For *G. s. marica*, group size ranges between 10 and 50 animals (Vesey-Fitzherald 1952), but 50 to 100 animals have also been reported previously in Saudi Arabia (Harrison and Bates 1991). Kingdon (1990) reports that the subspecies *G. s. marica* is commonly found in medium-sized groups of 6–12 animals throughout the year. While in the Harrat al-Harrah Protected Area in northern Saudi Arabia, seasonal patterns in group size are seen varying between 9 and 14 animals during the winter months, and 2–5 animals during the rest of the year (Cunningham and Wronski 2011). According to Habibi (1991), gregariousness is also evident amongst *G. s. marica* in captivity or when populations were still abundant on the Arabian Peninsula.

Despite the fact that living in large groups provides a number of important advantages for ungulates, especially regarding protection from predators (Berger 1978; Roberts 1996), most goitered gazelles prefer to stay in small groups (Zhevnerov 1984; Blank 1990). Life in small groups is typical not only for goitered gazelles, but also for many other gazelle species that live mainly in arid zones of Asia and Africa (Estes 1991). Jarman (1974) also showed that there is a direct relationship between antelopes' habitat and their behavior, including preference in group size. It is obvious that the preference of small groups in the goitered gazelles is also associated with the peculiarities of their feeding behavior.

The overall low productivity of desert pastures leads to the fact that the structure of the gazelle population is characterized primarily by a small group organization (Bugaev 1989). Goitered gazelles eat a wide variety of plant species in any inhabited environment, and almost all local plants are included in their diet, although some of them are eaten more often and others less often, depending on season and the taste characteristics of the plants themselves (Zhevnerov et al. 1983; Blank 1990; Kingswood and Blank 1996). Goitered gazelles belong to the intermediate eaters of ungulate, that is, they can behave like grazing species if environmental conditions allow, but they also can behave like typical browsers, when shrubs are their most abundant source of forage (Hofmann 1989). In Kazakhstan, where northern types of shrub deserts are typical for most of goitered gazelle habitats, they behave like typical browsers and selectively feed most of the year, when, as a rule, the whole plant is not eaten, but only the parts with the highest nutritional value and lowest fiber content (green shoots, flowers, buds). This kind of food does not occur in great amounts because of scarcity and being concentrated only a few places that are highly dispersed throughout the territory (Jarman 1974), and in order to find enough food, the gazelles need to constantly move around their environment to collect and compete for food from many different sources; therefore, they disperse widely by mostly small groups (Blank 1990). In these conditions, it is impossible for them to gather in large groups and still be able to collect sufficient amounts of food with high enough nutritional value for the whole herd (Jarman 1974). It is no coincidence that it is in winter, when there is no high-quality food, that the most deceased gazelles are observed, because low quality food cannot sustain them for very long (Zhevnerov 1984; Hofmann 1989).

Other gazelle species and their relatives are basically gregarious animals, and they will form groups ranging from a few to hundreds and even thousands of individuals, when conditions allow during the rainy season (for Thomson's gazelle—Walther et al. 1983; and Soemmerring's gazelles—Gebremedhin and Yirga 2005). Especially large herds are observed during migrations periods. In contrast, arid-dwelling species, such as the gerenuk (*Litocranius walleri*) and dibatag (*Ammodorcas clarkei*) form only small groups of less than ten animals, which is less than in any other Antilopinae species (Walther et al. 1983).

On Barsakelmes Island, goitered gazelles form into groups of different numbers and age and gender throughout the year. Groups of two to three animals are the most common, but there are also herds of ten or more individuals. The change in herding is largely determined by the gazelles' biological cycle, weather conditions, the availability of free places suitable for grazing and population size. Herds of up to 20 gazelles or more are rarely observed, and occur mainly in unfavorable winters, when gazelles are forced to concentrate in limited areas due to a lack of forage. Sometimes gazelles gather in places that are especially convenient for rest or protection from adverse weather, but these clusters of gazelles are temporary in nature, differ in their level of connectedness between individuals, and can only relatively be called a herd. Similar groupings can also be observed in early spring, where they gather in pastures with early green shoots, and in autumn, where they congregate in areas with their preferred forage of *Salsola* species. These herds are formed from groups of three to six individuals in close proximity to each other. In case of danger or during transitions to places of rest and overnight stays, they again divide into separate groups (Zhevnerov 1984).

In my study in the Ili Hollow, goitered gazelles change their group size most often based on their yearly biological cycle. The female and male groups are very changeable across months, but this does not apply to mixed-sex groups and yearlings. The female herds have minimal size in spring (May), while these groups are largest in autumn-winter (October through December). Male groups, in contrast, are largest in spring (April) and smallest in autumn-winter (November and December). Mixed-sex herds reach maximal size during spring and autumn migration periods, when goitered gazelles gather into large groups to migrate from winter habitats to spring-summer areas and back (March to April and October); rarely do group sizes increase in mid-summer (in July) (Kostin 1955; Blank 1990). These migration changes, however, are not very significant. Yearling groups fluctuate randomly over months without a significant difference. Most single females are observed when pregnant females go into seclusion to birth and hide their fawns in May and June, while single males are common all year round, though most numerous during the autumn rut, in November and December (Blank et al. 2012a; Table 11.1).

11.3 Social Composition Changes over Seasons

In goitered gazelles, there are adult female and male single sex groups, as well as mixed-sex and yearling groups (Zhevnerov 1984; Marmazinskaya 1999). The most numerous of all group types (90%) are female and male groups, while mixed herds and yearling groups are fairly uncommon (Table 11.2). The rarity of mixed groups suggests that females and males demonstrate sexual segregation for most of the year (Cransac et al. 1998; Shi et al. 2005). As for the yearlings, they prefer to join the female and male groups and only a small part of them form their own separate groups (Blank et al. 2012b).

Most gazelles stay in female groups (52%), whereas in male groups, there are noticeably fewer individuals (35%). Adult individuals of different sex and age gather in mixed groups which are largest compared to other group types but most variable in size. In contrast, only yearlings stay in the smallest yearling groups, with quite limited variability. Female groups consist mainly of adult females and fawns, which makes up 90% of individuals, while the remaining 10% are female-yearlings.

Table 11.1 Monthly mean group size (±standard deviation) of the different *Gazella subgutturosa* group types in the Ili Hollow goitered gazelle population (by Blank et al. 2012a)

Month	Female groups	Male groups	Mixed groups	Yearling groups
March	2.6 ± 0.25	2.02 ± 0.22	6.80 ± 2.85	2.5 ± 0.50
	n = 48	*n* = 149	*n* = 5	*n* = 4
April	2.69 ± 0.32	3.01 ± 0.34	5.34 ± 0.53	1.45 ± 0.17
	n = 298	*n* = 171	*n* = 45	*n* = 22
May	1.72 ± 0.05	2.43 ± 0.11	4.41 ± 0.30	1.63 ± 0.11
	n = 527	*n* = 429	*n* = 67	*n* = 80
June	2.37 ± 0.08	2.26 ± 0.11	4.75 ± 0.60	1.42 ± 0.08
	n = 699	*n* = 479	*n* = 37	*n* = 95
July	2.77 ± 0.06	2.06 ± 0.09	5.96 ± 0.71	1.41 ± 0.07
	n = 886	*n* = 358	*n* = 55	*n* = 90
August	2.71 ± 0.09	1.88 ± 0.11	4.25 ± 1.60	1±
	n = 253	*n* = 145	*n* = 4	*n* = 1
September	2.29 ± 0.11	1.71 ± 0.19	–	–
	n = 72	*n* = 31		
October	3.47 ± 0.06	1.85 ± 0.21	10.17 ± 2.66	1±
	n = 29	*n* = 34	*n* = 6	*n* = 1
November	3.28 ± 0.13	1.52 ± 0.06	5.40 ± 0.57	2.0 ± 0.21
	n = 252	*n* = 283	*n* = 42	*n* = 35
December	3.81 ± 0.34	1.59 ± 0.13	3.69 ± 0.40	1.80 ± 0.37
	n = 63	*n* = 201	*n* = 16	*n* = 5

Table 11.2 Observed number, proportion and size of the different *Gazella subgutturosa* group types in the Ili Hollow goitered gazelle population (sum of groups—6017 and sum of individuals—14,345) (by Blank et al. 2012b)

Group type	Number	Proportion (%)	Mean group Size ± SE	Group size Range (lim)	Standard deviation
Female					
Groups	3127	52.0	2.40 ± 0.03	1–22	1.68
Individuals	7505	52.3	*n* = 3127		
Male					
Groups	2280	37.9	2.14 ± 0.05	1–49	2.16
Individuals	4879	34.0	*n* = 2280		
Mixed					
Groups	277	4.6	5.29 ± 0.25	2–70	3.92
Individuals	1465	10.2	*n* = 277		
Yearling					
Groups	333	5.5	1.49 ± 0.05	1–6	0.85
Individuals	496	3.5	*n* = 333		

Male groups consist mainly of adult males (to 87%) and a small number of the male-yearlings (13%) (Table 11.3).

In the Barsakelmes Island population, the same composition of goitered gazelle herding also can be observed; for example, from 1971 to 1976, 1696 encounters

Table 11.3 Age and sex composition of gazelles in the different group types in the Ili Hollow goitered gazelle population (sum of groups is 6017) (by Blank et al. 2012b)

Group types	Females (%)	Young (%)	Male (%)	Yearling (%)	Total number
Female group	51.4	38.4		10.2	7505
Male group			86.9	13.1	4879
Mixed group	40.4	5.9	37.4	16.3	1465
Yearling				100.0	496
Singletons	38.0	1.3	51.3	9.4	2482

with groups of gazelles accounted for 5656 individuals. The most frequently observed group consisted of 2 individuals, and on average, groups of two make up 23.4% of all encounters over the period of a year. In the vast majority of cases, these are adolescents separated from their mothers, or 48.7% of all groups consisting of a female and fawn. In other cases, groups of two adult males or two females can be observed in almost equal numbers. Solitary gazelles make up 24.1% of the population and most singles occur in May and November (33.4% and 37.1%, respectively), and less in summer and winter (9.8% and 17.5%, respectively). In the spring, the increase in the number of singles occurs due to the separation of pregnant females. After birthing, the females continue to stay close to their newborns until they are old enough to start following their mothers, and since the females disperse throughout the island during the birthing period, there is an increase in the number of singles in May. In November, the number of singles is associated with the rut, which are mainly adult males and juvenile males (Zhevnerov 1984).

Groups of 3 individuals account for 18.1%. They usually consist of a female with fawns (45.9%), young gazelles under the age of 1 year (41.7%) or adult males (12.4%). Herds of 7–10 gazelles make up 7.3% of encounters. These large groups are relatively evenly distributed in all seasons of the year, except for the end April through June. In June, the young mostly follow their mothers creating an increase in the number of groups of two to three individuals compared to the spring period, and the number of singles decreases. Later, two to three family groups can merge into one herd. Male groups are most common from July through September. In October preceding the rut, most males begin to join female groups, and in early November, most individual males establish their rutting territories (Zhevnerov 1984).

Mean group size in goitered gazelles varies considerably over seasons, decreasing progressively from April, when average group sizes are largest, to minimal sizes in June (M ± SE = 1.73 ± 0.14 ind/gr, $n = 85$), after which mean group sizes started to increase up to a maximal values in February (6.35 ± 0.83 ind/gr, $n = 55$). These seasonal changes are most significant in female groups, less noticeable in mixed groups and yearling groups, with the size of male groups varying slightly (Blank et al. 2012a). In Indian blackbucks, the mean group size can be reach higher level, M ± SE = 13.84 ± 1.89 ind/gr (Rai 2019).

In March, females migrate in crowds and then in April, gradually split into smaller groups of usually two to three individuals. In May, most leave their groups to stay in seclusion during birthing and most of the hiding period, when it is possible to see both single females and females with their fawns. As a result, the number of female groups drops to their minimum. At the end of the hiding time in July, several mothers form a group from two or more families, when most of single females disappear, and all females again stay in female unisex groups. During August–October, female groups gradually increase the size of their groups and during the rut in November and December, they have the largest groups. Females continue to stay in these large groups through winter (January and February), before migrating to their birthing areas, splitting into smaller groups again in March (Blank et al. 2012a).

The size of goitered gazelle female groups are typically small and can change from two to six females with fawns (Blank et al. 2012a). Other gazelle species form groups of various size from 2 to 20 individuals depending on species and population density (Walther et al. 1983). In Grant's gazelles (*Nanger granti*—Walther 1972) and mountain gazelles (*Gazella gazella*—Grau 1974), the maximal size of a female group can be about 40 individuals. In blackbucks, it may range up to 80 members (Cary 1976). In Soemmerring's gazelle (*Nanger sommeringii*) groups of females from 11 to 29 ($M \pm SE = 18.3 \pm 1.5$ ind/gr) are most common in some areas, while in more favorable environments with rich grassy vegetation, they gather usually in groups of 5–196 individuals ($M \pm SE = 125.7 \pm 19.0$ ind/gr), though in certain conditions, single groups can reach up to as many as 250 individuals (Gebremedhin and Yirga 2005). In Serengeti population of Thomson's gazelles (*Eudorcas thomsonii*), female herds can reach 400 animals (Walther et al. 1983).

Male unisex groups also congregate to migrate in March, before also splitting into separate groups, which are larger compared to female groups. They stay in these groups the entire spring and summer (April through September), first increasing group size, then reducing their size to some extent. Male groups contain adult males and yearlings. In October, male groups migrate to rutting areas, where they are divided into single territorial males and bachelors that stay in their own bachelor groups (November and December). After the rut, males gather in larger groups and stay in male unisex and mixed groups with females through winter (January and February) until spring. Male group size can vary in a wide range, but generally they are larger than female groups in spring and summer and smaller in the rutting season. In the Ili Hollow, groups larger than ten individuals were exclusively rarely seen with most groups containing from two to six members (Blank et al. 2012b), and even when the population density was much higher in the Betpak-Dala Desert (Kazakhstan), the maximum group size was only up to a high of 12 males (Sokolov 1959). In Indian blackbuck (*Antilope cervicapra*), male groups of 20 individuals are common (Mungall 1978). In African Soemmerring's gazelles, male groups change in sizes between from 2 to 50 individuals, with average of 32 ind/gr (Gebremedhin and Yirga 2005). In Grant's and Thomson's gazelles, male unisex groups of up to 80 members can be seen occasionally (Walther et al. 1983).

Mixed-sex groups are the largest but also very unstable groups containing several males and females. Mixed-sex groups gather mainly during the spring migration to birthing areas in March and partly in April and during the autumn migration to rutting areas in October, though mixed-sex groups can sometimes be observed in other seasons, though very rarely. The rest of the time, adult males are usually segregated from females. The size of mixed groups can vary from three individuals up to more than 400 in Grant's and Thomson's gazelles, as well as in Indian blackbucks, where groups have been seen up to thousands of animals in the past (Walther et al. 1983). Thus, the average size of mixed groups is considerably larger than female and male unisex groups, particularly in open plain habitats (Walther 1972). Fairly frequently, the sexes form subgroups within the mixed herds (Walther et al. 1983). In Antilopinae, mixed herds will appear when the population reaches a certain density. During the 1980s, I observed goitered gazelle mixed-sex groups of usually less than 10 individuals, though once I did see a herd of 70 individuals once during spring migration (Blank et al. 2012b). During the 1940s, though, it was common to see groups numbering in the several dozens, and sometimes up to 150–200 individuals (Sokolov 1959). On the Mangyshlak Peninsula (W Kazakhstan) in the 1940s, according to winter survey data, goitered gazelles sometimes gathered in herds of up to 500 individuals. From a gazelle airplane survey on March 20, 1944, in the Syugatinsky Valley (SE Kazakhstan), large herds of gazelles were noted: seven herds of 100–150 and four herds of 150–200 gazelles and one herd of 300 individuals were seen (Shnarevich 1952). Mixed-sex herds are common in other gazelle species, except mountain gazelles in Israel, where the population density is not enough to form these groups; mixed-sex groups are also not typical in gerenuk (Walther et al. 1983).

Groups of adolescents and yearlings are the smallest and rarely observed in goitered gazelles. Single sub-adult males start to appear in November and December, when rutting males begin to chase and expel sub-adult males from female groups, while sub-adult females usually stay with their mothers until the females are ready to birth their new fawns. Many sub-adult males join adult males, but some form separate yearling groups, which also can join to adult males from time to time (Blank et al. 2012a). Groups of goitered gazelle yearlings are also uncommon and the smallest among other gazelle species (Walther et al. 1983).

Irrespective of group composition, throughout the year, there are two periods that differ sharply in goitered gazelle herd size: the spring-summer period, when the gazelles remain as singles or in small groups of two to five individuals, and the autumn-winter period, when the gazelles form into herds from a few tens to hundreds of individuals. In autumn through winter, gazelle groups are at their largest, with some kind of break for rutting period, when there are many single males and their small groups, while female groups reach their maximum size at this time. On average, herds during this time consist of several dozen individuals. However, in areas where gazelle density is high, herds of several hundred can form, especially when on good pastures (Heptner et al. 1961; Hosseini-Zavarei et al. 2010). After September in Mongolia, individuals and small herds of gazelles begin to unite in groups of 10–30 or more individuals (Bannikov 1954).

Similar seasonal changes have been observed for the *G.s. subgutturosa* population in the Barsakelmes Island (Zhevnerov 1984), for the *G.s. hillieriana* population in Xinjiang (Qiao et al. 2008), and for the *G.s. marica* in Saudi Arabia (Habibi et al. 1993; Cunningham and Wronski 2011). For China, Jiang (1996) argued that in spring and summer, goitered gazelles form herds of 2–6 individuals and in winter as many as 20 to 30 individuals, with temporary herds of as many as hundreds of gazelles gathering at watering places (Zhigang et al. 1996).

In *G.s. marica* in Saudi Arabia, the smallest groups are observed when forage is scarce and widely scattered, while the largest are during the rainy season (Cunningham and Wronski 2011). The same phenomenon of larger groups during the wet season and the smallest during the dry months was found for many ungulate species in the Crater Ngoro Ngoro (Leweri et al. 2022). However, in my study in the Ili Hollow, group size changes are more related to the yearly biological cycle, when adult females have a minimal group size for the birthing and hiding period (May into June), though high-quality forage is richest in these months (Blank et al. 2012a).

All these groups have variable characteristics and their content changes continuously. Groups exist from several minutes to several hours. Family groups, consisting of a female and her fawn, are the only stable kind of associations and exist in goitered gazelles from 4 to 5 months (April–October). Several family groups can join together, and barren females and young males also can associate with family groups. During August through October, sub-adults start to separate from the family groups, with sub-adult males leaving first, followed by sub-adult females (Marmazinskaya 1999).

Outside of periods of increased sexual activity, hierarchy is the main way of ordering relationships among conspecifics within goitered gazelle populations. Personal recognition among animals in unstable groupings and aggregations is impossible and there is no clear linear hierarchy. Mutual determination of rank occurs by age, body size, and degree of development of secondary sexual characteristics. Individual behavioral features also play a significant role in hierarchical interactions. Males dominate females (with the exception of juvenile males), and within sex classes, older individuals dominated young ones. Between animals of the same sex and age group, and between young of both sexes, the roles of active and passive partners in hierarchical contacts often change. Stable personalized relationships with a clear linear hierarchy among males and dominance of males over females are established in long-existing captive groups. Among females, several individuals occupying the highest and lowest ranks are clearly identified. The order of subordination between the females of the middle ranks often changes. The hierarchical relationship between adult females and young males is complex and depends on the individual characteristics of the animals and the nature of the interaction. The hierarchy within sexes is maintained through antagonistic interactions. The main parameters during which hierarchical interactions take place are not maintaining individual distances, competition for resting places, and females and territorial areas during the rut. The frequency of hierarchical contacts increases during periods of hunger, stressful situations, and increased sexual activity of males (Marmazinskaya 1999).

Antilopinae herds are open societies with their members coming and going when they like, and individuals splitting off and gathering together in various combinations. Sometimes herds may remain constant in size and composition over weeks and months. In spite of their gregariousness, gazelles are distance animals and keep a certain individual distance between each other (Walther et al. 1983). This distance can vary with sex, age, and activity, but usually does not come down to bodily touch

(Walther 1977). The solitary status of an individual is always temporary in gazelles and cannot be considered to be a special social class (Walther et al. 1983). In all Antilopinae species, individual isolation may be in three cases: first, fawns stay alone and hide during the first weeks of its life; second, females leave their herd and stay alone when giving birth; and third, a male leaves a herd for the establishment of his own individual territory (Walther et al. 1983).

It is possible to conclude that the social structure of goitered gazelles changes as a result of yearly biological cycles related to birthing and rutting activities, as well as spring and autumn migrations. It turns out that the social structure helps the adaptation of a species to external conditions and successful reproduction, expanding populations as a measure of the species welfare. Though seasons also have an impact on group size, when they are smaller during hot and dry seasons and larger during cold and rainy periods, their influence is not as powerful or determinant as the yearly biological cycle (Blank et al. 2012a).

Jarman (1974) found that the body size and social structure of antelopes and their feeding behavior, which is determined by the eatable plant distribution and the degree of selectivity, are closely related and adapted to each other. His conclusions were made on the basis of long-term observations of the group size and social behavior of only antelopes of different body size in the natural environment of the African continent and a comparison of their feeding behaviors under various conditions. Subsequently, additional comprehensive phylogenetic analyses were done of 250 ungulate species from 10 families to take into consideration Jarman's hypotheses and these studies support his ideas completely (Szemán et al. 2021). However, even though different gazelle species inhabit various ecosystems, from those rich in grasses on the African savanna to extra-arid areas of the Sahara and the cold deserts of Central Asia, they all have a surprisingly similar social structure, and in most cases, prefer to live in small groups. Brashares et al. (2000) decided to do additional analyses the Jarman (1974) hypothesis about mutual influence feeding selectivity on the group and body sizes using the most modern statistical techniques. These analyses have supported Jarman's idea in general, but in considering the phylogenetic analysis, this support is quite week. Despite the fact that the size of the groups and the entire social structure of different species are adapted to various local external conditions, the phylogenetic history of species imposes certain restrictions and restrains these differences, and as a result, goitered gazelles have many similarities in social structure and group size with other gazelle species (Blank et al. 2012b).

References

Bannikov AG (1954) Persian gazelle. In: Heptner VG (ed) Mammals of Mongolian People Republic. Publication of USSR Academy of Sciences, Moscow, pp 250–258. (in Russian)

Berger J (1978) Group size, foraging, and antipredator ploys: an analysis of bighorn sheep decisions. Behav Ecol Sociobiol:91–99

Berger J, Cunningham C (1988) Size-related effects on search times in north American grassland female ungulates. Ecology 69(1):177–183

Blank DA (1990) Persian gazelle. In: Kovshar AF (ed) Rare animals of desert regions. Nauka Press of Kazakh SSR, Alma-Ata, pp 56–80. (in Russian)

Blank DA, Ruckstuhl KE, Yang W (2012a) Social organization in goitered gazelle (*Gazella subgutturosa* Güldenstaedt, 1780). Ethol Ecol Evol 24:306–321

Blank D, Yang W, Xia C, Xu W (2012b) Grouping pattern of the goitered gazelle, *Gazella subgutturosa* (Cetartiodactyla: Bovidae) in Kazakhstan. Mammalia 76(2):149–155

Brashares JS, Garland TJ, Arcese P (2000) Phylogenetic analysis of coadaptation in behavior, diet, and body size in the African antelope. Behav Ecol 11(4):452–463

Bugaev KE (1989) Large mammals of the trans-Altai Gobi (ecology and conservation). Ph.D. thesis. Moscow State University, Moscow

Cary ER (1976) Territorial and reproductive behavior of the black buck antelope (*Antelope cervicapra*). Texas A&M University, Texas

Chu HJ, Jiang ZG, Ge Y, Jiang F, Tao Y, Wang C (2009) Population densities and number of khulan and goitered gazelle in Mt. Kalamaili ungulate nature reserve. Biodivers Sci 17:414–422

Cransac N, Gerard JF, Maublanc ML, Pepin D (1998) An example of segregation between age and sex classes only weakly related to habitat use in mouflon sheep (*Ovis gmelini*). J Zool 244:371–378

Creel S, Schuette P, Christianson D (2014) Effects of predation risk on group size, vigilance, and foraging behavior in an African ungulate community. Behav Ecol 25(4):773–784

Cunningham PL, Wronski T (2011) Seasonal changes in group size and composition of Arabian sand gazelle *Gazella subgutturosa marica* Thomas, 1897 during a period of drought in central western Saudi Arabia. Curr Zool 57:36–42

Dementiev GP (1935) Mongolian gazelle, goitered gazelle, saiga antelope. KOIZ Publication, Moscow, pp 15–30. (in Russian)

Dereviagin PYA (1947) On seasonal migrations of goitered gazelle from southern pre-Balkhash deserts into inter-mountain valley Kalkan-Matay- Mynbulack. Bull Acad Sci Kazakh SSR 4(25):33–38. (in Russian)

Dunbar RIM, Shultz S (2010) Bondedness and sociality. Behaviour 147:775–803

Estes RD (1974) In: The Behaviour of Ungulates and Its Relation to Management, Geist V, Walther F (eds) Social organization of the African Bovidae. IUCN, Morges, pp 166–205

Estes RD (1991) The behavior guide to African mammals. University of California Press, Berkeley-Los Angeles

Flint VE, Prisyazhnuk VE (1986) Population status and perspectives of goitered gazelle restoration in the USSR. In: Flint VE, Prisyazhnuk VE (eds) Conservation and perspectives of the number restoration of goitered gazelle in USSR. All-Union Scientific Institute of Nature Conservation and Nature Reserves, Moscow, pp 3–8. (in Russian)

Gajera N, Mahato AKR, Kumar VV (2014) Habitat preference and social composition of antelopes in arid region of Kachchh, Gujarat, India. Asian J Conserv Biol 3:164–169

Gao X, Xu K, Yao J, Jia Z (1996) The population structure of goitered gazelle in Xinjiang. Acta Theriologica Sinica 16:14–18. (in Chinese)

Gebremedhin B, Yirga S (2005) A study on abundance, group size and composition of Soemmering's gazelle (*Gazella soemmerringii*) in awash National Park and Alledeghi wildlife reserve, Ethiopia. SINET: Ethiopian J Sci 28(2):161–170

Gerard JF, Bideau E, Maublanc ML, Loisel P, Marchal C (2002) Herd size in large herbivores: encoded in the individuals or emergent? Biol Bull 202:275–282

Grau GA (1974) Behavior of mountain gazelle in Israel. Ph.D. dissertation. Texas A&M Univ, College Station

Groves C, Grubb P (2011) Ungulate taxonomy. Johns Hopkins University Press, Baltimore

Habibi K (1991) Arabian Gazelles. National Commission for Wildlife Conservation and Development Publication No 4, Riyadh

Habibi K, Thouless CR, Lindsay N (1993) Comparative behavior of sand and mountain gazelles. J Zool 229:41–53

Harrison DL, Bates PJJ (1991) The mammals of Arabia. Harrison Zoologicl Museum, Kent

Heptner G, Nasimovich AA, Bannikov AG (1961) Mammals of Soviet Union, vol. 1. Artiodactyla and Perissodactyla. Vysshaya Shkola, Moscow. (in Russian)

Heude PM (1894) Etudes odontologiques, etc. Memoires concernant l'histoire naturelle de l'Empire chinois par des peres de la Compagnie de Jesus. Impr. de la Mission Catholique, Chang-Hai, 2(3-4):117–247

Hillman JC (1987) Group size and association patterns of the common eland (*Tragelaphus oryx*). J Zool Lond 213:641–663

Hofmann RR (1989) Evolutionary steps of ecophysiological adaptation and diversification of ruminants: a comparative view of their digestive system. Oecologia 78:443–457

Hosseini-Zavarei F, Farhadinia MS, Hemami MR, Karami M, Daniali R, Omidi M (2010) Sex-age structure of bovids in Ghameshlou, Central Iran: (Mammalia: Bovidae). Zool Middle East 51(1):3–8

Isvaran K (2007) Intraspecific variation in group size in the blackbuck antelope: the roles of habitat structure and forage at different spatial scales. Oecologia 154(2):435–444

Jarman RJ (1974) The social organization of antelope in relation to their ecology. Behaviour 48:215–267

Jha RR, Isvaran K (2022) Antelope space-use and behavior indicate multilevel responses to varying anthropogenic influences in a highly human-dominated landscape. Ecol Evol 12(10):e9372

Jiang Z (1996) Current status of antelopes in China. J Northeast Forest Univ 7(1):58–62

Johnson CN (1983) Variations in group size and composition in red and western grey kangaroos, *Macropus Rufus* (Desmarest) and *M. Fuliginosus* (Desmarest). Aust Wildl Res 10:25–31

Kasozi H, Montgomery RA (2020) Variability in the estimation of ungulate group sizes complicates ecological inference. Ecol Evol 10:6881–6889

Kingdon J (1990) Arabian mammals: a natural history. Government Press, Ministry of Information, Isa Town

Kingswood SC, Blank DA (1996) Gazella subgutturosa. Mamm Species 518:1–10

Kostin VP (1955) Materials on ecology and commercial value of goitered gazelle. Proceedings of the institute of zoology, vol 4. Academy of Sciences of Kazakh SSR, Alma-Ata, pp 67–82. (in Russian)

Krause J, Ruxton GD (2002) Living in groups. Oxford University Press, Oxford

Lawes MJ, Nanni RF (1993) The density, habitat use and social organization of Dorcas gazelles (*Gazella dorcas*) in Makhtesh Ramon, Negev Desert, Israel. J Arid Environ 24:177–196

Leweri CM, Bartzke GS, Msuha MJ, Treydte AC (2022) Spatial and seasonal group size variation of wild mammalian herbivores in multiple use landscapes of the Ngorongoro conservation area, Tanzania. PLoS One 17(4):e0267082

Mallon DP, Kingswood SC (2001) Antelopes. Part 4. North Africa, the Middle East, and Asia. Global survey and regional plans. The IUCN/SSC Antelope Specialist Group, IUCN, Glands

Mambetzhumaev AM (1970) Goitered gazelle. FAN Publication, Tashkent. (in Russian)

Marmazinskaya NV (1999) Social behavior of goitered gazelle (*Gazella subgutturosa* /Guld., 1780/) during breeding period in the conditions ecocenter "Goitered gazelle". Ph.D. thesis in zoology. Institute of zoology of Uzbek Academy of Sciences, Tashkent, pp 1–20. (in Russian)

Mungall EC (1978) The Indian blackbuck antelope: A Texas view. Cesar Kleberg Studies in Natural Resources, No. 3, Texas A. and M. University 184 pp.

Qiao J, Yang W, Xu W, Liu W (2008) Social structure of goitered gazelles (*Gazella subgutturosa*) in Kalamaili Mountain nature reserve, Xinjiang. Acta Theriol Sin 28:353–357

Rai D (2019) Crowding, group size and population structure of the blackbuck *Antilope cervicapra* (Linnaeus, 1758) (Mammalia: Cetartiodactyla: Bovidae) in the semi-arid habitat of Haryana, India. J Threatened Taxa 11(9):14194–14203

Reiczigel J, Lang Z, Rozsa L, Tothmeresz B (2008) Measures of sociality: two different views of group size. Anim Behav 75:715–721

Roberts G (1996) Why individual vigilance declines as group size increases. Anim Behav 51:1077–1086

Ruckstuhl KE, Neuhaus P (2000) Sexual segregation in ungulates: a new approach. Behaviour 137:361–377

Ruckstuhl KE, Neuhaus P (2002) Sexual segregation in ungulates: a comparative test of three hypotheses. Biol Rev 77:77–96

Schmitt MH, Stears K, Wilmers C, Shrader AM (2014) Determining the relative importance of dilution and detection for zebra foraging in mixed-species herds. Anim Behav 96:151–158

Seebacher F, Krause J (2017) Physiological mechanisms underlying animal social behaviour. Philos Trans Royal Soc B Biol Sci 372:20160231

Shi J, Dunbar RIM, Bucjland D, Miller D (2005) Dynamics of grouping patterns and social segregation in feral goats (*Capra hircus*) on the isle of rum, NW Scotland. Mammalia 69(2):185–199

Shnarevich ID (1952) Quantitative accounting experience of some hunting animal numbers of the Alma-Ata nature reserve. In: Formozov AN (ed) Accounting methods of number and geographical distribution of terrestrial vertebrates. Academy of Sciences USSR, Moscow, pp 250–254. (in Russian)

Sludskiy AA (1956) Breeding of goitered gazelle. Proceedings of institute of zoology, Kazakh Academy of Sciences, vol 6. Nauka Press of Kazakh SSR, Alma-Ata, pp 78–108. (in Russian)

Sokolov II (1959) *Gazella subgutturosa* Guldenstaedt—Goitered gazelle. In: Pavlovskiy EN (ed) Fauna USSR, mammals: ungulates (orders Perissodactyla and Artiodactyla). Publication of the Academy of Sciences USSR, Moscow, pp 308–331. (in Russian)

Szemán K, Liker A, Székely T (2021) Social organization in ungulates: revisiting Jarman's hypotheses. J Evol Biol 34(4):604–613

Thomas O (1897) On a new gazelle from Central Arabia. The Annals and Magazine of Natural History, (6)19:162–163

Toïgo C, Gaillard JM, Michallet J (1996) La taille des groupes: un bioindicateur de l ' effectif des populations de bouquet in des Alpes (*Capra ibex ibex*)? Mammalia 60:463–472

Vesey-Fitzherald LDEF (1952) Wildlife in Arabia. Oryx 1(5):232–235

Walther FR (1972) Social grouping in Grant's gazelle (*Gazella granti* Brooke, 1872) in the Serengeti National Park. Z Tierpsychol 31:348–403

Walther FR (1977) Sex and activity dependency of distances between Thomson's gazelles (*Gazella thomsoni* Gunther 1884). Anim Behav 25:713–719

Walther FR, Mungall EC, Grau GA (1983) Gazelles and their relatives. A study in territorial behavior. Noyes Publication, Park Ridge

Wirtz P, Lorscher J (1983) Group size of antelopes in an east African national park. Behaviour 84:135–155

Worku EA, Bro-Jørgensen J, Evangelista PH, Bekele A, Atickem A, Stenseth NC (2023) Group size dynamics of the endangered mountain Nyala (*Tragelaphus buxtoni*) in protected areas of the Arsi and Ahmar Mountains, Ethiopia. Global Ecol Conserv 46:e02546

Zhevnerov VV (1984) Goitered gazelle of the Barsakelmes Island. Publishing of Nauka Kazakh SSR, Alma-Ata. (in Russian)

Zhevnerov VV, Bekenov AB, Sludskiy AA (1983) Goitered gazelle—*Gazella subgutturosa* Güldenstaedt, 1780. In: Gvozdev EV, Kapitonov VI (eds) Mammals of Kazakhstan, Bovidae, vol 3 (3). Nauka Publishing House of Kazakh SSR, Alma-Ata, pp 11–54. (in Russian)

Zhigang J, Zhongxin G, Yueqi S (1996) Current status of antelopes in China. J Northeast Forestry Univ 7:58–62

Impact of Population Density on Group Size 12

12.1 General Remarks

Group formation is a common behavior demonstrated by many ungulate species (Krause et al. 2002). Gathering ungulates in groups contributes to their improved survival and increased breeding success and many specific ecological hypotheses explain ways of improving population success in species which prefer to gather into groups of conspecifics (Ward and Webster 2016). Many of these hypotheses explain grouping behavior based on food availability—its distribution and quality, and the threat of predation (Delisle et al. 2023). Grouping behavior in relation to food resources that vary spatiotemporally can be explained by two strategies to assess suitable food. One strategy is the direct assessment of food distribution in relation to landscape features (resource dispersion hypothesis) and the other strategy is the observation of conspecifics to get information from their distribution on resource availability (conspecifics attraction hypothesis) (Peignier et al. 2019). All hypotheses on fear of predation, or an antipredator strategy, predict that group size enlarges with an increasing predation risk, with two main hypotheses: the many-eyes hypothesis, when ungulates increase the probability of predator detection but get a fitness advantage through more grazing while decreasing the vigilance rate needed per capita (Roberts 1996), and the dilution-of-risk hypothesis, which states that a predator kills a single animal per attack and the number of predation kills decrease per capita by increasing group size (Bertram 1978; Fortin et al. 2009).

Thus, ungulates gain a number of advantages by being in groups, and the hypotheses related to predation risk predict more benefits with increasing group size. However, group growth cannot be unlimited, as this increases the intra-group competition for food, since forage biomass declines with increasing population density

Blank et al. (2012a)—License Number—5477060538101

D. Blank, *The Behavioral Biology and Conservation of Goitered Gazelles (Gazella subgutturosa)*, https://doi.org/10.1007/978-3-031-83504-9_12

(Dalerum et al. 2008). As a result, ungulate species have to change their behavior, modifying group size and their entire social structure, distribution, proportion of feeding, and resting times (Jarman 1974). Mostly environmental parameters, as well as predation risk and social affinity, are involved to explain social structure and group size; therefore, ungulates that spread all over diverse habitats have various diets, feeding styles and body sizes, and as a consequence, they have different group sizes and social structures (Brashares et al. 2000). Each individual prefers to be in a group where it can satisfy its own essential physiological and social needs (Pepin and Gerard 2008).

Some mammals (elephants, primates, and whales) demonstrate an independence in their group size relative to population density (Lehmann and Boesch 2004; Wittemyer et al. 2005; Karczmarski et al. 2005), while in other species (ungulates, kangaroos), group sizes, though not stable, are very dependent on population density and are positively correlated with each other (Wirtz and Lorscher 1983; Lawes and Nanni 1993; Toigo et al. 1996; Mattiello et al. 2004). Many cervid species usually stay in social groups of three to six individuals but form even larger groups when population density increases (Stuwe and Hendrichs 1984; Barrette 1991; Thirgood 1996; Borkowski 2000; Hebblewhite and Pletscher 2002). Ungulate species show different trends in group sizes with an increase in population density, with female groups noticeably increasing in size and male groups showing a much weaker response, if any at all (for *Cervus canadensis*—Wal et al. 2013). Some antelope species (*Damaliscus*, *Connochaetes*, *Redunca*, *Kobus*) and mountain dwelling species, such as chamois (*Rupicapra pyrenaica*) and ibexes (*Capra ibex*), have large groups of up to ten individuals which increase with population density (Toïgo et al. 1996; Pepin and Gerard 2008). Other antelope species (*Hippotragus*) decrease group size with a declining population density (Owen-Smith et al. 2012). Social-dwelling ungulates such as peccaries (*Tayassu pecari*) can stay in really large groups of up to 300 individuals, but their group sizes decline with population density (Reyana-Hurtado et al. 2009).

The group size for moose (*Alces alces*) does not show a dependence on density, and their group sizes are quite stable in all subspecies of Eurasia and North America (Peek et al. 1974; Sweanor and Sandegren 1989). Steenbok (*Raphicerus camelus*), duikers (*Sylvicapra grimmia*) and gorals (*Nemorhaedus goral*) are mainly solitary-dwelling ungulates, and they maintain their group size without noticeable changes over seasons (Pendharkar and Goyal 1995; Bergstrom and Skarpe 1999; Mattiello et al. 2004). Territorial ungulates, such as oribi (*Oerebia ourebia*), demonstrate very limited changes of group sizes with changes in population density (Rowe-Rowe et al. 1992). Thus, population density has a significant impact on group sizes in highly social-dwelling species, having the ability to change group size across a very wide range and preferring to live in large groups, with huge crowds typical. In contrast, the influence of population density is very weak in solitary-dwelling or territorial species, preferring small groups with little variability (Blank et al. 2012a).

Goitered gazelles have the ability to gather into large groups and even crowds, but during most seasons, they prefer to stay in small groups, with a great number of singletons is typical for this species. Moreover, goitered gazelles are territorial but only during rutting season. Therefore, it will be interesting to consider if population density has an impact on goitered gazelle group size.

12.2 Interactions Between Population Density and Group Size in Goitered Gazelles

There is a whole range of information about the formation of certain groups by goitered gazelles, the number and composition of which varies throughout the year. At the same time, authors note an increase in group sizes during the autumn-winter period and a decrease in their group sizes in the spring-summer period (Vasenko 1950; Kostin 1955). With a high density, goitered gazelle herds in winter may consist of several tens or hundreds of individuals (Dereviagin 1947; Heptner et al. 1961), and in some cases, clusters of up to several thousand individuals have been observed (Dementiev 1935). However, this herd structure cannot be considered ordinary. It may be typical, but only for a short time, such as mass migrations and movements. In general, data from most authors show a weakly expressed herd instinct in the goitered gazelle (Sokolov 1959; Sludskiy 1956, 1977; Mambetzhumaev 1970a, b).

According to my study in Kazakhstan, it is clear that goitered gazelles prefer to live in small groups, usually of two to three individuals per group. During rutting season (November and December), each male has his own individual territory which he protects from other males; therefore, many single males are observed at this time. In contrast, outside of the rut many males stay in groups, and they roam freely all over their range (Blank 1998). Females also stay alone during the birthing period (May), but after several weeks postpartum, they gather into groups of several females again (Blank 1985). Large mixed-sex groups are quite rarely observed in goitered gazelles, though it is possible to find such groups during the spring (March and April) and autumn (September and October) migration periods (Blank et al. 2012b). Thus, goitered gazelles have mixed characteristics. On the one hand, many gazelles lead a solitary lifestyle, preferring to live in small groups, and males have territories only during the rut, as is typical for species whose group size does not depend on population density. On the other hand, goitered gazelles will gather in large aggregations for short periods, and for most of the year, males are not associated with territories and move widely around their territory, as is typical of social-dwelling species with group sizes sensitive to population density.

During my study of the goitered gazelle population in the Ili Hollow, I found that the number of gazelles and their population density increased consistently and significantly over my years of observations (1981–1989) (Table 12.1). Thus, I was able to analyze the effect of population density on the size of groups using this population as an example. I found that the mean group size varied significantly for most months except June. Despite the fact that the population density of goitered gazelles had been increasing all the time, the size of the groups changed differently in different months. In some cases, the size of the groups increased (April and December), in another decreased (May), and in other months, the size of the groups decreased

Table 12.1 Mean group size of goitered gazelle population in the Ili Hollow enlarging its population density (individuals per km^2) over years

Years-density ind/km^2	Months						
	April	May	June	July	September	November	December
1981- 1.53	–	–	–	–	–	$N = 230$ 3.23 ± 0.20	
1982 – 1.65	$N = 256$ 2.53 ± 0.15	$N = 211$ 2.26 ± 0.14	–	$N = 186$ 2.89 ± 0.17	$N = 53$ 2.30 ± 0.16	$N = 91$ 2.55 ± 0.17	
1983 – 1.74	–	$N = 211$ 2.23 ± 0.11	$N = 85$ 1.73 ± 0.14	–	–	–	
1984 – 1.69		$N = 286$ 2.14 ± 0.12	$N = 433$ 2.29 ± 0.10	$N = 162$ 2.09 ± 0.10	–	–	$N = 127$ 2.80 ± 0.21
1985 – 1.83	$N = 516$ 3.96 ± 0.23	$N = 412$ 2.45 ± 0.12	$N = 752$ 2.28 ± 0.08	$N = 760$ 2.71 ± 0.09		$N = 254$ 3.71 ± 0.25	$N = 313$ 2.88 ± 0.16
1986 – 2.15		–	$N = 328$ 2.26 ± 0.09	$N = 439$ 2.35 ± 0.08	–	$N = 429$ 2.85 ± 0.13	$N = 306$ 3.89 ± 0.25
1987 – 2.29		$N = 285$ 1.76 ± 0.08	$N = 853$ 2.39 ± 0.14	–	$N = 217$ 4.29 ± 0.39	–	–
1989 – 3.67		$N = 1037$ 1.83 ± 0.07	–	–	$N = 1517$ 3.29 ± 0.12	–	–
Significance of difference	T-test F = 13.537 df = 771 $P < 0.0001$	ANOVA F = 6.872 df = 5 $P < 0.0001$	ANOVA F = 0.998 df = 4 $P > 0.05$	ANOVA F = 6.633 df = 3 $P < 0.0001$	ANOVA F = 5.252 df = 2 $P = 0.005$	ANOVA F = 5.422 df = 3 $P = 0.001$	ANOVA F = 8.269 df = 2 $P < 0.0001$

Blank et al. (2012a)—License Number—5477060538101

and increased without much regularity (July, September, and November). Statistical analyses showed that neither population density nor seasonal differences had a significant effect on the mean group size. Unlike the mean group size, which varied significantly by months and years, the ratio of singletons to groups almost did not change by month in most cases (64% cases) and the population density practically did not affect the frequency of meetings of two kinds of groups (singletons vs groups of several individuals), with the exception of May and partly December (Tables 12.2 and 12.3) (Blank et al. 2012a). Since May is the birthing period for goitered gazelles, when most females leave their groups and stay alone for as much as a few weeks before birth and after postpartum during their fawns' hiding period, groups are smallest during this period, but in June, the proportion of singletons and groups was closer to parity compared to groups of mothers and fawns and the uniting of mothers with their yearlings. Since December is the rutting period, when many adult males stay alone within their individual territories, the portion of singletons increases. However, there was another trend of increasing sizes of female groups during this time; therefore, the difference between singletons and groups was absent at the beginning of the rut in November and was not very distinctive in December (Table 12.3).

Comparing the data on herd size of goitered gazelles in different geographical areas of their habitat, it can be noted that the average size of groups in certain seasons of the year is similar. Only the maximum sizes of groups vary significantly, which is determined by the total number within populations. In 1976, with the total number of gazelles on the Barsakelmes Island up to 400 individuals, groups of 15–20 individuals were much more common than in 1970 and 1971, when their total number was kept at the level of 120–170 individuals. So, in 1971, the maximum group of 20 gazelles was recorded only once for the year, and in 1970, groups of more than even 12 individuals were not observed. In January and February 1975, temporary accumulations of 36 to 40 gazelles were observed three times. In 1947 by comparison, with the number of gazelles on the island at 800–900 individuals, herds of up to 40–45 individuals were often found in the autumn and winter periods (Vasenko 1950).

In SE Turkmenistan, as a result of a sharp decline in the number of gazelles, the size of herds has also decreased. In winter of 1956 and 1957 in the Badkhyz Uplands, it was possible to see temporary herds of up to 100–200 individuals, while in 1964 and 1965, herds of 20 gazelles became rarely observed (Gorelov 1972). Long-term observations of gazelles show that the composition of groups are unstable. When two or more groups unite and then divide, individuals are exchanged. The residence time of individual animals in a particular group can range from several hours to several days or more. For example, tamed gazelles were united into a single group only on their feeding grounds, and the rest of the time they split into several separate groups. A longer stability in the number of individuals was observed in a group of young gazelles in 1975, from mid-February through April, with a group of 2 males and 2 females aged 10–11 months. The constancy of the main core of this group was not violated during this time, however, periodically the entire herd of four young gazelles joined one adult females, who, as a rule, kept apart from all others. The

Table 12.2 Proportion (%) of singletons to groups (2–4 individuals) of the goitered gazelle in the Ili Hollow population enlarging its population density (individuals per km^2) over years

	Months						
Years—density ind/km^2	April	May	June	July	September	November	December
1981–1.53						33/46	
1982–1.65	46/39	50/40		24/61	30/64	35/48	
1983–1.74		43/48	59/38				
1984–1.69		54/38	43/50	46/49			37/46
1985–1.83	28/48	42/45	49/42	32/57		30/46	40/44
1986–2.15			40/52	36/57		38/48	36/39
1987–2.29		60/36	58/33		27/50		
1989–3.67		70/23			41/43		
Chi-square (χ^2) df = 1; P	7.332 1 0.007	208.957-0.580 $P = 0.000$-0.446 Totally 12 of 15 cases were $P < 0.05$	96.481-0.662 $P = 0.000$-0.416 Totally 6 of 10 cases were $P < 0.05$	25.154-2.636 $P = 0.000$-0.104 Totally 4 of 6 cases Were $P < 0.05$	12.145-0.206 $P = 0.000$-0.650 Totally 2 of 3 cases Were $P < 0.05$	1.254-0.001 $P = 0.263$-0.980 Totally 6 cases Were $P > 0.05$	0.592-0.045 $P = 0.442$-0.832 Totally 3 cases Were $P > 0.05$

Blank et al. (2012a)—License Number—5477060538101

Table 12.3 Pearson Correlation index for proportion (singletons/groups of 2–4 gazelles) changing in the Ili Hollow population over years

Months	Group size	Pearson index, *N* and *P*.
May	Singletons Groups	0.855, $N = 6$, $P = 0.030$ −0.852, $N = 6$, $P = 0.031$
June	Singletons Groups	0.083, $N = 5$, $P = 0.894$ −0.265, $N = 5$, $P = 0.666$
July	Singletons Groups	−0.084, $N = 4$, $P = 0.916$ 0.117, $N = 4$, $P = 0.883$
September	Singletons Groups	0.680, $N = 3$, $P = 0.524$ −0.995, $N = 3$, $P = 0.061$
November	Singletons Groups	0.514, $N = 4$, $P = 0.486$ 0.471, $N = 4$, $P = 0.529$
December	Singletons Groups	−0.448, $N = 3$, $P = 0.704$ −1.000, $N = 3$, $P = 0.013$

Blank et al. (2012a)—License Number—5477060538101

mutual relations of females in the group were seen in the great tolerance shown before the beginning of the birthing season. Once this time began, the females left their group and retired to their chosen birthing areas, where upon their behavior noticeably changed and aggressiveness towards other individuals appeared (Zhevnerov 1984).

I conducted goitered gazelle surveys in different deserts of Kazakhstan, where the population density of goitered gazelles was diverse, but much lower compared to the Ili Valley population (Tables 12.4 and 12.5). Even though the difference in population density in those deserts was several times lower compared to the Ili Valley, only in a few cases did I observe a distinctive difference in mean group sizes, while in all other cases, there was no difference at all (Table 12.4). A comparison of these cases showed that only if the difference in population density was more than seven times, then there was also a difference in the mean group size. If the difference in population density was below this barrier, then the size of the groups did not change much. Similar uniformity was found in the frequencies of occurrence of different groups (proportion of singletons and gazelles in groups) and the group class difference also was significantly lower when population density was lower than seven times (Blank et al. 2012a). Such a non-linear increase in group sizes with an increase in population density also occurs in other species. For example, the size of groups of some antelopes (*Connochaetes*, *Tragelaphus*) begins to increase dramatically only if the density increases three times (Owen-Smith 2006), in logarithmic in Alpine ibex (*Capra ibex*—Toigo et al. 1996) and red kangaroo (*Macropus rufus*—Johnson 1983). Group size may depend of cubic root of population density (*Rupicapra pyrenaica*—Pépin and Gerard 2008). The bison (*Bison bison*) does not increase its group size in the case of a three-fold increase in population density (Jung 2020).

All these data suggest that in the goitered gazelle population, density has a rather weak effect on the size of groups. If such an influence was significant, then with an increase in population density, the size of the groups would noticeably increase

Table 12.4 The mean group size of goitered gazelle among populations in various deserts with different density

Populations	Density ind/km^2	Mean group size ind per group	*N*	T-test	*P*
Kapchagaj reserve - Aktau (12.1986)	2.15 0.30	3.55 ± 0.266 4.00 ± 0.712	221 34	F = 0.943 df = 243	0.333
Kapchagaj reserve - Aktau (06.1987)	4	2.39 ± 0.144 2.56 ± 0.287	853 54	F = 0.185 df = 905	0.667
Kapchagaj reserve - PanfilovKarakum	2.29 0.35	2.39 ± 0.144 1.33 ± 0.106	853 39	F = 4.259 df = 890	0.039
Aktau - PanfilovKarakum	0.54 0.35	2.56 ± 0.287 1.33 ± 0.106	54 39	F = 19.323 df = 91	0.000
Kapchagaj reserve - Boguty1 (05.1987)	2.29 0.23	2.39 ± +0.144 1.73 ± +0.080	853 234	F = 11.839 df = 1085	0.001
Kapchagaj reserve - Boguty1 (06.1987)	2.29 0.23	2.39 ± 0.144 1.41 ± 0.113	853 56	F = 5.261 df = 907	0.022
Kapchagaj reserve - Boguty2 (08.1989)	3.67 0.27	3.17 ± 0.119 2.81 ± 0.322	1506 48	F = 1.872 df = 1552	0.171
Kapchagaj reserve - Taukum (12.1982)	1.65 0.22	4.05 ± 0.266 2.62 ± 0.299	266 31	F = 6.763 df = 295	0.010
Taukum - (09.1988) Chu Muunkum	0.31 0.17	1.88 ± 0.147 1.90 ± 0.204	69 24	F = 0.372 df = 87	0.543
Taukum (07.1983) Saryishikotrau– Chu Muunkum	0.22 0.07 0.10	1.50 ± 0.107 1.70 ± 0.300 1.90 ± 0.246	46 19 20	ANOVA F = 2.770 df = 2	0.155
Taukum – Saryishikotrau – BetpakDala (09.1986)	0.18 0.08 0.05	1.73 ± 0.159 1.95 ± 0.223 2.00 ± 0.246	30 22 12	ANOVA F = 0.532 df = 2	0.591

Blank et al. (2012a)—License Number—5477060538101

almost linearly. In my case, an increase in population density caused an increase in the size of groups only in a few cases, and in most months, there were indefinite fluctuations in the size of groups or even a decrease. Although there was a decrease in the size of the groups if the difference in population density was seven or more times smaller, this result was not obvious enough to talk about as an unambiguous dependence of group size on population density. In May, due to the solitary lifestyle of females, the size of groups decreases sharply, regardless of any increase in population density, and during the rut, the number of single males increases dramatically, and the size of the male groups decreases. At the same time, the size of female groups increases, resulting in an uncertain trend during the rut. During migrations, the size of the group increases, and the group gathers in numerous clusters, and all the biological processes of the annual cycle of goitered gazelles change the size of groups so much that an increase in population density does not have a visible effect. During the rest of the year, most of the goitered gazelle population leads a solitary or in small group lifestyle. Similar group size changes also are observed in Mongolia (Kaczensky et al. 2015) and China (Peng et al. 2024). Obviously, therefore, the relationship between group size and population density in goitered gazelles is closer

Table 12.5 Group size classes of the goitered gazelle in various populations with different density

Populations	Density ind/km^2	Singles/groups	Chi-square (χ^2)
Kapchagaj reserve Aktau (12.1986)	2.15 0.30	32/44 44/29	18.021, df = 1, $P = 0.000$
Kapchagaj reserve Aktau (06.1987)	2.29 0.54	58/33 41/46	38.574, df = 1, $P = 0.000$
Kapchagaj reserve PanfilovKarakum	2.29 0.35	58/33 74/26	584.616, df = 1, $P = 0.000$
Aktau - PanfilovKarakum	0.54 0.35	41/46 74/26	18.062, df = 1, $P = 0.000$
Kapchagaj reserve Boguty1 (05.1987)	2.29 0.23	58/33 62/35	263.108, df = 1,$P = 0.000$
Kapchagaj reserve Boguty1 (06.1987)	2.29 0.23	58/33 71/27	518.423, df = 1, $P = 0.000$
Kapchagaj reserve Boguty2 (08.1989)	2.75 0.27	41/43 38/46	7.040, df = 1, $P = 0.008$
Kapchagaj reserve Taukum (12.1982)	1.65 0.22	29/40 39/42	11.481, df = 1, $P = 0.001$
Taukum - (09.1988) Muunkum	0.31 0.17	48/49 40/60	2.391, df = 1, $P = 0.122$
Taukum- Saryishikotrau (07.1983)	0.22 0.07	63/37 50/50	3.130, df = 1, $P = 0.077$
Taukum – Muunkum (07.1983)	0.22 0.10	63/37 45/50	2.000, df = 1, $P = 0.157$
Saryishikotrau – Muunkum	0.07 0.10	50/50 45/50	0.053, df = 1, $P = 0.819$
Taukum – Saryishikotrau (09.1986)	0.18 0.08	50/50 42/58	0.519, df = 1, $P = 0.471$
Taukum – BetpakDala (09.1986)	0.18 0.05	50/50 33/67	3.320, df = 1, $P = 0.068$
Saryishikotrau – BetpakDala (09.1986)	0.08 0.05	42/58 33/67	0.712, df = 1, $P = 0.399$

to that in solitary and territorial species than in social ungulates. In mountain gazelles (*Gazella gazella*), when the population density is very low, these gazelles are usually seen singly or in pairs; when population density increases, groups of several individuals are observed (Walther et al. 1983). In Israel, large groups of mountain gazelles are not seen at all in spite of their high population density (Grau 1974), even though they did occur there at the beginning of the twentieth century when herds of 500 to 600 individuals were common (Mendelssohn 1974). Other subspecies of goitered gazelles (*G. s. marica* and *G.s. farasani*—Cunningham and Wronski 2011a, b) have the same social structure with a large proportion of single individuals, which obviously demonstrates that their populations also react little to an increase in population density.

Thus, goitered gazelles, with mixed characteristics between species that lead a solitary or territorial lifestyle and purely social ungulates living in large groups, demonstrate a very small relationship between population density and group sizes, with any significant effect only noticeable when the density changes by seven or more times. The size of the groups and the entire social structure are closely related to the food resources of the habitat and its distribution as group sizes grow with food availability (Geist 1974; Borkowski 2000; Isvaran 2007). For example, Soemmerring's gazelles (*Nanger soemmerringii*) gather in groups of 6–10 individuals (mean group size 4.4 ind/km) in areas of poor forage, but groups of 20–50, and up to 250 individuals (16.8 ind/km), are common in environments with greater amounts of forage (Gebremedhin and Yirga 2005), but if these resources are severely limited or widely scattered throughout the territory, then ungulates cannot form too large of groups for which there are not enough food resources; instead, they need to observe high selectivity and collect high-quality food by dispersing small-sized groups all over the area. Therefore, diet selectivity is negatively correlated with group size (Brashares et al. 2000). This is why goitered gazelles increase the size of their groups only up to a certain limit, after which an increase in population density can only lead to an increase in the number of groups and their wider distribution over their territory (Krause et al. 2002). At the same time, the necessities of their annual biological cycle are very important for the successful reproduction and maintenance of the population, which are also adapted to the requirements and characteristics of the environment, including protection from predators. Therefore, biological factors play a much greater role in the size of groups and the goitered gazelles' entire social structure. In fact, a weak response to an increase in population density and a high sensitivity of their social structure and yearly biological cycle lead to the optimal size of groups for the specific local conditions, creating a favorable prerequisite for sufficient nutrition, protection from predators, successful reproduction and prosperity, or even expanding of the population.

References

Barrette C (1991) The size of axis deer fluid groups in Wilpattu National Park, Shi Lanka. Mammalia 55:207–220

Bergstrom R, Skarpe C (1999) The abundance of large wild herbivores in a semi-arid savanna in relation to seasons, pans and livestock. Afr J Ecol 37:12–26

Bertram BC (1978) Living in groups: predators and prey. In: Krebs JR, Davies NB (eds) Behavioural ecology: an evolutionary approach. Blackwell, Oxford, pp 64–96

Blank DA (1985) Peculiarities of social and reproductive behaviour of *Gazella subgutturosa* in the Ili Valley. Zool Zhurnal 64:1059–1070. (in Russian, English abstract)

Blank DA (1998) Mating behavior of the Persian gazelle *Gazella subgutturosa* Güldenstaedt, 1780. Mammalia 62(4):409–419

Blank DA, Ruckstuhl KE, Yang W (2012a) Influence of population density on group sizes in goitered gazelle (*Gazella subgutturosa* Guld., 1780). Europ J Wildl Res 58(6):981–989

Blank D, Yang W, Xia C, Xu W (2012b) Grouping pattern of the goitered gazelle, *Gazella subgutturosa* (Cetartiodactyla: Bovidae) in Kazakhstan. Mammalia 76(2):149–155

Borkowski J (2000) Influence of the density of a sika deer population on activity, habitat use, and group size. Can J Zool 78:1369–1374

Brashares JS, Garland TJ, Arcese P (2000) Phylogenetic analysis of coadaptation in behavior, diet, and body size in the African antelope. Behav Ecol 11(4):452–463

Cunningham PL, Wronski T (2011a) Population structure of Farasan gazelle. Mammalia 75:157–161

Cunningham PL, Wronski T (2011b) Seasonal changes in group size and composition of Arabian sand gazelle *Gazella subgutturosa marica* Thomas, 1897 during a period of drought in central western Saudi Arabia. Curr Zool 57:36–42

Dalerum F, Lange H, Skarpe C, Rooke T, Inga B, Bateman PW (2008) Foraging competition, vigilance and group size in two species of gregarious antelope. S Afr J Wildl Res-24-month delayed open access 38(2):138–145

Delisle ZJ, Sample RD, Swihart RK (2023) Fear, concealment, and time of day interactively predict group size of a common ungulate. Ecosphere 14(10):e4685

Dementiev GP (1935) Mongolian gazelle, goitered gazelle, saiga antelope. KOIZ Publication, Moscow, pp 15–30. (in Russian)

Dereviagin PYA (1947) On seasonal migrations of goitered gazelle from southern pre-Balkhash deserts into inter-mountain valley Kalkan-Matay- Mynbulack. Bull Acad Sci Kazakh SSR 4(25):33–38. (in Russian)

Fortin D, Fortin ME, Beyer HL, Duchesne T, Courant S, Dancose K (2009) Group-size-mediated habitat selection and group fusion–fission dynamics of bison under predation risk. Ecology 90(9):2480–2490

Gebremedhin B, Yirga S (2005) A study on abundance, group size and composition of Soemmering's gazelle (*Gazella soemmerringii*) in awash National Park and Alledeghi wildlife reserve, Ethiopia. SINET: Ethiopian J Sci 28(2):161–170

Geist V (1974) On the relationship of social evolution and ecology in ungulates. Am Zool 14(1):205–220

Gorelov UK (1972) Breeding of goitered gazelle (*Gazella subgutturosa* Güldenstaedt) and problem of restoring of its number in the Badkhyz (SE Turkmenistan). In: Theriology, vol 1. Nauka of Siberian Department of Academy of Sciences of USSR, Novosibirsk, pp 420–424. (in Russian)

Grau GA (1974) Behavior of mountain gazelle in Israel. Ph.D. dissertation. Texas A&M Univ, Texas

Hebblewhite M, Pletscher DH (2002) Effect of elk group size on predation by wolves. Can J Zool 80:800–809

Heptner G, Nasimovich AA, Bannikov AG (1961) Mammals of Soviet Union. In: Artiodactyla and Perissodactyla, vol 1. Vysshaya Shkola Publishing House, Moscow

Isvaran K (2007) Intraspecific variation in group size in the blackbuck antelope: the roles of habitat structure and forage at different spatial scales. Oecologia 154(2):435–444

Jarman RJ (1974) The social organization of antelope in relation to their ecology. Behaviour 48:215–267

Johnson CN (1983) Variations in group size and composition in red and western grey kangaroos, *Macropus Rufus* (Desmarest) and *M. Fuliginosus* (Desmarest). Aust Wildl Res 10:25–31

Jung TS (2020) Investigating local concerns regarding large mammal restoration: group size in a growing population of reintroduced bison (*Bison bison*). Global Ecol Conserv 24:e01303

Kaczensky P, Ganbaatar O, Altansukh N, Enksaikhan N, Kramer-Schadt S (2015) Monitoring of khulans and goitered gazelles in the Mongolian Gobi- potential and limitations of ground-based line transects. Open Ecol J 8:92–110

Karczmarski L, Wursig B, Gailey G, Larson KW, Vanderlip C (2005) Spinner dolphins in a remote Hawaiian atoll: social grouping and population structure. Behav Ecol 16:675–685

Kostin VP (1955) Materials on ecology and commercial value of goitered gazelle. Proceedings of the institute of zoology, vol. 4. Publication of Academy of Sciences of Kazakh SSR, Alma-Ata, pp 67–82. (in Russian)

Krause J, Ruxton GD, Ruxton G, Ruxton IG (2002) Living in groups. Oxford University Press, Oxford

Lawes MJ, Nanni RF (1993) The density, habitat use and social organization of Dorcas gazelles (*Gazella dorcas*) in Makhtesh Ramon, Negev Desert, Israel. J Arid Environ 24:177–196

Lehmann J, Boesch C (2004) To fission or to fusion: effects of community size on wild chimpanzee (*Pan troglodytes verus*) social organization. Behav Ecol Sociobiol 56:207–216

Mambetzhumaev AM (1970a) Goitered gazelle. FAN Publication, Tashkent, 196 pp. (in Russian)

Mambetzhumaev AM (1970b) On the number and distribution of goitered gazelles in the Ustyurt plateau and Kyzylkum Desert. Zool Zhurnal 49(3):431–434. (in Russian)

Mattiello S, Zanoni C, Du Plessis H, Heinzl E, Crimella MC (2004) Habitat use and group size of African wild ungulates in a Namibian game ranch. Game Wildlife Sci 21(4):735–745

Mendelssohn H (1974) The development of the populations of gazelles in Israel and their behavioral adaptations. Int Union Conserv Nat Natural Resour Publications New Series 24:722–743

Owen-Smith N (2006) Demographic determination of the shape of density dependence for the three African ungulate populations. Ecol Monogr 76(1):93–109

Owen-Smith N, Chirima GJ, Macandza V, Le Roux E (2012) Shrinking sable antelope numbers in Kruger National Park: what is suppressing population recovery? Anim Conserv 15(2):195–204

Peek JM, LeResche RE, Stevens DR (1974) Dynamics of moose aggregations in Alaska, Minnesota and Montana. J Wildl Manag 55:126–137

Peignier M, Webber QM, Koen EL, Laforge MP, Robitaille AL, Vander Wal E (2019) Space use and social association in a gregarious ungulate: testing the conspecific attraction and resource dispersion hypotheses. Ecol Evol 9(9):5133–5145

Pendharkar A, Goyal SP (1995) Group size and composition of the gray goral in Simbalbara sanctuary and Darpur reserved Forest, India. J Mammal 76(3):906–911

Peng D, Zhang Z, Chen J, Meng D, Liang Y, Hu T, Teng L, Liu Z (2024) The activity patterns and grouping characteristics of the remaining goitered gazelle (*Gazella subgutturosa*) in an isolated habitat of Western China. Animals 14(16):2338

Pepin D, Gerard JF (2008) Group dynamics and local population density dependence of group size in the Pyrenean chamois, *Rupicapra pyrenaica*. Anim Behav 75:361–369

Reyana-Hurtado RR, Naranjo E, Chapman CA (2009) Hunting and the conservation of a social ungulate: the white-lipped peccary *Tayassu pecari* in Calakmul, Mexico. Oryx 44(1):89–96

Roberts G (1996) Why individual vigilance declines as group size increases. Anim Behav 51:1077–1086

Rowe-Rowe DT, Everett PS, Perrin MR (1992) Group sizes of oribis in different habitats. South Afr J Zool 27:140–143

Sludskiy AA (1956) Breeding of goitered gazelle. Proceedings of institute of zoology, Kazakh Academy of Sciences, Vol. 6. Nauka Press of Kazakh SSR, Alma-Ata, pp 78–108. (in Russian)

Sludskiy AA (1977) Goitered gazelle. In: Rare animals of USSR. Publication of Lesnaya Promyshlennost, Moscow, pp 28–61. (in Russian)

Sokolov II (1959) *Gazella subgutturosa* Guldenstaedt—Goitered gazelle. In: Pavlovskiy EN (ed) Fauna USSR, mammals: ungulates (orders Perissodactyla and Artiodactyla). Publication of the Academy of Sciences USSR, Moscow, pp 308–331. (in Russian)

Stuwe M, Hendrichs H (1984) Organization of roe deer *Capreolus capreolus* in an open habitat. Z Sauget 49:359–367

Sweanor PY, Sandegren F (1989) Winter-range philopatry of seasonally migratory moose. J Appl Ecol 26:25–33

Thirgood SJ (1996) Ecological factors influencing sexual segregation and group size in fallow deer (*Dama dama*). J Zool 239:783–797

Toïgo C, Gaillard JM, Michallet J (1996) La taille des groupes: un bioindicateur de l ' effectif des populations de bouquetin des Alpes (*Capra ibex ibex*)? Mammalia 60:463–472

Vasenko EP (1950) Ecology and distribution of the saiga antelope. Proceedings of the Barsakelmes state nature reserve, issue 1. Kazakh State Publication, Alma-Ata, pp 38–115. (in Russian)

Wal V, Eric FM, Beest V, Brook RK (2013) Density-dependent effects on group size are sex-specific in a gregarious ungulate. PLoS One 8(1):e53777

Walther FR, Mungall EC, Grau GA (1983) Gazelles and their relatives. A study in territorial behavior. Noyes Publication, Park Ridge

Ward A, Webster M (2016) Sociality: the behaviour of group-living animals. Springer, New York

Wirtz P, Lorscher J (1983) Group size of antelopes in an east African national park. Behaviour 84:135–155

Wittemyer G, Douglas-Hamilton I, Getz WM (2005) The socioecology of elephants: analysis of the processes creating multitiered social structures. Anim Behav 69:1357–1371

Zhevnerov VV (1984) Goitered gazelle of the Barsakelmes Island. Nauka Publishing House of Kazakh SSR, Alma-Ata. (in Russian)

Segregation

13

13.1 General Remarks

There are ungulate species, in which males have considerably larger body sizes compared to females and outside of the rutting season, they typically stay separate from each other in unisex groups; this phenomenon is called sexual segregation, and it is quite widespread among ungulates (Main et al. 1996; Bleich et al. 1997; Ruckstuhl and Neuhaus 2002). Sexual segregation is a characteristic of mostly social ungulates, in which males and females have noticeable body size dimorphism, pronounced seasonality in reproduction, and free movement of females and males most of the year without individual male territories or monopolization of females for mating (Ruckstuhl and Neuhaus 2000). Sexual segregation is typical for ungulate species with a well-marked difference between the body sizes of the sexes, where males generally are at least 20% larger than females. Body size dimorphism of sexes plays a particularly important role in the emergence of sexual segregation, as the greater the dimorphism, the more likely it is that males and females prefer to be in separate groups (Mysterud 2000). Although the difference in body size is obviously the main factor leading to sexual segregation, additional factors also have impact on this process. For example, reproductive status, sex ratio, population density, and food abundance can influence the degree and duration of sexual segregation (Ruckstuhl et al. 2006). There is also social segregation, when males and females stay in separate groups most of year except for breeding season, and because of different environmental demands, males and females become segregated by habitat, as well (Clutton-Brock et al. 1982).

Sexual segregation can probably be explained by the activity budget hypothesis better than the predation-risk and the forage-selection hypotheses (Ruckstuhl 2007), although each of the hypotheses is partially confirmed by field research (Ciuti et al. 2004; Calhim et al. 2006; Loe et al. 2006) and they are not mutually exclusive (Main et al. 1996). With an increased difference in body size, males and females are increasingly showing a difference in their activity in order to fully satisfy their food

D. Blank, *The Behavioral Biology and Conservation of Goitered Gazelles (Gazella subgutturosa)*, https://doi.org/10.1007/978-3-031-83504-9_13

needs. Larger males need more food, but have relatively less energy requirements; therefore they are more tolerant of lower feed quality compared to smaller females (Mysterud 2000; Barboza and Bowyer 2000). Females, on the other hand, prefer higher-quality food, especially pregnant and lactating mothers, and have to graze for a longer time in order to collect more scattered high-quality food (Mellado et al. 2005). Males also need a longer time to chew and digest large amounts of lower-quality food, so they have a greater need to lie down and chew the cud more thoroughly. Whereas females walk more for grazing and do not need as long resting time to masticate (Ruckstuhl 1998). From all this, it can be seen that males and females have different needs for the duration of grazing and walking, as well as for the duration of rest. Therefore, being in the same herd, adult males and females must expend too much energy to synchronize the activity of all herd members and as a consequence they separate into unisex male and female groups, within which members have similar constraints and behaviors and a high synchronization of activities. That is why males and females in species with high body size dimorphism unite only for the rut, after which they quickly break up into separate unisex groups because of high energy costs and high instability of mixed-sex groups (Bon and Campan 1996; Ruckstuhl and Neuhaus 2002; Villerette et al. 2006).

Many species of caprids (*Capra ibex*, *Capra nubiana*, *Ovis canadensis*, *O. dalli*, *O. musimon*, *Rupicapra pyrenaica*) and cervids (*Cervus elaphus*, *C. nippon*, *Odocoileus hemionus*, *O. virginianus*, *Rangifer tarandus*) have a striking difference in body sizes, when males are 1.5 to 2 times larger than females and all have distinctive sexual segregation, when adult males and females live in separate unisex groups most of a year, gathering into their mixed-sex groups mostly during the rut. As for antelopes, some of them have actual body size dimorphism and related sexual segregation, including impala *Aepycerus melampus*, bontebok *Damaliscus dorcas*, gerenuks *Litocranius walleri*, *Tragelaphus spekei*, *T. strepsiceros* (Ruckstuhl and Neuhaus 2002). Small African antelopes do not have body size dimorphism at all, and they are also territorial, so sexual segregation is not found these species (*Madoqua kirkii*; *Oreotragus oreotragus*). In addition, many other antelopes have insignificant differences in body sizes across sexes, including *Antilocapra americana*, and *Oryx gazella*. Thomson's gazelles (*Eudorcas thomsonii*) and both species of wildebeests (*Connochaetes*) are assigned to a separate species category that does not reach the barrier of a 20% difference in body size between the sexes and can form temporary mixed groups together with territorial males. Goitered gazelles belong to antelopes which have body size dimorphism of up to 30% like impalas or bontebok, but can create temporaary mixed groups with territorial males similar to Thomson's gazelles.

13.2 Segregation in Goitered Gazelles

One of the most important indicators of the presence of sexual segregation in ungulate species is the proportion of mixed groups among all other categories of social groups. During my study in the Ili Hallow, mixed-sex groups were 4.6% of all groups observed and in which 10.2% of all individuals were contained. This is much lower compared to female unisex groups (53.8%, n = 6.017) and male groups (34.0%). The most numerous mixed-sex groups were found in April and October and most rarely observed in August and September. April is the time of the spring migration, when goitered gazelles gather in large herds and migrate to their spring pastures and birth sites, and in October is their autumn migration, when the reverse process occurs and goitered gazelles gather in large groups to migrate to their places for rutting and wintering. As for the rutting period, when females and males should be in mixed groups, the proportion of these groups was higher in the initial rutting period in November, than in other months except April and October. In the final period of the rut in December, the proportion of mixed groups was almost the same as outside the rut (Fig. 13.1).

At first glance, this phenomenon of a small number of mixed groups contradicts logic, since males and females must be together at least to breed. This contradiction is explained by the peculiarities of goitered gazelle reproduction, when female groups only cross the network of individual territories of males during diurnal migrations, but do not stay in each territory for very long. After 10–15 min, they leave one male's territory for the next, crossing the boundaries of the neighboring territory (Blank 1998). Such associations of a single male and a group of females are difficult to call a uniform group, so they were recorded separately as a single male and a group of females. Only at the end of the rutting period in December, when most females become ready to mate will they stay in a territory of a certain male for a longer, sometimes up to 3 days. Just during the mating period can a group of females and a territorial male be considered as a temporary group, while immediately after mating, the females leave the male's territory again and the percentage of mixed groups dropped again to its minimum. This is why the proportion of mixed

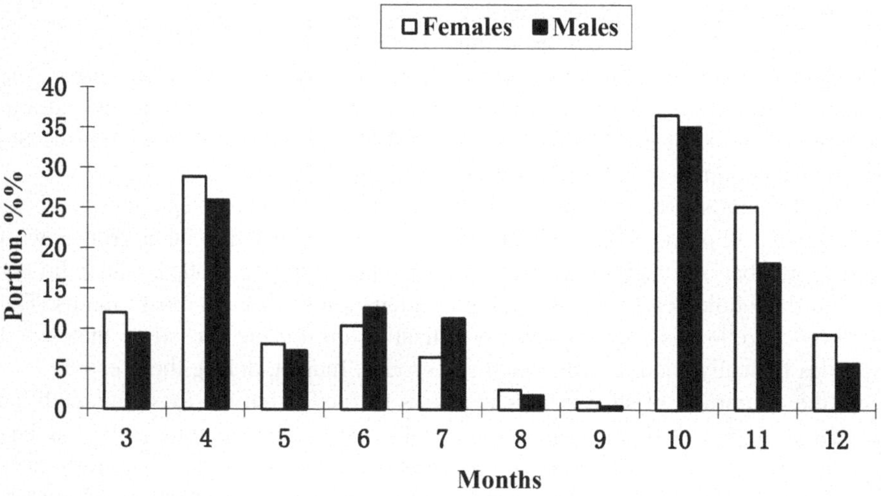

Fig. 13.1 Portions of females and males in mixed-sex groups

groups was unexpectedly lower compared to the periods of spring and autumn migrations.

Another important indicator by which the presence or absence of sexual segregation in ungulates can be determined is the proportion of adult females and males in mixed groups compared to those in unisex male or female groups. Similar to the proportion of mixed-sex groups, the number of adult males and females in mixed and unisex groups varied significantly over seasons. The maximal percentage of adult gazelles of both sexes in mixed-sex groups compared unisex groups was also higher in April and October, and the lowest percentage of males and females in mixed groups was observed in August and September without any significant difference in other seasons (Blank et al. 2012). At the beginning of the rut in November, the portion of females was significantly higher compared to other months except for the migration periods in April and October. However, the percentage of males in mixed-sex groups was only partially significant—four cases across 10 months (Fig. 13.2). At the end of the rut in December, the proportion of adult gazelles of both sexes in mixed groups vs unisex herds dropped significantly compared to November and was almost the same with other months at a lower level of 13%. Only during the spring and autumn migrations in April and October, and during the first part of the rut in November was this proportion higher than 13% (Blank et al. 2012).

Finally, the most complex and special Conrad's segregation coefficient (SC) (Conradt 1998), which is calculated by the formula.

$SC_{social} = 1—[(N—1)/X \cdot Y] \cdot \Sigma[x_i \cdot y_i /(n_i—1)]$, where $N (X + Y)$ is the total number of adult gazelles registered; X is the total number adult males; Y is the total number of adult females; x_i is the number of adult males in ith group; y_i is the number of females in ith geoup; n_i is the total number of adults in n_i group ($n_i = x_i + y_i$). SC can fluctuate between 0 (no segregation) and 1 (maximal segregation).

The SC as applied to the goitered gazelles in the Ili Hollow showed as a population they were completely segregated in August (0.98) and September (0.94). At the beginning of the rut in November, the segregation coefficient was lower (0.80), because more females were in the company of males, but segregation was higher in December with a higher segregation coefficient (0.86) compared to November. The minimal level of the segregation coefficient was observed during spring and autumn migrations, in April and October (0.71 and 0.67). In the rest of the year, the SC fluctuated at quite a high level, between 0.86 and 0.91 (Blank et al. 2012).

All three indicators show that goitered gazelles have sexual segregation. A surprisingly small proportion of mixed groups existed compared to other groups, with a great number of females and males in their separate unisex groups, which did not mix of the whole year; there was a high segregation coefficient most months. The goitered gazelle has an interesting organization during the rut, when males and females logically should form mixed-sex groups. Indeed, during the breeding season, the proportion of such groups increases markedly, but not as significantly as expected, since sexual segregation between the sexes remained quite high even during the rutting season. The point is that females and territorial males can form only temporary groups that break up within minutes, and only for a short period of mating (up to 3 days), when a female stays in a male's territory for mating. Such a short

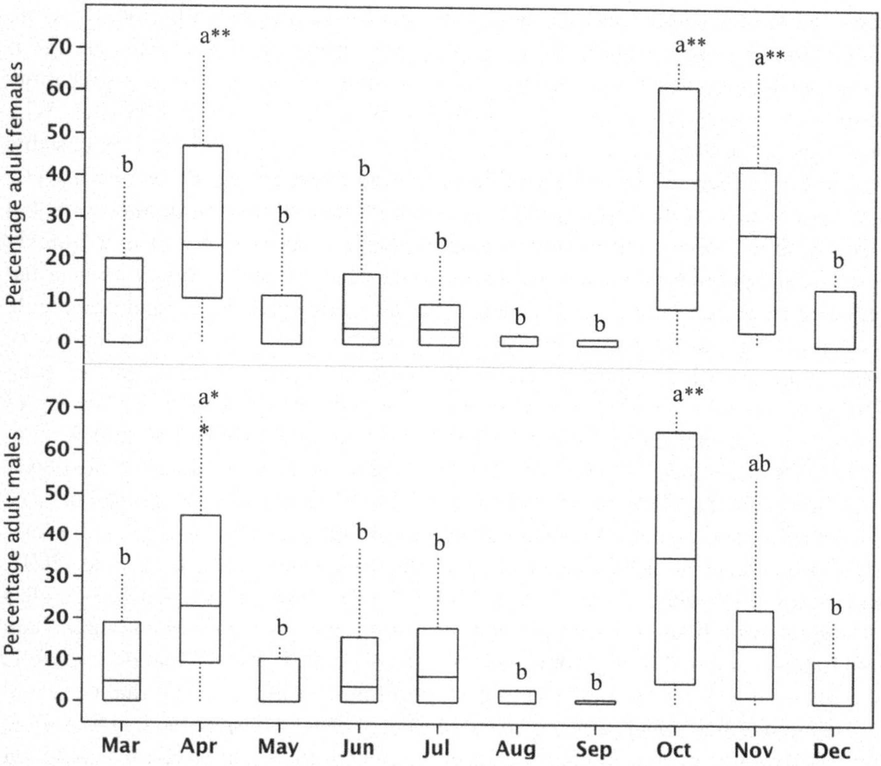

Fig. 13.2 Proportion (box plots with confidence intervals) of adult female and male goitered gazelles (*Gazella subgutturosa*) found in mixed-sex groups from March to December 1981–1987 in Ili Hollow, SE Kazakhstan. Means and standard errors are shown. Different letters above each box plot indicate significant differences in values, whereas both letters above the one box plot indicate partial significance, based on one-way ANOVA followed by a Fisher's least significant differences (LSD) post hoc test *, $P < 0.05$; **, $P < 0.001$ (Blank et al. 2012—License Number-)

period for the formation of a mixed-sex groups is associated with high synchronization of mating in most females, allowing for mass birthing to accommodate the short period of spring and forage abundance before the onset of the summer heat and drought. During the rut, there are also mixed groups of bachelors with females, but these groups appear and break up also very quickly.

Only during the spring and autumn migrations are there many mixed-sex groups that move to the birthing sites in spring and the rutting sites in autumn. Migrations last no more than 7–10 days, when goitered gazelles move in one direction in a continuous wave. It is difficult to say how long these mixed groups last or if they are even a group, since there is a strong impression that they are more casual fellow travelers, merging and separating, than members of the same group and they break up quite quickly.

Thus, it can be said that mixed groups are very short-lived in goitered gazelles and sexual segregation is observed almost all year round, and even during the rut, segregation between the sexes is quite high. As for the migration period, these are rather temporary aggregations of male and female groups than real mixed groups. It turns out that goitered gazelles have a very high degree of sexual segregation, despite the relatively low body size dimorphism between males and females. Why does this particular species have such a high degree of segregation? It is possible that not only the high synchronization of mating and birthing allows this gazelle species to successfully reproduce in the harsh climate of the northern deserts, but also the ability to give birth to twins and even triplets which allows them to quickly increase the population during favorable years, which do not occur every year but replaced by dry and hungry years and do not allow this gazelle to quickly increase its population continuously. Therefore, females are engaged in intensive feeding of their numerous (2–3) offspring during the short but most favorable time of year. This explains why females and males are separated during birthing and rearing of fawns, when sexual segregation is peaked (Bon et al. 2001; Ciuti et al. 2004).

Social or habitat segregation of adult males and females is typical for most body size dimorphic ungulate species, when sexes stay in separate unisex groups most of the year, but their gather into united mixed-sex groups during the rut season for mating (Miquelle et al. 1992; Main et al. 1996; Ruckstuhl 1998; Bon et al. 2001; Ruckstuhl and Neuhaus 2002; Bowyer 2004). Unlike other species, goitered gazelles exhibit fairly high rates of segregation even during rutting season, when females can join to territorial males only for a short time to mate. It is obvious that other gazelle species have a similar social system during rutting and likely a high rate of sexual segregation (e.g., *Procapra przewalskii*—Lei et al. 2001), despite the fact that some authors wrongly attribute all gazelle species to the group of ungulates, which do not have sexual segregation at all (Yearsley and Pérez-Barbería 2005), but other authors (Ruckstuhl and Neuhaus 2002) consider temporary associations of females with territorial males as temporary mixed groups and define them into a separate category of segregated species (for example, Thomson gazelles). I believe that goitered gazelle females remain separated from territorial males within their territories, since the visits of females are very short without synchronized activities with the male, which is necessary for group formation (Meldrum and Ruckstuhl 2009). This is why segregation of the goitered gazelle looks so unique.

References

Barboza PS, Bowyer RT (2000) Sexual segregation in dimorphic deer: a new gastrocentric hypothesis. J Mammal 81(2):473–489

Blank DA (1998) Mating behavior of the Persian gazelle *Gazella subgutturosa* Guldenstaedt, 1780. Mammalia 62(4):409–419

Blank DA, Ruckstuhl KE, Yang W (2012) Sexual segregation in goitered gazelles (*Gazella subgutturosa*). Can J Zool 90:955–960

Bleich VC, Bowyer RT, Wehausen JD (1997) Sexual segregation in mountain sheep: resources or predation? Wildl Monogr 134:1–50

Bon R, Campan R (1996) Unexplained sexual segregation in polygamous ungulates: a defense of an ontogenetic approach. Behav Proc 38:131–154

Bon R, Rideau C, Villaret JC, Joachim J (2001) Segregation is not only a matter of sex in alpine ibex, *Capra ibex ibex*. Anim Behav 62(3):495–504

Bowyer RT (2004) Sexual segregation in ruminants: definitions, hypotheses, and implications for conservation and management. J Mammal 85(6):1039–1052

Calhim S, Shi J, Dunbar RIM (2006) Sexual segregation among feral goats: testing between alternative hypotheses. Anim Behav 72(1):31–41

Ciuti S, Davini S, Luccarini S, Apollonio M (2004) Could the predation risk hypothesis explain large-scale spatial sexual segregation in fallow deer (*Dama dama*)? Behav Ecol Sociobiol 56:552–564

Clutton-Brock TH, Guinness FE, Albon SD (1982) Red deer: behavior and ecology of two sexes. University of Chicago press, Chicago

Conradt L (1998) Measuring the degree of sexual segregation in group-living animals. J Anim Ecol 67(2):217–226

Lei R, Jiang Z, Liu B (2001) Group pattern and social segregation in Przewalski's gazelle (*Procapra przewalskii*) around Qinghai Lake, China. J Zool 255(2):175–180

Loe LE, Irvine RJ, Bonenfant C, Stien A, Langvatn R, Albon SD, Mysterud A, Stenseth NC (2006) Testing five hypotheses of sexual segregation in an arctic ungulate. J Anim Ecol 75(2):485–496

Main MB, Weckerly FW, Bleich VC (1996) Sexual segregation in ungulates: new directions for research. J Mammal 77:449–461

Meldrum GE, Ruckstuhl KE (2009) Mixed-sex group formation by bighorn sheep in winter: trading costs of synchrony for benefits of group living. Anim Behav 77(4):919–929

Mellado M, Rodriguez A, Villarreal JA, Rodriguez R, Salinas J, López R (2005) Gender and tooth wear effects on diets of grazing goats. Small Rumin Res 57(2–3):105–114

Miquelle DG, Peek JM, Van Ballenberghe V (1992) Sexual segregation in Alaskan moose. Wildl Monogr 122:1–57

Mysterud A (2000) The relationship between ecological segregation and sexual body size dimorphism in large herbivores. Oecologia 124:40–54

Ruckstuhl KE (1998) Foraging behaviour and sexual segregation in bighorn sheep. Anim Behav 56(1):99–106

Ruckstuhl KE (2007) Sexual segregation in vertebrates: proximate and ultimate causes. Integr Comp Biol 47(2):245–257

Ruckstuhl KE, Neuhaus P (2000) Sexual segregation in ungulates: a new approach. Behaviour 137(3):361–377

Ruckstuhl KE, Neuhaus P (2002) Sexual segregation in ungulates: a comparative test of three hypotheses. Biol Rev Camb Philos Soc 77(1):77–96

Ruckstuhl KE, Manica A, MacColl ADC, Pilkington JG, Clutton-Brock TH (2006) The effects of castration, sex ratio and population density on social segregation and habitat use in Soay sheep. Behav Ecol Sociobiol 59:694–703

Villerette N, Helder R, Angibault JM, Cargnelutti B, Gerard JF (2006) Sexual segregation in fallow deer: are mixed-sex groups especially unstable because of asynchrony between the sexes? Comptes rendus. Biologies 329(7):551–558

Yearsley JM, Pérez-Barbería FJ (2005) Does the activity budget hypothesis explain sexual segregation in ungulates? Anim Behav 69(2):257–267

Sex Ratio

14

14.1 General Remarks

The sex ratio is one of the most important topics in the discussion of evolutionary mechanisms at the population level (Cameron 2004). From an evolutionary biology point of view, mothers of different physical conditions produce young of different sexes, according to their further contribution to the propagation of the population and the mothers' own reproductive success. According to the Trivers-Willard hypothesis (Trivers-Willard 1973), a mother in better physical conditions will produce sons, which demand more maternal investment to rear than daughters (Clutton-Brock et al. 1981; Wolff 1988); however, male offspring are more beneficial than females, because successful males in polygynous species can produce many more offspring than females (Clutton-Brock et al. 1982). In contrast, females in poor condition will birth daughters, because daughters are more viable than sons and will produce more offspring than unsuccessful sons who received insufficient maternal investment because of forage scarcity. This hypothesis has three assumptions: (a) phenotype quality of young at weaning closely depends on the mother's physical quality; (b) adult phenotypic quality of ungulates has a close relationship with their quality at weaning; and (c) the quality of adult males affects the breeding success more than females.

In some ungulate species, the sex of newborns can be different depending on the mother's social status, which has a strong effect on the maternal care. As a result, dominant mothers give births mainly to sons, which are heavier at birth and require a greater and earlier maternal investment. Young females of Saharan arrui (*Ammotragus lervia*), for example, produce a single female in their first birth, with subsequent yearly births following the sequence FF>M>MF>MM along with the

Blank and Yang (2013)—License Number—5477051490285

D. Blank, *The Behavioral Biology and Conservation of Goitered Gazelles (Gazella subgutturosa)*, https://doi.org/10.1007/978-3-031-83504-9_14

increasing social status of the mother (Cassinello and Gomendio 1996). Many studies on various mammals, including ungulates, have confirmed this hypothesis that in good conditions mothers produce more males, and in bad conditions, they produce more females: red deer *Cervus elaphus* (Clutton-Brock et al. 1984), American bison *Bison bison* (Rutberg 1986), reindeer *Rangifer tarandus* (Kojola and Eloranta 1989), mule deer *Odocoileus hemionus* (Kucera 1991), pronghorn (*Antilocapra americana*—Hewison and Gaillard 1999), snow goat *Oreamnos americanus* (Côté and Festa-Bianchet 2001). On the contrary, in ungulate species whose quality of care does not depend on a female's dominance in the herd hierarchy, the predictions of the Trivers-Willard's hypothesis (1973) turn out to be incorrect. For example, in bighorn sheep (*Ovis canadensis*), the lamb sex ratio is independent of the quality of a mother's care (Blanchard et al. 2004).

However, other authors believe that empirical results actually contradict the Trivers-Willard hypothesis both between species and among populations (Pélabon et al. 1995; Hewison and Gaillard 1999; Hewison et al. 1999), because changes in the adaptation of the sexual ratio also depend on many other environmental factors that were obviously not taken into account when developing the initial hypothesis (Kruuk et al. 1999; Rosenfeld and Roberts 2004; West et al. 2005) and assumptions of this hypothesis are found only for limited ungulate species (*Cervus elaphus*, *Rangifer tarandus*, *Dama dama*, *Ovis canadensis*, *O. aries* and *Equus caballus*—Hewison and Gaillard 1999; Blanchard et al. 2004; Cameron and Linklater 2007). In addition, there is an "advantaged daughter hypothesis" which is theoretically similar to the Trivers-Willard hypothesis but has opposite predictions: mothers in good conditions have daughters in which they invest more care and mothers in poor conditions birth more males, which receive poorer care (Hiraiwa-Hasegawa 1993; Hewison and Gaillard 1999). There is one more hypothesis for consideration: the "local resource competition hypothesis," which states that it is better for females in poor conditions to produce sons, because males ultimately disperse more than daughters and thus compete less with their mothers (Clark 1978; Silk 1983). Some researches of white-tailed deer (*Odocoileus virginianus*—Verme 1983; Caley and Nudds 1987), reindeer (Skogland 1986), roe deer (*Capreolus capreolus*—Hewison and Gaillard 1996; Hewison et al. 1999), and springbok *Antidorcas marsupialis* (Krüger et al. 2005) support this hypothesis.

In this regard, it would be interesting to consider fluctuations in the sexual ratio in goitered gazelles, which have a fission–fusion type of groups that quickly change their composition and, as a consequence, have an unstable hierarchy (Marmazinskaya 1999). In addition, ungulate species with less sexual dimorphism show a declining sex-bias ratio (Byers and Moodie 1990). All these suggest that gazelles are unlikely to be among the species that fall within the definition of the Triver-Willard hypothesis.

14.2 Sex Ratio in the Barsakelmes Island and Betpak-Dala Desert

The sex ratio of gazelles at birth is close to 1:1, with some predominance of females, so in Betpak-Dala in 1955, out of 59 large embryos and newborn fawns, 28 were found to be males (47.4%) and 31 females (52.6%), and in 1956, out of 12 embryos, 4 were male (33.3%) and 8 were female (66.7%), respectively. On the Barsakelmes Island, Zhevnerov (1984) found, out of 80 newborns, there were 42 (52.5%) females and 38 (47.5%) males. Of the 47 mothers of these newborns, 10 mothers (21.3%) had a male and female twins, 12 (25.5%) had 2 females twins, 11 (23.4%) had 2 male twins, 6 (12.8%) had a single male, and 8 (17.0%) had a single female, for a total of 46.8% males and 53.2% females. Twins are more often male and female (41.7%) than 2 females (22.3%). Among mature gazelles, this ratio is changing noticeably towards the predominance of females over males. This is facilitated by the earlier sexual maturity of females and increased mortality of males in winter. During the rut, males lose significant weight (up to 30%) and strength, becoming more sensitive to adverse environmental factors and a shortage of feed. The ratio of winter deaths of females and males among adults varies in different years from 1: 2.3 to 1: 6.2. On average, there are four dead males per one dead female. In the group of young animals, the gender deviation is closer to 1: 1 with a slight predominance of males (Zhevnerov 1984). Generally, goitered gazelle populations are 36–50% female, 3–36% male, and 15–51% young (Kingswood and Blank 1996).

Females are strongly predominant among adults. According to Sludskiy's (1956) records, in the Betpak-Dala Desert over a number of years (in May), gazelle males made up 5.4% to 21.1% (on average 8.1%) and gazelle females 78.9% to 94.6% (average 91.8%) of the population. For a number of years on the Barsakelmes Island, adult males made up on average of 13.4%, adult females 40.4% and young 46.2% of the population (Table 14.1). The small number of adult males is explained by their annually high death rate after the rut as a result of bad weather, such as dzhuts (unusual severe winter with a thick snow cover), and predation. The sharpest disproportion in a group of mature individuals is observed after winters with severe snow, so after the winters of 1971–1972 and 1975–1976, there were 12.8 and 9.4

Table 14.1 The sex and age content of observed groups in goitered gazelles in the Barsakelmes Island (1970–1976) (adapted from Zhevnerov 1984)

Year	Total number of gazelles	Including						Ratio females to males
		Adult males		Adult females		Fawns		
		Number	%	Number	%	Number	%	
1970	180	34	18.9	63	35.0	83	46.1	1.8
1971	250	54	21.6	79	31.6	117	46.8	1.5
1972	110	5	4.5	64	58.2	41	37.3	12.8
1973	154	22	14.3	41	26.6	91	59.1	1.9
1974	250	22	8.8	96	38.4	132	52.8	4.4
1975	364	72	19.8	141	38.7	151	41.5	1.9
1976	328	18	5.5	170	51.8	140	42.7	9.4
Average	–	32	13.9	93	40.0	107	46.1	4.8

females per male, respectively (Table 14.1) In the autumn period, there were from 1.4 to 12.8 females per adult male, or 4.8. on average (Zhevnerov 1984).

The dynamics in the number of gazelles on the Barsakelmes Island is determined by a number of interrelated factors, the main ones being isolation of the island, the island's meteorological influences during winter, and the availability of food. Barsakelmes Island has the characteristics of a northern (temperate) deserts of Eurasia. One of the features of this zone is the periodically recurring mass death of ungulates from winter starvation and climate: a high, dense snow cover, ice, frost, prolonged blizzards. The complex of these adverse weather conditions causes exhaustion and death among ungulates (Sludskiy 1963).

In a breeding season that experiences dzhut weather conditions, the overall gazelle population increases less than usual. For example, in August 1951 after the dzhut of 1950–1951, out of 203 gazelles counted in the Betpak-Dala, 110 (54.2%) were fawns, 86 (42.3%) were adult females, and only 7 (3.5%) were adult males (Zhevnerov et al. 1983).

Body size dimorphism in goitered gazelles is not so pronounced as in caprids (Kingswood and Blank 1996; Ruckstuhl and Neuhaus 2002; Cunningham et al. 2011) and a bias in the sex ratio is usually not high. Therefore, the sexual ratio at birth is close to parity in goitered gazelles (Zhevnerov 1984; Cunningham and Wronski 2011). However, males have higher mortality compared to females, which higher mortality starting from the fetus stage until adulthood, with an especially high mortality rate in unfavorable conditions (Cunningham and Wronski 2011).

The same mortality imbalance can be seen in Cuvier's gazelles, with the death rate among males during the first month of life is found to be significantly higher compared to females of the same age (Alados and Escós 1994), and populations of other gazelle species usually have a distinctive female bias as well: *Nanger granti* and *Gazella gazella* have up to two females per male (Walther 1972; Baharav 1983), and *Gazella bennetti* and *Gazella thomsonii* have up to 2.8 females per male (FitzGibbon 1990; Rahmani 1990). In some cases, such as in *G. thomsonii*, a six-fold disparity can exist between sexes (FitzGibbon and Lazarus 1995) and in *G. subgutturosa*, there can be even a 12-fold disparity (Zhevnerov 1984).

14.3 Sex Ratio in the Ili Hollow

During my 6-year observations of goitered gazelles in the Ili Valley (1981–1986) (Table 14.2), 38.2% were adult males and 34.8% were females. This means that the ratio of males to females was 49:51 averaged over the entire monitoring period. Despite the fact that the ratio of females to males fluctuated very widely from month to month, from 23% to 70%, these fluctuations had a certain pattern, repeating from year to year with the sex ratio turning out to be the same over the years. In most cases, deviations in the proportions of females to males ranged between 40% and 60%, and no statistical differences were found between the months. The only exception was May, when the ratio had a significant difference with other months except March and April, and if in most months, there was an excess in the number of

Table 14.2 The deviation of observed sex ratio (females: males, %) from the equal Proportion of sexes (50: 50) over months during different years and result of the Binomial test of the female male proportion (value of *P* and *N*)

Months	1981	1982	1983	1984	1985	1986
March			49:51 0.836 *N* = 210	33:67 0.000 *N* = 124		
April		51:49 0.587 *N* = 487			38:62 0.000 *N* = 582	
May		33:67 0.000 *N* = 392	41:59 0.000 *N* = 362	27:73 0.000 *N* = 429	40:60 0.000 *N* = 634	29:71 0.000 *N* = 906
June	56:44 0.111 *N* = 191		43:57 0.251 *N* = 92	59:41 0.000 *N* = 524	60:40 0.000 *N* = 765	54:46 0.074 *N* = 428
July	70:30 0.007 *N* = 50	50:50 0.000 *N* = 245		56:44 0.068 *N* = 220	62:38 0.000 *N* = 1030	61:39 0.000 *N* = 525
August		53:47 0.088 *N* = 634				
September		47:53 0.731 *N* = 76				
October	49:51 0.868 *N* = 145					
November	53:47 0.311 *N* = 389	61:39 0.019 *N* = 114			67:33 0.000 *N* = 290	61:39 0.000 *N* = 375
December				53:47 0.457 *N* = 181	59:41 0.067 *N* = 108	46:54 0.422 *N* = 155

females over males, to some extent, especially significant in November (up to 1:0.49—Fig. 14.1), then in March and April (1 of 2 observed years), and particularly in May (all years), there were noticeably more males than females (Table 14.2; Blank and Yang 2013).

Most social ungulate populations with a polygenic type of reproduction have more females than males, the so-called female-biased skew in the population. According to the consequences of the sex-allocation hypothesis (Trivers and Willard 1973; Hewison and Gaillard 1999), this is due to the fact that males have increased mortality at all ages compared to females and therefore the proportion of males decreases, especially many weakened males die after rutting periods or in years with unfavorable harsh winters. In favorable years, males have limited mortality and the ratio of males to females gradually evens out and becomes closer to equality (Blank and Yang 2013). Many polygamous ungulates have a male to female ratio

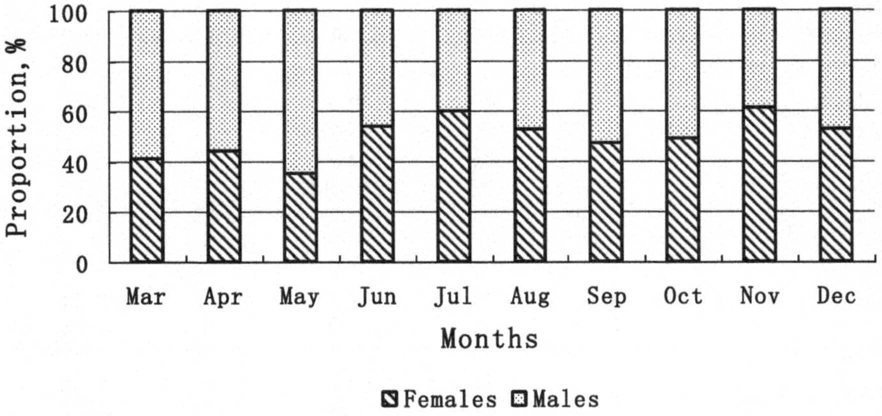

Fig. 14.1 Proportions of females and males of goitered gazelles during different months. January and February are absent because of lack of data. Other month data are average from different years with exception of August, September, and October, which have data of 1 year: March (1983, 1984), April (1982, 1985), May (1982–1986), June (1981, 1983–1986), July (1981–1982, 1984–1986), August (1982), September (1982), October (1981), November (1981–1982, 1985–1986), December (1984–1986) (Blank and Yang 2013)

close to parity (Berger and Gompper 1999). Thus, the sex ratio fluctuates from year to year from parity to a strong predominance of females in the population.

In my case, however, the sex ratio varied a little from year to year, and most months the proportion of females and males were close to parity, and only in May and partly in March and April, the number of females noticeably fall and males were found significantly more often. It is difficult to explain such monthly fluctuations using the sex-allocation hypothesis, but if we consider the annual biological cycle of goitered gazelles, then a certain likelihood is revealed. May is birthing month when most females leave their groups and stay alone. In addition, pregnant females try to lead a secretive, sedentary lifestyle, when they lie down and hide a lot along dry riverbeds and among the hilly landscape, avoiding open spaces. Males, on the other hand, do not change their lifestyle during the birthing period and often graze in open areas and gather in relatively large groups. Since surveys are carried out by car along the main dirt roads, which pass through open areas, it is much easier to see males from afar, since they run when a vehicle appears, while the females hide within the landscape and freeze rather than run away from any potential danger. Thus, the results of my surveys show not an actual change in the ratio of females to males, but rather a change in the behavior of females leading to them not being counted.

This experience shows that it is necessary to study the lifestyle of a particular species very thoroughly in order to correctly explain the results of a study. Therefore, observations of the behavior of ungulates must be carried out over a long period to understand what is happening in reality over time. Another interesting conclusion can be drawn from these data: that some females begin to lead an isolated lifestyle in preparation for birthing starting in April or even March, a month or a month and a half before actually giving birth, compared to others who leave their herds 1 or 2 weeks before birthing, as some researchers report. These are different females preparing for parturition in different ways, some earlier or others later, which impacts our understanding of this part of a female's parental behaviors. This point needs to be clarified.

References

Alados CL, Escós JM (1994) Variation in the sex ratio of a low dimorphic polygynous species with high level of maternal reproductive effort: Cuvier's gazelle. Ethol Ecol Evol 6:301–311

Baharav D (1983) Reroductive strategies in female mountain and Dorcas gazelles (*Gazella gazella* and *Gazella dorcas*). J Zool 200:445–453

Berger J, Gompper ME (1999) Sex ratios in extant ungulates: products of contemporary predation or past life histories? J Mamm 80(4):1084–1113

Blanchard P, Festa-Bianchet M, Gaillard JM, Jorgenson JT (2004) Maternal condition and offspring sex ratio in polygynous ungulates: a case study of bighorn sheep. Behav Ecol 16(1):274–279

Blank DA, Yang W (2013) Sex ratio in goitered gazelles (*Gazella subgutturosa* Güldenstaed, 1780). Acta Theriol 58:73–78

Byers JA, Moodie JD (1990) Sex-specific maternal investment in pronghorn, and the question of a limit on differential provisioning in ungulates. Beh Ecol Soc 26:157–164

Caley MJ, Nudds TD (1987) Sex-ratio adjustment in Odocoileus: does local resource competition play a role? Am Nat 129:452–457

Cameron EZ (2004) Facultative adjustment of mammalian sex ratios in support of the Trivers–Willard hypothesis: evidence for a mechanism. Proc R Soc B 271:1723–1728

Cameron EZ, Linklater WL (2007) Extreme sex ratio variation in relation to change in condition around conception. Biol Lett 3(4):395–397

Cassinello J, Gomendio M (1996) Adaptive variation in litter size and sex ratio at birth in a sexually dimorphic ungulate. Proc Royal Soc B 263:1461–1466

Clark AB (1978) Sex ratio and local resource competition. Science 201:163–165

Clutton-Brock TH, Albon SD, Guinness FE (1981) Parental investment in male and female offspring in polygynous mammals. Nature 289:487–489

Clutton-Brock TH, Guiness FE, Albon SD (1982) Red Deer. Behavior and Ecology of Two Sexes. University of Chicago Press, Chicago

Clutton-Brock TH, Albon SD, Guinness FE (1984) Maternal dominance, breeding success and birth sex ratios in red deer. Nature 308(5957):358–360

Côté SD, Festa-Bianchet M (2001) Birthdate, mass and survival in mountain goat kids: effects of maternal characteristics and forage quality. Oecologia 127:230–238

Cunningham PL, Wronski T (2011) Sex ratios of Arabian sand gazelle *Gazella subgutturosa marica* Thomas, 1897 in the Mahazat as Sayd protected area, Saudi Arabia. Mammalia 75:243–248

Cunningham PL, Sandoka M, Wronski T (2011) Some morphological characteristics of Arabian sand gazelle *Gazella subgutturosa marica* and its implications for management. Eur J Wildl Res 57(4):865–872

FitzGibbon CD (1990) Anti-predator strategies of immature Thomson's gazelles: hiding and the prone response. Anim Behav 40:846–855

FitzGibbon CD, Lazarus J (1995) Antipredator behavior of Serengeti ungulates: individual differences and population consequences. In: Sinclair ARE, Arcese P (eds) Serengeti II: dynamics, management, and conservation of an ecosystem. The University of Chicago Press, Chicago, pp 274–296

Hewison AJM, Gaillard JM (1996) Birth-sex ratios and local resource competition in roe deer, *Capreolus capreolus*. Behav Ecol 7(4):461–464

Hewison AJM, Gaillard JM (1999) Successful sons or advantaged daughters? The Trivers– Willard model and sex-biased maternal investment in ungulates. Trends Ecol Evol 14:229–234

Hewison AJM, Andersen R, Gaillard J-M, Linnell JDC, Delorme D (1999) Contradictory findings in studies of sex ratio variation in roe deer (*Capreolus capreolus*). Behav Ecol Sociobiol 45:339–348

Hiraiwa-Hasegawa M (1993) Skewed birth sex ratios in primates: should high-ranking mothers have daughters or sons? Trends Ecol Evol 8:395–400

Kingswood SC, Blank DA (1996) Gazella subgutturosa. Mammalian Species 518:1–10

Kojola I, Eloranta E (1989) Influences of maternal body weight, age, and parity on sex ratio in semidomesticated reindeer (*Rangifer t. tarandus*). Evolution 43(6):1331–1336

Krüger O, Radford AN, Anderson C, Liversidge R (2005) Successful sons or superior daughters: sex–ratio variation in springbok. Proc Royal Soc B Biol Sci 272(1561):375–381

Kruuk LEB, Clutton-Brock TH, Albon SD, Pemberton JM, Guinness FE (1999) Population density affects sex ratio in red deer. Nature 399:459–461

Kucera TE (1991) Adaptive variation in sex ratios of offspring in nutritionally stressed mule deer. J Mamm 72(4):745–749

Marmazinskaya NV (1999) Social behavior of goitered gazelle (*Gazella subgutturosa/*Guld., 1780/) during breeding period in the conditions ecocenter "Goitered gazelle". Ph.D. thesis in zoology. Institute of Zoology of Uzbek Academy of Sciences, Uzbekistan, pp 1–20. (in Russian)

Pélabon C, Gaillard J-M, Loison A, Portier C (1995) Is sex biased maternal care limited by total maternal expenditure in polygynous ungulates? Behav Ecol Sociobiol 37:311–319

Rahmani AR (1990) Distribution, density, group size and conservation of the Indian gazelle or chinkara *Gazella bennetti* (Sykes 1831) in Rajasthan, India. Biol Conserv 51:177–189

Rosenfeld CS, Roberts RM (2004) Maternal diet and other factors affecting offspring sex ratio: a review. Biol Reprod 71:1063–1070

Ruckstuhl KE, Neuhaus P (2002) Sexual segregation in ungulates: a comparative test of three hypotheses. Biol Rev Camb Philos Soc 77(1):77–96

Rutberg AT (1986) Lactation and fetal sex ratios in American bison. Am Nat 27:89–94

Silk JB (1983) Local resource competition and facultative adjustment of sex ratios in relation to competitive ability. Am Nat 121:56–66

Skogland T (1986) Sex ratio variation in relation to maternal condition and parental investment in wild reindeer. Oikos 46:417–419

Sludskiy AA (1956) Breeding of goitered gazelle. Proceedings of institute of zoology, Kazakh Academy of Sciences, vol 6. Nauka Press of Kazakh SSR, Alma-Ata, pp 78–108. (in Russian)

Sludskiy AA (1963) The mass death of wild ungulates in steppes and deserts of Europe and Asia. Proc Institute Zool Acad Sci Kazakh SSR 20:5–88. (in Russian)

Trivers RL, Willard DE (1973) Natural selection of parental ability to vary the sex ratio of offspring. Science 179:90–92

Verme LJ (1983) Sex ratio variation in Odocoileus: a crucial review. J Wildl Manag 47:573–582

Walther FR (1972) Social grouping in Grant's gazelle (*Gazella granti* Brooke, 1872) in the Serengeti National Park. Z Tierpsychol 31:348–403

West SA, Shuker DM, Sheldon BC (2005) Sex-ratio adjustment when relatives interact: a test of constraints on adaptation. Evolution 59(6):1211–1228

Wolff JO (1988) Maternal investment and sex ratio adjustment in American bison calves. Beh Ecol Sociobiol 23:127–133

Zhevnerov VV (1984) Goitered gazelle of the Barsakelmes Island. Nauka Publishing House of Kazakh SSR, Alma-Ata. 144 pp (in Russian)

Zhevnerov VV, Bekenov AB, Sludskiy AA (1983) Goitered gazelle—*Gazella subgutturosa* Güldenstaedt, 1780. In: Gvozdev EV, Kapitonov VI (eds) Mammals of Kazakhstan, Bovidae, vol 3 (3). Nauka Publishing House of Kazakh SSR, Alma-Ata, pp 11–54. (in Russian)

15 Evolution of Social Behavior

15.1 Introduction

Resource distribution, habitat structure, and predation pressure are thought to be the primary forces driving the evolution in the variations seen in the social organization within and between species (Brashares and Arcese 2002). The first bovids were small forest dwellers similar to the currently living Indian four-horned antelopes (*Tetracerus quadricornis*) and duikers (Cephalophinae) (Lumpkin and Kranz 1984; Estes 1974; Baskaran et al. 2011). Bovids' evolution developed as an adaptive dispersal from forests into open environments creating varied niches in the savannas, steppes, and deserts and providing a high herbivore biomass. Many behavioral adaptations to open habitats were exactly the opposite to ancestral forest lifestyles. The forest antelopes remain the least changed from their ancestors (Eisenberg 1966; Thenius 1968), while antelopes of open environments evolved extensively, adapting to new habitats (Pérez-Barbería et al. 2002; Estes 1974). The earliest bovids were assumed to have been typically solitary, small body-sized, sedentary, cryptically colored browsers with short, simple horns, and preferring concealment from danger. In contrast, species of open habitats are usually gregarious grazers, have medium to large sizes bodies often with distinctive marks, possess large and complex horns, greater mobility, and prefer to run from danger. The most socially advanced are highly gregarious species that tend to achieve the highest population densities. All bovids can be characterized as (1) solitary/territorial (Cephalophini, Neotragini, Redunca); (2) solitary/non-territorial (bushbuck, Sitatunga); (3) gregarious/ territorial (most species); or (4) gregarious/non-territorial (Tragelaphini, Bovini). Seemingly most bovids are territorial; the known exceptions all belong to the Bovinae species (Estes 1974; Bärmann 2014).

A grouping tendency is observed in many solitary antelopes when they are attracted to open environments. Such temporary groups were the likely antecedents of true herds. Individuals had to gather together in order to leave closed habitats, and the herd structure replaces the sense of cover, providing a feeling of

D. Blank, *The Behavioral Biology and Conservation of Goitered Gazelles (Gazella subgutturosa)*, https://doi.org/10.1007/978-3-031-83504-9_15

concealment for the individual. For example, forest-dwelling, solitary antelopes, such as reedbucks (*Redunca redunca*; *R. fulvorufula*), usually live alone or in pairs (Taylor and Skinner 2006). However, when they lose their cover in the dry season (through annual fires), they often come together in temporary herds of up to a dozen individuals, in some areas, it is possible to see herds of 100 to 200 and even 300 to 400 individuals, though such large herds are the exception (Estes 1974, 1991a, 1991b). Having joined into a large group, related individuals in reedbucks cease to associate with one another in preference to members of other families (Jungius 1970, 1971). Conversely, the presence of cover inhibits group formation, mainly because of the difficulty in keeping in visual contact (Einsenberg 1966). It is clear that species with a dispersed distribution pattern and concealing habits in closed environments, in other seasons, have the potential to form large social gatherings in open habitats. A tendency for group formation has been noted in all tribes of solitary antelopes: the least in duikers, in which females seemingly play an active part in territorial defense, and the greatest in the reedbucks, in which females show almost no sign of territoriality (Estes 1974, 1991a, 1991b).

Most gregarious species remain sedentary and form only small herds with a changeable composition (Estes 1976; Estes 1974; Marmazinskaya 1999). The migratory habits and aggregations of some plains' species have an alternate tendency to concentrate in the dry season and disperse in the wet season (Bolger et al. 2008; Walther 1984; Estes 1974). Most antelopes are gregarious/territorial. The whole range utilized by a population is divided into individual territories, each defended typically year-round in tropical areas (Thomson's gazelle *Eudorcas thomsoni* and Grant's gazelles *Nanger granti* (partly), gerenuk *Litocranius walleri*, mountain gazelles *Gazella gazella*—Walther 1984; Estes 1974; Walther et al. 1983; Habibi et al. 1993; Bärmann 2014), but only during relatively short rutting season in moderate, northern climates and elevated plateaus (goitered gazelles *Gazella subgutturosa*, Mongolian gazelles *Procapra gutturosa* and Tibetan gazelles *Procapra picticaudata*—Blank 1998; Schaller 1998; Vrba and Schaller 2000). Each territory is protected by an adult male, whose sexual and aggressive behavior has a great impact on the distribution of other conspecifics, females and males (Walther 1972; Estes 1974; Blank 1998).

For social ungulates, the size and composition of groups are the most basic elements of their social organization (Shi et al. 2005). In this chapter, I consider the social organization of goitered gazelles and discover the interrelationships between adaptations to their environment and social behavior. In addition, I will discuss the origins of the main patterns of social behavior, defining the place of goitered gazelles among other Antilopini species and the social development of goitered gazelle behavior among the evolutional trends of other antelope species. Goitered gazelles belong to the category of most numerous antelopes (at least in the past) that are gregarious and territorial at the same time (Mallon and Kingswood 2001).

15.2 Results and Discussion

The earliest bovids were in fact forest dwellers. *Eotragus,* the earliest known bovid, was a small antelope with low-crowned (brachydont) cheek teeth, wide-set, short, straight horns located above the orbits (in males only), and pronounced lachrymal fossa in both sexes, which presumably contained sizeable preorbital glands (Pilgrim 1939; Thenius 1969). Among living bovids there are several species mostly from the polyphyletic assemblage Neotragini (or Raphicerana) that show a similar horn morphology and social organization. All of them were apparently forest-dwelling browsers (Bärmann 2014; Estes 1974). Originating in forests, the first ruminants later penetrated into the grasslands and sub-deserts, becoming the ancestral forms for all Antilopini (Estes 1974; Kuznetsova and Kholodova 2003). The forest antelopes remain the least changed from their ancestors (Eisenberg 1966; Thenius 1968), and are monogamous, territorial browsers with a concealment lifestyle; these antelopes are inclined to solitary lives because dietary selectivity is negatively correlated with group size (Walther 1984; Estes 1991a, 1991b; Brashares et al. 2000). In contrast, the most progressive antelopes live in the open environments of savannas and arid zones and form herds of various sizes. Goitered gazelles have a similar social system compared to other gazelle species and are a typically gregarious, polygynous species of open arid environments. The discovery of a *Gazella* sp. in Upper Miocene deposits at Ft. Ternan, Kenya (Gentry 1970), shows that gazelles, at least, have roamed the arid plains of East Africa for the last 14 million years (Estes 1974). Goitered gazelles are migratory, territorial, and have female herds (family groups), male groups, mixed herds and solitary males, including territorial males during the rutting season only; they stay in large home ranges (Estes 1974; Blank et al. 2012a).

Goitered gazelles form true but loose herds which can change their group composition throughout the day (Zhevnerov 1984; Marmazinskaya 1999). The Arabian sand gazelle *Gazella subgutturosa marica* also show a tendency for daily group structure changes by grouping together (>10 individuals together) when resting and dispersing when foraging (Launay and Launay 1992). They generally avoid closed environments and reluctantly enter bushy thickets to reach watering places or to graze fresh vegetation during the hot summers. These visits happen sporadically, but since gazelles are afraid of heavy cover, they only occur along the edges of the shrubby thickets so the gazelles can avoid going too far inside the thickets. At the slightest alarm, they run out into the open environment to determine the degree of danger; if the danger is real, they run away exclusively in the open area. Goitered gazelles use saxaul forest (*Haloxylon aphylum*) as a protection from cold, windy weather, but they stay there only along the edge of the desert forest, without going deep inside (Zhevnerov et al. 1983).

The goitered gazelle is definitely a species of open environments, but flat landscapes are not among their most preferable. They like bumpy dunes covered by sparse shrubs, dry-river beds with steep banks, and the rugged terrain of foothills and intermountain valleys (Zhevnerov et al. 1983). This is not accidental. Goitered gazelles stay in areas with a thin snow cover during winter (not more than 10 cm), because they cannot reach vegetation under thick snow since their narrow hooves are not adapted to shoveling through snow like reindeer (*Rangifer tarandus*) or musk ox (*Ovibos moschatus*), with their large shovel-like hooves (Ihl and Klein 2001). Therefore, goitered gazelles live in southern areas, where snowfalls are not typical and snow covers are absent or usually thin and unstable during winter. In the northern parts of their habitat, goitered gazelles prefer areas where strong winter winds blow out the snow cover, making vegetation easily available. However, goitered gazelles need protection from cold winter winds, because their hair is not adapted to windy conditions (Zhevnerov et al. 1983). In this regard, they use deep dry-river beds and the more rugged environment of the foothills to protect themselves against unfavorable weather conditions (Blank 2020b). In addition, a rugged landscape promotes an uneven distribution of the snow cover. As a consequence, goitered gazelles always can find snowless sites with available forage (Blank 1990).

Goitered gazelles unquestionably have a gregarious type of social organization. It could not be any other way, because any antelopes in open environments have to band together and gather into herds when leaving a closed habitat, since the herd is a substitute for natural cover, providing concealment for the individual among the many (Estes 1974). However, goitered gazelles form mainly small herds of one to three individuals, though mainly during migrations, they can aggregate into herds of several tens of individuals (Blank et al. 2012a, c). In the past, when the population density was much higher than it is today, herds of several hundred and even thousands of gazelles were observed, but these numbers were most likely typical for only short migration periods. However, goitered gazelle groups are not very stable and can easily integrated and disintegrated within hours (Zhevnerov 1984; Marmazinskaya 1999). Eisenberg (1966) argues that loose and therefore flexible social systems are the rule for species adapted for unstable habitats (e.g., arid biomes), while species with more closed social systems are specialized for more predictable, stable habitats (e.g., wet savanna).

The point here is that browsing is related to highly selective feeding on widely dispersed plants, and as a result, browsers tend to have a more dispersed social organization, where antelopes have small groups of changeable composition that usually are spread widely over a large area. In contrast, non-selective grazers can graze side by side on suitable pastures in groups of mainly large numbers (Estes 1974; Jarman 1974). Goitered gazelles are likely mixed feeders but demonstrate typical browsing behavior in the northern deserts of Central Asia. Their main sources of food are shrubs and dwarf-shrubs, while grasses and herbs are not common, especially in gravel deserts. In contrast, these gazelles may be grazers in the southern ephemeral deserts, but only for a short period when the grasses are available in the winter to spring period (Cunningham 2013). Since goitered gazelles spend most of their time browsing, their social structure consists of mainly small, unstable groups that have a loose character, easily aggregating into large herds and splitting again into smaller groups over a short period (Walther et al. 1983; Marmazinskaya 1999).

In addition, a small body size is interrelated with browsing habits, since small-sized animals demand relatively more energy input than large-sized animals; therefore, their food has to be of a very high quality. This type of food is available in small amounts and usually spread widely; consequently, antelopes have to have a small enough body-size to feed themselves (Estes 1974; Jarman 1974). That means dietary selectivity is negatively correlated with body size (Brashares 2002; Brashares et al. 2000). Since goitered gazelles are mixed feeders, alternating browsing and grazing, depending on seasonal conditions (Cunningham 2013), they have a larger body size than most forest antelopes, such as duikers or four-horned antelopes, but smaller than highly gregarious grazing antelopes of open environments, such as blue wildebeests *Connohaetes taurinus* or hartebeests *Alcelaphus buselaphus* (Estes 1974; Baskaran et al. 2011). In addition, gregarious species of open environments have a large body size more often than forest antelopes, because group size is positively correlated with body size (Louys et al. 2013; Brashares et al. 2000).

Most of the intermediate to highly gregarious antelope species need a well-developed social structure and communication systems between individuals within their herds, so they need contrasting body markings. Indeed, many highly gregarious species have markings of moderate to strong contrast within their overall coloration. Most savanna species of open habitats have extensive white markings, which are conspicuous on dark animals. In contrast, solitary species usually have relatively non-contrasting markings on their hair-cover, because being solitary, they need concealment to a greater extent than communication (Estes 1974; Caro and Stankowich 2010). In goitered gazelles, there is a black tail on the background of a pure-white rump patch. Tail-flagging is often used as a signal to conspecifics about danger and the arousal status of the signaler. In a calm situation, the tail is lowered and splits the color of the rump patch which is a signal to other gazelles that everything is calm, but as soon as a gazelle raises its tail vertically, the rump patch becomes pure white and can even seem to glow, as the white hair of the rump patch stands on end with excitement and becomes visible at a great distance. This is a clear alarm signal, and when the gazelle starts to run away, the rump patch seems to move up and down which becomes a signal of very close danger, causing all gazelles to run without delay (Blank 2018).

There are also some facial black markings and stripes in goitered gazelles, while other markings are not as conspicuous. The black stripe on the sides of the body and the stripes around the rump patch, seen in Thomson's gazelles, are absent in goitered gazelles (Zhevnerov et al. 1983). Contrast markings are more conspicuous in highly gregarious species (large groups) than in less sociable species (small groups), because the larger the group the greater security to individuals than in small groups (Estes 1974). In comparing Thomson's gazelles and goitered gazelles, the first is a highly gregarious gazelle species, while the second has only small-sized groups; therefore, a Thomson's gazelle has more distinctive stripes and spots on its body, while a goitered gazelle has coloration that is more suitable for concealment (Zhevnerov et al. 1983).

In sexually dimorphic antelopes, only males have dark coloration, while females and young have more concealing coloration from tan or grey to medium brown which is important for secrecy and obscurity, especially for hiding young as a part of their anti-predator strategy (Estes 1974). Goitered gazelles do not have sexual dimorphism in coloration, though their young have really masking coloring and their unvarying dark-brown shade makes hiding young that stay motionless absolutely undetectable from the surrounding background, even from a near distance (Zhevnerov 1984; Blank 1997).

Usually, breeding antelope males avoid direct battles with other territorial males, and instead prefer to display horn threat postures or make ritualized intention movements and symbolic actions, as well as meet and repel his opponent's strikes with his horns (Walther 1974). Similar to other gazelle species, goitered gazelle males have well-developed lyre-shaped horns (Kingswood and Blank 1996). They are quite simple, but they are effective to show aggressive displays and defend their bodies against strikes from an opponent (Blank and Yang 2014; Blank et al. 2015b; Geist 1966). Males of moderately to highly gregarious species all have advanced (large and/or complex) horns and skeletal structures adapted to absorb severe impacts during intraspecific combat. Therefore, horn development has a direct relationship with social behavior (Geist 1974).

Gregarious antelopes have a more developed and intensive social behavior, essential for aggressive interactions to confront other males to determine and maintain their exact hierarchical position within a herd. In addition, the reproductive success in most advanced ungulates is based on absolute dominance rather than territorial dominance. Therefore, body-size dimorphism, when males may be up to two times a female's weight, appear in many gregarious ungulates having a polygyny social system, while it is absent in small, monogamous antelopes in closed habitats (forest or shrub thickets). Body-size dimorphism is explained by the sexual selection hypothesis that a bovine male's hierarchical rank is based on body size which provides a very important advantage in competition with opponents for breeding success: a larger and better-weaponed male has a much higher chance to win competitions for breeding rights (Pérez-Barbería et al. 2002). Therefore, a male of a dimorphic species experiences continuing horn and body growth during most his life (*Ovis* spp.—Geist 1971; *Capra sibirica*—Fedosenko and Blank 2001). In goitered gazelle males, social hierarchy in the herd is fairly weak (Blank et al. 2015b), because this species uses territorial dominance for mating success; so goitered gazelle males do not need a body size larger than their opponents for battle success, and hierarchical dominance in a herd is rather redundant. Therefore, goitered gazelle males cease their growth at the time of maturation and dimorphism between males and females is relatively small (Blank et al. 2012b; Kingswood and Blank 1996). The same phenomenon is observed in other gregarious antelopes of open habitats, such as impala *Aepyceros melampus*, Thomson's gazelles, blue wildebeests (*Connochaetes taurinus*) (Estes 1974).

The percentage of antelope females with well-developed horns also increases with the herd's level of gregariousness (up to 20% of species), though other species are hornless or have short spikes like duikers and small gazelles (Estes 1974). Goitered gazelle females usually do not have horns, though some have very short (up to 7 cm length) crooked horns (Zhevnerov et al. 1983; Blank 1997). In contrast in Arabian sand gazelles (*G. s. marica*), females have well-developed horns, though thinner than in males (Kingswood and Blank 1996). However, it has been recently determined that sand gazelles are not a goitered gazelle subspecies and likely belong to another, separate species (*G. marica*) (Wacher et al. 2011). Hornlessness in goitered gazelle females is a quite a primitive feature of this species, considering the fact that antelope evolution goes from hornless females to females with horns of similar size to males (Estes 1974). This means that the social behavior of goitered gazelle females is not as developed, and dominance hierarchy and social distance are not as clearly expressed in this species. The most developed social hierarchy needs a high level of individual aggression, since aggression between females is most pronounced in horned species (Estes 1974). This is not surprising considering the fact that females of goitered gazelles live mostly in small herds and during the several months of their young's hiding period they stay mostly alone; large female herds are not typical for goitered gazelles, which gather into larger groups only during the short rutting period and a few weeks during migrations (Blank et al. 2012b, 2012c).

Packer (1983) reasoned that horns of females are important not for social aggressive interaction, but rather for protection against predators, because their horns are straighter and thinner than in males and they are more specialized as defensive weapons against predators. Larger females protect themselves and their young more often than smaller body-sized females which prefer flight; therefore, larger females (more than 25 kg) typically have horns. Most goitered gazelle females are equal or smaller than 25 kg and consequently they are hornless (Packer 1983) and protect their young only against such small predators as foxes by using their front legs, but prefer flight in the case of large predators (wolves, dogs, humans) (Blank 2020a). However, Estes (1991b) holds a different opinion and thinks that female horns are not suitable for protection of their young against predators, but rather for the protection of their young from male aggression.

Most antelopes of open environments, including gazelles, are quite sedentary species similar to their forest ancestors (Estes 1991a). However, in many of them this is a seasonal habit, when antelopes settle into a region with rich forage, dispersing widely over the area in the wet season, concentrating again in the dry season to migrate to another location with more favorable conditions. Such migrations for forage and water are typical for many African Antilopini species (Thomson's gazelles, Grant's gazelles, springboks *Antidorcas marsupialis*) (Estes 1991a). Only, gerenuks (*Litocranius walleri*) and dibatags (*Ammodorcas clarkei*) are sedentary antelopes of open environments having a more ancestral lifestyle and meeting all their food requirements by browsing trees and bushes all year round (Leuthold 1978).

Goitered gazelles are also a sedentary species and, in most seasons, stay more or less in the same area, shifting slightly during their daily movements or moving around their home range (Blank 1990). Only during short periods in spring (end-March to early-April) and autumn (October) do they concentrate into large crowds and migrate from their wintering areas to their birthing grounds and summer regions where they stay until autumn before returning to their rutting and wintering grounds again. In some especially dry years, goitered gazelles also have summer migrations from dry areas to regions with more favorable conditions for forage and watering places (Zhevnerov et al. 1983). Social behavior of migrating goitered gazelles is very similar to other gazelles, when all groups gather into huge crowds and move as a continuous wave toward one direction. The difference between African gazelles and goitered gazelles is only in their motivation to migrate—seasonal droughts in Africa and seasonal snow covers and frosts in Central Asian deserts (Estes 1974; Zhevnerov et al. 1983).

During the rutting season, goitered gazelle males have a seasonal territorial social system once a year (in November and December—Blank 1998), when most males establish a network of individual territories along the routes female groups use during the day to move to and from their night pastures. The individual territories of rutting males are important for breeding, because mating opportunities are limited almost exclusively to males with individual territories (Estes 1974; Marmazinskaya 1999). These territories are maintained by ceaseless olfactory and vocal markings. In contrast to forest antelopes that never advertise their territories through vocalizations (Jarman 1974), territorial males in goitered gazelles do use roaring vocalizations to mark their territories and to display to other males their territorial status (Blank et al. 2014c). Other antelope species of open habitats also produce calls during rutting: Uganda's kob (*Kobus kob*) whistle as a territorial display; impala males roar, as do Mongolian gazelles *Procapra gutturosa* and springboks; while Grant's and Thomson's gazelles do not make loud calls during territorial displays (Jarman 1974; Estes 1991a; Blank et al. 2014c).

In breeding season, bachelor males stay in herds and continue to make their daily movements with or without the females. Females stay in their largest groups during the rut compared to other seasons and mixed herds and groups of adolescents are rarely observed. In contrast, single gazelles are very common all year round, both males that are most numerous during the rutting seasons (territorial males), and females that are especially in great numbers during birthing season (Blank et al. 2012a, 2012c). The territorial social system during breeding season is a fairly ancient feature of goitered gazelle behaviors, since all small ancestral forest antelopes also have a territorial social system, though these forest antelopes are landlords during all year round and not just during the rut (Estes 1974). Territorial males are common to both solitary and gregarious social systems. In fact, territoriality represents a direct holdover from a more primitive social system having originated in solitary ancestors with exclusive rights over a fixed home range. The difference is that gregarious antelopes have individual territories that do not include a male's entire home range and are not resource based, while solitary antelopes do protect their entire home range and defend their forage to attract females by the quality of forage within their range (Jarman 1983). To a very large extent, antelope social systems are built around male territoriality, which is not limited to mating with females, but also has a great impact on the distribution, density, and movements within the entire population (Estes 1974; Vrba and Schaller 2000).

Despite a number of similarities, goitered gazelles have a notable difference from small forest antelopes as groups of goitered gazelle females are not territorial and instead cross individual territories of rutting males almost without lingering. This is a great evolutionary advance in goitered gazelle behavior compared to females of small forest antelopes, which are sedentary and stay continuously within a male's territory, some even participating in border marking (Brotherton and

Manser 1997; Estes 1974; Brashares et al. 2000). At the same time, however, the territorial social system of goitered gazelles is somewhat antiquated compared to most advanced, large body-sized antelopes that, in social behavior, establish individual territories rather conditionally and without any clear boundaries or border markings (wildebeest), or breed by staying in male-female mixed groups without any sign of territoriality (oryx *Oryx beiza*, eland *Taurotragus oryx*) (Kiley-Worthington 1978).

Goitered gazelle males maintain their territories by marking the ground along their borders by leaving urine and feces marks, rubbing secretions from their preorbital glands on vegetation, and horning bushes and trees. In addition, goitered gazelles have inguinal glands, interdigital glands, and metatarsal glands, the functions of which are still not clear (Blank et al. 2014a, 2014b). External skin glands are closely related to the social behavior of ungulates, because secretions of these glands are a very important way of communication between conspecifics (Geist 1974; Blank et al. 2014a, 2014b). Their abundance unites goitered gazelles with their ancestral small forest antelopes, widely using skin glands for marking territory. Goitered gazelle females possess the same glands as males, but females do not have territories to mark (and their preorbital glands are not functional). They also do not leave urine or fecal pellets and cannot leave horning marks because they have no horns (Blank and Yang 2014; Blank et al. 2014a, 2014b). This is a progressive step in the development of this species' social behavior, considering that most evolutionary developments in antelopes do not have individual territories and do not mark with skin glands, even if they possess some of them (Estes 1974).

Bachelor herds are composed of yearlings to adult males, numbering from 2 individuals into hundreds in socially advanced antelopes (Estes 1974). Goitered gazelles usually have small bachelor herds from two to three individuals during the rut, while outside of rutting season, males can form bigger groups of up to eight to ten individuals, but rarely larger (Blank et al. 2012a). Presumably, male groups result from the same social attraction that brings females together, although male attachments appear weaker, since individual spacing in male herds is usually greater than in female herds. The main significance of bachelor herds in antelope society is that they provide a haven for males from the time they are expelled from nursery herds until they mature and become territorial. Accordingly, adults in bachelor herds could be considered as the most expendable members in a population because of their greater mortality. In advanced species with a high density, bachelor herds also provide sanctuary for adult males, which may include the unfit and old, as well as fit but supernumerary individuals and exhausted territorial males (Estes 1974; Bärmann 2014). The bachelor herds in goitered gazelles have likely the same significance (Blank et al. 2012a).

Apart from the real autumn rutting period in November and December, goitered gazelles also have a spring peak of sexual activity in April and May, when some adult males establish individual territories and increase their chasing female activity. This spring sexual activity has less intensity than the real rut in autumn. Mounting and copulation are not observed in the spring peak of sexual activity (though pregnancies have been known to occur). Nevertheless, the social composition during this spring pseudo rut is the same as during the real rutting season (Blank and Fedosenko 1983; Blank 1998). It has been suggested that this is spring peak is a fading remnant from ancestral populations during more favorable epochs, when breeding occurred more than once a year (Tsapluk 1972). Indeed, many African antelopes are territorial most of the year, because the tropical-savanna climate is capable of sustaining large, sedentary populations and year-round breeding. Species start to be seasonal breeders when the environment does not allow for continuous breeding (Estes 1991a). In addition, many gazelle species regularly exhibit two calving seasons a year (Loggers 1992). An Arabian subspecies of goitered gazelles, the sand gazelle (*G.s. marica*) (may be separate species *G. marica*), for example, has two peaks of birthing a year, February into April (62 offspring) and September into December (46 offspring) (Hague and Smith 1996). These data suggest that goitered gazelles can give births twice a year at least under favorable conditions (Sempéré et al. 2001).

Another behavioral holdover that is even more ancient than territoriality is the hiding instinct, which has been retained by the young of most antelope species except the most advanced. Generally, adult antelopes have lost the need for concealment, though in exceptional cases, even adult plain antelopes will still lie flat to the ground to escape detection. This behavior is seen rarely in Thomson's gazelles and Beiza oryxes (Walther 1969; Stewart 1963), but similar behavior is observed in adult pregnant females of goitered gazelles, though extremely rarely (Blank 1997); when the goitered gazelle population in Kazakhstan was significantly more numerous, Sludskiy (1956) observed several times during May observations how single adult pregnant females and groups of two to three females, noticing people in the distance, quickly lay down in the thickets of dwarf shrubs, trying to hide.

In contrast, such behavior is typical for antelope young, which continue to use it for their survival, especially immediately after birth. The young of both solitary and gregarious species display effective concealment in the same way. The young choose their own hiding places and take a sleeping posture, stretching the body with the lowest possible profile. In the fullest expression of this behavior, an infant remains entirely motionless with only slight respiratory movements and eye movements or an occasional ear flick to indicate the fawn is alive (Lent 1974). Fawns lie still until the mother calls to them for suckling. The mothers usually stay at a distance from their hiding young and only two to four times a day call their young from a close distance to suckle. While her fawn is suckling, the mother licks its genitalia and anus, stimulating it to eliminate metabolic wastes which she eats. After the brief activity and suckling, the fawn chooses a new hiding place and the mother leaves her young again (Estes 1974; Lent 1974). The main variation in this hiding system is the duration of the concealment period, which lasts anywhere from a few weeks (waterbuck—Hanks et al. 1969) to several months (reedbuck—Jungius 1970).

Ancestral traits are apparent in the ontogeny of other antelopes in open environments, notably in the natal coloration and instinct to hide displayed by most newborn antelopes (Estes 1974). Goitered gazelle fawns demonstrate their hiding over a two-month period (mid-May to mid-July), when fawns stay without their mothers most of the time. The body coloration of newborns is very different from the adults. Newborns have a solid light brownish solid color and is so consistent with the colors of the environment that it is impossible to see a motionless newborn, even from when close. After their hiding period, fawns start to change their body coloration through molting and their appearance becomes similar to the adults (Zhevnerov et al. 1983).

It is necessary to clarify that the entire hiding period (staying without mothers) in goitered gazelle lasts 2 months, but the concealment period, by itself, when fawns lie still and alone and in hiding, is significantly shorter and lasts up to a maximum of 2 weeks, after which fawns move from place to place and even graze independently in the absence of their mother. During this period, fawns meet the young from other families and form creches that move around together. This means that the social life of fawns starts earlier than joining a herd of adults at the age of 2 months (Blank and Yang 2015a, 2015b; Blank et al. 2015a). Since the young's concealment is lost at this time, the fawn is utterly dependent on the presence of conspecifics for concealment and protection, as a herd of young takes the place of cover for them, similar to gregarious adults. An individual becomes one of many and the more animals there are, the harder it becomes for a carnivore to single out any one fawn (Estes 1974). In a stress situation caused by a chasing predator, a fawn will follow the groups of adults without hesitation, even though it is unable to locate its mother before fleeing, and the reunion of mother and young can be delayed until the immediate danger is past (Lent 1974). In goitered gazelles, fawns behave the similarly and will also run after conspecifics from a predator attack. This means that socialization of goitered gazelle fawns starts from the moment fawns begin to walk independently from their mothers (Blank and Yang 2017). Although strictly speaking, the socialization of the young begins from their first days of life, when the newborn begins to communicate with its mother. Without this communication, fawns grow up to be socially inferior adult gazelles (Lent 1974; Blank and Yang 2017).

Natural selection thus favors the evolution of a short birthing season, when most females give births over a period of several days. A short birthing peak allows an advantage for the survival of the population by producing more young animals than a predator can handle at one time, essentially minimizing loss due to predation (Estes 1974). The goitered gazelle short mating and birthing periods, with most females giving birth within a few days, are an antipredator strategy of this species (Blank 1985; Blank et al. 2015a). In addition, females of gregarious antelope species either flee from large carnivores or counterattack predators that are smaller size than themselves, primarily in response to their young's distress calls (Estes 1974; Brashares et al. 2000), while small, solitary forest antelopes seek cover to hide (Underwood 1982; Ford and Goheen 2015; Brashares 2002 Brashares et al. 2000). Goitered gazelle females also attack foxes that enter the area of their hiding fawns. Sometimes a sub-adult female that accompanies an adult female also will get involved in the attack of a carnivore (Blank 2020a). On the other hand, cooperative defense by the herd, which is typical with the bovine social system, is not as developed in antelopes (Estes 1974) and is also absent in goitered gazelles (Blank 2020a). Another reason for a short birthing period is purely ecological, since it generally occurs during a season that coincides with the peak of forage availability, when moisture and temperature combine to produce intensive vegetation growth. These conditions of plentiful forage and the availability of watering places in arid areas are favorable for the survival of both fawns and mothers and promote the quick growth of young (Martin 2000). In the northern deserts, the time of year with the most favorable conditions in spring, between late the cold weather of winter and the early hot, dry months of summer, is very short. Therefore, goitered gazelle females birth most of their fawns at the beginning of this period, when it is warm enough but not hot and the green forage is of high quality and abundant for rearing young (Blank and Yang 2015a, 2015b). This similar behavior to tundra ungulates that have a very short favorable season for calving (6 to 9 weeks), so they also have very short rutting and calving cycles (Geist 1974).

Thus, gazelles, including goitered gazelles, occupy a median position in the evolution of antelopes and are a transitional form between forest dwarf antelopes and large highly gregarious antelopes of the savanna (wildebeest, hartebeest *Alcelaphus buselaphus*, eland), and other open arid habitats (*Oryx beiza*). In spite of the fact that all Antilopini species (gazelles and their allies) have many similarities with goitered gazelle social behavior, they are quite heterogeneous in its evolutionary development. Species of this taxon are one of the most diverse groups of bovids and have distinctive differences in the development of their social organizations: from rudimentary social behavior in sedentary gerenuk, where associations between females are casual and many females and males stay alone, to the socially advanced gazelles (*Nanger dama*, *Nanger granti*, *Eudorcas thomsonii*), forming large aggregations and having essentially migratory habits (Estes 1974; Bärmann 2014). In this variety of social organizations, goitered gazelles occupy an intermediate position, because they form relatively small herds, have a clumped, uneven distribution with relatively low density compared to most advanced gazelle species and generally have a sedentary lifestyle with small-scaled seasonal migrations (Blank et al. 2012a). Moreover, goitered gazelles also have some rudimentary elements of social behavior: they have many singletons within the population, possess small, loose groups, and fewer aggressive contacts between hornless females. Such social behaviors of goitered gazelles bring them closer to gerenuks, which differs by a rudimentary social system, than to large, highly gregarious, advanced gazelles.

References

Bärmann EV (2014) The evolution of body size, horn shape and social behaviour state in crowns Antilopini—an ancestral character state analysis. Zitteliana B 32:185–186

Baskaran N, Kannan V, Thiyagesan K, Desai A (2011) Behavioural ecology of four- horned antelope (*Tetracerus quadricornis* de Blainville, 1816) in the tropical forests of southern India. Mamm Biol 76:741–747

Blank DA (1985) Peculiarities of social and reproductive behaviour of *Gazella subgutturosa* in the Ili Valley. Zool Zhurnal 64:1059–1070. (in Russian, English abstract)

Blank DA (1990) Persian gazelle. In: Kovshar AF (ed) Rare animals of desert regions. Nauka Press of Kazakh SSR, Alma-Ata, pp 56–80. (in Russian)

Blank DA (1997) Social and reproductive behavior in the Persian gazelle. Ph.D. dissert. The University of Tel-Aviv, 189 pp

Blank DA (1998) Mating behavior of the Persian gazelle *Gazella subgutturosa* Guldenstaedt, 1780. Mammalia 62(4):409–419

Blank DA (2018) The use of tail-flagging and white rump-patch in alarm behavior of goitered gazelles. Behav Processes 151:44–53

Blank DA (2020a) Behavioral responses of goitered gazelles to potential threat. Mamm Res 65(1):141–149

Blank DA (2020b) Using microclimate of arid landscape as a resource in goitered gazelle comfort behavior. J Arid Environ 180:104201

Blank DA, Fedosenko AK (1983) The pseudo-rut in goitered gazelle. In: The behavior of animal in society. Materials of third all-union conference for animal behavior, vol 2. Nauka Press, Moscow, pp 258–259. (in Russian)

Blank DA, Yang W (2014) Does social status of males change their aggressive behavior repertoire in goitered gazelle (*Gazella subgutturosa* Güld., 1780)? Behav Processes 108:20–26

Blank DA, Yang W (2015a) Strategy of goitered gazelle suckling behavior and its adaption to the environment. Mamm Res 60(4):393–401

Blank DA, Yang W (2015b) Suckling behavior in goitered gazelle: do females invest more in twins or singletons? Zoology 118:348–356

Blank DA, Yang W (2017) Mother-young recognition in goitered gazelle during hiding period. Behav Processes 142:21–28

Blank DA, Ruckstuhl KE, Yang W (2012a) Social organization in goitered gazelle (*Gazella subgutturosa* Guldenstaedt, 1780). Ethol Ecol Evol 24:306–321

Blank DA, Ruckstuhl KE, Yang W (2012b) Sexual segregation in goitered gazelles (*Gazella subgutturosa*). Can J Zool 90:955–960

Blank DA, Ruckstuhl KE, Yang W (2012c) Influence of population density on group sizes in goitered gazelle (*Gazella subgutturosa* Guld., 1780). Eur J Wildl Res 58(6):981–989

Blank DA, Ruckstuhl K, Yang W (2014a) Secretion marking with preorbital glands in goitered gazelle, *Gazella subgutturosa* (Artiodactyla: Boidae). Folia Zool 63(2):127–136

Blank DA, Ruckstuhl KE, Yang W (2014b) The economics of scent marking with urine and feces in goitered gazelle (*Gazella subgutturosa*). Mamm Res 60(1):51–60

Blank DA, Ruckstuhl K, Yang W (2014c) Roaring function in male goitered gazelles. Behav Processes 106:152–159

Blank DA, Ruckstuhl K, Yang W (2015a) Seasonal dynamics of agonistic displays in territorial and non-territorial males of goitered gazelle. Zoology 118(1):63–68

Blank DA, Ruckstuhl K, Yang W (2015b) Antipredator strategy of female goitered gazelles (*Gazella subgutturosa* Guld., 1780) with hiding fawn. Behav Processes 119:44–49

Bolger DT, Newmark WD, Morrison TA, Doak DF (2008) The need for integrative approaches to understand and conserve migratory ungulates. Ecol Lett 11(1):63–77

Brashares JS (2002) The ecology and evolution of behavioral variation in the African antelope and the relevance of behavior for conservation. PhD diss. University of British Columbia

Brashares JS, Arcese P (2002) Role of forage, habitat and predation in the behavioural plasticity of a small African antelope. J Anim Ecol 1:626–638

Brashares JS, Garland TJ, Arcese P (2000) Phylogenetic analysis of coadaptation in behavior, diet, and body size in the African antelope. Behav Ecol 11(4):452–463

Brotherton PNM, Manser BM (1997) Female dispersion and the evolution of monogamy in the dik-dik. Anim Behav 54(6):1413–1424

Caro T, Stankowich T (2010) The function of contrasting pelage markings in artiodactyls. Behav Ecol 21(1):78–84

Cunningham PL (2013) Plants included in the diet of Arabian Sand Gazelle (Reem) from Saudi Arabia. J King Saud Univ Sci 25:167–173

Eisenberg JF (1966) The social organization of mammals. Handbuch des Zoologie 8(26):1–92

Estes RD (1974) Social organization of the African Bovidae. In: Geist V, Walther F (eds) The behaviour of ungulates and its relation to management. IUCN Publ. No. 24, International Union for the Conservation of ature and natural resources, Morges, Switzerland, pp 166–205
Estes RD (1976) The significance of breeding synchrony in the wildebeest. Afr J Ecol 14(2):135–152
Estes RD (1991a) The behavior guide to African mammals. University of California Press, Berkeley-Los Angeles
Estes RD (1991b) The significance of horns and other male secondary sexual characters in female bovids. Appl Anim Behav Sci 29(1–4):403–451
Fedosenko AK, Blank DA (2001) Capra sibirica. Mamm Species 675:1–13
Ford AT, Goheen JR (2015) An experimental study on risk effects in a dwarf antelope, *Madoqua guentheri*. J Mammal 96(5):918–926
Geist V (1966) The evolution of horn-like organs. Behaviour 27:175–214
Geist V (1971) Mountain sheep. University of Chicago Press, Chicago
Geist V (1974) On the relationship of ecology and behaviour in the evolution of ungulates: theoretical considerations. In: Geist V, Walther F (eds) The behaviour of ungulates and its relation to management. IUCN, Morges, Switzerland, pp 247–273
Gentry AW (1970) The Bovidae (*Mammalia*) of the Ft. Ternan fossil fauna. In: Leakey LSB, Savage RJG (eds) Fossil vertebrates of Africa. Academic Press, London, New York, pp 243–323
Habibi K, Thouless CR, Lindsay N (1993) Comparative behavior of sand and mountain gazelles. J Zool (London) 229:41–53
Hague MN, Smith TR (1996) Reintroduction of Arabian sand gazelle Gazella subgutturosa marica in Saudi Arabia. Biol. Conserv., 76:203–207
Hanks J, Price MS, Wrangham RW (1969) Some aspects of the ecology and behaviour of the defassa waterbuck (*Kobus defassa*) in Zambia. Mammalia 33(3):473–494
Ihl C, Klein DR (2001) Habitat and diet selection by muskoxen and reindeer in western Alaska. J Wildl Manag 65(4):964–972
Jarman RJ (1974) The social organization of antelope in relation to their ecology. Behaviour 48:215–267
Jarman P (1983) Mating system and sexual dimorphism in large, terrestrial, mammalian herbivores. Biol Rev 58:485–520
Jungius H (1970) Studies on the breeding biology of the reedbuck (*Redunca arundinum* Boddaert: 1785) in the Kruger N.P. Z Säugetierkunde 35:129–146
Jungius H (1971) The biology and behaviour of the reedbuck (*Redunca arundinum* Boddaert, 1785) in the Kruger National Park. Mammalia depicta. P. Parey-Verlag, Hamburg, Berlin, 106 pp.
Kiley-Worthington M (1978) The social organization of a small captive group of eland, oryx and roan antelope with an analysis of personality profiles. Behaviour 66(1–2):32–54
Kingswood SC, Blank DA (1996) Gazella subgutturosa. Mamm Species 518:1–10
Kuznetsova MV, Kholodova MV (2003) Revision of phylogenetic relationships in the Antilopinae subfamily on the basis of the mitochondrial rRNA and b-Spectrin nuclear gene sequences. Dokl Biol Sci 391:333–336
Launay F, Launay C (1992) Daily activity and social organization of the goitered gazelle *Gazella subgutturosa marica*. In: Spitz F, Janeau G, Gonzalez G, Aulagnier S (eds) Ongulates/ungulates 91. S.F.E.P.M. - I.R.G.M., Paris, Toulouse, pp 373–377
Lent PC (1974) Mother–infant relationships in ungulates. In: Geist V, Walther F (eds) The behaviour of ungulates and its relation to management. IUCN Publications, Morges, Switzerland, pp 14–55
Leuthold W (1978) On social organization of the gerenuk *Litocranius walleri* (Brooke, 1878). J Anim Ecol 47:471–490
Loggers CO (1992) Population characteristics of Dorcas gazelles in Morocco. Afr J Ecol 30:301–308
Louys J, Montanari S, Plummer T, Hertel F, Bishop LC (2013) Evolutionary divergence and convergence in shape and size within African antelope proximal phalanges. J Mamm Evol 20(3):239–248
Lumpkin S, Kranz KR (1984) Cephalophus sylvicultor. Mamm Species 225:1–7

Mallon DP, Kingswood SC (2001) Antelopes. Part 4. North Africa, the Middle East, and Asia. Global survey and regional plans. The IUCN/SSC antelope specialist group. IUCN, Glands and Cambridge, UK

Marmazinskaya NV (1999) Social behavior of goitered gazelle (*Gazella subgutturosa* /Güld., 1780/ during breeding period in the conditions ecocenter "Goitered gazelle". Ph.D. thesis in zoology, Institute of zoology of Uzbek Academy of Sciences, Uzbekistan, Tashkent, pp 1–20 (in Russian)

Martin L (2000) Gazelle (*Gazella* spp.) behavioural ecology: predicting animal behavior for prehistoric environments in south-west Asia. J Zool (London) 250:13–30

Packer C (1983) Sexual dimorphism: the horns of African antelopes. Science 221(4616):1191–1193

Pérez-Barberia FJ, Gordon IJ, Pagel M (2002) The origin of sexual dimorphism in body size in ungulates. Evolution 56:1276–1285

Pilgrim GE (1939) The fossil Bovidae of India. Mem Geol Surv India, Palaeont Indica 26(1):1–356

Schaller GB (1998) Wildlife of the Tibetan steppe. University of Chicago Press, Chicago, IL

Sempéré AJ, Brown N, Pereladova OB, Bahloul K, Lacroix A, Soldatova N (2001) Comparative analysis of reproductive cycles in female Persian gazelle (*Gazella subgutturosa subgutturosa*) (Central Asia) and sand gazelle (*Gazella subgutturosa marica*) (Arabian Peninsula). Gen Comp Endocrinol 121:57–65

Shi J, Dunbar RIM, Bucjland D, Miller D (2005) Dynamics of grouping patterns and social segregation in feral goats (*Capra hircus*) on the Isle of Rum, NW Scotland. Mammalia 69(2):185–199

Sludskiy AA (1956) Breeding of goitered gazelle. In: Proceedings of Institute of zoology, Kazakh Academy of Sciences, Vol. 6. Nauka Press of Kazakh SSR, Alma-Ata, pp 78–108. (in Russian)

Stewart DRM (1963) The Arabian oryx (*Oryx leucoryx* Pallas). East Afr Wildl J 1:103–117

Taylor WA, Skinner JD (2006) A review of the social organization of mountain reedbuck, *Redunca fulvorufula*, and grey rhebok, *Pelea capreolus*, in relation to their ecology. Trans R Soc S Afr 61(1):8–10

Thenius E (1968) In Grzimeks Tierleben, Enzyklopädie des Tierreches 13, p 552

Thenius E (1969) Stammesgeschichte der Säugetiere (einschliesslich der Hominiden). Hdb Zool 8(47–48): 722 pp.

Tsapluk OE (1972) Age and seasonal dynamics of sexual activity of goitered gazelle. In: Proceedings of Academy of Sciences of the Kazakh SSR, Biological series, vol 3, pp 39–46. (in Russian)

Underwood R (1982) Vigilance behaviour in grazing African antelopes. Behaviour 79(2):81–107

Vrba ES, Schaller GB (2000) Phylogeny of Bovidae based on behavior, glands, skulls, and postcrania. In: Vrba ES, Schaller GB (eds) Antelopes, deer, and relatives. Fossil record, behavioral ecology, systematics, and conservation. Yale University Press, New Haven and London, pp 203–222

Wacher T, Wronski T, Hammond RL, Winney B, Blacket MJ, Hundermark KJ et al (2011) Phylogenetic analysis of mitochondrial DNA sequences reveals polyphyly in the goitered gazelle (*Gazella subgutturosa*). Conserv Genet 12(3):827–831

Walther FR (1969) Ethologische Beobachtungen bei der kunstlichen Aufzucht eines Blessbockkalbes (*Damaliscus dorcas philippsi* Harper, 1939). Zool Gar 36:191–215

Walther FR (1972) Territorial behaviour in certain horned ungulates, with special reference to the examples of Thomson's and Grant's gazelles. Afr Zool 7(1):303–307

Walther FR (1974) Some reflections on expressive behavior in combats and courtship of certain horned ungulates. In: Geist V, Walther F (eds) The behaviour of ungulates and its relation to management. IUCN, Morges, Switzerland, pp 56–106

Walther FR (1984) Communication and expression in hoofed animals. Indiana University Press, Bloomington

Walther FR, Mungall EC, Grau GA (1983) Gazelles and their relatives. A study in territorial behavior. Noyes Publication, Park Ridge, NJ, USA

Zhevnerov VV (1984) Goitered gazelle of the Barsakelmes Island. Nauka Publishing House of Kazakh SSR, Alma-Ata. (in Russian)

Zhevnerov VV, Bekenov AB, Sludskiy AA (1983) Goitered gazelle—*Gazella subgutturosa* Güldenstaedt, 1780. In: Gvozdev EV, Kapitonov VI (eds) Mammals of Kazakhstan, Bovidae, vol 3, No. 3. Nauka Publishing House of Kazakh SSR, Alma-Ata, pp 11–54. (in Russian)

Defensive Behavior 16

16.1 Alarm Signals (Stotting, Hissing, Urination/Defecation) (License Number–5477010446224)

16.1.1 General Remarks

Alarm signals are very common for artiodactyls (Tilson and Norton 1981). Ungulates, being in a group, get many advantages and one of the most important is a more reliable protection from an unexpected attack by predators. Each individual in the group can safely graze, and when a predator appears, some pair of many eyes will notice it and give a signal to everyone else (the "many eyes effect"), and in a large group, predators will not necessarily catch you ("dilution effect") (Dehn 1990; Périquet et al. 2010). In this regard, predation plays an important role in the development of sociality (Caro et al. 2004). Ungulates do not run away immediately, even if they discover the source of danger, but carefully consider the threat itself and around it, performing a preflight checking and displaying auditory or visual alarm signals at the same time (Stankowish and Coss 2007). The wary behavior of one individual is transmitted to others and then everyone begins to explore the potential danger together. At the same time, everyone simultaneously monitors the reaction of others, and as soon as the timidest individual begins to run away, the whole group rushes after her, often without even seeing the danger itself (Fortin et al. 2004).

The use of various alarm signals in a dangerous situation and during escaping is widespread phenomenon among rodents, lagomorphs, and ungulates (Guthrie 1971). In goitered gazelles, I observed usage contrasting body coloration (white rump patch) and changes in the position and movement of the black tail, the position of the ears, extraordinary gaits (spotting), and sound signals (Blank 2018a, 2018b, 2018c, 2018d). These signals of a prey can be intended not only for conspecifics but also for predators and have to increase the survival likelihood of other group members at the expense of the alarming individual (Hirth and McCullough 1977).

D. Blank, *The Behavioral Biology and Conservation of Goitered Gazelles (Gazella subgutturosa)*, https://doi.org/10.1007/978-3-031-83504-9_16

Stotting also called pronking or pronging is very typical behavior for some cervids, antilocaprids, and bovids. Stotting is a very conspicuous behavior when a hoofed animal pushes off with all four legs and jumps as high as possible, and the stotting animal is very clearly visible from afar (Walther 1984). Influence of stotting intensifies by the white rump patch with erected hairs during this performance and this gait and white rump patch linked closely and their impact cannot be separated from each other (Estes 1967). Stotting behavior is especially remarkable for ungulate species of open habitats with sparse vegetation and firm soil. Bovid species are osteologyically better adapted to such a kind of environment compared to cervids and they have ability of fast run, long horizontal jumping (strides), and high stotting (Leinders 1979). Stotting is atypical for mountain ungulates, although some of them prefer to live on leveled plateaus, such as argali. The only exception is the mouflons (*Ovis vignei*), which inhabit leveled lowlands, although the only mention of stotting in alarm has been found in this species (Fedosenko 2002).

Stotting behavior has been known for many years and has been described in detail in many papers, but in fact this behavior was described in depth mostly for one species, Thomson's gazelle (Caro 1986b; FitzGibbon and Fanshawe 1988). There are 11 hypotheses about the purpose of this unusual behavior. Many review papers have also been written (Caro 1986a), but so far, the authors have not come to a consensus on the function of stotting; it is possible that this behavior has a different addressee and performs a different function in different species. Therefore, in this chapter I will consider the stotting behavior of the goitered gazelle and try to determine the main function of it in this species.

Many cervid species produce loud alarm barks in danger situations (Cap et al. 2008), bovid species snorts, and whistle in alarm (Walther 1984; Estes 1991). Springboks (*Antidorcas marsupialis*), for example, produce nasal sounds of a forceful, whistling snort of high pitches (Bigalke 1972). Obviously, vocalization in different species can perform a different function. The barking of roe deer (*Capreolus capreolus*), for example, is mainly a pursuit-deterrent signal to the predator and secondary using of this vocalization in males is guarding of their individual territories (Reby et al. 1999). Barking of muntjacs (*Muntiacus reversii*) also likely has a deterrent effect and can play the function of distress call alarming conspecifics (Yahner 1980). In contrast, alarm calls of Spanish ibex *(Capra pyrenaica)* are mostly warning to the conspecifics (Alados and Escos 1988). Snorting topi (*Damaliscus lunatus jimela*) warn conspecifics, but at the same time males can use this snort for retaining females in his territory for mating (Bro-Jørgensen and Pangle 2010). Regarding goitered gazelle alarm calls, nobody described them and firstly it was found that this gazelle hisses in the case of alarm (Blank 2018c). So, in this chapter I consider this behavior due to the discovery of the main function of hissing in goitered gazelles. Urination–defecation acts are well known as a pure physiological function of excretion. Using urine and pellets for marking borders of individual territories is also described by many authors (Walther 1984; Estes 1991). However, it is very rare for authors to mention urination and defecation in the context of defensive alarm behavior (Autenreith and Fitchter 1975; Alvarez et al. 1976). In this chapter, I will also consider this kind of alarm signals.

Table 16.1 Description of the recorded displays (ethogram) during escaping behavior in goitered gazelle within the study period according to Walther (1969), Blank (1998), Stankowich and Coss (2006), Blank et al. (2015) (Blank 2018c—License Number—5477010446224)

Behavioral patterns	Description
Galloping	Go at a rapid pace; fastest gait. This is a four-beat gait in which all four feet are off the ground at the same time during each stride and are extended as far as possible to cover ground. This gait evolved to allow ungulates to escape quickly from predators and relies on a burst of speed, rather than stamina. The goal of this gait is to get away from a dangerous location, rather than to keep up a sustained flight (Figs. 16.1 and 16.2)
Trotting	Go with speed as a trot is between a walk and a gallop. This is a natural jumping two-beat gait, when both pairs of diagonal legs alternate in touching down, with a pause in between. A trot is very energy-efficient and an ungulate can sustain this gait for long periods of time
Walking	A walking gait is the process that animals use to move themselves at their slowest pace. A natural walking gait in four equal movements: Each leg lifts and touches down diagonally in succession (Fig. 16.3)
Alarm in standing and laying position	Gazelle standing with its head and neck stiffly upright with ears pricked and turned forward and with muzzle and pinnae aimed directly at the predator (Fig. 16.4).
Flight	Movement away from the gazelle's original location, ranging from a slow walk to trotting and galloping, stotting. Flight necessarily followed alarm (alert) behavior, though flight occurred almost instantly (FID $\approx$ AD) when a predator was detected too close (too late). Career flight is the most rapid galloping gait (Fig. 16.5)
Sudden jump during grazing	Gazelle abruptly jumps aside while grazing, or starts to run but then stops after covering several meters and reassesses its situation
Alarm stop	Gazelle starts to run, stops, and turns around and look
Stotting	Stotting is a common behavior of gazelles, in which they leap off the ground high into the air with all four legs simultaneously held rigid and straight, and tail and head held up. They may run a little when they land and then stott again or performing a series of stotts (Fig. 16.6)
Hissing (very soft snort or sneezing)	These sounds were uttered by powerful air inflation through semi-closed pressed nostrils, when mouth was closed and pressed, as well. This signal sounds like fiut'(soft "t" like in "ti"), with very soft hoarse whistle or hissing soft whistle. At the beginning, throat (goiter) and cheeks swelled out with air that then pulled out through nose and may be partly through mouth—through teeth or closed lips, which did some kind of movements during this vocalization. Sometimes staying on the place, gazelle tensed body muscles and did jerks forward by the entire body, powerfully pushing air from its lungs and rising a tail up during this procedure (Fig. 16.7)
Alarm urination similar to ordinary urination procedure	Male: Stepping forward and stretching the body, lowering the belly toward ground, the hind legs sloping backward; tail is lifted. Female: Raising the tail, squatting close to the ground, and taking a crouched posture with the vulva only a few cm above the ground and urinating (Fig. 16.8)
Alarm defecation similar to ordinary defecation procedure	Male: Abrupt bringing of hind legs forward so that hooves are placed lateral to the stationary forelegs, squatting closer to the ground and taking a crouched posture, and defecating. Defecation posture is typically following urination act without pause. Female: Raising the tail and defecating while walking or grazing without a special posture. Defecation and urination acts are often not linked. Apart from the purely physiological excretion function, and for marking territory, urination–defection acts are observed in ungulates before or during escaping behavior from danger

16.1.2 Alarm Signals in Goitered Gazelles

By running away from a danger, goitered gazelles give a number of alarm signals. But beyond that, any type of fast gait acts as an alarm signal for them (Figs. 16.1, 16.2, 16.3, 16.4, 16.5). The most extraordinary and even bizarre of them is stotting (Fig. 16.6) (Table 16.1). During my study, goitered gazelles performed stotting most often (observed 146 cases), urination–defecation acts less often (104 cases), and hissing sounds were made least frequently (75 cases) (Figs. 16.7, 16.8, and 16.9). Besides, goitered gazelles perform urination–defecation acts mostly before running away and less often they do it during short stops, when the escaping running has already started. Adult females showed all these alarm signals most often among conspecifics, adult males did it less often, and fawns and especially sub-adults demonstrated these signals rarely (Fig. 16.9). Adult females demonstrated stotting 1.68 times more often than adult males did it (n = 64 vs 38, respectively); such a difference between females and males is less distinctive in the case of hissing sounds (1.52 times, n = 38 vs 25, respectively), and least difference was observed in the cases of urination–defecation alarm acts (1.10 times, n = 45 vs 41 cases) (Blank 2018c).

A walking human, which is perceived as a danger, triggers most often stotting and alarm urination–defecation acts before goitered gazelle escapes. In contrast, goitered gazelle's hissing is induced mostly by a standing human and more rarely

Fig. 16.1 Galloping in goitered gazelles: This is fast gait, the goal of which is to get away from a dangerous location as soon as possible. Galloping herd of goitered gazelles in Ustyurt Plateau area, some males open their mouth during running. W Kazakhstan. 29.04.2011. Photo by M. Pestov with kind permission

Fig. 16.2 Galloping with jumps: Sometimes goitered gazelles gallop with jumping, especially in shrubby areas. Drawing by U. Kenzhetaev with kind permission

Fig. 16.3 Walking: In calm situations goitered gazelles are walking by step. Drawing by U. Kenzhetaev with kind permission

by a walking human. In addition, stotting, urination–defecation acts, and hissing are displayed rarely as a reaction to other categories of a danger (Fig. 16.10).

In goitered gazelles, display of stotting is observed quite infrequently most of months (March–June and October–December), when frequencies change between 0.02 cases per hour up to 0.13 cases per hour. But this behavior is observed more often in July (0.29 cases per hour, n = 33) and even most frequently in August (0.71 cases per hour, n = 28) and September (0.62 cases per hour, n = 6) with a sharp decreasing rate thereafter. There are observations that African antelopes show stotting more often with an abundance of food and less often in dry seasons (FitzGibbon and Fanshawe 1988). In the case of the goitered gazelles, it could also be assumed that stotting is observed more often in spring during the rainy period with an abundance of food. But my data shows that this is not the case. In spring, goitered

Fig. 16.4 Alarm reaction of two goitered gazelle females to the potential danger. Issyk-Kul goitered gazelle nursery for reacclimatizing, 24.11.2022. Photo by D. Blank

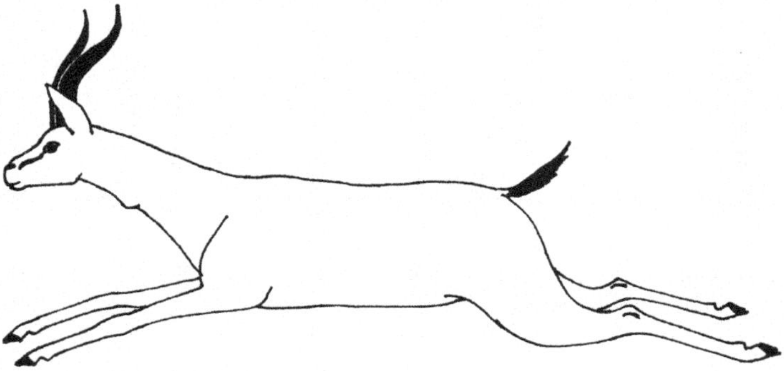

Fig. 16.5 Career flight in goitered gazelles is the most rapid galloping gait, which observed when the predator is too close. Drawing by U. Kenzhetaev with kind permission

gazelles show a minimal frequency of this behavior, whereas frequent stotting is observed during the hottest and driest period of the year, from July to September.

I observed stotting most frequently in adult females, whereas adult males perform this pattern two times less often; sub-adults demonstrate stotting 6 to 7 times less frequently than adult females. There are observations that all alarm signals including stotting are noted more often in kin groups than in herds of unconnected animals. White-tailed deer female groups (*Odocoileus virginianus*), which consist of closely related knit adult females, yearlings, and fawns, more readily perform

Fig. 16.6 Stotting is the most distinctive alarm behavior of goitered gazelles. Kalamali Ungulate Nature Reserve, Xinjiang, China. 06.07.2011. Photo by D. Blank

flight reactions with stotting displays than male herds, which represent accidently gathered adult males and yearlings (Hirth and McCullough 1977; Stankowich 2008). In my study of goitered gazelles, stotting is observed more frequently in July, which is the end of the fawns' hiding period, with the maximal rate in August–September, when fawns follow their mothers continuously and repeat all their behaviors, including escaping abilities. So, according to these investigations goitered gazelle mothers display their stotting behaviors mostly for their fawns, especially frequently during 2 to 3 months after the ending of hiding period. Afterward, stotting is demonstrated quite rarely (Blank 2018c).

Goitered gazelle stot most often as a reaction to a wolf approaching (11.1% cases, n = 63), while much less of this behavior is observed due to human disturbance (2.3%, n = 6269): moving car (0.06%) and walking human (1.4% cases, n = 6269). Thus, goitered gazelles have stotting response especially often during confronting with their natural coursing predator, the wolf. Among human disturbances, these gazelles stot more frequently during an encounter with a slowly walking human (a slow-speed "predator') than to a moving car (a high-speed "predator"). Observing the natural responses of Thomson's gazelles (*Eudorcas thomsonii*) to their natural predators, faster wild dogs and slower spotted hyena, Fitzgibbon and Fanshawe noted different reactions to different predators; these gazelles stot more readily to the faster than to the slower predators. In my study, goitered gazelles reacted more often to a slow "predator" than to a fast one. Obviously, natural predators and human disturbance are perceived as different categories of danger, and

Fig. 16.7 Alarm hissing: Goitered gazelle female is hissing in danger situation, puffing out the cheeks and letting air through the nose or half-closed mouth. Drawing by U. Kenzhetaev with kind permission

Fig. 16.8 Goitered gazelle female urinates before escaping from danger. Kalamali Ungulate Nature Reserve, Xinjiang, China. 06.09.2006. Photo by D. Blank

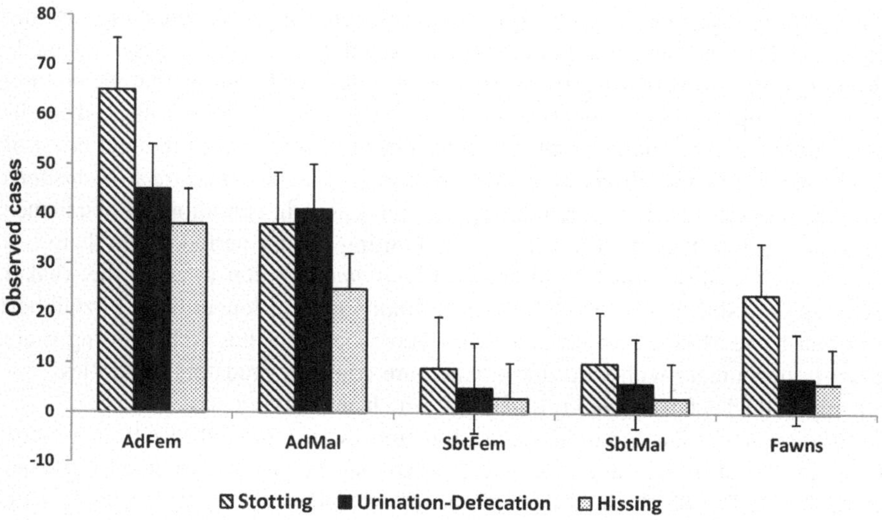

Fig. 16.9 Number of observed cases of stotting, alarm urination-defecation, and hissing, which were performed by goitered gazelles of different ages and sexes during the entire period of observations (AdFem—adult females, AdMal—adult males, SbtFem—sub-adult females, SbtMal—sub-adult males, and fawns) (Blank 2018c—License Number—5477010446224)

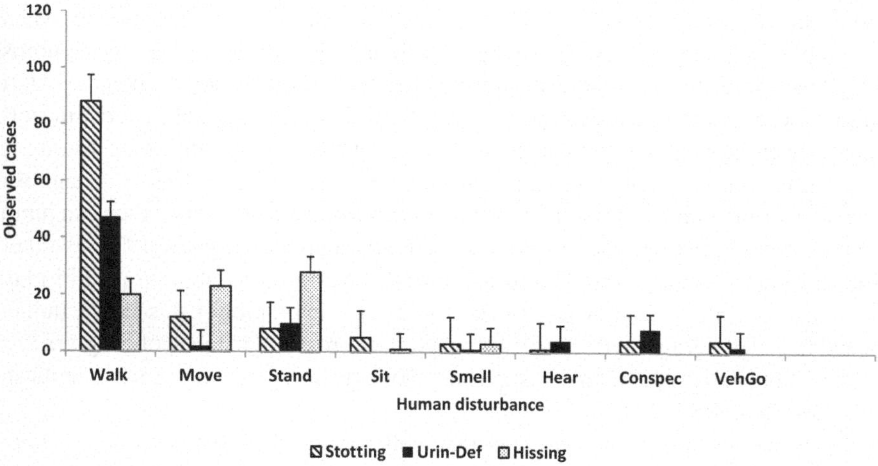

Fig. 16.10 Number of observed cases of stotting, alarm urination-defecation cases, and hissing, which were performed by goitered gazelles as a response to different kinds of human disturbance and conspecifics alarm behavior during the entire period of observations (Walk—walking human, Move—standing human making movements, Stand—standing immobile human, Sit—sitting immobile human, Smell—detection of human smell, Hear—hearing human noises, Conspec—response to alarm and escaping behavior of conspecifics, VehGo—moving vehicle, Unknw—far-distant or unknown danger source) (Blank 2018c—License Number 5477010446224)

these gazelles accept a person on foot as a much more dangerous threat than driving cars, which are not uncommon in these places and do not cause a great fear while driving on the road (Blank 2018c). There is a widespread opinion that the stotting display is employed as an honest signal for a predator that the prey is in good condition and has energy and endurance to outrun during escaping. An indirect proof of this statement is the observation that Thomson's gazelles (*Eudorcas thomsonii*) demonstrate stotting more often during the wet season in conditions of abundance of food, whereas they do it much less often during the dry period, when there is a lack of high-quality forage. Wild dogs (*Lycaon pictus*) prefer to chase the weaker gazelles for a shorter time than the stronger individual for longer time. FitzGibbon and Fanshawe (1988) also observed that Thomson's gazelles were stotting more often in encounters with wild dogs, which are coursing predators, while less frequently with cheetahs (*Acinonyx jubatus*), which are stalking predators.

Stotting likely has a partial testing function, when goitered gazelle may stot, jumping almost in the same place, as if provoking the predator to attack (Walther et al. 1983). The black-tailed deer (*Odocoileus hemionus*) can start running with stotting, but then they stop to check the predator reaction (Stankowich and Coss 2006). Caro (1986a) states that stotting has slower speed than usual galloping and decreases flight speed. This explains why stotting in springbok (*Antidorcas marsupialis*—Bigalke 1972) and Thomson's gazelles (Estes 1967; Walther 1984) and bounding in white-tailed deer (Hirth and McCullough 1977) are performed mainly in safe situations, when a predator unlikely has a chance to catch a stotting prey. White-tailed deer (*Odocoileus virginianus*), for example, do not bound when encountering a human at close distance (Hirth and McCullough 1977); pronghorns (*Antilocapra americana*) perform stotting only for a short distance when they start running and then switch to gallop (Kitchen 1974); Thomson's gazelles perform stotting only at the beginning of escaping from an attacking lion,; after that stotting is transformed into normal galloping if chasing continues (Pitcher 1979); springboks display stotting mainly during increasing or decreasing tension, when they can jump up to 2 m high, but they do not do it when really alarmed (Bigalke 1972; Walther 1981); Grant's gazelles and Thomson's gazelles demonstrate stotting also in play content; especially often fawns perform stotting during running (Estes 1967). Indian blackbucks (*Antelope cervicapra*) stot when they cross the patches of long grasses even without real predator attack; such behavior may be a precaution against ambush predator (Schaller 1967).

Gorbunov (1983) also mentioned that goitered gazelles display stotting more often, when a real danger is absent. Some behavior like that I found in goitered gazelle's behavior and they stot mainly in less dangerous situations, when they confront a walking human (a low-speed danger) and rarely they do it when they see a moving car (high-speed danger). From the other side, display stotting is performed for real danger (moving human) but not for potential danger and risk is very low or uncertain (sounds or smell from far-distant source, sitting or standing human).

Many cervid species utter loud, harsh calls and barks after detecting something dangerous (Cap et al. 2008); bovids species snort or whistle in alarm situation (Estes 1991). Alarm signals have two main functions, including pursuit-deterrent signal directed to the predator and warning addressed to conspecifics (Caro 1994; Cap et al. 2008). Roe deer (*Capreolus capreolus*—Reby et al. 1999) and some African bovids (Caro 1994) utter alarm calls (barks and snorts) not for alarming conspecifics, but for signaling to the predator that it was detected. In dangerous situations, goitered gazelles also produce quite soft hissing sounds which are heard from a close range. I try to figure out the main function of this vocalization in goitereed gazelles.

Hissing sounds in goitered gazelles are produced by forcibly blowing air through the nostrils without the participation of vocal ligaments. At the same time, air is first drawn into the mouth and the cheeks are inflated, and then it is passed through the nostrils and possibly partially through the semi-closed mouth. Sounds produced with a nose are typically soft and according to Walther (1969), such snorting sounds may be heard within 30 m. However, in the case of goitered gazelles, I heard alarm hissing at a distance of 50–70 m, sometimes 100–150 m. Alarm hissing of goitered gazelles is not observed in cases of facing wolf or other animals, but it is registered during human disturbances. Goitered gazelles produce hissing mostly during summer, growing from 0.10 cases per hour in June up to 0.20 cases per hour in August (n = 20 and 8, respectively), while these sounds are rarely heard in other seasons, varying from 0.01 cases per hour in spring, up to 0.08 cases per hour in autumn-winter.

Alarm hissing of goitered gazelles unlikely has pursuit-deterrent significance because they do not produce these sounds when confronting fast-speed threats having high risk (wolf and moving car). However, these gazelles produce hissing sounds mostly when they encounter a real danger, but when attacking risk is minimal (e.g., immobile or slowly walking human) and less often when the potential threat is uncertain or in a distance (e.g., catching smell without detecting a human). Marmazinskaya (2000) also found that hissing or "short staccato snorting" is produced in cases of a weak alarm, when goitered gazelles periodically emit such signals for a relatively long time (several minutes) and then move on to grazing or crossing the path of the threat, while sometimes, at the same time, they move away from the source of the alarm. In the case of an obvious and unexpected danger, gazelles immediately give loud, prolonged sounds before rushing to flight (Marmazinskaya 2000). In my study, I did not observe these loud alarm signals of goitered gazelles in dangerous situations (Blank 2018c). Many antelopes (Cephalophini, Neotragini, Antilopini tribes) in danger also use mostly nose soft sounds as alert signals for close distances (Estes 1991). It is quite another matter for saiga antelopes, which also make nasal sounds, but they are so greatly increased by the nasal cameras and passages of the enlarged nose that these sounds can be heard for several kilometers. When wolves get close to a thousandth herd of saigas and they notice it, the females gather in a dense group and begin to moo strongly. When several thousand females moo at the same time, the moo turns into a powerful rumble that scares the wolf so much that it refuses to attack. The alarm warning sounds of saigas gather them in a dense group and are definitely addressed at predators at the same time (pursuit-deterrent signal) (Sludskiy 1962).

Ungulate species (a number of European cervids and small African antelopes, golden takin *Budorcas taxicolor bedfordi*), preferring closed environment habitats (thickets, forest) or rugged mountain landscapes (chamois *Rupicapra rupicapra*; ibexes *C. sibirica*, *Capra pyrenaica*; klipspringer *Oreotragus oreotragus*), utter loud harsh barking or whistling sounds, producing by vocal cords of the larynx and having the ability to penetrate over long distances (Alados and Escos 1988; Fedosenko and Blank 2001; Cap et al. 2008; Baruzzi et al. 2017). In the situation of great level of alarm, some species can utter loud whistling sounds using their nose (impala—Schenkel 1966), but these are rare cases (Walther 1984). All these ungulates use their loud vocalization as a pursuit-deterrent signal in order to inform predators that they are found and identified. Ungulates of open habitats (gazelles, reindeer *Rangifer tarandus*, pampas deer *Ozotoceros bezoarticus*), from the other side, use mainly soft alarm sounds (hissing or snorting), producing by great exhalation and inhalation of air through the nose cavities. These ungulate species use these sounds mostly for warming of their conspecifics (Bigalke 1972; Estes 1991; Cap et al. 2008).

Loud sounds are intended for the long-range propagation and ungulates of closed environments and mountain rugged habitats are not afraid to be found by predators because they are under the shelter of forest or landscape. In contrast, ungulates of open environments have some kind of fear to attract the attention of more predators, because they do not have protection of the environment and only their defense is escaping in time, and in case of a danger, they transmit alarm signals for short distances primary to their conspecifics (Caro 1994). So, goitered gazelles utter soft hissing vocalization likely as a warning signal for their conspecifics (Blank 2018c). In addition to hissing, alarmed goitered gazelle produces another soft signal for the conspecifics by kicking of a foreleg hoof to the ground (Kostin 1955; Blank 1997).

In goitered gazelles, adult females perform alarm hissing more often than adult males and even much more than sub-adults and fawns (Fig. 16.10). This is a common phenomenon in ungulates when females snort more often than adult males (Caro et al. 2004). Most often adult females in goitered gazelles utter alarm hissing in July and August. This is the end of the hiding period in fawns, when they start to follow their mothers constantly and when managing the behavior of fawns is very important especially in alarm situations, when control of fawn's escaping behavior is particularly crucial for its surviving. So, adult females address their hissing vocalization likely for their fawns. The same is observed in feral goats (*Capra hircus*), kids of which react to alarm snorts much more readily than adult animals (Shank 1972). White-tailed deer also snort more often in female groups with their fawns, whereas males in their unisex groups do it quite rarely, escaping apart from others in complete silence (Hirth and McCullough 1977).

Goitered gazelles begin utter alarm snorting at the moment when they identify the potential danger (standing or moving human) and rarely in other kinds of danger. Similar to stotting, the gazelles produce hissing when the potential threat is real, but the possibility of attack is negligible. They do not utter hissing when the threat is uncertain or opposite, when the threat is close, and when conditions are too risky (fast moving car). White-tailed deer also snort frequently in the presence of humans when the risk is little, but do not vocalize and slip quietly without alarming conspecifics if the situation is too risky (Hirth and McCullough 1977).

Alarm sounds can contain several contexts depending on the situation and simultaneously are a warning signal for conspecifics and a pursuit-deterrent signal for predators (Caro et al. 2004; Yang et al. 2012). Alarm snorting in many ungulate species is rather information directed to predator that it is detected (for klipspringers *Oreotragus oreotragus*—Tilson and Norton 1981; topi *Damaliscus korrigum*—Caro 1994; muntjacs *Muntiacus reevesee*—Yahner 1980; roe deer *Capreolus capreolus*—Reby et al. 1999). In contrast to hissing and snorting, which are related to species of open environment and rather the warning signal, whistling is auditory signal of ungulates from closed habitat and addressed mainly to the predator (Caro et al. 2004). Some alarm sounds may be used without any threat but are employed in the context of mating, territorial, social, or agonistic behaviors in the case of highly exciting situations (Reby et al. 1999; Rossi et al. 2002; Cap et al. 2008; Bro-Jørgensen and Pangle 2010).

The main function of urination–defecation acts is a purely physiological need of the body to remove decomposition products from the body through urine and excrements. Another well-known function of urination–defecation acts is territorial marking when the territorial male defiantly marks the boundaries of his site in front of the enemy during territorial conflicts (Blank et al. 2014). However, there is another rather unknown function of urination–defecation: some kind of alarm signal during danger situations, when goitered gazelles urinate before escaping or mid-flight from a predator (Blank 2018c). Urination–defecation acts are displayed mainly in cases of human disturbances (walking human) (Fig. 16.8—5345), less often during meeting with other ungulates, such as kulan (*Equus hemionus*) and argali (*Ovis ammon*). All these species are slowly moving objects with low chances of attack. Goitered gazelles perform alarm urination–defecation rarely in the case of uncertain and far-distant danger and even in the case of a fast-moving car with potential high risk. The rate of the alarm urination–defecation acts varies between 0.04 and 0.23 cases per hour over months without any regularity. Females and males demonstrate alarm urination–defecation acts with almost the same frequency (Blank 2018c).

Walther (1984) thinks that alarm urination–defecation act is in fact a nervous excitement because of fright. Experiments of physiologists showed that the excretion frequency is closely related to emotional status especially when animal is frightened (Alvarez et al. 1976). Marmazinskaya (2000) also suggests that in a stressful, dangerous situation urination or defecation occurs as a reaction to stress. Observations after cattle behavior demonstrated that urine of stressful cow increases the fear in other cows (Boissy et al. 1998). Demonstration of urination or defecation posed by one individual of goitered gazelles stimulates the same behavior in other individuals of the group. Apparently, these behaviors and distinctive poses are used as a signal of danger for other conspecifics. It is also possible that the release of volatile substances with the smell of anxiety, along with urine and feces, is of no small importance, as well as the release of metabolites from the body before fleeing from danger. The urination–defecation posture per se is an alarm signal for all conspecifics and they urinate and defecate one after another (Marmazinskaya 2000). Apart from goitered gazelles, urination–defecation acts in alarm situations are described for pronghorns (*Antilocapra americana*—Autenreith and Fitchter 1975), impala (*Aepyceros melampus*—Schenkel 1966), fallow deer (*Dama dama*—Alvarez et al. 1976), and warthogs (*Phacochoerus aethiopicus*- Frädrich 1965).

There is another signal in goitered gazelles—this is stamping with the front foot, which is likely a visual signal, alerting conspecifics and typical for ungulate species of open habitats (Caro et al. 2004). As a rule, females are stamping twice more frequently than males (for white-tailed deer—Caro 2005). In my study, this behavior is observed quite rarely and only in females and it was noted in the relationship between the mother and the fawns during the hiding period, when the females attempted to lay their fawn. This behavior obviously had the effect of warning about danger for all conspecifics (Caro et al. 1995; Caro et al. 2004), but in the case of fawns, they almost instantly fell to the ground and froze. Mothers often stamped their forelegs, even without any danger, to put down a fawn that had been following its mother for too long. Other gazelle species (Thomson's gazelles and Grant's gazelles *Nanger granti*) also demonstrated front leg stamping in alarm situations and since stamping is usually accompanied by alarm snorting, Estes (1976) suggested that this signal has a warning effect. Though twitching of the skin flank is usual for Thomson's gazelles and observed (in less extent) in Grant's gazelles in dangerous situations, I did not observe this behavior in goitered gazelles.

In spite of the fact that hissing and stotting patterns are performed by all sexes and ages of goitered gazelles in dangerous situations, mostly females use these signals, especially after the ending of fawns' hiding period, to teach them to escape from any danger circumstances. Alarm urination–defecation acts arecused more for releasing stress in alarm situations, though these behaviors are contagious and have an important alarm effect for conspecifics.

16.2 Tail-Flagging and White Rump Patch as Alarm Signals

16.2.1 General Remarks

The morphological adaptations of any prey species, its behavior, and physical development are aimed at reducing the risk of predation as much as possible (Stankowich 2008). One of the main ways to escape predation in antelopes of open environments is running at a speed that exceeds the maximal speed of predators. For example, many African antelopes have abilities to increase their speed up to 80 km/h (Grant's gazelles *Nanger granti*, Thomson's gazelles *Eudorcas thomsonii*, wildebeest *Connochaetes gnou*) (Sludskiy 1962).

It is widely known that animal coloration has three main functions: physiological needs, protective coloration for concealment, and communication between individuals. To exchange information, ungulates rarely use a common coloration, but rather various stripes and spots located on the outward parts of the legs, tail, body, rump, head, and ears, which have contrasting coloration to the rest of the body, for example, a black tail contrasting with the white rump patch in goitered gazelles (Ortolani 1999; Caro 2005), which is often performed in their defensive behavior as an expression of alarm (Caro et al. 2004). Ungulates behave in such a way that these contrasting spots and stripes will be clearly visible. Species with contrasting-colored tails and a white rump patch are strongly socially gregarious and lead a diurnal lifestyle in open environments (Caro 2005).

Different authors define tail position and movements in different ways, including tail swaying, tail wagging, tail flicking, and tail-flagging, giving different names to the same behavior, or vice versa, calling the same behavior in different terms which leads to a lot of confusion in the description of ungulate behavior (Alvarez et al. 1976; Kiley-Worthington 1976; Alados 1986; LaGory 1981; Caro et al. 2004). Unlike other tail positions and movements, the vertical position of the tail plays an important role mostly in defensive behavior and has been called tail-flagging. This term is employed not only when the tail is in a vertical position, but also for movements of a vertical tail from side to side (Kiley-Worthington 1976; Hickman 1979; Stankowich 2008). I accept Caro et al.'s (2004) definition of a vertical tail and affirm that any vertical position of the tail is a strong alarm signal, even if the ungulate does not make any tail movements. In other words, my observations show that the position of the tail is noticeably more important for alarm behavior and straightly related to the level of fear than the tails' actual movements (Kiley-Worthington 1976). Therefore, it is quite acceptable to call the vertical position of a stationary tail as tail-flagging, although, as a rule, a vertical tail position is accompanied by movements from side to side. Many authors defined a vertical tail, or at least when it is raised higher than horizontal, by the term "tail-flagging" (Guthrie 1971; Alvarez et al. 1976; Bildstein 1983; Alados 1986; LaGory 1987; Caro et al. 1995). Smith (1991) has identified four main hypotheses that explain the adaptive nature of tail-flagging: conveying alarm (Estes and Goddard 1967), pursuit-deterrent function (Caro et al. 1995; Stankowich 2008), cohesive signal to conspecifics (Hirth and McCullough 1977), and a flash pattern (Edmunds 1974).

It is possible to define tail positions according to an ungulate's level of arousal: the tail is hanging down in a relaxed, neutral position with an absence of arousal; the tail is in a horizontal or lower position when in an intermediate arousal state; and the tail is held vertically or sometimes vaulted over the back under the greatest level of agitation (Figs. 16.11), especially high agitation is seen when tail-flagging combines with stotting (Fig. 16.12a, b). The tail moves up-down and/or side-to-side, when the danger is not so clear. A detailed ethogram of tail movements and positions is presented in Table 16.2.

When the relatively short tail rises vertically, it immediately exposes a snow-white rump patch in most cervids. When the tail is down, it is relatively wide enough to almost completely hide the white patch, which allows the animal to merge into the environment. Many deer species, such as swamp deer (*Cervus duvaucelii*), Indian hog deer (*Axis porcinus*), sambar deer (*Cervus unicolor*), and spotted deer chital (*Axis axis*) (Schaller 1967), are inhabitants of closed forest biotopes, and as a main antipredator behavior, are characterized by freezing in place which provides them concealment within their woodland environment. In contrast, when in danger, a deer's raised tail exposes the white rump patch making a more visible and a powerful, bright warning signal for conspecifics. Most antelope species, including

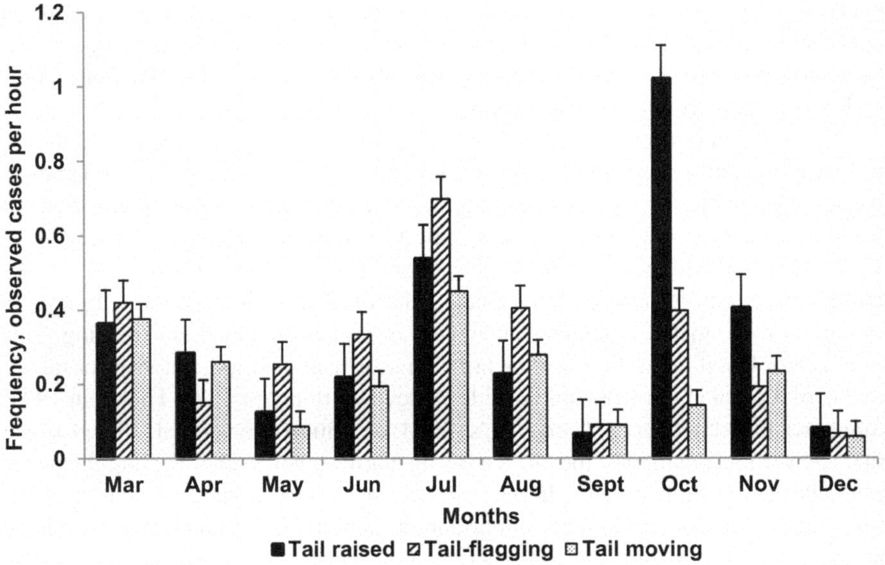

Fig. 16.11 Frequency fluctuations of different positions of the tail in goitered gazelles (tailed raised lower than back, tail-flagging, tail moving up-down and side to side) in alarm situations over months (by Blank 2018d—License Number—5477011106167)

goitered gazelles, live in open habitats and camouflage is not particularly useful to them, since they are clearly visible at great distances, and their most effective antipredator strategy is to run fast. Therefore, tails are quite narrow in antelopes and cannot hide the rump patch in calm situations, but instead divide the white rump patch into two equal halves. But as soon as the tail rises vertically, the rump patch is no longer divided and its hairs stand on end (piloerection) making the white spot especially enlarged, bright, and noticeable from a distance (Walther 1969). This is an emotional reaction to stress and that is an alarm signal of a real danger approaching for conspecifics (Guthrie 1971; Alvarez et al. 1976). Magnification through piloerection of an already large rump-patch area is found in many gazelle species, such as Grant's gazelle (*Nanger granti*—Estes 1967), Dorcas gazelle (*Gazella dorcas*—Alados 1986), Thomson's gazelle (*Eudorcas thomsoni*), springbok (*Antidorcas marsupialis*—Maberly 1963), and other antelopes, including pronghorn (*Antilocarpa americana*—Kitchen 1974), and *Tragelaphus* sp. (Kiley-Worthington 1976).

The function of flashing the white rump patch as a signal can be described through four main hypotheses: (a) a signal of warning of approaching danger; (b) a signal to "fallow me" to maintain cohesiveness among conspecifics or mostly for young to follow their mothers; (c) an important role in the submissive behavior to neutralize a dominant's aggressiveness; and (d) a pursuit-deterrent display to a predator (Guthrie 1971; Stankowich 2008). Moreover, Stankowich (2008) argues that the position of the tail is not as important in antipredator behavior of black-tailed

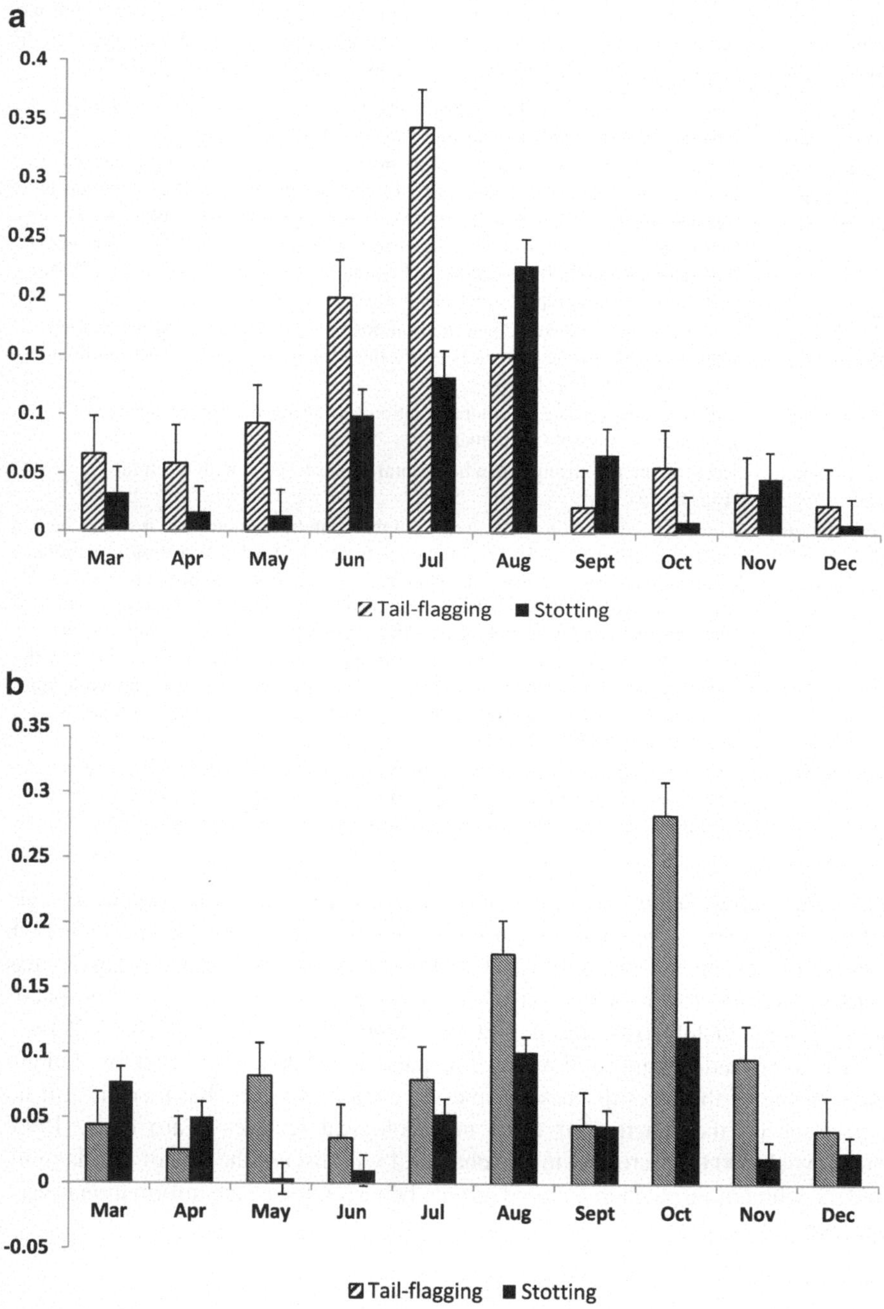

Fig. 16.12 Frequency fluctuations of tail-flagging and stotting observed in adult females (**a**) and adult males (**b**) of goitered gazelle in alarm situations over months (by Blank 2018d—License Number—5477011106167)

Table 16.2 Description of the recorded tail position in goitered gazelles during behavioral patterns (ethogram) while escaping based on research by Walther (1969), Kiley-Worthington (1976), Alados (1986), Stankowich (2008) (Blank 2018d—License Number—5477011106167)

Alarm for assessment	Standing or lying immobile. Gazelle erects its neck, puts its ears forward, lowers its croup, and tenses its muscles (Fig. 16.13).
Tail flicking (wagging, swaying)	Repeated lateral tail movements: The brief swaying of the tail from one side to the other or sometimes in a complete circular rotation, when the tail begins in a relaxed downward position and quickly rotates clockwise or counterclockwise to 90–180° before returning to a downward position. This pattern occurs mainly during relaxed conditions just at the beginning of a bout of locomotion. Rump patch is uncovered repeatedly (flashing) (Fig. 16.14).
Tail hanging down	The black tail is relaxed down (neutral position), vertically, in calm situations with minimum excitement and divides the white rump patch in the middle into two equal parts (Fig. 16.15).
Raised tail	Tail is erected up to a horizontal position (not above the spine) in cases of intermediate arousal (excitement)
Horizontal tail	Keeping the tail strongly in a horizontal position (level with the spine) (Fig. 16.16)
Tail-flagging	The tail is held vertically or erected higher than horizontal up to a strong vertical position (Fig. 16.17), or even putting it on the back in some situations of great agitation, exposing the white ramp patch to its maximum degree. In this position, the tail can make stereotypically repetitive motor lateral movements from left to right across the hindquarters and back again one or more times in succession. The white rump patch is completely uncovered, with a black spot only around the anus (Fig. 16.18). At the same time, the white hairs of the rump patch of the outline are projected outward, making rump patch more prominent (Fig. 16.19).
Tail moving	Making pendulous movements of the tail in side-to-side and/or up-down directions. During movements, the tail can be erected to various degrees: Vertically, horizontally, higher, and lower than horizontal (Fig. 16.20)

deer (*Odocoileus hemionus columbianus*), but rather the tail simply opens a bright white rump patch, which plays a major role in antipredator behavior, which is both a pursuit-deterrent and an alarm signal at the same time. I think that tail-flagging has a distinctive function in the defensive behavior of goitered gazelles, since conspecifics react immediately to the raised position and movements of the tail (Blank 2018d).

The use of tail-flagging and white rump patch in the defensive behavior of ungulates is one of the most discussed topics by various authors, but there is still no consensus on the function of these morphological features (Caro et al. 1995; Stankowich 2008). Therefore, in this chapter, I will discuss the role of tail-flagging and the white rump patch in goitered gazelle behavior, for which information is very limited.

16.2.2 Tail-Flagging and White Rump-Patch Displays in Goitered Gazelles

Among most distinctive morphological adaptations in goitered gazelles are the black tail and white rump patch beneath of it, which are used as alarm signals when

Fig. 16.13 Alarm for assessment: Goitered gazelle female is standing immobile and tries to assess the kind of potential danger. Kalamali Ungulate Nature Reserve, Xinjiang, China. 23.11.2020. Photo of F. Xu with kind permission

Fig. 16.14 Tail flicking in goitered gazelles: Repeated lateral tail movements from one side to the other. Issyk-Kul goitered gazelle nursery for reacclimatizing, 26.06.2022. Photo by D. Blank

Fig. 16.15 Tail hanging down in goitered gazelles: In calm situations the tail is relaxed in neutral position, vertically down. Issyk-Kul goitered gazelle nursery for reacclimatizing, 26.06.2022. Photo by D. Blank

Fig. 16.16 Horizontal tail in goitered gazelles: keeping the tail strongly in horizontal position. Issyk-Kul goitered gazelle nursery for reacclimatizing, 26.06.2022. Photo by D. Blank

Fig. 16.17 Resting goitered gazelles merge with the surrounding background and almost invisible. Southern Ustyurt Chink (Kaplankyr), SW Kazakhstan, 01.08.2019. Photo of the camera-trap of the Ustyurt State Nature Reserve team with kind permission

encountering a predator (Tilson and Norton 1981; Kingswood and Blank 1996). The black-colored tail of goitered gazelles, which contrasts against the background of a pure white rump patch, plays an important role in transmitting warning signals to their conspecifics or pursuit-deterrent signals to a predator. The position of the tail and its movements almost directly reflect the emotional state of the gazelle. The vertically erected position of the tail (or tail-flagging) relays a very strong excitement, if the danger is real and near, while tail-flagging is related to alarm and flight behaviors (Alados 1986). A tail lowered vertically down signals a calm relaxed state when goitered gazelles can graze relax or rest. When the tail is in an intermediate position close to horizontal, the animal is relaying an average level of arousal. When faced with danger, goitered gazelles also raise their tails to various levels depending on their level of excitement: I most often observed the vertical tail position (37.4%, n = 917); with almost the same frequency, though slightly less (35.2%), the tail was raised lower than horizontal level; noted less often were tail movements (25%), and least often, the tail was in an exact horizontal position (1.4%) or lowered down without movements (1.0%, n = 917). Sub-adults and fawns used their tails for signaling least often compared to adults, and adult males signaled significantly more than the young, while adult females showed tail signals more often than any other conspecifics (Table 16.3). The frequencies of tail signals in goitered gazelles were

Fig. 16.18 Tail flagging in goitered gazelle herd: The white rump-patch is very bright when completely uncovered, when tail is raised during escaping of goitered gazelles. Southern Ustyurt Chink (Kaplankyr), SW Kazakhstan. 01.08.2019. Photo of the camera-trap of the Ustyurt State Nature Reserve team with kind permission

observed very often in July and even more in October; in other months, tail signals were displayed significantly less often (Fig. 16.11) (Blank 2018d).

The tail-flagging frequency, or keeping tail in vertical position, by adult females was maximal in July and significantly less in June and August. In other months, this behavior was rarely found (Fig. 16.12). It turns out that goitered gazelle females display tail-flagging frequently in May through August when they stay alone or in small groups. In contrast, in autumn and winter, they use tail-flagging rarely, but they stay in larger groups, indicating that tail-flagging likely has an inverse relationship to the size of the groups (Blank et al. 2012). The monthly frequency dynamics in adult females of a tail raised lower than horizontal was different compared to tail-flagging, and similar to tail movements, three maxima were observed in October, April, and July (Table 16.4). Tail movements of fawns were most often observed in July with a minimum of movements seen before and after July. Adult males,vhowever, performed other seasonal fluctuations compared to adult females, where tail-flagging coupled with stotting changed randomly over months with a higher frequency in August and a maximum in October (Fig. 16.13). It was interesting that adult males had lower frequencies of tail-flagging during the rutting season (November and December) than in other months. Sub-adults performed tail-flagging rarely all year round (Blank 2018d).

Fig. 16.19 Tail flagging in goitered gazelles: The white hairs of the rump-patch of the outline are projected outwards, making rump patch more prominent. Kalamali Ungulate Nature Reserve, Xinjiang, China. 06.09.2006. Photo by D. Blank

Fig. 16.20 Tail movements in goitered gazelles: Making pendulous movements of the tail in side to side and/or up-down directions. Issyk-Kul goitered gazelle nursery for reacclimatizing. 25.06.2022. Photo by D. Blank

Table 16.3 Tail positions (Raised tail—raised tail lower than horizontal; Horizontal tail—keeping tail strongly in a horizontal position; Tail-flagging—keeping tail raised vertically; Tail moving—up-down or side-to-side movements of tail; Tail down—tail hanging down) as responses of goitered gazelles of various sexes and ages to potential threat (upper—number of observed cases, lower—proportions in %) (by Blank 2018d—License Number—5477011106167)

Behaviors	Adult Female	Adult Male	Sub-adult Female	Sub-adult Male	Fawns	Total
Number observed individuals	1654	1169	213	283	976	4295
Raised tail	140 43.2	78 24.1	31 9.6	35 10.8	40 12.3	324 100
Horizontal tail	7 53.8	1 7.7	1 7.7	–	4 30.8	13 100
Tail-flagging	168 49.0	84 24.5	19 5.5	17 5.0	55 16.0	343 100
Tail moving	101 44.3	66 28.9	17 7.5	11 4.8	33 14.5	229 100
Tail down	2 22.2	4 44.4	3 33.4	–	–	9 100
Total	418	233	71	63	132	917 100

Tail signals of goitered gazelles are noted frequently when encountering moving or even standing humans, while these signals are rarely observed when goitered gazelles encounter a moving car or hear a car engine, or some other far-distant potential danger (Table 16.5). When the danger is real, goitered gazelles typically show vertical tail-flagging and less often a raised tail. In contrast, when the gazelles identify a strange smell or sound or see an alarm response from other gazelles, they prefer to show a raised tail and rarely tail-flagging and usually move away from the area (Table 16.5).

According to my observations, when a goitered gazelle in a group perceives any danger, it freezes, looking attentively at the danger. The alarmed individual, as a rule, displays tail-flagging with its hindquarters directed to conspecifics, which see the erected tail very clearly, though the freezing posture by itself is enough for the group members to pay attention to the alarmed individual and then watch in the same direction (Walther 1969). If the group cannot find the source of danger, they watch the alarmed individual, and if it starts to run, all conspecifics follow or even run ahead. From these observations, it is clear that tail-flagging is an important warning signal for conspecifics (Blank 2018d).

Comparing functions of tail-flagging in 200 ungulate species demonstrated that in most cases tail-flagging is an important part of communication between conspecifics, though multiple functions of a vertical tail in the same species are also possible (Caro et al. 2004). Observations of white-tailed deer (*Odocoileus virginianus*) response behaviors in threat situations found that fawns have the most benefit from a mother's tail-flagging, which is displayed most often during intensive communication between a fawn and its mother, and in turn, reduces fawn mortality during a period when fawns are most vulnerable to predation.

Table 16.4 Monthly fluctuations of observed frequencies (acts per hour) of running with tail-flagging, tail raised, and tail moving in goitered gazelles over months (by Blank 2018d—License Number—5477011106167)

Patterns	March	April	May	June	July	August	Sept.	October	Nov.	Dec.
Observed hours	90.6	119.3	280.2	200.7	113.3	39.6	44.6	35.3	145.9	121.4
Tail-flagging										
Adult Females	6 0.07	7 0.06	26 0.09	40 0.20	39 0.34	6 0.15	1 0.02	2 0.06	5 0.03	3 0.02
Fawns	–	–	7 0.02	14 0.07	30 0.26	3 0.08	–	–	1 0.01	–
Adult Males	4 0.04	3 0.03	23 0.08	7 0.03	9 0.08	7 0.18	2 0.04	10 0.28	14 0.10	5 0.04
Sub-adult Females	2 0.02	1 0.01	4 0.01	3 0.01	–	–	1 0.02	2 0.06	6 0.04	–
Sub-adult Males	–	1 0.01	11 0.04	2 0.01	1 0.01	–	–	–	2 0.01	–
Tail raised										
Adult Females	8 0.09	27 0.23	29 0.10	18 0.09	28 0.25	3 0.08	2 0.04	9 0.26	15 0.10	1 0.01
Fawns	–	–	–	12 0.06	25 0.22	2 0.05	–	–	1 0.01	–
Adult Males	12 0.13	7 0.06	2 0.01	5 0.02	2 0.02	4 0.10	–	12 0.34	29 0.20	5 0.04
Sub-adult Females	3 0.03	–	4 0.01	1 0.0	3 0.03	–	1 0.02	3 0.09	12 0.08	4 0.03
Sub-adult Males	10 0.11	–	–	8 0.04	3 0.03	–	–	12 0.34	2 0.01	–
Tail moving										

(continued)

Table 16.4 (continued)

Patterns	March	April	May	June	July	August	Sept.	October	Nov.	Dec.
Adult Females	8 0.09	25 0.21	14 0.05	16 0.08	24 0.21	2 0.05	2 0.04	1 0.03	9 0.06	–
Fawns	–	–	–	11 0.05	20 0.18	1 0.03	–	–	1 0.01	–
Adult Males	12 0.13	6 0.05	5 0.02	4 0.02	4 0.04	8 0.20	1 0.02	4 0.11	17 0.12	5 0.04
Sub-adult Females	3 0.03	–	5 0.02	1 0.0	–	–	1 0.02	–	5 0.03	2 0.02
Sub-adult Males	11 0.12	–	–	7 0.03	2 0.02	–	–	–	2 0.01	–

Table 16.5 Different tail positions in goitered gazelles during their responses to the threat depending on the different kinds of danger source: human, vehicle and conspecifics (upper—number of observed cases, lower—proportion in %). Legend: Walk—human walking, Move—movements of standing human, Stand—standing human motionlessly, Sit—sitting human, Smell—arriving smell from motionless human, Hear—hearing some strange sounds, Conspec.—reaction to galloping conspecifics, Veh. moves—gazelles saw moving vehicle, Veh. Hear—gazelles heard car engine, but did not see it; Unknown—reaction to far-distant or unknown source of danger (by Blank 2018d—License Number—5477011106167)

Patterns	Walk	Move	Stand	Sit	Smell	Hear	Con-spec.	Veh.moves	Veh.hear	Un-known	Total
Tail Raised	90 30.9	35 12.0	89 30.6	5 1.7	10 3.4	8 2.8	13 4.5	6 2.1	3 1.0	32 11.0	291 100
Tail horizontal	–	–	3 0.4	2 0.2	–	–	–	–	–	3 0.4	8 100
Tail vertical	125 44.2	63 22.2	29 10.2	3 1.1	3 1.1	5 1.8	4 1.4	19 6.7	5 1.8	27 9.5	283 100
Tail moving	48 23.5	38 18.6	64 31.4	1 0.5	10 4.9	10 4.9	10 4.9	3 1.5	1 0.5	19 9.3	204 100
Tail down	5 0.5	–	2 0.2	3 0.3	–	–	–	–	–	–	10 100

I observed a similar situation in goitered gazelles, when mothers demonstrated tail-flagging behavior with increasing frequency from June to August. During the first weeks in May, mothers stayed apart from their fawns, which remained lying down alone most of the day (Blank et al. 2015); therefore, mothers' tail-flagging behavior was noted rarely during this month. Thereafter in June and July, mothers stayed with their fawns for longer and longer periods of time and the rate of tail-flagging increased significantly, reaching its maximum in July, when fawns followed their mothers most of the daytime and the hiding period was ending. This period is especially important for fawns as they shift their antipredator behaviors from concealment and freezing in threat situations to fleeing to outrun a predator. This time is quite stressful for mothers as they display tail-flagging and white rump-patch presentation signals most often, even in cases of minimal threats in order to teach their fawns the new strategy of how to not become a victim of predators.

In August, fawns stay with their mothers constantly and repeat all their movements and actions. As a result, tail-flagging displays alone begin to decrease and eventually lose their importance, replaced by maximal displays of the white rump patch, which is performed in conjunction with tail-flagging for communicating with each other in dangerous situations. Later, both tail-flagging and white rump-patch behaviors are observed rarely, likely losing their importance for intraspecific communication. During the same period (May through July), sub-adults and adult males rarely perform either tail-flagging or white rump-patch signals. Thus, it turns out that the main purpose of these displays is managing the movements of goitered gazelle fawns by their mothers so that they learn the defensive behaviors necessary to escape from predation (Blank 2018d).

A vertical tail (tail-flagging) can be a signal for all conspecifics, but the fact is that most often females demonstrate this behavior in the months when they are outside the group and communicating mainly with their fawns. My observations suggest that the tail-flagging and white rump patch are directed by females almost entirely for communications with their fawns during alarm situations. Therefore, tail-flagging and white rump-patch demonstrations do not depend on the size of a group and are observed more often from single females than in male groups. The females of white-tailed deer also display tail-flagging twice more often than adult males and use it frequently from the first days after giving birth and continue through their fawns first 6 weeks with a gradual decrease thereafter (Hirth and McCullough 1977; Smith 1991).

A white rump patch is one of the most characteristic color features of the goitered gazelle body. When a gazelle lies down, it practically merges with surrounding soil due to the sandy coloring of the upper body parts (Fig. 16.17). But as soon as the gazelle begins to run, the white rump patch swaying while running is visible from afar. At the same time, the black tail is raised vertically upward and emphasizes the contrast with the snow-white rump patch, making it especially noticeable (Fig. 16.18). Perhaps due to the bright white coloring of the rump patch, goitered gazelles can easily follow each other at night (Sokolov 1959). Hunters claim that Asiatic hounds can catch up to a single goitered gazelle, but not to a tight running group. Sludskiy (1962) believes that the gazelles in a group can run faster, but it is possible that the flickering effect of white rump patches in multiple gazelles distracts the attention of the dogs and prevents them from chasing their prey fast enough. Thus, a white rump patch in goitered gazelles has a pursuit-deterrent and distracting effect.

In highly stressful situations, the raising of the tail and presenting the white rump patch, as well as running away in a dense group, is a typical response for many species of gazelles and other antelopes, as well as cervids (Walther 1964; Bigalke 1972; Alvarez et al. 1976; Leuthold 1977, 1978; Estes 1991). From these many authors' observations, it can be assumed that the tail-flagging and white rump-patch presentation, exaggerated through piloerection, are closely related to each other, and in addition to being an alarm signal, also has a pursuit-deterrent effect for many ungulate species (Blank 2018d).

Both tail-flagging and white rum-patch presentation are usual for ungulate species which often perform stotting as part of their defensive behaviors (Caro et al. 2004; Caro 2009). Goitered gazelles demonstrate stotting is less frequently than tail-flagging. Frequencies of stotting and tail-flagging increased between March and June, reaching their maximum in July. Adult gazelles perform stotting, but tail-flagging is shown much more often than stotting, and only in August did stotting surpass tail-flagging in frequency to reach its yearly maximum. After that, the frequency of both drops sharply and in September to December they lose their value and are observed quite rarely without a significant difference between these behaviors (Fig. 16.12a). Adult males also displayed stotting less often than tail-flagging, but with other dynamics, with stotting reaching its maximum in August and October, fluctuating randomly in other months (Fig. 16.12b) (Blank 2018d).

Being more demonstrative to attract attention than to increase speed, stotting is demonstrated at safe distances when there is no threat of a direct predator attack (Walther 1977). Goitered gazelles and other gazelle species (Thomson's gazelles and Dorcas gazelles) also perform stotting and tail-flagging simultaneously, which enhances the communicative effect of each of them (Alados 1986; Estes 1991). During a normal gallop, the vertical lifting of the tail (tail-flagging) is usually not observed, as the tail is usually in a lower position, often below the horizontal (Alados 1986). The same behavior is possible to see in white-tailed deer, which do tail-flagging during alarm assessment, but lower the tail when running (Bildstein 1983), even though the cost of a tail-flagging display, as a physical act, is small (Caro et al. 1995). Other gazelle species (springboks and Grant's gazelles) even press their tails between their haunches during normal running, though they perform tail-flagging during alarm assessment (Walther 1968; Bigalke 1972). In addition, tail-flagging is very visible, maybe seen without stotting and therefore observed more frequently than stotting. So, tail-flagging has independent signaling effect, and mothers use mostly this behavior at the beginning for managing fawns' behavior; later the role of white rump-patch signal gradually increases reaching its maximum and most powerful warn signal in combination with tail-flagging for fawns occurs in August. Thus, three behavioral patterns, tail-flagging, white rump patch, and stotting, are closely linked to each other and when demonstrated together reach the maximal warning and pursuit-deterrent effects (Caro et al. 2004).

16.3 Escaping Behavior (License Number—5477020170232)

16.3.1 General Remarks

In order to increase their chances of survival, ungulates must change their behavior in such a way as to avoid the risk of becoming a victim of predator attacks. To do this, they have to change movements, foraging, daily activities (Verdolin 2006); reshape habitat use (Bonnot et al. 2012); lengthen vigilance time (Brown and Kotler 2004); modify group sizes (Caro 2005; Creel et al. 2014); find safer areas (Stankowich and Coss 2007); and enhance the entire antipredator strategy (Croes et al. 2007). All these actions have certain energetical costs, including consumption of food of lower quality (Christianson and Creel 2010), lessening feeding rate (Brown and Kotler 2004), or generating stress responses (Clinchy et al. 2013). Enlarging of group sizes and spending more time for vigilance are two of the most studied responses of ungulates covered by published papers (Underwood 1982; Prins and Iason 1989; FitzGibbon 1993; Hunter and Skinner 1998; Toïgo 1999; Caro 2005; Creel et al. 2005; Halofsky and Ripple 2008; Creel et al. 2014). Less papers devoted preflight assessment and ungulate responses when approaching a predator (Caro 2005). The decision of how and when to react to a danger and where to run is crucial for survival of ungulates; if there is a mistake in responding and selecting an optimal strategy or they wrongly retreat away from threat, ungulates may pay with their lives (Stankowich and Coss 2006).

Numerous articles on the study of prey defensive behavior have demonstrated a wide variety of responses of ungulates to predator attacks, including various escape variants, diversity visual and auditory signals, unusual gait during fleeing, using landscape environment, or even aggression of ungulates toward predatots (Reynolds 1983; Caro et al. 2004; Grovenburg et al. 2009; Lingle and Pellis 2002; Lingle et al. 2005). However, in any case, when a hoofed animal encounters a predator, it modifies its behavior according to the distance to the predator, the degree of risk, and the distance to the nearest refuge place (Stankowich and Coss 2007). The ungulate has to do a number of functions, including predator identification, degree of threat, where and when to run based on distance and predator's speed of approaching, and arranging the most optimal avoid behavior (Stankowich and Coss 2007). Ydenberg and Dill (1986) proposed the model of optimality which states that the cost of delaying on the place (e.g., be killed as a result) exceeds the costs of running away (losing feeding and breeding opportunities and energy waste) than the animal is escaping. Animals modify their behaviors as a result of a trade-off between different parameters of their life and show a great variability in antipredator actions (Frid 2003; Stankowich 2008). However, in any case the prey needs to enlarge the flight initiation distance when the risk is increasing and maintain both temporal and spatial resources of safety, making escape decisions (Cárdenas et al. 2005).

Properties of prey (experience, body size, and group size), predators (their size, directness, and speed of coming), and characteristics of habitat (distance of refuge, landscape roughness) are all important for the evaluation of threat degree (Stankowich and Blumstein 2005; Liley and Creel 2008; Baruzzi et al. 2017). The

Table 16.6 Description of the recorded displays (ethogram) during escaping behavior in goitered gazelle within study period according to Walther (1969), Blank (1998), Blank et al. (2015), Stankowich and Coss (2006) (by Blank 2018b—License Number—5477020170232)

Behavioral patterns	Description
Galloping	Go at a rapid pace, fastest gait. This is a four-beat gait in which all four feet are off the ground at the same time during each stride and they extended as far as possible to cover ground. This gait evolved to allow ungulate to escape quickly from predators and relies on a burst of speed, rather than stamina. This gait goal is to get away from a dangerous location, rather than to keep up a sustained flight (Figs. 16.1 and 16.2)
Trotting	Go with speed as the trot is between a walk and a gallop. This is natural jumping two beat gait, when both pairs of diagonal legs alternate in touching down, with a pause in between. Trot is very energy-efficient and an ungulate can sustain this gait for long periods of time
Walking	A walking gait is the process that animals use to move themselves at a slowest pace. Natural walking gait in four equal movements: each leg lifts and touches down diagonally in succession
Alarm in standing or laying position	Gazelle standing with its head and neck upright and ears erected and their heads oriented directly to the danger source once alerted
Flight	Movement away from the gazelle's original location, ranging from a slow walk to trotting and galloping, stotting. Flight necessarily followed alarm (alert) behavior, though flight occurred almost instantly (FID≈AD) when a predator was detected too close (too late)
Jump sudden during grazing	During grazing a gazelle abruptly jumped aside or starts to rush anywhere, but stops after covering several meters and does assessment. Such behavior was typically observed in cases of unexpected appearance conspecifics
Alarm stop during escaping	Gazelle starts to run, stops during escaping, turns round, and does additional assessment of approacher and then continues their running. Sometimes such stops repeat

flee distance grows with rising the threat, which depends on many determinants, such as behavioral features of prey and predator and also environmental characteristics (Ydenberg and Dill 1986; Stankowich and Blumstein 2005; Caro 2005). Prey animals have the ability to identify when the predator is hunting and when it is busy in other activities (resting, seeking a mate) (Jayakody et al. 2008). Moreover, ungulates are able to differentiate between non-predatory and predatory species having a similar outlook (Peckarsky and Dodson 1980; Coleman et al. 2008), and also between inactive and active predators (Dill 1974; Coss and Ramakrishnan 2000). The threat magnitude is the main factor influencing the escaping actions, and the prey use a quicker gait and will run for longer distances when predator shows increasing attack probability (Stankowich and Blumstein 2005; Stankowich and Coss 2006).

In this chapter, I will consider the features of the goitered gazelle's escaping and risk assessment behavior, observing several behavioral patterns; the ethogram is presented in Table 16.6.

16.3.2 Escaping Behavior in Goitered Gazelles

By running away from a predator or another threat, goitered gazelles of all age-sex groups prefer gallop and less frequently they walk away or trot while walking in the direction of a danger exclusively rarely. Only sub-adult females have different preferences in gaits compared to conspecifics, favoring trotting as often as gallop (Table 16.7). This is not surprising since gregarious ungulates of open habitats have the only protection, i.e., running away at as high a speed as possible so that no predator can catch up (FitzGibbon and Lazarus 1995). Being a species highly sensitive to any anxiety and excitable, goitered gazelles flee away with maximal speed (gallop) at once when they detect the real danger as any gregarious antelopes do (Estes 1991). High-speed galloping is effective in avoiding predation strategy for goitered gazelles, because they typically gain an advantage in speed over the wolves, which are main large stalking predators for the gazelles in Asia (Sludskiy 1962). The gregarious white-tailed deer has the same strategy of fleeing by gallop from wolves, which are also the most harmful predators for these deer (Lingle and Pellis 2002).

Before making a decision to flee, goitered gazelles usually stand still and assess the degree of potential danger and their response will be different for various kinds of danger. Alerted goitered gazelles, similar to mule deer (*Odocoileus hemionus*), direct their muzzles straightly toward the threat source, as a result of their identifying the location and kind of the danger more precisely than species like white-tailed deer (*O. virginianus*) that typically do not pinpoint the danger (Lingle and Wilson 2001). The quick identifying of the nature of threat is very important for goitered gazelles like other antelope species of open environments because they rescue themselves from predator attack in time. Such ability to define the kind of danger is not crucial for forest-dwelling deer and dwarf antelopes that are usually more affiliated in their antipredator behavior with freezing and concealment instead of escaping flight (Leinders 1979; Lingle and Wilson 2001), although sometimes forest-dwelling ungulates (duikers *Cephalophus monticola*) can change their habits and, instead of freezing, immediately run away from a predator (Croes et al. 2007) (Table 16.8).

Table 16.7 Gait preference during escape from danger in goitered gazelles of different sexes and ages (upper—number of observed cases, lower—proportion in %) (by Blank 2018b—License Number—5477020170232)

Gait	Adult female	Adult male	Sub-adult female	Sub-adult male	Fawns	Unknown	Total
Galloping	1028 72.1	763 71.6	72 42.6	150 66.9	639 77.5	2319 90.5	4971 79.3
Trotting	188 13.2	128 12.0	61 36.1	36 16.1	82 9.9	33 1.3	528 8.4
Walk from danger	189 13.3	166 15.6	33 19.5	36 16.1	103 12.5	209 8.2	736 11.8
Walk to danger	20 1.4	8 0.8	3 1.8	2 0.9	1 0.1	–	34 0.5
Total	1425 100	1065 100	169 100	224 100	825 100	2561 100	6269 100

Table 16.8 Alarm stop during escape vs running without stopping in goitered gazelles of different sexes and ages when escaping from danger; observed cases of alarm while in laying position and sudden jumps during grazing (upper—number of observed cases, lower—proportion in %) (Blank 2018b—License Number—5477020170232)

Behavioral patterns	Adult Female	Adult Male	Sub-adult Female	Sub-adult Male	Fawns	Unknown	Total
Alarm stop during escape	1856 78.8	1086 74.2	70 68.6	93 56.0	192 38.2	355 22.1	3652 59.0
Running without stop	476 20.2	360 24.6	29 28.4	71 42.8	306 60.8	1250 77.9	2492 40.2
Alarm while laying	20 0.9	9 0.6	2 2.0	1 0.6	4 0.8	–	36 0.6
Jump away suddenly	20 0.1	8 0.6	1 1.0	1 0.6	1 0.2	–	13 0.2
Total	2354 100	1463 100	102 100	166 100	503 100	1605 100	6193 100

In addition, during their escape by running, goitered gazelles stop after a short flight due to their looking back (Gorbunov 1983; Zhevnerov 1984), monitoring the predator actions for the correction of assessment regarding the speed and direction of their escaping fleeing and rarely they gallop without delay (Table 16.9). This correction of threat evaluation is incredibly important for decreasing ambiguity of how and where to flee and mitigate the cost of escaping (Blumstein 2008). During escaping from a predator, standstill after a short primary flight is observed as a typical behavior for Thomson's gazelles (*Eudorcas thomsonii*), pronghorns (*Antilocapra americana*), dwarf African antelopes (tribe Neotragini), duikers (tribe Cephalophini), and even for argali (Walther 1969; Autenreith and Fitchter 1975; Estes 1991; Fedosenko and Blank 2005). However, goitered gazelles do stops for additional assessment not in every case. They do it without fail, when they run from far-distant, immobile, or uncertain danger, but they do not stop when they are chased intensively by a predator and are in a situation of real threat to be victim. The gazelles do not stop, for example, when they are chased hardly by wolves or by cheetah (Sludskiy 1962; Walther 1969).

Black-tailed deer make a preflight assessment during significantly longer time than goitered gazelles, especially during the tourist season, when a number of deer species (white-tailed deer and red deer) estimate human appearance as low-risk threat (Grau and Grau 1980; de Boer et al. 2004; Jayakody et al. 2008). In contrast, during the hunting season the same deer conceives humans as great risk, increasing flight distance and making shorter assessment duration. In my observation area, walking humans, as well as moving cars, may do poaching actions; therefore, both of them are perceived as a great danger and cause a galloping escaping with an almost complete absence of scan assessment before and during fleeing. This pattern is observed also in caribou (*Rangifer tarandus*—Aastrup 2000), white-tailed deer (Behrend and Lubeck 1968), red deer (Jayakody et al. 2008), roe deer (*Capreolus capreolus*—de Boer et al. 2004), moose (Alces alces—Altmann 1958), impala (*Aepyceros melampus*—Matson et al. 2005).

Table 16.9 Gait preference during escaping from perceived danger in goitered gazelles depending on different actions presented by danger source: observed human, vehicle or conspecifics (upper—number of observed cases, lower—proportion in %). Legend: Walk—human walking, Move—movements of standing human, Stand—standing human motionlessly, Sit—sitting human, Smell—arriving smell from motionless human, Hear—hearing some strange sounds, Conspec.—reaction to galloping conspecifics, Veh. moves—gaselles saw moving vehicle, Veh. Hear—gazelles heard car engine, but did not see it; Unknown—reaction to far-distant or unknown source of danger (Blank 2018b—License Number—5477020170232)

Patterns	Walk	Move	Stand	Sit	Smell	Hear	Con-spec.	Veh.moves	Veh.hear	Un-known	Total
Galloping	2819 84.0	77 52.0	49 22.8	3 15.8	17 47.2	9 32.1	36 78.3	1757 85.8	1 2.4	70 61.4	4838 80.0
Trotting	306 9.1	29 19.6	50 23.2	10 52.6	10 27.8	11 39.3	2 4.3	48 2.3	14 33.3	25 21.9	505 8.3
Walk from danger	229 6.8	41 27.7	108 50.2	5 26.3	8 22.2	7 25.0	6 13.1	242 11.8	26 61.9	14 12.3	686 11.3
Walk to danger	1 0.1	1 0.7	8 3.8	1 5.3	1 2.8	1 3.6	2 4.3	1 0.1	1 2.4	5 4.4	22 0.4
Total	3355 100	148 100	215 100	19 100	36 100	28 100	46 100	2048 100	42 100	114 100	6051 100

Ungulates conceive less danger when they have protective areas in close distance, where they have low chances of being victim (Stankowich and Coss 2007). Many deer species (black-tailed deer *Odocoileus hemionus columbianus*), red deer (*Cervus elaphus*), and mule deer (*O.h.hemionus*) use woodland habitats and guanaco *Lama guanicoe* use undulating landscape as a protective cover, preferring dense vegetative areas when fleeing from predators (Geist 1981; Stankowich and Coss 2007; Jayakody et al. 2008; Taraborelli et al. 2012). In closed habitats, deer enable the pursuer to come closer than open-habitat ungulates allow to predator approach (Lingle and Wilson 2001), though they are more vigilant in closed environment (Altendorf et al. 2001). Goitered gazelles have no refuge cover during escaping from a predator, but they conceive relatively safe in the rough landscape and they can allow the threat to get closer similar to woodland-dwelling deer (Blumstein 2008).

The most forceful fleeing response (gallop) of goitered gazelles is observed after they detect walking human during approaching or moving vehicle or when conspecifics gallop away. Their response is less intensive when they see some movements of sitting human or feel his smell (Table 16.9), and the quietest reaction when the gazelles hear unusual sounds (car engine from afar) but have no ability to identify the cause or see an unmoved standing or sitting human; then the response will be walking away or trotting. Some behaviors of potential predator (e.g., people or car) are conceived by prey as a predation threat though people just travel or observe animals (Stankowich and Coss 2007; Jayakody et al. 2008). But in this case, the preys also perform the same behavioral actions if it is like they are being chased and they gallop away, lengthening the distance from a danger as soon as possible (Frid and Dill 2002; Stankowich and Coss 2006).

Goitered gazelles are constantly monitoring each other's behavior and if someone is alarmed, freezes, and stares attentively in some direction for a long time, then everyone starts to look at the alarmed gazelle first and then begins to look for the source of danger. If other gazelles cannot find it, they continue to follow the alarmed individual, and when the alarmed gazelle starts to run away, everyone follows it, even if they have not seen the danger, reacting keenly to the alarm and running of any member of the group. This is a universal response of social ungulates living in open environments, when run of one individual contagiously initiates the flight of the entire group (Sirot and Touzalin 2009). This reaction is seen in many deer species, such as caribou (*Rangifer tarandus* – Aastrup 2000), black-tailed deer (Stankowich and Coss 2006), and white-tailed deer (Lingle and Wilson 2001). In the case of a larger herd, it is more likely that any of them may be showing alarm even without any predator itself. It is a logical conclusion from here that a larger herd is more likely to become anxious earlier, stay in wariness longer time, and can start running sooner (Aastrup 2000; Lingle and Wilson 2001). Moreover, larger herds flee for longer distances escaping danger (Aastrup 2000, Taylor and Knight 2003).

Ungulates of closed environments are more reactive to sounds and smell because of low visibility of their habitat (Berger et al. 2001), and the presence of wolf scent, for example, can increase the frequency of vigilance in red deer (*Cervus elaphus*) from 22 to 46% and reduce their grazing time from 32% to 12% (Kuijper et al. 2014). Goitered gazelles live in open environments and when they hear extraneous sounds and feel an unusual scent without vision of their source, they flee away with smaller probability compared to cases when they detect a moving danger. This suggests that it is much more important for ungulates of open biotopes to see danger than to rely on hearing or smell of potential threat (for *Lama guanicoe*—Sarno et al. 2008). Similar to goitered gazelles, pronghorns (*Antilocapra americana*) and Thomson's gazelles (*Eudorcas thomsonii*), first of all, react to vision and only then on smell and hearing (Walther 1969; Autenreith and Fitchter 1975). Nevertheless, the smell of predators can have a certain effect on ungulate antipredator behavior (Apfelbach et al. 2005; Chamaillé-Jammes et al. 2014). When goitered gazelles detect uncertain danger, which does not move, without producing smell or sounds (e.g., immobile human), they have a calmer reaction and walk or trot away. I observed a similar behavior when goitered gazelles reacted on sounds of moving car from far distances; they walked away and less often trotted. Thus, goitered gazelles favor gallop only in the cases of a faster moving threat, in my case, when vision of moving car or at least a walking human (Blank 2018d).

In addition, more experienced adult females and males in most cases stop during escaping more frequently than sub-adults, while unpracticed fawns do the opposite and flee away without delay more often than stop in halfway (Table 16.10). This suggests that experience is crucial for the right evaluation of danger and proper actions due to survival at the end. Therefore, young ungulates still associated with their mother die less often from predators than those who already lead an independent life, since their experience is insufficient for effective protection from predation (White and Berger 2001). The goitered gazelles stop especially frequently when the situation is relatively safe and they run from immobile or far-distant danger, while in the case of real potential threat (approaching car or human in close range) they typically will not stop during their escaping running (Table 16.10) (Blank 2018d).

Compared to other angles, goitered gazelles favor a sideway escaping close to straight angle (90°) with regard to the trajectory of coming danger independently of predator behavior (freezing or walking human) and speed (vehicle vs human) (Table 16.11). Less preferable for them is running in a parallel path to overtake or traverse the predator vector: in some cases twice or more times. It is possible to assume that parallel flight is a first phase to overtake moving danger due to crossing of the way; then goitered gazelles desire crossing of the way at the same rate as sideway escaping. Rarely the gazelles flee along the same trajectory or in reverse direction to the pursuer. Goitered gazelles cross a way to walking human (slow threat) more frequently than to moving car (fast threat), which needs likely more efforts to outrun than walking human. All gazelles behave similarly irrespective of their age and sex, with the exception of sub-adult females which do not favor a sideway escape (Blank 2018d).

Table 16.10 Alarm stop during escape vs running without stopping and causes of observed cases of alarm in laying position and sudden jumps while grazing in goitered gazelles depending on different actions of danger source: observed human, vehicle, and conspecifics. (upper—number of observed cases, lower—proportion in %). Legend: Walk—human walking, Move—movements of standing human, Stand—standing human motionlessly, Sit—sitting human, Smell—arriving smell from motionless human, Hear—hearing some strange sounds, Conspec.—reaction to galloping conspecifics, Veh. moves—gazelles saw moving vehicle, Veh. Hear—gazelles heard car engine, but did not see it; Unknown—reaction to far-distant or unknown source of danger (Blank 2018b—License Number—5477020170232)

Patterns	Walk	Move	Stand	Sit	Smell	Hear	Con-spec.	Veh.moves	Veh.hear	Un-known	Total
Alarm stop during escape	574 25.2	103 81.7	125 91.9	5 100.0	17 73.9	21 87.5	71 89.9	315 30.5	9 64.3	2159 99.1	3399 57.5
Running without stop	1692 74.2	22 17.5	8 5.9	–	6 26.1	3 12.5	–	709 68.7	5 35.7	13 0.6	2458 41.6
Alarm while laying	14 0.6	1 0.8	2.2	–	–	–	3 3.8	8 0.8	–	1 0.04	30 0.6
Jump away suddenly	–	–	–	–	–	–	5 6.3	–	–	6 0.3	11 0.3
Total	2280 100	126 100	136 100	5 100	23 100	24 100	79 100	1032 100	14 100	2179 100	5898 100

Table 16.11 Different escaping angles in goitered gazelles of various sexes and ages (upper—number of observed cases, lower—proportion in %). Legend: < 90°—predominantly directly away; 90°— predominantly lateral escaping, > 90°—predominantly directly toward the predator; Parallel route—mostly in the same vector with predator; cross-way—cross vector of the predator moving (Blank 2018b—License Number—5477020170232)

Escaping angle	AdFemale	AdMale	SbtFemale	SbtMale	Fawns	Unknown	Sum
< 90°	35 8.6	48 13.6	13 25.5	–	9 4.2	268 16.0	373 13.5
90°	161 39.5	134 38.1	15 29.4	26 38.8	76 35.7	797 47.8	1209 43.8
> 90°	40 9.8	37 10.5	6 11.8	7 10.4	15 7.1	98 5.9	203 7.4
Parallel route	82 20.1	75 21.3	5 9.8	15 22.4	39 18.3	315 18.9	531 19.2
Cross-way	90 22.0	58 16.5	12 23.5	19 28.4	74 34.7	190 11.4	443 16.1
Sum	408 100	352 100	51 100	67 100	213 100	1668 100	2759 100

During observation of the escaping behavior of black-tailed deer, it was found that flight distance has an impact on angle escaping, and deer run with acute antgle when a predaor is close and escape sideways (directly away) when the pursuer is far (Geist 1981; Stankowich and Coss 2007). I did not observe this phenomenon in goitered gazelles and they prefer escaping flights sideways or crossway trajectories irrespective of the distance to pursuer. Zhevnerov (1984) observed that goitered gazelles run in one line in relatively calm situations, but this order disappears and changes into a chaotic panic run in cases of sudden and close danger. In addition, he noted several times that goitered gazelles can run toward danger in some cases; for example, a male can run to females even if the human comes his way.

During fleeing away from a danger, any ungulate is tending to arrive at the area behind of obstacles in relation to a predator (uneven landscapes, trees, shrubs) making some kind of protection which will interfere with the pursuit (Stankowich and Coss 2007). Therefore, goitered gazelles typically escape to rough terrain, which is preferable by the gazelles as a refuge cover (Blank 1990). The same behavior is seen in mule deer (*O. hemionus hemionus*) and black-tailed deer (*O. hemionus columbianus*), which always are escaping toward broken landscapes, where they are less often assaulted by predators than in plain areas (Lingle and Wilson 2001; Lingle and Pellis 2002; Stankowich and Coss 2007).

During fleeing, goitered gazelles have a tendency to escape upward to any high place in the surrounding landscape (uplands, foothills, banks of dry river beds); less preferable are horizontal surfaces and downward slopes are least chosen (Table 16.12). The gazelles of different ages and sexes behave mostly in a similar way in their desire to arrive at upland places with better points of view for checking the surrounding areas. Goitered gazelles prefer elevated places independently from threat, though most troubled events (moving car or just hearing car engine) force these ungulates to flee to the upland points more frequently, while in the cases of

Table 16.12 Different running directions (parallel, to, or away from elevated points or mountains) in goitered gazelles of various sexes and ages while escaping from danger (upper—number of observed cases, lower—proportion in %). Stop on elevated ground (upper—cases of goitered gazelle stops on tops of elevated points, lower—proportion of stops on elevations compared to general number of cases running upward in %) (Blank 2018b—License Number—5477020170232)

Running direction	Adult female	Adult male	Sub-adult female	Sub-adult male	Fawns	Unknown	Total
Running to (from) elevated points or in parallel direction							
Upward	501 68.4	326 68.8	53 73.6	58 60.4	331 66.1	758 83.3	2027 72.8
Downward	89 12.1	68 14.3	9 12.5	16 16.7	61 12.2	91 10.0	334 12.0
Horizontal	143 19.5	80 16.9	10 13.9	22 22.9	109 21.7	61 6.7	425 15.2
Total	733 100	474 100	72 100	96 100	501 100	910 100	2786 100
Stop on elevated tops	88 17.6	38 11.6	21 39.6	4 6.9	38 11.5	36 4.7	225 10.0
Running to mountains (upward), from mountains (downward) and parallel along mountains							
To Mount	264 75.6	192 72.4	27 75.0	32 64.0	122 71.3	469 88.0	1106 78.8
From mount	22 6.3	28 10.6	5 13.9	6 12.0	10 5.9	19 3.6	90 6.4
Parallel mount	63 18.1	45 17.0	4 11.1	12 24.0	39 22.8	45 8.4	208 14.8
Total	349 100	265 100	36 100	50 100	171 100	533 100	1404 100

distant threat or uncertain situations, goitered gazelles escape to higher and horizontal places with the same likelihood (Table 16.13).

Escaping from danger to the elevated places, goitered gazelles stop in the highest point and look back and rescan area around, assessing the progress of the pursuer, though sometimes they continue running without stopping. Adult females display assessment behavior in the highest points more often than other conspecifics excepting sub-adult females which do it more often than adults, though adult females run to elevated points much more often than sub-adult females (Table 16.12). In most ungulates, females devote more time to vigilance and are more careful, since they are responsible not only for their own safety, but for the survival of their offspring (Bøving and Post 1997). Goitered gazelles stop on the highest points relatively more frequently when the threat is perceived less dangerous (distant unknown danger, feel the smell without detecting human, standing human), while they stop less frequently when danger is more serious (hearing car and especially seeing the moving car, walking human) (Table 16.13) (Blank 2018d). Mule deer also usually stop in the upland points for additional assessment, but also they do not do it during serious threat, for example, attacking coyotes (Lingle and Wilson 2001).

When goitered gazelles encounter potential threat near the mountains, most of them flee toward them and ascend along the slope, less often they flee along the

Table 16.13 Different running directions (to, from, or parallel to elevated points or mountains) while escaping from danger in goitered gazelles depending on different actions of danger source: observed human, vehicle or conspecifics (upper—number of observed cases, lower—proportion in %). Legend: Walk—human walking, Move—movements of standing human, Stand—standing human motionlessly, Sit—sitting human, Smell—arriving smell from motionless human, Hear—hearing some strange sounds, Conspec.—reaction to galloping conspecifics, Veh. moves—gazelles saw moving vehicle, Veh. Hear—gazelles heard car engine, but did not see it; Unknown—reaction to far-distant or unknown source of danger. Stop on elevated (upper—cases of goitered gazelle stops on tops of elevated points, lower—proportion of stops on elevations compared to general number of cases running upward in %) (Blank 2018b—License Number—5477020170232)

Running direction	Walk	Move	Stand	Sit	Smell	Hear	Con-spec.	Veh.moves	Veh.hear	Un-known	Total
Running to (from) elevated points or in parallel direction											
Upward	1032 62.8	23 71.9	39 69.6	4	15 78.9	21 67.7	13 76.5	735 93.4	25 100	38 53.5	1945 72.4
Downward	289 17.6	2 6.2	17 30.4	–	1 5.3	10 32.3	4 23.5	4 0.5	–	2 2.8	329 12.2
Horizontal	323 19.6	7 21.9	–	–	3 15.8	–	–	48 6.1	–	31 43.7	412 15.3
Sum	1644 100	32 100	56 100	4 100	19 100	31 100	17 100	787 100	25 100	71 100	2686 100
Stop on elevated tops	84 8.1	9 39.1	16 41.0	2 50.0	11 73.3	4 19.1	3 23.1	8 6.5	6 24.0	32 84.2	215 11.0
Running to mountains (upward), from mountains (downward) and parallel along mountains											
To mount	511 73.3	7 77.8	12 32.4	2 50.0	10	4	4	411 93.0	72 100	16 45.7	1049 79.8
From mount	77 11.0	–	4 10.8	2 50.0	–	–	–	3 0.7	–	1 2.9	87 6.6
Parallel mount	109 15.6	2 22.2	21 56.8	–	–	–	–	28 6.3	–	18 51.4	178 13.5
Total	697 100	9 100	37 100	4 100	10 100	4 100	4 100	442 100	72 100	35 100	1314 100

mountain foothills and exclusively rarely they flee down from mountain slopes (Table 16.12). Similar to escaping to elevated points, goitered gazelles used mountains for escaping place more often in cases of real danger (approaching car) and less frequently when they encounter less threatening objects (distance of unknown threat, standing or sitting human), but instead they run parallel to the mountain slopes (Blank 2018d). Black-tailed deer have very similar behavior (Stankowich and Coss 2007). Goitered gazelles and black-tailed deer are not really ungulates living in mountains, but they have the same evading behavior typical for mountain-dwelling species, when they aim higher up the slope than the pursuer (Fedosenko and Blank 2001, 2005; Baruzzi et al. 2017). Coyotes, for example, have limited possibilities to be successful in hunting mule deer when they climbed higher up the slope and the predators stop the chasing group of mule deer staying in vantage place (Lingle 2002).

Thus, similar to any antelope of open environment, goitered gazelles' only defense from predator attack is high-speed running; therefore, they prefer evading gallop during real threat and slower gaits if the situation is not so dangerous. Before escaping and during fleeing, adult goitered gazelles stop and do risk assessment in obligatory order, sub-adults do it more rarely, and fawns run away without delay, suggesting that experience is very important for antipredator behavior. However, goitered gazelles perform assessment stopping if the situation is quite safe for them; otherwise, they run without delay. Running from threat, the gazelles flee sideways or cross the pursuer's path and always escape toward elevated places and mountain slopes, as well as they choose rough landscape as a refuge cover. Although goitered gazelles are plain-dwelling ungulates and mainly use speed flee as protection from predator attacks, they use also rough terrain as additional protection from persecutions.

16.4 Diversity of Risk Assessment (License Number—5477020890098)

16.4.1 General Remarks

Predators play an exceptional role in shaping all dimensions of a prey species existence, especially in changing ungulate behavior (Lingle 2001; Favreau et al. 2013). Ungulates of open biotopes, in order to resist the pressure from predators, many times gather in large groups to better detect an approaching danger in time for the prey to react ("many eyes effect"); this is the most important benefit of a social lifestyle (Childress and Lung 2003; Creel et al. 2014; Pays et al. 2012). To survive under the pressure from predation, ungulates have to achieve a balance between minimal vulnerability to predators and maximal time and conditions to cost-effectively forage (Hunter et al. 2007). There are two ways to improve antipredator behavior: first, a short-term reaction, when ungulates magnify their vigilance frequency in the high likelihood of a predator attack, especially an increase in the scanning rate in smaller groups (Caro et al. 2004), and second, a long-term response by expanding group size, where larger herds provide more safety through the "many eyes effect" (Lima and Dill 1990) and less risk because of the "dilution effect" with a decrease in danger to any single individual (Roberts 1996). Under a high predation pressure, the reaction may vary between different ungulate species: some species do not perform any visible response; others increase their vigilance rate, while still others increase or decrease their group sizes (Quenette 1990; Caro 2005; Creel et al. 2014; Beauchamp 2015).

When a threat has been identified, the prey animal then makes a risk assessment to evaluate the danger scale and choose where, when, and how to avoid the threat. The most vigilant and sensitive individuals watch the suspicious object and collect information which is then disseminated among conspecifics using visual or vocal signals (Walther 1984; Caro et al. 2004)—a wrong solution or a response that is too slow may lead to a loss of life (Stankowich and Coss 2006). The risk assessment is a very important part of antipredator behavior in ungulates, but it is poorly understood compared to other elements of defensive strategy and behaviors (Stankowich and Coss 2007). In this chapter, I will discuss the variability of risk assessments in goitered gazelles, their role, and importance in the antipredator strategy of this species. The main kinds of assessment behavior I observed in goitered gazelles (ethogram) are presented in Table 16.14.

16.4.2 Variety of Risk Assessment in Goitered Gazelles

Similar to pronghorns (*Antilocapra americana*), goitered gazelles know their environment and any prominent objects in it very well; any small visual changes in the landscape, such as new objects, attract their attention immediately (Autenreith and Fitchter 1975). As a result, an alerted gazelle freezes in place and looks attentively at the suspicious or unfamiliar object, sometimes for a long time. Through periodical routine acts of the short-span vigilance during grazing, goitered gazelles quickly recognize the alarm posture of any individual, which stands still without motions, ears raised with the auricular directed forward, muzzle oriented toward the potential threat, head and neck extended upward with muscles strained—and an alert display of one individual is an alarm signal for others. All gazelles in the group instantly stop their grazing and other activities, and the whole group assumes an alarm posture as a result of contagious effect and looks at alarmed gazelle intently and then looks in the same direction as alarmed individual, trying to identify the source of the threat, and after that watches alarmed individual again. If gazelles cannot determine the cause of the alarm, then they approach closely the excited individual and stand next to, behind, or in front of it and again look in the same direction they try to find the cause of the alarm from the viewpoint of alarmed gazelle. An alarm display has a low cost to ungulates, but it alerts a number of gazelles prepared to flee away from any possible danger (Caro et al. 2004). Thomson's gazelles (*Eudorcas thomsonii*) will gather in large herds that spread very widely when grazing, but when one gazelle shows alarm behavior, the signal propagates very quickly over the herd like a billow and the entire group instantly stares in the same direction (Walther 1969).

Table 16.14 Description of recorded displays (ethogram) observed during escaping behavior in goitered gazelles within study period according to Walther (1969), Autenreith and Fitchter (1975), FitzGibbon (1994), Stankowich and Coss (2006) (Blank 2019—License Number—5477020890098)

Behavioral patterns	Description
Alarm in standing or laying position (steady gaze)	The gazelle stood with its head and neck upright (long necked posture) with its ears erect, auricular oriented forward, head oriented directly in the direction of the disturbance source, croup lowered, and muscles tensed. This degree of attention with no interruption can be up to 30 min (Figs. 16.17 and 16.21)
Contagious effect of alarm posture and/or flight	If as few as just one gazelle spotted a potential danger and froze in an alarm posture, all conspecifics were provoked to immediately stop their activities and watch the alarmed individual attentively, and then shift their gaze in the same direction to look for the alarm source. In some cases, a conspecific moved very close to the alarmed individual and stood near or in front or behind of it to look from the same point in the same direction to find the threat. When an alarmed individual runs, other conspecifics either run immediately or shift their gaze to try to find the source of the threat and then either ran or searched further, depending on the alarm level (Figs. 16.4 and 16.18)
Exploratory getting closer or investigative approach	After an initial assessment, a gazelle took a few steps toward the suspicious object and continued to stare while in an alarm posture. After a while, the gazelle retreated and came back to the initial point, from where it continued staring at the same point. Such a maneuver was repeated one or twice more
Vigilance with maximum concealment	A gazelle gradually climbed to the top of an elevated area (hill, dune, or side of a dry river bed), initially raised just the top of its head to the level of its eyes, froze, and looked around attentively; then ascended further and put its head and part of its neck above the hill and looked around again; the gazelle arrived at the top and surveyed the landscape. While doing this, the gazelles froze in an alarm posture, with its head and neck upright, ears erected with the auricular turned forward; it turned its head to one side periodically with the tail erected horizontally and waving intensely from side to side. This behavior was typical for the first alert individual, usually a female, while conspecifics stood still from behind, watching the female and waiting until she finished checking the surroundings and arriving at the hill top; and only after the female started grazing, did others join her without any special vigilant acts while climbing to the elevated area. in a dry river bed, gazelles may do the opposite and after spotting a potential danger, they will descend lower into the river bed so that the whole body becomes invisible and only the head, or the top of the head to the eyes, peers over the side (Fig. 16.22). Often a goitered gazelle female watches a bizarre object, only sticking her head out from behind the edge of the hillock slope. Then the head disappears, but the female does not run away, but moves to another place behind the slope and sticks its head out again in another place and considers the potential danger from another angle
Monitoring a threat through the branches of a bush or saxaul tree	A gazelle stood behind a bush or saxaul tree and looked through the branches (Fig. 16.23) or only put its head out from behind branches to monitor the movements of the threat (Fig. 16.24)

(continued)

Table 16.14 (continued)

Behavioral patterns	Description
Volte	After looking at a potential danger for a time, a gazelle started to walk away with quick steps in excitement, shaking its lifted tail from side to side, but not leaving the area. Instead the animal walked in a small circle or walked from one side to the other, returning to its starting point to continue to look at the same point
Mobbing	A predator inspection or investigative behavior in which prey species approach and follow predators or a bizarre object, particularly when it is first encountered. In other cases, several individuals of a prey species mob a predator by cooperatively attacking or harassing it, often to protect offspring
Unexpected inspection	After a risk assessment by freezing in an alarm posture and staring at a suspicious object, the gazelle changed tactics and gave the appearance that it stopped its alarm and started to graze. Suddenly, though, it abruptly lifted its head and made an unexpected inspection, staring intently at the same point. This behavioral maneuver was repeated 2–3 times consecutively, alternating between grazing and unexpected inspections

Fig. 16.21 Alarm in standing position: Goitered gazelles can stand in alarm up to 30 min. Issyk-Kul goitered gazelle nursery for reacclimatizing, 25.06.2022. Photo by D. Blank

This behavior is a characteristic primarily seen in adult goitered gazelles, females and males almost equally (47% vs 53%, n = 55, respectively), while sub-adults and fawns pay noticeably less attention to alarmed conspecifics (Fig. 16.25). If a worried gazelle starts to flee, the entire group typically follows, fleeing instantly, though

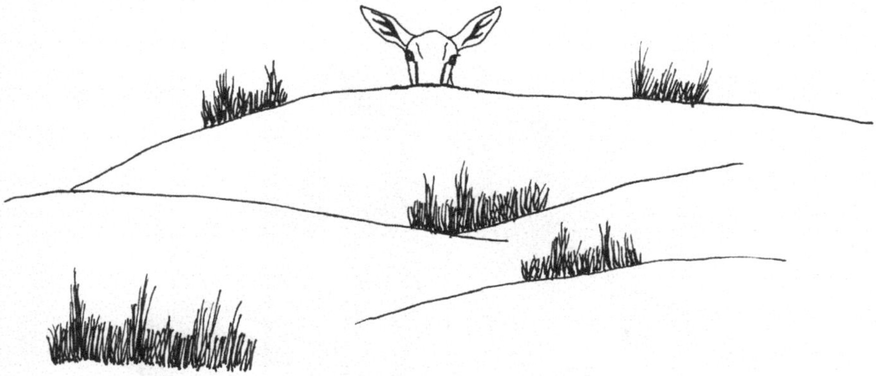

Fig. 16.22 Vigilance with maximum concealment: Goitered gazelles do assessment, sticking only his head out from behind the hillock. Drawing by U. Kenzhetaev with kind permission

Fig. 16.23 Concealment behind tree: Goitered gazelles do assessment the potential danger through shrub branches. Drawing by U. Kenzhetaev with kind permission

adult females do this 2.8 times less frequently than adult males (n = 49). Less often, other members of a group remain in place and continue searching for the source of the alarm before beginning to run, and adult females react this way more often than adult males. In general, the alert posture has a greater contagious effect for goitered gazelle adults, and this behavior is common for many ungulate species. The older, more experienced bighorn sheep, for example, are more sensitive to displays of

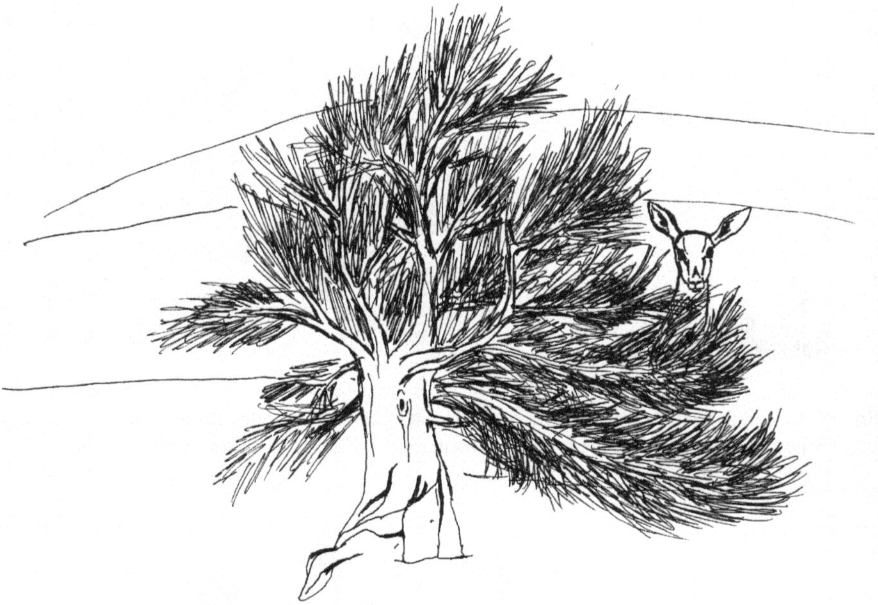

Fig. 16.24 Combination assessment and concealment: Goitered gazelles do assessment, putting its head out from behind branches. Drawing by U. Kenzhetaev with kind permission

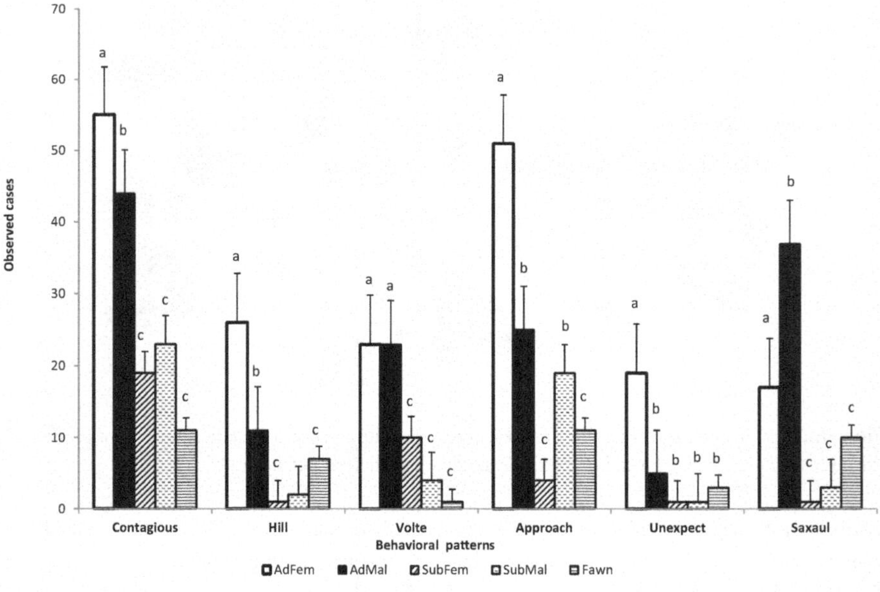

Fig. 16.25 Frequency of observed risk assessment displays in different age and sex groups of goitered gazelles (Legend: AdFem—adult females, AdMal—adult males, SubFem—sub-adult females, SubMal—sub-adult males, fawns; Behavioral patterns: Contagious—contagious effect of alarm posture, Hill—gradual climbing to hill and concealment behind hill, Volte—volte pattern, Approach—investigate approach, Unexpect—unexpected inspection, Saxaul—concealment behind saxaul tree). Bars with different letters were significant ($P < 0.05$) (Blank 2019—License Number—5477020890098)

alerted individuals and faster and more willing to replicate their behavior compared to young and sub-adult sheep (McDougall and Ruckstuhl 2018). Similar behavior is seen in the greater kudu (*Tragelaphus strepsiceros*—Pays et al. 2012) and red deer (*Cervus elaphus*—Lung and Childress 2006), but in fallow deer (*Dama dama*), this contagious response difference between adult deer and fawns is not observed (Schall and Ropartz 1985).

The trigger to alarm behavior of an entire goitered gazelle group is most often the first alarmed individual (56% cases, n = 169), to a lesser extent an uncertain, non-identifiable, or far-distant potential threat (25%), and much less often (13% to 15% cases), the source of anxiety is close and clear, such as a standing or approaching human. If an alarmed goitered gazelle group flees without a long risk assessment, the cause is usually a clear and present danger (56% cases, n = 62) and only limited cases are a far-distant, uncertain threat (9%); many cases of worried behavior decrease after a short period of time (35%), as most of these events are related to the behavior of group members that get excited about inconsequential things (Blank 2019).

Like in Thomson's gazelles (Walther 1969), an alerted goitered gazelle guides the entire group, as all group members have the same orientation, looking toward the same place. An alarm situation promotes the synchronization of all activities among group members, making them ready to flee. The contagious effect is an important behavior of social ungulates and the main way to transmit information about a threat to the whole group (McDougall and Ruckstuhl 2018; McDougall and Ruckstuhl 2018). The group of goitered gazelles generally does not run if they do not see the danger, but if an alerted gazelle flees, the group typically starts to follow even if they do not see the threat. Thomson's gazelles behave in exactly the same way and the entire group follows the first alarmed, running individual (Walther 1969). In pronghorns (*Antilocapra americana*), flight is so powerful a visual signal that it promotes the flight of even fawns as young as 3 months, even though their hiding period is not yet finished (Autenreith and Fitchter 1975). Similar actions are described for guanaco (*Lama guanicoe*—Taraborelli et al. 2012), bighorn sheep (*Ovis canadensis*—McDougall and Ruckstuhl 2018), and a number of deer species, including caribou (*Rangifer tarandus*—Aastrup 2000), mule deer (*Odocoileus hemionus*), and white-tailed deer (*O. virginianus*—Lingle and Wilson 2001).

The investigative approach of goitered gazelles to a potential threat is part of their antipredator behavior, when after looking at a potential threat for a long time, they either lose interest or go away or they continue their assessment and take several steps toward the object for a better assessment before returning back to their previous point. The periodical moving forward and back may be repeated several times. This is a quite common pattern and is seen two times more frequently than other behavioral patterns. Only the contagious effect of an alarm posture and flight is observed 1.3 times more often than moving closer. The frequency of use of the investigative approach is different between sexes, when adult females demonstrate this pattern most often among conspecifics (49% of all observed cases, n = 126) and adult males use this behavior half as often (22%). In contrast, sub-adults and fawns show this behavior rarely (4% to 15%, n = 126). Goitered gazelles come closer to a putative threat most frequently as a reaction to the approach of other ungulate species (kulan *Equus hemionus*, saiga *Saiga tatarica*, argali *Ovis ammon*—29%, n = 111) and noticeably fewer common addresses to a standing or walking human (18% and 16%, respectively). Other potential kinds of dangers generate an investigative approach much less often, between 3% and 9% of incidents, and are rarely observed in relation to the approach of a fox (n = 111). This investigative approach is not performed at all toward wolves, and 63 wolf/gazelle encounters were registered during this study.

Visual recognition in ungulates is more sensitive and better able to identify the threat potential of a moving object than a suspicious object that is stationary or only slightly moving (Walther 1969, 1984). In my study, goitered gazelles perform the investigative approach when they encounter something strange with unidentified intentions. The object was certainly perceived as being non-dangerous, although disturbing object. Therefore, these gazelles displayed this behavior as a reaction to the proximity of a large body-sized entity, such as another ungulate species (bizarre object) or when encountering an uncertain threat with incomplete information, a standing human for instance. However, when goitered gazelles were faced with a real, potential threat, such as a walking human or an object with a high likelihood to attack, wolves, or a horseman (a common poacher in the study area), then they did not use the investigative approach to risk assessment. The investigative approach is found in other ungulate species (Hinde 1970), like pronghorns (Autenreith and Fitchter 1975).

In contrast to the contagious alarm effect, which usually is a reaction to any even small cause of threat and appears because of too excitable conspecifics and their alarm displays, leading to running away, the investigative approach of some gazelles toward other ungulates does not have a contagious alarm effect and subsequent fleeing. Though significant differences are not found between the mechanisms of the investigative approach and the contagious alarm effect, the absence of a correlation between them suggests that the triggers for these two distinct behaviors are different (Blank 2019).

While simultaneously observing potential danger, goitered gazelles often demonstrate concealment behind landscape features (undulating relief, hills, banks of dry river beds) by sticking out only part of their heads from behind a slope, making themselves as unseen as possible. When they do this, they do not immediately go to the top of a slope, but rise gradually, with 2 to 3 stops until they reach the top point, where they also gradually stick their heads out to carefully inspect the area from the other side of the hill while remaining hidden. This behavior is observed 5 to 10 times more often in adult gazelles than in fawns and sub-adults, though fawns perform this pattern a bit more often than sub-adults. Adult males perform this means of vigilance by slowly climbing to the upland area using the relief for concealment three times less often (in 25% cases) than adult females (75% cases, n = 51; Fig. 16.25), for whom such behavior is most common. Another more rarely observed variant is the gazelle going down into a deep wadi where they also observe a potential threat while hiding most of the body behind the riverbank and sticking out only their heads for vigilance. However, such concealment is seen 7 to 8 times less often than the slow climb to an elevated point and adult females demonstrate this behavior 3 times more often than adult males. Vigilance with the hiding component is seen in cases when goitered gazelles observe other ungulate species (argali, saiga) (50% cases, n = 92) and less frequently as a reaction to a standing observer (17%) or far-distant, uncertain anxieties (17%); this behavior is seen very rarely with other disturbances (1% to 5%, n = 92) (Blank 2019).

In addition, there is another kind of hiding using terrain features while observing an unusual object. This is when a goitered gazelle sticks only its head out from behind a slope and watches for possible danger for a long time. Then the head disappears, and the gazelle moves to another place behind the slope and the gazelle's head appears again, observing the same object from a different angle. Obviously, this behavior is due to a lack of information about a bizarre object and the desire to conduct a deeper risk assessment through observation. This pattern was observed as a reaction to a sitting, motionless human, and is more likely due to curiosity than fear, when the gazelle is not in immediate danger.

Goitered gazelles not only use irregularities in the landscape, but also large vegetation (shrubs, saxaul trees) behind which to remain hidden while observing a suspicious object in case of a possible threat (Figs. 16.23 and 16.24). Goitered gazelles stay behind saxaul trees or large shrubs and look through the branches or stick only their heads out of the tree or shrub crown. Concealment behind vegetation is seen in these gazelles 1.5 more frequently than the concealment behind a landscape features, and adult females use this kind of hiding half as often than adult males, who do it most often (Fig. 16.25). The fawns and sub-adults hide behind vegetation quite rarely, with adults using vegetation for hiding 7 to 10 times more often than young gazelles, though fawns do it slightly more often than sub-adults. The concealment behind large vegetation is noted in goitered gazelles as a reaction to a walking human (64%, n = 58), significantly less frequently to human gestures (14%), and infrequently to other disturbances (2% to 7%, n = 58). Thus, gradually moving up a slope and hiding behind large vegetation looks like similar behaviors when goitered gazelles are watching a disturbing object, remaining relatively invisible to the source of the alarm. Adult gazelles exhibit both these behaviors significantly more often than young individuals, although gradually climbing a slope is more typical for females, and hiding behind vegetation is more typical for males. The gradual climb is mostly related to the proximity of other large sized ungulate species or the unusual behavior of a familiar object (standing or sitting motionless human) and both are more likely related to curiosity than fear. In contrast, concealment behind trees/shrubs is more likely tied to insufficient information on a suspicious object with real potential of danger (walking human). So, different triggers promote gradual climbing and hiding behind vegetation in goitered gazelles (Fig. 16.25) (Blank 2019). Hiding behind vegetation is also described for the goitered gazelle population in the Barsakelmes Island (W Kazakhstan) and the Bukhara Breeding Center (Uzbekistan) (Zhevnerov 1984; Marmazinskaya 1996).

Forest-dwelling small African antelopes and caprid and especially cervid species, inhabiting closed environments, commonly freeze in place behind trees/shrubs for concealment, hiding through their immobility and cryptic coloration (Jarman

1974; Estes 1991; Takada et al. 2018). In contrast, concealment for goitered gazelles is not as practical as an antipredator strategy, though even open habitat species like goitered gazelles do everything to detect predators as soon as possible and recognize their intentions while trying to be as invisible as possible using the landscape and vegetation for hiding (Caro 2005; Langerhans 2007). Pronghorns and Thomson's gazelles are species of open environments, but they use the foothills, hilly valleys, and dry river beds, as well as other landscape features for concealment during risk assessments (Walther 1969; Autenreith and Fitchter 1975; O'Gara 1978). Whatever method, prey animals attempt to decrease the possibility of detection by a predator during risk assessments (Langerhans 2007).

Volte behavior is observed in goitered gazelles after they have looked at a disturbing place for a period of time; after that they leave a point of vigilance, but then return to its previous point and continue to observe the same disturbing object. Volte behavior is observed as often as hiding behind vegetation. Both adult females and males perform this defensive behavior with parity between sexes, but adults do it significantly more frequently than fawns and sub-adults (Fig. 16.25). The returning (volte) behavior is a common reaction to the close approach of large ungulates (29%, n = 48) or a human (standing—23% and walking human—19%), while other distressing factors cause this behavior to be noticeably less frequent (8% to 10%, n = 48). Circling back (volte) for further observation represents a behavior of contradictory desire in the goitered gazelle between escaping immediately or continuing a risk assessment of the disturbing object. This pattern is displayed in the cases of uncertainty between curiosity and fear (other ungulates proximity and moving human) and evidently in the absence of an immediate threat and related to a scarcity of information and/or curiosity. The volte display is observed because of the same causes as gradual slope climbing, though with a higher risk of a real threat (walking human) (Blank 2019). Thomson's gazelles also perform volte display in a state of alarm. After looking at a predator for some time, the gazelle unexpectedly makes a small circle at a rapid pace, returning to its initial point to continue watching (Walther 1969).

Mobbing is a rarely observed assessment behavior in goitered gazelles, when the ungulates follow a predator. This is the least often used defensive behavior, when usually only 1 or 2 gazelles are involved. Mainly adult females demonstrate mobbing (62%, n = 13), while adult males do it 2.7 time less frequently (23%), and sub-adults rarely (15%). Foxes (465, n = 13) and argali (46%) are the most common recipients of mobbing, but rarely saiga antelopes (8%). And mobbing was never addressed to humans or wolves (n = 63). When detecting a fox, especially coming to an area with a hiding fawn, goitered gazelle mothers, and sometimes with the help of a sub-adult daughter, gallop toward the fox and try to strike it with their front hooves and/or head-butt it. Goitered gazelle males usually do not join the attacking females and instead just watch the female chase the fox. This behavior of males is typical for many ungulate species (Thomson's gazelle *Eudorcas thomsoni* - Wyman 1967; Grant's gazelle *Nanger granti*—Costelloe 2017; pronghorn—Marion and Sexton 1979; Lipetz and Bekoff 1980; Berger 2005; white-tailed deer—Grovenburg et al. 2009). There is only one known case when a pronghorn sub-adult male protected a fawn from predation (Berger 2005). Thomson's gazelles in large groups identify a predator approaching (to 100 to 200 m) and continue to follow it, keeping a safe distance. The gazelles stay in a state of high alert so the predator does not have the opportunity for a sudden attack.

Mobbing is defined as an ungulate group movement following a predator or other disturbing object, including another offbeat ungulate species or even an individual of the same species perhaps acting strangely. Thomson's gazelles have been seen to follow (mob) a stalking predator (lion—*Panthera leo* or cheetah—*Acinonyx jubtus*) for more than an hour, though they do not mob a coursing predator (African wild dogs—*Lycaon pictus*) (FitzGibbon 1994). Goitered gazelles also do not mob wolves (*Canis lupus*—coursing predator), but will follow foxes (coursing predator), as well as harmless argali (*Ovis ammon*) and rarely saiga antelope (*Saiga tatarica*). Wolves are large-sized predators and live and hunt in packs that chase their prey and are harmful to both adult goitered gazelles and their fawns. In contrast, foxes are solitary predators, hunting alone, and dangerous to newborn fawns and not adult gazelles (Sludskiy 1962). Other ungulate species do not present any danger to goitered gazelles of any age. Mobbing a fox can change into an aggressive chase with the intention to hurt it with a kick from front hooves or head-butt, but goitered gazelles make no aggressive actions when mobbing a large-sized ungulate. Thomson's gazelles display mobbing almost every day, following predators around and using about 4% of their daily time budget. In addition, Thomson's gazelles can gather into herds of several thousand with large numbers participating in mobbing (FitzGibbon 1994). Goitered gazelles, on other hand, perform mobbing rarely, maybe only several times a year and usually with only 1 to 2 gazelles. This difference between these two species is likely related to the low population density of goitered gazelles, as well as predators, in my study area, while in African environments, where prey–predator encounters can happen several times a day (Pays et al. 2012).

Thomson's gazelles show mobbing when the herd is large enough and vegetation is quite short and the predator does not have the ability to hide for a sudden attack. This behavior is demonstrated to a predator that is resting or busy with other activities that are not related to hunting, thus insuring that the gazelles will not be attacked unexpectedly. Goitered gazelle fawns and sub-adults rarely participate in adults' mobbing, while immature Thomson's gazelles do it more frequently than adults, though they are more vulnerable to predation (FitzGibbon 1994). Pronghorns are also known to mob predators (Berger et al. 1983; Reynolds 1983).

Making unexpected inspections is rarely noted as part of assessment behavior, when goitered gazelles suddenly raise their heads demonstrating vigilance when feeding. Examining a suspicious object, the gazelle tries to determine what has attracted its attention, alive or not alive, trying to catch movement. If the object is alive, then it carries a potential danger. Once identified, the gazelle can estimate its degree of risk. If the initial assessment did not help to decide whether this object is alive (did not move), then the gazelle appears to graze again to demonstrate a peaceful behavior. A gazelle assuming the behavior of a victim stimulates a frozen predator to try to creep closer for an attack, and in so doing has to move. This is what the gazelle is trying to achieve, catching any movement of the object by suddenly raising its head from grazing for an abrupt examination of the disturbing object. According to Walther (1969), moving objects are much easier for gazelles to identify for making a risk assessment. This pattern is most typical for adult females (70% of cases, n = 27; Fig. 16.25), with adult males taking this action nearly 5 times less frequently (15%), and fawns and sub-adults performing it rarely (11% and 4%, n = 27, respectively). The unexpected raising of the head to look at a potential threat occurred mostly with a standing human (74%, n = 27), while other disturbances promoted this behavior very rarely (Blank 2019).

All actions described above are in fact different kinds of risk assessment behavior which is a very important part of any antipredator strategy (Stankowich and Coss 2006). The model of optimality states that a prey animal will continue to stay in one place until the potential cost of delaying will be less than the minimal cost of escaping (Ydenberg and Dill 1986). Risk assessment is so important for ungulates that they will continuously assess the risk until they no longer have an alternative except to run (Stankowich and Coss 2006).

The sex and age of ungulates are significant factors that have an impact on antipredator behavior (Frid 1997; Hunter and Skinner 1998). Unexperienced in threat evaluation and escaping strategy, goitered gazelle fawns and sub-adults perform vigilance and risk assessment behavior considerably less frequently than adults; often young individuals cannot identify predators after detection, are weakly coordinated, and are unlikely to start running in time (FitzGibbon and Lazarus 1995; Caro 2005). As a result, young gazelles are the main targets of hunting predators, because they are more vulnerable and easily caught (Barrett 1984; Smith et al. 2000; Childress and Lung 2003). Therefore, mothers typically dedicate more of their time to vigilance, scanning longer and being more responsive, and allowing their young to graze more so it will grow stronger and faster and be prepared for survival. Hence, females—especially mothers—are the most vigilant faction in a population (FitzGibbon and Lazarus 1995; Childress and Lung 2003; Lung and Childress 2006). It is clear why adult females are active with risk assessment more often and longer than adult males, and this is a common trait in many ungulate species besides goitered gazelles, also Thomson's gazelles (FitzGibbon 1990), caribou (Bøving and Post 1997), red deer (*Cervus elaphus*—Childress and Lung 2003), and guanacos (Marino and Baldi 2008).

Risk assessment for goitered gazelles involves various kinds of displays, each used for a specific condition and addressed to different levels of a potential threat. The main task of a risk assessment is to detect a predator and its intention, remain unnoticeable for as long as possible, and find an optimal time and manner for making a successful escape while minimizing any cost at the same time (Blank 2019).

16.5 Vigilance, Staring and Escape Running

16.5.1 General Remarks

Ungulates change their activity budget in accordance with pressure from predators, paying more attention to vigilance at the expense of time for other activities (FitzGibbon 1989; Burger and Gochfeld 1994; Toïgo 1999; Li et al. 2009). Predation pressure is a prevalent determinant forming feeding behavior of ungulate species and it is the main reason for the decrease in the foraging rate, altering of the daily food ration, and habitat preference (Brown 1999; Altendorf et al. 2001; Brown and Kotler 2004; Christianson and Creel 2008). Ungulates demonstrate a great variability in antipredator behavior, but all species behavior can be divided into two categories: to reduce the chances of detection by a predator and to improve the chances of surviving after being seen by a predator (Langerhans 2007).

Ungulates quite often routinely stop grazing and look around the surrounding area for possible danger; this pattern is called vigilance (Walther 1969). By making frequent stops for vigilance, ungulates try to discover and identify a predator before being found themselves. When prey detects danger, it demonstrates variable responses to escape predation (Langerhans 2007). Various species perform vigilance while standing or lying down, as well as during feeding with different rates depending on their security status and habitat conditions (Walther 1969; Bøving and Post 1997). The more vigilant ungulate has the ability to discover danger early enough to leave the area in time to escape and also to transfer this information to less cautious conspecifics (FitzGibbon 1989; Cresswell et al. 2003). Social antelope species inhabit mainly open environments and their only protection against predators is vigilance and flight (Brashares et al. 2000). In large groups, there is always at least one individual with its head raised at any one time for vigilance and then transferring information about a possible threat immediately to conspecifics through communicative signals, including stotting, snorting, and tail-flagging (Caro et al. 2004).

Individual vigilance declines as group size enlarges because there are more ears and eyes to detect a predator earlier and at farther distances (the "many eyes" effect of collective detection). This sharing of vigilance time, in turn, allows each individual in the group longer time to feed (Pulliam 1973; Lima 1995). In addition, the more animals in a herd the lower the predation risk to any one individual through the dilution effect ("safety in numbers") (Foster and Treherne 1981; Roberts 1996). This negative correlation between vigilance frequency and group size is not evident (Beauchamp 2008). Various ungulate species have different preferences when faced with dangerous situations. While some raise their vigilance rate without enlarging group size, others have a strong preference for enlarging their group size without modifying their vigilance rate; these are complementary reactions (Creel et al. 2014).

The herd is led continuously by an adult female. Usually, she walks 40 to 50 m ahead of the rest of the head, often stopping to look around and listen. If the leader stops, other members of the group also stop. In a calm environment, the gazelles are in a relaxed state, but when danger is detected, they tense and assume an alert pose with head raised and staring in one direction for an extended period of time. Gazelles have been known to stand completely motionless for 5 to 8 min, observing a

Table 16.15 Description of the recorded displays (ethogram) during escaping behavior in goitered gazelle within the study period according to Walther (1969); Kiley-Worthington (1976); Stankowich and Coss (2006), Blank et al. (2015) (by Blank 2018a—License Number—5477021276455)

Behavioral patterns	Description
Routine vigilance or looking around	After keeping the head down for minutes and sometimes even for a quarter of an hour, a gazelle rises its head up and checks surroundings around
Freezing for assessment or staring	Standing or lying immobile. Gazelle erects its neck, puts its ears forward, lowers its croup, and tenses its muscles. In this long necked posture of freezing, a gazelle looks for potential danger or conspecifics at a far distance or something unusual, when it is not able to find out at once whether the thing is dangerous or not, a gazelle shows a strong tendency to obtain more exact information by optical means alone
Tail flagging (vertical position)	Tail is held vertically or erected higher than horizontal up to strongly vertical position, with putting it on the back in some situations of great agitation, exposing its white ramp patch in greatest degree. In this position, tail can make stereotypical repetitive motor lateral movements from left to right across the hindquarters and back again one or more times in succession. The white rump patch is completely uncovered, and black spot is around the anus only. At the same time, the white hair of rump patch of the outline is projected outwards, making rump patch more prominent

suspicious object (Zhevnerov et al. 1983). In ungulate species of open habitats, vision is significantly more important for recognizing danger than scent or sound, which are used more by species of closed environments (Walther 1969; Estes 1991). Interactions between group size and vigilance rate are the most considered topics in antipredator strategy in ungulates, while much less understood are the factors influencing the interrelationship of these two alternatives (Creel et al. 2014). Therefore, in this chapter, I consider the different behavioral reactions of goitered gazelles (vigilance, staring, and escape running) to their threat of danger and their choices in antipredator behavior. Observed displays of escape behavior (ethogram) in goitered gazelles are described in Table 16.15.

16.5.2 Routine Vigilance and Avoidance Behavior in Goitered Gazelles

In a calm environment, goitered gazelles mostly graze and rest, but even in these relatively safe situations, these ungulates perform routine vigilance, by raising their heads occasionally and looking around while chewing their collected vegetation in their mouth. In case of appearance of some suspicious moving object, the gazelles immediately stop grazing and stare attentively without moving sometimes for a long time to determine if the disturbing object is dangerous. If not determined to be dangerous, the gazelles continue grazing; otherwise the whole group gets alarmed and

Table 16.16 Different responses (escape running, staring, and routine vigilance) of goitered gazelles of various sexes and ages (upper—number of observed cases, lower—proportions in %) (by Blank 2018a—License Number—5477021276455)

Behaviors	Adult female	Adult male	Sub-adult female	Sub-adult male	Fawns	Total
Escape running	1225 38.0	936 29.0	151 4.7	209 6.5	703 21.8	3224 100
Staring	686 46.3	406 27.4	79 5.3	102 6.9	210 14.1	1483 100
Routine vigilance	1509 64.9	781 33.6	2 0.1	3 0.1	30 1.3	2325 100

watch the object and, when necessary, the entire group flees. More experienced adult gazelles demonstrate routine vigilance more frequently than sub-adults and fawns. Adult females are typically initiators of these patterns (routine vigilance, staring, and escape fleeing) and perform these behaviors most often when among conspecifics (Table 16.16). Adult males, in contrast, display elements of defensive behavior more rarely compared to adult females. The sub-adults perform these elements of defensive behaviors rarely, while fawns demonstrate antipredator behavior at an intermediate rate. Similarly, more active antipredator behavior and higher vigilance are found for the greater kudu (Pays et al. 2012) and red deer (*Cervus elaphus*—Lung and Childress 2006), while other authors have not found a difference in vigilance rate based on age. In fallow deer, for example, age differences are not noted (Schall and Ropartz 1985).

Goitered gazelle mothers are especially susceptible to threats of danger when raising fawns, which have no experience or ability to spot or escape from predators. Therefore, mothers significantly increase their vigilance rate during this time. This behavior is typical for many ungulate species, when females increase the frequency of inspections, especially during postpartum, and do it more often than males and sub-adults (Clutton-Brock et al. 1982; FitzGibbon 1989; Burger and Gochfeld 1994; Li et al. 2009). In blue wildebeest (*Connochaetes taurinus*), impala (*Aepyceros melampus*) (Hunter and Skinner 1998), chamois (*Rupicapra rupicapra*—Ferretti et al. 2014), and fallow deer (*Dama dama*—Schall and Ropartz 1985), females also show a significantly higher and longer vigilance rate than adult males. In Spanish ibex (*Capra pyrenaica*), however, males are more vigilant than females (Alados 1985) and the greater kudu (*Tragelaphus strepsiceros*—Pays et al. 2012) does not show a difference in vigilance between sexes. The same contradictory results are found for Przewalski's gazelle (*Procapra przewalskii*—Li et al. 2009), where males have the same or even higher vigilance rate as females. These contradictions are most likely due to the fact that these authors do not distinguish between routine antipredator vigilance and social vigilance related to rutting activities (Blank 2018a).

Goitered gazelles stare and flee away most frequently when they come across a walking human or moving car, while in many other cases, the gazelles do this significantly less often (Table 16.17). Routine vigilance is showed for different reasons compared to staring and running and is most often as a reaction to an uncertain or

Table 16.17 Causes of observed cases of escape running, staring, and routine vigilance in goitered gazelles depending on different actions of danger source: observed human, vehicle, and conspecifics. (upper—number of observed cases, lower—proportion in %). Legend: Walk—human walking, Move—movements of standing human, Stand—standing human motionlessly, Sit—sitting human, Smell—arriving smell from motionless human (assumption), Hear—hearing some strange sounds (assumption), Conspec.—reaction to galloping conspecifics, Veh. moves—gazelles saw moving vehicle, Veh. Hear—gazelles heard car engine, but did not see it; Unknown—reaction to far-distant or unknown source of danger (by Blank 2018a—License Number—5477021276455)

Patterns	Walk	Move	Stand	Sit	Smell	Hear	Con-spec.	Veh.moves	Veh.hear	Un-known	Total
Escape running	3210 58.9	110 2.0	74 1.4	26 0.5	64 1.2	22 0.4	62 1.1	1729 31.8	28 0.5	122 2.2	5447 100
Staring	644 43.0	89 5.9	179 11.9	20 1.3	25 1.7	20 1.3	2 5.5	353 23.6	31 2.1	56 3.7	1499 100
Routine vigilance	7 0.3	1 0.0	4 0.2	–	2 0.1	11 0.5	201 8.7	1 0.0	–	2093 90.2	2310 100

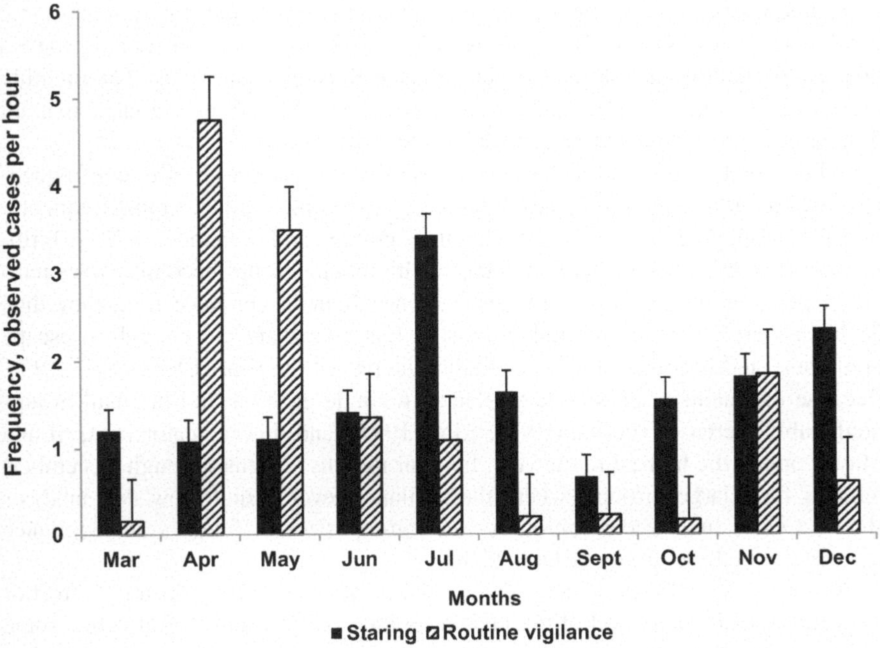

Fig. 16.26 Frequency (observed cases per hour observations) of staring and routine vigilance in goitered gazelles over months (by Blank 2018a—License Number—5477021276455)

far-distant danger with an unidentified source and significantly less often as a response to the alert behavior of conspecifics; other sources of anxiety are noted rarely (Blank 2018a).

Females change groups in different ways over months compared to male herds. The smallest female groups are seen in April and May, when most females leave their groups when birthing young (Table 11.1). Female groups enlarge noticeably in July as they stay in larger-sized groups for the rest of the year. The smallest male groups are found in November and December, as well as in April, because of rutting activities. The size of mixed groups is relatively small and changes randomly over months, while largest groups are observed in March, April, and slightly less in December. The expected negative correlation between the rate of routine vigilance rate and group size is found in female groups, while it is not detected in male or mixed groups. The rate of staring and running fluctuates over months independently of group size and regardless of group composition (Blank 2018a).

Goitered gazelles display the maximal rates of routine vigilance and staring in different months: the highest rate of routine vigilance is found in April and May, while staring is noted most commonly in July (Fig. 16.26). Routine vigilance was demonstrated three times less often (0.8 cases per hour in average) in August through December compared to March through July (2.4 cases per hour), while staring has the opposite allocation, and in July through December, it was seen two times more frequently (2.3 cases per hour in average) than in March through June (1.2 cases per hour) (Table 16.18). The monthly variations of escape running have a greater correlation with staring displays than with routine vigilance. The monthly changes in the rate of staring had a weak correlation with routing vigilance in adult females and were almost absent in adult males (Blank 2018a).

Adult females and adult males perform routine vigilance with different frequencies over months (Fig. 16.27). Adult females use routine vigilance most frequently in April (when pregnant females leaving their groups) and even more in May (birthing month with highly vulnerable young), with female groups becoming the smallest of the year with usually only one member. Female vigilance frequency then declines sharply in June through August (when young are old enough to escape predators), and becomes almost negligible the rest of the year (Blank et al. 2012). Because the intensity of routine vigilance in single gazelles and in small groups predictably increases (Pulliam 1973; Lima 1995), and this indicator in April and May is one of the highest in the year. In other months (August through December) females form larger groups, where the vigilance level is quite low. My analysis shows a close inverse correlation between group sizes of females and vigilance frequencies in these groups (Blank 2018a).

Adult males demonstrate acts of vigilance in April and rarely during the rest of the year. April is the period of the false or pseudo rut (Tsapluk 1972), when some males occupy their own territorial areas where they chase females. Since there are more single males, male groups decrease in size and, as a result, the male rate of vigilance increases. There is an even greater reduction in male group size during the autumn rutting season in November and December, with many single territorial males; however, the frequency of vigilance does not increase, remaining almost at the same level. This is an unexpected result, which is difficult to explain in terms of theory which proposes that the effect of increasing vigilance frequency is a decrease in group size (Lima 1995; Roberts 1996). My analysis demonstrates a lack of group size effect on the vigilance rate in male herds, except for the slight bump in April. The same vigilance behaviors are observed in Przewalski's gazelles, where the female groups perform the inverse correlation between the size of groups and vigilance rate, while the same phenomenon is not found for male groups (Li et al. 2009, 2012). Investigations of vigilance behavior of some ungulate species (impala—Dunham 1979; Defassa waterbuck *Kobus ellipsiprymnus defassa* and wildebeest—Burger and Gochfeld 1994) during rut have shown that the frequency of scanning the environment in males is higher than in females. I imagine the differences from my results are related to different approaches for collecting and analyzing data. In my study, I defined alarm vigilance of males as related to social vigilance, when rutting males just monitor the activities of other males. In addition, I calculated staring and routing vigilance as separate patterns, while other authors calculated all behavioral patterns together, getting a general scan frequency part of which was related to rutting activities rather than alarm behavior (Blank 2018a).

Frequencies of staring displayed by adult females and males also vary over months. Adult males show an almost a steady rate growth in staring rate from May to August with a noticeable peak in October, and a gradual decline after that. Adult

Table 16.18 Observed frequencies of escape running, staring, and routine vigilance in goitered gazelles over months (by Blank 2018a—License Number—5477021276455)

Patterns	March	April	May	June	July	August	Sept.	October	Nov.	Dec.
Observed hours	90.6	119.3	280.2	200.7	113.3	39.6	9.6	35.3	145.9	121.4
Escape running	373	359	481	992	1283	239	254	57	698	862
	4.12	3.01	1.72	4.94	11.32	6.03	26.46	1.61	4.78	7.10
Freezing	109	129	312	284	389	65	30	55	266	288
	1.20	1.08	1.11	1.41	3.43	1.64	3.13	1.56	1.82	2.37
Routine vigilance	14	567	979	272	124	8	10	6	270	75
	0.15	4.75	3.49	1.35	1.09	0.20	1.04	0.17	1.85	0.62

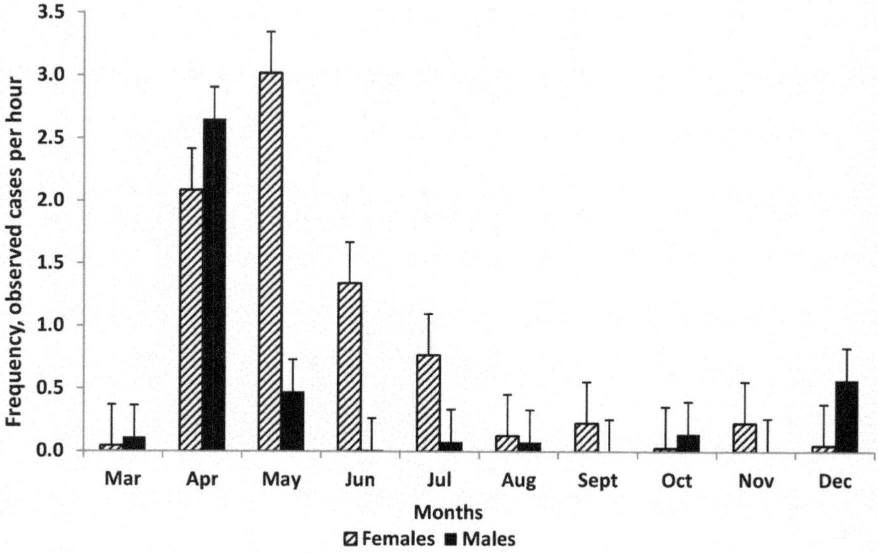

Fig. 16.27 Frequency (observed cases per hour observations) of the routine vigilance in adult females and males of goitered gazelles over months (by Blank 2018a—License Number—5477021276455)

females enhance their staring frequencies during March through June, with the highest rate in July, and decrease the rate thereafter (Fig. 16.28). The group size effect is not found for monthly changes in staring frequency of females, which is more related to the period of hiding fawns (May–July) and the enhanced sensitivity and disturbance of mothers when monitoring their hiding fawns from a distance. Not by accident does the staring rate reach its maximum frequency at the end of hiding period, when the mobility of the independent movement of fawns is maximal (Blank 2017). After the hiding period, fawns stay with their mothers continuously, and staring related to fawn's control and monitoring loses its significance, declining considerably (Blank 2018a). The staring rate of adult males changes randomly over months, with some increase in the importance of staring in October and part of November, and at a higher rate than in females (Fig. 16.23). Randomness of this behavior suggests that staring in adult males is more related to real threats, than depending on yearly biological cycles, though there is still a small impact during the rutting period (Blank 2018a).

Regarding running away, goitered gazelle adult females do it much frequently compared to males in July, when difference between sexes is maximal, while this behavior is seen in both sexes almost at the same rate in other months. Adult males and females show the highest rate of escaping running in July and to a lesser extent in August and rarely the rest of year (Fig. 16.29). A close positive correlation is found between staring and escape running, since the goitered gazelles often flee

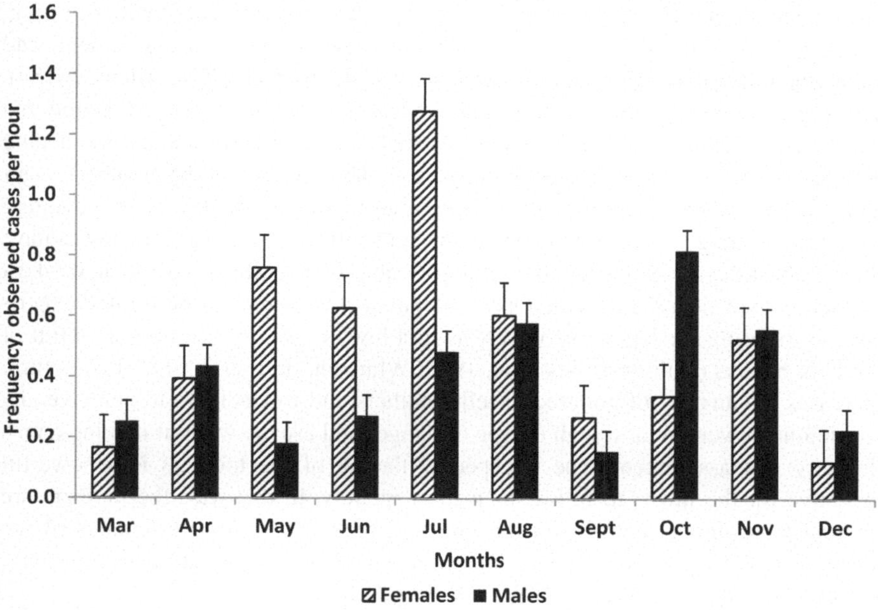

Fig. 16.28 Frequency (observed cases per hour observations) of the staring pattern in adult females and males of goitered gazelles over months (Blank 2018a—License Number—5477021276455)

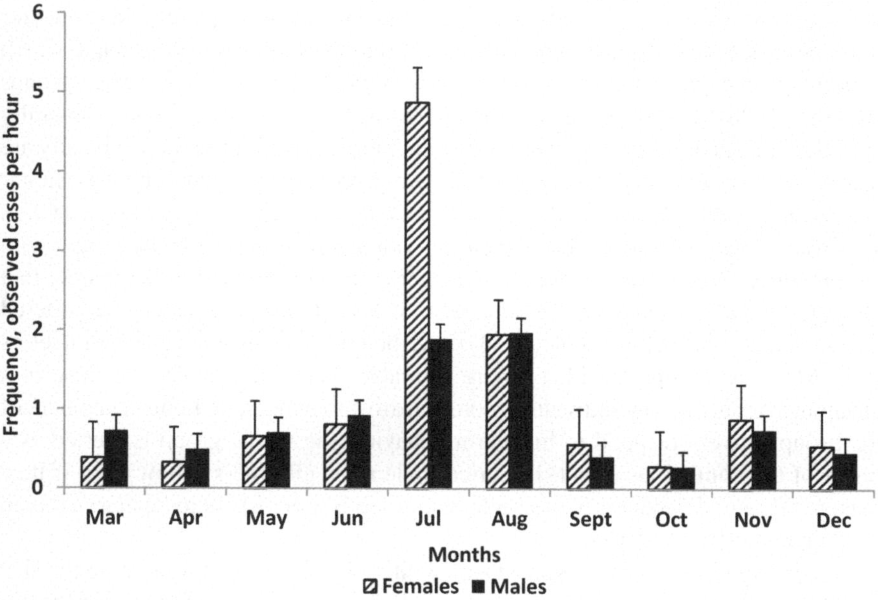

Fig. 16.29 Frequency (observed cases per hour observations) of the escape running pattern in adult females and males of goitered gazelles over months (by Blank 2018a—License Number—5477021276455)

away after staring. However, adult females flee away more frequently than staring, which indicates that they escape run without previous preflight assessment, and stare when they perceive a high threat (Stankowich and Coss 2006; Blank 2018b). This occurs especially often in July, and to a lesser extent in August, when goitered gazelle adult females' threat discrepancy between staring cases and escape fleeing is the greatest. It means that, during this period, the females feel the greatest danger and it occurs when the hiding period is coming to an end and fawns are changing from hiding strategy to following behavior. At this time, goitered gazelle fawns have their highest degree of independent mobility and their mothers have their weakest impact on their fawns' behavior. Open independent movements and limited experience at avoiding predators expose fawns to a higher risk of becoming a victim of predator attacks (Clutton-Brock et al. 1982; White and Berger 2001). This, in turn, increases the anxiety of goitered gazelle mothers and forces them to perceive any suspicious movement as a high degree of danger and escape without making a preliminary risk assessment. The unexpected fleeing of the mothers is a powerful incentive for the fawns to follow them immediately. In this way, the mothers are obviously regaining control over the safety of their fawns. Moose females (*Alces alces*) also show a higher level of vigilance when their young are active and less anxiety when they are hiding (White and Berger 2001).

Escape running in adult males is seen most frequently in July and August, with substantially lower frequencies in other months. These monthly changes in the fleeing rate show a weak correlation with the group size effect (Lima 1995; Roberts 1996). Like with females, adult males also usually perform escape fleeing usually more often than staring, signifying that in some cases they perceive a great danger and run away without a preflight danger assessment. Most frequently this behavior is found in July and August, when they form into their largest male groups, which contradicts the basic theory claiming prey animals feel safer in larger groups (Roberts 1996). Maybe there is another reason for this phenomenon. Generally speaking, males of many ungulate species, including goitered gazelles, typically are less sensitive to potential danger and have a weaker response than females (Hunter and Skinner 1998; Ferretti et al. 2014). Especially often territorial males show less sensitivity when they do not leave their individual territories even when faced with a direct threat, fleeing only from close danger at the last moment. It is possibly that all gazelle species behave in this manner; at least I saw this pattern in goitered gazelle males and Walther et al. (1983) described it for Thomson's gazelle males as well. Males in groups are likely more excitable than single males, because in a group there is always an individual more sensitive to danger, and other members of the group actively respond to his alarm behavior; the whole group is alerted as a result of the contagious effect. If alerted male runs, all others follow even if they have not detected danger. Thus, a male in a group is more likely to be alarmed than a single male (Blank 2018a).

According to my study in Kazakhstan, goitered gazelles do not increase the size of their groups as a reaction to a threat, even during periods when females and males prefer solitary life. Females enhance their vigilance frequency during a fawn's hiding period when they are more sensitive to danger. In contrast, adult males do not increase vigilance or group size, but paradoxically they are most vigilant in large groups. Thus, goitered gazelle females and males have divergent antipredator strategies. Similar behavior is found in the Mediterranean mouflon (*Ovis gmelini*—Benoist et al. 2013) and chamois (*Rupicapra rupicapra*—Ferretti et al. 2014), which is consistent with the hypothesis that females and males of dimorphic ungulates make divergent trade-offs between food consumption amounts and vigilance rate (Barnier et al. 2016).

The conception of group size effect argues that there is a straight inverse correlation between group size and vigilance: a prey animal expends less time on vigilance and more time on feeding in larger groups ("many eyes" effect and safety in number effects—Pulliam 1973; Foster and Treherne 1981; Lima 1995; Roberts 1996) and, conversely, spends more time on vigilance in smaller groups and less on forage. However, the most investigations have yet to show a straight and clear correlation between group size and vigilance, as this correlation maybe negative, positive, or absent. In actual fact, a correlation between vigilance and group size is not always evident or simple and often nonlinear (Elgar 1989; Quenette 1990; Beauchamp 2008). The size of groups depends not only on the pressure of predation, but also on environmental conditions: for example, ungulate herds are commonly larger in open habitats and their reactions to the danger of predation intensify there (Valeix et al. 2009; Periquet et al. 2012). In addition, some ungulates cannot modify or enlarge their group size because of the limitations of food resources (distribution, quantity, and quality) (Jarman 1974; Valeix et al. 2009).

Beyond that, a prey's antipredator response may be divergent depending on other factors, including the individual's peripheral or central location in the group, body size and social status of a prey, proximity to a refuge area, and even time of day, season, and temperature, as well as the distance to the nearest individual (Eby and Ritchie 2013; Underwood 1982; Quenette 1990). In addition, a predator's hunting method (stalker or courser) and its size can cause various defensive reactions (Thaker et al. 2011). Unfortunately, all these factors are not taken into account when studying a prey/predator relationship and authors of most studies pay special attention to some factors, but completely ignore others; as a result, most studies by various authors are inconsistent and very difficult to compare (Eby and Ritchie 2013; Quenette 1990; Underwood 1982).

Although there are two main responses to an increase in danger—increase vigilance frequency and enlarge group size—different species demonstrate preferences for one method over another. Some ungulates resist predators by increasing their concentration through increasing the size of a group, while others cannot do this because of environmental limitations, increasing their frequency of vigilance instead (Quenette 1990). The greater kudu, zebras (*Equus quagga*), Grant's gazelles (*Nanger grati*), and Przewalski's gazelles enhance the frequency of their vigilance in the presence of a predator without enlarging their group sizes, but impala do not do this; wildebeest (*Connochaetus taurinus*) will modify their vigilance rate according to the specifics of their habitat (Li et al. 2009; Pays and Fritz, 2012; Periquet et al. 2012; Creel et al. 2014). Other ungulate species can enlarge (zebra) or decrease their group sizes (blue wildebeest) in the predator's presence without raising their rate of vigilance (Creel et al. 2014).

According to my study, goitered gazelles prefer to modify their vigilance frequency in reaction to a possible danger rather than enlarge their group size, because they live in an arid environment with very poor and scattered vegetation, meaning they have to stay in small groups to collect enough forage for all individuals (Blank et al. 2012). My study also demonstrates that goitered gazelles of the same sex and age will change their vigilance depending on the physiological status of their biological cycle (rut for males and fawn's hiding period for females) (Blank 2018a).

16.6 Behavioral Responses to Threats

16.6.1 General Remarks

Before running from a predator, ungulates usually make a preflight risk assessment to identify a predator, estimate its danger level, and then take actions to decrease the chances of being caught. There can be a variety of antipredator actions, such as enlarging group size, raising the vigilance rate for more quickly spotting and identifying of threat, changing location, and/or reducing their activity (Caro 2005; Apfelbach et al. 2005). There are a number of factors that have an impact on an ungulates' sense of risk and what actions to take, like to where and when to flee, but the most important are the morphological characteristics and features of prey and predator and their behaviors; environmental factors also play a role in an ungulates' risk assessment, though these influences are smaller than the traits of the prey and predator (Stankowich and Coss 2006; Liley and Creel 2008; Creel et al. 2014). Ungulates are evolutionarily adapted to react quickly and effectively to the presence and attack of predators and choose the most optimal way to avoid capture (Stankowich and Coss 2007).

It is known that ungulates perform alarm behavior as soon as they see any unknown object in their environment irrespective of whether the threat is real or not (Walther 1969). However, ungulates have the ability to differentiate between dangerous and safe circumstances, between predatory and nonpredatory animals, and between hunting and inactive predators to determine the most optimal way to escape when necessary (Peckarsky and Dodson 1980; Caro 2005; Coss and Ramakrishnan 2000; Stankowich and Coss 2006). In open spaces, goitered gazelles can identify danger from a distance of up to 2 km (Zhevnerov 1975). Using alarm signals, ungulates share information among conspecifics to decrease the predation risk for the entire herd (Yahner 2012).

There is limited information about the variability of ungulate responses of different kinds of threats and factors determining the perception of a safe or dangerous situation and the magnitude of danger presented by various predators (Liley and Creel 2008). Moreover, many studies devoted to ungulate antipredator behavior used simulated artificial risks as threat factors, while studies of antipredator responses of a prey animal to natural predators and other ungulate species in the wild are much more limited (Liley and Creel 2008). In this chapter, I consider the variability of antipredator behavior in goitered gazelles and their responses to the presence of different predators and ungulate species. Indices of body sizes in ungulate and predator species that interact with goitered gazelles are presented in Table 16.19.

16.6.2 Responses of Goitered Gazelles to Various Kinds of Threat

During my study of the Ili Hollow population, the goitered gazelles encountered various other ungulates, predators, and humans. I described the gazelles' responses to each of these species, analyzing which behavioral features of the other animals were perceived by goitered gazelles as more dangerous and which ones were less frightening. To measure the magnitude of fear in goitered gazelles to various potential threats, I used a ratio of some antipredator responses to others (Table 16.20). This ratio can be called an index of fear. During encounters with other ungulate species, goitered gazelles were least afraid of kulans (*Equus hemionus*) and ibexes (*Capra sibirica*) were the scariest for them. Their responses to domestic cows, saiga antelopes, and argali (in decreasing order) were in between that range. Among predators, goitered gazelles fled most often (as compared to freezing in place) from a horseman with an escorting dog and least frequently from foxes, while wolves and a lone dog fell in between (Table 16.20).

In order to get away from danger as quickly as possible, gazelles most often used their fastest gait of a gallop and much less often a trot. Galloping is a more common reaction toward predators than other ungulates (ratio of trotting to galloping), and trotting is a more common response toward other ungulates. Goitered gazelles trot most frequently from kulans and most rarely after facing argali, with the domestic cow, ibex, and saiga (in increasing order) in the middle. Among predators, gazelles rarely trot from wolves and horsemen and do not trot at all from a fox or dog (Table 16.20). Galloping away in panic with no preliminary risk assessment is typical for dogs and especially horsemen, while not characteristic for most encounters with other predators or ungulate species. Among predators, goitered gazelles walk away most often from foxes and wolves, rarely from dogs, and never from a horseman; among ungulates, they walk from saiga antelope and domestic cows and rarely from kulan and argali. Sometimes gazelles walk toward a suspicious animal, though this occurs much more rarely than walking away. Among predators, goitered gazelles walk mostly toward foxes, sometimes turning it into a running chase, while they never walk toward other predators. Among ungulates, goitered gazelles walk mostly toward saiga antelopes and kulans, rarely toward argali, and never toward ibexes or domestic cows (Blank 2020).

Vigilance in goitered gazelles when in a recumbent position is a rare phenomenon and is observed in encounters with foxes and argali (Table 16.20). Alarm urination and stotting are found more often than recumbent vigilance, and stotting is a more common pattern than alarm urination when facing other ungulates, while the

Table 16.19 Size indices in observed ungulate and predator species that were encountered by goitered gazelles (Body length—120–125 cm, wither height—79 cm, weight—29–34 kg) according to Solomatin (1973), Fedosenko and Savinov (1983), Fedosenko and Kapitonov (1983), Zhevnerov et al. (1983), Fadeev and Sludskiy (1983) (Blank 2020—License Number—5477030865611)

Sizes indices	Kulan	Argali	Saiga	Cow	Ibex	Fox	Wolf	Dog	Hs-man
Body length (cm)	206–240	160–168	125–150	180–220	161–171	60–80	105–140	65–75	200–220
Withers height (cm)	116–136	105–110	66–78	120–140	95–104	25–30	0–85	55–60	150–170
Weight (kg)	250–350	122–125	37–49	350–450	83–114	5–7	35–40	30–40	350–500

Table 16.20 Different escape rates of goitered gazelles depending on perceived threat (Kulan—*Equus hemionus*; Argali—*Ovis ammon*; Saiga—*Saiga tatarica*; Cow—domestic cow; Ibex—*Capra sibirica*; Fox—*Vulpes vulpes*; Wolf—*Canis lupus*; Dog—domestic dog; Hs-man—horseman with a dog; WalkFr—walking from danger; WalkTo—walking toward danger) (Blank 2020—License Number—5477030865611)

Escape rate	Kulan	Argali	Saiga	Cow	Ibex	Fox	Wolf	Dog	Hs-man
Encounter cases	36	241	33	31	81	27	112	60	27
Running/ Freezing	3/26 0.115	33/175 0.189	6/16 0.375	8/16 0.5	59/22 2.682	1/14 0.071	33/63 0.524	27/29 0.931	29/8 3.625
Galloping/ Freezing	1/26 0.038	26/175 0.149	6/16 0.375	6/16 0.375	37/22 1.682	1/14 0.071	31/63 0.492	27/29 0.931	27/8 3.375
Trotting/ Freezing	2/26 0.077	7/175 0.04	6/16 0.375	2/16 0.125	22/22 1	–	1/63 0.016	–	2/8 0.25
NoAssess/ Galloping	–	7/33 0.212	–	–	–	–	–	2/27 0.074	25/27 0.926
Trotting/ Galloping	2/1 2	7/33 0.212	6/6 1	2/6 0.333	22/37 0.595	–	1/37 0.027	–	2/27 0.074
WalkFr/ Freezing	3/26 0.115	23/175 0.131	9/16 0.562	7/16 0.438	–	4/14 0.286	16/63 0.254	4/29 0.138	–
WalkTo/ Freezing	4/26 0.154	8/175 0.046	2/16 0.125	–	–	2/14 0.143	–	–	–
AttackTo/ Freezing	–	–	–	–	–	6/14 0.429	–	–	–
Laying/ Freezing	–	2/175 0.011	–	–	–	–	–	–	–
Urinating/ Running	1/3 0.333	9/33 0.273	–	–	–	–	1/33 0.03	1/27 0.037	6/29 0.207
Stotting/ Running	1/3 0.333	3/33 0.091	–	4/16 0.25	–	–	7/33 0.212	–	–

opposite is observed when goitered gazelles encounter predators. The gazelles display alarm urination more common when facing kulan, less frequently with argali and horsemen, rarely with wolves, and even less frequently with dogs. In contrast, alarm urination is not displayed by goitered gazelles at all with many ungulate species, including the ibex, cow, and saiga antelope. Stotting is demonstrated by goitered gazelles when facing wolves (Table 16.20—Blank 2020).

When encountering ibexes, wolves, dogs, and horsemen, goitered gazelles strive to raise their altitude during escape by running up along a mountain slope. Much less frequently, they flee without changing elevation (cases with domestic cows) and exclusively rarely they descend altitude (only in encounters with argali) (Table 16.21). Goitered gazelles prefer to flee toward the mountains as a response to ibexes and horsemen, as well as argali and dogs.

In danger environments, alarmed goitered gazelles typically keep their tails typically in a vertical position; less frequently the tail is raised but in a lower position than horizontal. Gazelles display a vertical tail more commonly to predators than to ungulates and hold their tails vertically mainly as a response to horsemen, a little

Table 16.21 Escaping direction in goitered gazelles (Argali—*Ovis ammon*; Saiga—*Saiga tatarica*; Cow—domestic cow; Ibex—*Capra sibirica*; Wolf—*Canis lupus*; Dog—domestic dog; Hs-man—horseman with dog; Up—climbing up to mountains; Horizon—no change to elevation; Down—descending from mountains; ToMount—running toward mountains; ParalMount—running along mountains; FromMount—running opposite side from mountains) (Blank 2020—License Number—5477030865611)

Escaping direction	Argali	Cow	Ibex	Wolf	Dog	Hs-man
Up	12/17 0.706	1/5 0.2	59/59 1	25/31 0.806	7/10 0.7	29/29 1
Horizon	–	4/5 0.8	–	6/31 0.194	3/10 0.3	–
Down	5/17 0.294	–	–	–	–	–
ToMount	12/15 0.8	–	59/59 1	4/31 0.129	7/10 0.7	29/29 1
ParalMount	–	5/5 1	–	27/31 0.871	3/10 0.3	–
FromMount	3/15 0.2	–	–	–	–	–

Table 16.22 Tail position during escaping behavior of goitered gazelle depending on perceived threat (Argali—*Ovis ammon*; Saiga—*Saiga tatarica*; Cow—domestic cow; Ibex—*Capra sibirica*; Fox—*Vulpes vulpes*; Wolf—*Canis lupus*; Dog—domestic dog; Hs-man—horseman with a dog) (Blank 2020—License Number—5477030865611)

Tail position	Argali	Cow	Fox	Wolf	Dog	Hs-man
Tail vertical	9/28 0.321	8/15 0.533	7/14 0.5	6/7 0.857	12/13 0.923	29/29 1
Tail raised	18/28 0.643	7/15 0.467	7/14 0.5	1/7 0.143	1/13 0.077	–
Tail down	1/28 0.036	–	–	–	–	–
Tail waving/ vertical+raised	17/27 0.63	5/15 0.333	7/14 0.5	1/7 0.143	–	–

less frequently to kulans, wolves, and dogs (in increasing order), more rarely to cows and foxes, and least often to argali, where the tail is rather raised lower than horizontal (Table 16.22). A relaxed tail (down position) is found rarely (only once as a reaction to argali) and atypically during defensive behavior with other animals. Goitered gazelles wave their tails when they cross the path of an argali or fox, more rarely with cows, and least frequently with wolves and kulans; tail waving is not seen with a dog or horseman (Blank 2020).

If I compare reactions of goitered gazelles to other ungulates, gazelles most often gallop away (never walk) from ibexes and often without a preflight risk assessment, and least often from kulans. In contrast, they walk or run from kulans with the same probability but follow kulans even more frequently. Ibexes do not promote stotting in goitered gazelles, while kulans promote stotting more frequently than any other ungulate species. It is widely known that stotting is demonstrated by prey animals when in a safe environment (Blank 2018c). All this suggests that goitered gazelles perceive their highest magnitude of fear level from ibexes and their least danger is experienced from kulans, in spite of the fact that ibexes are 1.5 to 2 times smaller than kulans. In general, a prey species feels a greater threat from larger body animals, but in this case, this perception is reversed. But it is not only the body size that matters, but also the behavior of the potential threat and habitat characteristics that influence the risk evaluation of prey (Ydenberg and Dill 1986; Stankowich and Blumstein 2005; Kuijper et al. 2014). Since kulans are a species of open habitats favoring almost completely flat areas (Bekenov and Fadeev 1984), they can be seen from a great distance. The combination of the kulans' large body size and a preference for grazing on flat areas, their unhurried movements without changing speed or trajectory, and their habit of standing motionless for hours make unexpected encounter of goitered gazelles with kulans almost an incredible event. Ibexes, on the other hand, behave opposite to kulans by preferring heavily broken terrain, a higher level of maneuverability, sharper movements, and more sudden advent. In addition, ibexes produce great noises when walking along slopes, because many stones often fall down when they cross over the scree-covered stony terrain (for Siberian ibexes—Fedosenko and Blank 2001; for kulans—Bekenov and Fadeev 1984).

Also, since kulans prefer similar smooth terrain and stay in the same areas with goitered gazelles, the two species encounter each other frequently, and goitered gazelles are familiar with the kulans' presence and habits; therefore, this species is identified by gazelles as "not dangerous" (Ydenberg and Dill 1986). In contrast, goitered gazelles see ibexes very rarely, since they live in different landscapes and only encounter each other along the border between the mountain slopes and arid plains. For this same reason, gazelles are more afraid of domestic cows, which they see rarely, and thus more frequently gallop away from them than from saiga antelopes. In addition, larger cows are scarier to gazelles than smaller saiga antelopes. Gazelles have a calmer reaction to saiga antelopes, from which the gazelles usually trot or walk away slowly, and sometimes will even walk toward them. Walther (1969) observed a similar flight in Thomson's gazelle escaping after meeting black rhinos (*Diceros bicornis*), which are rarely found on the African continent and therefore unfamiliar to the gazelles. Thus, the degree of familiarity is very important for the perception of threat to goitered gazelles and prey animals in general (Blank 2020).

Regarding predators, goitered gazelles sense various magnitude of risk, depending on their size and hunting manner (Walther 1969; Ydenberg and Dill 1986; Stankowich and Blumstein 2005; Chamaillé-Jammes et al. 2014). In my study, goitered gazelles feel the most risk from herdsmen (especially a horseman with a dog) and least fear from foxes. The dog itself causes additional fear and ungulates are more sensitive to dogs than even tourists they occasionally encounter, promoting more vigilance and enlarging of group size (for cattle—Welp et al. 2004; for *Cervus elaphus*—Jayakody et al. 2008; for domestic sheep—MacArthur et al. 1982).

Foxes are small carnivores that hunt alone and not in packs and are only harmful to gazelle fawns and only during their first few days after birth; foxes are harmless to adult gazelles (Blank 1990). For this reason, goitered gazelle adults do not fear foxes and usually walk away from them in a relaxed manner and never flee away. However, the behavior of a mother with hiding young is quite different: as soon as the fox approaches her hidden fawn, she runs directly toward it and tries to hit it with her front hooves while chasing it away. Goitered gazelles attack only the small-sized foxes, but flee from larger predators, such as wolves, dogs, and humans. Other prey species behave in the same way, chasing small predators and avoiding large ones. For example, females of mule deer (*Odocoileus hemionus*—Lingle and Pellis 2002), red deer (*Cervus elaphus*—Gese 1998), and pronghorns (*Antilocapra americana*—Byers 1997) will chase away coyotes (*Canis latrans*) but flee from larger predators.

Stotting and the vertical position of the tail are not demonstrated by goitered gazelles when meeting foxes, but instead demonstrate a lower level of excitement with the tail no higher than horizontal, which indicates that the gazelles do not see foxes as a dangerous predator. In contrast, mounted herdsmen, especially when accompanied by dogs, are seen as the most harmful threat because of their high and sudden maneuverability that gives them the ability to chase prey at any instant (Estes and Goddard 1967). Therefore, goitered gazelles enter a state of panic as soon as they see this "predator" and, with a vertical tail, gallop away as far as 1 to 2 km without a preflight risk assessment; goitered gazelles very infrequently trot and never walk away from herdsmen (Blank 2020).

Interestingly, goitered gazelles perceive both the wolf and dog as a medium-risk danger, second only after a herdsman; moreover, the dog is seen as a greater threat than the wolf. Both carnivores have large bodies compared to foxes, with great potential to be dangerous for both goitered gazelle adults and fawns. And in my study area, wolves are the most common predator of gazelles (Blank 1990). For this reason, goitered gazelles have a more active response to wolves than the fox. However, goitered gazelles are less reactive to a wolf than a herdsman. The difference is that a wolf cannot overtake a galloping gazelle during a chase, but a horseman can (Sludskiy 1962). In addition, gazelles have a more intense response to herdsmen, who are often armed, since many have survived after previous encounters with this predator or take their cues from conspecifics that have (Liley and Creel 2008). Comparing the reaction between a wolf and ibex, goitered gazelles usually make a risk assessment toward wolves, but flee immediately from ibexes. Here, the suddenness of the ibexes' appearance generates greater fear. Thomson's gazelles behave in a similar way and flee immediately as if facing their most harmful predator, when they detect the sudden, unexpected appearance of any animals, including birds, small mammals, and even conspecifics, which are not perceived as hazardous in other situations (Walther 1969). The unexpected appearance of wolves can only be seen during the collective hunting of the pack, when they chase their prey, while others wait in ambush to meet their victim when it approaches. In most observed cases, goitered gazelles encounter wolves when they are not busy hunting, but instead leisurely jogging across the plain. In these cases, the gazelles see wolves from afar, and groups that are closer to approaching wolves warn of potential danger, and most gazelles are alerting and already waiting for the predators' appearance; therefore, the gazelles delay their escaping flee and do a preflight risk assessment (Blank 2020).

It is also an interesting point as to why goitered gazelles are more afraid of a dog more than a wolf, even though these carnivores have a similar body size, appearance, and predation skills. The distinction is related to the disparity in their behavior, creating an impression of differences in their threat potential Blumstein 2008). Wolves live by natural laws and start hunting only when they have a sufficient chance for success; otherwise, the hunt does not begin even if they are hungry; wolves never randomly run after any potential prey that they encounter but identify their prey in advance and "assign" roles within the pack during the hunt. How they agree among themselves remains a mystery. Goitered gazelles, therefore, are chased by wolves from time to time, but most encounters with these predators do not result in being hunted. For this reason, goitered gazelles do not flee immediately, but almost always do a preflight risk assessment to identify the wolves' intention to attack or not. In addition, goitered gazelles perform stotting for wolves, like Thomson's gazelles, which display this pattern to any coursing predator to check its intentions (FitzGibbon and Lazarus 1995). A dog is a completely different matter, and goitered gazelles flee away from it in real panic. Since dog's food and protection are fully provided by a human, a dog does not need to save its energy in order to evaluate its potential for hunting success before attacking, so a dog will always chase an animal with the intent of killing it, even if catching it is hopeless. That's why, goitered gazelles gallop away immediately after detecting a dog; doing a risk assessment and stotting would be unhelpful and dangerous in this situation (Blank 2020).

The environment and especially the terrain also influence an escape strategy in prey (Liley and Creel 2008; Barja and Rosellini 2008). Goitered gazelles prefer to run from any threat toward a mountain feature and escape upward along the slope from any threat, though this preference is most common when they face their greatest threats (herdsmen and ibexes). For a closer, less intense danger, a rugged landscape creates some kind of hiding protection and allows them to flee shorter distances. Similar behavior is found in mule deer (*Odocoileus hemionus*) and caribou (*Rangifer tarandus*), which can use dense forest and shrubby vegetation and a rugged landscape for protection against any threat (Lent 1966; Lingle and Pellis 2002; Stankowich and Coss 2006).

Thus, this study suggests that, in general, goitered gazelles perceive more danger from predator species (wolves, dogs, herdsmen) than from other ungulates, though their response may be different depending on the predator's size and its hunting traits. Goitered gazelles also perform defensive behaviors toward other ungulates (argali, kulan, ibex, saiga, and cow), though their reaction to different species varies depending on the size, behavioral traits, approaching speed, directness, and suddenness of appearance. In most cases, goitered gazelles make a preflight risk assessment to identify a potential threat and determine its intentions, while they flee away immediately when a threat is perceived as highly dangerous irrespective of any real harm it may present.

References

Aastrup P (2000) Responses of West Greenland caribou to the approach of humans on foot. Polar Res 19:83–90

Alados CL (1985) An analysis of vigilance in the Spanish ibex (*Capra pyrenaica*). Z Tierpsychol 68:58–64

Alados CL (1986) Aggressive behaviour, sexual strategies and their relation to age in male Spanish ibex (*Capra pyrenaica*). Behav Process 12:145–158

Alados CL, Escos J (1988) Alarm calls and flight behavior in Spanish ibex (*Capra pyrenaica*). Biol Behav 13:11–21

Altendorf KB, Laundre JW, Lopez Gonzales CA, Brown JS (2001) Assessing effects of predation risk on foraging behavior of mule deer. J Mammal 82(2):430–439

Altmann M (1958) Social integration of the moose calf. Anim Behav 6(3–4):155–159

Alvarez F, Braza F, Norzagaray A (1976) The use of the rump patch in the fallow deer (*D. dama*). Behaviour 56(3–4):298–308

Apfelbach R, Blanchard CD, Blanchard RJ, Hayes RA, McGregor IS (2005) The effects of predator odors in mammalian prey species: a review of field and laboratory studies. Neurosci Behav Rev 29:1123–1144

Autenreith RE, Fitchter E (1975) On the behavior and socialization of pronghorn fawn. Wildl Monogr 42:1–111

Barja I, Rosellini S (2008) Does habitat type modify group size in roe deer and red deer under predation risk by Iberian wolves? Can J Zool 86:170–176

Barnier F, Duncan P, Fritz H, Blanchard P, Rubenstein DI, Pays O (2016) Between-gender differences in vigilance do not necessarily lead to differences in foraging-vigilance tradeoffs. Oecologia 181:757–768

Barrett MW (1984) Movements, habitat use, and predation on pronghorn fawns in Alberta. J Wildl Manag 48(2):542–550

Baruzzi C, Lovari S, Fattorini N (2017) Catch me if you can: antipredator behavior of chamois to wolf. Ethol Ecol Evol 1:1–10

Beauchamp G (2008) What is the magnitude of the group-size effect on vigilance? Behav Ecol 19:1361–1368

Beauchamp G (2015) Animal vigilance: monitoring predators and competitors. Academic Press, Oxford

Behrend DF, Lubeck RA (1968) Summer flight behavior of white-tailed deer in two Adirondack forests. J Wildl Manag 32:615–618

Bekenov AB, Fadeev VA (1984) Kulan. In: Gvozdev EV, Strautman EI (eds) Mammals of Kazakhstan, vol 3, No. 4. Nauka Kaz SSR, Alma-Ata, pp 189–217. (in Russian)

Benoist S, Garel M, Cugnasse JM, Blanchard P (2013) Human disturbances, habitat characteristics and social environment generate sex-specific responses in vigilance of Mediterranean mouflon. PlosOne 8(12):e82960

Berger KM (2005) Defense of pronghorn fawns by adult male pronghorn against coyotes. West North Am Nat 65(2):267–268

Berger J, Daneke D, Johnson J, Berwick SH (1983) Pronghorn foraging economy and predator avoidance in a desert ecosystem: implications for the conversion of large mammalian herbivores. Biol Conserv 25(3):193–208

Berger J, Swenson JE, Persson IL (2001) Recolonizing carnivores and naïve prey: conservation lessons from Pleistocene. Science 291:1036–1039

Bigalke RC (1972) Observations on the behavior and feeding habits of the springbok, *Antidorcas marsupialis*. Zool Afr 7:333–359

Bildstein KL (1983) Why white-tailed deer flag their tails. Am Nat 121:709–715

Blank DA (1990) Persian gazelle. In: Kovshar AF (ed) Rare animals of desert regions. Nauka Press of Kazakh SSR, Alma-Ata, pp 56–80. (in Russian)

Blank DA (1997) Social and reproductive behavior in the Persian gazelle. Ph.D. dissert. The University of Tel-Aviv, 189 pp

Blank DA (1998) Mating behavior of the Persian gazelle Gazella subgutturosa Guldenstaedt, 1780. Mammalia 62(4): 409–419

Blank DA (2017) Antipredator tactics are largely maternally controlled in goitered gazelle, a hider ungulate. Behav Processes 136:28–35

Blank DA (2018a) Vigilance, staring and escape running in antipredator behavior of goitered gazelle. Behav Processes 157:408–416

Blank DA (2018b) Escaping behavior in goitered gazelle. Behav Processes 147:38–47

Blank DA (2018c) Alarm signal in goitered gazelle with special reference to stotting, hissing, and alarm urination-defecation. Zoology 131:29–35

Blank DA (2018d) The use of tail-flagging and white rump-patch in alarm behavior of goitered gazelles. Behav Processes 151:44–53

Blank DA (2019) Diversity of risk assessment pattern in antipredator behavior of goitered gazelles. Behav Processes 158:211–218

Blank DA (2020) Behavioral responses of goitered gazelles to potential threat. Mammal Res 65(1):141–149

Blank DA, Ruckstuhl KE, Yang W (2012) Social organization in goitered gazelle (*Gazella subgutturosa* Güldenstaedt, 1780). Ethol Ecol Evol 24:306–321

Blank DA, Ruckstuhl KE, Yang W (2014) The economics of scent marking with urine and feces in goitered gazelle (*Gazella subgutturosa*). Mammal Res 60(1):51–60

Blank DA, Ruckstuhl K, Yang W (2015) Antipredator strategy of female goitered gazelles (*Gazella subgutturosa* Güld., 1780) with hiding fawn. Behav Processes 119:44–49

Blumstein DT (2008) Fourteen security lessons from antipredator behavior. In: Natural security: a Darwinian approach to a dangerous world, p 147e158

Boissy A, Terlouw C, Le Neindre P (1998) Presence of cues from stressed conspecifics increases reactivity to aversive events in cattle: evidence for the existence of alarm substances in urine. Physiol Behav 63(4):489–495

Bonnot N, Morellet N, Verheyden H, Cargnelutti B, Lourtet B, Klein F, Hewison AJM (2012) Habitat use under predation risk: hunting, roads and human dwellings influence the spatial behavior of roe deer. Eur J Wildl Res 59(2):185–193

Bøving PS, Post E (1997) Vigilance and foraging behavior of female caribou in relation to predation risk. Rangifer 17(2):55–63

Brashares JS, Garland TJ, Arcese P (2000) Phylogenetic analysis of coadaptation in behavior, diet, and body size in the African antelope. Behav Ecol 11(4):452–463

Bro-Jørgensen J, Pangle WM (2010) Male Topi antelopes alarm snort deceptively to retain females for mating. Am Nat 176(1):E33–E39

Brown JS (1999) Vigilance, patch use, and habitat selection: foraging under predation risk. Evol Ecol Res 1:49–71

Brown JS, Kotler BP (2004) Hazardous duty pay and the foraging cost of predation. Ecol Lett 7:999–1014

Burger J, Gochfeld M (1994) Vigilance in African mammals—differences among mothers, other females, and males. Behaviour 131:153–169

Byers JA (1997) American pronghorn: social adaptation and the ghost of predators past. University of Chicago Press, Chicago, IL

Cap H, Deleporte P, Joachin J, Reby D (2008) Male vocal behavior and phylogeny in deer. Cladistics 24:1–15

Cárdenas YL, Shen B, Zung L, Blumstein DT (2005) Evaluating temporal and spatial margins of safety in galahs. Anim Behav 70(6):1395–1399

Caro TM (1986a) The functions of stotting in Thomson's gazelles: some tests of the predictions. Anim Behav 34:663–684

Caro TM (1986b) The function of stotting: a review of the hypotheses. Anim Behav 34:649–662

Caro TM (1994) Ungulate antipredator behavior: preliminary and comparative data from African bovids. Behaviour 128(3):189–228

Caro TM (2005) Antipredator defenses in birds and mammals. University of Chicago Press, Chicago, IL
Caro TM (2009) Contrasting coloration in terrestrial mammals. Philos Trans R Soc B 364:537–548
Caro TM, Lombardo L, Goldizen AW, Kelly M (1995) Tail-flagging and other antipredator signals in white-tailed deer: new data and synthesis. Behav Ecol 6(4):442–450
Caro TM, Graham CM, Stoner CJ, Vargas JK (2004) Adaptive significance of antipredator behavior in artiodactyls. Anim Behav 67:205–228
Chamaillé-Jammes S, Malcuit H, Le Saout S, Martin JL (2014) Innate threat-sensitive foraging: black-tailed deer remain more fearful of wolf than of the less dangerous black bear even after 100 years of wolf absence. Oecologia 174(4):1151–1158
Childress M, Lung MA (2003) Predation risk, gender and the group size effect: does elk vigilance depend upon the behaviour of conspecifics? Anim Behav 66:389–398
Christianson D, Creel S (2008) Risk effect in elk: sex-specific responses in grazing and browsing due to predation risk from wolves. Behav Ecol 19:1258–1266
Christianson D, Creel S (2010) A nutritionally mediated risk effect of wolves on elk. Ecology 91(4):1184–1191
Clinchy M, Sheriff MJ, Zanette LY (2013) Predator-induced stress and the ecology of fear. Funct Ecol 27:56–65
Clutton-Brock TH, Guiness FE, Albon SD (1982) Red Deer. Behavior and ecology of two sexes. University of Chicago Press, Chicago, IL
Coleman A, Richardson D, Schechter R, Blumstein DT (2008) Does habituation to humans influence predator discrimination in Gunther's dik-diks (*Madoqua guentheri*)? Biol Lett 4:250–252
Coss RG, Ramakrishnan U (2000) Perceptual aspects of leopard recognition by wild bonnet macaques (*Macaca radiata*). Behaviour 137:315–335
Costelloe BR (2017) Participation by a male Grant's gazelle in the defense of a fawn. Afr J Ecol 56(2):378–380
Creel S, Winnie JJ, Maxwell B, Hamlin K, Creel M (2005) Elk alter habitat selection as an antipredator response to wolves. Ecology 86(12):3387–3397
Creel S, Schuette P, Christianson D (2014) Effects of predation risk on group size, vigilance, and foraging behavior in an African ungulate community. Behav Ecol 25(4):773–784
Cresswell W, Lind J, Kaby U, Quinn JL, Jakobsson S (2003) Does an opportunistic predator preferentially attack nonvigilant prey? Anim Behav 66:643–648
Croes BM, Laurance WF, Lahm SA, Tchignoumba L, Alonso A, Lee ME, Campbell P, Buji R (2007) The influence of hunting on antipredator behavior in Central African monkeys and duikers. Biotropica 39(2):257–263
De Boer HY, Van Breukelen L, Hootsmans MJ, Van Wieren SE (2004) Flight distance in roe deer *Capreolus capreolus* and fallow deer *Dama dama* as related to hunting and other factors. Wildl Biol 10(1):35–41
Dehn MM (1990) Vigilance for predators: detection Anim. Behav and delution effects. Behav Ecol Sociol 26:337–342
Dill LM (1974) The escape response of the zebra danio (*Brachydanio rerio*). I. The stimulus for escape. Anim Behav 22:711–722
Dunham K (1979) The feeding ecology of impala (*Aepyceros melampus*) in the Sengwa region of Rhodesis. M. Phil. Thesis. University of Rhodesia
Eby S, Ritchie ME (2013) The impacts of burning on Thomson'd gazelles' *Gazella thomsoni* vigilance in Serengeti National Park, Tanzania. Afr J Ecol 51(2):337–342
Edmunds M (1974) Defense in Animals. Longman, Essex
Elgar MA (1989) Predator vigilance and group size in mammals and birds: a critical review of the empirical evidence. Biol Rev 64:13–33
Estes RD (1967) The comparative behaviour of grants and Thomson's gazelles. J Mamm 48:189–209
Estes RD (1976) The significance of breeding synchrony in the wildebeest. Afr J Ecol 14(2):135
Estes RD (1991) The behavior guide to African mammals. University of California Press, Berkeley-Los Angeles

Estes RD, Goddard J (1967) Prey selection and hunting behavior of the African wild dog. J Wildl Manag 31:52–70

Fadeev VA, Sludskiy AA (1983) Saiga. In: Gvozdev EV, VI Kapitonov (eds) Mammals of Kazakhstan. Nauka of Kazakh SSR, Alma-Ata, Kazakhstan, pp 54–92

Favreau F-R, Pays O, Goldizen AW, Fritz H (2013) Short-term behavioural responses of impalas in simulated antipredator and social contexts. PLosOne 8(12):e84970. https://doi.org/10.1371/journal.pone.0084970

Fedosenko AK (2002) Urial: conservation status, ecology, behavior, conservation and economical using. Centrokhotcontrol Press, Moscow (in Russian)

Fedosenko AK, Blank DA (2001) Capra sibirica. Mamm Species 675:1–13

Fedosenko AK, Blank DA (2005) Ovis ammon. Mamm Species 773:1–15

Fedosenko AK, Kapitonov VI (1983) Argali. In: Mammals of Kazakhstan (Gvozdev EV and Kapitonov VI, eds), Vol 3, part 3. Nauka of Kazakh SSR, Alma-Ata, Kazakstan: 144–209 pp

Fedosenko AK, Savinov EF (1983) The Siberian ibex. In: Mammals of Kazakhstan (Gvozdev EV and Kapitonov VI, eds). Vol. 3, Part 3. Nauka of Kazkh SSR, Alma-Ata, Kazakhstan: 92–143 pp

Ferretti F, Costa A, Corazza M, Pietrocini V, Cesaretti G, Lovari S (2014) Males are faster foragers than females: intersexual differences of foraging behavior in the Apennine chamois. Behav Ecol Sociobiol 68:1335–1344

FitzGibbon CD (1989) A cost to individuals with reduced vigilance in groups of Thomson's gazelles hunted by cheetahs. Anim Behav 37:508–510

FitzGibbon CD (1990) Anti-predator strategies of immature Thomson's gazelles: hiding and the prone response. Anim Behav 40:846–855

FitzGibbon CD (1993) Antipredator strategies of female Thomson's gazelles with hidden fawns. J Mammal 74(3):758–762

FitzGibbon CD (1994) The costs and benefits of predator inspection behaviour in Thomson's gazelles. Behav Ecol Sociobiol 34:139–148

FitzGibbon CD, Fanshawe JH (1988) Stotting in Thomson's gazelles: an honest signal of condition. Behav Ecol Sociobiol 23(2):69–74

FitzGibbon CD, Lazarus J (1995) Antipredator behavior of Serengeti ungulates: individual differences and population consequences. In: Sinclair ARE, Arcese P (eds) Serengeti II: dynamics, management, and conservation of an ecosystem. The University of Chicago Press, Chicago and London, pp 274–296

Fortin D, Boyce MS, Merrill EH, Fryxell JM (2004) Foraging costs of vigilance in large mammalian herbivores. Oikos 107:172–180

Foster WA, Treherne JE (1981) Evidence for the dilution effect in the selfish herd from fish predation on a marine insect. Nature 293:466–467

Frädrich H (1965) Zur Biologie and Ethologie des Warzenschweins (*Phacochoerus aethiopicus* Pallas), unter Berücksichtigung des Verhaltens anderer Suiden. Z Tierpsychol 22:328–393

Frid A (1997) Vigilance by female Dall's sheep: interactions between predation risk factors. Anim Behav 53:799–808

Frid A (2003) Dall's sheep responses to overflights by helicopter and fixed-wing aircraft. Biol Conserv 110:387–399

Frid A, Dill L (2002) Human-caused disturbance stimuli as a form of predation risk. Conserv Ecol 6(1):11

Geist V (1981) Behavior: adaptive strategies in mule deer. In: Wallmo OC (ed) Mule and black-tailed deer of North America. University of Nebraska Press, Lincoln, NE, pp 156–223

Gese EM (1998) Threat of predation: do ungulates behave aggressively towards different members of a coyote pack? Can J Zool 77:499–503

Gorbunov AV (1983) About ecology and behavior of goitered gazelles in the Krasnovodsk peninsula. Proc Turkmen Acad Sci, Ser Biol Sci 3:66–68. (in Russian)

Grau GA, Grau BL (1980) Effects of hunting on hunter efforts and white-tailed deer behavior. Ohio J Sci 80:150–156

Grovenburg TW, Jenks JA, Jacques CN, Klaver RW, Swanson CC (2009) Aggressive defensive behavior by free-ranging white-tailed deer. J Mammal 90(5):1218–1223

Guthrie RD (1971) A new theory of mammalian rump patch evolution. Behaviour 38:132–145
Halofsky JS, Ripple WJ (2008) Fine-scale predation risk on elk after wolf reintroduction in Yellowstone National Park, USA. Oecologia 155:869–877
Hickman GC (1979) The mammalian tail: a review of functions. Mamm Rev 9:143–157
Hinde RA (1970) Animal behaviour. A synthesis of ethology and comparative psychology. McGraw-Hill Book Co, New York
Hirth DH, McCullough DR (1977) Evolution of alarm signals in ungulate with special reference to white-tailed deer. Am Nat 111(977):31–42
Hunter LTB, Skinner JD (1998) Vigilance behavior in African ungulates: the role of predation pressure. Behaviour 135(2):195–211
Hunter LT, Durant SM, Caro TM (2007) To fleee or not to flee: predator avoidance by cheetahs at kills. Behav Ecol Sociobiol 61:1033–1042
Jarman RJ (1974) The social organization of antelope in relation to their ecology. Behaviour 48:215–267
Jayakody S, Sibbald AM, Gordon IJ, Lambin X (2008) Red deer Cervus elaphus vigilance behavior differs with habitat and type of human disturbance. Wildl Biol 14(1):81–91
Kiley-Worthington M (1976) The tail movements of ungulates, canids and felids with particular reference to their causation and function as displays. Behaviour 56:69–114
Kingswood SC, Blank DA (1996) Gazella subgutturosa. Mamm Species 518:1–10
Kitchen DW (1974) Social behavior and ecology of the pronghorn. Wildl Monogr 38:3–96
Kostin VP (1955) Materials on ecology and commercial value of goitered gazelle. In: Proceedings of the Institute of zoology, vol. 4. Academy of Sciences of Kazakh SSR, Alma-Ata, pp 67–82. (in Russian)
Kuijper DPJ, Verwijmeren M, Churski M, Zbyryt A, Schmidt K, Jędrzejewska B, Smith C (2014) What cues do ungulates use to assess predation risk in dense temperate forests? PLOSOne 9(1):e84607
LaGory KE (1981) The possible communicative role of tail-flicking in white-tailed deer. Anim Behav 29:966
LaGory KE (1987) The influence of habitat and group characteristics on the alarm and flight response of white-tailed deer. Anim Behav 35(1):20–25
Langerhans RB (2007) Evolutionary consequences of predation: avoidance, escape, reproduction, and diversification. In: Elewa AMT (ed) Predation in organisms: a distinct phenomenon. Springer, Berlin, pp 177–220
Leinders JJM (1979) On the osteology and function of the digits of some ruminants and their bearing on taxonomy. Z Säugetierkunde 44:305–319
Lent PC (1966) Calving and related social behavior in the barren-ground caribou. Z Tierpsychol 6:701–756
Leuthold W (1977) African ungulates. A comparative review of their ethology and behavioral ecology. Springer, Berlin
Leuthold W (1978) On social organization of the gerenuk *Litocranius walleri* (Brooke, 1878). J Anim Ecol 47:471–490
Li Y, Xu W, Qiao J, Yang W (2009) The spatio-temporal distribution and habitat selection of *Gazella subgutturosa* in Kalamaili Mountain nature reserve in four seasons. Arid Land Geogr 32:261–267. (in Chinese with English abstract)
Li C, Jiang Z, Li L, Li Z, Fang H, Li C, Beauchamp G (2012) Effect of reproductive status, social rank, sex and group size on vigilance patterns in Przewalski's gazelle. PLoSOne 7(2):e32607. https://doi.org/10.1371/journal.pone.0032607
Liley S, Creel S (2008) What best explains vigilance in elk: characteristics of prey, predators, or the environment? Behav Ecol 19(2):245–254
Lima SL (1995) Back to the basics of anti-predatory vigilance: the group size effect. Anim Behav 49:11–20
Lima SL, Dill LM (1990) Behavioral decisions made under the risk of predation –a review and prospectus. Can J Zool 68:619–640

Lingle S (2001) Antipredator strategies and grouping patterns in white-tailed deer and mule deer. Ethology 107:295–314
Lingle S (2002) Coyote predation and habitat segregation of white-tailed deer and mule deer. Ecology 83:2037–2047
Lingle S, Pellis S (2002) Fight or flight? Antipredator behavior and the escalation of coyote encounters with deer. Oecologia 131:154–164
Lingle S, Wilson WF (2001) Detection and avoidance of predators in white-tailed deer (*Odocoileus virginianus*) and mule deer (*O. Hemionus*). Ecology 107:125–147
Lingle S, Pellis SM, Wilson WF (2005) Interspecific variation in antipredator behaviour leads to differential vulnerability of mule deer and white-tailed deer fawns to coyote predation early in life. J Anim Ecol 74:1140–1149
Lipetz VE, Bekoff M (1980) Possible functions of predator harassment in pronghorn antelope. J Mammal 61:741–743
Lung MA, Childress MJ (2006) The influence of conspecifics and predation risk on the vigilance of elk (*Cervus elaphus*) in Yellowstone National Park. Behav Ecol 18:12–20
Maberly CTA (1963) The game animals of southern Africa. T. Nelson and Sons, Johannesburg
MacArthur RA, Geist V, Johnston RH (1982) Cardiac and behavioral responses of mountain sheep to human disturbance. J Wildl Manag 46:351–358
Marino A, Baldi R (2008) Vigilance patterns of territorial guanacos (*Lama guanicoe*): the role of reproductive interests and predation risk. Ethology 114:413–423
Marion KR, Sexton OJ (1979) Protective behavior by male pronghorn, *Antilocapra americana* (Artiodactyla). South Nat 24(4):709–710
Marmazinskaya NV (1996) Territorial and marking behavior of goitered gazelle *Gazella subgutturosa* (Artiodactyla, Bovidae) in the Bukhara ecological center. Zool Zhurnal (Moscow) 75(11):1737–1751. (in Russian, English summary)
Marmazinskaya NV (2000) Development of social behavior in goitered gazelle (*Gazella subgutturosa subgutturosa* Gueldenstaedt, 1780). In: Biodiversity conservation in the specially protected areas of Uzbekistan. Chinor ENK, pp 87–94. (in Russian)
Matson TK, Goldizen AW, Putland DA (2005) Factors affecting the vigilance and flight behaviour of impalas. S Afr J Wildl Res-24-month delayed open access 35(1):1–11
McDougall PL, Ruckstuhl KE (2018) Doing what your neighbor does: neighbor proximity, familiarity and postural alignment behavioural mimicry. Anim Behav 135:177–185
O'Gara BW (1978) Antilocapra americana. Mamm Species 90:1–7
Ortolani A (1999) Spots, stripes, tail tips and dark eyes: predicting the function of carnivore colour patterns in carnivores using the comparative method. Biol J Linn Soc 67:433–476
Pays O, Fritz H (2012) The effect of simulated African wild dog presence on antipredator behavior of kudu and impala. Ethology 118(10):1018–1027
Pays O, Sirot E, Fritz H (2012) Collective vigilance in the greater kudu: towards a better understanding of synchronization patterns. Ethology 118:1–9
Peckarsky BL, Dodson SI (1980) Do stonefly predators influence benthic distributions in streams? Ecology 61(6):1275–1282
Périquet S, Valeix M, Loveridge AJ, Madzikanda H, Macdonald DW, Fritz H (2010) Individual vigilance of African herbivores while drinking: the role of immediate predation risk and context. Anim Behav 79:665–671
Periquet S, Todd-Jones L, Valeix M, Stapelkamp B, Elliot N, Wijer M, Pays O, Fortin D, Madzikanda H, Fritz H et al (2012) Influence of immediate predation risk by lions on the vigilance of prey of different body size. Behav Ecol 23:970–976
Pitcher TJ (1979) He who hesitates, lives. Is stotting anti-ambush behavior? Am Nat 113(3): 453–456
Prins HHT, Iason GR (1989) Dangerous lion and nonchalant buffalo. Behaviour 108:262–296
Pulliam HR (1973) On the advantages of flocking. J Theor Biol 38:419–422
Quenette PY (1990) Functions of vigilance behavior in mammals: a review. Acta Ecol 11(6):801–818
Reby D, Cargnelutti B, Hewison AJM (1999) Contexts and possible functions of barking in roe deer. Anim Behav 57:1121–1128

Reynolds TD (1983) Pronghorn responses to hunting coyotes. Great Basin Nat 43(1):7. https://scholarsarchive.byu.edu/gbn/vol43/issl/7
Roberts G (1996) Why individual vigilance declines as group size increases. Anim Behav 51:1077–1086
Rossi I, Mauri L, Laficara S, Apollonia M (2002) Barking in roe deer (*Capreolus capreolus*); seasonal trends and possible functions. Hystrix 13(1–2):13–18
Sarno RJ, Grigione MM, Arvidson LD (2008) Lack of response of an open-habitat ungulate to the presence of predator urine. Rev Chil Hist Nat 81:179–183
Schall A, Ropartz P (1985) Le comportement de surveillance chez le daim (*Dama dama*). In: C.R. Acad. Sc. Paris 16 301, series III, pp 731–736
Schaller GB (1967) The deer and the Tiger. University of Chicago Press, Chicago
Schenkel R (1966) On sociology and behavior in impala (*Aepyceros melampus suara* Matschie). Z Säugetierk 31:177–205
Shank CC (1972) Some aspects of social behavior in a population of feral goats (*Capra hircus* L.). Z Tierpsychol 30:488–528
Sirot E, Touzalin F (2009) Coordination and synchronization of vigilance in groups of prey: the role of collective detection and predators' preference for stragglers. Am Nat 173(1):47–59
Sludskiy AA (1962) Interrelation between predators and prey. In: Proceedings of the Institute of zoology, vol 17. Publication of Academy of Sciences KazSSR, pp 24–143. (in Russian)
Smith WP (1991) Ontogeny and adaptiveness of tail-flagging behavior in white-tailed deer. Am Nat 138(1):190–200
Smith DW, Mech LD, Meagher M, Clark WE, Jaffee R, Phillips MK, Mack JA (2000) Wolf- bison interactions in Yellowstone National Park. J Mammal 81:1128–1135
Sokolov II (1959) *Gazella subgutturosa* Güldenstaedt—Goitered gazelle. In: Pavlovskiy EN (ed) Fauna USSR, mammals: ungulates (Orders Perissodactyla and Artiodactyla). Publication of the Academy of Sciences USSR, Moscow and Leningrad, pp 308–331. (in Russian)
Solomatin AO (1973) Kulan, Nauka Press: Moscow (in Russian)
Stankowich T (2008) Ungulate flight responses to human disturbance: a review and meta-analysis. Biol Conserv 141:2159–2173
Stankowich T, Blumstein DT (2005) Fear in animals: a review and meta-analysis of the risk assessment. Proc R Soc Bull 25:139–145
Stankowich T, Coss RG (2006) Effects of predator behavior and proximity on risk assessment by Columbian black-tailed deer. Behav Ecol 17:246–254
Stankowich T, Coss RG (2007) Effect of risk assessment, predator behavior, and habitat on escape behavior in Columbian black-tailed deer. Behav Ecol 18:358–367
Takada H, Nakamura K, Takatsuki S, Minami M (2018) Freezing behavior of the Japanese serow (*Capricornis crispus*) in relation to habitat and group size. Mamm Res 63(1):107–112
Taraborelli P, Gregorio P, Moreno P, Novado A, Carmanhahi P (2012) Cooperative vigilance: the guanaco's (*Lama guanicoe*) key antipredator mechanism. Behav Processes 91:82–89
Taylor AR, Knight RL (2003) Wildlife responses to recreation and associated visitor perceptions. Ecol Appl 13:951–963
Thaker M, Vanak AT, Owen CR, Ogden MB, Niemann SM, Slotow R (2011) Minimizing predation risk in a landscape of multiple predators: effects on the spatial distribution of African ungulates. Ecology 92:398407
Tilson RL, Norton PM (1981) Alarm duetting and pursuit deterrence in an African antelope. Am Nat 118:455–462
Toïgo C (1999) Vigilance behavior in lactating female alpine ibex. Can J Zool 77:1060–1063
Tsapluk OE (1972) Age and seasonal dynamics of sexual activity of goitered gazelle. In: Proceedings of Academy of Sciences of the Kazakh SSR, Biological series, vol 3, pp 39–46. (in Russian)
Underwood R (1982) Vigilance behaviour in grazing African antelopes. Behaviour 79(2):81–107
Valeix M, Loveridge AJ, Chamaillé-Jammes S, Davidson Z, Murindagomo F, Fritz H, Macdonald DW (2009) Behavioral adjustments of African herbivores to predation risk by lions: spatiotemporal variations influence habitat use. Ecology 90:23–30

Verdolin JL (2006) Meta-analysis of foraging and predation risk trade-offs in terrestrial systems. Behav Ecol Sociobiol 60:457–464

Walther FR (1964) Verhaltensstudien an der Gattung Tragelaphus De Blainville, 1816 in Gefangenschaft, unter besonderer Berucksichtigung des Sozialverhaltens. Z Tierpsychol 21:393–467

Walther FR (1968) Verhalten der Gazellen. A. Ziemsan Verlag, Wittenberg, Lutherstadt

Walther FR (1969) Ethologische Beobachtungen bei der kunstlichen Aufzucht eines Blessbockkalbes (*Damaliscus dorcas philippsi* Harper, 1939). Zool Gart 36:191–215

Walther FR (1977) Sex and activity dependency of distances between Thomson's gazelles (*Gazella thomsoni* Günther 1884). Anim Behav 25:713–719

Walther FR (1981) Remarks on behavior of springbok, *Antidorcas marsupialis* Zimmermann, 1780. Zool Gart 2:81–103

Walther FR (1984) Communication and expression in hoofed animals. Indiana University Press, Bloomington

Walther FR, Mungall EC, Grau GA (1983) Gazelles and their relatives. A study in territorial behavior. Noyes Publication, Park Ridge, NJ, USA

Welp T, Rushen J, Kramer DL, Festa-Bianchet M, de Passille AMB (2004) Vigilance as a measure of fear in dairy cattle. Appl Anim Behav Sci 87:1–13

White KS, Berger J (2001) Antipredator strategies of Alaskan moose: are maternal trade-offs influenced by offspring activity? Can J Zool 79:2055–2062

Wyman J (1967) The jackals of the Serengeti. Animals 10:79–83

Yahner RH (1980) Barking in a primitive ungulate, *Muntiacus reevesi*: function and adaptiveness. Am Nat 116(2):157–177

Yahner RH (2012) Communication. In: Wildlife and conservation. Springer, New York, NY, pp 101–111

Yang C, Xiao Z, Guo Y, Xiong Y, Zhang X, Yue B (2012) Alarm signals of the Sichuan Sika deer *Cervus nippon sichuanicus*. Zool Sci 29:423–427

Ydenberg RC, Dill LM (1986) The economics of fleeing from predators. Adv Study Behav 16:229–249

Zhevnerov VV (1975) The ethology and ecology of goitered gazelle during the rutting period in the Barsa-Kelmes Island (Aral Sea). In: Proceedings of the Academy of Sciences of Kazakh SSR, Biological series, vol 1, pp 26–33. (in Russian)

Zhevnerov VV (1984) Goitered gazelle of the Barsakelmes Island. Nauka Publishing House of Kazakh SSR, Alma-Ata. (in Russian)

Zhevnerov VV, Bekenov AB, Sludskiy AA (1983) Goitered gazelle – *Gazella subgutturosa* Güldenstaedt, 1780. In: Gvozdev EV, Kapitonov VI (eds) Mammals of Kazakhstan, Bovidae, vol 3, No. 3. Nauka Publishing House of Kazakh SSR, Alma-Ata, pp 11–54. (in Russian)

Playing Behavior 17

17.1 General Remarks

Play is a very important behavior in human babies allowing them to undergo normal development to shape future adult behavior. Play is also part of the cultural traditions. Playing behavior is a well-known and very common phenomenon for both bird and mammal species. Less is known about game playing in octopuses, fishes, and reptiles (Beach 1945; Chick 2008). Many studies of animal and human play behavior have tried to define the function of play (Fagen 1981).

The traditional surplus energy hypothesis states that play is the way to use up surplus energy, which appears in particularly favorable and optimal conditions of an abundance of plant forage (Chick 2008). However, during play, young develop not only motor skills, but also their cognitive abilities (optimal arousal model) to achieve better results through their entire lifespan (foraging, social success, breeding) (Allen and Bekoff 1997; Chick 2008). As the outcome of the surplus energy hypothesis was created, the locomotor-ontogeny of play hypothesis, stating that this play behavior occurs especially often during the suckling period when fawns get milk rich of a variety of nutrients; as they get older, they receive less and less milk nutrition and the frequency of plays gradually declines, decreasing sharply after weaning (Muller-Schwarze et al. 1982; Newberry et al. 1988; Jensen and Kyhn 2000).). According to the practice-exercise hypothesis, during play, ungulate young develop mostly their physical motor skills, which can be used in various contexts, including fighting, mating, or antipredator behavior, and can even enhance their arousal level (Bekoff 2001; Chick 2008). Fagen (1974) supposed that new behaviors appear during play and spread among conspecifics thereafter and start to be adaptive for the whole population. Finally, the social play hypothesis proposed that both young and adults involve into social interactions and hierarchical fights with conspecifics when ungulates, especially the young, learn to connect with each other and behave or even fight to establish their own hierarchical status. During social play, animals learn the acceptable rules for communication with conspecifics, how to

(Open Access)

D. Blank, *The Behavioral Biology and Conservation of Goitered Gazelles (Gazella subgutturosa)*, https://doi.org/10.1007/978-3-031-83504-9_17

play fairly, trust partners, and how to solve conflict situations (Dugatkin and Bekoff 2003). Spinka et al. (2001) thinks that through playing, offspring get training for the unexpected events, while Bekoff (2004) argues that play is the "foundation of fairness" and prepares the young for adult life.

Play behavior is one of the most often explored subjects of research in the psychical development of humans, while considerably less attention is paid to animals (Bekoff 1995, 2004; Chick 2008). At the same time, among animals, most investigations have been devoted to play behavior in primates under laboratory environments (Tacconi and Palagi 2009) and laboratory rodents or domestic animals (Jensen and Kyhn 2000; McDonnell and Poulin 2002). Play behavior in wild animals in natural habitats, however, is limited to only a few species and is very fragmentary (Bekoff 2004). Therefore, in this chapter, I will describe playing behavior in goitered gazelles in detail and consider the main hypotheses of playing behavior. The description of various kinds of playing activities (ethogram) recorded during my observations is presented in Table 17.1.

17.2 Playing Behavior in Goitered Gazelle

In the wild goitered gazelle population in Ili Hollow, playing behavior is often observed in fawns (61% of all observed cases, n = 558). Adult males play 3 times less often than fawns (20%), while adult females and subadults show playing behavior rarely (8% in adult females, 7% in subadult males and 4% in subadult females). Most often, goitered gazelle fawns play in the presence of their mothers (98.6%, n = 143), much less often in a group of fawns, often independent from their mothers (1.4%), and fawns do not play alone. The most frequent kind of play in goitered gazelles of any age and sex is running after each other, sometimes making stotting jumps or tossing of their hind legs (72.7%, n = 334), much more rarely will gazelles make several different kinds of jumps, butting, and head waving (Table 17.2). Fawns exhibit the largest variety of games compared to adults and subadults, while adult females perform only running, sometimes with stotting; the variability of playing behavior in adult males and subadults is between fawns and adult females (Blank and Yang 2012).

Goitered gazelles typically gallop during play (49% cases, n = 33), less frequently they trot (33%) and least often (18%), they run at their highest speed (careering). During play fleeing, gazelles usually keep their tails in a vertical position (76%, n = 26), much less often they hold their tails in middle or low position (3.8%). Usually, goitered gazelles keep their tails in vertical position and ears pressed to their heads when galloping or careering, while tails are in a lower position and their ear tips are higher or erect during slower gaits.

In most cases, goitered gazelle games start suddenly and last for only a short time, from a few seconds to a half minute (54%); further, longer term games become less frequent the longer they are: 14% cases of 0.5 to 1 min duration, 1% to 7% cases of 1.5 to 7 min duration. All adults, subadults and fawns chose to play for a shorter time rather than a longer time (Blank and Yang 2012). Goitered gazelles

Table 17.1 Description of the recorded playing patterns (ethogram) in goitered gazelle (Blank and Yang 2012—Open Access)

Play type	Description
Locomotor play	All acts connected with movements of any body part or the whole animal body in the presence or absence of partner and no involve physical contact.
Running: Run chasing or lone-rushing	Running with partners (chasing) or alone (lone-rushing) in straight trajectory to some direction and back or by circle around the mother or group of adult gazelles or among other gazelles in zigzag or loop manner. Tail is erected and ears pressed to the head. Movement forward.
Run + signal jumps	Running with high jumps: all four legs are lifted from the ground and the whole body is elevated 0.5–0.7 m above ground surface. Tail is erected and ears pressed to the head. Movement forward-upward.
Run + tossing hind legs	Gazelle runs and tosses the hindquarter in the same time: Two hind legs are lifted from the ground and hindquarter of the body is elevated (tossed) and waved from side to side. Both hind legs are kicked in a posterior direction. Tail and ears are in different position. Movement forward, spinning and reverse direction.
Gaits: Trot	It is a two-beat gait between a walk and a run, in which the legs move in diagonal pairs, but not quite simultaneously.
Gallop	It is a four-beat gait, which is faster than trot.
Career	It is very fast gallop, when front legs are stretching forward and hind legs are stretching backward for maximum possible extent.
Jumping: Jump + spinning	All four legs are lifted and the body is elevated above ground surface. Spinning (until 360°) movements in the air.
Jump + moving forward-backward	The same like former, only movement forward and backward.
Jump + kick up	The same like former, only kicking up by hind or all 4 legs in the air.
Head shaking-waving	Head is shaken or waved up-down or from side to side in playful irregular manner or sometimes rotates widely.
Social play:	All acts connected with movements directed to the partner.
Horn-forehead threat	One gazelle walks toward other one with lowered head and horns or forehead directed forward.
Butting	Two gazelles are standing front to front, butting head (horns, forehead) against head (horns, forehead) of partner in a playful manner, no chasing.
Muzzle pushing	The same like butting, only gazelles fight with the muzzle side.
Mount	One gazelle is mounting another staying on the place or following him by walk in vertical two-pedal position like male doing posture and movements during mating with female.
Invitation for play	The body is ascended from front to back, the head is lowered, legs are stretched forward, tail is twitched from side to side. The posture is similar to that of dog making invitation for play ("bow" for dogs).

play most frequently in May and June, with play behavior difficult to observe in other months (Fig. 17.1). Goitered gazelle fawns begin their playing activities in May, though play is still relatively rare during first days after birth, but by June, the playing frequency rises by three-fold compared to May, with a sharp reduction during July. Further, in August and September, fawns' games become more and more rarely noted (Fig. 17.2). Games of adult gazelles have a different dynamic, as they

Table 17.2 Proportion of various playing patterns, events, and number of playing individuals (Blank and Yang 2012—Open Access)

Play patterns	Cases number	Portion, %	Number of gazelles	Portion, %
Run, chasing or lone-rushing	178	53.3	330	52.6
Run + signal jumps	42	12.5	83	13.2
Run + tossing hind legs	23	6.9	36	5.7
Run totally	243	72.7	449	71.5
Jumps + spinning	9	2.7	16	2.6
Jumps + moving forward-back	4	1.2	6	1.0
Jumps + kick up	8	2.4	15	2.4
Jumps totally	21	6.3	37	6.0
Butting	22	6.6	59	9.4
Head shaking-waving	26	7.8	44	7.0
Mounting each other	8	2.4	15	2.4
Horn-forehead threat	6	1.8	13	2.1
Muzzle pushing	1	0.3	3	0.5
Invitation for play	4	1.2	4	0.6
Horning plants and ground	3	0.9	3	0.5
Total	334	100	627	100

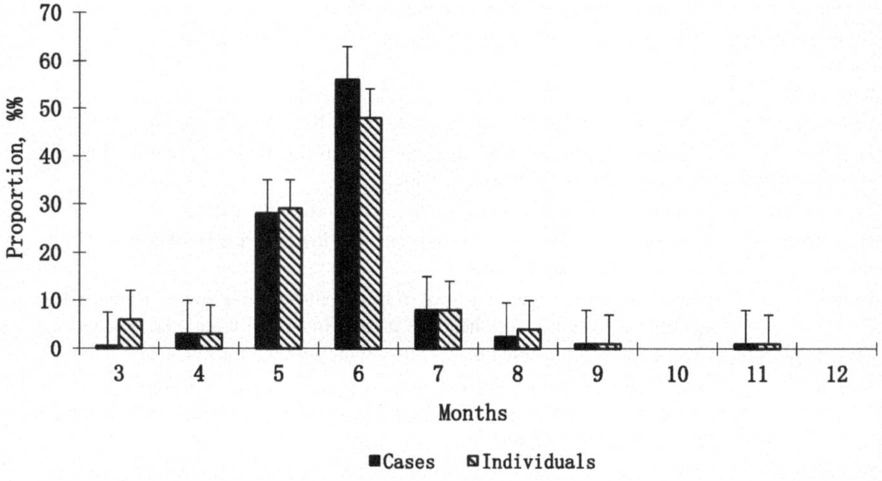

Fig. 17.1 Proportion of playing pattern over months (cases: number of observed cases, n = 200 and individuals: number of playing individuals, n = 396 (Blank and Yang 2012—Open Access)

play most often in May, with a decline of frequency in June. The rest of the year, adult gazelles play rarely and never in winter (October through December). The play intensity of adults and young gazelles has a significant difference in dynamics in May and June, but this difference is absent in other months. Adult males and subadults perform playing behavior most often in May, while adult females play most often in June (Fig. 17.3a, b).

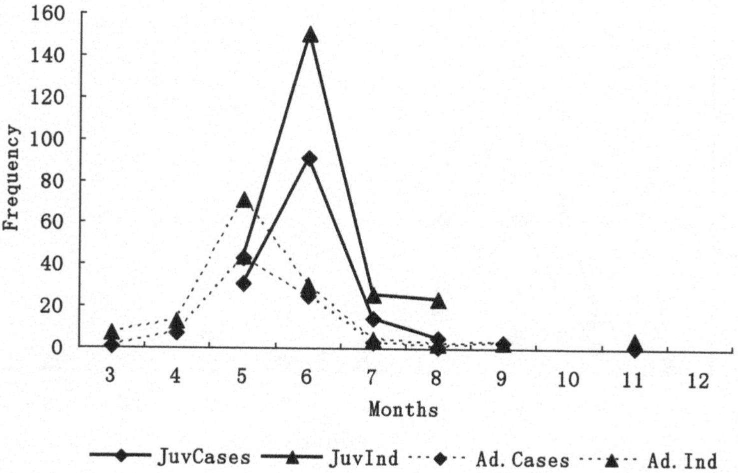

Fig. 17.2 Frequency of playing patterns of young and adult gazelles over months (frequency of play cases for young—JuvCases, n = 141 and for adults—Ad.Cases, n = 84; and frequency of playing individuals number—JuvInd, n = 245 and Ad.Ind, n = 134) (Blank and Yang 2012—Open Access)

Generally, goitered gazelles demonstrate gaming activities in full accordance with their regular activities with peaks in the morning (7:00–8:00) and evening (19:00–21:00); the highest rate is noted in the evening, after 19:00, and the lowest rate occurs midday (Fig. 17.4), but daytime play activity is still different to some extent in adults and young gazelles. Play activity in fawns is observed every 2 to 3 h within the mornings (at 5:00, 7:00–8:00 and 10:00 a.m.), declining after midday, then playing often again after 19:00. Adult males usually play in the afternoons (after 16:00), showing a maximum of playing activities at 19:00. Adult females most often play in the evening hours, after 20:00 (Fig. 17.4), while fawns end their plays after 20:00; adult males rarely play at this time (Blank and Yang 2012).

Goitered gazelle fawns mostly perform locomotor play (running after each other) and start playing from their first days of life in May (22%); the amount of playing increases significantly, reaching its peak in June (63%), with a drastic decline playing activities in July and August (Table 17.3). Social play (collaboratively interacting with others) in fawns begins in June when it reaches its maximal frequency, with a decline in July and rare cases in August. Object play (playing with an object) is observed exclusively rarely in goitered gazelle fawns. Adult males, females, and subadults show locomotor play with approximately the same frequency (36%, 31%, and 33%, respectively, n = 101), though subadult males play more frequently (20%, n = 101) than subadult females (13%). Adult males and subadult males demonstrate social play more rarely (up to 3 times and 9 times, respectively) than locomotor play, while adult females and subadult females do not perform social play at all (Table 17.4). Adult males and subadult males generally show locomotor play a bit more often than adult females and subadult females, while social play is typical only for males.

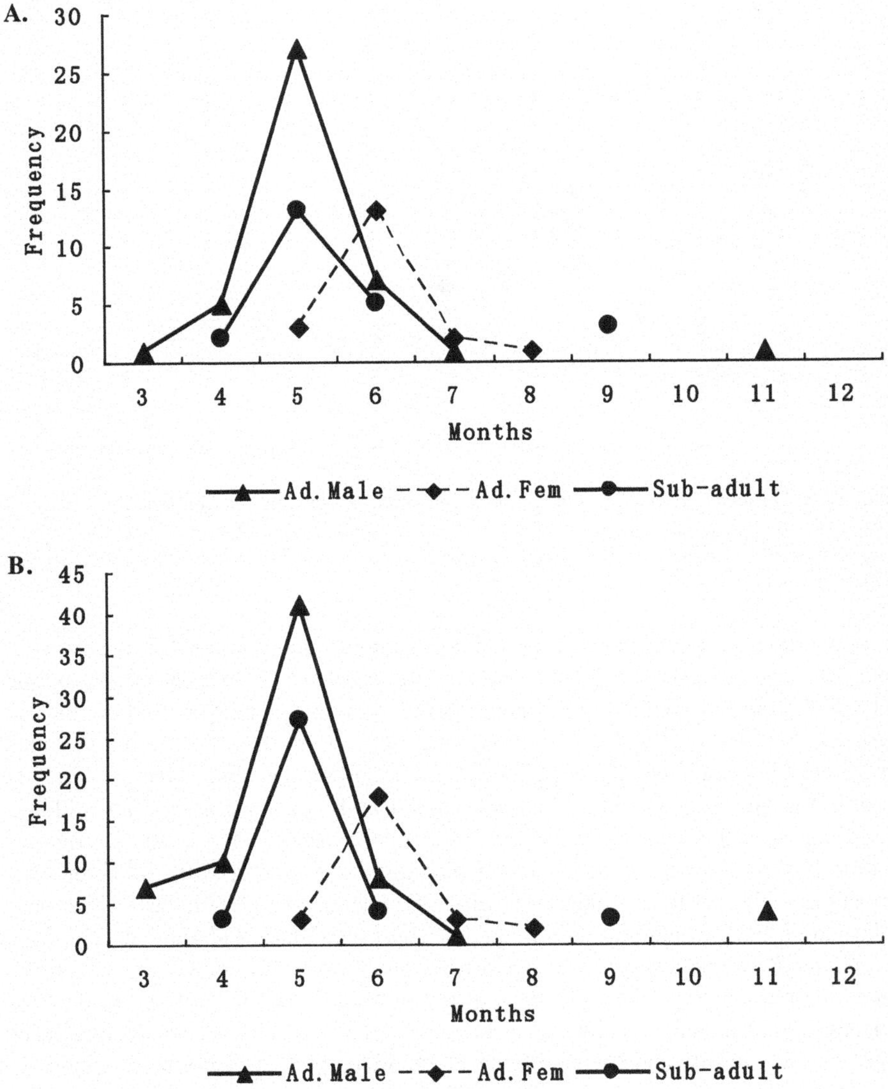

Fig. 17.3 Frequency of playing patterns in adult and sub-adult gazelles over months (**A**)—frequency of cases: Ad.Male, n = 42; Ad.Fem., n = 19; Sub-adult, n = 23 and (**B**)—frequency of playing individuals number: Ad.Male, n = 71; Ad.Fem., n = 26; Sub-adult, n = 37) (Blank and Yang 2012—Open Access)

In a captive population of goitered gazelles, the transition to play activity could be stimulated by the following situations: the beginning of rain, a gust of wind, the movement of a person around the enclosure, an unexpected reduction in the minimum individual distance between gazelles, transfer to a new enclosure, or after an opponent demonstrates a threat (Marmazinskaya 2000). In the wild, fawns prefer to

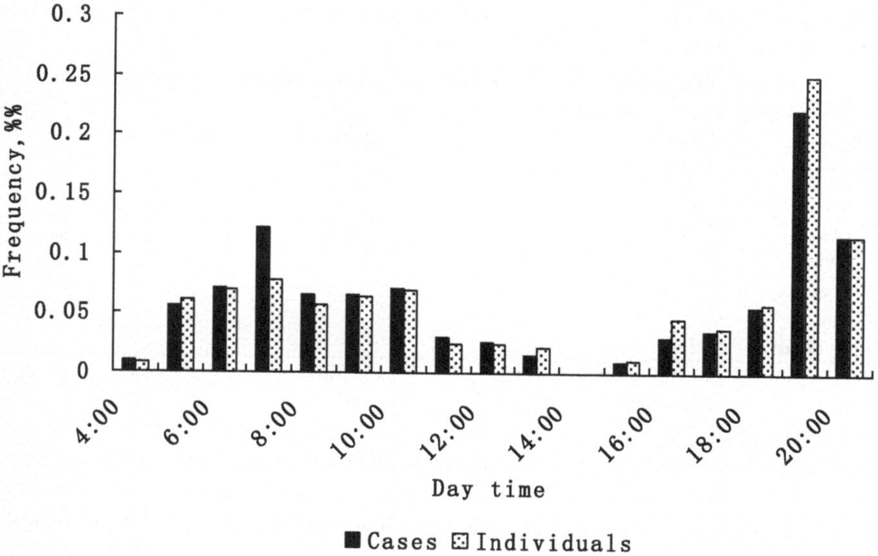

Fig. 17.4 Frequency of playing pattern during a day time (for observed cases, n = 198 and number of playing individuals, n = 371) (Blank and Yang 2012—Open Access)

Table 17.3 Proportion of locomotor and social play over months (young birth period was noted at mid-May in our study area) (Blank and Yang 2012—Open Access)

	April	May	June	July	August	Total
Locomotor play (number/portion, %)						
Cases	1/0.6	39/21.9	112/62.9	20/11.2	6/3.4	178/100
Individuals	1/0.4	57/19.5	176/60.3	41/14.0	17/5.8	292/100
Social play (number/portion, %)						
Cases	–	–	20/76.9	5/19.2	1/3.9	26/100
Individuals	–	–	40/74.1	12/22.2	2/3.7	54/100
Object play (number/portion, %)						
Cases	–	–	2/	–	–	2/
Individuals	–	–	2/	–	–	2/

Explanation: Cases, number of registered cases of play; Individuals, number of playing individuals

play in the company of other fawns, but usually with the audience of their mothers (39% cases, n = 459). In cases when a fawn plays alone, the attendance of its mother is necessary (24%). Quite frequently, playing of a fawn alone or with other fawns occurs during suckling bouts (20%). Rarely will fawn play with adult males or subadults together (3.1% and 2.1%, respectively, n = 459). Games of adults are seen during feeding (8.2%) or after ending head-butting bouts (5.2%), when fleeing from strong wind (2.1%), during a fox chasing or after resting (1%) (Blank and Yang 2012).

Playing adult males and fawns always stay in groups, while adult females (one case) and subadults (two cases) sometimes can play alone. Games of adult males are

Table 17.4 Proportion of locomotor and social play for different age-sex classes during the whole period of study (Blank and Yang 2012—Open Access)

	Adult males	Adult females	Yearling males	Yearling females
Locomotor play (number/portion, %)				
Cases	37/67.3	31/	20/90.9	13/
Individuals	89/69.5	35/	29/90.6	23/
Social play (number/portion, %)				
Cases	15/27.3	–	2/9.1	–
Individuals	36/28.1	–	3/9.4	–
Object play (number/portion, %)				
Cases	3/5.4	–	–	–
Individuals	3/2.4	–	–	–
Totally:	55/100	31/100	22/100	13/100
Cases	100/100	35/100	32/100	23/100
Individuals				

Explanation: Cases, number of registered cases of play; Individuals, number of playing individuals

observed in large male groups of more than 3 individuals (86%, n = 73). Subadults are engaged in games when they stay in groups of 2 to 3 individuals (57%, n = 30), much less often in larger groups. Fawns often play when they are in the company of three individuals (mother and her twin fawns) and rarely in larger groups. Fawns, playing alone, are more often in a group of two—a fawn with its mother (45%, n = 112), less frequently in groups of three (29%), and least in larger herds (Blank and Yang 2012).

Marmazinskaya (2000) observed similar playing behavior of goitered gazelles in the semi- captive conditions of the Bukhara Breeding Center. Short playing activity is seen in fawns as early as 2 days old, though their movements are still poorly coordinated. Play behavior includes running accompanied by other behaviors. When running, the fawn's ears are parallel and directed backward, the mouth is slightly open, the head is shaking, the hind legs are bucking, mounts are attempted, threat displays made during jumps, butting of other individuals, and running up to other gazelles for nasal sniffing. During their games, fawns make three types of sound signals: danger, summoning, and short noises without opening their mouths. In the semi-captive population, synchronous play in fawns occurs both in the absence of their mother as well as while following their mothers. There are individual differences in the frequency of play activity, and bursts of play in young gazelles have been recorded until they reached the age of 12 months. Adults, including dominant males and pregnant females, also join in synchronous games with the young (Marmazinskaya 2000).

It is clear from these data that goitered gazelle fawns are playing much more frequently than adults and subadults, sustaining the practice-exercise hypothesis, and it seems reasonable to define playing's main function as the way to quickly develop muscles and bones and connective tissues for the fawn to be able to run from predators as soon as possible (Brownlee 1954; Gazzola et al. 2005). Following strength skills, playing is used to develop "motor training," meaning that play activities lead to the growth and strengthening of all organs related to motor functions, as well as the peripheral nervous system, brain, cardiovascular system, and any other physical functions (Bekoff and Byers 1981; Byers and Walker 1995; Byers 1999). This hypothesis describes a certain dynamic of play behavior that starts with a low frequency of activity from the first days of life and increases over time to reach the animal's maximum stage of development with decreasing growing rate at the end (Byers and Walker 1995). These dynamics are typical for many ungulate species, including giraffes (*Giraffa camelopardalis*—Pratt and Anderson 1979), pronghorns (*Antilocapra americana*—Miller and Byers 1991), bighorn sheep (*Ovis canadensis*—Hass and Jenni 1993), and reindeer (*Rangifer tarandus*—Espmark 1971). Playing behavior is seen especially often, literally daily, during the most vigorous growing time and commonly during certain hours of the day (Byers 1984; Miller and Byers 1991). In my study, goitered gazelles played most often in June, when their physical development and growth are at their highest rate (Zhevnerov et al. 1983). Thus, the dynamics of playing behavior in goitered gazelle fawns is completely explained by the practice-exercise hypothesis (Blank and Yang 2012).

Another a well-known and long held hypothesis is the release of surplus energy, which argues that when an animal is in optimal condition with plenty of high-quality forage and low predation threat and all physiological needs are met, excessive energy is generated which is spent through playing activities. This is a quite logical hypothesis, although it is difficult to test (Bekoff and Byers 1981). Fawns of black-tailed deer (*Odocoileus hemionus*—Müller-Schwarze 1968) and kids of domestic goats (*Capra hircus*), though, deprived of play did not demonstrate a noticeable difference in muscle development with individuals which played; however, later they had deviations from normal behavior. In addition, cattle calves that were deprived of play increased their playing activities when able (Jensen 2001). In my study, goitered gazelle fawns had their longest and most frequent suckling bouts in May and June, with a decline in July; correspondingly, play frequencies gradually rose in May, reaching the maximum in June, with a sharp decline in July. Fawns often played within suckling bouts in the mornings, and many games by most fawns also occurred again in the evenings after getting their evening meals from their mothers. Thus, playing behavior of goitered gazelle fawns also match by the surplus energy hypothesis (Blank and Yang 2012).

Adults of other ungulate species perform play behaviors with very low frequencies compared to young (Byers and Walker 1995). Goitered gazelle adult males and subadults play primarily in May, which is the most productive month for vegetation growth (high precipitation rate and warm weather) when high quality forage is in abundance (Zhevnerov et al. 1983). Adult males perform their play activities in the mornings and especially often in the evenings, preferring to play when the weather is cooler and they are well fed. In contrast to conspecifics, goitered gazelle adult females are more playful in June, when they quite frequently join to their fawns in play. Adult females of aoudad (*Ammotragus lervia*) often participate in the games of their young (Habibi 1977). In addition, adult females play frequently in late evening when they are well-fed after their evening pasturing and when they have finish all the suckling of their fawns. Thus, playing behavior of goitered gazelle adults also match the surplus energy hypothesis (Blank and Yang 2012).

The third hypothesis, locomotor play ontogeny, states that an intensive milk supply is available during the most vigorous physical development of ungulate young, with a decline thereafter. Indeed, locomotor play starts with a low frequency in goitered gazelles fawns immediately after birth, then quickly increases (in May), reaching its maximum when their most intensive milk supply is available (in June) and then declines as they near weaning (in July). This is a common phenomenon for locomotor play development in young of most ungulate species (Byers and Walker 1995), including cattle calves (Jensen and Kyhn 2000; Krachun et al. 2010), pigs (Newberry et al. 1988), and black-tailed deer (*Odocoileus hemionus*—Müller-Schwarze et al. 1982), when locomotor play sharply declines with the start of weaning. So, my data for goitered gazelle fawns also matches the locomotor play ontogeny hypothesis (Blank and Yang 2012).

The social function of play states that young males more often participate in fighting or sexual play, while young females do it rarely, if at all. This phenomenon is found in young of many ungulates: Cuvier's gazelles (*Gazella cuvieri*), bighorn sheep (*Ovis canadensis*—Hass and Jenni 1993), Siberian ibex (*Capra sibirica*—Byers 1980), and in domestic ungulates (pigs, cattle and domestic sheep—Berger 1979; Dobao et al. 1985; Vitale et al. 1986). In fact, this phenomenon is a part of the practice-exercise hypothesis, claiming that animals enhance those abilities that are necessary for their future well-being as adults (Bekoff and Byers 1981). My study of goitered gazelles showed that adult and subadult males were involved frequently in fighting and sexual activities (social plays), while females usually do not participate at all in social plays. Thus, my data also matches the definition of the social function hypothesis (Blank and Yang 2012).

In studying play behavior of ungulates, different categories of games appeared and disappeared at different developmental ages. Locomotor games were characteristic of Cuvier's gazelle fawns in the first days of their life, reached their highest frequency at the age of 2 weeks, and then gradually disappeared. Social plays, in contrast, began significantly later compared to locomotor play and lasted until its maximal level in June at the age of 4 to 6 weeks, decreasing thereafter (Gomendio 1988). Goitered gazelle fawns had a different dynamic: their maximal rate of locomotor play occurred later than Cuvier's gazelle fawns, at the age of 4 to 6 weeks (in June), with a sharp decrease in July at the age of 7 to 8 weeks. This disparity between these two gazelle species is likely related to their different body sizes and weights, with goitered gazelles larger antelopes (up to 43 kg in males—Heptner et al. 1961) than Cuvier's gazelles (up to 35 kg in males—Kingdon 1997), and larger sized ungulate young experience longer developing periods than smaller species (Blueweiss et al. 1978). Goitered gazelle fawns began to engage in social play at 3-weeks of age, similar to Cuvier's gazelle fawns (Gomendio 1988), but unlike Cuvier's gazelle fawn, goitered gazelle fawns reached their maximal levels of social play during the same period as locomotor play in June, while in July and August, the frequencies of both social and locomotor play declined almost at the same time. So, in goitered gazelle fawns, I did not find any differences in development progress dynamics of these two kinds of plays. Thus, my data for goitered gazelle fawns are conflicting with the social play ontogeny hypothesis.

In contrast to locomotor plays, objects play was seen very rarely in both Cuvier's gazelle fawns (Gomendio 1988) and goitered gazelle fawns. Indeed, locomotor play is much more important for developing skills for future life of all ungulate species than their object playing (Byers and Walker 1995), which is more typically seen in young primates (Fagen 1981).

Thus, my data for goitered gazelle play behavior generally matches all hypotheses except the specific stages of fawn's locomotor play and social play which develop in the same time in contrast to the playing behavior of the Cuvier's gazelles, where locomotor play occurs much earlier fawn's age, while social play appear several weeks later.

References

Allen C, Bekoff M (1997) Species of mind: the philosophy and biology of cognitive ethology. MIT Press, Cambridge, MA

Beach FA (1945) Current concepts of play in animals. Am Nat 79(785):523–541

Bekoff M (1995) Play signals as punctuation: the structure of social play in canids. Behaviour 132:419–429

Bekoff M (2001) Social play behaviour. J Conscious Stud 8(2):81–90

Bekoff M (2004) Wild justice and fair play: cooperation, forgiveness, and morality in animals. Biol Philos 19:489–520

Bekoff M, Byers JA (1981) A critical reanalysis of the ontogeny of mammalian social and locomotor play: an ethological hornet's nest. In: Immelmann K, Barlow GW, Petrinovich L, Main M (eds) Behavioural development: the Bielefeld interdisciplinary project. Cambridge University Press, New York, pp 296–337

Berger CD (1979) Weaning conflict in desert and mountain bighorn sheep (*Ovis canadensis*): an ecological interpretation. Z Tierpsychol 50:188–200

Blank DA, Yang W (2012) Play behaiour in goitered gazelle, *Gazella subgutturosa* (Artiodactyla: Bovidae) in Kazakhstan. Folia Zool 61(2):161–171

Blueweiss L, Fox H, Kudzma V, Nakashima D, Peters R, Sams S (1978) Relationship between body size and some life history parameters. Oecologia 37:257–272

Brownlee A (1954) Play in domestic cattle in Britain: an analysis of its nature. Br Vet J 110:48–68

Byers JA (1980) Play partner preferences in Siberian ibex, *Capra ibex sibirica*. Z Tierpsycho 53:23–40

Byers JA (1984) Play in ungulates. In: Smith PK (ed) Play in animals and humans. Blackwell, Oxford, pp 43–65

Byers JA (1999) The distribution of play behaviour among Australian marsupials. J Zool (London) 247:349–356

Byers JA, Walker C (1995) Refining the motor training hypothesis for the evolution of play. Am Nat 146(1):25–40

Chick G (2008) Altruism in animal play and human ritual. World Cult eJ 16(2):1–17

Dobao MT, Rodriganez J, Silio L (1985) Choice of companion in social play in piglets. Appl Anim Behav Sci 13:259–266

Dugatkin LA, Bekoff M (2003) Play and the evolution of fairness: a game theory model. Behav Process 60:209–214

Espmark Y (1971) Individual recognition by voice in reindeer mother-young relationship. Behaviour 40:295–301

Fagen R (1974) Selective and evolutionary aspects of animal play. Am Nat 108(964):850–858

Fagen R (1981) Animal play behavior. Oxford University Press, Oxford

Gazzola A, Bertelli I, Avanzinelli E, Tolosano A, Bertotto P, Apollonio M (2005) Predation by wolves (*Canis lupus*) on wild and domestic ungulates of the western Alps, Italy. J Zool (London) 266:205–213

Gomendio M (1988) The development of different types of play in gazelles and functions of play. Anim Behav 36:825–836

Habibi K (1977) The mammals of Afghanistan: their distribution and status.FO:DP/AFG/74/016. Field Document No.1. FAO, Rome

Hass CC, Jenni DA (1993) Social play among juvenile bighorn sheep: structure, development, and relationship to adult behavior. Ethology 93:105–116

Heptner G, Nasimovich AA, Bannikov AG (1961) Mammals of Soviet Union, Vol. 1. Artiodactyla and Perissodactyla. Vysshaya Shkola Publishing House, Moscow. (in Russian)

Jensen MB (2001) A note of the effect of isolation during testing and length of previous confinement on locomotor behaviour during open-field test in dairy calves. Appl Anim Behav Sci 70:309–315

Jensen MB, Kyhn R (2000) Play behaviour in group-housed dairy calves, the effect of space allowance. Appl Anim Behav Sci 67:35–46

Kingdon J (1997) The kingdom field guide to African mammals. Academic Press, London and New York

Krachun C, Rushen J, de Passille AM (2010) Play behaviour in dairy calves is reduced by weaning and by low energy intake. Appl Anim Behav Sci 122:71–76

Marmazinskaya NV (2000) Development of social behavior in goitered gazelle (*Gazella subgutturosa subgutturosa* Güldenstaedt, 1780). Biodiversity conservation in the specially protected areas of Uzbekistan. Chinor ENK, pp 87–94. (in Russian)

McDonnell SM, Poulin A (2002) Equid play ethogram. Appl Anim Behav Sci 78:263–290

Miller MN, Byers JA (1991) The energetic cost of locomotor play in pronghorn fawns. Anim Behav 41:1007–1013

Müller-Schwarze D (1968) Play deprivation in deer. Behaviour 31(1–2):144–162

Müller-Schwarze D, Stagge B, Muller-Schwarze C (1982) Play behaviour: persistence, decrease, and energetic compensation during food shortage in deer fawns. Science 215:85–87

Newberry RC, Wood-Gush DGM, Hall JW (1988) Playful behaviour of piglets. Behav Process 17:205–216

Pratt DM, Anderson V (1979) Giraffe cow-calf relationships and social development of the calf in the Serengeti. Z Tierpsychol 51:233–251

Spinka M, Newberry RC, Bekoff M (2001) Mammalian play: training for the unexpected. Q Rev Biol 76:141–168

Tacconi G, Palagi E (2009) Play behavioural tactics under space reduction: social challenges in bonobos, *Pan paniscus*. Anim Behav 78:469–476

Vitale AF, Tenucci M, Papini M, Lovari S (1986) Social behaviour of the calves of semi-wild Maremma cattle, *Bos primigenius taurus*. Appl Anim Behav Sci 16:217–231

Zhevnerov VV, Bekenov AB, Sludskiy AA (1983) Goitered gazelle—*Gazella subgutturosa* Güldenstaedt, 1780. In: Gvozdev EV, Kapitonov VI (eds) Mammals of Kazakhstan, Bovidae, vol 3, No. 3. Nauka Publishing House of Kazakh SSR, Alma-Ata, pp 11–54. (in Russian)

Comfort Behavior 18

18.1 Using Microclimate in Comfort Behavior

Blank (2020)—License Number—5477040274237.

18.1.1 General Remarks

Northern deserts are very cold during winter and very hot during summer. The high deficit of water and the high solar radiation with limited cover are typical for this habitat; therefore, saving body water loss and maintaining efficient thermoregulation under any conditions, including abrupt daily and seasonal drops of temperature, are the most important phenomena for ungulates of arid areas (Feldhamer et al. 1999; Mader et al. 2010). In the case of overheating, there are many negative consequences: inability to fertilization, increased likelihood of miscarriage, and loss of milk production (Souza et al. 2005). The physiological incapacity to cope with dehydration and the increase in body temperature lead to death (Curtis 1983). In the cold of winter, ungulates also need to maintain their body temperature using morphological adaptations (thremo-isolation) and physiological mechanisms (thermoregulation). In addition, ungulates can use behavioral mechanisms, employing relief irregularities, such as shrub thickets and trees, high dunes and dry riverbanks, and deep and narrow gorge, to find a more suitable microclimate to mediate the effects from high heat or cold winds (Zhevnerov et al. 1983; Blank and Li 2022).

Ungulates need to maintain the fluctuations of their body temperature within a narrow range (Fuller et al. 2005). There are three main mechanisms, including morphological, physiological, and behavioral, which provide this thermal and water balance between the body and ambient environment (Cain et al. 2006). Physiological mechanisms, however, have a high energetic cost (Souza et al. 2005); therefore, ungulates use these mechanisms only in critical situations, which usually apply to the behavioral mechanism, such as finding a shelter with more favorable conditions,

D. Blank, *The Behavioral Biology and Conservation of Goitered Gazelles (Gazella subgutturosa)*, https://doi.org/10.1007/978-3-031-83504-9_18

which supply better protection from the harsh environment (Fuller et al. 2005; Holcomb 2017). In addition to the selection of more suitable thermally habitats, ungulates will change, if needed, their daily activities, breeding schedule, migration, diet, social behaviors, and/or body orientation (Bartholomew 1987; Huey 1991).

The focus of this chapter is the behavioral responses of goitered gazelles in natural conditions to hot, sunny conditions and cold, windy periods of weather and using the microclimate features of their landscape, including shady and windy sites during the hot months and wind-protected, sunny places during the cold months. Fluctuations in the monthly average temperature (°C) and wind speed (m/s) in my study area during March–November are presented in Fig. 18.1 (Blank 2020).

18.1.2 Comfort Behavior in Goitered Gazelles

During cold seasons, goitered gazelles avoid the use of plateaus and hills for their grazing and resting, while they often stay at elevated points for their daytime resting period during warm months. It is possible to see resting gazelles on the high parts of the terrain most often in June and they continue to use elevated areas until the end of the hot season in September. Less frequently, gazelles can be seen on higher terrain in April and May, when it is warm, but not hot yet, and almost never are they found on elevations in November and March (Fig. 18.2). In contrast, goitered gazelles are seen in the lowered parts of their landscape (deep river beds, narrow gorges between foothills) for shelter from winds most frequently in March, when cold, strong winds occur frequently, with some decline in these locations in April and May, which are warmer and less windy; least often they use these lower sites in June and July (hottest months), and in other months, their preference in lowland, windless places is observed only occasionally. The significant raise in their

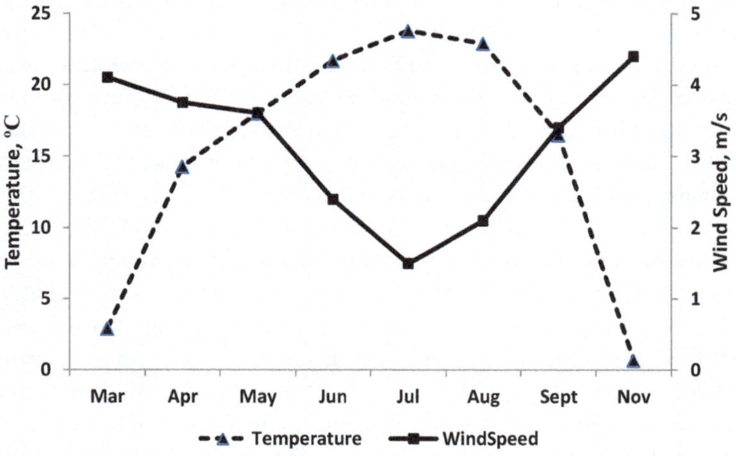

Fig. 18.1 Fluctuations of monthly average temperature (°C) and wind speed (m/s) in the study area during March–November (Blank 2020—License Number—5477040274237)

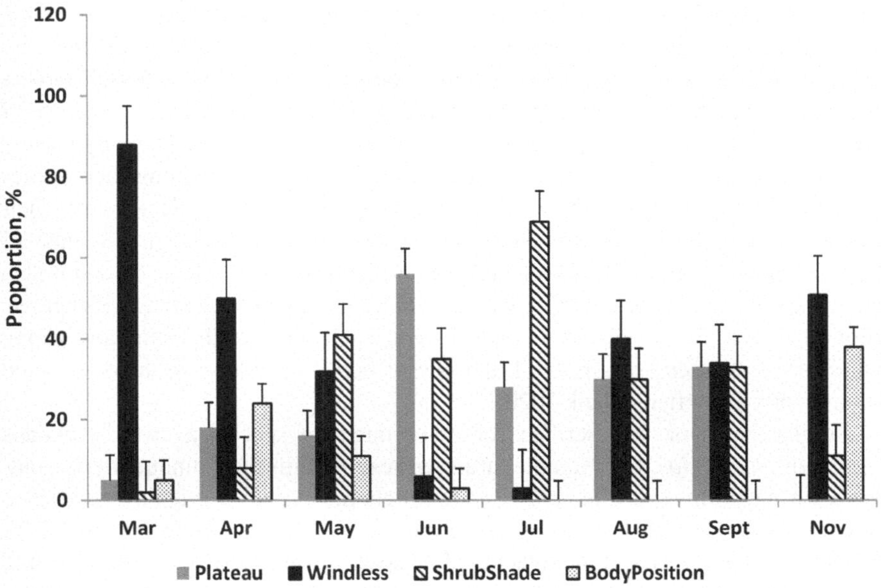

Fig. 18.2 Proportions of using elevated points (Plateau), windless depressions (Windless), shade of shrubs/trees (ShrubShade), and body orientation against wind (BodyPosition) by goitered gazelles over months (Blank 2020—License Number—5477040274237)

preference to use deep riverbeds during November is due to the return of the strong cold winds (Figs. 18.1 and 18.2). In my study area, the spring months (March through May) are windier than summer months (June and July) (Fig. 18.1), so it is perfectly logical that the goitered gazelles would seek out windless places in spring more frequently than in summer. The deep depressions protect the gazelles from the winds, but in the open areas, they will also turn their hindquarters toward the wind, which is seen mostly in April and May, as well as in November and rarely during the rest of the year (Fig. 18.2) (Blank 2020).

Orienting the body along in the direction of the wind is seen significantly less often compared to staying in windless places, and this disparity is maximal in March (95%, n = 56) and less in November, April, and May (42%, 33%, and 26%, n = 33, 55, and 39, respectively). Why did goitered gazelles prefer to use deep depressions significantly more often in March than the body orientation toward the wind? The winds are very strong in March, and fur conductivity rises with higher-force winds (Lentz and Hart 1960), while the goitered gazelle coat of hair is only weakly resistant, providing low effectivity against strong winds (Zhevnerov et al. 1983); therefore, windless places like deep dry riverbeds are much more effective and preferable as a shelter. In November, the winds are also quite strong and lower ambient temperatures are common compared to March, but at this time, goitered gazelles use body orientation much more frequently, demonstrating the yearly maximum of this behavior. November is also the rutting season for goitered gazelles, when males establish their individual territories mostly in the open plain areas, since highly rugged terrain, with its closed characteristics, is completely inappropriate for rutting activities. During this period, rutting males cannot escape the winds inside dry riverbeds even for short time, because they need to protect their territorial borders constantly and chase females that pass through; all these actions are necessary or territorial males would risk losing their territory to another male. The only protection from cold winds for territorial males in this situation is orienting their bodies with their hindquarters into the wind. Therefore, goitered gazelles demonstrate their maximal rate of this behavior in November. Another, but smaller, high frequency of body orienting along winds is seen in April during the pseudo or false rut and is also at the expense of territorial males, who are forced to stay in open areas for their territorial actions. Since only some males become territorial at this time and strong winds are less often seen in April, the rate of body orientation behavior is noted significantly less often (Blank 2020).

Similar behavior is described for other ungulate species as well. Aoudads (*Ammotragus lervia*), for example, are seen resting with their hindquarters windward during storms (Schaller 1977), and pronghorns (*Antilocapra americana*) also orient their bodies with their rumps toward the winds during winters (Bruns 1977). However, in most cases, numerous ungulate species use landscape features to find windless sites for protection against strong winds. Siberian roe deer (*Capreolus pygargus*), saiga antelopes (*Saiga tatarica*), and kulans (*Equus hemionus*) are known to search for protection from cold winds within thickets of large shrubs, behind and between sand dunes, inside deep wadis and ravines and other lowlands. During strong sandstorms and snowstorms, ungulate species of open environments that usually avoid forests and shrub thickets, enter inside of coniferous forests (Mongolian gazelles *Procapra gutturosa*), and tugai forests along large rivers and reed thickets around lakes and riverbanks (kulans and saiga antelopes) (Sludskiy 1963).

In addition to using the windier parts of their topography, goitered gazelles utilize any shady place they can find (banks of dry river beds, hills and cliffs), as well as tall shrubs and trees for resting (Fig. 18.3) during the hottest midday hours in May through September, with a peak in these cases in July (Fig. 18.2). Besides, adult females with their fawns prefer to stay in the shade of shrubs and trees: 73% cases in females and only 27% in adult males, n = 67, while they use upland areas less often: 24% cases in females vs 76% cases in adult males, n = 89). Adult males, in contrast, demonstrate their preference for elevated windy sites and stay in the shade of vegetation less frequently and subadults use the shade of vegetation the least (Fig. 18.4). The avoidance to using uplands in adult females and especially those with fawns, could be related more to avoidance of predation than comfort behavior. Within the first weeks after birth, when fawns are especially vulnerable to predation, mothers and especially their fawns have to be undetectable to predators, as fawns hide alone for long periods of time and invisibility is their only protection from predation. Fawns lying in high locations would be more conspicuous and easier for predators to detect than those resting between shrubs in the lower areas of a wadi. Since mothers have to stay close to the lower area to monitor their hiding fawns and protect them from small carnivores, it is unfeasible for the mothers to move higher up to rest in upland areas (Blank et al. 2015).

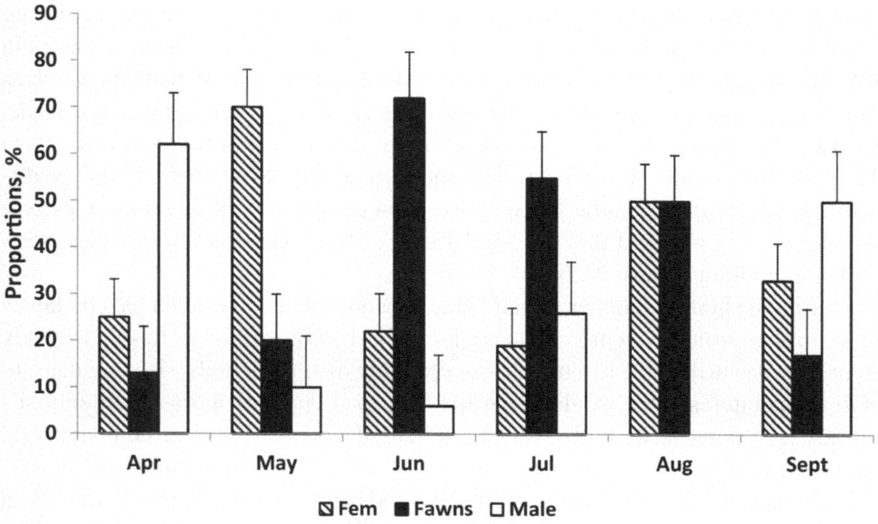

Fig. 18.3 Proportions of observed cases of staying in the shade by females, males, and fawns over months (Blank 2020—License Number—5477040274237)

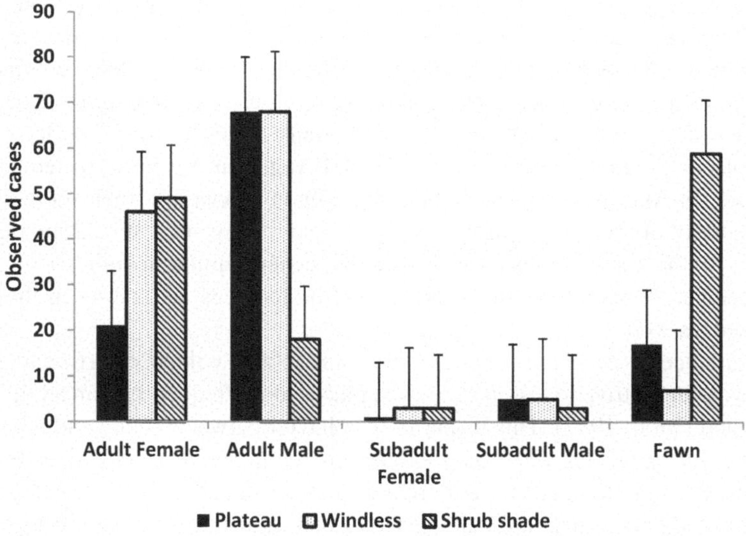

Fig. 18.4 Observed cases of using elevated points (Plateau), windless depressions (Windless) and shade of shrubs and trees (Shrub shade) by goitered gazelles of different sex and age (Blank 2020—License Number—5477040274237)

Adult females use the shade from trees and shrubs most frequently in May, less in August and rarely during the rest of the year. In contrast, adult males prefer shady places in April, a bit less often in September and rarely in other months. Moreover, during the hottest summer months, fawns show a higher precedence for shade from shrubs and trees compared to both male and female adults. Most often, fawns are found in shady locations in June, with a slow decline in the use of these places in July and August (Fig. 18.3). Trees are rarely found in the gazelle habitats, whereas shrubs are common vegetation, and thus, more readily available to provide shade. Fawns prefer shade from vegetation even more than their mothers and since most desert shrubs rise low above the ground and provide very little shade, it may be difficult for adults to hide in the shade of low-growing vegetation; therefore, they prefer larger shady places if they are available or at least open and more windy areas for better cooling (Blank 2020).

During the hottest summer period (June through August), goitered gazelle fawns strive to shady sites even more than adults. This objective of fawns to stay in shady areas may be attributable to a higher susceptibility of smaller body-size ungulates to high temperatures. This would mean the heat load depends more on an animal's surface area than on its body weight, and smaller animals have a comparatively larger body surface area than larger species (Heleski and Murtazashvili 2010). As a consequence, small body-sized animals cool and heat faster than larger ones (Peters 1983). During the first weeks of a goitered gazelle fawns' life in May, there was a low preference to be in shady locations compared to adult females and males. May is a relatively cool month, with hot day relatively rare. Most importantly, a fawns' resting site during this time is almost completely dependent on their mother's anti-predator strategy and typically do not choose their own hiding places (Blank 1985), but as June becomes hotter, fawns are much more independent in choosing their own hiding sites and demonstrate a higher preference for shady places compared to conspecifics, because of being more sensitive to ambient temperatures. Due to the rapid growth of body size (Kingswood and Blank 1996), a fawns' preference for shady places gradually decreases in July and August in response to temperature decreases. In August, when the hiding phase ends, fawns join their mothers who regain control of their offspring's movements and resting sites, synchronizing their activities so that the preference for shady sites became similar in both (Blank 2017). In September, temperatures drop further and preferences for resting in the shade decreased, as well.

In hartebeest (*Alcelaphus buselaphus*) and black wildebeest (*Connochaetes gnou*), yearlings strive to rest in shadowed places more frequently than adults (Ben-Shahar and Fairall 1987). This disparity in behavior between young and adult individuals is likely greater when the more difference in age and body sizes between them and young individuals have more divergence from adults than yearlings.

In my study area, there are many dry riverbeds, most of which are shallow with soft sandy soil which is attractive for gazelles for resting all year round. Therefore, in general, all goitered gazelles irrespective of their sex or age, more often rest inside the dry wadis than in stony areas outside of them. They show a higher preference for deep, dry river beds in the colder months of spring (March) and winter (November and December) (82–93% cases, n = 60 and 95, respectively) than during hot months (April through July), when the preference fluctuated between 51% and 73% (n = 33 and 96). Adult gazelles also rest in wadis most frequently (70% cases, n = 141 in females and 71% cases, n = 118 in males); subadults do it a bit more rarely (69% cases, n = 13 in females and 67% cases, n = 18 in males); and fawns resting in wadis is noted least frequently (64% cases, n = 55) (Blank 2020).

Domestic cows favor softer bed sites, where they stay longer than in harder bed sites (Chaplin et al. 2000; Manninen et al. 2002; Tucker and Weary 2004; Rushen et al. 2007). In the Ili Hollow, most of the ground is covered by gravel and stony deserts and only the numerous dry riverbeds that run in different directions through this undulating plain, provide goitered gazelles with soft sandy soils, which gazelles scrap with their front legs additionally before lying down for making bed sites especially soft. In addition, this soft, sandy soil was discovered to be the best material for maintaining the health and cleanliness of ungulates, protecting them from lameness, claw diseases, hock lesions, enhancing hoof health (Cook 2003; Espejo et al. 2006; Norring et al. 2008). Strong and healthy legs and hooves are incredibly important for goitered gazelles, so they are able to survive in the wild with fast running escapes from predators (Blank 2018).

In conclusion, goitered gazelles are well adapted to their desert habitat, using landscape features with various microclimates and preferring to hide inside deep depressions to avoid strong winter winds, and going to elevated windy sites or locations with shade from landscape features or tall vegetation during hot summers. Orienting the rump toward the winds is also used by goitered gazelles, though quite rarely and only during rutting period (Blank 2020).

18.2 Bedding Behavior

18.2.1 General Remarks

The bedding is closely related to resting behavior which is a very important part of maintenance behavior. Comfortable resting when lying down promotes a healthy body and legs, which are especially important for ungulates of open environments for running escaping predators. Therefore, the selection of comfortable bedding sites is a behavior of high priority (Munksgaard et al. 2005). Animals that are deprived of good, beneficial rest demonstrate unhealthy behaviors, such as increased stress level, and unhealthy physical problems, such as claw health (Singh et al. 1994; Fisher et al. 2002). The choice of bed-site placement is also crucial for successful rest: the bedding location must provide protection from severe weather (strong winds, heavy rains, strong insolation) to maintain essential thermoregulation, and most importantly, to provide security from unexpected predator attacks (Smith et al. 1986; Ockenfels and Brooks 1994). These bed-site requirements are especially important for fawns that still have imperfect thermoregulation and cannot effectively cope with extreme weather conditions. In addition, fawns are much more vulnerable to predation and their resting location needs to provide concealment, particularly to hiding species (Gerlach and Vaughan 1991; Alldredge et al. 1991; Olson et al. 2005; Bongi et al. 2008; Moorter et al. 2009; Lehman et al. 2009).

It is necessary for the animal to not only choose the right bed-site for resting, but also to make additional preparations before lying down. A number of mountain-dwelling species do not scratch solid ground, but readily scratch at soft soils (Schaller 1977). Therefore, the softness of the ground at a resting site is particularly important, which animals increase through pawing and ripping at the hard surface to loosen it. Sandy soils are in great demands not only because of their softness, but also because the sandy soil helps ungulates to keep their legs and hooves in a healthy state and protect them from lameness (Cook 2003; Espejo et al. 2006). Bed-sites that are especially comfortable may be used many times and many ungulates clean bed-sites from gravel and stone debris, tree branches, and other solid objects. Many ungulate species clean resting sites from snow cover in snowy winters, to avoid resting above the snow cover due to the imperfect insulation characteristics of their hair coat. Though pawing of bedding places is a common behavior of ungulates, specific studies of this pattern are quite limited, chiefly in wild environments, despite an abundance of fragmentary information in published papers (Kitchen 1974; Alldredge et al. 1991; Bongi et al. 2008; Lehman et al. 2009; Kjellander et al. 2012). Therefore, in this chapter, I consider the bedding behavior of goitered gazelles, which also has not had special research. Descriptions of the recorded displays (ethogram) of bedding behavior in goitered gazelles within my study period are presented in Table 18.1.

18.2.2 Bedding Behavior in Goitered Gazelles

For rest and protection from the weather, goitered gazelles arrange bedding places that are small oval, egg-shaped depressions in the ground (on average 60 × 30 cm in size). The beds of gazelles are usually located 5 to 30 m from each other. Antelopes that walk together within a small group do not lie down all together, but 100 to 300 m away from each other. During the day, animals rest, as a rule, at the feeding site with a good view of the surrounding terrain. In bad weather, especially with strong winds and cold, they bed down in bushes and ridge sands. The most convenient beds are used repeatedly and are deepened to 20 cm. For bed-sites, animals choose areas without vegetation, which are slightly loosened with their front hooves and there can be several beds under one favorite bush, which gazelles use serially, as the shadows move. In winter, gazelles rake the snow with their hooves and arrange beds in depressions or on the leeward side of bushes and hills. The location of the beds is determined by the season of the year and a number of microclimate factors. Gazelles can use convenient and safe places for resting repeatedly (Sokolov 1959; Zhevnerov et al. 1983).

According to Mambetzhumaev (1970), nighttime summer resting places of goitered gazelles can be found on the open plains with sagebrush growth (*Artemisia* sp.), regardless of the season of the year: from here, Mambetzhumaev thought, it would be easier for them to spot danger and escape. In contrast, Zhevnerov (1984) believes that the determining factor for gazelles in the choice of places for night rest in winter is the degree of protection from adverse weather, and with excellent hearing, goitered gazelles rely more on sound at night than sight. Thus, of the 83 nighttime beds he found in winter, 47 were located at the base of dunes, 9 were in

Table 18.1 Description of the recorded displays (ethogram) of bedding behavior in goitered gazelles within the study period (by Geist 1963; Schaller 1977; Walther 1984; Walther et al. 1983; Blank 2021)

Behavioral patterns	Description
Searching bed-site	The gazelle lowers its muzzle closely to the ground and moves from one to another place exploring the ground surface (Fig. 18.5).
Preparation of resting place: Scratching (pawing) the bed-site	Before lying down, a gazelle scratches the ground with a front leg several times, or alternately with both front legs, in a series of 2–3 acts and periodically lowers its muzzle to the area being prepared, probing if the place is ready for lying down (Fig. 18.6).
Turning above bed-site	Before lying down, a gazelle turns round in the place above the bed-site mainly to 180°, but sometimes from 90° to 360°.
Changing body side during resting	After lying on one side for some time, a gazelle stands periodically (ca. Per half an hour), sometimes standing still for a short period of time, followed by a roughly 180° turn, then lies down in the same bed-site on its other side, its head pointing in the opposite direction to its previous orientation; sometimes the gazelle makes turns from 90° to 360° but rarely makes several turns in the same place.
Stretching after resting in lying position Two types of stretches: a. Stretching in a length b. Stretching (arching) the entire body	A gazelle puts its front legs far forward from its hind legs, the spine is bent down slightly and one hind leg is raised and stretched far backward. The tail is erected upward to its extreme vertical position or even lies down on the back (Fig. 18.7). A gazelle keeps its hind legs not far from its front legs, the spine is bent down or alternatively bent upward, when the front legs are put closer to the hind legs and spine is bent upward; the tail is erected upward to its extreme vertical position; the muzzle top is pressed to the neck, which is bent upward; the ears are spread apart or sometimes backward at the tops (Fig. 18.8; Fig. 18.9).
Resting in a shady place	During hot periods, a gazelle tries to find shady places along high dry river beds near saxaul trees or large shrubs. Trying to get fully in the shade, or as much of its body as possible shaded (at least the head), the gazelle snuggles into the river bank as close as possible (Fig. 18.10).

depressions on the border with ridge sands, 19 were in thickets of bushes, and only 8 were on the open plain. Moreover, most of these beds were located on the leeward side of the hills. During frosts with strong winds and snowstorms, bumpy sands with thickets of shrubs become the main place of shelter for gazelles. At the same time, when gazelles feel the approach of a blizzard in advance, they will go into hiding among shrub thickets or deep dry river beds 4 to 8 h before it begins (Mambetzhumaev 1970).

In winter, the beds are arranged near the ridges of dunes on the leeward side or under the protection of bushes. Gazelles lay down and rest on windless days, but with strong winds and snowfalls in their places of shelter, they can be seen standing in close proximity to each other. For example, on January 23, 1972, with an air temperature of—13 °C and a wind force of 16 m/s, most of the gazelles encountered stood in groups of three to five individuals on the leeward side of sand ridges with overhanging bushes. In the shelter areas, the wind force at a height of 70 cm did not exceed 2 m/s. The heads and rumps of many gazelles were covered with frozen

Fig. 18.5 Searching bed-site: Goitered gazelle explore the ground surface with muzzle lowered downward. Issyk-Kul goitered gazelle nursery for reacclimatizing. 26.06.2022. Photo by D. Blank

Fig. 18.6 Bedding in goitered gazelles: A female scratches the bed-site by her front leg. Issyk-Kul goitered gazelle nursery for reacclimatizing. 25.11.2022. Photo by D. Blank

Fig. 18.7 Stretching in length in goitered gazelles: After rest, goitered gazelle stretches his body, when the spine is bent down slightly and one hind leg is raised and stretched far backward. Issyk-Kul goitered gazelle nursery for reacclimatizing. 26.06.2022. Photo by D. Blank

Fig. 18.8 Stretching with an upward bend, when the front legs are put closer to the hind legs and spine is bent upward; the tail is erected up to vertical position. Drawing by U. Kenzhetaev with kind permission

Fig. 18.9 Arching the entire body: a goitered gazelle bends its spine slightly, a muzzle top pressed to the neck and tail raise in vertical position. Issyk-Kul goitered gazelle nursery for reacclimatizing. 25.11.2022. Photo by D. Blank

Fig. 18.10 During hot daytime goitered gazelles prefer to rest in the shade. Issyk-Kul goitered gazelle nursery for reacclimatizing. 26.06.2022. Photo by D. Blank

snow. Some of them, mostly youngsters, stood huddled with their necks retracted and their backs hunched. When confronted with the threat of danger, the gazelles left their hiding places reluctantly, and instead of going out onto the plain, they went deep into the ridge sands. In winter, the animals spent their nights in most cases in permanent places. The beds were often arranged in twos or threes next to each other. From "repeated use," the beds were deepened into the slope of the dune by up to 80 cm and had a diameter of about 1.2 to 1.3 m, resembling a niche. Often the links between two neighboring niches were destroyed, and the animals found themselves together on the same bed. Two young males were observed for 8 days, and they always used the same one niche. The animals came to their nighttime places to sleep at dusk, and before lying down, they loosened up the soil again for a few seconds. Of the gazelles that fell from exhaustion that winter, 80% were found in or near these shelters. In the northwestern part of the fixed sands, 6 niches with dead gazelles were found on a 200-meter strip. In the average climatic conditions of low-snow winters, such beds were not observed in gazelles. In ordinary low-snow winters, such beds were not observed. Of the 83-night beds we discovered in the winters of 1970 through 1975 on the island of Barsakelmes, 47 (56.5%) were located at the base of dunes, 9 (10.8%) were in depressions on the border with the ridge sands, 19 (22.9%) were in the thickets of shrubs, and only 8 (9.8%) were on the open plain but even then near the hills (Zhevnerov et al. 1983).

During summer days, goitered gazelles prefer to lie with at least their heads in the shade of even just a small bush. In pistachio plantations, as well as in deserts where single trees (*Ulmus* spp.) grow, gazelles lie down in the shade of trees during the day. The beds, which are used many times, have the appearance of oval holes of up to 70 to 90 cm long. At the edge of the hole, there is almost always a pile of goitered gazelle droppings. The beds, as a rule, are located singly, a few dozen or more meters from each other. At night, gazelles willingly lie on smooth takyrs, while during the day, if possible, they lie under steep cliffs; sometimes, from frequent use, small niches form in the cliff wall (Heptner et al. 1961).

During summer days on Barsakelmes Island, gazelles lie down where it is cooler: on the seashore, desirable plump salt marshes, as well as in the shade of shrubs. Sometimes several beds are visible around one bush, arranged in a circle, so the gazelle can follow the shadow as it moves. In summer, in the depths of the fixed sands, salt marshes and open inter-dune spaces, well blown by the wind, goitered gazelles are found resting, mainly early in the morning. At this time, their beds are arranged on the open slopes of the dunes, where the soil is well cooled overnight. As the air temperature increases, the animals move to the borders of the sands, to places blown by the wind (usually to the first pile of dunes from the sea). During the warm nights of the year, the gazelles choose to rest in the open flat areas and on hills and other elevations. Less often, gazelles stay overnight in the ridge sands, arranging their beds on the crests on the sandy ridges. Sometimes they lie down at the very edge of shrub thickets or go deep into them, settling down in clearings and salt marshes. In winter, they prefer to stay in places protected from the wind, regardless of the degree of visibility of the terrain. The vast majority of beds are located in ridge sands, as well as in depressions and depressions on the border of thickets of shrubs and open spaces. In the ridge sands, the beds are arranged at the base of sandy mounds with overhanging bushes. Gazelles lie sideways or backwards to the wind in winter, and head to the wind in summer (Zhevnerov 1984).

Typically, goitered gazelles rest in recumbent position and can be found primarily within dry river beds all year round. Mostly adult gazelles favor staying inside of dry river beds (more than 70% of cases, n = 259), without a great disparity between sexes. Their attraction to sandy wadis maybe explained partly by searching for shelter from strong winds, but such shelter goitered gazelles can found mainly in the deep river beds, but they prefer to rest in wadis not only during the cold and windy seasons, but also during hot summers. In addition, the gazelles will rest in riverbeds of various depths, including the very shallow ones. All this suggests that the most attractive thing in the dry riverbeds for gazelles is the soft sandy soil, on which they have an especially comfortable setting to rest (Blank 2021). Observations of domestic cows under experimental conditions confirm this idea that animals always prefer the softest materials for long periods of reclining (Manninen et al. 2002; Rushen et al. 2007). The longer a resting place is used, the more the sand becomes compacted, losing its softness and resulting in it being used less and less often (Drissler et al. 2005). Among other things, a sandy bed layer is good for maintaining the health and cleanliness of hooves and legs, which are extremely important as their main protection for escaping running from predators (Norring et al. 2008).

Since different spots in a wadi may be more attractive than others, the more dominant gazelles will expel subordinates from the more comfortable beds in order to lie down themselves (Zhevnerov 1984; Blank 2019). This displacement behavior of dominants over subordinates for the most attractive place for bedding down is very typical and frequently seen in many ungulate species, such as pronghorn (Kitchen 1974), Siberian ibex (*Capra sibirica*—Fedosenko and Blank 2001), and argali (*Ovis ammon*—Fedosenko and Blank 2005). The competition for better bed sites may be over more shady spots in hot summers to avoid overheating (Hass 1991) or warmer, sunnier or snowless places with better protection from strong winds during winters, when ungulates struggle against hypothermia (Sludskiy 1963; Mysterud and Østbye 1995).

Goitered gazelles can rest in the same bedding place multiple times (40% cases, n = 432), with adult females doing it less frequently (22%, n = 138) than adult males (54%, n = 203), while the frequency of fawns is in-between (37%, n = 54). Seeking the most suitable lying place, gazelles will search their immediate area with their muzzles close to the soil; this behavior is more common for adult males (95%, n = 100) than for adult females (47% cases, n = 81) and is very rarely seen in fawns (15%, n = 46). After locating a convenient place, the gazelles commonly scrape the ground with their front hooves before resting (one third of all observed cases). Adult gazelles prepare their bed-sites by scratching comparatively frequently (33% cases, n = 300) with no difference disparity between females and males. Fawns prepare their resting places by scraping the ground longer than conspecifics (7.7 s in average, n = 21; up to 20 successive movements by front legs), with adult females doing it for the shortest time (5.9 s in average, n = 461) and adult males in between (7.1 s in average, n = 697) making up to 53 scratching movements per session (Blank 2021).

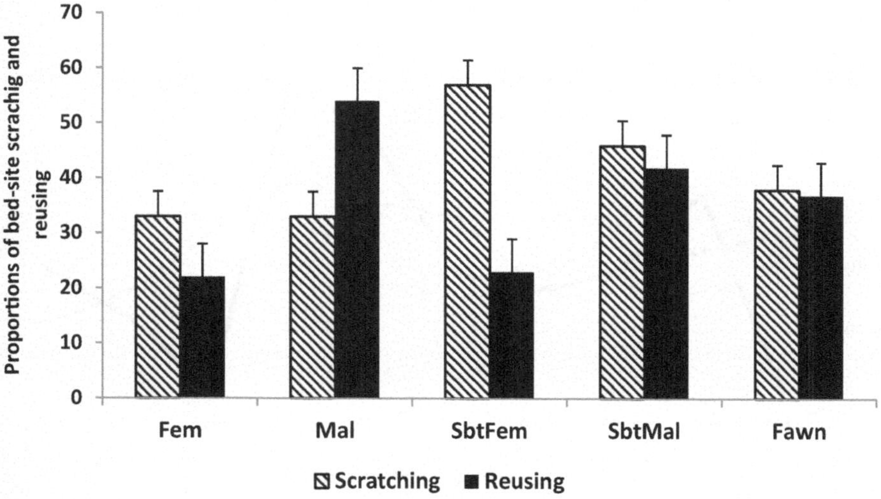

Fig. 18.11 Proportions (%) of bed-site scratching and rest place reusing compared to all observed cases in different age and sex classes of goitered gazelles. Legend: Scratching—proportion of scratching cases, reusing—proportion of bed-sites reusing; Fem—adult females, Mal—Adult males, SbtFem—subadult females, SbtMale—subadult males (by Blank 2021)

When scrapping the ground, goitered gazelles can use one front hoof several times consecutively (63%, n = 192) or more rarely both front hooves in turns (37%, n = 71), but they keep their muzzles lowered closely to the soil during the entire process of the bed-site preparation. In many cases, adult males do not scrap previously used resting places, preparing only new sites. In contrast, adult females and fawns scrap both old and new resting places (Fig. 18.11). There are also seasonal fluctuations in the rate of ground scratching and reuse of bed sites. Ground scratching declines from March through June, and in July through September, the rate starts to increase to its yearly maximal level and then drops to its yearly minimum in December (Fig. 18.12). Bed-site reuse dynamics fluctuate in a similar way, reaching its maximal level in September and dropping to its minimum in June and October (Blank 2021).

During the process of lowering to a lying position, gazelles commonly spin around in place from 180° (69% cases, n = 80) up to a complete circle (360°) (19%), but rarely (12%, n = 80) make 2 to 3 circles successively. Spinning before lying down is more common for adult males (52%, n = 73), less often for females (43%, n = 63), and least frequently for fawns (24%, n = 41). The purpose of this behavior is not clear, but it gives the impression that the goitered gazelles are seeking the most convenient placement within the bed site. Among ungulates, this bedding behavior feature is described only for musk deer (*Moschus moschiferus*—Sokolov et al. 1981). However, canids, especially dogs and wolves, display spinning in the place before lowering to a lying position quite frequently. It is assumed that this

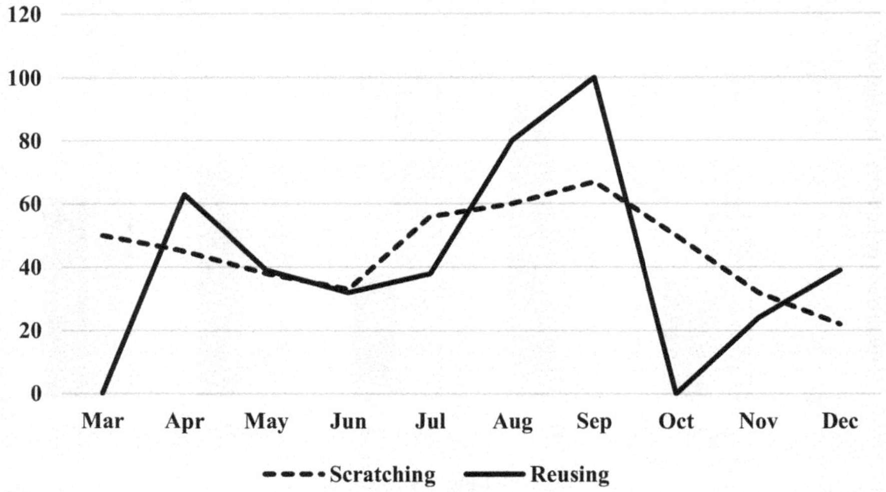

Fig. 18.12 Proportions (%) of bed-site scratching and rest place reusing compared to all observed cases over months. Legend: Scratching—proportion of scratching cases, reusing—proportion of bed-sites reusing (by Blank 2021)

behavior is a way that animals flatten the vegetation around their resting place to make it more convenient (Davis 2012).

Many cervid and bovid species demonstrated this same routine behavior while preparing a bed for resting, excepting of spin pattern, which is rarely mentioned. Some ungulates display the bed site preparation behavior sporadically and not everywhere and only in some seasons, while others do it willingly, often and all year round (Schaller 1977). Mountain-dwelling bovids can establish resting platforms on slopes with steep grades pawing the dense soil with their front hooves repeatedly or commonly they remove debris from bed sites (more than ten scraping movements per session in bharal *Pseudois nayaur*) (Geist 1971; Schaller 1977; Smith et al. 1986). The musk ox (*Ovibos moschatus*—Schaefer and Messier 1995) and reindeer (*Rangifer tarandus*—Kojola 1989) typically rest lying down directly on the snow cover, because of the excellent thermal insulation of their hair coat; they rarely clean their bed sites of snow, but commonly scratch through the snow cover when foraging. Red deer (*Cervus elaphus*—Chen et al. 1999), moose (*Alces alces*—Geist 1963), pronghorns (*Antilocapra americana*—Kitchen 1974), and saiga antelopes (*Saiga tatarica*—Sludskiy 1963; Pohle 1974) also rarely clean their resting places of the snow. In contrast, the Asiatic ibex (*Capra sibirica*), Bezoar ibex (*C. aegagrus*), markhor (*C. falconeri*), and musk deer (*Moschus moschiferus*—Sokolov et al. 1981), clean snow off their resting places and will not lie down on snow covered beds, because the bare earth is a better means of sustaining a body temperature balance (Schaller 1977; Sokolov et al. 1981; Sargeant et al. 1994).

Goitered gazelles avoid resting on snowy ground, because their hair coat is not adapted to cold and windy weather, making it poorly equipped to lie above the snow; therefore, they always clean their bed sites of snow or select resting places on bare, snowless ground (Sludskiy 1963). During the very snowy winter of 1971/1972, Zhevnerov (1984) checked 30 gazelle bed sites on Barsakelmes Island, and all of them were cleaned of snow. These observations led to the logical conclusion that in my study population in the Ili Hollow, the goitered gazelles also would scratch at their bed sites more often in winter than summer due to the presence of snow. However, my observations showed the opposite behavior; they scrapped their resting places more frequently in summer than in winter. This discrepancy is most likely due the differences in study areas, since the snow in my study area only stays for a short time and quickly melts or blows away by strong winds, and there are always many areas devoid of snow and suitable for resting without scratching off snow (Blank 1990). According to my data, goitered gazelles scratch their bed site places more likely to enhance their comfort level, particularly during warm months (Blank 2021).

During long-time midday rests, goitered gazelles do not lie on one side for very long time, and periodically stand up and stay in a standing position for up to several minutes (10% of cases, n = 39 in males and 32%, n = 28 in females). After that, they change their body orientation, turning for 90 to 180° before laying down again, only on the other side of the body. Changing body sides while lying down is a compulsory procedure for all resting gazelles, but this happens at different intervals. In some cases, they change positions after only 1 to 5 min of resting, while in other cases, it happens every hour, though rarely longer: one female was observed to stay in a lying in a recumbent position on the same side for 3 h and 25 min without getting up, but most often, the interval for goitered gazelles to change body side lasts about a half an hour on average (28 to 30 min, n = 863) for all age and sex classes. This is associated with the formation of pressure sores if an animal lies in the same position for a long time (Blank 2021). This phenomenon is widely known from human medical care, as well as considered for domestic ungulates, but periodically changing body sides during long resting periods has not been discussed for wild ungulates in nature (Bailey et al. 1983; Kierney et al. 1998; Caron and Mehler 2009; Cushing and Phillips 2013).

During investigations of the goitered gazelle population on Barsakelmes Island, Zhevnerov (1984) observed this same phenomenon. The animal does not lie motionless on its bed for long. After 40 to 50 min, it rises and stands for 1 to 2 min or moves from one place to another, and then having loosened the ground of the bed, it lies down again. This also helps to protect it from overheating, since the air temperature at the height of a standing animal is usually several degrees lower than on the surface of the ground. On July 7, 1970, at 11 o'clock on Barsakelmes Island, the difference in temperature at different locations was 7 °C on the plump salt marsh, and 18 °C on the sand by the sea. In the heat, gazelles like to lie on their backs with their light belly facing the sun, and in the cold, they are usually "curled up in a ball" (Zhevnerov 1984). I also observed this phenomenon in the Issyk-Kul Breeding Center for gazelle reacclimatization (Blank 2021, unpublished report). In the hot summer, goitered gazelles laid in a more open posture with their legs stretched out to the side and removed from under the body (Fig. 18.13), while in cold winter, the goitered gazelles preferred to take the shape of a ball (Fig. 18.14). In addition,

Fig. 18.13 Bedding postures: During summer, goitered gazelles lie in more open posture, when legs removed from under the body. Issyk-Kul goitered gazelle nursery for reacclimatizing. 26.06.2022. Photo by D. Blank

Fig. 18.14 Bedding posture in goitered gazelles: During winter, the gazelles prefer to take the shape of a ball. Issyk-Kul goitered gazelle nursery for reacclimatizing. 25.11.2022. Photo by D. Blank

Fig. 18.15 Bedding posture in goitered gazelles during summer: During sleeping, goitered gazelles like to put their head forward on the ground. Issyk-Kul goitered gazelle nursery for reacclimatizing. 26.06.2022. Photo by D. Blank

during both summer (Fig. 18.15) and winter (Fig. 18.16), they put their heads forward on the ground.

Adult gazelles get up from their resting places to change body sides more rarely compared to fawns. Zhevnerov (1984) observed on July 6, 1973, this behavior in the Barsakelmes Island population (western Kazakhstan): in a group of 2 females and 3 fawns during their two-hour resting period, the fawns got up per every 15 to 30 min, for a total of 5 times, while the females got up only twice. According to 18 observations, the average time interval between the fawns' movements when resting in an open areas was 27 min (Zhevnerov et al. 1983).

After the gazelles get up from resting, they are stretching for 1 to 8 s (3.7 ± 0.3 s on average, n = 27). Adult goitered gazelles perform the stretching behavior almost necessarily, especially after ending their rest (98% of cases, n = 182), skipping this procedure is rare. In contrast, fawns fulfill stretch significantly less frequently (26% of cases, n = 182). Two types of body stretching can be distinguished: in some cases, the gazelles stretch forward for the entire length of their body, with the front legs stretched far forward and the hind legs backwards (stretching body in length); in other cases, they arch their spines downward (rarely is the back bent up), with a highly raised head and the chin pressed chin to the neck, neck and tail are lifted, as well (stretching body in height). In most situations, goitered gazelles prefer stretching in length (79% to 83% in females and fawns, n = 64 and 24, respectively; and 59% to 62% in males, n = 77). Stretching in body length is especially preferable by adult females (83% of cases, n = 58) and their body arching is rarely observed

Fig. 18.16 Bedding postures: Even in winter, goitered gazelles like to have a sleep, putting the head forward on the ground. Issyk-Kul goitered gazelle nursery for reacclimatizing. 25.11.2022. Photo by D. Blank

(17%). Adult males perform the body arching significantly more frequently (38%, n = 77) and stretching in length less often (62%) than adult females; fawns prefer more stretching in body length (77%, n = 24) than arching (21%) (Blank 2021).

Other ungulate species, such as moose (Geist 1963), musk deer (Sokolov et al. 1981), and pronghorn (Kitchen 1974), also display stretching, usually immediately after recumbent rest. As a result of stretching, an exercise can be seen of alternating muscle tension with muscle relaxation, which supports muscle tonus and a readiness for the ungulate to engage in walking, running or other mobile activities (Kitchen 1974). Different ungulate species may perform stretching in various ways, but the main function of this behavior is to maintain tonus through increasing blood circulation and "waking up" the muscles to be ready for movement and escape, if necessary (Byers and Bekoff 1981; Biewener 2005). Similar behavior is performed by humans by using stretching gymnastics, particularly before heavy physical work or other activities requiring an intensive muscle load, such as running; such exercises keep high adaptability of the muscles to protect the body from injuries to elbows, shoulders, and knees (Anderson 2010).

Thus, bedding behavior of goitered gazelles is similar to other ungulate species, though turning around before lying down and periodically changing the side of the body they are lying on is not described for other hoofed mammals. However, I do not think that this is a behavior specific to goitered gazelles but it is rather the absence of research of these interesting behaviors in other ungulates.

References

Alldredge AW, Deblinger RD, Peterson J (1991) Birth and fawn site selection by pronghorns in a sagebrush-steppe community. J Wildl Manag 55(2):222–227

Anderson B (2010) Stretching. Shelter Publications, Inc, Bolinas, CA

Bailey JV, Badtram G, Farrow CS (1983) Unusual bone healing in a white-tailed deer. Can Vet J 24(11):335

Bartholomew GA (1987) Interspecific comparison as a tool for ecological physiologists. In: Feder ME, Bennett AF, Burggren WW, Huey RB (eds) New directions in ecological physiology. Cambridge University Press, Cambridge, pp 11–37

Ben-Shahar R, Fairall N (1987) Comparison of the diurnal activity pattern of blue wildebeest and red hartebeest. S Afr J Wildl Res 17(2):49–54

Biewener AA (2005) Biomechanical consequences of scaling. J Exp Biol 208(9):1665–1676

Blank DA (1985) Peculiarities of social and reproductive behaviour of *Gazella subgutturosa* in the Ili Valley. Zool Zhurnal (Moscow) 64:1059–1070. (in Russian, English abstract)

Blank DA (1990) Persian gazelle. In: Kovshar AF (ed) Rare animals of desert regions. Nauka Press of Kazakh SSR, Alma-Ata, pp 56–80. (in Russian)

Blank DA (2017) Antipredator tactics are largely maternally controlled in goitered gazelle, a hider ungulate. Behav Processes 136:28–35

Blank DA (2018) Escaping behavior in goitered gazelle. Behav Processes 147:38–47

Blank DA (2019) Female-female aggression in goitered gazelles: the desire for isolation. Behav Processes 164:186–192

Blank DA (2020) Using microclimate of arid landscape as a resource in goitered gazelle comfort behavior. J Arid Environ 180:104201

Blank DA (2021) Bedding behavior in goitered gazelles. Ethol Ecol Evol 33(2):170–183

Blank D, Li Y (2022) Antelope adaptations to counteract overheating and water deficit in arid environments. J Arid Land 14(10):1069–1085

Blank DA, Ruckstuhl K, Yang W (2015) Antipredator strategy of female goitered gazelles (*Gazella subgutturosa* Güld., 1780) with hiding fawn. Behav Processes 119:44–49

Bongi P, Ciuti S, Grignolio S, Del Frate M, Simi S, Gandelli D, Apollonio M (2008) Anti- predator behavior, space use and habitat selection in female roe deer during the fawning season in wolf area. J Zool (London) 276(3):242–251

Bruns EH (1977) Winter behavior of pronghorns in relation to habitat. J Wildl Manag 41(3):560–571

Byers JA, Bekoff M (1981) Social, spacing, and cooperative behavior of the collared peccary, *Tayassu tajacu*. J Mammal 62(4):767–785

Cain JW, Krausman PR, Rosenstock SS, Turner JC (2006) Mechanisms of thermoregulation and water balance in desert ungulates. Wildl Soc Bull 34(3):570–581

Caron JP, Mehler SJ (2009) Laparoscopic mesh incisional hernioplasty in five horses. Vet Surg 38(3):318–325

Chaplin SJ, Tierney G, Stockwell C, Logue DN, Kelly M (2000) An evaluation of mattresses and mats in two dairy units. Appl. Anim. Behav. Sci., 66:263–272

Chen H, Li F, Luo L, Wang H, Ma J, Li F (1999) Winter bed-site selection by red deer *Cervus elaphus xanthopygus* and roe deer *Capreolus capreolus bedfordi* in forests of northeastern China. Acta Theriol 44(2):195–206

Cook NB (2003) Prevalence of lameness among dairy cattle in Wisconsin as a function of housing type and stall surface. J Am Vet Med Assoc 223:1324–1328

Curtis SE (1983) Environmental Management in Animal Agriculture. Iowa State Univ. Press, Ames, IA

Cushing CA, Phillips LG (2013) Evidence-based medicine: pressure sores. Plast Reconstr Surg 132(6):1720–1732

Davis T (2012) Why dogs do that: a collection of curious canine behaviors. Willow Creek Press, Minocqua, WI

Drissler M, Gaworski M, Tucker CB, Weary DM (2005) Freestall maintenance: effects on lying behavior of dairy cattle. J Dairy Sci 88(7):2381–2387
Espejo LA, Endres MI, Salfer JA (2006) Prevalence of lameness in high-producing Holstein cows housed in freestall barns in Minnesota. J Dairy Sci 89:3052–3058
Fedosenko AK, Blank DA (2001) Capra sibirica. Mamm Species 675:1–13
Fedosenko AK, Blank DA (2005) Ovis ammon. Mamm Species 773:1–15
Feldhamer GA, Drickamer LC, Vessey SH, Merritt JF (1999) Mammalogy: adaptation, diversity, and ecology. WCB McGraw-Hill, Boston, Massachusetts, USA
Fisher DO, Blomberg SP, Owens IPF (2002) Convergent maternal care strategies in ungulates and macropods. Evolution 56(1):167–176
Fuller A, Kamerman PR, Maloney SK, Matthee A, Mitchell G, Mitchell D (2005) A year in the thermal life of a free-ranging herd of springbok Antidorcas marsupialis. J Exp Biol 208:2855–2864
Geist V (1963) On the behaviour of the North American moose (*Alces alces andersoni*, Peterson 1950) in British Columbia. Behaviour 20(3–4):377–416
Geist V (1971) Mountain sheep. University of Chicago Press, Chicago
Gerlach TP, Vaughan MR (1991) Mule deer fawn bed site selection on the Pinyon Canyon Maneuver site, Colorado. Southwest Nat 36:255–258
Hass CC (1991) Social status in female bighorn sheep (*Ovis canadensis*): expression, development and reproductive correlates. J Zool (London) 225:509–523
Heleski CR, Murtazashvili I (2010) Daytime shelter-seeking behavior in domestic horses. J Vet Behav 5(5):276–282
Heptner G, Nasimovich AA, Bannikov AG (1961) Mammals of Soviet Union, Vol.1. Artiodactyla and Perissodactyla. Vysshaya Shkola Publishing House, Moscow. (in Russian)
Holcomb KE (2017) Is shade for horses a comfort resource or a minimum requirement? J Anim Sci (Sofia) 95:4206–4212
Huey RB (1991) Physiological consequences of habitat selection. Am Nat 137:91–112
Kierney PC, Engrav LH, Isik FF, Esselman PC, Cardenas DD, Rand RP (1998) Results of 268 pressure sores in 158 patients managed jointly by plastic surgery and rehabilitation medicine. Plast Reconstr Surg 102(3):765–772
Kingswood SC, Blank DA (1996) Gazella subgutturosa. Mamm Species 518:1–10
Kitchen DW (1974) Social behavior and ecology of the pronghorn. Wildl Monogr 38:3–96
Kjellander P, Startholm I, Bergvall UA, Jarnemo A (2012) Habitat use, bed-site selection and mortality rate in neonate fallow deer *Dama dama*. Wildl Biol 18(3):280–291
Kojola I (1989) Mother's dominance status and differential investment in reindeer calves. Anim Behav 38(2):177–185
Lehman CP, Hartland JD, Keller BJ, Millspaugh JJ, Brundige GC (2009) Bed site selection of fawn pronghorn in Custer State Park, South Dacota. Prairie Nat 41:100–109
Lentz CP, Hart JS (1960) The effect of wind and moisture on heat loss through the fur of new-born caribou. Can J Zool 38:679–688
Mader T, Johnson L, Gaughan JB (2010) A comprehensive index for assessing environmental stress in animals. J Anim Sci 88:2153–2165
Mambetzhumaev AM (1970) Goitered gazelle. FAN Publishing House, Tashkent. (in Russian)
Manninen E, de Passiller AM, Rushen J, Norring M, Saloniemi H (2002) Preferences of dairy cows kept in unheated buildings for different kind of cubicle flooring. Appl Anim Behav Sci 75:281–292
Moorter BV, Gaillard JM, McLoughlin PD, Delorme D (2009) Maternal and individual effects in selection of bed sites and their consequences for fawns survival at different spatial scales. Oecologia 159:669–678
Munksgaard L, Jensen MB, Pedersen LJ, Hansen SW, Matthews L (2005) Quantifying behavioural priorities — effects of time constraints on behaviour of dairy cows, *Bos taurus*. Appl Anim Behav Sci 92(1–2):3–14
Mysterud A, Østbye E (1995) Bed-site selection by European roe deer (*Capreolus capreolus*) in southern Norway during winter. Can J Zool 73(5):924–932

Norring M, Manninen E, De Passillé AM, Rushen J, Munksgaard L, Saloniemi H (2008) Effects of sand and straw bedding on the lying behavior, cleanliness, and hoof and hock injuries of dairy cows. J Dairy Sci 91(2):570–576

Ockenfels RA, Brooks DE (1994) Summer diurnal bed sites of Coues white-tailed deer. J Wildl Manage 58:70–75

Olson KA, Fuller TK, Schaller GB, Lhagvasuren B, Odonkhuu D (2005) Reproduction, neonatal weights, and first-year survival of Mongolian gazelles (*Procapra gutturosa*). J Zool (London) 265:227–233

Peters RH (1983) The ecological implications of body size. Cambridge University Press, Cambridge

Pohle C (1974) Haltung und Zucht der saiga-Antelope (*Saiga tatarica*) im Tierpark Berlin. Zool Garten 61:263–266

Rushen J, Haley D, de Passiller AM (2007) Effect of softer flooring in tie stalls on resting behavior and leg injuries of lactating cows. J Dairy Sci 90:3647–3651

Sargeant GA, Eberhardt LE, Peek JM (1994) Thermoregulation by mule deer (*Odocoileus hemionus*) in arid rangelands of southcentral Washington. J Mammal 75:536–537

Schaefer JA, Messier F (1995) Winter foraging by muskoxen: a hierarchical approach to patch residence time and cratering behaviour. Oecologia 104(1):39–44

Schaller GB (1977) Mountain monarchs. Wild Sheep and Goats of the Himalaya. The University of Chicago Press, Chicago/London

Singh SS, Ward WR, Murray RD (1994) An angiographic evaluation of vascular changes in sole lesions in the hooves of cattle. Br Vet J 150(1):41–52

Sludskiy AA (1963) The mass death of wild ungulates in steppes and deserts of Europe and Asia. Proc Inst Zool Acad Sci KazSSR, Alma-Ata 20:5–88. (in Russian)

Smith HD, Oveson MC, Pritchett CL (1986) Characteristics of mule deer beds. Great Basin Nat 46:542–546

Sokolov II (1959) *Gazella subgutturosa* Guldenstaedt—Goitered gazelle. In: Pavlovskiy EN (ed) Fauna USSR, mammals: ungulates (Orders Perissodactyla and Artiodactyla). Publication of the Academy of Sciences USSR, Moscow and Leningrad, pp 308–331. (in Russian)

Sokolov VE, Prihodiko VI, Smirin VM (1981) Postures and distinctive movements in behavior of musk deer (*Moschus moschiferus*). Part 2. Zool Zhurnal (Moscow) 60(10):1553–1561. (in Russian)

Souza ED, Souza BB, Souza WH (2005) Determinação dos parâmetros fisiológicos e gradiente térmico de diferentes grupos genéticos de caprinos no Semi-Árido. Revista Ciência e Agrotecnologia Lavras 29:177–184

Tucker CB, Weary DM (2004) Bedding on geotextile mattresses: how much is needed to improve cow comfort? J Dairy Sci 87:2889–2895

Walther FR (1984) Communication and Expression in Hoofed Animals. Indiana University Press, Bloomington

Walther FR, Mungall EC, Grau GA (1983) Gazelles and their relatives. A study in territorial behavior. Noyes Publication, Park Ridge, N.J., USA, 239 pp

Zhevnerov VV (1984) Goitered gazelle of the Barsakelmes Island. Nauka Publishing House of Kazakh SSR, Alma-Ata. (in Russian)

Zhevnerov VV, Bekenov AB, Sludskiy AA (1983) Goitered gazelle—*Gazella subgutturosa* Güldenstaedt, 1780. In: Gvozdev EV, Kapitonov VI (eds) Mammals of Kazakhstan, Bovidae, vol 3, No. 3. Nauka Publishing House of Kazakh SSR, Alma-Ata, pp 11–54. (in Russian)

Main Causes of Mortality and Mass Death 19

19.1 Mortality Causes

19.1.1 Young Mortality

In many mammalian species, fertility is often greatly reduced as a result of resorption and abortions of embryos, which are the result of insufficient or incomplete nutrition of females during pregnancy because of unfavorable environmental conditions, such as during dzhuts. Some abortions can also be caused by infection or invasion of parasites. Resorption and abortion of embryos sometimes become a mass phenomenon during dzhuts (Sludskiy 1953). In years with favorable conditions for wintering gazelles, resorption is observed only in isolated cases. For example, in the Betpak-Dala Desert in 1955, out of 32 pregnant and females at once after giving birth, resorption was observed in only one yearling female pregnant for the first time. Of her two embryos, one was almost thoroughly resorbed, while the second embryo developed normally and weighed 1100 g (male). In general, only one of the 60 embryos examined was resorbed (1.66%) (Sludskiy 1956).

Infant mortality is high in gazelles. If an average of 1.8 fawns (100%) are born per 1 female (data for 3 years in the Betpak-Dala Desert), then by early autumn on average only 1.1 are left, or 61.1% of the spring fawns. In June and July, on Barsakelmes Island, 1.2 fawns were born per female in 1970, 0.88 per female in 1971, 1.6 per female in 1972, and 1.8 in 1973. In the summer, the mortality of fawns on this island was small at about 3%, but in winter, the mortality increased sharply from 4.3% to 28.4% in different years for an average of 14.8%. Consequently, the loss of young gazelles from birth to one-year of age can be 27.3% to 51.4%, or 37% on average (Zhevnerov 1984).

As can be seen in Table 19.1, the fertility indicators decreased monthly due to the death of fawns in the first months of life. By the end of September, about 40% of the fawns born alive had died. In Kazakhstan during favorable years, the mortality of young is low, with 70% to 80% of young surviving until winter. In the seasons

D. Blank, *The Behavioral Biology and Conservation of Goitered Gazelles (Gazella subgutturosa)*, https://doi.org/10.1007/978-3-031-83504-9_19

Table 19.1 Mortality of young in the Betpak-Dala Desert (adapted from Sludskiy 1956)

Age and months	Number of observed families	Total number of fawns	Average number of fawns per mother	Average proportion (%) of embryo and fawns	Mortality of young to month end, %
Embryo and newborns	32	59	1.8	100	–
May	10	16	1.6	88.9	11.1
June	40	60	1.5	83.3	16.7
July	87	108	1.2	66.6	33.4
August	67	110	1.6	88.9	11.1
September	30	33	1.2	61.1	38.9

Table 19.2 The goitered gazelle mortality of different sex and age in the Barsakelmes Island during winter (adapted from Zhevnerov 1984)

Winter season	Total found corpses	Including							
		Adult males		Adult females		Fawn males		Fawn females	
		Number	%	Number	%	Number	%	Number	%
1971/72	95	59	62.2	8	8.4	14	14.7	14	14.7
1972/73	9	5	55.6	–	–	3	33.3	1	11.1
1973/74	8	3	37.5	1	12.5	2	25.0	2	25.0
1974/75	32	14	43.8	1	3.1	10	31.2	7	21.9
1975/76	103	42	40.8	18	17.5	23	22.3	20	19.4
Average	–	22	43.4	7	12.8	10	23.7	8	20.1

following dzhut, the increase in the number of gazelles is less significant. For example, after the dzhut of 1950 to 1951 in the south of Kazakhstan, 203 gazelles were counted in August of 1951 in the Betpak-Dala Desert, of which only 100 individuals (or 54.2%) were young; among the 86 adult gazelles, 42.3% were adult females and only 3.5% were adult males (Sludskiy 1956).

On Barsakelmes Island, the mortality of fawns is about 20% in the first days of life (Table 19.2). Among the 80 newborns born during the period from 1971 through 1976, a total of 15 (18.7%) were found dead. The mortality of fawns in summer and autumn in favorable years does not exceed 3%. During 1971 through 1976, 597 goitered gazelles were seen in July, 279 of them were fawns (46.9%); in October 384 gazelles were found with 170 (44.2%) of them fawns. There is a close relationship between the condition of pastures and the reproductive capacity of gazelles. The smallest number of single females and the largest number of females with fawns were observed in the years with the best pasture conditions (1973–1975—Table 7.3) (Zhevnerov 1984).

Predators, especially foxes, are the main cause of newborn mortality (Sludskiy 1956; Gorelov 1972), with many young dying during the first several days after birth. This is evidenced by the number of remains of dead fawns near fox holes and at birthing sites. According to observations of Zhevnerov (1984), of the 80 newborn gazelles, 18.7% were found dead from fox predation. Of 12 fawns born to tame gazelles, three were destroyed by foxes on the third or fourth day of life (Zhevnerov 1984). In fact, the deaths of young animals during birth are somewhat higher than can be seen, since some of the bodies remain unaccounted for. Therefore, it is possible to assume that in the first days the mortality rate among newborns is more than 20%. In the future, the deaths of stronger fawns during summer in normal years are about 3%. A lack of drinking water may be also a serious factor in the mortality of young gazelles. After the onset of summer heat and the drying of ephemeral vegetation, young gazelles, left for one reason or another without a mother, suffer due to the inability to regulate their water balance. Often, at the same time, inexperienced grazers will eat food that is too dry and die from a lack of water (a semi-dry lump of undigested food can be found in their stomach), or finding a source of too salty water will drink and die from salt poisoning (traces of hemorrhage on the inner walls of the stomach is found) (Gorelov 1972). Apparently, the ability to choose only plants that contain enough water or determine the taste of the suitable water for drinking is not immediately developed in gazelles. The death of young gazelles from thirst and/or water poisoning occurs mainly in the years with early and dry summers (Gorelov 1972). Thus, goitered gazelle populations have increased mortality of young during their first months of life, with usually 40% of newborns dying by the end of September (Table 19.3).

In a captive herd of *G.s. marica*, the greater the weight at birth and sex of a fawn are good predictors of the fawn's survival over the first 90 days of life; with female fawns having a higher survival rate than male fawns. The weight of fawn-singletons is greater than the weight of twins, so in general, singletons have better chances of survival than each of the twins. Therefore, the weight of the fawns and offspring number are negatively interrelated with each other. This contradiction is compensated by favorable conditions of environments, when females give birth to more twins with minimal mortality in the spring (March–April and partly May) and more singles in other seasons of the year. The female pedigree and/or individual differences in postnatal care have limited effect on survival during the first year of life. Although pedigree can explain the variances in the survival among individual offspring, it is less important for first-year survival than during the first 90 days of life. Survival to the age of reproduction in females was predicted by a single trait: weight at birth (Riesch et al. 2013).

Table 19.3 Proportion of adult females to sub-adult females in November–December in the Barsakelmes Island population of goitered gazelles (adapted from Zhevnerov 1984)

			Portion sub-adults	
Year	Month	Total number of encountered females	Number	Proportion, %
1969	November	48	17	35.4
	December	89	32	35.9
1970	November	36	11	32.2
	December	54	21	37.1
Totally		227	81	35.6

19.1.2 Adult Gazelle Mortality

On the whole, annual mortality rates are 9% to 18% for females, 27% to 58% for males, and 3% to 50% for neonates. Mortality is generally lowest during summer and highest during winter (Kingswood and Blank 1996).

Let's consider the changes in the mortality rate of goitered gazelles on the example of population on the Barsakelmes Island. The goitered gazelles were released there in 1929. By the autumn of 1948, the population of this species reached a maximum (about 2 thousand individuals—Vasenko 1950). This period of time can be considered as the first phase of the goitered gazelle acclimatization without competitors and predators leading to a quick increase in the population number. In the harsh winter of 1948–1949, a high mortality was noted among the gazelles. In January alone, 500 carcasses were found. After heavy rains in early August, the death of animals due to necrobacillosis was observed. As a result of these natural disasters, by the end of 1949, only 60 gazelles remained, and the number of the gazelle population for the next 28 years did not approach its maximum of 1948. This period can be considered the second stage of acclimatization and also as a phase of depression (Zhevnerov 1977a, 1977b).

Since the establishment of the Barsakelmes Reserve in 1939, severe winters were ten times until 1970s, alternating with breaks from a year to 4 years. Of the ungulates that live on the island—kulan, saiga and gazelle, the gazelle is the morphologically least adapted to the harsh snowy winters. The critical depth and density of snow for the gazelle is much less than for the saiga, and gazelles stop feeding on flat pastures earlier than saiga and move into ridge sands, which is a much more limited area. Saiga antelopes, whose numbers remained at the level of 1 to 3 thousand for the last two decades (1950s through 1970s), also moved to the ridge sands as the snow depth increased, creating serious food competition for the gazelles (Zhevnerov 1984).

The conditions in which gazelles fall in harsh winters were clearly demonstrated by the winter of 1972–1973. The duration of the stable snow cover was 83 days. The average height of the snow cover was 12.5 cm. During the winter period, there were 39 days with strong winds (12 to 18 m/s) and blizzards. The average monthly air temperature for January–February was −14 °C, an absolute minimum of −29.7 °C. Under the influence of frequent strong winds, snow is rapidly compacted, and by the second week of January in the lowlands on the plain, the snow density reached 0.24 g/cm^3, and on the plateau 0.30 g/cm^3. Due to drought during the growing season, the vegetation of lowland pastures was suppressed, and the height of the plant life did not exceed 8 to 10 cm; therefore, the forage in the pasture was mostly inaccessible to ungulates. Saigas in mid-January, and gazelles much earlier, moved to the ridge sands, where it was easier to get to food. On lowland pastures, goitered gazelles were rare and mainly located in areas with an unevenly snow cover. A significant concentration of ungulates on the limited area of fixed sands (1369 ha) led to a large pasture load on this plant complex. During such winter, saxaul's annual twigs, especially seeds and galls, were eaten almost completely by both gazelles and saigas to a reachable height. Near the saxaul bushes, the gazelles lingered for 5 to

10 min, often falling into the snow up to their bellies. The same behavior was observed with *Ephedra* shrubs. On February 12 in the early 1970s, along a 5-kilometer route, fresh shoveling of snow was observed in 70% of all *Ephedra* bushes encountered. The role of plants containing alkaloids, such as *Anabasis aphylla* and *Eremosparton,* has significantly increased in the diet of the gazelle. In favorable winters, 37 species of plants are noted in the gazelle diet. However, in the winter of 1971–1972, the species number decreased by 56.8%, amounting to only 16 plant species (Zhevnerov 1984).

Of the unfavorable factors that adversely affect the state of gazelles in winter, in addition to the depth and density of the snow cover, are low air temperatures and strong winds. The complex of these three factors can be both the direct cause of the death of gazelles, and indirect cause by disrupting the rhythm of grazing animals. Frequent strong winds combined with low air temperatures force gazelles to leave open pastures for shelter. Under such conditions, along with a short period of day light, gazelles do not have time to satisfy their need for food leading to a severe energy depletion. The weight of a fallen adult male is in the range of 18 to 23 kg, which is more than 30% of a normal weight loss. The color of the muscle tissue is pale blue and fat deposits are completely absent. Some animals also have darkening in the lungs (Zhevnerov 1984).

Prolonged starvation apparently can cause problems with the thermoregulation of the body in some individuals, since cases of death of gazelles from hypothermia have been recorded. So, after severe frosts that lasted from January 27 to January 30 of 1972, when the air temperature dropped to −29 °C, two dead adult males were found lying in a sleeping animal position curled up in a ball. The first cases of death of gazelles were noted in the second decade (10–20 January) and a mass death of gazelles began in February. Most of those encountered gazelles that were still alive were noticeably exhausted: when fleeing, many had their tails lowered and their movements were sluggish. All the bodies of the gazelles that fell in January and February, with rare exceptions, were found in the ridge sands. On flat pastures, the dead animals began to appear in mid-March after the snow cover disappeared (Zhevnerov et al. 1983).

In the following spring, when counting fallen gazelles, 102 were found dead. Since the bodies were quickly dragged away by foxes, this number of dead does not fully characterize the overall winter loss. A comparison of the autumn population of 1971 and the spring of 1972 showed that the death rate among the gazelles was 68% of the population. The death rate of kulans and saigas was 2.5% and 30%, respectively. Thus, with the example of the winter of 1971–1972, it can be seen that among the ungulates of the Barsakelmes Island, the goitered gazelles are most susceptible to adverse factors in severe winters (Zhevnerov 1984).

The damage inflicted on the gazelle population in severe winters is determined not only by the actual death of gazelles. In females, after dzhut, embryo resorption and abortion are often noted, and as a result, the number of offspring for the year decreases and female barrenness increases. In the summer of 1972, the birth rate was 0.88 fawns per female, and 24.6% of the females were barren. The rapid recovery of the population affected by dzhut begins only in the second or third year after

Table 19.4 Dynamics of the goitered gazelle number in the Barsakelmes Island during 1970–1976 (adapted from Zhevnerov 1984)

Year	Season of survey	Gazelle number	Winter mortality, %	Population growth, %
1970	Autumn	180		
1971	Spring	170	5.5	–
	Autumn	250		47.0
1972	Spring	80	68.0	
	Autumn	110		37.5
1973	Spring	100	9.0	
	Autumn	158		58.0
1974	Spring	150	5.0	
	Autumn	250		66.6
1975	Spring	210	15.0	
	Autumn	364		70.3
1976	Spring	236	35.2	
	Autumn	330		41.3
Average		–	22.7	53.4

its end (Sludskiy 1963), and the increase in the number of gazelles on the island actually occurs between severe winters (Table 19.4). From Table 19.4, it can be seen that the increase in the number of goitered gazelles by autumn due to young animals was increased for 37.5% to 70.3% (average 53.4%), and mortality for the winter was on average 22.7%. The annual increase in the number of gazelles was on average 30.7% individuals. In favorable winters, mortality does not exceed 15%, and in snowy winters it can reach 35.2% to 68%. Consequently, under favorable conditions, goitered gazelles can restore their numbers to the level preceding the harsh winter during years. The frequency of snowy winters is on average every 4 years. In addition, further population growth is strongly hindered by predators and unfavorable feeding conditions. In dry years, combined with severe winters, the population number decline is especially sharp, since goitered gazelles are deprived of the opportunity to avoid unfavorable conditions by migrations or sufficiently wide movements, in which case, it can take at least 3 to 4 years to restore the population (Zhevnerov 1984).

In contrast to Persian goitered gazelles, Arabian sand gazelles (*G. subgutturosa marica*) die most often during severe summer droughts, as seen when about 300 skulls were collected during a 1-year period (March 2008—March 2009). Winters in Saudi Arabia are more favorable seasons, with warm and relatively rainy weather with more green food (Cunningham and Wacher 2009). Most skulls of the dead were males, and the natural mortality provided an imbalanced sex ratio skewed toward males (1.39:1), though the sex ratio of living gazelle population was skewed toward adult females (1:1.07). Young males in the age of 18 to 24 months had a mortality rate of 70.6%, and adult males of 18 months to 7 years had a mortality of 88.8%. Young females of 18 to 24 months had mortality rate of 31.1% and females older than 2 years old died at a rate 43.2% (Cunningham and Wronski 2011).

Causes of mortality in captive animals (number of deaths in parentheses) include stress or trauma from predation (54), capture operations (49), fence injuries (7), fighting (6), unspecified factors (74), or unknown causes (118—Griner 1983; Rietkerk et al. 1992). Digestive disorders (20), malnutrition (18), respiratory ailments (3), and urinary stasis induced by genital suckling (10) were also a cause of death, particularly in neonates (Griner 1983; Rietkerk et al. 1992). In a sample of 785 births, 13 were stillbirths (Griner 1983; Rietkerk et al. 1992). Mortality causing pathogens include *Carynebacterium pyogenes* (45), *Mycobacterium* sp. (5), and *Cryptosporidium* sp. in combination with *Escherichia coli* (1) (Rietkerk et al. 1992).

19.2 Dzhut (Mass Death Due to Starvation)

Recurrent deaths of ungulates and other animals from winter starvation and unfavorable meteorological factors are one of the specific features of the steppes, semi-deserts, and northern temperate deserts of Eurasia. This is so common and frequent a phenomenon that it has received a special name among many peoples: the Mongols call it dzud or tsagan-dzud (white dzud), Kalmyks call it zud, Kazakhs call it jute and the Russian version is dzhut. In the first extant written sources of the ancient Romans, Greeks, Slavs, and other European and Asian peoples, there are often indications of the mass death of domestic and wild ungulates in the steppes and deserts in the most severe winters, which indicates how widespread this phenomenon is and its importance to humans. Thus, the Roman poet Virgil (70–19 BC) described this phenomenon in the steppes of the Black Sea region: "…snow is falling all over the air; herds are dying. Large bodies of bulls are surrounded by snow and deer herds in dense snow get bogged down in new snow drifts, from which the tops of their horns barely protrude. Here hunters don't set dogs on them, they don't set nets, they don't put them to flight…but they are hit at close range with iron, while they are trying in vain to break through the standing mountains of snow with their breasts" (Sludskiy 1963).

Buchinskiy (1954) summarized information about the severe winters that were on the Russian plains that were described in the chronicles. According to his data, severe and snowy winters in the south of the European part of modern Russia were repeated periodically (every few years).

Dzhut is a direct consequence of a number of meteorological factors acting independently or in combination: unusually deep, dense snow cover, ice, icy snow cover, prolonged blizzards or a strong cold wind after a thaw. The complex of these unfavorable weather conditions often creates a lack of food due to thick an ice cover on the snow, which ungulates cannot penetrate, or a thick snow cover, making pasture food inaccessible to ungulates. The inaccessibility of food leads to exhaustion and death, and excessively low temperatures during blizzards kill animals outright (Sludskiy 1963).

In the Mongolian steppe and deserts, winter precipitation is negligible. Only occasionally does snow cover the northern steppe part of this country, while its southern half rarely receives snow. In the Gobi Desert, snow that falls there only in some years usually disappears with the first rays of sun, and evaporates without wetting even the surface layer of the soil. In Inner Mongolia (China), winter is dry and harsh. The coldest month is January, when the temperature can drop to −30 °C to −40 °C. The low snow cover here, if it occurs, evaporates quickly, as in Mongolia. However, in both parts of Mongolia, some winters are very snowy that lead to dzhut. The mass death of livestock in these countries may also be a consequence of cold, northwesterly winds reaching storm strength. Having a low and unstable snow cover in the steppes and semi-deserts of Mongolia and Inner Mongolia would be quite favorable for wintering ungulates, but the instability of this climate feature often acts as a negative environmental factor. With dry food, very low temperatures, and the absence of snow, ungulates lose the source of drinking water they need and, in order to quench their thirst, they ironically are forced to look for areas with a snow cover. In search of watering places, ungulates often migrate to the foothills or go north until they find snow, often as far north as the steppes of Transbaikalia (Sludskiy 1963).

Dzhuts are also characteristic of Mongolia, where over the past 90 years, they have been observed almost 20 times in various regions. During these dzhuts, both domestic ungulates (up to 3/4 of the total livestock population) and wild ungulates died across large territories. In contrast to dzhuts in Eastern Europe, Transcaucasia, southern Kazakhstan, and other Central Asian republics where a dzhut is the result of high snow cover accompanied by ice crusts and blizzards, dzhuts in eastern Kazakhstan, Mongolia, and northern China, are the result of an unusually high snow cover occurring in calm weather. In Kazakhstan, dzhut most often begins in January or February and at the end of April, or sometimes even in May. On average, a dzhut lasts about 75 days. In some winters, dzhuts cover huge territories, occupying up to half of the entire area of a particular natural zone or even several zones located from north to south. When the number of wild ungulates is high, and their ranges are large and continuous, dzhuts, sometimes destroying all animals in even large regions, could not lead them to the complete extermination of the species throughout their distribution area. Ten to twenty-five years after a strong dzhut, the number and ranges of affected ungulates are usually restored. At present, when the number and ranges of many species have greatly decreased and broken into separate fragments, dzhuts can lead to the complete extinction of a species in certain parts of its range and sometimes to the complete extinction of the species. Among wild ungulates, goitered gazelles die first, then all other ungulate species of the deserts and semi-deserts. The study of factors affecting the movement of the number of ungulates in the steppes, semi-deserts and deserts of Eurasia has shown that the first cause belongs to dzhuts, which causes fluctuations in the number of ungulates in significant territories. The increase in the number of wild ungulates occurs between dzhuts (Sludskiy 1963).

Winters in the north of the desert zone are severe. The northern part of the Aral Sea freezes annually for 4 to 5 months. The average temperature in January in the lower reaches of the Syr Darya is −12 °C. In some winters, temperatures drop in the Betpak-Dala Desert to −49 °C and in the Alakul Basin to −50 °C. In clay ephemeral deserts, winter is much warmer. The average January temperature in Tashkent is

−1.3 °C, although sometimes the temperature drops to −33 °C. It is very warm in winter in the south of Turkmenistan, where every year in January there are days when the air temperature rises to +20 °C. In the low-lying deserts in the eastern part of Central Asia (Mongolia, Inner Mongolia, and Xinjiang), winter is harsh. The average January temperature in the Gobi is—16 °C to −18 °C, with an absolute minimum is −40 °C to −43 °C (Sludskiy 1963).

In Kazakhstan's deserts of the northern type, the depth of the snow cover ranges from 4 cm to 15 cm, or 3 to 5 times less than in the steppes. In some winters, especially in Uzbekistan and Turkmenistan, there is absolutely no stable snow cover, whereas in Kazakhstan the snow cover usually lasts between 25 and 100 days. Due to strong winds blowing in the deserts, terrain irregularities (bumpy sands, hills) and solar insolation, the snow cover in this zone is very uneven, and large areas devoid of snow are found here all winter. Kazakhstan and especially the more southern deserts are characterized by frequent thaws, quickly followed by frosts, which lead to the formation of an ice crust on the surface of the snow cover. In some years, the thickness of the ice crust on the soil varies from 2 cm to 10 cm or more. Sometimes grasses and shrubs are also covered with 1 cm to 2 cm of thick ice. Due to the low and short duration snow cover, the desert zone, especially its southern half, serves as a wintering place for several ungulate species, both sedentary and migrating there from the steppes and semi-deserts. Ungulates avoid wintering only in the eastern parts of Central Asia because of the negligible snow cover, going to areas where there is some snow to quench their thirst (Sludskiy 1963).

Often in winter, masses of cold Arctic air invade the desert zone from the Kara Sea and the Taimyr Peninsula, unusually establishing a high snow cover in the deserts and frequent severe blizzards and frosts. During large outflows of Arctic air to the south, wintering wild ungulates end up in uncharacteristic environmental conditions, which leads to dzhut—mass exhaustion and death. Due to the different wintering conditions in different areas of the steppes, semi-deserts and deserts, the directions and length of seasonal migrations of ungulates are different. In the autumn in Kazakhstan, most species of ungulates, especially goitered gazelles, go far to the south every year, completely leaving the semi-desert zone for wintering to desert areas. In the first half of some snowy, frosty winters, ungulates will go north to the snow line, while some stay wintering in the semi-desert and others remain near the desert and later move with the snow line to the south (Sludskiy 1963).

Goitered gazelles are less adapted to low temperatures than other desert ungulates. In contrast to saiga, they try to hide from the wind more often by sheltering under a cliff, behind a dune or bushes. When lying down to rest, goitered gazelles usually tear up the snow to the ground before lying down. By winter, all ungulates, including gazelles, have put on a good layer of fat. For example, on the sides of some gazelles, the fat layer was measured between 1.5 cm and 2 cm, with up to 4.5 kg of fat per individual. During the rutting season, however, which occurs in goitered gazelles in November into January, rutting adult males lose all or most all of their fat. Thus, during a dzhut year, they tolerate cold and lack of food worse than females and are the first to die from starvation (Sludskiy 1963).

During dzhuts, goitered gazelles, fleeing from cold winds, enter the dunes, tugai forests in ravines, and/or reeds near rivers and lakes, even though they avoid most of these shelters in normal winters. During the dzhut of 1945–1946, thousands of gazelles appeared in the tugai forests of the lower reaches of the Ili River, and in the winter of 1953–1954, many of them came from Ustyurt and other places to the thickets of the Amu Darya delta. These gazelles during the subsequent dzhuts remained in the coastal thickets of the Syr Darya, Chu, Ili, and other rivers (Sludskiy 1963). Ungulates also suffer from low temperatures if a thaw is suddenly replaced by a severe frost and/or blizzard. In blizzards, goitered gazelles do not graze; instead, they pack into bushes to escape the wind, where they are swept by snow. There are cases where weakened gazelles, caught in snowdrifts, were found frozen standing upright or in the pose of a resting animal. A similar observation was made on Barsakelmes Island in the Aral Sea: On March 14, 1945, a thick fog hung over the morning and a weak wind blew. By the middle of the day, wet snow began to fall in large flakes. Gradually, the strength of the wind increased and reached the severity of a stormy speed. The snowfall was so thick that nothing could be seen two steps away and it lasted for 2 days. After the blizzard on the plains, the snow depth reached 70 cm, and where shrubs grew, snowdrifts up to 2 m to 2.5 m were piled up. When the blizzard ended, about 20 corpses of goitered gazelles were found under one of the cliffs on the shore. Apparently, they were hiding there from the strong wind and froze. In addition, more than 50 dead gazelles were found in other various parts of the island. Some, in groups of four to five individuals, were lying under bushes growing in sand mounds, while the rest of the dead bodies of gazelles were scattered singly. The frozen bodies of some individuals were standing upright in the bushes, where they fled and huddled from the blizzard, got stuck in the deep, wind-compacted snow and froze to death. Among the individuals who died during the blizzard, adult males predominated, weakened from the rut (Sludskiy 1953).

The ungulates of the deserts of Central Asia have a large weight load (491 g/cm^2) on the supporting surface of their hooves. This is much more than that of the reindeer—140 to 180 g/cm^2. Due to the heavy load on the track of the gazelle, with weak compaction of snow and the formation of fragile layers on it that cannot withstand the gravity of gazelles, their movement becomes even more difficult. Walking on the fragile surface, the animals fall through constantly and cut the skin on their front legs on the icy crust. Leg injuries increase if ungulates are pursued by predators or poachers.

Less often than other species of semi-deserts and deserts, gazelles shovel snow with their feet. Poor adaptability to shoveling snow can be explained by the fact that gazelles originate from regions where the snow cover is usually low and only lasts for a short time. Therefore, where the thick snow cover appears in southern regions, goitered gazelles are forced to migrate far distances. So, during the dzhut of 1945–1946, many gazelles appeared in the lower reaches (the floodplain) of the Ili River in December. Their footprints in the snow were often marked by spots of blood oozing from wounded legs. In most gazelles harvested in January and February, the skin on the metatarsals was cut with inclusion; in some gazelles harvested in January and February, the skin on the metatarsals was cut; in some gazelles, the cut legs festered and swelled. In Badkhyz in southeastern Turkmenistan, during the dzhut of 1956–1957, some of the dead gazelles had the skin on their metatarsals cut off by an icy crust of snow cover. In the same area, in January 1959, there was again an ice crust on the snow cover, which led to leg injuries. There were also individuals whose blood circulation in the extremities was disrupted as a result of injuries. In such gazelles, the front legs were affected by gangrene, and they lost the ability to move. Some animals with wounded legs lay down under bushes of *Halimodendron halodendron* and the prickly branches and seed pods slowly overtook them, causing the deaths of some individuals. Some ungulates died on the bed like they are sleeping (Sludskiy 1963).

In snowy winters, during the establishment of the ice crust, the death of goitered gazelles from wolves increases significantly. The load on a snowy trail of a desert wolf is less than that of a goitered gazelle by 2.5 to 3 times. Consequently, wolves can run on thin ice without falling through, whereas the ice crust will not support gazelles. In such winters, gazelles are killed by wolves by the hundreds and even the thousand individuals, and during some dzhuts, wolves can almost completely exterminate goitered gazelles over vast territories (Sludskiy 1963).

In winter, most ungulates wintering in desert and semi-desert zones feed mainly on wormwood (*Artemisia*), *Salsola* and *Anabasis* and less often with small species of cereals, and most willingly they eat green leaves of wormwood that appear back in the fall on the root part of the bush, then *Anabasis salsa*, twigs of wormwood, *Salsola arbuscula* and *S. laricifolia* and other plants. When the snow cover is 10 cm high, the green shoots of sagebrush and part of *Anabasis salsa* are unobtainable, and with a snow depth of 20 to 25 cm, other stunted vegetation is covered. Only large shrubs and semi-shrubs, such as *Haloxylon aphylla* and *Salsola arbuscula*, remain above the surface, but dense thickets of bushes are often completely covered with snow, as well. Snow in semi-deserts and northern deserts lasts for a long time. This has led to the fact that most of the ungulates wintering here have adapted to foraging under the snow. If the snow cover is loose, they can push the snow apart with their lips and muzzles during grazing, exposing the bush, the animal can tear off leaves and branches. Gazelles with loose snow cover, getting to the food, push the snow with their muzzles even when the snow cover reaches 30 cm. On open plains, fallen snow usually condenses quickly. When ungulates can no longer push the dense snow with their muzzles, they are forced to shovel the snow with their front hooves. Horses, argali, roe deer, deer, and even saiga can shovel snow well, but goitered gazelles cannot. These animals begin to shovel snow when a compacted snow cover is established at a depth that exceeds 4 to 8 cm, as well as an ice crust appears on the snow. Ungulates shovel snow in those areas where the tops of grasses stick out above the snow surface; however, if they are completely covered with snow, the ungulates stop feeding on them (Sludskiy 1963).

In Kazakhstan, the deaths of large numbers of gazelles have been noted many times: in southeastern Kazakhstan, the dzhut of 1945–1946 was especially disastrous, and in southern Kazakhstan, the mass death of gazelles was noted in 1950–1951 (Sludskiy 1953). In January 1946, in the Balkhash region, the height of the snow cover in places reached 50 cm, and as a result of thaws, an ice crust formed, causing the gazelles to severely injured their legs. Gazelles kept to the thickets of bushes along the Ili River, many moved with difficulty. The weight of adult animals was below normal by 10 kg and a mass death in February and March led to the loss of several thousand gazelles in the lower reaches of the river (Heptner et al. 1961). A large case of gazelle deaths resulted when an ice crust formed in western Kazakhstan in 1938 and in western Mongolia in 1935. High snows of up to 40 cm fell in Badkhyz in 1934 and 1957, which also led to the death of gazelles (Heptner et al. 1961). Since gazelles are not able to make such long-distance migrations like the saiga, after dzhut winters, the number of goitered gazelles drops severely and in some areas these ungulates disappear for a long time (Heptner et al. 1961; Sludskiy 1953).

Thus, as the height of the snow cover increases, the list of the most important winter foods for ungulates rapidly decreases, with one forage plant falling out of the diet after another. Summing up all the information about the dzhuts that have occurred in the southern parts of deserts of Central Asia over the past 100 years, there were 16 harsh winters during this period, of which nine were especially severe for wild ungulates. In the semi-deserts and steppes of western Kazakhstan, dzhuts occurred more often than in the southern regions. Forty-four dzhuts have been recorded at various intervals over the past 150 years, of which 19 were particularly heavy. They were also repeated at various intervals from 1 year to 22 years, though most often were observed every 5 to 7 years. Especially large dzhuts usually occurred every 10 to 20 years, with smaller ones also occur in the intervals between them. Most severe winters followed severe droughts. Most ungulate deaths were caused by thick snow and rarely the appearance of an ice crust on the surface of the snow cover (Sludskiy 1963).

With heavy snowfalls, the gazelles tend to go to snowless areas; they often accumulate in the foothills, where the snow is blown away by the winds and quickly melts (Sludskiy 1953). There are cases when, like in the winter of 1924–1925 in the Shirvan and Milskaya steppes of Transcaucasia, during heavy snowfalls, gazelles joined flocks of sheep (Heptner et al. 1961). In Transcaucasia, snowy winters occur once every 15 to 20 years, and the number of gazelles in such years drops sharply due to the high death rate (Heptner et al. 1961). Fleeing from the cold winds and strong snowstorms, all hoofed animals go sometimes hundreds of kilometers, moving with the wind and often against it. At the same time, they die by the thousands by falling into viscous salt marshes, ravines covered with snow, from cliffs, or going far from the shore into a sea of ice (Sludskiy 1963).

Individual dzhuts carry away tens of thousands to millions of wild and domestic ungulates, which die of exhaustion and freezing, but their destructive role is not limited to this. During a severe drought, which usually precedes a dzhut and the harsh winter following it, the normal course of various physiological processes in the animal's body is disrupted. As a rule, during dzhut and in the first months after its completion, ungulates are severely depleted. Beriberi is widely spread, causing various diseases of the skin, digestive tract and other internal organs. The body's resistance to infectious and invasive diseases sharply decreases. Dzhut is often accompanied by mass abortion, embryo resorption and still births. The activity of the sex glands lowers significantly, to the point of complete cessation of function. It takes a long time for the body of an animal that has survived a dzhut to recover. Some surviving ungulates do not recover completely, with bareness increasing sharply to where some individuals become infertile for the rest of their lives. Some ungulates reach such a state of exhaustion that they cannot stand up on their own and some domestic animals in this condition die quickly. The lack of plant foods and their dryness during drought reduce milk production in lactating females. In dry years, the lactation of goitered gazelles stops earlier by 1–1.5 months than in usual years. During a drought, young ungulates do not develop well, as they do not receive enough fully nutritious milk and are forced to feed on dry, low-nutrient plants, and weak young animals do not tolerate heavy winters (Sludskiy 1963).

In some droughts, ungulates suffer greatly from lack of water, which depletes an animal as well as malnutrition. Therefore, some goitered gazelles looked for the remains of water at wells for watering livestock, and after rare, brief thunderstorms, the roads were trampled by wild hooves, quenching thirst from the puddles left after the rain. Such unusually difficult living conditions forced ungulates to migrate over considerable distances in search of pastures and watering holes. Exhaustion and death of ungulates during droughts are known in Kazakhstan and Mongolia. Thus, droughts, causing exhausted and weakened animals, greatly contribute to deaths during dzhut. Starving ungulates during dzhut feeding on coarse, dry food and even inedible objects led to a digestive disorders and they quickly died. Dzhut in the south of Kazakhstan (1928–1929) was accompanied by a continuous infection with scabs and lice, as well as intestinal diseases. During another dzhut, ungulates suffered from scabies and eczema on the basis of vitamin deficiency and large areas of the body were deprived of hair. It was characteristic that all the goitered gazelles that fell during this dzhut had stomachs filled with coarse fodder (branches of *Halimodendron halodendron*, dry *Salsola*, reeds *Phragmites communis*), which were glued together into hard and durable balls, which were the result of an upset stomach (Sludskiy 1963).

During fasting, the body is initially supported by previously accumulated fat reserves, so well-fed individuals are able to endure hunger longer than thin ones. During dzhut, males exhausted during the rut are the first to die in large numbers, following by the young, then still growing individuals, and least are adult females. During dzhut, in long-starving gazelles and other ungulates, the mezdra is greatly thinned, becomes dry and loose, and darken in color. Among tanners, the skins acquired during this phenomenon are known as hungry sheepskin and result in very low marketable qualities. In long-starved ungulates, the muscles lose their normal pinkish color and become yellowish. The meat of a severely starved individual has an unpleasant smell of decomposing urea and, at times, even dogs refuse to eat it. In long-starved gazelles, the bone marrow in the tubular bones almost completely disappears (Sludskiy 1963).

From the point of view of veterinary science, hunger in ungulates is associated not only with an insufficient intake of food, but also especially with its lack of quality. Eating even a small amount of non-nutritious, hard-to-digest food, often with a lack of mineral substances, is unable to protect against severe starvation. The inevitable companions of fasting are vitamin deficiencies, which cause a number of disorders in the body, creating the basis for the development of secondary severe pathologies. The main clinical and patho-anatomical sign of starvation is the emaciation of the animal—a sharp decrease in body volume and weight due to the loss of biological matter. Emaciation is mainly the result of a sharp predominance of dissimilation processes that are not covered by the assimilation of fiber substances that do not arrive in sufficient quantities. Violation of the water-salt and protein regime in cells also leads to a qualitative change in protoplasm, to a decrease in the dispersion of its colloids, and the overall decrease in body weight which can be very significant at up to three to four times the initial average weight. In a severely starving animal, muscles and glands lose 40% to 70% of their original weight (Sludskiy 1963).

First of all, there is a complete absence of fat in subcutaneous and intermuscular tissue and in internal fat depots. The omentum and mesentery remain in the form of a thin, transparent, dryish film. The so-called serous atrophy of epicardial fat is very characteristic. In this case, the fatty tissue of the epicardium is enlarged and gelatinous, and when it is incised, a clear liquid flows down. The same phenomenon is observed in the perinephrine tissue. The disappearance of fat causes about 25% of the total weight loss of the animal's body. The second most important sign is skeletal muscle atrophy. It accounts for about 50% of all weight loss. The muscles become pale, dry, and greatly reduced, which is why the outlines of the skeleton clearly appear in the animal. Atrophy, for example, of the intercostal muscles often reaches the point that they turn into completely transparent, with only slightly reddish plates. The bone apparatus also changes dramatically. Histological examination shows a significant reduction and resorption of bone marrow and content of tubular bones, both in spongy and compact matter. Due to this, fractures easily occur, with very poorly healing in an emaciated animal. Yellow bone marrow undergoes serous atrophy, red bone marrow decreases, which leads to a sharp reduction in hematopoiesis, creating condition similar to anemia. The skin is thinning, especially the epidermis, which often has only one to two layers instead of many layers of cells. The hair follicles also atrophy, often causing the formation of areas of baldness (Sludskiy 1963).

Of the internal organs, the liver is most atrophied. Weight can drop to half of normal. The liver becomes dense, with sharpened edges, and takes on a dark color. The hepatic girders thin out and turn out to be widely spread by dilated capillaries or, in more chronic cases, overgrown connective tissue. The spleen is significantly reduced, the kidneys are somewhat lessened. The heart resists atrophic changes longer than other organs, but with very prolonged exhaustion, the heart muscle can also decrease, which causes the coronary arteries to become serpentine as in human sclerosis. The sex glands sharply atrophy, especially the ovaries, in which the germinal layer disappears and atresia of developing follicles occurs. In the testes, degenerative and necrobiotic changes occur in the cells of the seminal tubules, up to the complete emptying of the tubules. This is sometimes preceded by phagocytosis (absorption) of spermatozoa by sertoli cells. The stroma is sharply sclerosed. The endocrine tissue of the testes and ovaries is preserved for a long time. Since prolonged fasting is usually accompanied by the adoption of coarse, unusual food, then in the last stages, secondary severe lesions of the digestive tract develop, in which mechanical and chemical effects on the mucosa are accompanied by activation of microflora that has no pathogenic value in the normal state of the body, but pathogenic in starvation. As a result, catarrhal or even hemorrhagic inflammatory processes develop in the stomach and intestines, accompanied by diarrhea. These phenomena aggravate overall starvation. A very frequent companion of starvation is focal catarrhal or catarrhal-necrotic mnemonias (pneumonia), from which animals quickly die. A general decrease in the body's resistance in depleted animals contributes to outbreaks of infectious and parasitic diseases. First of all, with the young and old individuals dying first from exhaustion (Sludskiy 1963).

Most of the changes associated with starvation are reversible. With the cessation of starvation and the gradual transition to a normal diet, the structure and functions of the organs are restored. However, with a very long, gradual progressive fasting, changes in the organs remain irreversible for a very long time and do not recover with the normalization of nutrition. Such individuals often die with a completely normal diet, especially during their first time of transition to it. It takes a very long time for the genitals to recover, and even complete sterility can develop in young females (Sludskiy 1963).

It is very difficult to carry out, on a large scale, any event aimed at improving the wintering conditions of wild ungulates in an unorganized territory. Therefore, the first and most important will be the organization of a system of sanctuaries, reserves, and complex commercial hunting areas in the areas of mass wintering of wild ungulates. All these organizations have to do the following work (Sludskiy 1963).

1. In case of the appearance of crust, ice layers, heavy compaction of snow and ice, which complicate the movement and grazing of ungulates, it will be necessary to use special rippers to free the snow cover. A harrow with a wooden frame and iron teeth designed for loosening snow cover and ice crust up to 10 cm thick is also proposed (Sakharov 1948). The destruction of the snow cover will be best carried out in narrow strips of 20 to 40 m, placing a footprint of 2 to 3 m. In every large wintering area of wild ungulates, it will be necessary to have 5 to 10 tractors and 10 to 30 rippers for snow by January.
2. In wet years, it will be necessary to harvest hay where it grows and make insurance stocks of feed from it in pressed form. In order for animals to get used to having it available and know the places where it will be laid out, this effort should be practiced on a small scale from the beginning of winter.
3. Do not allow cattle in places of winter concentrations of wild ungulates.
4. Arrange special watering holes for wild ungulates and do not let cattle to use them.
5. Conduct an intense fight against wolves using whatever methods are available.
6. Strengthen the fight against poachers in wintering grounds. It will be necessary to oblige the population to keep dogs on leashes during dzhuts.
7. In hunting farms, ungulate hunting should be stopped by December 15, so as not to disperse ungulates from their best wintering grounds.

8. If a dzhut is expected after a severe spring-summer drought, and the number of ungulates is high, it will be necessary to increase the ungulate extraction plan in order to significantly reduce their number before the start of winter to reduce starvation and prevent useless death.

The recovery in the numbers of the goitered gazelles is much slower than that of the saiga. This is determined by less mobility of the gazelle, lower fertility, and greater mortality from predators, including humans. During dzhuts, gazelles die relatively more often than saigas. For example, after the dzhut winters of 1945–1946 and 1950–1951 in Kazakhstan, the goitered gazelle, which disappeared from a number of areas, did not recover there for 5 to 7 years, while the number of saiga reached its initial level after 2 to 3 years (Sludskiy 1953). A similar phenomenon took place in the basin of the Great Western Lakes of Mongolia, where the number of goitered gazelles that disappeared because of ice covered snow of 1935 did not recover fully after 10 years, while the number of saiga recovered after 3 to 4 years (Bannikov 1954). The fecundity of the gazelle is such that due to the gain of young animals, under very favorable conditions, the goitered gazelle numbers may increase by 70 to 80% by autumn. However, in the years after severe winters, by August only 50% to 55% of the young were found in the population (Sludskiy 1956).

The death of goitered gazelle fawns is significant and amounts to 40% to 50% by autumn (Sludskiy 1956; Heptner et al. 1961). The number of barren females is two to three times more than that of the saiga. A significant death of young animals, a relatively large number of barren females, combined with a relatively late puberty in most young females (a year later than in saiga), determine the slow recovery of the goitered gazelle population. It is possible that the uneven deaths between the sexes in the years after the dzhut also play a negative role in this regard. In normal years, the sex ratio of the goitered gazelles is such that adult males make up about 20% of the population; in the years after a dzhut, during which males are the first to die, they make up only 3.5 to 5.5% of the entire population (Sludskiy 1956).

19.3 Diseases

Diseases of goitered gazelles have been studied for a long time. Spontaneous tularemia disease is documented (Drobinskiy and Klimukhin 1948). In the 1940s, in the Dangara district of the Tajik SSR, goitered gazelles (and sheep) were sick and died in large numbers from foot-and-mouth disease. In the same region, according to the veterinarian Polotov, during World War II gazelles died from sheep pox (Sokolov 1959). In zoos, for example, in Moscow, gazelles often died from necrobacillosis (Kovalenko Ya 1948), and in 1935, two goitered gazelles were sick with mycosis there (Tsvetaeva 1941). Coccidia were found in the intestines of gazelles from the Betpak-Dala Desert (*Eimeria elegans*), but apparently were not pathogenic for them. In the same location, spirochaetes were found in the gazelle, but their species was not defined. The mass death of gazelles from epizootics is rare and has had only a local distribution (Zhevnerov et al. 1983). The identical coccidia (*Eimeria gazella*) was found in *G. s. marica* in Saudi Arabia (Kasim et al. 1991). Normal bacterial flora was described for *G. s. marica* in Saudi Arabia (Mohammed et al. 2011).

About 150 papers have reported on viral diseases of goitered gazelles from across the many different parts of their vast range (Sarukhanova 2021). Viral diseases infecting goitered gazelles include the plague of small ruminants, bluetongue, epizootic hemorrhagic disease, parainfluenza-3, foot-and-mouth disease, malignant catarrhal fever, parainfluenza-3, and sheep pox (Doyle and Heuschele 1983; Heuschele et al. 1982; Zhevnerov et al. 1983). Doyle and Heuschele (1983) did not find antibodies to bovinae virus diarrhea and infectious bovine rhinotracheitis in goitered gazelles in spite of their presence in several other species of ruminants. Goitered gazelles are susceptible to bacterial infection (*Arizona* sp. and *Salmonella* sp.), omphalophlebitis, poliocephalomalacia, and ulcerative entries (Griner 1983; Rietkerk et al. 1992).

Sometimes the gazelles are affected by infectious diseases from domestic animals through pastures, watering places and the air. In the 1940s, in the Dangara region of Tajikistan, goitered gazelles were affected and died in large numbers from foot-and-mouth disease, which was common among domestic sheep. During World War II, the gazelles were also sick and died from sheep pox (Sokolov 1959).

Peste des petits ruminants (PPR) is a highly infectious disease of small ruminants and caused by small ruminant morbillivirus (SRMV), formerly called peste-des-petits-ruminants virus (PPRV). These viruses (PPRVs) re-emerged in China at the end of 2013 and then spread rapidly into 22 provinces through movement of live goats and sheep (Liu et al. 2018). All samples from wild species, including goitered gazelles were found to be positive for these viruses (Li et al. 2017). However, 94 goitered gazelles were checked for a number of diseases and *Brucella* spp. (significant zoonotic disease) was found in two gazelles (2.6%, n = 94), *Chlamydophyla abortus* (intracellular gram-negative bacteria) in 16 gazelles (16%), *Coxiella burnetii* (causes Q fever) in 8 gazelles (8.5%), while *Taxoplasma gondii* (a ubiquitous apicomplexan parasite) was not detected in any of the 94 gazelles. These five zoonotic pathogens are commonly found in domestic livestock (Wu et al. 2020). PPR (Peste des petits ruminants) was found also in Iran, where from 1500 up to 2500 wild goats and goitered gazelles died with signs of PPR (Marashi et al. 2017).

Rift valley fever was not found in any goitered gazelle in Turkey's analyses made during 2000–2006. This is a vector-borne zoonotic disease, is caused by a phlebovirus (Family Bunyaviridae). All 82 gazelle samples were negative (Gür et al. 2017). However, Peste des petits ruminants virus was found in 10 gazelles (12%) (Gür and Albayrak 2010). Deerpox virus from the subfamily Chordopoxirinae was found in goitered gazelles in the zoologic park in Minnesota (USA) (Bracht et al. 2013).

In addition, in Iran listeriosis (*Listeria monocytogenes*) infections (6.09%, n = 82) and toxoplasmosis (*Taxoplasma gondii*) (28.04%, n = 82) were found but no seropositive gazelles were tested for *Brucella* antibodies. There is no significant difference between male and female gazelles in the seroprevalences of toxoplasmosis and listeriosis (Gökçen et al. 2007). However, on Kish Island (Iran), brucellosis was detected in goitered gazelles, which were infected from local domestic cattle. As a result, most outside females having contact with cattle aborted their young, while females in captivity without the brucellosis viruses gave birth successfully (Memarian et al. 2015).

Sarcoptes scabiei infestation was diagnosed in four freshly dead and three net-captured gazelles while ranging freely. The captured animals presented with an alopecic pruritic skin disease with signs of crusted skin lesions, numerous small nodules which first appeared on the lips or nostrils and then extended towards the eyelids, around the ears, and in some cases, over the entire face, neck, trunk, and legs. Skin over the affected area gradually became bald, thick, and hard being dry and doughy to the touch, and serous fluid or sometimes blood oozed from the lesions which had a severe malodor. Skin scrapings confirmed the presence of the mite *S. scabiei*. Histopathology of lesions demonstrated marked acanthosis, hyperkeratosis, and parakeratosis. Microscopical examination also revealed all stages of *S. scabiei*, which were located mainly in the stratum corneum and also in the stratum granulosum. During the capture and sampling of the animals, four persons ranging in age from 25 to 62 years were exposed to scabies. Two relatives of one of them were also affected by familiar contact. Clinical signs appeared within 9 days of exposure, and they developed several pruritic erythematous papules with intense itching (Bazargani et al. 2007).

Bluetongue is an infectious but non-contagious disease, characterized mainly with fever, hyperemia and ulceration of oral mucosa, coronitis, lameness, abortion, and death. The Bluetongue virus (BTV) is a type species of the *Orbivirus* genus, family Reoviridae, having double stranded RNA genome which consists of ten segments. Twenty-four serotypes have been recognized so far (Roy 2002). BTV has a very wide host range including sheep, cattle, goats, gazella spp., bighorn sheep, antelope, and various types of artiodactiles (Hourrigan and Klingsporn 1975; Hoff and Hoff 1976; Bender et al. 2003).

Epidemiology of BTV infection is vector dependent, especially *Culicoides* spp. (Mellor et al. 2000). The main species carrying the virus are *C. imicola* and *C. obselotus* (Dik 1989; Yilmaz 1994), but *C. imicola* is the more important species in Turkey (Mellor et al. 1995). Distribution of *Culicoides* spp. lie in a band of latitudes between 35°S and 40°N; this could be extended up to 50°N (Mellor et al. 2000).

In addition, the seropositive rate of goitered gazelles for the epizootic hemorrhagic disease virus reached 2.43% after checking 82 individuals from Anatolia (Turkey) (Albayrak et al. 2010). However, Gür (2008) in Turkey (Eastern Anatolia) showed that 40.2% of the animals were identified as seropositive to the Bluetongue virus (sheep catarrhal fever). Despite the fact that almost no animals have been found to have obvious clinical signs of the disease, researchers note that the virus can affect the reproductive capabilities of animals, which in turn can affect the overall dynamics of the population (Gür 2008).

In captive goitered gazelles, *Trueperella pyogenes* was found which is associated with interdigital necrobacillosis. Four male and two female gazelles presented with weight loss, front limb swelling, and persistent lameness despite antibiotic treatment. The animals were reluctant to bear weight on the affected limbs, and the diagnosis of interdigital necrobacillosis was made based on physical exam, bacteriologic evaluation, and radiographic imaging. In all cases, the interdigital skin and subcutaneous tissues were affected in both forelimbs. Exungulation (loss of the hoof) occurred in one female gazelle. Despite aggressive topical and systemic treatment, the two female gazelles died. Gross pathology detected abscesses in the heart, lung, and liver. *T. pyogenes* was isolated as the major pathogen. This is the first report of interdigital necrobacillosis caused by *T. pyogenes* in *Gazella subgutturosa* (Çeçen et al. 2018).

An outbreak of *Salmonella enterica* serotype *Choleraesuis enteritis* occurred in two juvenile goitered gazelles over a 5-week period at the Minnesota Zoo. Diagnosis was made postmortem on one gazelle, and a second gazelle was diagnosed via a fecal culture. Salmonellosis was considered to be a contributing factor besides ostertagiasis for the death of one goitered gazelle and for the diarrhea of another goitered gazelle. A third gazelle became ill in the same time period, but *Salmonella* infection was not confirmed by culture. All exhibited the clinical signs of profuse, watery diarrhea. The gazelles developed a protein-losing enteropathy (Wolf et al. 2011).

References

Albayrak H, Ozan E, Gur S (2010) A serologic investigation of epizootic hemorrhagic disease virus (EHDV) in cattle and *Gazella subgutturosa subgutturosa* in Turkey. Trop Anim Health Prod 42(8):1589–1591

Bannikov AG (1954) Persian gazelle. In: Heptner VG (ed) Mammals of Mongolian People Republic. Publication of USSR Academy of Sciences, Moscow-Leningrad, pp 250–258. (in Russian)

Bazargani T, Hallan JA, Nabian S, Rahbari S (2007) Sarcoptic mange of gazelle (*Gazella subgutturosa*) and its medical importance in Iran. Parasitol Res 101:1517–1520

Bender LC, Li H, Thomson BC, Morrow PC, Valdez R (2003) Infectious disease survey of gemsbok in New Mexico. J Wildl Dis 39:772–778

Bracht AJ, Armién AG, Carrillo C, O'Hearn ES, Fabian AW, Moran KE, Lu Z, Ariyakumer DS, Rasmussen JM, Metwally SA (2013) Isolation and characterization of a Cervidpoxvirus from a goitered gazelle (*Gazella subgutturosa*) from a zoologic park in Minnesota. J Zoo Wildl Med 44(3):589–595

Buchinskiy IE (1954) The essays of the Russian plain climate during historical epoch. Hidrometeorological Publishing House, Leningrad, pp 1–89. (in Russian)

Çeçen G, Buyukcangaz EK, Çalışkan ÜG, Cangul TI, Akdesir E (2018) Interdigital necrobacillosis associated with *Trueperella pyogenes* in goitered gazelles (*Gazella subgutturosa*). J Zoo Wildl Med 49(2):429–434

Cunningham PL, Wacher T (2009) Changes in the distribution, abundance and status of Arabian sand gazelle (*Gazella subgutturosa marica*) in Saudi Arabia: a review. Mammalia 73:203–210

Cunningham PL, Wronski T (2011) Sex ratios of Arabian sand gazelle *Gazella subgutturosa marica* Thomas, 1897 in the Mahazat as Sayd protected area, Saudi Arabia. Mammalia 75:243–248

Dik B (1989) Konya ve çevresinde bulunan Culicoides (Diptera:Ceratopogonidae) türleri üzerine arastirmalar. Doktora Tezi, Ankara Üniversitesi Saglik Bilimleri Enstitüsü, Ankara. (in Turkish, with English abstract), pp. 1–85

Doyle LG, Heuschele WP (1983) Prevalence of antibody to major viral diseases in captive exotic ruminants. Am Assoc Zoo Vet Annu Proc 1983:76–82

Drobinskiy VV, Klimukhin VK (1948) Tularemia in mountain areas. J Microbiol Epidemiol Immunol 1. (in Russian) (cited by Zhevnerov et al. 1983)

Gökçen A, Atlaş MG, Sevgili M, Babür C, Çelebi B, Kılıç S (2007) Detecting toxoplasma, listeria and Brucella antibodies in goitered gazelles in Turkey. Medycyna Wet 63(9):1064–1066

Gorelov UK (1972) Breeding of goitered gazelle (*Gazella subgutturosa* Güldenstaedt) and problem of restoring of its number in the Badkhyz (SE Turkmenistan). In: Theriology, vol 1. Nauka of Siberian Department of Academy of Sciences of USSR, Novosibirsk, pp 420–424. (in Russian)

Griner LA (1983) Pathology of zoo animals. A review of necropsies conducted over a fourteen-year period at the San Diego zoo and San Diego Wild Animal Park. Zoological Society of San Diego, 608 pp.

Gür S (2008) A serologic investigation of blue tongue virus (BTV) in cattle, sheep and *Gazella subgutturosa subgutturosa* in southeastern Turkey. Trop Anim Health Prod 40(3):217–221

Gür S, Albayrak H (2010) Seroprevalance of peste des petits ruminants (PPR) in goitered gazelle (Gazella subgutturosa subgutturosa) in Turkey. Journal of wildlife diseases, 46(2):673–677

Gür S, Kale M, Erol N, Yapici O, Mamak N, Yavru S (2017) The first serological evidence for Rift Valley fever infection in the camel, goitered gazelle and Anatolian water buffaloes in Turkey. Tropical animal health and production, 49(7):1531–1535

Heptner G, Nasimovich AA, Bannikov AG (1961) Mammals of Soviet Union, Vol.1. Artiodactyla and Perissodactyla. Vysshaya Shkola Publishing House, Moscow. (in Russian)

Heuschele WP, Doyle LG, Hooker EA, Gottling KL, Kawanabe PS (1982) Current status of some important viruses of domestic ruminants in captive wild ruminants in the U.S.A. Am Assoc Zoo Vet Annu proc 1982:94–121

Hoff GL, Hoff DM (1976) Bluetongue and epizootic hemorrhagic disease. A review of these diseases in nondomestic artiodactyls. J Zoo Anim Med 7:26–30

Hourrigan JL, Klingsporn AL (1975) Bluetongue: the disease in cattle. Aust Vet J 51:170–174

Kasim AA, Alyousif MS, Al Shawa YR (1991) Redescription of *Eimeria gazella* Musaev, 1970 (Apicomplexa: Eimeriidae) from the Arabian gazelle, *Gazella gazella arabica* (Artiodactyla: Bovidae) in Saudi Arabia. Parasitologia 33(2–3):107–109

Kingswood SC, Blank DA (1996) Gazella subgutturosa. Mamm Species 518:1–10

Kovalenko Ya R (1948) Necrobacillosis of farm animals. Publication of Selkhozogiz, Moscow, pp 1–271. (in Russian)

Li J, Li L, Wu X, Liu F, Zou Y, Wang Q, Liu C, Bao J, Wang W, Ma W, Lin H (2017) Diagnosis of peste des petits ruminants in wild and domestic animals in Xinjiang, China, 2013–2016. Transbound Emerg Dis 64(6):e43–e47

Liu F, Li J, Li L, Liu Y, Wu X, Wang Z (2018) Peste des petits ruminants in China since its first outbreak in 2007: a 10-year review. Transbound Emerg Dis 65(3):638–648

Marashi M, Masoudi S, Moghadam MK, Modirrousta H, Marashi M, Parvizifar M, Dargi M, Saljooghian M, Homan F, Hoffmann B, Schulz C (2017) Peste des petits ruminants' virus in vulnerable wild small ruminants, Iran, 2014–2016. Emerg Infect Dis 23(4):704

Mellor PS, Jennings DM, Hambling C, Burgu I, Urman HK, Akca Y, Hazýroglu R, Alkan F, Yonguc AD, Ozkul A, Eren H, Altýnsaat S (1995) Control of Akabane disease and surveillance of Bluetongue and Ephemeral fever. United Nations Development Programme Food and Agriculture Organisation of the United Nation, Rome

Mellor PS, Boorman J, Baylis M (2000) Culicoides biting midges: Their role as arbovirus vectors. Annual Review of Entomology, 45:307–340

Memarian I, Moghani F, Chegini S, Shahdari A, Hamidi A (2015) Brucellosis and stormy abortion in Persian goitered gazelle (*Gazella subgutturosa subgutturosa*). Dis Zoo Wild Anim:68

Mohammed SAS, Sawsan AO, Abdulla SK, Abdulaziz NA, Osama BM (2011) Normal bacterial flora from idmi (*Gazella gazella*) and reem gazelles (*Gazella subgutturosa marica*) in Saudi Arabia. Afr J Microbiol Res 5(28):5090–5096

Riesch R, Martin RA, Lerp H, Plath M, Wronski T (2013) Size and sex matter: reproductive biology and determinants of offspring survival in *Gazella marica*. Biol J Linn Soc 110(1):116–127

Rietkerk FE, Lindsay N, Tatwany H, Mubarak S, Badri O, Williamson DT (1992) Population dynamics of a captive herd of Arabian sand gazelles. Ongulés/Ungulates 91:379–382

Roy P (2002) Orbivirus. In: Tidonna CD, Darai G (eds) The springer index of viruses. Springer-Verlag, Berlin Germany, pp 957–963

Sakharov IV (1948) Testing of the simplest devices for the destruction of the snow-ice crust. In: Proceedings of the Agricultural Institute, engineering series, vol 1. Kazgosizdat, Alma-Ata, pp 96–103. (in Russian)

Sarukhanova S (2021) The study level of the goitered gazelle (Artiodactyla, Bovidae) in Azerbaijan. Bull Sci Pract 7(9):42–58. (in Russian)

Sludskiy AA (1953) The mass death of wild ungulates in the Kazakhstan deserts and its impact on the animal numbers. In: Proceedings of the institute of zoology, vol. 2. Publication of Academy of Sciences KazSSR, Alma-Ata, pp 3–30. (in Russian)

Sludskiy AA (1956) Breeding of goitered gazelle. In: Proceedings of institute of zoology, Kazakh Academy of Sciences, Vol. 6. Nauka Press of Kazakh SSR, Alma-Ata, pp 78–108. (in Russian)

Sludskiy AA (1963) The mass death of wild ungulates in steppes and deserts of Europe and Asia. In: Proceedings of the Institute of Zoology, Academy of Sciences of the Kazakh SSR, 20. Nauka of the KazSSR, Alma-Ata, pp 5–88. (in Russian)

Sokolov II (1959) *Gazella subgutturosa* Güldenstaedt—Goitered gazelle. In: Pavlovskiy EN (ed) Fauna USSR, mammals: ungulates (Orders Perissodactyla and Artiodactyla). Publication of the Academy of Sciences USSR, Moscow and Leningrad, pp 308–331. (in Russian)

Tsvetaeva NP (1941) Animal diseases in the Moscow zoo. In: Proceedings of the Moscow zoo, vol 2, No. 2. Publication of Moscow Zoo, Moscow, pp 1–153. (in Russian)

Vasenko EP (1950) Ecology and distribution of the saiga antelope. In: Proceedings of the Barsakelmes State Nature Reserve, No. 1, Alma-Ata, pp 38–115. (in Russian)

Wolf TM, Wünschmann A, Morningstar-Shaw B, Pantlin GC, Rasmussen JM, Thompson RL (2011) An outbreak of *Salmonella enterica* serotype Choleraesuis in goitered gazelle (*Gazella subgutturosa subgutturosa*) and a Malayan tapir (Tapirus indicus). J Zoo Wildl Med 42(4):694–699

Wu JY, Li JJ, Wang DF, Wei YR, Meng XX, Tuerxun G, Bolati H, Liu KK, Muhan M, Shahan A, Dilixiati D (2020) Seroprevalence of five zoonotic pathogens in wild ruminants in Xinjiang, Northwest China. Vector-Borne Zoonotic Dis 20(12):882–887

Yılmaz B, Demircioğlu İ, Bozkaya F, İnce NG (2020) Three-dimensional tomographic reconstruction and morphometric analysis of skull in gazelles (Gazella subgutturosa). Ankara Üni. Vet. Fak. Derg., 67:161–168

Zhevnerov VV (1977a) Outcome and perspectives of the goitered gazelle acclimatization in the Barsakelmes Island. In: Gvozdev EV (ed) Rare animals and birds of Kazakhstan. Nauka Publishing House of KazSSR, Alma-Ata, pp 78–80. (in Russian)

Zhevnerov VV (1977b) Breeding of goitered gazelle under human control. In: Sokolov VE (ed) Rare species of mammals and their conservation, Moscow, pp 32–33. (in Russian)

Zhevnerov VV (1984) Goitered gazelle of the Barsakelmes Island. Nauka Publishing House of Kazakh SSR, Alma-Ata. (in Russian)

Zhevnerov VV, Bekenov AB, Sludskiy AA (1983) Goitered gazelle—*Gazella subgutturosa* Güldenstaedt, 1780. In: Gvozdev EV, Kapitonov VI (eds) Mammals of Kazakhstan, Bovidae, vol 3, No. 3. Nauka Publishing House of Kazakh SSR, Alma-Ata, pp 11–54. (in Russian)

Conservation Efforts 20

20.1 Conservation Status

Until fairly recently, goitered gazelle numbers were very large. In the past, goitered gazelles traveled vast distances during the fall and winter to avoid deep snow, and during the summer months in search of water (Heptner et al. 1961). In the early 1900s, it was estimated that there were approximately 1 million goitered gazelles in Soviet Central Asia alone. However, since then, their range has significantly diminished, and their population has been decreasing due to illegal hunting and the loss of suitable habitat. As a result, sedentary populations that currently exist have small home ranges and small numbers (Martin 2000). Goitered gazelles became extinct in Kuwait, Georgia, and Kyrgyzstan (Kingswood and Blank 1996; Mallon and Kingswood 2001). During the late-1990s, the entire goitered gazelle population was estimated at 120,000 to 140,000 (Mallon and Kingswood 2001), while the most recent survey in 2016 (IUCN 2017), estimated the world population of mature gazelles at 42,000 to 49,000. This ongoing decrease in numbers is observed across all regions inhabited by the goitered gazelles (Karlstetter and Mallon 2014) and is the result of two main causes: habitat loss primary due to human economic activity, and hunting and poaching, which are primarily motivated by the illegal meat trade, and to a lesser degree, skins and horns (Clark et al. 2006; Svizzero 2019).

Goitered gazelles have been intensively hunted for centuries. During the Early Bronze Age, from the Arabian Peninsula to the Caucasus and Central Asia, the hunting pressure was particularly intense. Thousands of large stone traps, known as desert kites, were used to capture and kill large numbers of goitered gazelles as they migrated (Bar-Oz et al. 2011; Zeder et al. 2013). In the mid to late 1300s, Muslim armies under the command of Timur Leng hunted goitered gazelles, killing about 40, 000 each year (Svizzero 2019). Despite this intense human predation, the goitered gazelle population was able to survive until recent times. However, modern economic development of arid regions, the penetration of human beings into the most remote areas of the deserts, and the widespread use of more and more accurate

D. Blank, *The Behavioral Biology and Conservation of Goitered Gazelles (Gazella subgutturosa)*, https://doi.org/10.1007/978-3-031-83504-9_20

firearms have led to the catastrophic decline of this species all over its range and fragmentation of most gazelle populations (Svizzero 2019).

Despite this devastation, the number of gazelles within the former USSR was still significant into the mid-1980s and was estimated to reach up to 30 thousand individuals. At the same time, the gazelles in this region showed a distinct tendency to decrease in both number and area. In the mid-1980s in Kazakhstan, their range was cut almost in half and their numbers declined by 50%. During this time, the area of distribution of goitered gazelles in Karakalpakstan (western Uzbekistan) decreased by 3 times, and their numbers dropped by 10 to 12 times. In the rest area of Uzbekistan, the area inhabited by goitered gazelles also shrank significantly and those still left have a mosaic structured territory. The gazelles' main migration routes were also disrupted (Marmazinskaya 1999). Over the past 25 years, the gazelles' area in Turkmenistan has decreased by 6 times, and their numbers by 2.5 times. Only in Azerbaijan has there been a noticeable increase in the goitered gazelle population. In a vast part of what was once the goitered gazelles total range, the state of the remaining populations is alarming. Even in the most viable regions, the position of this species is far from safe or stable. The fragmentation of the area into separate populations, the steady tendency to further reduce these areas, their decrease in the number of isolated groupings, and populations that have all arisen at the same time are all typical signs of degradation of the species. The main reasons for the steady deterioration of populations and habitats of the goitered gazelle in their natural desert environments are the large-scale economic development of these areas, the generally low levels of protective measures and thriving poaching (Figs. 20.1—M029 and 20.2—7685) (Flint and Prisyazhnuk 1986).

To the present time, experience has been accumulated showing that, under certain conditions, goitered gazelles largely lose their fear of humans and can inhabit densely populated landscapes that have been heavily transformed by humans. In the protected areas of the Shirvan steppes (Azerbaijan), gazelles graze in close proximity to flocks of sheep and busy highways. Examples of the rapid habituation of gazelles to humans are known from the reserve on Barsakelmes Island (Kazakhstan) and the Bukhara Breeding Center (Uzbekistan). Therefore, it can be argued that, with sufficient reason, goitered gazelles can be restored to their desert landscapes experiencing moderate anthropogenic stress, if the poaching pressure is completely removed and some biotechnical measures are carried out (Flint and Prisyazhnuk 1986). The areas of these territories in Kazakhstan and throughout Central Asia are huge, and it is key to first restore the goitered gazelle numbers and then explore the future economic use of this species. It is essential to create a thorough conservation strategy in order to preserve the goitered gazelle as a wild and free species. Among these measures is to:

1. Effectively stop motorized and other types of poaching altogether. Hunting gazelles was once traditional for the peoples of Central Asia (Fig. 20.2), and according to local beliefs, goitered gazelle meat has healing properties, which means it can bring a high price on the "black market." The pursuit of a gazelle on a motorcycle is not difficult, but despite the mass nature of poaching, it is soli-

Fig. 20.1 Despite the fact that goitered gazelle hunting is strictly prohibited by law in all Central Asian republics, goitered gazelle poaching for meat is a very common phenomenon. Poachers usually take only gazelle meat and leave the skins and legs at the crime scene. This is still fresh gazelle skin found in a small niche between the stones. Ustyurt, Kazakhstan. 28.04.2011. Photo by Mark Pestov with kind permission

Fig. 20.2 Hunting goitered gazelles was traditional business for local people for a long time. Iron traps near rare watering holes is traditional poaching way for local people in Ustyurt now. SW Kazakhstan. 25.04.2011. Photo by Mark Pestov with kind permission

tary. Therefore, fighting against poaching, especially in sparsely populated areas, presents great challenges, which are exacerbated by the high dispersion of animals throughout the desert terrain. The basis of the fight should be the purposeful activity of the republican and regional hunting inspections, manifested in the form of special, previously planned raids to the places of the greatest concentrations of gazelles attractive for poachers. Of great importance in the fight against poaching is the new scale of penalties for hunting of animals listed in the Red Book of the USSR, which set a fine of 800 rubles for each gazelle killed (Flint and Prisyazhnuk 1986). Now the fine is around 230,000 rubles (up to 3000$) and maybe a sentence of imprisonment of up to 5 years as well.

2. Create protected areas. For the preservation of goitered gazelles, as experience has shown, it is not necessary to completely remove protected land from the sphere of economic use. Also of consideration, the creation of special reserves is a difficult and long task. Therefore, the main form of these reservations for the restoration of goitered gazelles should be recognized as a provincial reserve, on territory where livestock grazing is allowed under certain conditions (e.g., the absence of weapons and dogs), but hunting is completely prohibited. Provincial nature reserves must have a staff to enforce protections, and at the same time, their organization needs to work with local communities to minimize resistance from local land users against creating nature reserves. Sanctuaries should be of significant size and created in places potentially suitable for gazelles. It is also necessary to carry out some biotechnical measures in the area of nature reserves, with the first step to create artificial watering holes. The creation of these nature reserves should be included in the state perspective plan for the development of protected areas (Flint and Prisyazhnuk 1986).
3. Create a network of nurseries for breeding goitered gazelles. Specialized gazelle nurseries can perform several important functions: genetic banks, providing diversity in the population and a scientific basis for studying the biology of the gazelle to develop sustainable plans for its economic use; a place to create a reserve of goitered gazelle stock for reintroduction and creation of new populations in protected areas—it is because of nurseries that it is possible to maintain wildlife populations in territories specified as nature reserves and without the presence of local animals, and nurseries can be considered as experimental ranges for the partial domestication of gazelles and also used as public information centers. Recent attempts to belittle the importance of nurseries as a means of preserving gazelles have no scientific justification, and the network of nurseries should expand (Flint and Prisyazhnuk 1986).
4. Strengthen public information and education for the protection of goitered gazelles. Promotion of the species protections and explanations of its biological and economic value are important links in the general strategy of maintaining successful safeguards. Information should be widely distributed by the mass media (newspapers, television, radio, cinema), as well as through the release of special posters with text in all languages of the local peoples of the countries on whose territory the goitered gazelles live. It is very important to promote all the efforts and successes in the restoration of the goitered gazelle, the work of the

nurseries, the need to fight against poaching, and explanations of the content of legislative acts on the protection of wildlife, as well as measures and penalties used within the responsibility for violations. It is important to emphasize that the strategy of restoring the number of goitered gazelles can be effective only if all these measures are thoroughly and successfully implemented regardless of the complexity or duration of the work (Flint and Prisyazhnuk 1986).

20.2 Status in Some Regions

20.2.1 Kazakhstan

The goitered gazelles once had a continuous distribution throughout the arid regions of Kazakhstan, with their northern borders extending to 46°–48°N. In the 1930s, their population was estimated to be over 200,000 individuals. However, their numbers steadily decreased, and by the late of the 1970s, solely around 10,000 remained (Afanasiev et al. 1953; Sludskiy 1977; Zhevnerov et al. 1983). The primary contributors to this decrease were excessive hunting activities and human economic activities, including overgrazing as a result of the pastoral livestock breeding system, which occupied all watering holes with domestic ungulates (Fig. 20.3—M0513) and used guard dogs for protecting livestock. At the same time, sheepherders and guard dogs hunted wildlife, while the shepherds' yurts served as shelters, aiding

Fig. 20.3 Domestic livestock occupy rare found watering holes in deserts and goitered gazelles forced to migrate or die. Kyzylkum Desert. 20.06.2018. Photo of Mark Pestov with kind permission

visiting poachers. In the decades of the 1940s and 1950s, new areas were rapidly opened up for the promotion of various industries. With the continual expansion of cultivated areas, the amount of available lands decreased, while the number of domestic livestock grew. This led to overgrazing, resulting in 63–64% of pastures becoming degraded (Kurochkina and Karibaeva, 1986). In addition, goitered gazelles are particularly vulnerable to the effects of winter dzhuts (lack of food due to icy pastures), which are caused by a lack of food due to icy pastures. This causes periodic mass mortality, which occurs every 7 to 8 years on average (Bekenov et al. 2001). Another powerful negative factor is the industrial development of deserts, such as mines, road construction, and geological exploration projects.

However, the most negative factor is illegal hunting or poaching, which leads to the mass extermination of goitered gazelles from all arid areas, even those that are the most remote and difficult to access. The government harvested gazelle meat intensively between 1942 and 1950; in the Ili Valley, it was killed by soldiers 14,000 gazelles at least during 1944 alone (Sludskiy 1963; Zhevnerov et al. 1983). In addition to the legal harvest, poaching was completely uncontrolled. Apart from the legal hunting, poaching was left unchecked. This led to a rapid decline in the population of goitered gazelle, with their numbers disappearing from most areas of their natural habitat.

In the 1970s, there was an improvement in the animal protection system which contributed to the progressive restoration of the population of goitered gazelles in Kazakhstan. During the 1980s, their numbers ranged from 30,000 to 50,000 individuals. However, the population which previously had a continuous range, split into many small populations as a result of fragmentation. The largest numbers were found in the Mangistau District in Western Kazakhstan and in the Altyn-Emel National Park in the Ili Valley (Southeastern Kazakhstan). The goitered gazelle population trends during the 1980s were diverse, with a slow rise in their number in the two large populations and a decline in the smaller, fragmented parts. Factors that contributed to the growth of the larger populations included a low human population in these areas, control of oil and gas production development, the establishment of new conservation areas in Mangistau, enhanced protection in Altyn-Emel National Park, and a decrease in grazing pressure by domestic livestock. On the other hand, causes that negatively affected the smaller populations included an increase in illegal hunting, the establishment of agricultural enterprises without considering environmental conservation issues, and natural habitat transformation (Bekenov et al. 2001).

In the early 1990s, the former USSR collapsed and five Central Asiatic republics gained independence. As a result, the former system of governmental protection organizations was eliminated, and new conservation efforts and organizations only began to form. In addition, open borders with surrounding countries allowed for the free flow of humans and control of the region by a central government under a trans-border exchange of goods pact weakened. Moreover, the free, unregulated selling of hunting rifles was permitted and the rate of poaching started to increase at a great pace and became a common at all times and everywhere, even within areas of nature reserves (Bekenov et al. 2001).

The basic challenges facing to the conservation of natural wildlife in Kazakhstan are the escalating rates of poaching and natural habitat devastation. To address these issues, the legal framework governing these activities must be enhanced, and current legislation must be strictly amplified. The hunting business has to be granted an independent economic status comparable to that of other sectors, such as industry and agriculture. This will facilitate the establishment of new areas for the conservation of wild species, including goitered gazelles and other ungulates. The authority of hunting inspectors must be enhanced to reach a status of police officers, and more severe punishment for poaching must be implemented, including the confiscation of vehicles and weapons used for illegal hunting. The abilities of state and regional conservation services must be improved through the provision of better mobility, reliable connection with colleagues, and other tools. Given the critical need to increase protection against poaching, additional funding must be allocated to the hunting inspectorate to ensure its effective operation. New protected territories of various ranks are necessary to create for saving present populations from vanishing. First of all, new nature reserves in the Taukum Desert, Kyzylkum Desert, and Buzachi Peninsula, each with an area of 30,000 to 50,000 hectares at least, are necessary to establish; a large new national park should be organized in the left bank of the Ili Valley and connect it to the existing Altyn–Emel National Park on the right bank of the Ili River. Additionally, wildlife sanctuaries should be founded in Muyunkum, Zhabaushkan, and Besbulak, each with an area of at least 100,000 hectares (Bekenov et al. 2001).

During the 2010s, a new issue was introduced in Kazakhstan that negatively impacts conservation efforts: constructing barbed wire fences along the entire border between Kazakhstan and Uzbekistan, as well as between Kazakhstan and Turkmenistan in the southern Ustyurt. This is a standard fence of eight horizontal rows of barbed wire fixed on posts, supplemented with diagonal rows. The fence height is around 150 cm, with a distance between posts of 4 m and a gap between wire rows of 20 cm. It is obvious that such barriers are a significant obstacle for ungulates, and after the installation of these border barriers, there is serious difficulty in the traditional cross-border migrations of ungulates (Pestov et al. 2018). Moreover, there are two passages with a length of 6.5 km through the fence from the Kazakhstan side, but no passageways exist in the border fence from the Turkmenistan side. As a result, goitered gazelles penetrate through this passage into the area between the two fences, and when the border car appears, the gazelles cannot find their way out and run from the cars until they fall from fatigue and die. In the southern Ustyurt Plateau, goitered gazelles drink from puddles after the spring snow melt until the paddles dry out, and then they start searching for water. In the huge waterless spaces of the Usturt Plateau, artificial water sources are very rarely found (Fig. 20.4—8300), so most goitered gazelles try to migrate to the Uzbekistan side, where the Sarykamysh depression contains great amounts of water, but the border fence blocks their way. Goitered gazelles walk along the fence for hours but cannot find passage. They try to penetrate the fence by force and die on the barbed wire (Fig. 20.5–6206). Some goitered gazelles try to jump over the fence, but it is too high. Their hind legs get entangled in the rows of wire, where they hang helplessly

Fig. 20.4 Artificial water sources are very rarely found in the Kyzylkum Desert, while most areas of the region are waterless. Kyzylkum Desert, Kazakhstan part. 10.05.2021. Photo by Mark Pestov with kind permission

Fig. 20.5 The desire to overcome the barbed fence is so strong among the goitered gazelles that they walk along it for hours, trying to find a passage and because of repeated attempts to overcome the barbed fence by force, they rub their necks and die from blood loss. A dead female goitered gazelle at the Kazakhstan-Uzbekistan border fence. Ustyurt Plateau, W Kazakhstan. 16.09.2019. Photo by Mark Pestov with kind permission

Fig. 20.6 Some goitered gazelles tried to jump over the fence, but they cannot do it and they get entangled in the fence and die on the wire fence, and only the legs and torn bones of the skeleton remain hanging on the fence at the end. Limbs in the fence. Southern Ustyurt, W Kazakhstan. 18.09.2019. Photo by Mark Pestov with kind permission

on the wire fence until they die, leaving their corpses to be eaten by various predators, with only the legs and torn bones of the skeleton hanging on the fence (Fig. 20.6—M6308). This fence is barbaric and has an impact on goitered gazelles much crueler than intensive poaching, because many gazelles die trapped in what can only be considered great thirst deprived of the opportunity to get to the water (Pestov et al. 2018).

20.2.2 Mongolia

The Gobi Desert in Mongolia is a vast expanse of arid terrain that serves as a crucial sanctuary for several threatened large steppe herbivores. The most disastrous feature for gazelles in this habitat is the snowy winters, which periodically occur in deserts. A sharp drop in the number of goitered gazelles was observed after snowy winters in all parts of their range (Sludskiy 1963), which is due to the relatively inability of gazelles to migrate over long distances compared to saiga antelope. The comparatively low reproduction rates and the special type of use of their territory by goitered gazelles determine the great vulnerability of populations to the effects of natural and anthropogenic factors. At present, in the context of a free market economy, socioeconomic transformations have impacted all regions of Mongolia. Infrastructure advancements, particularly in the context of mining operations, lead

Fig. 20.7 The Mongols have been hunting goitered gazelles for centuries. However, with the advent of modern hunting weapons and high passable cars, poaching is emptying vast spaces in the Mongolian Gobi deserts. The same sign in different scales (general view - "**a**" and close-up view - "**b**") demonstrating that gazelle hunting is prohibited was used as a target for which a gun was fired, which demonstrates the real disregard of the local population for animal protection. 16.06.2012. Great Gobi B, Mongolia. Photo by P. Kaczensky with kind permission

to fragmentation and destroyed habitats (Mallon and Jiang 2009; Kaczensky et al. 2015). Overstocking with livestock results in competition for continuously degrading pastures (Sheehy et al. 2010; Berger et al. 2013). The huge demand for wildlife products leads to high levels of illegal hunting (Pratt et al. 2004; Wingard and Zahler 2006), and widespread poaching with motor vehicles plays a decisive role in reducing the range and population (Fig. 20.7–5738, 5738) (Bugaev 1989). The Gobi Desert supports the largest population of goitered gazelles, which was estimated at 60,000 in the early 1990s, but has now declined to less than 30,000 at present (Buuveibaatar et al. 2017).

The Mongolian Gobi Desert is a home to the greatest existing population of migrated ungulates in the world, including the largest remaining population of goitered gazelles, with an estimated number of 60,000 individuals in the early 1990s (Ito et al. 2017). The railway connecting Russia and China cuts through this ecosystem, and there is a worry that it could hinder the movements of wild hoofed animals. Research has shown that most of the animals monitored on either side of the Ulaanbaatar–Beijing Railway never cross the railway. The construction of additional railways to support mining projects in the region poses an additional threat to the ungulates' important migration patterns across Mongolia. The railway, serving as an obstacle to gazelle migrations, directly and indirectly contributes to their mortality. The Mongolian harsh winters compel the Mongolian gazelles and other large herbivores to undertake long-range migrations in search of food. The presence of the railway acts as a hindrance, blocking their access to certain necessary food resources (Ito et al. 2017).

Thus, poaching is the primary reason for the decline in population size and distribution at present. The construction of railway lines will lead to the fragmentation of populations, resulting in a significant decrease in population numbers in the near future. The exploitation of natural resources, such as mining, the increasing presence of humans and their activities, and the intensification of livestock husbandry in most important habitats for wildlife, such as occupancy of oases and natural springs, the construction of fences along railway lines, and other human activities, all contribute to the limitation of migrations and movements. This, in turn, limits access to pasturing areas and water sources (Clark et al. 2006). As livestock populations grow in these habitats and habitat range areas, overgrazing becomes a serious issue, leading to the degradation of pastures and competition for remaining forage and water (Lhagvasuren and Milner-Gulland 1997).

According to the research conducted by Clark et al. in 2006, the goitered gazelle is classified as a species of concern at the international and regional levels. The species is also listed in the Convention on the Conservation of Migratory Species of Wild Animals (CMS) Appendix II, which provides legal protection for the species. In Mongolia, the goitered gazelle is legally protected by the Mongolian Laws on Fauna and Hunting, which prohibit hunting. The species is also listed as a critically endangered species in the second edition of the Mongolian Red Book, published by Shiirevdamba et al. in 1997.

20.2.3 China

Goitered gazelles were once common in western and northern China, in Xinjiang, northern Qinghai, Gansu, and Inner Mongolia. Isolated populations of separate subspecies (*G.s. yarkadensis*) are found in the Tarim Basin. In Inner Mongolia, goitered gazelles were once abundant. These gazelles can still be observed mainly close to the Mongolian border, while many other populations were exterminated. And an isolated population in the Qaidam Basin is close to extinction. The main threat is poaching and habitat destruction. It is crucial to intensify conservation efforts and establish protected areas, particularly in the Tarim and Qaidam basins, as suggested by Jiang and Sung in their 2001 study. In addition, goitetered gazelle have been insufficiently studied in this region, with most populations not studied at all.

20.2.4 Caucasus

In Caucasus, goitered gazelles lived in Azerbaijan, Georgia, and Armenia (Heptner et al. 1961). In Azerbaijan, gazelles were widely distributed in the semi-desert plains and foothill areas of the country (Sarukhanova 2021). Former distribution in the South Caucasus covered the Karabakh-Milsk Steppe, Mugan Steppe and KuraAraks lowlands, extending northwest to the Shiraki and Eldari steppes of

Georgia. In the northeast, they occurred on the Apsheron peninsula and up to about 41^0 N. In the nineteenth century they probably also existed in Armenia, in the Araz (Araks) Valley between Nakhchivan and Erevan (Heptner et al. 1961).

Range and numbers declined rapidly in the twentieth century, especially from the 1930s onwards, as a result of intensive hunting and wide-scale cultivation of the steppes. In the 1930s, hunting with the help of cars and trucks became a very large industry and the goitered gazelle population was destroyed in many regions. At the Karasu station and another small area, poachers extracted more than 500 to 600 gazelles per year by car. The same number was extracted by shepherds during the wintering of their cattle. By 1937, the spread of goitered gazelles in Azerbaijan was confined to eight isolated areas of semi-desert and poorly developed areas of the steppes and hummocks, which significantly decreased the population. In subsequent years, the range and number of goitered gazelles continued to decline rapidly with the development of motor transport, poaching and the agricultural expansion of semi-desert areas for growing cotton (Vereshchagin 1959).

An estimated 50,000 to 60,000 goitered gazelles were present at the start of the twentieth century but by the end of the 1940s only 5000 remained, all in Azerbaijan. By 1961 they had dwindled to only 130 in an area near the mouth of the Kura River on the Caspian coast. Byandovan Sanctuary (zakaznik) was established in 1961 to conserve this relict population (Heptner et al. 1961) followed by the adjacent Shirvan State Reserve in 1969. By 1972, the population of the gazelle had grown to 1500. Ten years later, in the 1980s, the number had increased to approximately 3500. In 2003, an area of 54,000 hectares was designated as the Shirvan National Park, with the primary objective of protecting the goitered gazelle. As a result, the population increased steadily to 4500 by 2003 and an estimated 6000 in 2012, with the current population in the park now staying around 6000 (Svizzero 2019). This represents >95% of the South Caucasus population. Small groups are still found elsewhere in Azerbaijan (Askerov et al. 2021) including sites in the NW of the country not far from the Georgian border (Korchay Sanctuary, Palandeken mountains) (Mallon et al. 2012).

In Georgia, their former distribution covered the steppes of the semiarid zone in the southeast, and in the 1870s, their range extended up the Kura/Mtkvari Valley as far as Avchal, about 10 km west of Tbilisi (Radde 1899). The species disappeared from Georgia by the 1930s due to uncontrolled hunting, encouraged in part by government policy designed to maximize livestock production. A few individual gazelles were reported on winter grazing grounds in the 1980s that had presumably wandered there from small relict populations in Azerbaijan (Mallon et al. 2012).

20.2.5 Turkmenistan

Until relatively recently, goitered gazelles were found almost all over the territory of Turkmenistan (76% of the country). In the nineteenth century, the number of gazelles was high within the entire habitat area, even close to human settlements and oases. Although the gazelles were a part of traditional hunting practices for humans, the gazelle population managed to survive. A noticeable decrease in the number of goitered gazelles and a reduction in their habitat in Turkmenistan began to be observed at the end of the nineteenth century, due to the construction of the Transcaspian railway, bringing a human increase in population in the cities and the arrival of new settlers. In many developed regions, gazelles were completely exterminated by the beginning of the twentieth century. In subsequent years, the decline of the gazelle population has continued along with the development of new lands (Kucheruk 1995).

During the 1930s, state procurements of wildlife began and in 1932, the state organization hunted 13 thousand gazelles. Then over the next five-year period (1933–1937), Turkmenistan hunted a total of 150 thousand gazelles (Laptev 1935), though during 1940s, goitered gazelles were still numerous and common (Gorbunov and Zarkhidze 1985). During that time, a new method of hunting gazelles with cars was invented, allowing humans to hunt both day and night. Gazelles were hunted during the days by being chased with cars that pushed them to exhaustion, and during the nights by being stunned with strong headlights that made them easy prey. Goitered gazelle hunting occurred all year round everywhere a car could travel without limitation. In addition to poaching, the habitat in western Turkmenistan has deteriorated. In these areas, saxaul (*Haloxylon persicum*) and large shrubs have long been devastated and shrubs of tasbyurgun (*Nanophyton erinaceum*) have been uprooted. In the 1980s, the cutting of wormwood, biyurgun (*Anabasis salsa*) thickets, saxaul, and *Salsola* thickets reached dangerous quantities.

Goitered gazelles were completely exterminated in most parts of Turkmenistan and the last refuge was the Turkmen Ustyurt, where it was still possible to see gazelles in significant numbers. However, even in those areas, the goitered gazelles were poached and became rarely observed during the early 1970s, and the Turkmen population split into a few small, isolated ones (Kucheruk 1995). Table 20.1 demonstrates the gradual decrease in the number of goitered gazelles in W Turkmenistan and the southern Ustyurt Plateau. Now only 3% of the area from their former natural habitat is still occupied by goitered gazelles (Gorelov 2001).

Table 20.1 The number of goitered gazelles and its changes during 1955–1983 according to the data of car surveys in western Turkmenistan and southern Ustyurt Plateau (adapted from Gorbunov and Zarkhidze 1985)

	Routes length (km) in different areas and average number of goitered gazelles (per 100 km of the route)									
Years	Krasnvodskoe Plateau		Zauzbokskiy district		Southern Ustyurt		Steppe Mangyshlak		Southwestern Turkmenistan	
1955–1959	4645	68.9	1009	101.3	1120	71.0	710	60.0	590	95.6
1960–1964	3005	14.6	707	49.6	620	66.0	610	10.5	595	40.1
1965–1969	725	6.9	695	37.3	590	56.5	–	–	1850	8.2
1970–1974	490	0.6	1960	1.4	1900	8.1	–	–	1210	0.7
1975–1979	688	0.7	695	4.0	1500	4.4	510	0.1	–	–
1980–1983	520	0.8	6250	0.4	3540	3.4	–	–	–	–

The main reason for the catastrophic decline in the number of goitered gazelles was widespread poaching, especially hunting them from under the headlights of a car, as well as chasing them on motorcycles, with the numbers of both types of vehicle increasing among the local population very quickly during 1950–1960s. On moonless nights, it was always possible to see the reflections of the headlights of cars hunting gazelles. In the field of view, as many as 10 to 12 light reflections or more could be seen at one time. Whole poaching "expeditions" were organized to the southern Ustyurt Plateau with some cars in pursuit and hunting and others used for stacking up the dead. In the 1960s, it was a common occurrence to see cars filled to the brim, stacked with their maximum load of goitered gazelle corpses drive into settlements. At that time, a single carcass could be bought for 5 to 10 rubles. Often the heads and skins were wastefully thrown away when the corpses were butchered. In the Depme Gorge, for example, it was possible to photograph large piles of gazelle legs that accumulated during the taking of several thousand skins. This persecution of goitered gazelles was conducted all year round, and neither pregnant females nor fawns were spared. With such a senselessly cruel and wasteful shooting of all gazelles that got in the way, the number of these once numerous animals decreased very quickly (Gorbunov and Zarkhidze 1985).

A number of measures have been taken to preserve the goitered gazelle in Turkmenistan. An important task is the need to create new watering holes near places of stable concentrations of gazelles, at crossings and in areas of migration, near wells, settlements, economic facilities, and weather stations. If proper protections are observed, the proximity of people is not an obstacle to gazelles, even visiting reservoirs that are located near populated areas. Gazelles can use artificial watering holes from the 1978 pipeline and its branches that run for 180 km from the Karakum Canal to the central areas of the Karakum Desert. In 1983, gazelles were noted to use these watering holes at night. Other potential sources of water for gazelles are reservoirs and transit channels with discharge collection and drainage water. In the republic, more than 5 km^3 of poorly mineralized drainage waters are discharged into the desert annually, entering the Sarykamysh depression and a number of other reservoirs. The main discharge of drainage waters of the Murghab oasis forms a chain of lakes and spillways with dense thickets of reeds and tugai-type vegetation with an area of 500 hectares. The conduits of large collector channels passing in desert areas constitute guaranteed, reliable watering holes in large areas (Rustamov et al. 1986).

The strategy for the protection of gazelles should be established considering the possible consequences of anthropogenic transformation of the natural environment. Turkmenistan has plans to expand the existing road network (about 800 km) and to build a Turkmen drainage channel from the collector of about 900 km to divert the discharged drainage waters from a whole group of oases. The collector will be laid across the entire republic in a latitudinal direction. It will dissect some concentration areas and cross the migration routes of goitered gazelles. Conditions for watering gazelles in the future should also be improved in connection with the proposed watering of pastures and the transfer of part of the flow of northern rivers to the south of the country. Significant changes in the appearance of the new landscape of the republic will also occur in connection with the development of irrigated agriculture and industrial development. Therefore, it is advisable to develop measures for the protection and restoration of gazelles and other species in conjunction with plans for the development of various natural resources for supporting growing populations (Rustamov et al. 1986).

To preserve gazelle populations, the Government of Turkmenistan has taken a number of measures. In 1982, work began on breeding gazelles in a semi-free state in the Gyaursky special nursery of the Nature Protection Society of Turkmenistan. Since February 1982, a special reserve has been established on Ogurchinskiy Island (Caspian Sea) to create a new population of gazelles. The absence of a single proven method of trapping goitered gazelles, however, has significantly hindered the systematic resettlement effort. In Turkmenistan, three methods of trapping are used: net trapping, immobilization, and the manual trapping of newborns. Analyzing the accumulated experiences of catching gazelles with the help of immobilizing agents and nets, the most effective and bloodless method was net trapping. In 1982, when catching gazelles for settlement on Ogurchinsky Island, 5 individuals (22%) died out of 23 caught using immobilization, and in 1983, only 2 died out of 12 captured gazelles (12%) (Rustamov et al. 1986).

When netting gazelles in Badkhyz, a shaded trailer with a specially equipped side of the car (Ural-675) was used a temporary detention pen. When after detention in this car for 7 days, the gazelles were well fed and willingly ate the food offered. Transportation the gazelles to the Gyaur nursery was carried out in the back of a truck, covered on the inside with slabs of reeds and a bottom of hay, the top the sides was covered with a tarpaulin. The journey took 11 h. With this method of transportation, all gazelles were delivered to the nursery without injury (Rustamov et al. 1986).

To restore the number of gazelles in Turkmenistan at the present time and in the future, it is advisable to:

1. Provide and implement the protection of migration routes, passages, local concentration sites, seasonal pastures, watering holes, birthing areas; the main part of the funds should be used to ensure reliable protection of natural populations.
2. Stop taking animals from natural populations; if it is necessary to create new nurseries, it is advisable to complete their breeding stock at the expense of individuals raised in previously established nurseries.
3. Create sanctuaries and other protected areas at the expense of ministries and departments that develop and exploit various natural resources of the republic.
4. Recommend the construction of watering sites to relevant production organizations, considering existing and planned sources (waterways for watering pastures, freshwater channels for irrigation, large collector channels, reservoirs, wells).
5. Have the Ministries of Forestry and Agriculture propose the establishment of feeding stations.
6. Improve technical equipment for environmental inspections (high-traffic vehicles, communications, technical equipment).

7. Combat the current alarming situation of the goitered gazelle requiring the creation of an "Operation Goitered Gazelles" project with the participation of many republican and local organizations, whose representatives will be responsible for the restoration and on-going conditions of goitered gazelles in Turkmenistan (Rustamov et al. 1986).

In Turkmenistan, efforts were made to establish captive breeding centers and launch species reintroduction programs. In the 1980s, attempts were made to establish gazelle breeding centers in Western Kopetdagh and Gyaurs. However, the most successful results were achieved by relocating animals to Ogurchinsky Island in the Caspian Sea, where the sea served as a natural barrier. Several dozen animals were released on the island, and the population grew to around 800–1000 individuals, which is the island's carrying capacity. When the population reached this number, some animals are relocated in other areas of Turkmenistan (Svizzero 2019).

20.2.6 Uzbekistan

The goitered gazelle once inhabited all desert and semi-desert areas of Uzbekistan. However, their population and distribution have significantly declined in the last 30–40 years. The distribution of the goitered gazelle is steadily decreasing and becoming fragmented, with isolated populations struggling to survive. At present there are seven sub-populations in various parts of the Kyzylkum Desert and the Ustyurt Plateau (Marmazinskaya and Mardonov 2004). The decrease in the number of goitered gazelles is the result of both anthropogenic (cattle grazing, poaching, irrigation) and natural (snowy winters, ice crust covering the snow) factors (Marmazinskaya 1999), though gazelles were able to cope with natural factors for hundreds of years and eventually restore their population numbers. Overgrazing by livestock and developing an irrigation system also played a significant role in their decrease, but goitered gazelles are a very adaptive species to the presence of humans and could learn to cope with the onset of economic activity. However, uncontrolled poaching occurring all over the gazelle's natural habitat, which is accompanied by the development of the deserts, does not leave the goitered gazelles a chance to survive in places where they once did (Blank 1997). As a result, the goitered gazelle has been included on the list of endangered species and listed in the Red Data Book of the former USSR (1984), as well as in the Red Books of the countries of Transcaucasia and Central Asia. In addition, this species is listed in category 2 on the IUCN scale as a species whose numbers are low and rapidly declining and under threat of extinction (Marmazinskaya 1999).

20.2.7 Tajikistan

Goitered gazelle is protected species in Tajikistan and is listed in the national Red Data Book (Abdusalyamov 2001). A few of these graceful creatures still reside in the semi-desert area of the Tigerbulk Nature Reserve, which covers approximately 50,000 hectares. Efforts to establish a captive breeding center at Tigerbulk were made, but were halted in 1992 due to the economic crisis in the country. The Karatau Nature Sanctuary, covering 14,400 hectares, is located in the southwest of the country within the gazelles' natural habitat. To ensure the survival of this species in Tajikistan, it is essential to enhance protection measures for the remaining gazelles by providing transportation and necessary equipment for wildlife protection. However, the only way to ensure the survival of goitered gazelles in Tajikistan is through captive breeding due to release gazelles into areas of their natural habitat range. It is crucial to upgrade the Karatau Nature Sanctuary into a fully functional nature reserve and enlarge its area to incorporate the entire Karatau Range. Goitered gazelles can then be released into Tuyuntau, Aruktau, and the Tereklitau Hills, as well as other locations in the southwest. In northern Tajikistan, a captive breeding center with an area of 5000 hectares should be organized in the Akbel'tau area. Additionally, a public awareness campaign should be launched throughout Tajikistan to highlight the importance of gazelle conservation. This campaign aims to engage the local population in the conservation of goitered gazelles, as described by Abdusalyamov in 2001.

20.2.8 Turkey

It was reported by Güldenstaedt (1780) that the species *Gazella subgutturosa* was native to Anatolia. Goitered gazelles, as they commonly known, live in a wide-open area in Turkey. (It is very possible that *G.s. marica* lives there.) Goitered gazelles are found in Turkey but only in the very limited area of the Ceylanpinar state-farm near the Syrian border. This population consisted of 1500 to 2000 goitered gazelles in 1968. Erkan (1978) stated that in the past, herds of approximately 100 gazelles were not uncommon in the vicinity of Ceylanpinar. However, by the mid-1960s, the size of these herds had dwindled to 40 to 50 individuals. Despite the fact that gazelles have been protected by law in Turkey for an extended period, hunting, capturing gazelle fawns for pets, and the use of agricultural chemicals remain the primary causes of their decline (Erkan 1978). During the 1970s, the goitered gazelle population decreased drastically, and its estimated number in Turkey was around 200–300 gazelles (Ogurlu 1992). About 300 gazelles were estimated to survive in the wild during end-1990s (Ölçer 2001). However, in the past, there were large herds of gazelles in the wild around the towns of Viransehir and Akcakale, as well as in the Karacadag area in Southeastern Anatolia. A survey conducted between 2001 and 2002 by the GAP Biodiversity Research Project, which was part of the UNDP's Southeast Anatolia Project, revealed that there were no viable populations of gazelles in the region.

In Turkey, as elsewhere, poaching constitutes a significant threat to large mammals, including the goitered gazelles. This is one of the primary reasons of the decline in the population of this species. Despite a ban on hunting, a significant portion of the goitered gazelle species has been affected by poaching. The capture of newborns during the breeding period has also played a role in the decline. Certain people in the community are drawn to the notion of adopting fawns as pets, attracted by their charm and the relative ease of their taming. The use of pesticides by farmers in the 1950s has been another major factor in the decline of gazelles in Turkey. These chemicals can harm not only pests but also nontarget organisms living near the areas where pesticides are applied. They can be dispersed into the soil and water, affecting many herbivores and carnivores (Mallon and Kingswood 2001; Mallon 2008; Durmuş 2010).

20.2.9 Iran

In Iran, the goitered gazelle populations are spread widely all over the country and once can be found in areas with steppe or semi-desert landscapes. They do not inhabit mountain ranges, the Caspian Sea region, or the far southern parts of the country (Karami et al. 2002). The almost continuous distribution all over the country was fragmented into many small populations, living mainly within protected areas. In many Iranian areas, goitered gazelles have become extinct. The number of goitered gazelles has decreased due to illegal hunting, the destruction of habitats caused by excessive grazing and the eradication of bushes, and abandonment of protected areas and transformation them into agricultural land, construction sites, mining sites, and military installations (Karami 1992).

The goitered gazelle population in Iran has suffered two phases of decrease. The first decline was due to the rapid introduction of semi-automatic rifles and off-road vehicles in the late 1940s and early 1950s. In the year 1967, the Game and Fish Department of Iran took action to safeguard goitered gazelles and their natural habitats through legislation. This resulted in the recovery of the populations of this species. The second downturn of the gazelle number and range occurred during and shortly after the Islamic Revolution in Iran in 1979, when the established framework of conservation organizations came to an end. The eight-year period of conflict and the Islamic Revolution had a severe detrimental impact on the environment and wildlife conservation. About 70% of protected areas and 50% of wildlife sanctuaries have become completely ineffective in providing habitat for wildlife. In summary, between 50% and 80% of wildlife species have suffered damage, and 80% of protected areas have been affected. The population of gazelles, which once numbered in the thousands, has shrunk by a factor of ten over the past two decades. The population of goitered gazelles has been significantly reduced in 10 protected areas, and in other areas they are on the verge of extinction. The reform of the Department of Environment in Iran during the 1980s enabled the regulation of illegal hunting and the restoration of some gazelle populations. However, smaller populations were wiped out within the protected areas. The Mooteh Wildlife Sanctuary now houses more than 2000 gazelles, and four other sanctuaries now have more than 1000 goitered gazelles (Hamami and Groves 2001).

The Iranian population of goitered gazelles appears to have experienced a loss of genetic diversity due to the limitations imposed by their restricted range and their small population size. The observed poor genetic makeup and limited genetic variation among populations are likely the result of a combination of human-induced and natural landscape characteristics that have contributed to the isolation of these populations. At present, the majority of gazelle populations are restricted to protected areas that are bordered by urban areas and extensive road infrastructure. The answer is uncertain if there is a low gene pool among the assumed to be isolated populations, or, what might be the consequences of altering the landscape are on the genetic diversity of the gazelles. The nomadic lifestyle of goitered gazelles is currently constrained by human encroachment on their habitat, and this can have a significant impact on the health of their gene pool, because of restriction of the gene circulation between populations. In addition, the breakup of the landscape resulting from the fragmentation of the goitered gazelles' previously continuous distribution across the central Iranian plateau also may have impact on the genetic variability of the populations. During the mid-1970s, there were numerous goitered gazelles, but hunting activities have drastically reduced their numbers outside of protected areas. Currently, only about 20% of the original population remains (Firouz 2005). In years to come, however, human actions can play a substantial role in reducing richness of the gene pool. Consequently, it is essential to restore degraded habitats, reintroduce extinct populations to available suitable areas, and create suitable corridors for genetic exchange between remaining wild populations. These measures are crucial for preserving genetic diversity and ensuring the long-term survival of the goitered gazelle population in Iran (Khosravi et al. 2017, 2018). In some areas, reintroduction projects are underway in the Moghan Plains in northwestern Iran and in the Ghazvine Plains in the central part of the country. The Sohrein Plain had ca. 1000 to 1500 goitered gazelles in 2002 and is thus an important refuge for this species in the country (Zachos et al. 2010).

Comprehensive national studies are necessary to evaluate the current states and numbers of goitered gazelle populations. Additional research into the classification, genetic makeup, and environmental ecological interactions of Iranian gazelles is essential to ensure the successful conservation of existing populations. Establishing breeding centers, particularly for populations of goitered gazelles in the western Zagros, would significantly contribute to the recover former goitered gazelle populations in Iran. Gazelles are a reliable barometer for the health of drylands, and in suitable environments, they can thrive in large numbers, posing minimal competition to domestic livestock. The preservation of healthy gazelle populations in Iran is contingent upon the prevention of poaching within protected areas and the implementation of strict hunting regulations outside of these areas (Hamami and Groves 2001).

20.2.10 Afghanistan

The preservation of wildlife in Afghanistan began with the creation of royal hunting preserves in the early period of the previous century. Nevertheless, since that time, there has been no legal framework in place to regulate and manage protected areas or safeguard wildlife. Certain regions were designated by royal or presidential edicts, and in the 1970s, attempts were made to curb unregulated hunting. The Ministry of Agriculture's Department of Forests and Range was tasked with preserving wildlife and maintaining protected areas, while the Afghan Tourist Organization was responsible for managing commercial hunting activities. In the 1970s, the goitered gazelles were safeguarded by imposing restrictions on large-scale hunting. However, it was challenging to enforce the ban due to a lack of effective monitoring. In the 1970s, several protected areas were created, as documented by Habibi in 2001. Afghanistan joined the World Heritage Convention in 1979, as reported by the International Union for Conservation of Nature (IUCN) in 1992.

The armed conflict with the former Soviet Union from 1979 to 1991, and the subsequent civil strife, have significantly harmed the natural environment of Afghanistan, particularly its forests (Habibi 2001). With the outbreak of civil unrest, data on the current state of antelopes is scarce and hard to come by. It is to be hoped that the Afghan goitered gazelles can continue to be inhabitants of the country. Due to the abundance of goitered gazelles in low-lying areas, they were at risk of being targeted by large-scale hunting activities conducted by expatriate groups engaged in development projects in the western deserts of Afghanistan. With the advent of four-wheel drive vehicles, this form of massacre was perpetuated by the local population, leading to widespread devastation of the population. Due to the current political climate, there is limited information available about the status of the species at this time (Habibi 2001).

Goitered gazelles had no lawful safeguarding in Afghanistan before the war. It is unknown if they still exist in any protected areas. Study the current conservation status is needed to know if the species still exists in Afghanistan and evaluate the status of any remaining populations. If they are found, it is essential to protect them from hunting (Habibi 2001).

20.2.11 Jordan (*G.s. marica*)

The previous distribution of antelopes in Jordan is not well-documented, but it is clear that their populations have significantly decreased in both range and numbers over the past century. Similar to other Arabian countries, *G. subgutturosa* is currently regarded as endangered. There is a dearth of data on antelopes in the eastern and southeastern deserts of Jordan, which may be due to a lack of field researchers rather than a true absence of these species. Antelopes have been the target of hunting in Arabia for centuries, but two elements have been particularly damaging: the use of "desert kites" in the past, which were dry stone wall structures, and the introduction of an all-terrain vehicle for hunting in recent decades. In the eastern lava deserts, where gazelles frequent, kites were built in numerous locations. These kites were also constructed along the paths of gazelle migration routes. The use of kites for capturing gazelles resulted in the capture of hundreds of individuals, leading to a significant decline in the gazelle population in these regions (Mendelssohn 1974; Hatough-Bouran and Disi 1991) Lately, advancements in conservation measures, heightened public consciousness regarding wildlife preservation, and the establishment of a network of protected areas have contributed to a more favorable conditions for wildlife in Jordan (Kiwan et al. 2001). Within this short timeframe (Jan 2015 to Jan 2016), at least 23 Arabian sand gazelles were hunted and killed. The population of these gazelles in Jordan is currently in danger of extinction (Eid and Handal 2018).

Antelopes are legally protected, and hunting regulations have been in place since 1973. Jordan has three types of protected areas: wildlife sanctuaries, pasture reserves, and national parks. Preserves for wildlife are the sole category whose primary purpose is to safeguard the natural world, particularly endangered species and their habitats. The Shaumari Wildlife Reserve is home to a group of *Gazella subgutturosa marica* that are kept in captivity. However, the breeding program has not been particularly successful.

A semi-captive herd of Arabian sand gazelles is also present in the fenced Shaumari Wildlife Reserve, but they, too, have not bred well. The group by number of 22 in 1990–1991 and 14 in 1994. Furthermore, Quemsiyeh et al. (1996) suggested that the original captive population of ten animals donated to the RSCN in 1978 possibly had a hybrid between *G. subgutturosa* and *G. dorcas*. The Arabian sand gazelle is listed as Vulnerable by the International Union for Conservation of Nature (IUCN) in 1996. The idea of introducing Arabian sand gazelles (Gazella marica) into the eastern desert from the Shaumari reserve has been put forward (Kiwan et al. 2001).

It is essential to determine the taxonomic position of the captive populations and to maintain captive breeding populations of pure *G. s. marica*. The sand gazelles from these herds should eventually be relocated to protected areas with suitable habitats. It is necessary to conduct surveys in the eastern desert in order to determine the current state of remnant populations and to ensure their protection. Arabian sand gazelles have been observed in the area of the proposed Burqu Reserve in the northeast (Kiwan et al. 2001).

20.2.12 Saudi Arabia

Large creatures like ungulates have been pursued with the aid of stone kites, which were likely employed in Neolithic times as walled enclosures for herding, trapping, and slaying gazelles (Rosen and Perevolotsky 1998), and many hunting drawings have been found in Stone Age petroglyphs all over Saudi Arabia. The increase in the number of people and associated land use for agriculture and livestock have created significant problems for wildlife survival. The historical record of the original gazelle populations is scarce. Since World War II, motorized water tankers and supplemental fodder for domestic livestock have resulted in strong competition for food, and motorized hunting has taken significant numbers of gazelles as well as other species (Williamson and Tatwany 1996). A significant decline in the population of the G.s. marica species in Saudi Arabia's natural environment was observed in the 1970s and 1980s (Nader 1989).

Following the creation of the Al-Khunfah conservation zone in 1988 (northern Saudi Arabia), the Arabian sand gazelle numbers were relatively stable between 200 and 400 individuals in the mid-1990s (Cunningham and Wacher 2009). Declines were mainly due to increased poaching since 1995, when 150 sand gazelles were still estimated to survive (Dunham et al. 2001). By 2005, only a small number remained and since then no official data were collected (Cunningham and Wacher 2009). In the Harratal-Harrah protected area (northern Saudi Arabia), the sand gazelles increased in number after protections were established in the late-1980s, increasing to almost 1000 individuals. In 2002, Wacher and Al Toum estimated a population of 240 to 500 gazelles, but after that the population decreased dramatically and apparently was close to extinction. Since 2005, no official data have been received from that area (Cunningham and Wacher 2009). In the Mahazat as-Sayd protected area, Arabian sand gazelle have increased exponentially during years with good rainfall, but mass mortalities occurred during 1999 (939 individuals were found dead) related to severe drought during 1998 and 1999 (Chassot et al. 2005). Good rainfall was observed between 2001 and 2004 again followed by a mass die-off of at least 679 animals (more than 50% of the population) during 2006 (Islam et al. 2007). By 2007, Islam et al. (2007) estimated the population number at ca. 1200 animals. The most recent estimates based on three counts range from 500 to 650 individuals. In the Uruq Bani Ma'arid protected area, sand gazelles have had a relatively stable population of around 300 individuals (Cunningham and Wacher 2009).

The number of free-roaming gazelles has significantly decreased, particularly in the northern protected areas. The population has shrunk by more than 30% in the past decade. The unregulated hunting and competition with domestic livestock are the primary factors contributing to the decline. In 2009, the total population was estimated at less than 10,000 individuals in the entire area of Saudi Arabia (Cunningham and Wacher 2009).

In conclusion, goitered gazelles are protected by law in most countries. However, environmental protections in most cases are mostly declarative in nature and generally not fulfilled. Poaching is poorly controlled everywhere, because governments do not believe that this is an important undertaking. The leaders of these countries are more concerned with the development of mining and other industries to increase revenues from exporting raw materials, and for this, they are intensively developing plans for expansion and increasing the networks of highways and railways lines. But in time, they will discover that they have actually destroyed the most valuable thing they have—an irreplaceable ecosystem and its varied and distinctive wildlife. Goitered gazelles are the most widely distributed ungulate, but soon, as a result of human efforts, this species may well become extinct. No matter how big the revenues from big economic projects are, it will quickly be squandered; but even more than financial loses, these countries will lose what makes them unique, their wild and free natural world and with it, in fact, a part of their future. For example, Mongolia is the last refuge for goitered gazelles and many other wild ungulates of steppes and deserts, and with its modern development it will turn into "scorched earth," and people will be left without a future in fact.

20.3 Re-acclimatization Projects

20.3.1 Kazakhstan

Barsakelmes Island (Aral Sea, Kazakhstan). Goitered gazelles (two males and seven females) were caught in the Karakalpak Autonomous Republic (Uzbekistan) and released on the Barsakelmes Island in 1929, where this species was not found previously (Zhevnerov 1984). The gazelles were very well acclimated to this island and began to increase their numbers rapidly to around 2300 individuals by 1947 and 4650 individuals by 1948. During the harsh winter of 1948 to 1949, heavy snowfalls that were followed by warm weather created an ice crust on the surface of the snow cover (dzhut). As a result of this ice cover, goitered gazelles could not break through to reach food, resulting in the main population dying because of starvation (1807 corpses of gazelles were counted) and leaving only 64 goitered gazelles alive by the autumn of 1949; by the end of this year, there were even fewer. The number of gazelles stayed mainly static for at least a decade, with their number not exceeding 2 to 3 dozen in 1957 (Sokolov 1959). In the autumn of 1960, an additional group of gazelles was taken from the Kara-Bugut Valley (S Kazakhstan) to Barsakelmes Island: 6 gazelles, of which 2 were males and 3 females at the age of 4 months and 1 female at the age of 16 months. The adult female died a day after her arrival; however, the remaining 5 adapted well to the new conditions, and during the 1970s, the population was around 200 gazelles (Zhevnerov 1984).

20.3.2 Kyrgyzstan

In October of 2021, 15 goitered gazelles that had been artificially fed were transferred from the Bukhara Breeding Center (Uzbekistan) to the enclosures of the Issyk-Kul Breeding Center (Kyrgyzstan). There were 5 males and 9 females, all at the age of 5 months except for one female of 15 months. Acclimatization was successful with no deaths (Fig. 20.8), and all were in good condition in November 2022, when the males actively courted the females. In April and May of 2023, 7 females gave birth and 8 fawns was born in the enclosure (Fig. 20.9). In December 2022, 12 more goitered gazelles were transferred from the Bukhara Breeding Center at the age of 6 or 7 months. At present, these gazelles are still in the stage of acclimatization to their new habitat (Fig. 20.10) (Blank D. 2022. Unpublished report).

20.3.3 Turkmenistan

The Ogurchinskiy sanctuary is located on Ogurchinskiy Island, which is situated close to the southeastern coast of the Caspian Sea. In the early 1980s, the area of this site was 4500 hectares, but this figure has been reduced due to rising sea levels. In 1982, a population of gazelles was established when 12 goitered gazelles were relocated from various locations in Turkmenistan and Uzbekistan. Two gazelles died

Fig. 20.8 First group of goitered gazelles during additional feeding in the enclosure. Issyk-Kul Breeding Center, Kyrgyzstan. 25.06.2022. Photo by David Blank

Fig. 20.9 Seven females from the first group gave birth eight fawns after 2 years keeping in the enclosures. Issyk-Kul Breeding Center, Kyrgyzstan. 06.05.2023. Photo by David Blank

Fig. 20.10 Second group of goitered gazelles during feeding by additional fodder (Lucerne hay). Issyk-Kul Breeding Center, Kyrgyzstan. 05.05.2023. Photo by David Blank

during transportation, while 10 others (2 males and 8 females) survived and bred. Protection of the sanctuary is helped by its island location; therefore, the acclimatization was successful, and 400 individuals were found there in 1994 (Flint and Prisyazhnuk 1986).

20.3.4 Uzbekistan

Goitered gazelle numbers decreased drastically in Uzbekistan. To re-establish goitered gazelles in their once native areas, a special Breeding Center was created for growing the gazelle population in a controlled area (Pereladova et al. 1998). The area of this Breeding Center offers a diverse environment: the northern section is characterized by a lack of rainfall, while the southern region boasts a richer variety of plant and animal life, with several lakes connected by underground waterways. The Bukhara Breeding Center was established in May 1977, when 44 goitered gazelles were released into a large, fenced area by size of 5126 ha in size. All 44 gazelles (23 females and 2 males) aged 1 to 2 years (20%), 3 years (40%) and 4 to 5 years (40%) were collected from hand reared individuals in different areas of Uzbekistan. Five of them were injured during transportation and apparently did not survive after release. A few weeks later, the flight distance of the gazelles increased from some dozens of meters to almost 500 m. The population at the center started to grow and peaked at 1224 gazelles in 1989, but 6 years later (1995), the population had dropped to 580 individuals. Later the population fluctuated in different years from 700 to 900 individuals (Pereladova et al. 1998).

20.3.5 Azerbaijan

In the Caucasus, goitered gazelles were widely distributed all over the semi-arid and arid plains and foothills of Azerbaijan, Georgia, and Armenia. The range and number of gazelles has declined rapidly since the 1930s due to intensive poaching and fragmentation of habitat as a result of agricultural development and increasing livestock numbers. In the late 1950s, only 200 gazelles were found in Azerbaijan (Safarov 1961) and the gazelle populations in Armenia and Georgia had gone extinct. The government of Azerbaijan established the Shirvan Protected Area (up to 640 km^2 today) to save the last gazelle population in the Caucasus. The number of goitered gazelles steadily increased up to 7000 in the Shirvan steppe region (Sarukhanova 2016); however, this large, isolated population remains under threat do to such things as outbreaks of diseases and climate disasters. The only way of restoring the goitered gazelles in the Caucasus, therefore, is through a reintroduction program (Askerov et al. 2021).

In Azerbaijan, six possible locations for the release were identified, and in Georgia, four sites were determined (Mallon 2008; Mallon et al. 2012). Two attempts to release captive-bred individuals were made in Georgia in 1989 and 1990, originating from Central Asia, and in 2009–2012, originating from Turkey. However, both efforts were unsuccessful, as reported by Mallon et al. in 2012. Consequently, the strategy for reintroducing the species was modified to a more aggressive approach, known as a «wild-to-wild» release, which does not involve confining the animals in a temporary enclosure prior to release (Askerov et al. 2021).

In early 2010s, the gazelles were caught during the night using expansive cotton fishing nets. The gazelles were captured at night using large cotton fish nets, without applying tranquilizers. From 2013 through 2016, captured gazelles were kept for a week in a small enclosure before releasing. In October 2013, the first reintroduction project was performed by releasing the gazelles directly into the Vashlovani National Park (Georgia), but it was not successful because the gazelles initially stayed 18 h into their transportation boxes, followed by three gazelles being poached and six others disappeared. Consequently, the reintroduction method was changed and the second project was implemented in the Eldari plain (Azerbaijan). A total of 174 individuals were moved and released immediately following transportation (referred to as hard release) in two locations. Currently, the combined population of the two groups stands at approximately 312 individuals, including young adults. It is anticipated that with future growth of the population of these released gazelles, a new, stable gazelle population will be created and survive (Askerov et al. 2021).

The results of re-acclimatization demonstrated that in the first years of the goitered gazelles' reintroduction to protected or fenced areas with a semi-captive method of animal maintenance and an absence of predators, the number of gazelles grow at an exponential rate. Conversely, in the presence of predators, population growth is severely limited or the population completely exterminated (Sarukhanova 2021).

20.3.6 Georgia

In Georgia, an unsuccessful reintroduction attempt took place in 1988 when 10 gazelles were taken to Vashlovani National Park from the Bukhara Breeding Centre in Uzbekistan; all died within 2 years due to disease and wolf predation. Flaws in the fence design enabled wolves to gain access to the pre-release enclosure which was sited in an area of *Pistacia* woodland, an unsuitable habitat for gazelles, and one that optimized predation possibilities for wolves while hindering vigilance and escape on the part of the gazelles (Mallon et al. 2012).

In 2006, the World Bank's Georgia Protected Area Development Project provided funding for another reintroduction. Preliminary agreement was reached with the Azerbaijan authorities to bring gazelles from the Shirvan State Reserve, but that operation could not be finalized before the funding deadline ran out. In 2008 the Ministry of Environment Protection and Natural Resources of Georgia and the Agency of Protected Areas made an agreement with the Ministry of Environment and Forestry of the Republic of Turkey, to assist Georgia in its efforts to reintroduce *Gazella subgutturosa* to its once native habitat (Mallon et al. 2012).

20.3.7 Turkey

In the 1940s and 1950s in Turkey, the range of goitered gazelles became more restricted from Suruç (Şanlıurfa) to Cizre, near the border with Syria. In recent years, the number of gazelles has continued to decrease, and now they are only found in Şanlıurfa. The factors contributing to this decline include poaching, the capture of young animals for sale, the expansion of agricultural land, and the use of pesticides (Ölçer 2001; Can and Lise 2004; Durmuş 2010). A breeding farm was founded at Ceylanpinar in 1977 in order to breed animals for re-introduction. Breeding started in 1978 with five gazelles—two from Savur (Mardin) and two from Tarsus (Mersin), where they had been reared by farmers (Erkan 1978). In 1982, 24 animals were found in the enclosures at Ceylanpinar, and by the 1980s, 80 animals are found there. In addition, 5 gazelles are kept in the zoo in Ankara (Durmuş 2010).

20.3.8 Saudi Arabia

It is considered that Arabian sand gazelles (*G.s. marica*) are found mainly in the Arabian Peninsula, where it is thought that up to 35% of the global population lives (Dunham et al. 2001). In the wild of Saudi Arabia, possibly two small populations still exist, one near the Jordanian border at Al Khunfah and the other at Harrat al-Harrah (Hague and Smith 1996). This information is based on fairly old data from the 1990s and it is unknown whether these populations have survived until now, because of the absence of recent information.

As a consequence of the drastic decline in the number of many sand gazelle populations, and the disappearance of many of them, Saudi Arabia established the National Commission for Wildlife Conservation and Development (NCWCD) in 1986. The prime targets were the improvement of the protection of the last two known wild populations of sand gazelles, and the start to a captive breeding program at the King Khalid Wildlife Research Center (KKWRC), located close to Riyadh, and the creation of a large stock of captive sand gazelles for reintroduction into the wild, and in fact, all sand gazelles later chosen for the reintroduction projects were obtained from the KKWRC. During 1990–1993, a total of 135 sand gazelles were translocated to the fenced area of Mahazat as-Sayd and, in 1994, several groups totaling 164 individuals were released in the area surrounded by a protective fence. The released population increased its size up to 300 gazelles and this reintroduction project was considered successful, as the new population that was established is considered viable, with a good potential for future breeding (Hague and Smith 1996).

20.3.9 Conservation Measures

The rapid decline in the number of gazelles in the late 1940s worried hunting organizations, and hunting of this species was banned everywhere (in Tajikistan—since 1946, in Turkmenistan and Uzbekistan—since 1950, in the Kazakhstan—since 1951 and in Kyrgyzstan—since 1952). For illegal hunting of goitered gazelles, according to the "Regulations on hunting in the Kazakh SSR" approved by the Council of Ministers on September 8, 1978, a guilty person was required to pay a fine of 500 rubles. However, despite this, poaching continued, especially at night using a car's headlights. The fight against poachers was still being poorly conducted, and those caught red-handed often avoided punishment due to the inertia of the local authorities. The goitered gazelle was listed in the Red Data Book of the Soviet Union on 1 January, 1970, and any hunting of this species was forbidden in the former USSR (Zhevnerov et al. 1983).

To make conservation of goitered gazelles effective, here are some steps follow (Blank 1990):

1. Make drivers of cars, all-terrain vehicles, tractors, or motorcycles even those belonging to government organizations, used for poaching face criminal responsibility and unconditionally are deprive them of their driver's license for a period of 1-year, malicious poachers for a period of 10 years. The heads of organizations and farms, with the knowledge or involvement of transport equipment used by poachers, should be administratively censured, and in the case of continued activity also made criminally responsible.

2. Forbid shepherds from having more than two dogs with one flock of sheep and keep the dogs under tight control when working during the day to drive cattle to seasonal pastures and on leashes at night to keep them from hunting gazelles.
3. Provide hunting inspectors and hunting personnel with walkie-talkies and a high-speed method of transport, including helicopters equipped with night vision devices. In recent years, the use of helicopters for patrolling hunting grounds in Kazakhstan has yielded excellent results.
4. Decrease the number of foxes everywhere in the desert zone by increasing the existing procurement prices for fox pelts by at least three times and continue paying bonuses for the extermination of wolves without reducing the payment amount.
5. Create a number of new nature reserves in the desert zone with an area of at least 250 to 300 thousand hectares each. The first places to establish reserves should be in "Ustyurt" on the western part of Ustyurt Plateau, as well as in Kugusem for the protection not only of the gazelles but also the Ustyurt mouflon; interstate "Kaplankyrskiy" nature reserve in the southern part of Ustyurt between Kazakhstan, Uzbekistan and Turkmenistan; a reserve in "Betpak-Dala" with branches in the central and eastern parts of the desert; and "Kyzylkum Nature Reserve" in the central part of the Kyzyl Kum, in the northwest Mynbulak Mountains.

During a dzhute (mass death because of ice crust) in nature reserves, hunting farms and nature sanctuaries, the following support work will be required:

1. When there is an ice crust, ice layers or other hindrances to movement and grazing of animals, the snow cover will need to be broken up and loosened. Destruction of the snow cover or ice would be best done in strips 20 to 40 m wide, placing the next strip 2 to 3 m from the previous one. By the month of January, in every large nature reserve or farm, it will be necessary to have 5 tractors and 10 to 30 plows for clearing away ice and snow.
2. During winter, disallow cattle grazing in areas with goitered gazelles concentrations.
3. During dzhutes, maximize the protection of distressed gazelles by requiring the local population to keep their dogs on a leash and especially eliminate the threat from street dogs.
4. It is essential to systematically make educational information available and work with local people to establish protecting gazelles as a priority: publish brochures and posters; make speeches in the press, on radio and television; educate children in the present and future importance of nature and the wildlife that inhabits it (Zhevnerov et al. 1983).

20.3.10 Immobilization

20.3.10.1 Experience Trapping and Immobilization in Turkmenistan

The capture of goitered gazelles in the Badkhyz Reserve was carried out over a number of years during the 1970s. The process consisted of collecting newborns and caring and bottle-feeding them in a zoo nursery up to a certain age. This method of trapping is quite laborious and requires a significant financial investment to feed young animals. There is also a significant loss (30% to 35%) through this process (Bozhko 1981).

In the Badkhyz Nature Reserve, the method of catching goitered gazelles at night with nets was also tested during the 1970s, but this method also proved ineffective, since it required the involvement of a large number of people and took many hours (up to 2 or 3 days to catch one adult gazelle) (Bozhko 1981).

A trial catch of goitered gazelles was run with neutralizing substances from a distance, using a hollow, a small-caliber Komarov bullet filled with ditilin. However, this method was not workable, since the bullets had a high killing power. Passing through the muscles, they penetrated to a depth of 15 cm, damaging vital organs, and ultimately killing the animal. Therefore, for the immobilization process, Gerasimov Yu A made injection arrows designed for shooting from 16 caliber smoothbore guns. The design of the arrow means it can be shot from a distance of 1 to 50 m (Bozhko 1981).

In the spring of 1979, experiments were conducted in the Badkhyz Nature Reserve on catching adult gazelles using injection arrows according to the Gerasimov method. The capture was carried out at night with the help of a GAZ-66 car equipped with a powerful headlight; a method used among poachers when hunting gazelles or tagging young Mongolian gazelles in Mongolia. From a moving car, with the help of the headlight, it was possible to approach to a distance of 30 to 50 m and make aimed shots. To immobilize goitered gazelles, a 50% sernylene paste (200 mg each) was used. Seven adult gazelles (five males and two females) were injected in the experiment using this method. Their trance-like state occurred 8 to 10 min after firing, causing the animal lose its balance, sway and fall. After being shot with the arrow and before its capture, the gazelle ran no more than 1 km, during which the gazelle was monitored with a headlight at a distance of 350 to 400 m. The trance-like effect in the captured gazelles from the moment of injection to transportation to the enclosure lasted 3 to 4 h, then the goitered gazelle completely returning to normal (Bozhko 1981).

The tested method of immobilizing goitered gazelles with arrows is relatively safe for the animals, since the arrow penetrates the muscles by only 3 to 3.5 cm and does not damage internal organs. This method allows trapping gazelles at any time of the year by a group of three people (driver, shooter, and catcher). The main advantage of this method is the ability to selectively remove individuals from a group and collecting from 5 to 8 individuals per night (Bozhko 1981).

20.3.10.2 Experience Immobilization in Saudi Arabia

Combinations of xylazine and ketamine have been used widely for anesthetizing ungulates. The combination produces a tranquil sedation, good muscle relaxation, and a smooth and uncomplicated recovery (Boever and Paluch 1974; Hall and Clarke 1991). Intravenous doses of 6.1 mg/kg xylazine and 4.9 mg/kg ketamine have been used successfully for captive goitered gazelles (Foster 1999). The detailed methodology and doses of the constituent components are described in Foster (1999) and available to an English-speaking reader.

20.4 Conservation and Breeding on Barsakelmes Island: Biotechnology

The goitered gazelle is well-adapted to desert environment but poorly adapted to inhabit areas with harsh winters and deep snow cover. This is evidenced by the very high weight load of this species on the supporting surface of their hooves, the small angle of the hooves' front wall, which makes it difficult and mostly ineffective to get food from under the snow. In addition, the high thermal conductivity of the coat does not provide adequate protection from low temperatures and cold wind (Zhevnerov 1984).

One of the ecological adaptations of ungulates to the extreme winter conditions is the change of habitats through migration. Gazelles of mainland populations stay in winter only in areas with an average snow cover height of no more than 10 cm (Sludskiy 1977). The isolation of the territory of Barsakelmes Island excludes this possibility; therefore, the main factor constraining the stable growth in the number of gazelles on the island is the periodically recurring deep snow and the associated lack of food. Adult males, weakened during the rutting season, and young individuals are particularly severely affected. The increase in the number of goitered gazelles on Barsakelmes Island actually occurs in the period between severe winters (Zhevnerov 1984).

Long-term stationary observations convincingly show the need for winter feeding. The timing and intensity of it should be regulated in accordance with the meteorological conditions of winter. Moreover, the availability of additional feed should be started before the snow cover is established, so that gazelles get used to visiting feeding grounds in advance of the snowfall. For additional forage, it is best to provide oats, barley and mixed feed, using compound feed at the first stage of additional forage in November and December before the start of stormy winds, and then alternate this with other forage components on windless days during the second half of winter. With the establishment of a high, and dense snow cover, when herbaceous vegetation in pastures becomes inaccessible, it will be necessary to lay out hay in small piles or hang sheaves on shrubs. Gazelles readily eat alfalfa and cane imported to the island. In the winters following the wet seasons, natural pasture forage, for example, *Artemisia terrae-albae*, can be partially used for additional forage, harvesting it in autumn from gullies and deep depressions that have no practical value in winter due to snow drifts (Zhevnerov 1984).

It is advisable to carry out additional forage from mid-November to March or April, before the snow cover melts. In the beginning, the food should be spread in small portions 1 to 2 times a week; then, with the increase in attendance of the feeding grounds by goitered gazelles, the amount should be increased and the intervals between should be reduced. Based on observations of tamed gazelles, the daily feeding ration to maintain a normal existence in winter should be 1 to 2 kg per individual. Feeding grounds should be placed with consideration of the terrain, the height of the snow cover and the direction of the prevailing winds. The most optimal fixed ridge sands are the main habitat of goitered gazelles in winter, and the sites should be dispersed evenly (from 1 to 1.5 km) along the entire line of the coastal zone, in places excluding snow drifts and with a good overview of the terrain. They can be placed directly in the fixed sands, in the open inter-row spaces. To maximize the conservation of the gazelle population in severe winters, it is necessary to have at least 30 permanent feeding points, the locations of which should remain unchanged from year to year to support familiarly (Zhevnerov 1984).

To eliminate interspecific competition during the feeding period, it is necessary to enclose feeding grounds. Using fences should ensure free access of goitered gazelles to the sites and exclude the possibility of penetration by kulans. To do this, the lower bar of the fence should be installed at a height of 80 cm from the ground, with the total height of the fence at 1.8 to 2 m, and the area no less than 7x7 m. The competitive relationship between gazelles and saigas at feeding grounds is not tense, since saigas visit feeding sites irregularly, and often individuals of these two species feed together. When a large group of saigas approach a feeding area at the same time, they displace gazelles, and even though this phenomenon is temporary, it still largely affects the overall course of feeding the gazelles. More important is the fact that the abundance of saigas undermines the main food base of the gazelles in winter. Apparently, at present, the habitation of these two species of animals in the limited territory of the island is impractical. In any case, the number of saigas should not exceed 150 to 200 individuals (Zhevnerov 1984).

Observations of tamed gazelles show the relationships between individuals of different sexes and ages and the relationship between groups on feeding grounds are changing. The most frequent cases of antagonism are among adult males. Aggressive interactions between females are rarer, and individuals who regularly visit feeding grounds dominate in this respect. Outside the feeding areas, gazelles gather in naturally formed groups of three to four individuals. On the feeding grounds, they once united and fed side by side, and there was rarely even the slightest manifestation of aggression between them. The number of aggressive contacts, however, increased with the appearance of one of the groups after a long absence, especially if it included a male. Uniform dispersal of feeding grounds in the territory excludes the accumulation of goitered gazelles in a few places. In fact, two to three permanent groups of the gazelles that live in this area and have private contacts with each other visit individual feeding locations. This helps to reduce the number of conflicts between individual groups of gazelles and creates a calmer environment during feeding (Zhevnerov 1984).

On Barsakelmes Island, the only permanent source of watering for gazelles is the sea. A sharp drop in the level of the Aral Sea and the associated increase of salt concentration in the water make it necessary to create artificial watering holes. For this purpose, a deep well was drilled on the island, the waters of which filled a natural depression with an area of about 3 hectares. The water is quite suitable for watering gazelles (salt concentration of 4 g/l); however, this did not completely solve the problem. The water source is located in the central part of the island, in close proximity to people, which significantly restricts the free access of gazelles to water. It is visited mainly by kulans which have become accustomed over many years to being around people. Gazelles and saigas use this watering hole in rare cases and at dusk. The main watering place for them, as before, remains the sea (Zhevnerov 1984).

References

Abdusalyamov IA (2001) Chapter 29. Tajikistan. In: Mallon DP, Kingswood SC (eds) Antelopes. Part 4. North Africa, the Middle East, and Asia. Global survey and regional plans. The IUCN/SSC antelope specialist group. IUCN, Gland, Switzerland and Cambridge, UK, pp 148–149

Afanasiev AV, Bazhanov VS, Karelov MN, Sludskiy AA, Strautman EN (1953) Goitered gazelle. In: Animal of Kazakhstan. Publication of Kazakh Academy of Sciences, Alma-Ata, pp 484–488. (in Russian)

Askerov E, Mallon DP, Zazanashvili N, Kochiashvili V, Sarukhanova S, Muradov A (2021) An attempt to restore the population of the Goitered Gazelle (*Gazella subgutturosa*) in its indigenous range in the Caucasus (Mammalia: Bovidae). Zool Middle East 67(3):189–197

Bar-Oz G, Nadel D, Avner U, Malkinson D (2011) Mass hunting game traps in the southern Levant: the Negev and Arabah "desert kites". Near East Archaeol 74(4):208–215

Bekenov AB, Blank DA, Grachev YA, Plakhov KN (2001) Chapter 26. Kazakhstan. In: Mallon DP, Kingswood SC (eds) Antelopes. Part 4. North Africa, the Middle East, and Asia. Global survey and regional plans. The IUCN/SSC antelope specialist group. IUCN, Gland, Switzerland and Cambridge, UK, pp 134–140

Berger J, Buuveibaatar B, Mishra C (2013) Globalization of the cashmere market and the decline of large mammals in Central Asia. Conserv Biol 27:679–689

Blank DA (1990) Persian gazelle. In: Kovshar AF (ed) Rare animals of desert regions. Nauka Press of Kazakh SSR, Alma-Ata, pp 56–80. (in Russian)

Blank DA (1997) Social and reproductive behavior in the Persian gazelle. The University of Tel-Aviv, Ph.D. dissert, 189 pp

Blank DA (2022) Behavior of goitered gazelles in the Issyk-Kul Breeding Center, Kyrgyzstan. Unpublished Report: May, 2022

Boever WJ, Paluch H (1974) Injectable anesthetics in wild ruminants. Vet Med Small Anim Clin 69:548–551

Bozhko VE (1981) Experience of the goitered gazelle immobilization in the Badkhyz. In: Proceedings of the Academy of Sciences of the Turkmen SSR, Series of Biological Sciences, vol 4, pp 92–93. (in Russian)

Bugaev KE (1989) Large mammals of the trans-Altai Gobi (ecology and conservation). Ph.D. thesis. Moscow State University, Moscow

Buuveibaatar B, Strindberg S, Kaczensky P, Payne J, Chimeddorj B, Naranbaatar G, Amarsaikhan S et al (2017) Mongolian Gobi supports the world's largest populations of khulan *Equus hemionus* and goitered gazelles *Gazella subgutturosa*. Oryx 51(4):639–647

Can OE, Lise Y (2004) Large mammal survey in southeastern Anatolia and recommendation for conservation (in GAP biodiversity research project 2001–2003-final report) DHKD (Turkish Society for the Conservation of nature), Istanbul, Turkey

Chassot P, Messochina P, Ostrowski S (2005) Re-introducing of Arabian oryx in the Kingdom of Saudi Arabia: up-date on population size in two protected areas. Re-introduction News 24:17–19
Clark EL, Mukhbat J, Dulamtseren S, Baillie JEM, Batsaikhan N, Samiya R, Stubbe M (2006) Mongolian red list of mammals. Zoological Society of London, London, UK
Cunningham PL, Wacher T (2009) Changes in the distribution, abundance and status of Arabian sand gazelle (*Gazella subgutturosa marica*) in Saudi Arabia: a review. Mammalia 73:203–210
Dunham KM, Williamson DT, Joubert E (2001) Chapter 8. Saudi Arabia. In: Mallon DP, Kingswood SC (eds) Antelopes. Part 4: North Africa, the Middle East, and Asia. Global survey and action plans.—SSC antelope specialist group. IUCN, Gland, Switzerland and Cambridge, UK, pp 55–62
Durmuş M (2010) Determination of home range size and habitat selection of gazelles (*Gazella subgutturosa*) by GPS telemetry in Şanliurfa (Master's thesis, Middle East Technical University)
Eid E, Handal R (2018) Illegal hunting in Jordan: using social media to assess impacts on wildlife. Oryx 52(4):730–735
Erkan F (1978) Ceylanpinar Seyahati. Av Dergisi 11(5):14–15. Ankara (in Turkish)
Firouz E (2005) The complete Fauna of Iran. I. B. Tauris, London, New York
Flint VE, Prisyazhnuk VE (1986) Population status and perspectives of goitered gazelle restoration in the USSR. In: Flint VE, Prisyazhnuk VE (eds) Conservation and perspectives of the number restoration of goitered gazelle in USSR. All-Union Scientific Institute of Nature Conservation and Nature Reserves, Moscow, pp 3–8. (in Russian)
Foster CA (1999) Immobilization of goitered gazelles (*Gazella subgutturosa*) and Arabian mountain gazelles (*Gazella gazella*) with xylazine-ketamine. J Zoo Wildl Med 30(3):448–450
Gorbunov AB, Zarkhidze VA (1985) Problem of the number restoration ungulates in the western Turkmenistan. Proceedings of the Academy of Sciences of the Turkmen SSR. Ser Biol Sci 6:41–46. (in Russian)
Gorelov YK (2001) Goitered gazelle. Chapter 30. Turkmenistan. In: Mallon DP, Kingswood SC (eds) Antelopes. Part 4. North Africa, the Middle East, and Asia. Global survey and regional plans. The IUCN/SSC antelope specialist group. IUCN, Gland, Switzerland and Cambridge, UK, pp 150–154
Güldenstaedt AI (1780) Antelope subgutturosa discripta. Acta Academiae Scientiarum Imperialis. Petropilitanae, for 1778, vol 1, pp 251–274
Habibi K (2001) Chapter 22. Afghanistan. In: Mallon DP, Kingswood SC (eds) Antelopes. Part 4. North Africa, the Middle East, and Asia. Global survey and regional plans. The IUCN/SSC antelope specialist group. IUCN, Gland, Switzerland and Cambridge, UK, pp 119–121
Hague MN, Smith TR (1996) Reintroduction of Arabian sand gazelle *Gazella subgutturosa marica* in Saudi Arabia. Biol Conserv 76:203–207
Hall LW, Clarke KW (1991) Veterinary Anaesthesia, 9th edn. Balliere Tindall, London, England
Hamami MR, Groves CP (2001) Chapter 21. Iran. In: Mallon DP, Kingswood SC (eds) Antelopes. Part 4. North Africa, the Middle East, and Asia. Global survey and regional plans. The IUCN/SSC antelope specialist group. IUCN, Gland, Switzerland and Cambridge, UK, pp 114–118
Hatough-Bouran A, Disi AM (1991) History, distribution and conservation of large mammals and their habitats in Jordan. Environ Conserv 18(19–32):44
Heptner G, Nasimovich AA, Bannikov AG (1961) Mammals of Soviet Union, Vol. 1. Artiodactyla and Perissodactyla. Vysshaya Shkola Publishing House, Moscow. (in Russian)
Islam MZ, Boug A, Anagariyah S, Ismail K, Robinson ER, Mohammed OB (2007) Catastrophic drought-induced die-off of reintroduced animals in Mahazat as-Sayd protected area in arid Central Saudi Arabia. National Wildlife Research Centre, King Khalid Wildlife Research Center and National Commission for Wildlife Conservation and Development, Saudi Arabia
Ito TY, Lhagvasuren B, Tsunekawa A, Shinoda M (2017) Habitat fragmentation by railways as a barrier to great migrations of ungulates in Mongolia. In: Railway Ecology, pp 229–246
IUCN (2017) Guidelines for using the IUCN Red List categories and criteria, ver. 13. – IUCN Standards and Petitions Subcommittee, IUCN, Gland, Switzerland and Cambridge, UK

Jiang Z, Sung W (2001) Chapter 33. China. In: Mallon DP, Kingswood SC (eds) Antelopes. Part 4. North Africa, the Middle East, and Asia. Global survey and regional plans. The IUCN/SSC antelope specialist group. IUCN, Glands and Cambridge, UK, pp 168–177

Kaczensky P, Ganbaatar O, Altansukh N, Enksaikhan N, Kramer-Schadt S (2015) Monitoring of khulans and goitered gazelles in the Mongolian Gobi- potential and limitations of ground-based line transects. Open Ecol J 8:92–110

Karami M (1992) Nature reserves in the Islamic Republic of Iran. Species 19:11

Karami M, Hemami MR, Groves CP (2002) Taxonomic, distributional and ecological data on gazelles in Iran. Zool Middle East 26:29–36

Karlstetter M, Mallon D (2014) Assessment of gaps and needs in migratory mammal conservation in Central Asia. Report prepared for the convention on the conservation of migratory species of wild animals (CMS) and the deutsche Gesellschaft für Internationale Zusammenarbeit (GIZ) mbH. Financed by the ecosystem restoration in Central Asia (ERCA) component of the European Union Forest and biodiversity governance including environmental monitoring project (FLERMONECA)

Khosravi R, Adavoudi R, Malekian M, Hemami MR (2017) Measuring environmental niche isolation between genetically management units of Goitered gazelle, *Gazella subgutturosa* (Güldenstadt, 1970) in Iran. J Wildl Biodivers 1(2):60–68

Khosravi R, Hemami MR, Malekian M, Silva TL, Rezaei HR, Brito JC (2018) Effect of landscape features on genetic structure of the goitered gazelle (*Gazella subgutturosa*) in Central Iran. Conserv Genet 19(2):323–336

Kingswood SC, Blank DA (1996) Gazella subgutturosa. Mamm Species 518:1–10

Kiwan K, Boef J, Boudari A (2001) Chapter 18. Jordan. In: Mallon DP, Kingswood SC (eds) Antelopes. Part 4. North Africa, the Middle East, and Asia. Global survey and regional plans. The IUCN/SSC antelope specialist group. IUCN, Glands and Cambridge, UK, pp 102–106

Kucheruk VV (1995) Mammals of Turkmenistan, Vol. 1. Carnivores, pinnipeds, ungulates. Publication of Ylym, Ashgabat, 317 pp (in Russian)

Kurochkina LYA, Karibaeva KN (1986) Forage plants of Kazakhstan deserts. Publication of Kainar, Alma-Ata, pp 1–206. (in Russian)

Laptev MK (1935) Vertebrate animals of Turkmenistan and their using. Proceedings of Turkmen Interdepartmental Committee for nature conservation and development of natural resources, vol 2. Publication of Turkmen State Publication, Ashkhabad, pp 87–114. (in Russian)

Lhagvasuren B, Milner-Gulland EJ (1997) The status and management of the Mongolian gazelle *Procapra gutturosa* population. Oryx 31:127–134

Mallon DP (2008) Rehabilitation of historical range of Goitered gazelles in Azerbaijan: rapid assessment survey, September 2008. WWF Caucasus, Project Report #GE0037, Tbilisi

Mallon DP, Jiang Z (2009) Grazers on the plains: challenges and prospects for large herbivores in Central Asia. J Appl Ecol 46:516–519

Mallon DP, Kingswood SC (eds) (2001) Antelopes. Part 4. North Africa, the Middle East, and Asia. Global Survey and Regional Plans. The IUCN/SSC Antelope Specialist Group. IUCN, Glands and Cambridge, UK

Mallon D, Woodfine T, Gilbert T (2012) Reintroduction of Goitered gazelles to the Iori plateau-Vashlovani landscape, Georgia. NACRES and FFI, Tbilisi

Marmazinskaya NV (1999) Social behavior of goitered gazelle (*Gazella subgutturosa* /Güld., 1780/ during breeding period in the conditions ecocenter "Goitered gazelle"). Ph.D. thesis in zoology, Institute of zoology of Uzbek Academy of Sciences, Tashkent, Uzbekistan, pp 1–20. (in Russian)

Marmazinskaya NV, Mardonov BK (2004) The current status of populations and conservation goitered gazelle (*Gazella subgutturosa subgutturosa* Güldenstaedt, 1780) in Uzbekistan. Proceedings of nature reserves of Uzbekistan (4–5), Tashkent: 124–133 (in Russian)

Martin L (2000) Gazelle (*Gazella* spp.) behavioural ecology: predicting animal behavior for prehistoric environments in south-west Asia. J Zool (London) 250:13–30

Mendelssohn H (1974) The development of the populations of gazelles in Israel and their behavioral adaptations, vol 24. International Union for Conservation of Nature and Natural Resources Publications New Series, pp 722–743

Nader IA (1989) Rare and endangered mammals of Saudi Arabia. In: Abuzinada AH, Goriup PD, Nade IA (eds) Wildlife conservation and development in Saudi Arabia. NCWCD, Riyadh, pp 220–233

Ogurlu I (1992) Wild ungulates of Turkey. In: Ongulés/Ungulates, vol 91, pp 575–577

Ölçer SY (2001) Goitered gazelle. Chapter 20. Turkey. In: Mallon DP, Kingswood SC (eds) Antelopes. Part 4. North Africa, the Middle East, and Asia. Global survey and regional plans. The IUCN/SSC antelope specialist group. IUCN, Gland, Switzerland and Cambridge, UK, pp 112–113

Pereladova OB, Bahloul K, Sempere AJ, Soldatova NV (1998) Influence of environmental factors on a population of goitred gazelles (*Gazella subgutturosa subgutturosa* Güldenstaedt, 1780) in semi-wild conditions in an arid-environment: a preliminary study. J Arid Environ 39:577–591

Pestov MV, Dieterich T, Teretyev VA, Nurmuhambetov ZE, Muhashov AT, Denisov DA (2018) Problems and solutions regarding the border fence that mitigates the migration of ungulates on the territory of the Mangystau oblast of the Republic of Kazakhstan. Selevinia 26:92–98. (on Russian, English abstract)

Pratt DG, MacMillan DC, Gordon IJ (2004) Local community attitudes to wildlife utilization in the changing economic and social context of Mongolia. Biodivers Conserv 13:591–613

Quemsiyeh MB, Amr ZS, Budari AM (1996) Status and conservation of Artiodactyla (Mammalia) in Jordan. Mammalia 60:417–430

Radde GI (1899) Collections of Caucasus Museum, 1. Zoology. Caucasus Museum, Tiflis, 520 pp. (in Russian)

Rosen B, Perevolotsky A (1998) The function of "desert kites"—hunting or livestock husbandry? Paleurient 24(1):7

Rustamov AK, Atamuradov KI, Sopyev OS, Kurbanov OP, Koshunov MA (1986) Goitered gazelles in Turkmenistan: population status, resettlement and breeding? Conservation measures. In: Flint VE, Prisyazhnuk VE (eds) Conservation and perspectives of the number restoration of goitered gazelle in USSR. All-Union Scientific Institute of Nature Conservation and Nature Reserves, Moscow, pp 17–21. (in Russian)

Safarov MA (1961) Results of airplane survey of goitered gazelle number in Azerbaijan. In: Proceedings of Academy of Sciences of Azerb SSR, Biological series, vol 8, pp 51–55. (in Russian)

Sarukhanova SA (2016) The historical and current status of Goitered gazelle (*Gazella subgutturosa* Güldenstaedt, 1780) in the South-Eastern Caucasus. Proceedings of the Azerbaijan Institute of Zoology dedicated to the 80th anniversary of the Institute of Zoology, Baku 34 N1: 119–126

Sarukhanova S (2021) The study level of the goitered gazelle (Artiodactyla, Bovidae) in Azerbaijan. Bull Sci Pract 7(9):42–58. (in Russian)

Sheehy DP, Sheehy CM, Johnson DE, Damiran D, Fiemengo M (2010) Livestock and wildlife in the Southern Gobi region, with special attention to wild ass Mongolia discussion papers East Asia and Pacific sustainable development department. World Bank, Washington, DC

Shiirevdamba TS, Shagdarsuren O, Erdenejav G, Amgalan L, Tsetsegmaa TS (eds) (1997) Mongolian red book. Ulaanbaatar, Mongolia

Sludskiy AA (1963) The mass death of wild ungulates in steppes and deserts of Europe and Asia, vol 20. Proceedings of the Institute of Zoology, Academy of Sciences of the Kazakh SSR, Alma-Ata, pp 5–88. (in Russian)

Sludskiy AA (1977) Goitered gazelle. In: Rare animals of USSR. Publication of Lesnaya Promyshlennost', Moscow, pp 28–61. (in Russian)

Sokolov II (1959) Gazella subgutturosa Guldenstaedt – Goitered gazelle. In: Pavlovskiy, E.N. (ed.) auna USSR, Mammals: Ungulates (Orders Perissodactyla and Artiodactyla). Publication of the Academy of Sciences USSR: Moscow and Leningrad, 308-331 pp. (in Russian)

Svizzero S (2019) Issues and challenges in the conservation of the goitered gazelle (*Gazella subgutturosa* Güldenstädt, 1780). Int J Zool Res 3(3):1–9

Vereshchagin NK (1959) Mammals of Caucasus. History of fauna evolution. Publication of the Academy of Sciences USSR, Leningrad. (in Russian)

Williamson D, Tatwany H (1996) Gazelle distribution and status in the Arabian peninsula. In: Greth A, Williamson D, Abuzinada AH (eds) Gazelle conservation in the Arabian peninsula. Proceedings of the symposium—Establishing priorities for gazelle conservation in the Arabian Peninsula, 31 October to 3 November 1992. National Commission for Wildlife Conservation and Development, Riyadh, pp 40–44

Wingard JR, Zahler P (2006) Silent steppe: the illegal wildlife trade crisis. Mongolia discussion papers, East Asia and Pacific environment and social development department. World Bank, Washington, DC

Zachos FE, Karami M, Ibenouazi Z, Hartl GB, Eckert I, Kirsching J (2010) Mamm Biol 75:277–282

Zeder M, Bar-Oz G, Rufolo S, Hole F (2013) New perspectives on the use of kites in mass-kills of Levantine gazelle: a view from northeastern Syria. Quat Int 297:110–125

Zhevnerov VV (1984) Goitered gazelle of the Barsakelmes Island. Nauka Publishing House of Kazakh SSR, Alma-Ata. (in Russian)

Zhevnerov VV, Bekenov AB, Sludskiy AA (1983) Goitered gazelle—*Gazella subgutturosa* Güldenstaedt, 1780. In: Gvozdev EV, Kapitonov VI (eds) Mammals of Kazakhstan, Bovidae, vol 3, No. 3. Nauka Publishing House of Kazakh SSR, Alma-Ata, pp 11–54. (in Russian)

Unexplored Phenomena in the Behavioral Biology of Goitered Gazelles

21

21.1 Taxonomy of *Gazella subgutturosa* Subspecies

In this book, I have adhered to the relatively old subspecies taxonomy of the goitered gazelle, according to which this species has four subspecies: (1). Persian gazelle (*Gazella subgutturosa subgutturosa*), living all over Central Asia, Afghanistan, Pakistan, Iran until the Zagros Mountains (Fig. 1.3); (2). Yarkand gazelle (*G. s. yarkadensis*), living only around the Takla-Makan Desert between the Eastern Tien Shan Mountains (Tengritag Mountains) and the Tibet Plateau (southern Xingjian, China); (3). Dzhungarian goitered gazelle (*G.s. hillieriana*), living in the Dzhungar Depression (China) and Gobi Desert (southern Mongolia); (4). Arabian sand gazelle (*G. s. marica*), living in Saudi Arabia, Jordan, Suria, Turkey, and southwestern Iran (to the west of the Zagros Mountains—Fig. 1.4) (Kingswood and Blank 1996; Hayatgheib et al. 2011). However, a number of genetic and morphological studies of the last 10 to 15 years (Wacher et al. 2011; Abduriyim et al. 2018; Fadakar et al. 2020) introduced some doubts about the accuracy of the former systematics, destroying the established old taxonomy, but not adding clarity to the new system.

Firstly, Wacher et al. (2011) stated that the Arabian sand gazelle is not a subspecies of goitered gazelles (*G.s. marica*), but a separate species (*G. marica*). This conclusion was made on the grounds that, according to the results of his genetic studies (sequence variation of a mitochondrial marker (Cytochrome b) of samples), Arabian sand gazelle is systematically closer to the *Gazella leptoceros* and *G. cuvieri* from northern Africa than to the goitered gazelle from Asia. This statement, however, completely contradicts the morphological and zoogeographical data. Can the Arabian sand gazelle be a separate species? Yes, maybe, but this must be proved through a large sampling of genetic materials taken from wild individuals of different populations to alleviate doubts about genetic purity and to eliminate hybrids that may have come from past captivity. Can the results of genetic studies contradict morphological data? This is very doubtful. Can the Arabian sand gazelle be

D. Blank, *The Behavioral Biology and Conservation of Goitered Gazelles (Gazella subgutturosa)*, https://doi.org/10.1007/978-3-031-83504-9_21

systematically closer to the African gazelles (*Gazella leptoceros* and *G. cuvieri*) compared to goitered gazelles? Here, there is a problem with the genetic purity of individuals from captivity, from whom genetic materials have been taken for research. Many gazelle species were kept together in enclosures in Saudi Arabia and Oman which produced interspecific hybrids. Therefore, the genetic material taken from these gazelles will always arouse suspicion of the genetic purity of the species. Therefore, any taxonomic affinity of Arabian sand gazelles to African *G. leptoceros* and *G. cuvieri* causes a very big doubt. Morphological and zoogeographical data speak against such proximity. From all that has been said, it can be concluded that additional studies of the taxonomic status of the Arabian sand gazelle are needed with more numerous materials from other individuals from different parts of the goitered gazelle natural habitats to clarify the taxonomy of these gazelles.

Secondly, a genetic investigation is necessary in the Euphrates-Tigris basin and the Zagros Mountains of Iran, which is considered the (so-called) hybrid zone between Arabian sand gazelles and goitered gazelles and get genetic confirmation that different populations with horned or hornless females are the result of hybridization of two subspecies or species. Turkey's populations need the same genetic investigations to confirm that Arabian sand gazelles live there or if this is also a hybrid zone between Arabian sand gazelles and goitered gazelles, because Turkey's gazelles have both horned and hornless females.

Third, Sorokin et al. (2011) found that there are two different populations in Azerbaijan with some goitered gazelles are genetically similar to the Persian goitered gazelles from Iran (*G.g. subgutturosa*), while others belong to a genetically unique group. Is it *G.s. marica* that was migrated from Turkey during former epochs or it is something specific; this should be define.

Fourths, Abduriyim et al. (2018) found that the Dhungarian Xinjiang populations of goitered gazelles (*G.s. hellieriana*) in China are genetically distinctive from the southern Takla Makan goitered gazelles (*G.s. yarkadensis*). However, he found a gene flow between the Takla-Makan population of goitered gazelles and the Uzbekistan and Turkmenistan populations (*G.s. subgutturosa*), indicating migrations between them, though such migrations are impossible in fact through the high ridges of Tien Shan and Pamir mountains, so this point needs additional clarification.

Fifth, it is necessary to establish genetic conformity in the taxonomy status of the Qaidam basin population of goitered gazelles (China): if they are the same *G.s. hillieriana* or belong to a separate subspecies (*G.s. reginae*) (Abduriyim et al. 2018).

Sixth, it is essential to clarify the taxonomy status of the goitered gazelle population in Tajikistan and answer the question of if *G.s. glacilicornis* (Stroganov 1956) is a real and separate subspecies of goitered gazelle or if this is just a population variability of the same nominative subspecies (*G.s. subgutturosa*), which is typical for other countries of Central Asia.

21.2 Social Structure of Goitered Gazelles and Alternative Breeding Tactics

During my observations in the Ili Hollow (SE Kazakhstan), I found that goitered gazelle males establish individual territories during the rutting season similar to other gazelle species (Blank 1985, 1998). Later, this observation was confirmed by investigations of this species in the semi-captive conditions of the Bukhara Nursery in Uzbekistan by Marmazinskaya (1996, 1999). Strictly speaking, I was not the first to find this territorial social structure in this species. The first observations of these behaviors in captive goitered gazelle males were made during the 1960s and described first in German by Fritz R. Walther (1968) and then in English (Walther et al. 1983). I was, however, the first who described the territorial social structure of these gazelles in wild populations. In contrast, zoological literature from the former Soviet Union reported the well-established opinion that goitered gazelle males had a harem social structure, where a male has a group of several females during the rutting season (Heptner et al. 1961; Zhevnerov 1984). Marmazinskaya (1996) did not see any contradiction in this and proposed the idea of an alternative breeding system: goitered gazelles had a territorial structure with a high population density and a harem social structure with a low population density.

Despite the attractiveness and logic of this hypothesis, it is most likely wrong. The fact is that all antelopes and gazelles do not have harems in the way they are described for horses and zebras (Estes 1974; Walther et al. 1983), but what happens if the population density is so low that a male no longer has to protect his territory from other males? This question was answered by Walther et al. (1983) as the male begins to travel along with the females, without becoming attached to a territory. At first glance, it seems to turn out to be a harem structure, but this is not the case. The main difference between a group of females with a male and a harem is that the harem is a permanent, very stable group of the same females, and the male returns any female to the harem if she tries to leave. The group composition of goitered gazelle females, however, is typically unstable and changes very quickly, with some females entering the group and others leaving it (Zhevnerov 1984; Marmazinskaya 1999).

Thus, even with a low population density, goitered gazelles do not have harems. Nevertheless, for full confidence, it would be good to conduct directed studies on the social structure of goitered gazelles during the rutting season to determine whether goitered gazelles have alternative breeding tactics apart from territorial in various population density and different habitat characteristics.

21.3 The Problem of Water for Territorial Males Living in Snowless Areas During the Rut

Goitered gazelles have a territorial social structure during the rut. During the principle phase of the rut when territorial males show the most intensive courtship behaviors toward females, rutting males do not leave their territories, during both day and night (Blank 1985, 1998). At this time, they get very thirsty because of the intense chasing of females and male intruders (Sludskiy 1956). In the northern deserts of Kazakhstan and Xinjiang, goitered gazelle males quench their thirst by eating snow, so they do not need to leave their territories (Blank 1998). It is not entirely clear, however, how territorial males solve the problem of thirst-quenching during the rut in southern habitats (Uzbekistan, Turkmenistan, Iran, Iraq, and Saudi Arabia), where snow rarely falls or does not fall at all. Perhaps territorial males are forced to locate their rutting territories near watering sources, or they leave their territories for a while to visit watering holes—this remains unknown. It would add significantly to the knowledge of goitered gazelles for a special study to be conducted on this issue on goitered gazelle populations living in completely snowless areas.

21.4 Urination-Defecation Territorial Marks Versus Latrines

Males in all gazelle species, including goitered gazelles, mark their individual rutting territories by urination-defecation spots (Walther et al. 1983). Periodically, only landlord-males or territorial neighbors renew their own markings (Blank 1998). In goitered gazelles, urination-defecation marks exist only during the rutting season; therefore, these marks do not increase by size significantly. They are relatively small in size and contain the excrement of one or two males. The function of rutting urination-defecation marks is to protect an individual's territorial borders from male-intruders (Walther et al. 1983). Latrines, in contrast, appear out of rutting season and are used by all males living in the area, all of whom leave their pellets in the latrines. As a result, latrines look like bare ground filled with large amounts of droppings of different sizes from different individuals. Latrines are usually fairly large in size (up to 1–1.7 m in diameter) and are likely a point of communication between males within their home range. Thus, in goitered gazelles, there are two kinds of scent marks: territorial urination-defecation marks during the rutting season and latrines outside of rutting season (Blank et al. 2015).

During my observations of goitered gazelles in the Ili Hollow, I noted numerous, but small-sized, territorial urination-defecation marks during the rutting season (in November and December), but a latrine was found only once in summer (Blank 1998). Pavlov et al. (2013) found numerous latrines of goitered gazelles in Tigrovaya Balka Nature Reserve in Tajikistan. Marmazinskaya (1996) in the Bukhara nursery and Zhevnerov (1984) in the Barsakelmes Island population found both urination-defecation marks during the rutting season and latrines outside of the rut. However, these studies do not distinguish between territorial urination-defecation marks and latrines. Therefore, it is difficult to understand what Marmazinskaya (1996) or Zhevnerov (1984) had in mind in any particular case. In this regard, it is necessary to conduct a special study of the differences between territorial rutting marks and latrines. In addition, it would be worth studying the social functions of latrines more closely with consideration of their daily use and impact on goitered gazelle behavior.

Regarding other gazelle species, specifically *Gazella arabica* (formerly *Gazella gazella cora*), Wronski and Plath (2010) and Wronski et al. (2013) called both types of scent marking locations of this species latrines, though these authors did distinguish different functions of scent marks, defining the function of rutting urination-defecation marks as a territorial protection and latrines as a means for communication. *Gazella arabica* males have individual territories, but not just seasonal like goitered gazelles, instead they hold and protect them for a long time. Is it possible that *Gazella gazella* and *Gazella arabica* have only large-sized urination-defecation territorial marking locations (latrines)? During my observations of *Gazella gazella acaciae* in Israel, I saw only small-sized urination-defecation marks and did not see latrines. Is it possible that *Gazella arabica* has two kinds of urination-defecation marks even though males have year-round territories? Does *G.s. marica* (or *G. marica*) also have two kinds of urination-defecation marks or just one kind (only latrines) with two different functions? It would be important to clarify this point for all species and subspecies of the gazelles.

21.5 Skin Scent Glands and Their Functions

There are four main skin scent glands in goitered gazelles (Pocock 1910; Sokolov 1959; Sokolov 1982). These are inguinal, carpal, pedal, and preorbital glands. Even though the histological structures of the skin glands are described quite well (Sokolov 1982), their function and role in the goitered gazelle behaviors are not as clear. Preorbital glands are located below and in front of each eye (in both males and females) and used by territorial males for marking shrubs and trees in defense from territorial male-intruders (Walther et al. 1983; Blank 1998). In this book, I consider preorbital makings in detail in Chap. 5 (sub-section 5.3.). Marking with preorbital glands). All gazelle species possess preorbital glands and many of them use these glands for marking and maintaining their individual territories (Walther et al. 1983).

Regarding other skin glands, the data are not so clear. The function of the inguinal glands has not been investigated enough and what information exists is completely incomprehensible. Zhevnerov (1984) found that the secretory activity of inguinal glands starts to increase simultaneously with the onset of the swelling of the udder. Their secretions, a yellowish liquid, begin to accumulate in the pockets of the glands and, as the birth approaches, these accumulated secretions are released onto the udder. Zhevnerov (1984) supposed that during the suckling by the fawn, the secretions get on the fawn's muzzle giving it an individual smell that helps the mother distinguish her own fawn from others. Except for Zhevnerov (1984), no other researchers have observed this phenomenon in goitered gazelles. It would be helpful to the understanding of this process to specifically study the function of the inguinal glands and their role in the individual recognition of fawns by their mothers. In addition, it would be interesting to check if the inguinal glands are active in males and what function they have in males.

Both male and female goitered gazelles possess interdigital (carpal) glands on all four legs. It can be assumed that these glands serve as a source for scent markings, and indeed, during herd movements, some goitered gazelles lowered their muzzles to the ground and seemed to be following the trail of conspecifics. But this was not always observed. It is also unclear that if these glands are used for marking, how do the secretions from the interdigital gland located above the hoofs get to the ground. Therefore, interdigital glands need to be the topic of a special study to define the function and seasonal activities in the life of goitered gazelles.

Goitered gazelles also have pedal glands located below the knee junction. In contrast to carpal glands, secretions of which no one noted, these glands produce a yellow greasy secretion. It is possible to assumes that these glands are used for marking resting sites, when goitered gazelles stand on their knees when laying down or standing up. However, it is not clear why resting sites are marked and how other individuals use this information for their life. Therefore, it is necessary to study seasonal activities and functions of these glands.

21.6 Scratching and Digging Ground with Front Hoofs

Gorelov (1972) found in the Badkhyz Reserve (SE Turkmenistan) that goitered gazelles were often digging the ground searching for juicy roots of plants, especially during the dry seasons. With gazelles in the wild in Ili Hollow, I noted this behavior very rarely. However, in the semi-captive herd of the Issyk-Kul Breeding Center, I saw goitered gazelles digging the sandy ground regularly (Figs. 21.1 and 21.2–7142, 7349). Do goitered gazelles dig the ground only to find juicy roots or are there other reasons for this behavior? Why do gazelles rarely scrape the ground in the gravelly deserts, but more often do in sandy soil; is this due to the hardness of

Fig. 21.1 Goitered gazelle female often scratch the ground by front legs during pasturing. Issyk-Kul Breeding Center, Kyrgyzstan. 06.05.2023. Photo by David Blank

Fig. 21.2 After scratching female eats lower parts of a plant and its root. Issyk-Kul Breeding Center, Kyrgyzstan. 07.05.2023. Photo by David Blank

the soil or there are other reasons for this difference in goitered gazelle behavior? A special study could answer these questions.

21.7 Triplets and Four Fawns in Goitered Gazelle Females

A small number of goitered gazelles give births of triplets, and there even have been cases of females birthing four fawns. How do mothers suckle three or four fawns? After all, goitered gazelles have only two active nipples (Sokolov 1959; Zhevnerov et al. 1983). Strictly speaking, females have two additional nipples, but it is believed that those nipples do not in function. Does this mean that the two other nipples begin to function when a mother needs to feed more than two fawns? Do all fawns born in threes or fours survive, or do one or two of them die of hunger soon after birth? Are there any cases of survival of all fawns from these more than normal multiple births? It would be very interesting to make these observations of mothers and their many newborns for a long enough period of times, in the wild and captivity, to answer these questions.

21.8 Milk Stealing and Allosuckling

During my study in the Ili Hollow, I saw many attempts of fawns trying to suckle from alien females (not mothers). Invariably, though, females suckled only their own fawns and refused others. However, in the Issyk-Kul Breeding Center, I observed allosuckling in a female, where one female started to suckle the fawn and then another female approached and suckled the same fawn again, licking his anal area. It this some kind of unique behavior of only semi-captive females or this phenomenon also found in the wild? How common is this phenomenon among goitered gazelles and under what conditions does it appear? It would be worth pursuing this issue of possible allosuckling by alien mothers when studying breeding behaviors of goitered gazelles in both captivity and the wild.

21.9 Horn Length

Looking at the photos of goitered gazelle males sent from Xinjiang and the Ustyurt Plateau region in western Kazakhstan, it struck my eye that the horns of adult males from Ustyurt are noticeably longer, while the horns of Xinjiang males are much shorter. In addition, the horns of young males develop significantly slower in the eastern parts of the range (Chinese Xinjiang and Mongolia) and develop more quickly in the western Kazakhstan. It would be useful to conduct a study of the length of the male horns in different parts of their natural habitat, specifically in the west of the natural habitats and east to find a reason of this phenomenon, if it turns out to real.

21.10 Mineral Consumption

I did not observe goitered gazelles using natural mineral sites during my many-year study in the wild. It is known that mineral consumption is more typical for ungulates living in mountainous areas, where the environment contains few salts. However, in arid environments, goitered gazelles eat salty plants and drink salty water, which likely supplies them with enough minerals, eliminating the need for additional salts. Nevertheless, some authors indicated that goitered gazelles observed in Kazakhstan used artificial salt stones left after domestic sheep were moved from winter desert pastures to summer mountain pastures (Zhevnerov et al. 1983), and Dauletbakov (in litt.) stated that salt consumption from the artificial salt stones was observed in goitered gazelles in the Altyn-Emel Nature Reserve (Ili Hollow). In addition, I also saw how females, after giving birth, readily licked a salt stone while living under semi-voluntary maintenance of goitered gazelles in the Issyk-Kul Breeding Center (Fig. 21.3–7046; Blank 2023). Because of these observations, I believe it would be worthwhile to conduct a special study on whether only goitered gazelle females after birthing need additional salts or do all individuals occasionally require additional minerals in their diet? Do goitered gazelles use salts more during certain seasons (molting, birthing) or do they need additional salts in all seasons? Do goitered gazelles have natural locations to find salt?

Surely there are many more interesting phenomena in the life of goitered gazelles that are still not identified or poorly studied. The goitered gazelle is the only gazelle species that over time moved far to the north and adapted to life in the northern deserts with their very cold, snowy winters and very hot, dry summers, but we may

Fig. 21.3 Goitered gazelle female readily licks salt-stone during birthing season. Issyk-Kul Breeding Center, Kyrgyzstan. 06.05.2023. Photo by David Blank

not have enough time left to study this species to its fullest, because it may be driven to extinction if countries responsible for goitered gazelle habitat continue to ignore the interests of this species and all wildlife under their care, when opening up and developing their wild lands. New economic and political-based projects, such as building border fences, new highways and railways, new oil and gas pipelines, developing more new mines, and expanding and driving huge herds of livestock into previously wild spaces, while at the same time removing all wildlife access to watering holes, will make the beauty and tranquility of our wild, natural spaces nonexistent for future generations. This loss, being witnessed all over the world, will change life for everyone forever.

References

Abduriyim S, Zibibulla G, Eli S, Ismayil Z, Halik M (2018) Phylogeny and genetic structure of the goitered gazelle (Artiodactyla, Bovidae) in North-Western China indicated by the hypervariable mitochondrial control region. Syst Biodivers 16:527–537

Blank DA (1985) Peculiarities of social and reproductive behaviour of *Gazella subgutturosa* in the Ili Valley. Zool Zhurnal (Moscow) 64:1059–1070. (in Russian, English abstract)

Blank DA (1998) Mating behavior of the Persian gazelle *Gazella subgutturosa* Güldenstaedt 1780. Mammalia 62(4):409–419

Blank DA (2023) Birthing season in goitered gazelles in conditions of the Issyk-Kul Breeding Center. Unpublished report for the Institute of Biology, Kyrgyz Academy of Sciences

Blank DA, Ruckstuhl K, Yang W (2015) The economics of scent marking with urine and feces in goitered gazelle (*Gazella subgutturosa*). Mamm Res 60(1):51–60

Estes RD (1974) Social organization of the African Bovidae. In: Geist, V., Walther, F. (eds) The behaviour of ungulates and its relation to management, IUCN Publ. No. 24, International Union for the Conservation of Nature and Natural Resources, Morges, Switzerland, pp. 166–205

Fadakar D, Barmann EV, Lerp H, Mirzakhah M, Nasari MN, Rezaei HR (2020) Diversification and subspecies patterning of the goitered gazelle (*Gazella subgutturosa*) in Iran. Ecol Evol 10:5877–5991

Gorelov UK (1972) Breeding of goitered gazelle (*Gazella subgutturosa* Güldenstaedt) and problem of restoring of its number in the Badkhyz (SE Turkmenistan). In: Theriology, vol 1. Nauka of Siberian Department of Academy of Sciences of USSR, Novosibirsk, pp 420–424. (in Russian)

Hayatgheib D, Karami M, Farahmand H, Mehrabani YH, Farhadinia MS (2011) Morphological study and comparison of western and eastern goitered gazelle (*Gazella subgutturosa*: Güeldenstaedt, 1780) populations in Iran. Int J Environ Res 5(1):225–232

Heptner G, Nasimovich AA, Bannikov AG (1961) Mammals of Soviet Union, Vol. 1. Artiodactyla and Perissodactyla. Vysshaya Shkola Publishing House, Moscow. (in Russian)

Kingswood SC, Blank DA (1996) Gazella subgutturosa. Mamm Species 518:1–10

Marmazinskaya NV (1996) Territorial and marking behavior of goitered gazelle *Gazella subgutturosa* (Artiodactyla, Bovidae) in the Bukhara ecological center. Zool Zhurnal (Moscow) 75(11):1737–1751. (in Russian, English summary)

Marmazinskaya NV (1999) Social behavior of goitered gazelle (*Gazella subgutturosa* /Güld., 1780/ during breeding period in the conditions ecocenter "Goitered gazelle". Ph.D. thesis in zoology, Institute of zoology of Uzbek Academy of Sciences, Tashkent, Uzbekistan, pp 1–20 (in Russian)

Pavlov VN, Vanisova EA, Nikolskii AA (2013) The vegetation on the signaling points of goitered gazelle *Gazella subgutturosa*. Bull Moscow Soc Nat 118(3):44–48. (in Russian, English abstract)

Pocock RI (1910, December 1) On the specialized cutaneous glands of ruminants. In: Proceedings of the Zoological Society of London, vol 80, No. 4. Blackwell Publishing Ltd, Oxford, UK, pp 840–845

Sludskiy AA (1956) Breeding of goitered gazelle. Proceedings of Institute of zoology, Kazakh Academy of Sciences, Vol. 6. Nauka Press of Kazakh SSR, Alma-Ata, p 78–108 (in Russian)

Sokolov II (1959) *Gazella subgutturosa* Güldenstaedt—Goitered gazelle. In: Pavlovskiy EN (ed) Fauna USSR, mammals: ungulates (Orders Perissodactyla and Artiodactyla). Publication of the Academy of Sciences USSR, Moscow and Leningrad, pp 308–331. (in Russian)

Sokolov VE (1982) Mammal skin. University of California Press, Berkeley

Sorokin PA, Soldatova NV, Lukarevskiy VS, Kholodova MV (2011) Genetic diversity and relations of the goitered gazelle (*Gazella subgutturosa*) groups from Uzbekistan, Turkmenistan, and Azerbaijan: analysis of the D-loop of mitochondrial DNA. Biol Bull 38(6):585–590

Stroganov SU (1956) Materials for knowledge of the theriological fauna of the Soviet Union. Proceedings of the Biological Institute, West Siberian Branch of the Academy of Sciences, zoological issue 1, Novosibirsk: 15–19 (in Russian)

Wacher T, Wronski T, Hammond RL, Winney B, Blacket MJ, Hundermark KJ et al (2011) Phylogenetic analysis of mitochondrial DNA sequences reveals polyphyly in the goitered gazelle (*Gazella subgutturosa*). Conserv Genet 12(3):827–831

Walther FR (1968) Verhalten der Gazellen. A. Ziemsen Verlag, Wittenberg Lutherstadt

Walther FR, Mungall EC, Grau GA (1983) Gazelles and their relatives. A study in territorial behavior. Noyes Publication, Park Ridge, NJ, USA

Wronski T, Plath M (2010) Characterisation of the spatial distribution of latrines in mountain gazelles (*Gazella gazella*): do latrines demark female group home ranges? J Zool Lond 280(1):92–101

Wronski T, Apio A, Plath M, Ziege M (2013) Sex-difference in the communicatory significance of localized defecation sites in Arabian gazelles (*Gazella arabica*). J Ethol 31(2):129–140

Zhevnerov VV (1984) Goitered gazelle of the Barsakelmes Island. Nauka Publishing House of Kazakh SSR, Alma-Ata. (in Russian)

Zhevnerov VV, Bekenov AB, Sludskiy AA (1983) Goitered gazelle—*Gazella subgutturosa* Güldenstaedt, 1780. In: Gvozdev EV, Kapitonov VI (eds) Mammals of Kazakhstan, Bovidae, vol 3, No. 3. Nauka Publishing House of Kazakh SSR, Alma-Ata, pp 11–54. (in Russian)

22 Instead of Conclusion

Around 100 years ago, the goitered gazelle was one of the most widespread ungulates, having a vast habitat area in Asia, where clusters of thousands of gazelles were often found, especially during migrations. The total number of the goitered gazelles in the former Soviet Central Asia alone was estimated at 1 million during the 1900s. Several decades have passed and with the advent of powerful firearms, goitered gazelles began to be exterminated in large numbers throughout their range. High-traffic vehicles appeared, and poachers began to hunt and kill gazelles in formerly remote and impassable areas. This was followed by intensive watering and agricultural development of the deserts, and then the influx of livestock that replaced goitered gazelles first in the best pasture lands and watering holes and then from everywhere. With the development, poaching penetrated into places where it was not found previously. After agricultural expansion, industrial development began, bringing people into the most arid desert regions, continuing to exterminate goitered gazelles whenever they were found.

All Asian countries have gone through this process, some earlier, and some later. As a result, there was a catastrophic decrease in the number of gazelles. The almost continuous natural habitat of goitered gazelles in Central Asia broke up into many small populations, each of which continued to slowly die out. People realized the problem when the goitered gazelle had disappeared from several countries, while in others, they have become so rare that this species can hardly be seen in the wild. Almost all countries have little control over poaching and almost completely do not limit the offensive invasion of the economic development of arid zones. Once the problem of a dying species was acknowledged, countries began to build fenced nature reserves in order to somehow limit poaching. Almost all countries have adopted laws prohibiting goitered gazelle hunting, but almost none one making serious efforts to actually preserve what gazelle populations that still exist. A number of projects were organized to restore wild populations, but any positive effect was noted only in the fenced reserves or on islands with limited access. The only exception is Azerbaijan, which has reintroduced several populations in former habitats.

D. Blank, *The Behavioral Biology and Conservation of Goitered Gazelles (Gazella subgutturosa)*, https://doi.org/10.1007/978-3-031-83504-9_22

The leaders of most other countries have almost no interest in preserving their biodiversity, so the last remnants of their former wild splendor are still being exterminated, such as in Mongolia, where the world largest goitered gazelle population will decrease drastically soon as a result of intensive industrial development and enlarging of the railway net all over the country.

If this trend continues, we will not have time to study goitered gazelles well enough to understand their biology. This species will disappear completely if countries, without consideration of the needs of goitered gazelles, continue to develop their new human-centric projects: building border fences, new highways and railways, new oil and gas pipelines, exploring more and more new mines, driving huge herds of livestock, and eliminating all access to available watering holes and pastures for goitered gazelles. At the same time, governments tend to look indifferently at the rampant poaching and destruction of thousands of gazelles. Goitered gazelles have lived in many countries throughout Central Asia and the Near and Middle East. All these countries, with the exception Azerbaijan, ignore the fact that goitered gazelles are disappearing from their once vast territories, and the remaining small groups are doomed to gradual extinction. These countries, both rich and poor, do not recognize the need to make the necessary efforts to protect their real wealth and futures—goitered gazelles as a part of their natural wildlife. Azerbaijan is only exception, whose country's leaders have made great efforts to restore its goitered gazelle populations. Unfortunately, other countries do not follow this example and are willing to destroy the great gifts that nature has given them. Soon it will only be possible to see goitered gazelles in museums and study them only from bone fragments (Fig. 22.1).

Fig. 22.1 Horns of goitered gazelle male found on takyr in the Southern Ustyurt Chink area (Kaplankyr). SW Kazakhstan. 22.05.2019. Photo by Mark Pestov with kind permission

The manufacturer's authorised representative in the EU is Springer Nature Customer Service Centre GmbH, Europaplatz 3, 69115 Heidelberg, Germany. If you have any concerns regarding our products, please contact ProductSafety@springernature.com

Printed and bound by CPI Group (UK) Ltd, Croydon, CR0 4YY

30/09/2026

02225562-0018